Geometric Algebra for
Computer Science

Geometric Algebra for Computer Science

An Object-Oriented Approach to Geometry

LEO DORST

DANIEL FONTIJNE

STEPHEN MANN

AMSTERDAM • BOSTON • HEIDELBERG • LONDON
NEW YORK • OXFORD • PARIS • SAN DIEGO
SAN FRANCISCO • SINGAPORE • SYDNEY • TOKYO
Morgan Kaufmann Publishers is an imprint of Elsevier

Morgan Kaufmann Publishers is an imprint of Elsevier
30 Corporate Drive, Suite 400, Burlington, MA 01803, USA

Library of Congress Cataloging-in-Publication Data
Application submitted

ISBN: 978-0-12-374942-0

For information on all Morgan Kaufmann publications,
visit our Web site at www.mkp.com or www.books.elsevier.com

Contents

PART III   IMPLEMENTING GEOMETRIC ALGEBRA

List of Figures

List of Tables

List of Programming Examples

Preface

Geometric algebra is a powerful and practical framework for the representation and solution of geometrical problems. We believe it to be eminently suitable to those sub-fields of computer science in which such issues occur: computer graphics, robotics, and computer vision. We wrote this book to explain the basic structure of geometric algebra, and to help the reader become a practical user. We employ various tools to get there:

- Explanations that are not more mathematical than we deem necessary, connecting algebra and geometry at every step
- A large number of interactive illustrations to get the "object-oriented" feeling of constructions that are dependent only on the geometric elements in them (rather than on coordinates)
- Drills and structural exercises for almost every chapter
- Detailed programming examples on elements of practical applications
- An extensive section on the implementational aspects of geometric algebra (Part III of this book)

This is the first book on geometric algebra that has been written especially for the computer science audience. When reading it, you should remember that geometric algebra is fundamentally simple, and fundamentally simplifying. That simplicity will not always be clear; precisely because it is so fundamental, it does basic things in a slightly different way and in a different notation. This requires your full attention, notably in the beginning, when we only seem to go over familiar things in a perhaps irritatingly different manner. The patterns we uncover, and the coordinate-free way in which we encode them, will all pay off in the end in generally applicable quantitative geometrical operators and constructions.

We emphasize that this is not primarily a book on programming, and that the subtitle "An Object-oriented Approach to Geometry" should not be interpreted too literally. It is intended to convey that we finally achieve clean computational "objects" (in the sense of object-oriented programming) to correspond to the oriented elements and operators of geometry by identifying them with "oriented objects" of the algebra.

AUDIENCE

The book is aimed at a graduate level; we only assume basic linear algebra (and a bit of calculus in Chapter 8). No prior knowledge of the techniques of computer graphics or robotics is required, though if you are familiar with those fields you will appreciate how much easier things are in geometric algebra. The book should also be well suited for self-study at the post-graduate level; in fact, we tried to write the book that we would have wanted ourselves for this purpose. Depending on your level of interest, you may want to read it in different ways.

- If you are a seasoned user of geometry and well versed in the techniques of casting geometry in linear algebra, but don't have much time, you will still find this book worthwhile. In a comfortable reading, you can absorb what is different in geometric algebra, and its structure will help you understand all those old tricks in your library. In our experience, it makes many arcane techniques comprehensible, and it helped us to learn from useful math books that we would otherwise never have dared to read. You may never actually use geometric algebra, but you will find it enlightening all the same. And who knows—you may come back for more.

- If you are currently writing code using the coordinate-based linear algebra, a background study of the techniques in this book will be helpful and constructive. The advantages for the previous category will apply to you as well. Moreover, you may find yourself doing derivations of formulas you need to program in the compact geometric algebra manner, and this will clarify and improve your implementations, even if you continue writing those in the old linear algebra vocabulary. In particular, the thinking behind your code will be more geometrical, less coordinate-based, and this will make it more transparent, more flexibly applicable (for instance, in higher dimensions), and ready to be translated into geometric algebra after the revolution.

- If you are starting out in geometric programming, take the time to absorb this book thoroughly. This geometric algebra way of thinking is quite natural, and we are rather envious that you can learn it from scratch, without having to unlearn old methods. With study and practice you will be able to write programs in geometric algebra rather fluently, and eventually contribute actively to its development.

Our style in this book is factual. We give you the necessary mathematics, but always relate the algebra to the geometry, so that you get the complete picture. Occasionally, there is a need for more extensive proofs to convince you of the consistency of aspects of the framework. When such a proof became too lengthy and did not further the arguments, it was relegated to an appendix. The derivations that remain in the text should be worth your time, since they are good practice in developing your skills. We have attempted to avoid the "pitfall of imprecision" in this somewhat narrative presentation style by providing the fundamental chapters with a summary of the essential results, for easy consultation via the index.

HISTORY

We do not constantly attribute all results, but that does not mean that we think that we developed all this ourselves. By its very nature, geometric algebra collates many partial results in a single framework, and the original sources become hard to trace in their original context. It is part of the pleasure of geometric algebra that it empowers the user; by mastering just a few techniques, you can usually easily rediscover the result you need.

Once you grasp its essence, geometric algebra will become so natural that you will wonder why we have not done geometry this way all along. The reason is a history of geometric (mis)representation, for almost all elements of geometric algebra are not new—in hindsight. Elements of the quantitative characterization of geometric constructions directly in terms of its elements are already present in the work of René Descartes (1595–1650); however, his followers thought it was easier to reduce his techniques to coordinate systems not related to the elements (nevertheless calling them Cartesian, in his honor). This gave us the mixed blessing of coordinates, and the tiresome custom of specifying geometry at the coordinate level (whereas coordinates should be relegated to the lowest implementational level, reserved for the actual computations). To have a more direct means of expression, Hermann Grassmann (1809–1877) developed a theory of extended quantities, allowing geometry to be based on more than points and vectors. Unfortunately, his ideas were ahead of their time, and his very compact notation made his work more obscure than it should have been. William Rowan Hamilton (1805–1865) developed quaternions for the algebra of rotations in 3D, and William Kingdon Clifford (1845–1879) defined a more general product between vectors that could incorporate general rigid body motions.

All these individual contributions pointed toward a geometric algebra, and at the end of the 19th century, there were various potentially useful systems to represent aspects of geometry. Gibbs (1839–1903) made a special selection of useful techniques for the 3D geometry of engineering, and this limited framework is basically what we have been using ever since in the geometrical applications of linear algebra. In a typical quote from his biography "using ideas of Grassmann, Gibbs produced a system much more easily applied to physics than that of Hamilton." In the process, we lost geometric algebra. Linear algebra and matrices, with their coordinate representations, became the mainstay of doing geometry, both in practice and in mathematical development. Matrices work, but in their usual form they only work on vectors, and this ignores Grassmann's insight that extended qualities can be elements of computation. (Tensors partially fix this, but in a cumbersome coordinate-based notation.)

With the arrival of quantum physics, convenient alternative representations for spatial motions were developed (notably for rotations), using complex numbers in "spinors." The complex nature of spinors was mistaken for an essential aspect of quantum mechanics, and the representations were not reapplied to everyday geometry. David Hestenes (1933–present) was perhaps the first to realize that the representational techniques in relativity and quantum mechanics were essentially manifestations of a fundamental "algebra of spatial relationships" that needed to be explored. He rescued the half-forgotten geometric algebra (by now called Clifford algebra and developed in nongeometric directions), developed it into an alternative to the classical linear algebra–based representations, and started advocating its universal use. In the 1990s, his voice was heard, and with the implementation of geometric algebra into interactive computer programs its practical applicability is becoming more apparent.

We can now finally begin to pick up the thread of geometrical representation where it was left around 1900. Gibbs was wrong in assuming that computing with the geometry of 3D space requires only representations of 3D points, although he did give us a powerful system to compute with those. This book will demonstrate that allowing more extended quantities in higher-dimensional representational spaces provides a more convenient executable language for geometry. Maybe we could have had this all along; but perhaps we indeed needed to wait for the arrival of computers to appreciate the effectiveness of this approach.

SOFTWARE

There are three main software packages associated with this book, each written with a different goal in mind (interaction, efficiency and illustration of algorithms, respectively). All three were developed by us, and can be found on the web site:

http://www.geometricalgebra.net

for free downloading.

- GAViewer is an interactive program that we used to generate the majority of the figures in this book. It was originally developed as a teaching tool, and a web tutorial is available, using GAViewer to explain the basics of geometric algebra. You can use GAViewer when reading the book to type in algebraic formulas and have them act on geometrical elements interactively. This interaction should aid your understanding of the correspondence between geometry and algebra considerably. The (simplified) code of the figures provides a starting point for your own experimentation.

- Gaigen 2 is geometric algebra implementation in C++ (and Java), intended for applications requiring more speed and efficiency than a simple tutorial. The GA sandbox source code package used for the programming examples and exercises in this book is built on top of Gaigen 2. To compile and run the programming examples in Part I and Part II, you only have to download the sandbox package from the web site.

- Our simplistic but educational "reference implementation" implements all algorithms and techniques discussed in Part III. It is written in Java and intended to show only the essential structure; we do not deem it usable for anything that is computationally intensive, since it can easily be 10 to 100 times slower than Gaigen 2.

If you are serious about implementing further applications, you can start with the GA sandbox package, or other available implementations of geometric algebra, or even write your own package.

ACKNOWLEDGMENTS

Of those who have helped us develop this work, we especially thank David Hestenes, not only for reinvigorating geometric algebra, but also for giving Leo an early introduction to the conformal model at a half-year sabbatical at Arizona State University. We are grateful to Joan Lasenby of Cambridge University for her detailed comments on the early chapters, and for providing some of the applied examples. We are also indebted to Timaeus Bouma for his keen insights that allowed our software to be well-founded in mathematical fact.

We gratefully acknowledge the support of the University of Amsterdam, especially professor Frans Groen; NWO (Netherlands Organization for Scientific Research) in project 612.012.006; and NSERC (Natural Sciences and Engineering Research Council of Canada).

Ultimately, though, this book would have been impossible without the home front:

Leo Dorst's parents and his wife Phyllis have always utterly supported him in his quest to understand new aspects of math and life; he dedicates this book to them. This second printing is also dedicated to his wondrous daughter Mia, fortuitously born one year after our book first appeared.

Daniel Fontijne owes many thanks to Yvonne for providing the fun and artistic reasons to study geometric algebra, and to Femke and Tijmen for the many refreshing breaks while working at home.

Stephen Mann would like to thank Jeanette, Mei, and Lilly for their support during the writing of this book.

SECOND CORRECTED PRINTING

This is the second printing of the book, correcting a number of errors found in the first printing by ourselves and alert readers and originally reported on the book's web site. We are grateful to all contributors (for full attribution, see the web site), especially Gregory Grunberg, Philip J. Kuntz, Ron Goldman, Mark McLaughlin, and Jeroen Spandaw, who each found several errors. But by far the greatest contributor was Allan Cortzen, who reported many errors, some quite subtle.

1 WHY GEOMETRIC ALGEBRA?

This book is about geometric algebra, a powerful computational system to describe and solve geometrical problems. You will see that it covers familiar ground—lines, planes, spheres, rotations, linear transformations, and more—but in an unfamiliar way. Our intention is to show you how basic operations on basic geometrical objects can be done differently, and better, using this new framework.

The intention of this first chapter is to give you a fair impression of what geometric algebra can do, how it does it, what old habits you will need to extend or replace, and what you will gain in the process.

1.1 AN EXAMPLE IN GEOMETRIC ALGEBRA

To convey the compactness of expression of geometric algebra, we give a brief example of a geometric situation, its description in geometric algebra, and the accompanying code that executes this description. It helps us discuss some of the important properties of the computational framework. You should of course read between the lines: you will be able to understand this example fully only at the end of Part II, but the principles should be clear enough now.

Suppose that we have three points c_1, c_2, c_3 in a 3-D space with a Euclidean metric, a line L, and a plane Π. We would like to construct a circle C through the three points, rotate it around the line L, and then reflect the whole scene in the plane Π. This is depicted in Figure 1.1. Here is how geometric algebra encodes this in its conformal model of Euclidean geometry:

1. **Circle**. The three points are denoted by three elements c_1, c_2, and c_3. The oriented circle through them is

$$C = c_1 \wedge c_2 \wedge c_3.$$

   The $\wedge$ symbol denotes the *outer product*, which constructs new elements of computation by an algebraic operation that geometrically connects basic elements (in this case, it connects points to form a circle). The outer product is antisymmetric: if you wanted a circle with opposite orientation through these points, it would be $-C$, which could be made as $-C = c_1 \wedge c_3 \wedge c_2$.

2. **Rotation**. The rotation of the circle C is made by a sandwiching product with an element R called a rotor, as

$$C \mapsto R\,C/R.$$

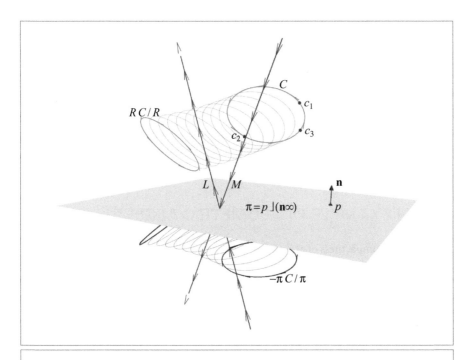

Figure 1.1: The rotation of a circle C (determined by three points c_1, c_2, c_3) around a line L, and the reflections of those elements in a plane Π.

The product involved here is the *geometric product*, which is the fundamental product of geometric algebra, and its corresponding division. The geometric product multiplies transformations. It is structure-preserving, because the rotated circle through three points is the circle through the three rotated points:

$$R(c_1 \wedge c_2 \wedge c_3)/R = (Rc_1/R) \wedge (Rc_2/R) \wedge (Rc_3/R).$$

Moreover, any element, not just a circle, is rotated by the same rotor-based formula. We define the value of the rotor that turns around the line L below.

3. **Line.** An oriented line L is also an element of geometric algebra. It can be constructed as a "circle" passing through two given points a_1 and a_2 and the point at infinity ∞, using the same outer product as in item 1:

$$L = a_1 \wedge a_2 \wedge \infty.$$

Alternatively, if you have a point on L and a direction vector $\mathbf{u}$ for L, you can make the same element as

$$L = a_1 \wedge \mathbf{u} \wedge \infty.$$

This specifies exactly the same element L by the same outer product, even though it takes different arguments. This algebraic equivalence saves the construction of many specific data types and their corresponding methods for what are geometrically the same elements.

The point at infinity ∞ is an essential element of this operational model of Euclidean geometry. It is a finite element of the algebra, with well-defined algebraic properties.

4. **Line Rotation.** The rotor that represents a rotation around the line L, with rotation angle ϕ, is

$$R = \exp(\phi L^*/2).$$

This shows that geometric algebra contains an exponentiation that can make elements into rotation operators. The element L^* is the *dual* of the line L. Dualization is an operation that takes the geometric complement. For the line L, its dual can be visualized as the nest of cylinders surrounding it.

If you would like to perform the rotation in N small steps, you can interpolate the rotor, using its logarithm to compute $R^{1/N}$, and applying that N times (we have done so in Figure 1.1, to give a better impression of the transformation). Other transformations, such as general rigid body motions, have logarithms as well in geometric algebra and can therefore be interpolated.

5. **Plane.** To reflect the whole situation with the line and the circles in a plane Π, we first need to represent that plane. Again, there are alternatives. The most straightforward is to construct the plane with the outer product of three points p_1, p_2, p_3 on the plane

and the point at infinity ∞, as $\Pi = p_1 \wedge p_2 \wedge p_3 \wedge \infty$. Alternatively, we can instead employ a specification by a normal vector $\mathbf{n}$ and a point p on the plane. This is a specification of the dual plane $\pi \equiv \Pi^*$, its geometric complement:

$$\pi = p \rfloor (\mathbf{n}\infty) = \mathbf{n} - (\mathbf{p} \cdot \mathbf{n}) \infty.$$

Here $\rfloor$ is a contraction product, used for metric computations in geometric algebra; it is a generalization of the inner product (or dot product) from vectors to the general elements of the algebra. The duality operation above is a special case of the contraction.

The change from p to $\mathbf{p}$ in the equation is not a typo: p denotes a point, $\mathbf{p}$ is its location vector relative to the (arbitrary) origin. The two entities are clearly distinct elements of geometric algebra, though computationally related.

6. **Reflection**. Either the plane Π or its geometric complement π determine a reflection operator. Points, circles, or lines (in fact, any element X) reflect in the plane in the same way:

$$X \mapsto -\pi X / \pi.$$

Here the reflection plane π, which is an oriented object of geometric algebra, acts as a reflector, again by means of a sandwiching using the geometric product. Note that the reflected circle has the proper orientation in Figure 1.1.

As with the rotation in item 2, there is obvious structure preservation: the reflection of the rotated circle is the rotation of the reflected circle (in the reflected line). We can even reflect the rotor to become $R' \equiv \pi \exp(\phi L^*/2)/\pi = \exp(-\phi (-\pi L^*/\pi)/2)$, which is the rotor around the reflected line, automatically turning in the opposite orientation.

7. **Programming**. In total, the scene of Figure 1.1 can be generated by a simple C++ program computing directly with the geometric objects in the problem statement, shown in Figure 1.2. The outcome is plotted immediately through the calls to the multivector drawing function `draw()`. And since it has been fully specified in terms of geometric entities, one can easily change any of them and update the picture. The computations are fast enough to do this and much more involved calculations in real time; the rendering is typically the slowest component.

Although the language is still unfamiliar, we hope you can see that this is geometric programming at a very desirable level, in terms of quantities that have a direct geometrical meaning. Each item occurring in any of the computations can be visualized. None of the operations on the elements needed to be specified in terms of their coordinates. Coordinates are only needed when entering the data, to specify precisely which points and lines are to be operated upon. The absence of this quantitative information may suggest that geometric algebra is merely an abstract specification language with obscure operators that merely convey the mathematical logic of geometry. It is much more than that: all expressions are quantitative prescriptions of computations, and can be executed directly. Geometric algebra is a programming language, especially tailored to handle geometry.

```
// l1, l2, c1, c2, c3, p1 are points, n is a direction vector
// OpenGL commands to set color are not shown
line L; circle C; dualPlane p;

L = unit_r(l1 ^ l2 ^ ni); // ni represents the point at infinity
C = c1 ^ c2 ^ c3;
p = p1 << (n ^ ni);

draw(L); // draw line (red)
draw(C); // draw cicle (green)
draw(p); // draw plane (yellow)

draw(−p * L * inverse(p)); // draw reflected line (magenta)
draw(−p * C * inverse(p)); // draw reflected circle (blue)

// compute rotation versor:
const float phi = (float)(M_PI / 2.0);
TRversor R;
R = exp(0.5f * phi * dual(L));

draw(R * C * inverse(R)); // draw rotated cicle (green)

// draw reflected, rotated circle (blue)
draw(−p * R * C * inverse(R) * inverse(p));

// draw interpolated circles
pointPair LR = log(R); // get log of R
for (float alpha = 0; alpha < 1.0; alpha += 0.1f)
{
  // compute interpolated rotor
  TRversor iR;
  iR = exp(alpha * LR);

  // draw rotated circle (light green)
  draw(iR * C * inverse(iR));

  // draw reflected, rotated circle (light blue)
  draw(−p * iR * C * inverse(iR) * inverse(p));
}
```

Figure 1.2: Code to generate Figure 1.1.

You may be concerned about the many different products that occurred in this application. If geometric algebra needs a new product for every new operation, its power would be understandable, but the system would rapidly grow unwieldy. This is perhaps the biggest surprise of all: *there is only one product that does it all*. It is the *geometric product* (discovered by William Kingdon Clifford in the 1870s), which we used implicitly in the example in the sandwiching operations of rotation and reflection. The other products ($\wedge$, $\rfloor$, *, sandwiching) are all specially derived products for the purposes of spanning, metric projection, complementation, and operating on other elements. They can all be defined in terms of the geometric product, and they correspond closely to how we think about geometry classically. That is the main reason that they have been given special symbols. Once you get used to them, you will appreciate the extra readability they offer. But it is important to realize that you really only need to implement one product to get the whole consistent functionality of geometric algebra.

Because of the structural properties of geometric algebra, this example can be extended in many ways. To name a few:

- **Spherical Reflection**. If we had instead wanted to reflect this situation in a sphere, this is done by

$$X \mapsto -\sigma X / \sigma.$$

Here σ is the dual representation of a sphere (it encodes a sphere with center c passing through p as the representational vector $p \rfloor (c \wedge \infty)$). We depict this in Figure 1.3.

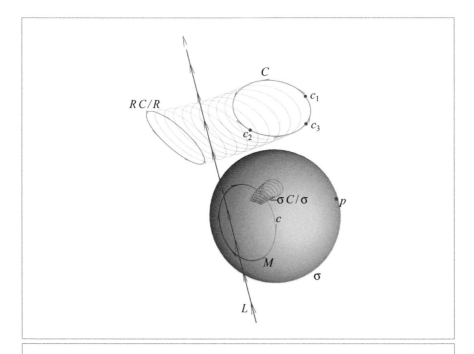

Figure 1.3: The rotation of a circle C (determined by three points c_1, c_2, c_3) around a line L, and the reflections of those elements in a sphere σ.

The only thing that is different from the program generating Figure 1.1 is that the plane π was replaced by the sphere σ, not only geometrically, but also algebraically. This generates the new reflection, which reflects the line L to become the circle $M = -\sigma L/\sigma$. It also converts the reflected rotor around M into the operation $\sigma R/\sigma$, which generates a scaled rotation around a circle, depicted in the figure. The whole structure of geometric relationships is nicely preserved.

- **Intersections.** The π-based reflection operator of item 6 takes the line L and produces the reflected line $-\pi L/\pi$, without even computing the intersection point of the line and the plane. If we had wanted to compute the intersection of line and plane, that would have been the point $\pi \rfloor L = \Pi^* \rfloor L$. This is another universal product, the `meet`, which computes the intersection of two elements Π and L.

- **Differentiation.** It is even possible to symbolically differentiate the final expression of the reflected rotated circle to any of the geometrical elements occurring in it. This permits a sensitivity analysis or a local linearization; for instance, discovering how the resulting reflected rotated circle would change if the plane π were to be moved and tilted slightly.

1.2 HOW IT WORKS AND HOW IT'S DIFFERENT

The example has given you an impression of what geometric algebra can do. To understand the structure of the book, you need a better feeling for what geometric algebra is, and how it relates to more classical techniques such as linear algebra.

The main features of geometric algebra are:

- **Vector Spaces as Modeling Tools.** Vectors can be used to represent aspects of geometry, but the precise correspondence is a modeling choice. Geometric algebra offers three increasingly powerful models for Euclidean geometry.

- **Subspaces as Elements of Computation.** Geometric algebra has products to combine vectors to new elements of computation. They represent oriented subspaces of any dimension, and they have rich geometric interpretations within the models.

- **Linear Transformations Extended.** A linear transformation on the vector space dictates how subspaces transform; this augments the power of linear algebra in a structural manner to the extended elements.

- **Universal Orthogonal Transformations.** Geometric algebra has a special representation of orthogonal transformations that is efficient and universally applicable in the same form to all geometric elements.

- **Objects Are Operators.** Geometric objects and operators are represented on a par, and exchangeable: objects can act as operators, and operators can be transformed like geometrical objects.

- *Closed Form Interpolation and Perturbation*. There is a geometric calculus that can be applied directly to geometrical objects and operators. It allows straightforward interpolation of Euclidean motions.

In the following subsections, we elaborate on each of these topics.

1.2.1 VECTOR SPACES AS MODELING TOOLS

When you use linear algebra to describe the geometry of elements in space, you use a real vector space $\mathbb{R}^m$. Geometric algebra starts with the same domain. In both frameworks, the vectors in an m-dimensional vector space $\mathbb{R}^m$ represent 1-D directions in that space. You can think of them as denoting lines through the origin. To do geometry flexibly, we want more than directions; we also want *points* in space. The vector space $\mathbb{R}^m$ does not have those by itself, though its vectors can be used to represent them.

Here it is necessary to be more precise. There are two structures involved in doing geometrical computations, both confusingly called "space."

- There is the physical 3-D space of everyday experience (what roboticists call the task space). It contains the objects that we want to describe computationally, to move around, to analyze data about, or to simply draw.
- Mathematics has developed the concept of a vector space, which is a space of abstract entities with properties originally inspired by the geometry of physical space.

Although an m-dimensional vector space is a mathematical generalization of 3-D physical space, it does not follow that 3-D physical space is best described by a 3-D vector space. In fact, in applications we are less interested in the *space* than in the *geometry*, which concerns the objects residing in the space. That geometry is defined by the motions that can freely move objects. In Euclidean geometry, those motions are translations, rotations, and reflections. Whenever two objects differ only by such transformations we refer to them as the same object, but at a different location, with a different orientation, or viewed in a mirror. (Sometimes *scaling* is also included in the permitted equivalences.)

So we should wonder what computational model, based in a vector space framework, can conveniently represent these natural motions of Euclidean geometry. Since the motions involve certain measures to be preserved (such as size), we typically use a metric vector space to model it. We present three possibilities that will recur in this book:

1. *The Vector Space Model*. A 3-D vector space with a Euclidean metric is well suited to describe the algebra of *directions* in 3-D physical space, and the operation of rotation that transforms directions. Rotations (and reflections) are orthogonal linear transformations: they preserve distances and angles. They can be represented by 3×3 orthogonal matrices or as quaternions (although the latter are not in the *linear* algebra of $\mathbb{R}^3$, we will see that they are in the *geometric* algebra of $\mathbb{R}^3$).

2. *The Homogeneous Model*. If you also want to describe translations in 3-D space, it is advantageous to use *homogeneous coordinates*. This employs the *vectors* of a

4-D vector space to represent *points* in physical 3-D space. Translations now also become linear transformations, and therefore combine well with the 3-D matrix representation of rotations.

The extra fourth dimension of the vector space can be interpreted as the *point at the origin* in the physical space. There is some freedom in choosing the metric of this 4-D vector space, which makes this model suitable for projective geometry.

3. ***The Conformal Model***. If we want the translations in 3-D physical space represented as *orthogonal* transformations (just as rotations were in the 3-D vector space model), we can do so by employing a 5-D vector space. This 5-D space needs to be given a special metric to embed the metric properties of Euclidean space. It is expressed as $\mathbb{R}^{4,1}$, a 5-D vector space with a Minkowski metric.

The vectors of the vector space $\mathbb{R}^{4,1}$ can be interpreted as dual spheres in 3-D physical space, including the zero-radius spheres that are points. The two extra dimensions are the *point at the origin* and the *point at infinity*.

This model was used in the example of Figure 1.1. It is called the conformal model because we get more geometry than merely the Euclidean motions: all conformal (i.e., angle-preserving) transformations can be represented as orthogonal transformations. One of those is inversion in a sphere, which explains why we could use a spherical reflector in Figure 1.3.

Although these models can all be treated and programmed using standard linear algebra, there is great advantage to using geometric algebra instead:

- Geometric algebra uses the subspace structure of the vector spaces to construct extended objects.

- Geometric algebra contains a particularly powerful method to represent orthogonal transformations.

The former is useful to all three models of Euclidean geometry; the latter specifically works for the first and third. In fact, the conformal model was invented before geometric algebra, but it lay dormant. Only with the tools that geometric algebra offers can we realize its computational potential. We will treat all these models in Part II of this book, with special attention to the conformal model. In Part I, we develop the techniques of geometric algebra, and prefer to illustrate those with the more familiar vector space model, to develop your intuition for its computational capabilities.

1.2.2 SUBSPACES AS ELEMENTS OF COMPUTATION

Whatever model you use to describe the geometry of physical space, understanding vector spaces and their transformations is a fundamental prerequisite. Linear algebra gives you techniques to compute with the basic elements (the vectors) by using matrices. Geometric algebra focuses on the *subspaces of a vector space as elements of computation*. It constructs these systematically from the underlying vector space, and extends the matrix techniques to transform them, even supplanting those completely when the transformations are orthogonal.

The outer product $\wedge$ has the constructive role of making subspaces out of vectors. It uses k independent vectors $\mathbf{v}_i$ to construct the computational element $\mathbf{v}_1 \wedge \mathbf{v}_2 \wedge \cdots \wedge \mathbf{v}_k$, which represents the k-dimensional subspace spanned by the $\mathbf{v}_i$. Such a subspace is *proper* (also known as *homogeneous*): it contains the origin of the vector space, the zero vector 0. An m-dimensional vector space has many independent proper subspaces: there are $\binom{m}{k}$ subspaces of k dimensions, for a total of 2^m subspaces of any dimension. This is a considerable amount of structure that comes for free with the vector space $\mathbb{R}^m$, which can be exploited to encode geometric entities.

Depending on how the vector space $\mathbb{R}^m$ is used to model geometry, we obtain different geometric interpretations of its outer product.

- In the vector space model, a vector represents a 1-D direction in space, which can be used to encode the direction of a line through the origin. This is a 1-D proper subspace. The outer product of two vectors then denotes a 2-D direction, which signifies the attitude of an oriented plane through the origin, a 2-D proper subspace of the vector space. The outer product of three vectors is a volume. Each of those has a magnitude and an orientation. This is illustrated in Figure 1.4(a,b).

- In the homogeneous model, a *vector* of the vector space represents a *point* in the physical space it models. Now the outer product of two vectors represents an oriented line in the physical space, and the outer product of three vectors is interpreted as an oriented plane. This is illustrated in Figure 1.4(c,d). By the way, this representation of lines is the geometric algebra form of Plücker coordinates, now naturally embedded in the rest of the framework.

- In the conformal model, the points of physical space are viewed as spheres of radius zero and represented as vectors of the vector space. The outer product of three points then represents an oriented circle, and the outer product of four points an oriented sphere. This is illustrated in Figure 1.4(e,f). If we include the point at infinity in the outer product, we get the "flat" elements that we could already represent in the homogeneous model, as the example in Section 1.1 showed.

It is very satisfying that there is one abstract product underlying such diverse constructions. However, these varied geometrical interpretations can confuse the study of its algebraic properties, so when we treat the outer product in Chapter 2 and the rest of Part I, we prefer to focus on the vector space model to guide your intuitive understanding of geometric algebra. In that form, the outer product dates back to Hermann Grassmann (1840) and is the foundation of the *Grassmann algebra* of the extended quantities we call proper subspaces. Grassmann algebra is the foundation of geometric algebra.

In standard linear algebra, subspaces are not this explicitly represented or constructed. One can assemble vectors $\mathbf{v}_i$ as columns in a matrix $[\![V]\!] = [\![\mathbf{v}_1 \ \mathbf{v}_2 \cdots \mathbf{v}_k]\!]$, and then treat the image of this matrix, $\mathrm{im}([\![V]\!])$, as a representation of the subspace, but this is not an

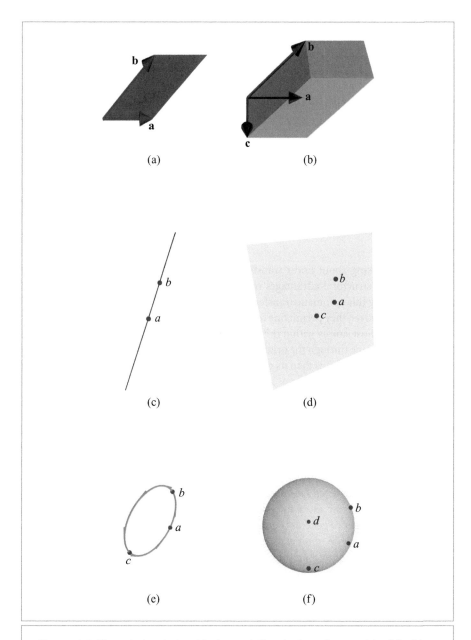

Figure 1.4: The outer product and its interpretations in the various models of Euclidean geometry. (a,b): the vector space model; (c,d): the homogeneous model; and (e,f): the conformal model.

integral part of the algebra; it is not a product in the same sense that the dot product is. If the matrix is square, we can take the determinant det($[\![V]\!]$) to represent the amount of area or volume of the subspace and its orientation, but if it is not square, such measures are less easily represented. Subspaces are simply not well represented in standard linear algebra.

1.2.3 LINEAR TRANSFORMATIONS EXTENDED

Linear transformations are defined by how they transform vectors in the vector space $\mathbb{R}^m$. As these vectors transform, so do the subspaces spanned by them. That fully defines how to extend a linear transformation to the subspace structure.

If one uses a matrix for the representation of the linear transformation on the vector space level, it is straightforward and automatic to extend this to a matrix that works on the subspace levels. You just take the outer product of its action on the basis vectors as its definition on the basis for subspaces. Now you can perform the same linear transformation on any subspace.

This way of thinking about linear transformations, with its use of the outer product, already provides structural advantages over the usual coordinate-based methods. Programs embedding this automatic transference of a vector space mapping to its subspaces are simpler. Moreover, they permit one to choose a representation for geometric elements that transforms most simply within this framework. An example is the representation of the attitude of a plane through the origin; its representation by a normal vector has more complicated transformations than its equally valid representation by an outer product of two vectors.

Within the subspace representation, a general product can be given for the intersection of subspaces (the `meet` product), which also transform in a structure-preserving manner under the extended linear transformations (the transform of an intersection is the intersection of the transforms). This uses more than the outer product alone; it also requires the contraction, or dualization.

The resulting consistent *subspace algebra* is good to understand first. Its subspace products are the algebraic extensions of familiar techniques in standard linear algebra. Seeing them in this more general framework will improve the way you program in linear algebra, even if you do not make explicit use of the extended data structures that the subspaces provide. Therefore we begin our journey with the treatment of this subspace algebra, in Chapters 2 to 5.

1.2.4 UNIVERSAL ORTHOGONAL TRANSFORMATIONS

In the vector space model and the conformal model, *orthogonal transformations* are used to represent basic motions of Euclidean geometry. This makes that type of linear transformation fundamental to doing geometry in those models. (General linear transformations are still useful to represent deformations of elements, on top of the basic motions of the geometry, but they are not as crucial).

Geometric algebra has a special way to represent orthogonal transformations, more powerful than using orthogonal matrices. These are *versors*, and the example in Section 1.1 showed two instances of them: a rotor and a reflector. A versor V transforms any element X of the geometric algebra according to the versor product:

$$X \mapsto (-1)^{xv}\, V X/V,$$

where the sign factor depends on the dimensionality of X and V, and need not concern us in this introduction. This operator product transcends matrices in that it can act directly on arbitrary elements: vectors, subspaces, and operators.

The product involved in the sandwiching of the versor product is the geometric product; as a consequence, subsequent operators multiply by the geometric product. For instance, $R_2\, (R_1\, X/R_1)/R_2 = (R_2\, R_1)\, X/(R_2\, R_1)$. This product is linear, associative, and invertible, but not commutative. That matches its geometric interpretation: orthogonal transformations are linear, associative, and invertible, but their order matters.

The two-sidedness of the versor product of an operator may come as a bit of a surprise, but you probably have seen such two-sided products before in a geometrical context.

- When the vectors of a space transform by a motion represented by $[\![M]\!]$ (so that $[\![\mathbf{x}]\!]$ becomes $[\![M]\!]\,[\![\mathbf{x}]\!]$), a matrix $[\![A]\!]$ transforms to become $[\![M]\!]\,[\![A]\!]\,[\![M]\!]^{-1}$. Note that in linear algebra, vectors and operators transform differently, whereas in geometric algebra they transform in the same manner.

- Another classical occurrence of the two-sided product is the quaternion representation of 3-D rotations. Those are in fact rotors and, therefore, versors. In the classical representation, you need to employ three imaginary numbers to represent them properly. We will see in Chapter 7 how geometric algebra simply uses the real subspaces of a 3-D vector space to construct quaternions. Quaternions are not intrinsically imaginary! Moreover, when given this context, they become universal operators, capable of rotating geometric subspaces (rather than only being applicable to other quaternions).

The versor form of an orthogonal transformation automatically guarantees the preservation of algebraic structure (more technically known as covariance). Geometrically, this implies that *the construction of an object from moved components equals the movement of the object constructed from the original components*. Here, "construction" can be the connection of the outer product, the intersection of the `meet`, the complementation of the duality operation, or any other geometrically significant operation.

You have seen in the example how this simplifies constructions. In traditional linear algebra, one can only transform *vectors* properly, using the matrices. So to move any construction one has built, one has to move the vectors on which it was based and rebuild it from scratch. With geometric algebra, it is possible to move the construction itself: the lines, circles, and other components, and moreover *all* of these are moved by the same versor construction with the same versor representing the motion.

You need no longer to be concerned about the type of element you are moving; they all automatically transform correctly. That means you also do not need to invent and build special functions to move lines, planes, or normal vectors and can avoid defining a motion method for each data structure, because all are generic. In fact, those differing methods may have been one of the reasons that forced you to distinguish the types in the first place. Now even that is not necessary, because they all find their place algebraically rather than by explicit construction, so fewer data types are required. This in turn reduces the number of cases in your program flow, and therefore may ultimately simplify the program itself.

1.2.5 OBJECTS ARE OPERATORS

In geometric algebra, operators can be specified directly in terms of geometric elements intrinsic to the problem.

We saw in Section 1.1, item 6, how the dual plane π (i.e., an object) could be used immediately as the reflector (i.e., an operator) to produce the reflected line and circles. We also constructed the rotor representing the rotation around the line L by exponentiating the line in item 4.

Geometric algebra offers a range of constructions to make versors. It is particularly simple to make the versors representing basic motions as ratios (i.e., using the division of the geometric product): the ratio of two planes is a rotation versor, the ratio of two points is a translation versor, and the ratio of two lines in 3-D is the screw motion that turns and slides one into the other. These constructions are very intuitive and satisfyingly general.

As you know, it is much harder to define operators in such direct geometrical terms using linear algebra. We summarize the usual techniques:

- There are several methods to construct rotation operators. Particularly intricate are various kinds of standardized systems of orientating frames by subsequent rotations around Euler angles, a source of errors due to the arbitrariness of the coordinate frames. One can construct a rotation matrix from the rotation axis directly (by Rodrigues' formula), and this is especially simple for a quaternion (which is already an element of geometric algebra). Unfortunately, even those are merely rotations at the origin. There is no simple formula like the $\exp(\phi L^*/2)$ of geometric algebra to convert a general axis L into a rotation operator.

- Translations are defined by the difference of vectors, which is simple enough, but note that it is a different procedure from defining rotations.

- A general rigid body motion converting one frame into another can be artificially split into its rotational aspects and translational aspects to produce the matrix. Unfortunately, the resulting motion matrix is hard to interpolate. More rewarding is a screw representation, but this requires specialized data structures and Chasles' theorem to compute.

The point is that these linear algebra constructions are specific for each case, and apparently tricky enough that the inventors are often remembered by name. By contrast, the geometric algebra definition of a motion operator as a ratio is easily reinvented by any application programmer.

1.2.6 CLOSED-FORM INTERPOLATION AND PERTURBATION

In many applications, one would like to apply a motion gradually or deform it continuously (for instance, to provide smooth camera motion between specified views). In geometric algebra, interpolation of motions is simple: one just applies the corresponding versor V piecemeal, in N steps of $V^{1/N}$. That N^{th} root of a motion versor V can be determined by a logarithm, in closed form, as $\exp(\log(V)/N)$. For a rotor representing a rotation at the origin, this retrieves the famous "slerp" interpolation formula of quaternions, but it extends beyond that to general Euclidean motions. Blending of motions can be done by blending their logarithms.

By contrast, it is notoriously difficult to interpolate matrices. The logarithm of a matrix can be defined but it is not elementary, and not in closed form. A straightforward way to compute it is to take the eigenvalue decomposition of the rigid body motion matrix in the homogeneous coordinate framework, and take the Nth root of the diagonal matrix. Such numerical techniques makes the matrix logarithm expensive to compute and hard to analyze.

Perturbations of motions are particularly easy to perform in geometric algebra: the small change in the versor-based motion VX/V to any element X can be simply computed as $X \times B$, the commutator product of X with the bivector logarithm of the perturbing versor. This is part of *geometric calculus*, an integrated method of taking derivatives of geometric elements relative to other geometric elements. It naturally gets into differential geometry, a natural constituent of any complete framework that deals with the geometry of physical space.

1.3 PROGRAMMING GEOMETRY

The structural possibilities of the algebra may theoretically be rich and inviting, but that does not necessarily mean that you would want to use it in practical applications. Yet we think you might.

1.3.1 YOU CAN ONLY GAIN

Geometric algebra is backwards-compatible with the methods you already use in your geometrical applications.

Using geometric algebra does not obviate any of the old techniques. Matrices, cross products, Plücker coordinates, complex numbers, and quaternions in their classical form are all included in geometric algebra, and it is simple to revert to them. We will indicate these connections at the appropriate places in the book, and in some applications we actually revert to classical linear algebra when we find that it is more efficient or that it provides numerical tools that have not yet been developed for geometric algebra. Yet seeing all these classical techniques in the context of the full algebra enriches them, and emphasizes their specific geometric nature.

The geometric algebra framework also exposes their cross-connections and provides universal operators, which can save time and code. For example, if you need to rotate a line, and you have a quaternion, you now have a choice: you can convert the quaternion to a rotation matrix and apply that to the positional and directional aspects of the line separately (the classical method), or you view the quaternion as a rotor and apply it immediately to the line representation (the geometric algebra method).

1.3.2 SOFTWARE IMPLEMENTATION

We have made several remarks on the simpler software structure that geometric algebra enables: universal operators, therefore fewer data types, no conversions between formalisms, and consequently a simpler data flow.

Having said that, there are some genuine concerns related to the size of geometric algebra. If you use the conformal model to calculate with the Euclidean geometry of 3-D space, you use a 5-D vector space and its $2^5 = 32$ subspaces. In effect, that requires a basis of 32 elements to represent an arbitrary element. Combining two elements could mean 32×32 real multiplies per geometric product, which seems prohibitive.

This is where the actual structure of geometric algebra comes to the rescue. We will explain these issues in detail in Part III, but we can reassure you now: geometric algebra can compete with classical approaches if one uses its algebraic structure to guide the implementation.

- Elements of geometric algebra are formed as products of other elements. This implies that one cannot make an arbitrary element of the 32-dimensional framework. Objects typically have a single dimensionality (which is three for circles and lines) or a special structure (all flats contain the point at infinity ∞). This makes the structure of geometrically significant elements rather sparse. A good software implementation can use this to reduce both storage and computation.

- On the other hand, the 32 slots of the algebra are all used somehow, because they are geometrically relevant. In a classical program, you might make a circle in 3-D and would then have to think of a way to store its seven parameters in a data structure. In geometric algebra, it automatically occupies some of the $\binom{5}{3} = 10$ slots of 3-vector elements in the 5-D model. As long as you only allocate the elements you need, you are not using more space than usual; you are just using the pre-existing structure to keep track of them.

- Using linear algebra, as you operate on the composite elements, you would have to invent and write methods (for instance, to intersect a circle and a plane). This would require special operations that you yourself would need to optimize for good performance. By contrast, in geometric algebra everything reduces to a few basic products, and their implementation can be optimized in advance. Moreover, these are so well-structured that this optimization can be automated.

- Since all is present in a single computational framework, there is no need for conversions between mathematically different elements (such as quaternions and rotation matrices). Though at the lower level such conversions may be done for reasons of efficiency, the applications programmer works within a unified system of geometrically significant elements of computation.

Using insights and techniques like this, we have implemented the conformal model and have used it in a typical ray-tracing application with a speed 25 percent slower than the optimized classical implementation (which makes it about as costly as the commonly used homogeneous coordinates and quaternion methods), and we believe that this overhead may be reduced to about 5 to 10 percent. Whether this is an acceptable price to pay for a much simpler high-level code is for you to decide.

We believe that geometric algebra will be competitive with classical methods when we also adapt algorithms to its new capabilities. For instance, to do a high-resolution rendering, you now have an alternative to using a much more dense triangulation (requiring many more computations). You could use the new and simple description of perturbations to differentially adapt rays of a coarse resolution to render an ostensibly smoother surface. Such computation-saving techniques would easily compensate for the slight loss of speed per calculation.

Such algorithms need to be developed if geometric algebra is to make it in the real world. We have written this book to raise a new generation of practitioners with sufficient fundamental, intuitive, and practical understanding of geometric algebra to help us develop these new techniques in spatial programming.

1.4 THE STRUCTURE OF THIS BOOK

We have chosen to write this book as a gradual development of the algebraic terms in tandem with geometric intuition. We describe the geometric concepts with increasing precision, and simultaneously develop the computational tools, culminating in the conformal model for Euclidean geometry. We do so in a style that is not more mathematical than we deem necessary, hopefully without sacrificing exactness of meaning. We believe this approach is more accessible than axiomatizing geometric algebra first, and then having to discover its significance afterwards.

The book consists of three parts that should be read in order (though sometimes a specialized chapter could be skipped without missing too much).

1.4.1 PART I: GEOMETRIC ALGEBRA

First, we get you accustomed to the outer product that spans subspaces (and to the desirability of the "algebraification of geometry"), then to a metric product that extends the usual dot product to these subspaces. These relatively straightforward extensions from linear algebra to a multilinear algebra (or subspace algebra) already allow you to extend linear mappings and to construct general intersection products for subspaces. Those capabilities will extend your linear algebra tool kit considerably.

Then we make the transition to true geometric algebra with the introduction of the geometric product, which incorporates all that went before and contains more beyond that. Here the disadvantage of the approach in this book is momentarily annoying, since we have to show that the new definitions of the terms from the earlier chapters are "backwards compatible." But once that has been established, we can rapidly move on from considering objects (the subspaces) to operators acting on them. We then easily absorb tools you may not have expected to encounter in real vector spaces, such as complex numbers and quaternions. Both are available in an integrated manner, real in all normal senses of the word, and geometrically easily understood.

Part I concludes with a chapter on geometric differentiation, to show that differential geometry is a natural aspect of geometric algebra (even though we will use it only incidentally in this book).

1.4.2 PART II: MODELS OF GEOMETRY

In Part II the new algebra will be used as a tool to model aspects of mostly Euclidean geometry. First, we treat directions in space, using the *vector space model*, already familiar from the visualizations used in Part I to motivate the algebra.

Next, we extend the vector embedding trick of homogeneous coordinates from practical computer science to the complete *homogeneous model* of geometric algebra, which includes Plücker coordinates and other useful methods.

Finally, in Chapter 13 we can begin to treat the *conformal model* of the motivating example in Section 1.1. The conformal model is the model that has Euclidean geometry as an intrinsic part of its structure; all Euclidean motions are represented as orthogonal transformations. We devote four chapters to its definition, constructions, operators, and its ability to describe general conformal transformations.

1.4.3 PART III: IMPLEMENTATION OF GEOMETRIC ALGEBRA

To use geometric algebra, you will need an implementation. Some are available, or you may decide to write your own. Naïve implementations run slow, because of the size of the algebra (32-D for the basis of the conformal model of a 3-D Euclidean space).

In the third part of this book, we give a computer scientist's view of the algebraic structure and describe aspects that are relevant to any efficient implementation, using its multiplicative and sparse nature. We end with a simple ray tracer to enable comparison of computational speeds of the various methods in a computer graphics application.

1.5 THE STRUCTURE OF THE CHAPTERS

Each regular chapter consists of an explanation of the structure of its subject. We explain this by developing the algebra with the geometry, and provide examples and illustrations. Most of the figures in this book have been rendered using our own software package, GAViewer. This package and the code for the figures are available on our web site,

http://www.geometricalgebra.net.

We recommend that you download the software, install it, and follow the instructions to upload the figures. You can then interact with them. At the very least you should be able to use your mouse for 3-D rotation and translation to get the proper spatial impression of the scene. But most figures also allow you to interactively modify the independent elements of the scene and study how the dependent elements then change.

You can even study the way we have constructed them[1] and change them on a command line; though if you plan to do that we suggest that you first complete the GAViewer tutorial on the web site. This will also allow you to type in formulas of numerical examples.

If you really plan to use geometric algebra for programming, we recommend doing the drills and programming exercises with each chapter. The programming exercises use a special library, the *GA sandbox*, also available from the web site. It provides the basic data structures and embedding of the products so that you can program directly in geometric algebra. This should most closely correspond to how you are likely to use it as an extension of your present programming environment.

We also provide structural exercises that help you think about the coherence of the geometric algebra subject in each chapter and ask you to extend some of the material or provide simple proofs. For answers to these exercises, consult the web site.

Historically, geometric algebra has many precursors, and we will naturally touch upon those as we develop our concepts and terms. We do not meticulously attribute all results and thoughts, but occasionally provide a bit of historic perspective and contrast with traditional approaches. At the end of each chapter, we will give some recommended literature for further study.

---

1   But be warned that some illustrative figures of the simpler models may use elements of the conformal model in their code, since that is the most natural language to specify general elements of Euclidean geometry.

PART I
GEOMETRIC ALGEBRA

2 SPANNING ORIENTED SUBSPACES

After many attempts at formalizing space and spatial relationships, the concept of a *vector space* emerged as the useful framework for geometrical computations. We use it as our point of departure, and use some of the standard linear algebra governing its mappings. Yet already we will have much to add to its usual structure. By the end of this chapter you will realize that a vector space is much more than merely a space of vectors, and that it is straightforward and useful to extend it computationally.

The crucial idea is to make the subspaces of a vector space explicit elements of computation. To build our algebra of subspaces, we revisit the familiar lines and planes through the origin. We investigate their geometrical properties carefully, and formalize those by the aid of a new algebraic *outer product*, which algebraically builds subspaces from vectors.

We consider the structure it gives us for the Grassmann space of subspaces of a vector space $\mathbb{R}^n$, and define many terms to describe its features.

Throughout this chapter, we consider a real n-dimensional vector space $\mathbb{R}^n$, but have no need for a metric; and we only treat its *homogeneous subspaces* (i.e., subspaces containing the origin).

2.1 VECTOR SPACES

We start with an n-dimensional vector space. However, the definition of a vector space in linear algebra is more general than what we need in this book, being defined over arbitrary fields of scalars. Since we are interested in computing with spatial elements, we will immediately narrow our focus and consider only n-dimensional vector spaces over the real numbers $\mathbb{R}$. Such a vector space $\mathbb{R}^n$ consists, by definition, of elements called *vectors* and an addition and multiplication by reals (called *scalars*), such that

$$
\begin{aligned}
&(1) & \mathbf{x} + \mathbf{y} \; &\in \; \mathbb{R}^n & &\forall\, \mathbf{x}, \mathbf{y} \in \mathbb{R}^n \\
&(2) & (\mathbf{x} + \mathbf{y}) + \mathbf{z} \; &= \; \mathbf{x} + (\mathbf{y} + \mathbf{z}) & &\forall\, \mathbf{x}, \mathbf{y}, \mathbf{z} \in \mathbb{R}^n \\
&(3) & \exists\, \mathbf{0} \in \mathbb{R}^n \; : \; \mathbf{x} + \mathbf{0} \; &= \; \mathbf{x} & &\forall\, \mathbf{x} \in \mathbb{R}^n \\
&(4) & \exists\, \mathbf{y} \in \mathbb{R}^n \; : \; \mathbf{x} + \mathbf{y} \; &= \; \mathbf{0} & &\forall\, \mathbf{x} \in \mathbb{R}^n \\
&(5) & \mathbf{x} + \mathbf{y} \; &= \; \mathbf{y} + \mathbf{x} & &\forall\, \mathbf{x}, \mathbf{y} \in \mathbb{R}^n \\
&(6) & \alpha\,(\mathbf{x} + \mathbf{y}) \; &= \; \alpha\,\mathbf{x} + \alpha\,\mathbf{y} & &\forall\, \alpha \in \mathbb{R}, \mathbf{x}, \mathbf{y} \in \mathbb{R}^n \\
&(7) & (\alpha + \beta)\,\mathbf{x} \; &= \; \alpha\,\mathbf{x} + \beta\,\mathbf{x} & &\forall\, \alpha, \beta \in \mathbb{R}, \mathbf{x} \in \mathbb{R}^n \\
&(8) & (\alpha\,\beta)\,\mathbf{x} \; &= \; \alpha\,(\beta\,\mathbf{x}) & &\forall\, \alpha, \beta \in \mathbb{R}, \mathbf{x} \in \mathbb{R}^n \\
&(9) & 1\,\mathbf{x} \; &= \; \mathbf{x} & &\forall\, \mathbf{x} \in \mathbb{R}^n
\end{aligned}
$$

Properties (1-4) make the vector space into a group, property (5) even into a commutative group, and properties (6-9) define how scalar multiplication works on its elements. All this shows that scalars are considered separate from vectors, and that no other elements are part of the usual definition of a vector space.

Such a vector space can contain subspaces that are also vector spaces. The dimensionality of a (sub)space is the maximum number of independent vectors in it (i.e., vectors that cannot be expressed as a scalar-weighted sum of other vectors). These subspaces must obviously contain the element $\mathbf{0}$, and are sometimes called *homogeneous subspaces* (or proper subspaces).

In standard linear algebra, specific homogeneous subspaces are defined implicitly, typically by sets of vector equations, or explicitly by parameterized expressions. That is workable, but it is really too cumbersome for such important algebraic features of any vector space. We will first turn subspaces into direct elements of computation, following the pioneering work by Grassmann (in 1844) that unfortunately did not make it into the mainstream linear algebra texts.

To do so, we revisit the familiar homogeneous subspaces of a vector space in their geometrical interpretation. We list their properties, defining terms that unify those across dimensions. This uncovers an algebraic product that can span them, thus making those subspaces and their properties elements of computation.

The concept of a subspace is independent of any *metric* properties a vector space might have. In this chapter, we therefore avoid using a metric and the inner product that defines it. This also implies that we cannot use orthonormal bases in our examples, which may make them look a bit less specific than they could be. Of course, the concepts still work when you do have a metric, and some of the exercises bring this out. This is the only chapter in which we avoid using a metric, since that is the cleanest way to show that the results actually hold for a vector space with any kind of metric.

2.2 ORIENTED LINE ELEMENTS

To develop our thinking about subspaces, we consider the homogeneous subspaces of a 3-D space, sketched in Figure 2.1. We skip over the 0-D subspace of Figure 2.1(a) for the moment, and start with lines through the origin.

2.2.1 PROPERTIES OF HOMOGENEOUS LINES

A line through the origin is a 1-D homogeneous subspace of the vector space $\mathbb{R}^n$. It is characterized by any nonzero vector $\mathbf{a}$, in the sense that any vector $\mathbf{x}$ denoting a point on the line is a multiple of $\mathbf{a}$. This gives a correspondence between the geometry and the algebra of a line:

$$\mathbf{x} \text{ on line determined by } \mathbf{a} \quad \Leftrightarrow \quad \mathbf{x} = \lambda\mathbf{a}, \text{ for some } \lambda \in \mathbb{R}, \tag{2.1}$$

see Figure 2.1(b). If you think of a line as a set of points, then any scalar multiple of $\mathbf{a}$ determines the same line. For many applications, this is enough, but in others you need more detailed properties, such as its heading (opposite for $\mathbf{a}$ and $-\mathbf{a}$) and its speed (twice as much for $2\mathbf{a}$ as for $\mathbf{a}$). The characterization of a line by a vector allows it to have those extra properties. In preparation for more general subspaces, we attempt to find descriptive

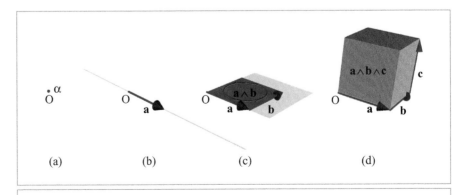

(a)   (b)   (c)   (d)

Figure 2.1: Spanning homogeneous subspaces in a 3-D vector space.

terms for those features that can transcend vectors. At first, these terms will seem merely intuitive, but we can give them more exact definitions when we are done, in Table 2.1.

- A line has an *attitude* (or stance) in the surrounding space; it is characterized by $\mu\mathbf{a}$, for any nonzero μ. We will use the term attitude purely in the sense of the subspace occupied by the line; lines characterized by $\mathbf{a}$ and $-\mathbf{a}$ have the same attitude. They both determine the same linear line-like carrier stretching to infinity in both directions.

- We can give the line an *orientation*; this means that we care about the sign of λ in (2.1). Then $\mathbf{a}$ and $-\mathbf{a}$ represent lines of different orientation (but $2\mathbf{a}$ has the same orientation as $\mathbf{a}$).[1] We will reserve the term *direction* for a combination of attitude and orientation; so the set of all vectors $\lambda\mathbf{a}$ with the same $\mathbf{a}$ and the same sign of λ represent lines with the same direction.

- We also care about a distance measure along the line, which is quantified in the magnitude of $\mathbf{a}$. Here we should be careful in choosing our term, since "magnitude" suggests a metric measure by which we can compare different lines. For now, we cannot, since we are still working in a nonmetric vector space. We choose to use the term *weight* (because the term speed does not transfer well to higher-dimensional subspaces). A line with twice the weight could be said to pass through its points twice as fast for the same change in λ. We will allow the weight to be negative, in which case the line is oriented oppositely to a standard direction for that 1-D subspace.

Those three properties of an oriented line through the origin are all part of its specification by a vector.

2.2.2 VISUALIZING VECTORS

As is customary, we visualize vectors as *arrows*. The straightness of the arrow indicates its nature of representing a 1-D subspace; the length is its weight, the attitude is indicated by the shaft, and the orientation by the ordering from tail to arrowhead.

The addition of arrows can be represented by the familiar parallelogram construction: place the two arrows to be added with their tails together, complete the parallelogram to obtain a fourth location, and the result is an arrow from the tails to the opposite point, as in Figure 2.2.

This and all similar figures have been generated using our software GAViewer, and you can download these figures to view and change them interactively (see Section 1.5).

The visualizations serve an important purpose in this book, and the interactive software is essential to get a good feeling for geometric algebra. Vector addition is a good example of the principle. If you would work on paper, in coordinates, you might think that vector addition in 3-D is the addition of three scalars. It is implemented in that way, of course,

---

1   This use of the term oriented line is common in oriented projective geometry (see [60]).

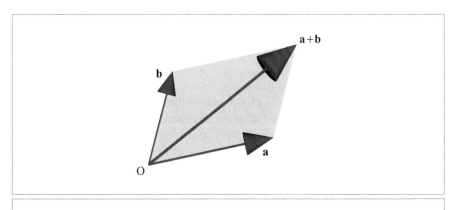

Figure 2.2: Imagining vector addition.

to generate the figures. But you should think and feel that vector addition is geometric: it completes the parallelogram of **a** and **b**, and algebraically there is no need to go lower than the notation **a** + **b**, which has all the properties you need. After you have loaded Figure 2.2 in GAViewer, you can drag **a** and **b**, and see the resulting change in **a** + **b**, in 3-D. That conveys precisely this coordinate-free feeling. Playing with this and the other figures will aid your intuition. It will help you dare to leave coordinates behind, and think about the algebra at the proper level of geometric primitives.

2.3 ORIENTED AREA ELEMENTS

2.3.1 PROPERTIES OF PLANES

A plane through the origin is a 2-D homogeneous subspace of the vector space $\mathbb{R}^n$. It may be determined by two linearly independent vectors, **a** and **b**, in the sense that any vector **x** in the plane can be written as

$$\mathbf{x} = \lambda\mathbf{a} + \mu\mathbf{b}.$$

This specification is not unique. For instance, replacing **a** and **b** by $(\mathbf{a} + \mathbf{b})/\sqrt{2}$ and $(\mathbf{a} - \mathbf{b})/\sqrt{2}$ gives the same set of vectors, as do many other linear combinations on **a** and **b**. We should think about replacing the specification with something more appropriate, which will make it easier to verify whether two specifications refer to the same plane.

Moreover, just like the homogeneous line, a homogeneous plane has more properties than just being a set of vectors. We list those properties, illustrated in Figure 2.1(c):

- A homogeneous plane has an *attitude* in the surrounding *n*-dimensional space $\mathbb{R}^n$. This is its "subspace aspect" of being a particular planar carrier. In 3-D, this is the property traditionally characterized by a normal vector for the plane, with any weight or orientation. But such a characterization by a vector is insufficient for a planar subspace of an n-dimensional space. (Another defect is that it characterizes the nonmetric concept of a particular 2-D subspace by a metric construction involving perpendicularity.)

- A plane may be considered to have an *orientation*, in the sense that the plane determined by two vectors **a** and **b** would have the opposite orientation of a plane determined by the vectors **b** and **a**. We use this often when we specify angles, speaking of the angle from **a** to **b** as being of opposite sign to the angle from **b** to **a**. The sign of the angle should be referred to more properly as relative to the orientation of the plane in which its defining vectors reside.

- A plane has a *measure of area*, which we shall call its *weight*. The plane determined by the vectors 2**a** and **b** has twice the weight (or double the area measure) of the plane determined by the vectors **a** and **b**. As with vectors, this weight is for now only a relative measure within planes of the same attitude. (We would need a metric to compare areas within different planes.)

In linear algebra, the orientation and the area measure are both well represented by the determinant of a matrix made of the two spanning vectors **a** and **b** of the plane: the orientation is its sign, the area measure its weight (both relative to orientation and area measure of the basis used to specify the coordinates of **a** and **b**). In 2-D, this specifies an *area element* of the plane. In 3-D, such an area element would be incomplete without a specification of the attitude of the plane in which it resides. Of course we would prefer to have a single algebraic element that contains all this geometric information about the plane.

2.3.2 INTRODUCING THE OUTER PRODUCT

We now introduce a product between vectors to aid in the specification of the plane containing the two vectors **a** and **b**. Its definition should allow us to retrieve all geometrical properties of the plane. We denote this algebraic product by $\mathbf{a} \wedge \mathbf{b}$.

The algebraic consequence of our geometrical desire to give the plane an orientation is that $\mathbf{a} \wedge \mathbf{b}$ should be opposite in sign to $\mathbf{b} \wedge \mathbf{a}$, so that $\mathbf{a} \wedge \mathbf{b} = -\mathbf{b} \wedge \mathbf{a}$. When **b** coincides with **a**, this would give the somewhat unusual algebraic result $\mathbf{a} \wedge \mathbf{a} = -\mathbf{a} \wedge \mathbf{a}$. This suggests that the square of **a**, using this product, must be zero. Geometrically, this is very reasonable: the vector **a** does not span a planar element with itself, and we may encode that as a planar element with weight zero.[2]

When we decrease the angle between **a** and **b**, the area spanned by **a** and **b** gets smaller as they become more parallel. In fact, in a space with a Euclidean metric you would expect the measure of area associated with the planar element $\mathbf{a} \wedge \mathbf{b}$ to be $\|\mathbf{a}\| \|\mathbf{b}\| \sin \phi$, with ϕ the angle between them. However, we should not make such an explicit property part of the definition of $\mathbf{a} \wedge \mathbf{b}$—it involves just too many extraneous concepts like norm and

---

2   We use "span" here informally, and different from the use in some linear algebra texts, where the span of two identical vectors would still be their common 1-D subspace rather than zero. That span is not very well-behaved; it is not even linear. In Chapter 5, its geometry will be encoded by the `join` product.

angle, which are moreover intrinsically metric. Instead, we try to define the product so that this metric area formula is a consequence of more basic axioms. We note that the area measure increases linearly in the magnitude of each of the vector factors of the product. So let us at least make the product bilinear, giving it proper scaling and distributivity relative to the constituent vectors.

Bilinearity and antisymmetry are already enough to define the product $\mathbf{a} \wedge \mathbf{b}$ completely. We call the result the *outer product* of $\mathbf{a}$ and $\mathbf{b}$. In view of the above, its defining properties are:

$$\text{Antisymmetry:} \quad \mathbf{a} \wedge \mathbf{b} = -\mathbf{b} \wedge \mathbf{a}$$
$$\text{Scaling:} \quad \mathbf{a} \wedge (\beta \, \mathbf{b}) = \beta \, (\mathbf{a} \wedge \mathbf{b})$$
$$\text{Distributivity:} \quad \mathbf{a} \wedge (\mathbf{b} + \mathbf{c}) = (\mathbf{a} \wedge \mathbf{b}) + (\mathbf{a} \wedge \mathbf{c})$$

We pronounce $\mathbf{a} \wedge \mathbf{b}$ as "a wedge b." The outcome of the outer product $\mathbf{a} \wedge \mathbf{b}$ of two vectors is called a *bivector* or, more properly, *2-blade* (we explain the difference between these terms in Section 2.9.3). It is an element of the algebra we are developing that is different from the scalars and vectors we have seen so far. Since the outer product is linear, its outcomes are elements of a linear space, the "bivector space." If we denote that space by $\bigwedge^2 \mathbb{R}^n$, the outer product is a mapping $\wedge : \mathbb{R}^n \times \mathbb{R}^n \to \bigwedge^2 \mathbb{R}^n$.

You may think of $\mathbf{a} \wedge \mathbf{b}$ as the span of $\mathbf{a}$ and $\mathbf{b}$, in a quantitative manner, or as an oriented area element of a particular homogeneous plane. Let us verify that that geometric interpretation is indeed consistent with its algebraic properties. If we take a basis $\{\mathbf{e}_1, \mathbf{e}_2\}$ in the subspace with the same attitude as $\mathbf{a} \wedge \mathbf{b}$, we can write $\mathbf{a} = a_1 \, \mathbf{e}_1 + a_2 \, \mathbf{e}_2$ and $\mathbf{b} = b_1 \, \mathbf{e}_1 + b_2 \, \mathbf{e}_2$. Then, using the definition, we develop the outer product to a recognizable form:

$$\begin{aligned}
\mathbf{a} \wedge \mathbf{b} &= (a_1 \, \mathbf{e}_1 + a_2 \, \mathbf{e}_2) \wedge (b_1 \, \mathbf{e}_1 + b_2 \, \mathbf{e}_2) \\
&= a_1 \, b_1 \, \mathbf{e}_1 \wedge \mathbf{e}_1 + a_1 \, b_2 \, \mathbf{e}_1 \wedge \mathbf{e}_2 + a_2 \, b_1 \, \mathbf{e}_2 \wedge \mathbf{e}_1 + a_2 \, b_2 \, \mathbf{e}_2 \wedge \mathbf{e}_2 \\
&= (a_1 \, b_2 - a_2 \, b_1) \, \mathbf{e}_1 \wedge \mathbf{e}_2
\end{aligned} \tag{2.2}$$

since the antisymmetry implies that $\mathbf{e}_1 \wedge \mathbf{e}_1 = -\mathbf{e}_1 \wedge \mathbf{e}_1$, so it must be equal to 0—as is any outer product of parallel vectors.

We can write the final result in terms of a determinant by introducing the matrix $[\![\mathbf{a} \ \mathbf{b}]\!]$ with, as its columns, the coefficients of $\mathbf{a}$ and $\mathbf{b}$ on the $\{\mathbf{e}_1, \mathbf{e}_2\}$-basis. This yields

$$\mathbf{a} \wedge \mathbf{b} = \det([\![\mathbf{a} \ \mathbf{b}]\!]) \, \mathbf{e}_1 \wedge \mathbf{e}_2.$$

This determinant you know from linear algebra as a relative measure for the oriented area spanned by $\mathbf{a}$ and $\mathbf{b}$ relative to the area spanned by the basis vectors $\mathbf{e}_1$ and $\mathbf{e}_2$. Its value is what we called the relative weight of a plane element. The other part of the result, $\mathbf{e}_1 \wedge \mathbf{e}_2$, can then be consistently interpreted as the geometrical unit in which area in this plane is measured (i.e., the amount of standard area with the correct attitude, in the plane spanned by the basis vectors $\mathbf{e}_1$ and $\mathbf{e}_2$). The orientation of the plane, and hence the relative orientation of $\mathbf{a} \wedge \mathbf{b}$, is specified by the order of $\mathbf{e}_1$ and $\mathbf{e}_2$.

The outcome of $\mathbf{a} \wedge \mathbf{b}$ neatly contains all three geometric properties in one algebraic element of computation, as we had hoped. If you want to see what happens when you have a Euclidean metric in the plane, do structural exercise 1.

In summary, we have the following algebraic representation of the geometry of homogeneous planes:

> $\mathbf{a} \wedge \mathbf{b}$ is a weighted, oriented area element of the 2-D subspace spanned by $\mathbf{a}$ and $\mathbf{b}$ (or other vectors producing the same attitude)

(see Figure 2.1(c), which also denotes a piece of the infinite homogeneous plane of which $\mathbf{a} \wedge \mathbf{b}$ is a part). This element $\mathbf{a} \wedge \mathbf{b}$ may become zero when $\mathbf{a}$ and $\mathbf{b}$ are parallel, or when either vector has zero norm, in agreement with our geometrical intuition that no planar element is spanned in those cases.

When we have multiple area elements in different planes in space, we cannot choose independent bases for each of them. Yet it is still possible to decompose $\mathbf{a} \wedge \mathbf{b}$ on a bivector basis for the whole space. We demonstrate this in a 3-D space $\mathbb{R}^3$ with a totally arbitrary basis $\{\mathbf{e}_1, \mathbf{e}_2, \mathbf{e}_3\}$ (not necessarily orthonormal). Let the coefficients of $\mathbf{a}$ and $\mathbf{b}$ on this basis be a_i and b_i, respectively. Then we compute, using the definition of the outer product:

$$
\begin{aligned}
\mathbf{a} \wedge \mathbf{b} &= (a_1\mathbf{e}_1 + a_2\mathbf{e}_2 + a_3\mathbf{e}_3) \wedge (b_1\mathbf{e}_1 + b_2\mathbf{e}_2 + b_3\mathbf{e}_3) \\
&= a_1 b_1(\mathbf{e}_1 \wedge \mathbf{e}_1) + a_1 b_2(\mathbf{e}_1 \wedge \mathbf{e}_2) + a_1 b_3(\mathbf{e}_1 \wedge \mathbf{e}_3) + \\
&\quad\, a_2 b_1(\mathbf{e}_2 \wedge \mathbf{e}_1) + a_2 b_2(\mathbf{e}_2 \wedge \mathbf{e}_2) + a_2 b_3(\mathbf{e}_2 \wedge \mathbf{e}_3) + \\
&\quad\, a_3 b_1(\mathbf{e}_3 \wedge \mathbf{e}_1) + a_3 b_2(\mathbf{e}_3 \wedge \mathbf{e}_2) + a_3 b_3(\mathbf{e}_3 \wedge \mathbf{e}_3) \\
&= (a_1 b_2 - a_2 b_1)\,\mathbf{e}_1 \wedge \mathbf{e}_2 + (a_2 b_3 - a_3 b_2)\,\mathbf{e}_2 \wedge \mathbf{e}_3 + (a_3 b_1 - a_1 b_3)\,\mathbf{e}_3 \wedge \mathbf{e}_1
\end{aligned}
\tag{2.3}
$$

This cannot be simplified further. We see that an outer product of two vectors in 3-D space can be written as a scalar-weighted sum of three standard elements $\mathbf{e}_1 \wedge \mathbf{e}_2$, $\mathbf{e}_2 \wedge \mathbf{e}_3$, $\mathbf{e}_3 \wedge \mathbf{e}_1$. Their weighting coefficients are obviously 2-D determinants, which we know represent directed area measures, now of the components of the original plane on the coordinate planes of the basis. The formula is then consistent with the interpretation of these three elements as *standard area elements* for the coordinate planes of the basis vectors.

It is a pleasant surprise that area elements in 3-D have such a decomposable structure as a weighted sum over a basis. Mathematically, this means that they reside in their own bivector space of $\bigwedge^2 \mathbb{R}^3$, of three dimensions, with basis elements $\mathbf{e}_{12} \equiv \mathbf{e}_1 \wedge \mathbf{e}_2$, $\mathbf{e}_{23} \equiv \mathbf{e}_2 \wedge \mathbf{e}_3$, $\mathbf{e}_{31} \equiv \mathbf{e}_3 \wedge \mathbf{e}_1$.[3]

---

3   This bivector space satisfies the mathematical axioms of a vector space, but it would be geometrically confusing to call its elements vectors. We will reserve the term vector exclusively for the elements of $\mathbb{R}^n$.

Any area element of the form $\mathbf{a} \wedge \mathbf{b}$ can be decomposed onto a basis of standard area elements.

In 3-D space, the converse also holds: a weighted sum of basis 2-blades is an area element that can be factorized by the outer product in the form $\mathbf{a} \wedge \mathbf{b}$. But that property does not hold in n-dimensional space; you cannot in general make factorizable 2-blades of the form $\mathbf{a} \wedge \mathbf{b}$ simply by adding basis bivectors with arbitrary weights. We get back to this important issue in Section 2.9.

With the outer product, we can generate additional structure from our initial 3-D vector space $\mathbb{R}^3$ and its real scalars $\mathbb{R}$. You already know from linear algebra that determinants of 2×2 matrices with 2-D vectors as columns are scalar areas, and that determinants of 3×3 matrices of 3-D vectors are scalar volumes; now we also have 2-D-oriented areas spanned by 3-D vectors as a 3-D linear space of 2-D determinants. Subspaces and their measures are beginning to fall into a new pattern.

2.3.3 VISUALIZING BIVECTORS

The algebraic properties of a bivector in 3-D are equivalent to those of a directed area element. It is good to get a mental picture of such area elements, and of their interaction using addition, that stays close to their algebraic properties. This will help us to "see" bivectors in problems and their solutions.

A first attempt might be Figure 2.3(a): $\mathbf{a}\wedge\mathbf{b}$ as the parallelogram spanned by $\mathbf{a}$ and $\mathbf{b}$. Since that does not convey the orientation (it looks commutative rather than anticommutative), we need to denote this orientation by something extra, such as by the oriented circular arc inside. We could also use a representation like Figure 2.3(b), in which the orientation follows naturally from the order of the arrows. We can extend this to show some more of the anticommutativity of the construction, which implies $\mathbf{a} \wedge \mathbf{b} = (-\mathbf{b} \wedge \mathbf{a}) = \mathbf{b} \wedge (-\mathbf{a}) = (-\mathbf{a})\wedge(-\mathbf{b})$ by showing all of those four vectors along the border, as in Figure 2.3(c). This shows that none of the vertices of the parallelogram can be seen as an anchor. Therefore the depiction of the bivector is free to move translationally within its carrier plane, though we will prefer drawing it near the origin. Also, since $\mathbf{a} \wedge \mathbf{a} = 0$, we have to remember that drawing the corresponding parallelogram is equivalent to drawing nothing at all, rather than some flattened shape.

Not only can we imagine sliding the bivector out of the origin, but we can also reshape it in some ways without changing its algebraic value. Anticommutativity of the outer product implies, for instance, that $(\mathbf{a} - \frac{1}{2}\mathbf{b}) \wedge \mathbf{b} = \mathbf{a} \wedge \mathbf{b} = (\mathbf{a} - \frac{1}{2}\mathbf{b}) \wedge (\frac{1}{2}\mathbf{a} + \frac{3}{4}\mathbf{b})$. Therefore, we are allowed to reshape the parallelogram geometrically by sliding the arrows along any of its parallel sides (Figure 2.3(d, e)), obtaining another faithful depiction of the same bivector $\mathbf{a} \wedge \mathbf{b}$. If you have a coordinate system, there may be an advantage to redrawing the bivector to have sides to be aligned with the coordinate vectors. But you should realize that there is no unique way of doing this; since $\mathbf{a}\wedge\mathbf{b} = (2\mathbf{a})\wedge(\mathbf{b}/2)$, and so on, the magnitudes of the components are adjustable (as long as area and orientation remain the same).

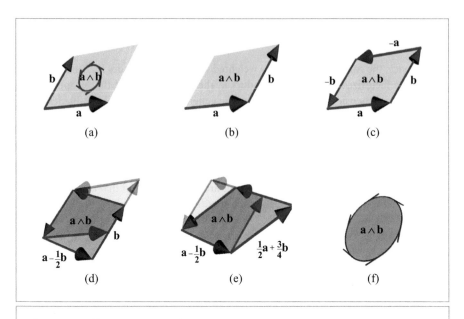

Figure 2.3: Bivector representations.

In fact, as soon as we have computed a bivector quantity, we have lost the identity of the vectors that generated it. We may therefore prefer to denote it by a circular area in the plane, as in Figure 2.3(f) (as long as we realize that even this circular shape is arbitrary).

We cannot emphasize enough that these redrawings all represent the *same* element. The algebraic bivector is not specific on shape; geometrically, it is an amount of oriented area in a specific plane, that's all. Initially, this may appear too vague to be useful; but we will soon see that this "reshapeability" is a strength, not a weakness.

2.3.4 VISUALIZING BIVECTOR ADDITION

We can also make a geometrical representation of *bivector addition*. In the same plane, addition of bivectors can be done by reshaping them until they can be added visually, preserving both magnitude and orientation of their area during the reshaping, (see Figure 2.4). To add $c \wedge d$ to $a \wedge b$ (assumed to be in the same plane!), first reshape $c \wedge d$ to be of the form $(\gamma c) \wedge (-b)$; see Figure 2.4(b). Place them side to side so that the $(-b)$ side of the first bivector matches the b side of the second. Then reshape the area again, so that $c \wedge d$ is finally of the form $\alpha a \wedge b$, as in Figure 2.4(c). Now the result is obviously $(1 + \alpha)a \wedge b$. But of course, you could change the order of the arguments and do not really need to do the reshaping at all. It is just a matter of putting oriented areas together and algebraically it is handled automatically by adding their coefficients relative to a bivector basis.

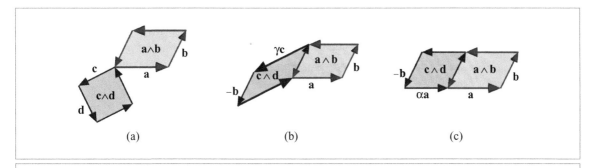

Figure 2.4: Imagining bivector addition in 2-D space.

Now consider the addition of two bivectors in $\mathbb{R}^3$. In 3-D space, two homogeneous planes intersect in at least a homogeneous line, so two bivectors must have some vector in common. Calling this vector $\mathbf{e}$, we can then reshape the two addends to have $\mathbf{e}$ as a factor. Then they are both aligned with $\mathbf{e}$, and the bivector addition reduces to vector addition. For example, we compute $(3\mathbf{e}_1 \wedge \mathbf{e}_2) + (\mathbf{e}_3 \wedge \mathbf{e}_2)$ (as illustrated in Figure 2.5(a)) by factoring out $\mathbf{e} = \mathbf{e}_2$ and adding $3\mathbf{e}_1$ and $\mathbf{e}_3$ (Figure 2.5(b)). You have to be a bit careful, since you may have to change the sign of one of the vectors before adding (Figures 2.5(c, d)) due to the antisymmetry of the outer product.

In higher-dimensional spaces, this geometric construction is likely to fail, since in general you will not be able to find a common 1-D direction for two planes. Algebraically this corresponds to the fact that not all bivectors can be factored as the outer product of two vectors; we call the ones that can 2-blades. The algebraic difference between bivectors and 2-blades will be treated in detail in Section 2.9.3.

2.4 ORIENTED VOLUME ELEMENTS

2.4.1 PROPERTIES OF VOLUMES

We now consider representations of oriented volumes. By analogy with homogeneous lines and planes, volumes can be treated as 3-D homogeneous subspaces of a vector space $\mathbb{R}^n$. They have the geometric properties we found for lines and planes, illustrated by analogy in Figure 2.1(d).

- Although it is somewhat hard to visualize for us 3-D beings, volumes have an *attitude* in spaces of more than three dimensions, denoting the 3-D subspace that contains them. In a 3-D space there is, of course, only one choice—all volumes are proportional to the volume of the unit cube.

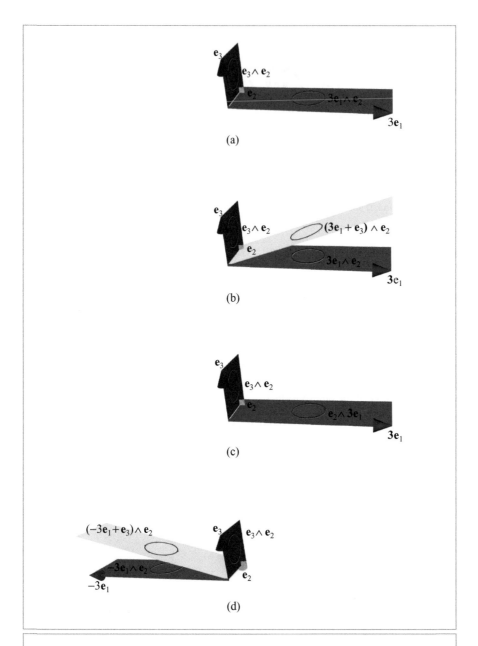

Figure 2.5: Bivector addition in 3-D space: orientation matters. (a),(b): $(3\mathbf{e}_1 \wedge \mathbf{e}_2)$ $+ (\mathbf{e}_3 \wedge \mathbf{e}_2) = (3\mathbf{e}_1 + \mathbf{e}_3) \wedge \mathbf{e}_2$; (c),(d): $(\mathbf{e}_2 \wedge 3\mathbf{e}_1) + (\mathbf{e}_3 \wedge \mathbf{e}_2) = (-3\mathbf{e}_1 + \mathbf{e}_3) \wedge \mathbf{e}_2$, which is a different bivector.

- The volume has an *orientation*, usually referred to as handedness. In the 3-D space of our example, the volume spanned by e_1, e_2, e_3 (in that order), has opposite orientation from that spanned by $e_1, e_2, -e_3$ (in that order); use a mirror in the $(e_1 \wedge e_2)$ plane to see this. The latter has the same orientation as the volume spanned by e_2, e_1, e_3 (in that order—use two more mirrors), and of any odd permutation of (e_1, e_2, e_3). Orientation of volumes is thus an antisymmetric property, and therefore there exist only two different orientations (a glove can be right-handed or left-handed—that's all).

- Volume has a scalar *weight*. It is well known from linear algebra that in 3-D space the signed magnitude of the volume spanned by a, b, and c is proportional to the determinant of the coefficient matrix $[\![a\,b\,c]\!]$ with a, b, and c as columns. It is therefore an antisymmetric linear function of the vectors.

We should try to find an algebraic product to represent these geometric properties of volumes.

2.4.2 ASSOCIATIVITY OF THE OUTER PRODUCT

The antisymmetry we signaled in the classical characterization of volume measures is a clue to its representation in our new algebra. We simply attempt to extend the "span" operation of the outer product to more than two terms. Algebraically, the most natural way is to define the outer product to be associative:

$$\text{associativity:} \quad (a \wedge b) \wedge c = a \wedge (b \wedge c).$$

We can thus write the volume element as $a \wedge b \wedge c$ without ambiguity. (You should realize that the outer product is still pairwise antisymmetric, so that $a \wedge b \wedge c = -a \wedge c \wedge b = c \wedge a \wedge b$, etc.) Geometrically, this property would imply that we can span the same oriented volume in different ways as the span of a planar element and a vector; if that were not true, this algebraic formalization would be inappropriate. Fortunately, Figure 2.6 confirms the geometric validity of the algebraic associativity. It visualizes the volume $a \wedge b \wedge c$ in several equivalent ways, all leading to the same oriented amount of 3-space.

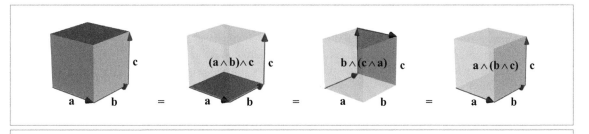

Figure 2.6: The associativity of the outer product.

We call the element formed as the outer product of three vectors a *trivector* or *3-blade* (the difference is explained in Section 2.9.3). To verify its interpretation, we may introduce an arbitrary basis $\{e_1, e_2, e_3\}$ in $\mathbb{R}^3$. Then the defining properties of the outer product yield

$$\mathbf{a} \wedge \mathbf{b} \wedge \mathbf{c} =$$
$$= (a_1\mathbf{e}_1 + a_2\mathbf{e}_2 + a_3\mathbf{e}_3) \wedge (b_1\mathbf{e}_1 + b_2\mathbf{e}_2 + b_3\mathbf{e}_3) \wedge (c_1\mathbf{e}_1 + c_2\mathbf{e}_2 + c_3\mathbf{e}_3) \quad (2.4)$$
$$= (a_1b_2c_3 - a_1b_3c_2 + a_2b_3c_1 - a_2b_1c_3 + a_3b_1c_2 - a_3b_2c_1)\, \mathbf{e}_1 \wedge \mathbf{e}_2 \wedge \mathbf{e}_3.$$

Therefore, any trivector in space $\mathbb{R}^3$ is a multiple of the trivector $\mathbf{e}_1 \wedge \mathbf{e}_2 \wedge \mathbf{e}_3$. The proportionality is the determinant of the spanning vectors on the basis $\{e_1, e_2, e_3\}$, which we recognize as the (relative) volume and its sign as the orientation relative to $\mathbf{e}_1 \wedge \mathbf{e}_2 \wedge \mathbf{e}_3$. In linear algebra, we would write that scalar as the determinant $\det([[\mathbf{a}\,\mathbf{b}\,\mathbf{c}]])$ of the matrix, with $\mathbf{a}$, $\mathbf{b}$, and $\mathbf{c}$ as columns. So the properties of a volume in the 3-D space $\mathbb{R}^3$ may be characterized by a scalar, at least if we have agreed upon some convention about the order of the basis elements (i.e., a handedness of the basis). That is indeed how it is done classically.

In subtle difference, our algebra of the outer product permits us to treat a volume element as a single trivector. This is tidier, since it requires no bookkeeping of an extraneous convention: as the handedness of the basis changes, the trivector coefficient automatically changes its orientation appropriately. It is also clearly different from a scalar, even in 3-D space, because it has the geometrical unit volume as part of its value. We can therefore clearly express what happens when that is chosen differently (for instance, with a left-handed rather than right-handed orientation). Moreover, this representation of volume elements as the outer product of three vectors carries over unchanged to spaces of arbitrary dimensionality, and those will be surprisingly important, even for computations on 3-D space.

2.4.3 VISUALIZATION OF TRIVECTORS

In our geometric verification of associativity, we met the obvious visualization of a 3-blade $\mathbf{a} \wedge \mathbf{b} \wedge \mathbf{c}$ in Figure 2.6, relating it to its vector factors $\mathbf{a}$, $\mathbf{b}$, and $\mathbf{c}$.

Of course, this amount of oriented volume may be sheared (similar to what we did to bivectors) as long as you do not change magnitude or orientation. If you have a basis, it is sometimes convenient to align it with the basis vectors. If there are no vectors available, just a trivector, you could draw it as a spherical volume, with some convention on how to denote its orientation.

In $\mathbb{R}^3$, the algebraic addition of trivectors is just a matter of adding their signed scalar magnitudes. Since that has no aspects of geometrical attitude, explicit visualization of trivector addition in 3-D is not really useful.

2.5 QUADVECTORS IN 3-D ARE ZERO

In the 3-D space $\mathbb{R}^3$ the outer product of any four vectors $\mathbf{a}$, $\mathbf{b}$, $\mathbf{c}$, $\mathbf{d}$ is zero. This is an automatic consequence of the outer product properties, and easily shown.

In $\mathbb{R}^3$, only three vectors can be independent, and therefore the fourth ($\mathbf{d}$) must be expressible as a weighted sum of the other three:

$$\mathbf{d} = \alpha\,\mathbf{a} + \beta\,\mathbf{b} + \gamma\,\mathbf{c}.$$

Associativity, distributivity, and antisymmetry then make the outer product of these four vectors zero:

$$\begin{aligned}
\mathbf{a} \wedge \mathbf{b} \wedge \mathbf{c} \wedge \mathbf{d} &= \mathbf{a} \wedge \mathbf{b} \wedge \mathbf{c} \wedge (\alpha\,\mathbf{a} + \beta\,\mathbf{b} + \gamma\,\mathbf{c}) \\
&= \mathbf{a} \wedge \mathbf{b} \wedge \mathbf{c} \wedge (\alpha\,\mathbf{a}) + \mathbf{a} \wedge \mathbf{b} \wedge \mathbf{c} \wedge (\beta\,\mathbf{b}) + \mathbf{a} \wedge \mathbf{b} \wedge \mathbf{c} \wedge (\gamma\,\mathbf{c}) \\
&= 0.
\end{aligned}$$

So the highest-order element that can exist in the subspace algebra of $\mathbb{R}^3$ is a trivector. It should be clear that this is not a limitation of the outer product algebra in general: if the space had more dimensions, the outer product would create the appropriate hyper-volumes, each with an attitude, orientation, and magnitude.

It is satisfying that the geometric uselessness of the construction of elements of higher dimension than n in $\mathbb{R}^n$ is reflected in the automatic algebraic outcome of 0. Geometrically, we should interpret that element 0 as the *empty subspace* of any dimensionality. So this one element 0 is the zero scalar, the zero vector, the zero bivector, and so on. There is no algebraic or geometric reason to distinguish between those, for the empty subspace has no attitude, orientation, or weight.

2.6 SCALARS INTERPRETED GEOMETRICALLY

We extend the new pattern of constructing subspaces downwards as well, to the lowest dimensional subspaces. In a consistent view, the set of points at the origin should be considered as a 0-D homogeneous subspace (just as lines through the origin were 1-D, and homogeneous planes were 2-D). The previous sections then suggest that it might be represented algebraically by an outer product of zero vector terms. The most general such element is a scalar (since only the scaling property of the outer product remains). So *scalars are 0-blades* and can be used to represent homogeneous points (i.e., points at the origin). Treating scalars as homogeneous points keeps our algebra and its geometrical interpretation nicely consistent, even though you may initially feel that we are stretching analogies a bit too far.

A 0-blade is depicted in Figure 2.1(a). As a subspace, it has the properties of attitude, orientation, and magnitude.

- **Attitude**. The attitude (the locational aspect of the point) is not very interesting; all homogeneous points sit at the origin.

- **Orientation**. The orientation of the point is the sign of the scalar that represents it. This will be useful. For instance, in $\mathbb{R}^3$, the point of intersection of a homogeneous line and a homogeneous plane can be assigned a different sign depending on whether the line enters the plane from the back or the front.

- **Weight**. A point has a weight, which can for instance be used to indicate the intersection strength of a line and a plane in $\mathbb{R}^3$ (we will see that a line almost parallel to a plane leads to a weaker intersection than one perpendicular to it).

Note that this inclusion of scalars among the subspaces reduces the artificial distinction between scalars in a field and vectors in a space that we had in the traditional vector space definition of Section 2.1. They are merely subspaces, just like all other subspaces distinguished by their dimensionality.

Striving towards a complete and consistent mathematical structure, we would like to have the outer product defined between any two elements, including scalars. So we extend the definition of the outer product to include scalars, in a straightforward manner, by defining

$$\alpha \wedge \mathbf{x} = \mathbf{x} \wedge \alpha = \alpha \mathbf{x}, \quad \text{and} \quad \alpha \wedge \beta = \alpha \beta \ \text{ for } \alpha, \beta \in \mathbb{R}. \tag{2.5}$$

In this view, the usual scalar multiplication in the vector space $\mathbb{R}^n$ definition of Section 2.1 is really the outer product in disguise. In this chapter, there has therefore in fact been only a single product in use.

By the way, beware of assuming that the outer product for scalars should have been antisymmetric to be consistent with the outer product for vectors. The outer product is not even antisymmetric for bivectors or 2-blades, for the property of associativity enforces symmetry. This is easily demonstrated:

$$
\begin{aligned}
(\mathbf{a}_1 \wedge \mathbf{a}_2) \wedge (\mathbf{b}_1 \wedge \mathbf{b}_2) &= \mathbf{a}_1 \wedge \mathbf{a}_2 \wedge \mathbf{b}_1 \wedge \mathbf{b}_2 \\
&= -\mathbf{a}_1 \wedge \mathbf{b}_1 \wedge \mathbf{a}_2 \wedge \mathbf{b}_2 \\
&= \mathbf{b}_1 \wedge \mathbf{a}_1 \wedge \mathbf{a}_2 \wedge \mathbf{b}_2 \\
&= -\mathbf{b}_1 \wedge \mathbf{a}_1 \wedge \mathbf{b}_2 \wedge \mathbf{a}_2 \\
&= \mathbf{b}_1 \wedge \mathbf{b}_2 \wedge \mathbf{a}_1 \wedge \mathbf{a}_2 \\
&= (\mathbf{b}_1 \wedge \mathbf{b}_2) \wedge (\mathbf{a}_1 \wedge \mathbf{a}_2).
\end{aligned}
$$

The general rule is given in Section 2.10: only for two elements of odd dimensionality is the outer product antisymmetric.

By (2.5), the algebraic fact that all scalars are a multiple of the number 1 can be interpreted geometrically: the 0-vector 1 represents the standard point at the origin, and all other weighted points at the origin are a multiple of it. Visualization of a scalar is therefore simply as a weighted point at the origin, as in Figure 2.1(a), and addition of such elements merely results in the addition of their weights. There is very little geometry in scalars, but they are part of the general pattern.

2.7 APPLICATIONS

Having subspaces as elements of computation is only the beginning of geometric algebra, but it already allows us a slightly different perspective on the solutions to some common problems in linear algebra.

2.7.1 SOLVING LINEAR EQUATIONS

If you have a basis $\{\mathbf{a}, \mathbf{b}\}$ in $\mathbb{R}^2$, any vector $\mathbf{x} \in \mathbb{R}^2$ can be written as

$$\mathbf{x} = \alpha\,\mathbf{a} + \beta\,\mathbf{b},$$

for some α and β, which need to be determined. With the outer product, you can solve such equations explicitly and draw a picture of the solution.

We proceed by taking the outer product of both sides with $\mathbf{a}$ and with $\mathbf{b}$, respectively. This yields two equations that simplify due to the antisymmetry of the outer product:

$$\mathbf{x} \wedge \mathbf{a} = \alpha\,\mathbf{a} \wedge \mathbf{a} + \beta\,\mathbf{b} \wedge \mathbf{a} = \beta\,\mathbf{b} \wedge \mathbf{a}$$
$$\mathbf{x} \wedge \mathbf{b} = \alpha\,\mathbf{a} \wedge \mathbf{b} + \beta\,\mathbf{b} \wedge \mathbf{b} = \alpha\,\mathbf{a} \wedge \mathbf{b}.$$

Since the 2-blades occurring on both sides must reside in the same homogeneous plane, they are both scalar multiples of the same basic 2-blade characterizing that plane. Therefore, their ratio is well defined. We write it as a division (symbolically, for now; when we have the full geometric algebra in Chapter 6, this will be algebraically exact). Using this ratio, β and α are immediately computable in terms of $\mathbf{x}$, $\mathbf{a}$, and $\mathbf{b}$. This gives the decomposition

$$\mathbf{x} = \frac{\mathbf{x} \wedge \mathbf{b}}{\mathbf{a} \wedge \mathbf{b}}\,\mathbf{a} + \frac{\mathbf{x} \wedge \mathbf{a}}{\mathbf{b} \wedge \mathbf{a}}\,\mathbf{b} \tag{2.6}$$

Figure 2.7 shows the geometric interpretation of this decomposition. The blue parallelogram representing the bivector $\mathbf{x} \wedge \mathbf{a}$ is algebraically equivalent to various reshapings of the same amount of area. We reshape it by sliding the vector pointing to $\mathbf{x}$, along a line parallel to $\mathbf{a}$, until it points in the direction $\mathbf{b}$. That gives the other blue parallelogram, computationally fully identical to the first. Now the ratio of area of this blue parallelogram to the area of $\mathbf{b} \wedge \mathbf{a}$ (also indicated) is precisely the stretch of $\mathbf{b}$ required. The same holds for the green parallelogram $\mathbf{x} \wedge \mathbf{b}$, which should be divided by $\mathbf{a} \wedge \mathbf{b}$. Notice the implicit sign change between the two cases, inherent in the denominator.

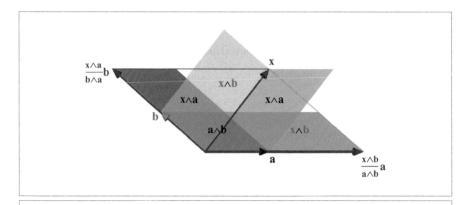

The geometrical reshapeability of the bivector is precisely the feature that gives us the correct algebraic formula. Practice "seeing" this, as it gives you a new tool for finding compact expressions in a coordinate-free manner. Play with the interactive figure in GAViewer to convince yourself that this works in all cases and relative directions of $\mathbf{a}$, $\mathbf{b}$, and $\mathbf{x}$ as long as they are coplanar (except, of course, when $\mathbf{a} \wedge \mathbf{b} = 0$).

In elementary linear algebra, we would view the expression for $\mathbf{x}$ as two equations for α and β, in components:

$$a_1 \alpha + b_1 \beta = x_1$$
$$a_2 \alpha + b_2 \beta = x_2.$$

Then *Cramer's rule* specifies the solutions as the ratio of two determinants:

$$\alpha = \frac{\det\left(\begin{bmatrix} x_1 & b_1 \\ x_2 & b_2 \end{bmatrix}\right)}{\det\left(\begin{bmatrix} a_1 & b_1 \\ a_2 & b_2 \end{bmatrix}\right)}, \quad \beta = \frac{\det\left(\begin{bmatrix} x_1 & a_1 \\ x_2 & a_2 \end{bmatrix}\right)}{\det\left(\begin{bmatrix} b_1 & a_1 \\ b_2 & a_2 \end{bmatrix}\right)}.$$

When you realize that the ratio of bivectors in a common plane is simply the ratio of their weights, which can be expressed as determinants, you realize that our earlier solution is in fact Cramer's rule. Yet Cramer's rule is usually explained as algebra, not as the geometrical ratio of areas it really is. The fact that it uses coordinates in its formulation easily makes one lose sight of the geometry involved.

To solve equations involving n basis vectors in $\mathbb{R}^n$, the same technique applies. You then need to take more outer products to get the parameters (see structural exercise 4).

2.7.2 INTERSECTING PLANAR LINES

Again in $\mathbb{R}^2$, consider the problem of intersecting a line L with position vector $\mathbf{p}$ and direction vector $\mathbf{u}$, with a line M with position vector $\mathbf{q}$ and direction vector $\mathbf{v}$. This is depicted in Figure 2.8.

Its similarity to Figure 2.7 is clear. To find the intersection point $\mathbf{x}$, we need to add an amount of $\mathbf{u}$ and $\mathbf{v}$, so we are effectively decomposing $\mathbf{x}$ in that basis. We might wish to use (2.6), but that would contain the unknown $\mathbf{x}$ on the right-hand side in the bivectors $\mathbf{x} \wedge \mathbf{u}$ and $\mathbf{x} \wedge \mathbf{v}$, so this approach appears to fail.

However, we have another way of making precisely those bivectors, for they are reshapeable: $\mathbf{x} \wedge \mathbf{u} = \mathbf{p} \wedge \mathbf{u}$, and $\mathbf{x} \wedge \mathbf{v} = \mathbf{q} \wedge \mathbf{v}$. Therefore, the same geometrical method now solves the intersection point as

$$\mathbf{x} = \frac{\mathbf{q} \wedge \mathbf{v}}{\mathbf{u} \wedge \mathbf{v}} \mathbf{u} + \frac{\mathbf{p} \wedge \mathbf{u}}{\mathbf{v} \wedge \mathbf{u}} \mathbf{v}. \tag{2.7}$$

This procedure shows that the flexibility in the geometric reshapeability of the bivectors gives an enjoyable amount of algebraic freedom. Once you learn to see bivectors in your problems, solutions can become fairly immediate.

This was merely meant to be an illustrative example of the use of the outer product. Much later, in Section 11.7.1, we will revisit lines and their representation, and the above computation will be a specific case of the `meet` operation, which will remove its somewhat arbitrary nature. We will then also treat skew lines in 3-D.

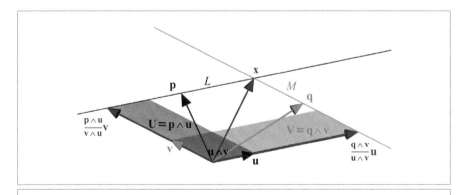

Figure 2.8: Intersecting lines in the plane.

2.8 HOMOGENEOUS SUBSPACE REPRESENTATION

Apart from these initial applications, we can also use this new algebraic instrument of the outer product to formalize some useful geometrical properties.

2.8.1 PARALLELNESS

The outer product of two vectors **a** and **b** forms a 2-blade proportional to their spanned area in their common plane. When you keep **a** constant but make **b** increasingly more parallel to it (by turning it in the common plane), you find that the weight of the bivector becomes smaller, for the area spanned by the vectors decreases. When the vectors are parallel, the bivector is zero; when they move beyond parallel (**b** turning to the other side of **a**) the bivector acquires the opposite orientation.

A 2-blade may thus be used as a measure of parallelness: $\mathbf{a} \wedge \mathbf{b}$ equals zero if and only if **a** and **b** are parallel (i.e., lie on the same 1-D subspace). Therefore $\mathbf{x} \wedge \mathbf{a} = 0$, considered as an equation in **x** for a given **a**, defines the vectors **x** of the homogeneous line determined by **a**. We cannot solve such equations yet (we will need a division for that, which we will only discuss in Chapter 6), but it is already easy to verify that $\mathbf{x} = \lambda \mathbf{a}$ is a solution (and in fact, the general solution). So we may surmise:

$$\mathbf{x} \text{ on line determined by } \mathbf{a} \quad \Leftrightarrow \quad \mathbf{x} = \lambda \mathbf{a} \quad \Leftrightarrow \quad \mathbf{x} \wedge \mathbf{a} = 0$$

Therefore "2-blades being zero" produces homogeneous line equations. When used in this way only the attitude of the line characterization matters, since both orientation and weight are scalar factors that can be divided out.

To obtain an equation for a plane, recognize that a 3-blade is zero if and only if the three vectors that compose it lie in the same plane (2-D subspace). Geometrically, we would say that they do not span a volume; the corresponding algebraic expression conveys that they are linearly dependent. It follows that we can use a 2-blade $\mathbf{a} \wedge \mathbf{b}$ to represent a plane through the origin, by using it to detect the vectors that do not span a volume with it:

$$\mathbf{x} \text{ on plane determined by } \mathbf{a} \text{ and } \mathbf{b} \quad \Longleftrightarrow \quad \mathbf{x} = \lambda \mathbf{a} + \mu \mathbf{b} \quad \Leftrightarrow \quad \mathbf{x} \wedge \mathbf{a} \wedge \mathbf{b} = 0$$

Using both **a** and **b** is in fact a bit too specific, as we mentioned at the beginning of this chapter; the same plane could have been characterized by different vectors. It is better to talk only about the 2-blade **B** of the plane:

$$\text{vector } \mathbf{x} \text{ in plane of } \mathbf{B} \quad \Longleftrightarrow \quad \mathbf{x} \wedge \mathbf{B} = 0.$$

This 2-blade **B** even represents a *directed plane*, for we can say that a point **y** is at the positive side of the plane if $\mathbf{y} \wedge \mathbf{B}$ is a positive volume (i.e., a positive multiple of the standard volume $\mathbf{I}_3 = \mathbf{e}_1 \wedge \mathbf{e}_2 \wedge \mathbf{e}_3$ for some standard basis $\{\mathbf{e}_i\}_{i=1}^{3}$ of the space).

2.8.2 DIRECT REPRESENTATION OF ORIENTED WEIGHTED SUBSPACES

These constructions are easily extended. In general, if we have a k-dimensional homogeneous subspace $\mathcal{A}$ spanned by k vectors $\mathbf{a}_1, \cdots, \mathbf{a}_k$, we can form the k-blade $\mathbf{A} = \mathbf{a}_1 \wedge \cdots \wedge \mathbf{a}_k$. We call this blade $\mathbf{A}$ the *direct representation* of the homogeneous subspace $\mathcal{A}$ (as opposed to the dual representation, which we will meet later). By that we mean that any vector in $\mathcal{A}$ satisfies $\mathbf{x} \wedge \mathbf{A} = 0$, and that, vice versa, any vector $\mathbf{x}$ satisfying this equation is in $\mathcal{A}$.

$$\mathbf{A} \text{ is a direct representation of } \mathcal{A} \quad : \quad (\mathbf{x} \in \mathcal{A} \iff \mathbf{x} \wedge \mathbf{A} = 0)$$

This will become such a useful construction that we will often identify the subspace $\mathcal{A}$ with the blade $\mathbf{A}$, and say that a vector is contained in a blade. Of course, there is an orientation and a weight involved in $\mathbf{A}$ that is not usually assumed for $\mathcal{A}$, so $\mathbf{A}$ is a more precise characterization of the geometry, with the properties defined on Table 2.1.

It is also convenient to lift this concept of containment from vectors to the level of blades. We will say that $\mathbf{A}$ *is contained in* $\mathbf{B}$, denoted $\mathbf{A} \subseteq \mathbf{B}$, if all vectors in $\mathbf{A}$ are also in $\mathbf{B}$. In formula:

$$\mathbf{a}_1 \wedge \cdots \wedge \mathbf{a}_k = \mathbf{A} \subseteq \mathbf{B} \iff \mathbf{a}_i \wedge \mathbf{B} = 0, \quad i = 1, \ldots, k. \tag{2.8}$$

Beware that this is *not* the same as $\mathbf{A} \wedge \mathbf{B} = 0$, since that would already hold if only one of the vectors in $\mathbf{A}$ was contained in $\mathbf{B}$.

2.8.3 NONMETRIC LENGTHS, AREAS, AND VOLUMES

Nowhere in this chapter have we used a metric for our computations. The outer product, which is the product of spanning and weighting, does not need one. Yet the lengths, areas, and volumes that can be computed using the outer product appear to have a metric feeling to them. We must emphasize that the lengths measurable by the outer product are always length *ratios* along the same line through the origin, that the areas are *ratios* of areas in

---

Table 2.1: Algebraic definition of the terms we use to denote the geometrical properties of a subspace as represented by a blade $\mathbf{A}$.

Term	Definition
Attitude	The equivalence class $\lambda \mathbf{A}$, for any $\lambda \in \mathbb{R}$
(Relative) weight	The value of λ in $\mathbf{A} = \lambda \mathbf{I}$
	(where $\mathbf{I}$ is a selected standard subspace with the same attitude)
(Relative) orientation	The sign of the weight relative to $\mathbf{I}$

the same plane through the origin, and that volumes similarly are *ratios* of volumes in the same space. For such ratios of full-grade elements within the same subspace you do not need a metric measure.

It is only when comparing lengths, areas, and volumes from different homogeneous subspaces that you need to introduce a metric. The metric permits you to rotate one vector onto another to check that the lengths are identical or what their ratio is. We will do that in the next chapter, using a Euclidean metric.

Having said that, we already have some useful instruments. In an n-dimensional space $\mathbb{R}^n$ we can compare arbitrary hypervolumes. If we have n vectors $\mathbf{a}_i$ $(i = 1, \dots, n)$, then the hypervolume of the parallelepiped spanned by them is proportional to a unit hypervolume in $\mathbb{R}^n$ by the magnitude of $\mathbf{a}_1 \wedge \dots \wedge \mathbf{a}_n$. The hypervolume of a *simplex* in that space, which is the convex body containing the origin and the n endpoints of the vectors $\mathbf{a}_i$ $(i = 1, \cdots, n)$, is a fraction of that:

$$\frac{1}{n!} \, \mathbf{a}_1 \wedge \cdots \wedge \mathbf{a}_n.$$

The same formula also applies to a simplex in each k-dimensional subspace of the n-dimensional subspace, as

$$\frac{1}{k!} \, \mathbf{a}_1 \wedge \cdots \wedge \mathbf{a}_k.$$

This one formula computes the relative oriented volume of a tetrahedron in 3-D, of the relative oriented area of a triangle in 2-D, of the relative oriented length of a vector in 1-D, and it even works for the relative oriented weight of a scalar in 0-D.

2.9 THE GRADED ALGEBRA OF SUBSPACES

We have introduced the geometric algebra of the outer product step by step. This section makes an inventory of the general patterns we have uncovered and introduces descriptive terms for its algebraic aspects.

2.9.1 BLADES AND GRADES

The outer product of k vectors is called a *k-blade*. This name (from Hestenes [33]) reflects its higher-dimensional nature and its flatness when used as a representation of a k-dimensional homogeneous subspace in a vector space model. A vector is a 1-blade, a bivector is a 2-blade (in 3-space), and a scalar may be referred to as a 0-blade.

The number k is called the *grade* of the k-blade (though you may find the term *step* in some literature). Algebraically, it is the number of vector factors in a nonzero k-blade, and we denote it by the grade() symbol:

$$\text{grade}(\mathbf{a}_1 \wedge \cdots \wedge \mathbf{a}_k) = k.$$

Geometrically, the grade is the dimensionality of the subspace that the k-blade represents, in the manner of Section 2.8.2. But we prefer to reserve the term *dimension* for the dimensionality n of the space $\mathbb{R}^n$ that we are considering. In terms of blades and grades, the dimension is the highest grade that a nonzero blade may have in a chosen (sub)space.

The outer product construction connects blades and their grades:

$$\text{grade}(\mathbf{A} \wedge \mathbf{B}) = \text{grade}(\mathbf{A}) + \text{grade}(\mathbf{B}).$$

Since the outcome of the product $\mathbf{A} \wedge \mathbf{B}$ may be zero for arbitrary $\mathbf{A}$ and $\mathbf{B}$, *the element 0 must be allowed to have any appropriate grade.* There is no algebraic reason to discriminate among the zero scalar, zero vector, and so on. Geometrically, the element 0 of the subspace algebra represents the *empty subspace*—which can clearly be of any grade.

2.9.2 THE LADDER OF SUBSPACES

We have seen that the blades can be decomposed as a weighted sum of basis blades. If we construct k-blades in an n-dimensional space, because of the antisymmetry of the outer product for vectors there are $\binom{n}{k}$ elements in this basis. The k-blades therefore reside in a $\binom{n}{k}$-dimensional linear space $\bigwedge^k \mathbb{R}^n$. But we will soon see that not all elements of this linear space are k-blades. Plotting the numbers of basis k-blades for various n, we obtain an inventory of the ladder of k-blades of different grades as in Table 2.2; you may recognize Pascal's triangle.

A blade of highest possible dimension is often called a *pseudoscalar* of the space. It obtained this name in algebraic recognition that it is like a scalar, in that it defines a 1-D hypervolume space in which all hypervolumes are multiples of each other. According to Table 2.2, it has a basis consisting of a single blade, as we saw in Section 2.4.2 for 3-blades in $\mathbb{R}^3$.

Table 2.2: Pascal's triangle of the number of basis k-blades in n-dimensional space.

	subspace grade k					
n	0	1	2	3	4	5
0	1					
1	1	1				
2	1	2	1			
3	1	3	3	1		
4	1	4	6	4	1	
5	1	5	10	10	5	1
$\cdots$		$\cdots$				

It is often sensible to appoint one n-blade as the unit pseudoscalar, both in magnitude and in orientation, relative to which the other volumes are measured. This is especially possible in a vector space with a nondegenerate metric, where we can introduce an orthonormal basis $\{e_i\}_{i=1}^{n}$, and the natural choice is $I_n \equiv e_1 \wedge \cdots \wedge e_n$.

When we are focused on a specific subspace of the full n-dimensional space, we will often speak of the pseudoscalar of that subspace—again meaning the largest blade that can reside in that subspace. We will use I_n for the chosen unit pseudoscalar of $\mathbb{R}^n$, and I for the pseudoscalar of a subspace, or another I-like symbol.

2.9.3 k-BLADES VERSUS k-VECTORS

We have constructed k-blades as the outer product of k vector factors. By derivations like we did for 2-blades in (2.3), it is easy to show that the properties of the outer product allow k-blades in $\mathbb{R}^n$ to to be decomposed on an $\binom{n}{k}$-dimensional basis.

One might be tempted to reverse this construction and attempt to make k-blades as a weighted combination of these basis k-blades. However, this does not work, for such sums are usually not factorizable in terms of the outer product. The first example occurs in $\mathbb{R}^4$. If $\{e_1, e_2, e_3, e_4\}$ is a basis for $\mathbb{R}^4$, then the element $A = e_1 \wedge e_2 + e_3 \wedge e_4$ simply can *not* be written as a 2-blade $a \wedge b$. We ask you to convince yourself of this in structural exercise 5.

We have no geometric interpretation for such nonblades in our vector space model. They are certainly not subspaces, for they contain no vectors; the equation $x \wedge (e_1 \wedge e_2 + e_3 \wedge e_4)$ $= 0$ can be shown to have no vector solution (other than 0). The geometrical role of such elements, if any, is different.

Yet it is very tempting to consider the linear space $\bigwedge^k \mathbb{R}^n$ spanned by the basis k-blades as a mathematical object of study, in which addition is permitted as a construction of new elements. A typical element constructed as a weighted sum of basis blades is called a *k-vector*; its grade aspect is often called *step*. You will find much mathematical literature about the algebraic properties of such constructions—though necessarily little about its geometric significance. Within the context of k-vectors, the blades are sometimes known as *simple k-vectors* (or some other term reflecting their factorizability).

The k-blades are elements of this space (remember this: *k-blades are k-vectors*), but it is not elementary to specify the necessary and sufficient conditions for a k-vector to be a k-blade. (This problem has only recently been solved in [20], and the outcome is not easily summarized.) Only 0-vectors, 1-vectors, $(n-1)$-vectors, and n-vectors are *always* also blades in n-dimensional space. As a consequence, in 3-D space all k-vectors are k-blades, but already in 4-D space one can make 2-vectors that are not 2-blades. Since we need 4-D and even 5-D vector spaces to model 3-D physical space, the distinction between k-blades and k-vectors will be important to us.

Because of the bilinear nature of the outer product, it is quite natural to extend it from k-blades to k-vectors by distributing the operation over the sum of blades. Following established mathematical tradition, it is tempting to give the most general form of any theorem, and supplant k-blades by k-vectors wherever we can, whether the result is geometrically meaningful or not. We do this every now and then, when we find a geometric reason for doing so. Yet some theorems we will encounter later are true for blades only, so we need a way to distinguish them notationally from the k-vectors.

In Part I of this book, we denote k-vectors by nonbold capital letters and k-blades by bold capital letters, possibly denoting the grade as a subscript (in Part II, we will need the distinction between bold and nonbold to denote something else). So $\mathbf{A}_k$ is a k-blade, but A_k is a k-vector (which is not necessarily a blade). Vectors are always blades, and denoted by bold lowercase, such as $\mathbf{a}$. Scalars are blades, and denoted by lowercase Greek, such as α. These conventions are summarized in Table 2.3.

2.9.4 THE GRASSMANN ALGEBRA OF MULTIVECTORS

The construction of k-vectors as a sum of k-blades makes sense algebraically, but since its elements are not necessarily subspaces, we cannot be assured of the geometrical

Table 2.3: Notational Conventions for Blades and Multivectors for Part I of This Book.

α, β, etc.	Scalar
a_i, b^i, etc.	Scalar components of vectors
$\mathbf{a}$, $\mathbf{b}$, etc.	Vector
$\mathbf{e}_i$	Basis vector, typically in an orthonormal basis
$\mathbf{b}_i$	Basis vector, nonorthonormal basis
$\mathbf{A}$, $\mathbf{B}$, etc.	General blade
$\mathbf{A}_k$, $\mathbf{B}_k$, etc.	General blade of grade k
A_k, B_k, etc.	A k-vector, not necessarily a blade
A, B, etc.	General multivector, not necessarily a k-vector or a blade
$\mathbf{I}_n$, $\mathbf{I}$, etc.	(Unit) pseudoscalars
$\mathbb{R}^n$	n-Dimensional vector space over the field $\mathbb{R}$
$\bigwedge^k \mathbb{R}^n$	The linear space of k-vectors, in which k-blades also reside
$\bigwedge \mathbb{R}^n$	The linear space of multivectors (Grassmann space)
grade($\mathbf{A}$)	The grade of $\mathbf{A}$
$\langle A \rangle_k$	The k-grade part of a multivector A
$\widetilde{\mathbf{A}}$	The reverse of $\mathbf{A}$, equal to $(-1)^{\text{grade}(\mathbf{A})(\text{grade}(\mathbf{A})-1)/2}\mathbf{A}$
$\widehat{\mathbf{A}}$	The grade involution of $\mathbf{A}$, equal to $(-1)^{\text{grade}(\mathbf{A})}\mathbf{A}$

significance. We will therefore exclude it from geometric algebra—at least until we know the corresponding geometry. Still, it is a well-studied structure with useful theorems, so we will discuss it briefly to give you access to that literature.

If we allow the addition of k-blades to make k-vectors, we obtain what mathematicians call a *graded algebra*, since each element has a well-defined grade (even though not each element is a product of k vector factors). But they do not stop there. When they also allow the addition between elements of *different* grades, they obtain the most general structure that can be made out of addition $+$ and outer product $\wedge$. This results in a linear space of elements of mixed grade; these are called *multivectors*.

It is simple to extend the outer product to multivectors, using its linearity and distributivity. For instance:

$$(1 + e_1) \wedge (1 + e_2) = 1 \wedge 1 + 1 \wedge e_2 + e_1 \wedge 1 + e_1 \wedge e_2 = 1 + e_1 + e_2 + e_1 \wedge e_2.$$

Mathematicians call the structure thus created the *Grassmann algebra* (or exterior algebra) for the *Grassmann space*, $\bigwedge \mathbb{R}^n$. The name pays homage to Hermann Grassmann (1809–1877), who defined the outer product to make subspaces into elements of computation. It is a somewhat ironic attribution, as Grassmann might actually have preferred not to admit the k-vectors (let alone the multivectors) in an algebra named after him, since they cannot represent the geometrical subspaces he intended to encode formally.

The Grassmann algebra of a 3-D vector space with basis $\{e_1, e_2, e_3\}$ is in itself a linear space of $2^3 = 8$ dimensions. A basis for it is

$$\left\{ \underbrace{1}_{\text{scalars}}, \underbrace{e_1, e_2, e_3}_{\text{vector space}}, \underbrace{e_1 \wedge e_2, e_2 \wedge e_3, e_3 \wedge e_1}_{\text{bivector space}}, \underbrace{e_1 \wedge e_2 \wedge e_3}_{\text{trivector space}} \right\}. \qquad (2.9)$$

In an n-dimensional space, there are $\binom{n}{k}$ basis elements of grade k. The total number of independent k-vectors of any grade supported by the vector space $\mathbb{R}^n$ is

$$\sum_{k=0}^{n} \binom{n}{k} = 2^n.$$

Therefore the Grassmann algebra of an n-dimensional space requires a basis of 2^n elements. This same basis is of course also useful for the decomposition of k-blades, so an algebra for blades also has $\binom{n}{k}$ basis blades of grade k, for a total of 2^n over all the blades. But we reiterate that when we list that basis for blades, we should only intend it for decomposition purposes, not as a linear space of arbitrary additive combinations. There is unfortunately no standard notation for the submanifold of the k-vector space $\bigwedge^k \mathbb{R}^n$ that contains the blades, so we will use the slightly less precise k-vector notation even when we mean k-blades only, and let the context make that distinction.

When we have multivectors of mixed grade, it is convenient to have the *grade operator* $\langle \ \rangle_k : \bigwedge \mathbb{R}^n \to \bigwedge^k \mathbb{R}^n$, which selects the multivector part of grade k (note that this is not necessarily a k-blade). With that, we can express the outer product between arbitrary elements of the Grassmann algebra as

$$A \wedge B = \sum_{k=0}^{n} \sum_{l=0}^{n} \langle A \rangle_k \wedge \langle B \rangle_l.$$

The grade-raising property of the outer product can be stated in terms of the grade operator as

$$\langle A \wedge B \rangle_k = \sum_{i=0}^{k} \langle A \rangle_i \wedge \langle B \rangle_{k-i}. \tag{2.10}$$

The result of (2.10) may be 0, so as before 0 is a multivector of any grade. Within a Grassmann algebra, the multivector 0 has the properties of the zero element of both outer product and addition:

$$0 \wedge A = 0 \quad \text{and} \quad 0 + A = A.$$

These sensible correspondences should be enough to "translate" the literature on Grassmann algebra, and peruse it for geometrically meaningful results.

2.9.5 REVERSION AND GRADE INVOLUTION

In future computations, we often need to reverse the order of the vectors spanning a blade, and other monadic operations on blades. We introduce notations for them now, since we have all the necessary ingredients, and give some useful properties.

Define the *reversion* $\tilde{}$ as an operation that takes a k-blade $\mathbf{A} = \mathbf{a}_1 \wedge \mathbf{a}_2 \wedge \cdots \wedge \mathbf{a}_k$ and produces its *reverse*:

$$\textit{reversion:} \quad \widetilde{\mathbf{A}} \equiv \mathbf{a}_k \wedge \mathbf{a}_{k-1} \wedge \cdots \wedge \mathbf{a}_1, \tag{2.11}$$

which just has all vectors of $\mathbf{A}$ in reverse order. The notation by the tilde is chosen to be reminiscent of an editor's notation for an interchange of terms. (Some literature denotes it by $\mathbf{A}^\dagger$ instead, since it is related to complex conjugation in a certain context.) This definition appears to require a factorization of $\mathbf{A}$, but its consequence is that a reversion just leads to a grade-dependent change of sign for $\mathbf{A}$. In (2.11), we can restore the terms in $\widetilde{\mathbf{A}}$ to their original order in $\mathbf{A}$. This requires $\frac{1}{2}k(k-1)$ swaps of neighboring terms, which provides an overall sign of $(-1)^{k(k-1)/2}$. So we may equivalently define, for a k-blade $\mathbf{A}_k$:

$$\textit{reversion:} \quad \widetilde{\mathbf{A}}_k = (-1)^{\frac{1}{2}k(k-1)} \mathbf{A}_k. \tag{2.12}$$

This is the preferred definition in computations. Note that this sign change exhibits a $+ + - - + + - - \cdots$ pattern over the grades, with a periodicity of four.

By extension, we can apply a similar definition to the reversion of a general multivector consisting of a sum of elements of different grades in a Grassmann algebra. This is then simply defined as the monadic operation $\tilde{}: \bigwedge \mathbb{R}^n \to \bigwedge \mathbb{R}^n$

$$\widetilde{A} \equiv \sum_k (-1)^{\frac{1}{2}k(k-1)} \langle A \rangle_k.$$

The reversion has the following useful structural properties:

$$(\widetilde{A})^{\sim} = A \quad \text{and} \quad (A \wedge B)^{\sim} = \widetilde{B} \wedge \widetilde{A}.$$

which, together with its action on scalars and vectors, actually may be used to define it algebraically. In mathematical terms, the reversion is often called an *anti-involution* because of these properties; it is an involution since doing it twice is the identity, and it is anti since it reverses order of the reverses.

By contrast, the useful *grade involution* is defined as the monadic operation $\hat{} : \bigwedge^k \mathbb{R}^n \to \bigwedge^k \mathbb{R}^n$ that swaps the parity of the grade:

$$\text{grade involution: } \widehat{\mathbf{A}}_k = (-1)^k \mathbf{A}_k.$$

The grade involution is easily extended to multivectors, and has the properties

$$(\widehat{A})^{\wedge} = A \quad \text{and} \quad (A \wedge B)^{\wedge} = \widehat{A} \wedge \widehat{B}.$$

These properties for the grade involution and the reversion above have nonbold A and B, so these are formulated for multivectors. Of course they also hold for the special case of blades. You can always specialize the theorems of Grassmann algebra in this manner. But do not reverse this process carelessly: only the *linear and distributive* properties of blades extend to multivectors!

2.10 SUMMARY OF OUTER PRODUCT PROPERTIES

Assembling the elements of the definitions throughout this chapter, we now have enough to specify the outer product in the full generality of our needs in this book. We have compacted them a little; the scaling law now follows from associativity and the outer product with scalars, and so on.

In this list, α is a scalar, $\mathbf{a}$ and $\mathbf{b}$ are vectors, and A and B general multivectors of possibly mixed grade.

The outer product is a dyadic product $\wedge : \bigwedge \mathbb{R}^n \times \bigwedge \mathbb{R}^n \to \bigwedge \mathbb{R}^n$, with the following properties:

$$\text{associativity: } A \wedge (B \wedge C) = (A \wedge B) \wedge C$$
$$\text{distributivity: } A \wedge (B + C) = (A \wedge B) + (A \wedge C)$$
$$\text{distributivity: } (A + B) \wedge C = (A \wedge C) + (B \wedge C)$$
$$\text{antisymmetry: } \mathbf{a} \wedge \mathbf{b} = -\mathbf{b} \wedge \mathbf{a}$$
$$\text{scalars: } \alpha \wedge \mathbf{a} = \mathbf{a} \wedge \alpha = \alpha \mathbf{a}$$

We will often use αA as the natural notation for $\alpha \wedge A$.

This set of rules enables computation of any outer product: the distributivity laws allow expansion to outer products of blades, associativity reduces those to outer products of vectors, and antisymmetry and scalar multiplication then permit full simplification to some standard form.

The antisymmetry property for vectors can be lifted to blades. For a k-blade $\mathbf{A}_k$ and an l-blade $\mathbf{B}_l$, it becomes

$$\mathbf{A}_k \wedge \mathbf{B}_l = (-1)^{kl}\, \mathbf{B}_l \wedge \mathbf{A}_k. \tag{2.13}$$

It is good practice in the properties of the outer product to prove this for yourself. Note that (2.13) implies that the outer product is only antisymmetric for two blades of odd grade (which of course includes vectors).

We repeat the important grade-raising property of the outer product:

$$\text{grade}(\mathbf{A} \wedge \mathbf{B}) = \text{grade}(\mathbf{A}) + \text{grade}(\mathbf{B}).$$

This explicitly shows how the outer product constructs the ladder of blades from mere vectors, and is in fact a mapping $\wedge : \bigwedge^k \mathbb{R}^n \times \bigwedge^l \mathbb{R}^n \to \bigwedge^{k+l} \mathbb{R}^n$ connecting blades of specific grades.

2.11 FURTHER READING

With only the outer product defined, it is a bit early to refer you to useful and productive literature. If you are interested in a bit of history, the idea to encode subspaces originated with Hermann Grassmann. There has been some recent recognition of the debt we owe him, as well as embarrassment at the historical neglect of his ideas in mainstream linear algebra [59, 28]. This makes for frustrating reading, for it makes you realize that we could have had these techniques in our tool kit all along.

2.12 EXERCISES

2.12.1 DRILLS

1. Compute the outer products of the following 3-space expressions, giving the results relative to the basis $\{1, \mathbf{e}_1, \mathbf{e}_2, \mathbf{e}_3, \mathbf{e}_1 \wedge \mathbf{e}_2, \mathbf{e}_2 \wedge \mathbf{e}_3, \mathbf{e}_3 \wedge \mathbf{e}_1, \mathbf{e}_1 \wedge \mathbf{e}_2 \wedge \mathbf{e}_3\}$. Show your work.

   (a) $(\mathbf{e}_1 + \mathbf{e}_2) \wedge (\mathbf{e}_1 + \mathbf{e}_3)$
   (b) $(\mathbf{e}_1 + \mathbf{e}_2 + \mathbf{e}_3) \wedge (2\mathbf{e}_1)$
   (c) $(\mathbf{e}_1 - \mathbf{e}_2) \wedge (\mathbf{e}_1 - \mathbf{e}_3)$
   (d) $(\mathbf{e}_1 + \mathbf{e}_2) \wedge (0.5\,\mathbf{e}_1 + 2\mathbf{e}_2 + 3\mathbf{e}_3)$
   (e) $(\mathbf{e}_1 \wedge \mathbf{e}_2) \wedge (\mathbf{e}_1 + \mathbf{e}_3)$
   (f) $(\mathbf{e}_1 + \mathbf{e}_2) \wedge (\mathbf{e}_1 \wedge \mathbf{e}_2 + \mathbf{e}_2 \wedge \mathbf{e}_3)$

2. Given the 2-blade $\mathbf{B} = \mathbf{e}_1 \wedge (\mathbf{e}_2 - \mathbf{e}_3)$ that represents a plane, determine if each of the following vectors lies in that plane. Show your work.

   (a) $\mathbf{e}_1$
   (b) $\mathbf{e}_1 + \mathbf{e}_2$
   (c) $\mathbf{e}_1 + \mathbf{e}_2 + \mathbf{e}_3$
   (d) $2\mathbf{e}_1 - \mathbf{e}_2 + \mathbf{e}_3$

3. What is the area of the parallelogram spanned by the vectors $\mathbf{a} = \mathbf{e}_1 + 2\mathbf{e}_2$ and $\mathbf{b} = -\mathbf{e}_1 - \mathbf{e}_2$ (relative to the area of $\mathbf{e}_1 \wedge \mathbf{e}_2$)?

4. Compute the intersection of the nonhomogeneous line L with position vector $\mathbf{e}_1$ and direction vector $\mathbf{e}_2$, and the line M with position vector $\mathbf{e}_2$ and direction vector $(\mathbf{e}_1 + \mathbf{e}_2)$, using 2-blades. Does the basis $\{\mathbf{e}_1, \mathbf{e}_2\}$ have to be orthonormal?

5. Compute $(2 + 3\mathbf{e}_3) \wedge (\mathbf{e}_1 + \mathbf{e}_2 \wedge \mathbf{e}_3)$ using the grade-based defining equations of Section 2.9.4.

2.12.2 STRUCTURAL EXERCISES

1. The outer product was defined for a vector space $\mathbb{R}^n$ without a metric, but it is of course still defined when we do have a metric space. In $\mathbb{R}^2$ with Euclidean metric, choose an orthonormal basis $\{\mathbf{e}_1, \mathbf{e}_2\}$ in the plane of $\mathbf{a}$ and $\mathbf{b}$ such that $\mathbf{e}_1$ is parallel to $\mathbf{a}$. Write $\mathbf{a} = \alpha \mathbf{e}_1$ and $\mathbf{b} = \beta(\cos\phi\, \mathbf{e}_1 + \sin\phi\, \mathbf{e}_2)$, where ϕ is the angle from $\mathbf{a}$ to $\mathbf{b}$. Evaluate the outer product. Your result should be:

$$\mathbf{a} \wedge \mathbf{b} = \alpha\beta \sin\phi\, (\mathbf{e}_1 \wedge \mathbf{e}_2). \qquad (2.14)$$

   What is the geometrical interpretation?

2. Reconcile (2.14) (which uses lengths α and β and an angle ϕ) with (2.2) (which uses coordinates).

3. The anticommutative algebra has unusual properties, so you should be careful when computing. For real numbers $(x + y)(x - y) = x^2 - y^2$, and for the dot product of two vectors (in a metric vector space) this corresponds simply to $(\mathbf{x} + \mathbf{y}) \cdot (\mathbf{x} - \mathbf{y}) = \mathbf{x} \cdot \mathbf{x} - \mathbf{y} \cdot \mathbf{y}$. Now for comparison compute $(\mathbf{x} + \mathbf{y}) \wedge (\mathbf{x} - \mathbf{y})$ and simplify as far as possible. You should get $-2\mathbf{x} \wedge \mathbf{y}$, which is a rather different result than the other products give! Verify with a drawing that this algebraic result makes perfect sense geometrically in terms of oriented areas.

4. Solve a 3-D version of the problem in Section 2.7.1:

$$\mathbf{x} = \alpha\mathbf{a} + \beta\mathbf{b} + \gamma\mathbf{c},$$

   using an appropriate choice of outer products to selectively compute α, β, γ. What is the geometry of the resulting solution?

5. Consider $\mathbb{R}^4$ with basis $\{\mathbf{e}_i\}_{i=1}^4$. Show that the 2-vector $\mathbf{B} = \mathbf{e}_1 \wedge \mathbf{e}_2 + \mathbf{e}_3 \wedge \mathbf{e}_4$ is not a 2-blade. (i.e., it cannot be written as the outer product of two vectors). (Hint: Set $\mathbf{a} \wedge \mathbf{b} = \mathbf{B}$, develop $\mathbf{a}$ and $\mathbf{b}$ onto the basis, expand the outer product onto the bivector basis, and attempt to solve the resulting set of scalar equations.)

6. Show that $\mathbf{B} = \mathbf{e}_1 \wedge \mathbf{e}_2 + \mathbf{e}_3 \wedge \mathbf{e}_4$ of the previous exercise does not contain any vector other than 0 (see Section 2.8.2 for the definition of *contain*).

7. (The general case of the previous exercises.) Show that a non-zero A contains precisely k independent vectors if and only if A is of the form $A = \mathbf{a}_1 \wedge \mathbf{a}_2 \wedge \cdots \wedge \mathbf{a}_k$ (i.e., if and only if A is a k-blade). This shows that among the multivectors, *only k-blades represent k-dimensional subspaces.*

8. In some literature on Grassmann algebras, one defines the *Clifford conjugate* $\overline{\mathbf{A}}_k$ as

$$\text{Clifford conjugate: } \overline{\mathbf{A}}_k \equiv \widehat{\widetilde{\mathbf{A}}}_k.$$

Is it an involution or an anti-involution? Derive the sign-change for $\mathbf{A}_k$ as an alternative definition of the Clifford conjugate.

9. Prove (2.13): $\mathbf{A}_k \wedge \mathbf{B}_l = (-1)^{kl} \mathbf{B}_l \wedge \mathbf{A}_k$.

2.13 PROGRAMMING EXAMPLES AND EXERCISES

At the end of nearly every chapter in Part I and Part II, we provide some C++ programming examples to make the material less abstract. Some examples simply provide interactive versions of figures that you may also find in the book, as alternatives to the `GAViewer` versions, that are closer to the way you would program them yourselves. Other examples illustrate some important concept that is introduced in the chapter. A few examples go further and actually compute something useful and applicable, like singularity detection or external camera calibration. Yet other examples benchmark the performance of certain techniques and compare efficiency of the solutions of geometric algebra with the classical way. In all, we have intended the examples to be a helpful starting point for your own programming work.

The source code package for the examples can be downloaded from the web site:

http://www.geometricalgebra.net

The package contains projects or makefiles for Windows (Visual Studio .NET) and Linux, Mac OS X, and Solaris (GCC, autotools). We refer you to the instructions provided with the package on how to install it. Our solutions to the programming exercises are provided in the package in a separate directory to help you when you are stuck.

The package comes with a library that we have entitled *GA Sandbox*. This library should make it easy to play around with geometric algebra as used in this book. All basic operations are implemented for the various models of geometry, along with several useful algorithms. The implementation is based on our geometric algebra implementation `Gaigen 2`.

Below is a list of peculiarities of the GA sandbox implementation to bear in mind when reading the source code listings. To learn more about using the sandbox implementation, see the documentation that comes with the package. To learn more about implementation of geometric algebra in general, see Part III of this book.

Models

We use various models of geometry, and the sandbox provides an implementation for each of them:

    e2ga : The vector space model of 2-D Euclidean geometry (2-D algebra).
    e3ga : The vector space model of 3-D Euclidean geometry (3-D algebra).
    h3ga : The homogeneous model of 3-D Euclidean geometry (4-D algebra).
    c3ga : The conformal model of 3-D Euclidean geometry (5-D algebra).

Note that the number in the name of each implementation refers to the physical space that is modeled, not to the dimension of the vector space of algebra. Also note that using a particular algebra is only a matter of taste and simplicity. The conformal model c3ga embeds all the other models, so in principle we could use it to do everything; but as the book builds up to this model, so do the examples.

General Multivectors Versus Specialized Multivectors

Gaigen 2 allows you to use both *general multivectors* and *specialized multivectors*. Examples of specialized multivector classes are vector, bivector, and rotor. These classes can store only the coordinates that are required to represent those types. The other coordinates are assumed to be 0 (for example, assigning a vector value to a bivector variable always results in 0, because the bivector cannot hold a vector value). Using specialized multivectors saves memory, and—more importantly—allows the implementation to be highly optimized.

However, sometimes you may need a variable that can hold *any* multivector value, because you may not know whether the variable will be a vector, a rotor, or any other value. For this purpose, the mv class is provided. It is slower than the specialized classes, but more generic.

Underscore Constructors

Due to the internals of Gaigen 2, there are conversion functions that we have named "underscore constructors." The underscore constructor is not really a constructor in the C++ sense, but rather a regular function that converts an arbitrary multivector value to a specialized multivector variable. Using underscore constructors is required under certain conditions. For example, if you want to assign general multivector to a vector, you should write:

```
mv X = e1;
vector v = _vector(X); // <- note the underscore constructor
```

We trust that you do not find the underscore constructors to be very distracting when reading the source code, since they actually help remind you of the type of function arguments. For example:

```
mv X = e1;
mv Y = e2 ^ e3;
// call function float foo(vector v, bivector b):
float result = foo(_vector(X), _bivector(Y));
```

Operator Bindings

The C++ operator bindings are listed in Table 2.4. Most of these will be defined only later in the book, but it is good to have them all in one table. The geometric product is denoted by a half-space in the book, but this would obviously cause confusion in the code, so we use $*$. This resembles the notation for the scalar product $*$ of Chapter 3. Since the scalar product is used only rarely in both book and code, this should not be a problem. The symbols for the contractions of Chapter 3 have been chosen to point at the elements of lowest grade (i.e., the left contraction points left). We will see how the vector inner product coincides with the contractions and the scalar product, so we need no special symbol for it. That is convenient, since ".." is reserved for referencing the members of a class or structure in C++.

Coordinates

Although geometric algebra is coordinate-free at the level of application programming, its implementations make heavy use of coordinates internally. While we can avoid exposing

Table 2.4: C++ Operator Bindings.		
Code Symbol	Functionality	Book Symbol
$+$	Addition	$+$
$-$	Subtraction (binary), negation (unary)	$-$
$*$	Geometric product	
^	Outer product	$\wedge$
$<<$	Left contraction	$\rfloor$
$>>$	Right contraction	$\lfloor$
%	Scalar product	$*$
$<<$ or % or $>>$	Vector inner product	$\cdot$

ourselves to coordinates most of the time (and you should practice this!), occasionally we
need to access them:

- The most common reason is to transfer multivector values to a library that is not
  based on geometric algebra. For example, to send a vertex to OpenGL, we would
  use:

```
vector v = ...;
glVertex3f(v.e1(), v.e2(), v.e3());
```

  The functions `e1()`, `e2()`, and `e3()` return the respective coordinates of the vector. You can also retrieve all coordinates as an array using the `getC()` function, so
  that you may act on them with matrices:

```
vector v = ...;
const float *C = v.getC(vector_e1_e2_e3);
glVertex3fv(C); // <- this line is just an example of
               //    how to use 'C'
```

  Constants such as `vector_e1_e2_e3` must be passed to `getC()` for two paternal-
  istic reasons:

  1. The constant improves readability: the reader immediately knows how many
     coordinates are returned, and what basis blades they refer to.
  2. `Gaigen 2` *generates* its implementation from specification. Someone
     could decide to reorder the coordinates of `vector` (e.g., `e2` before `e1`), and
     regenerate the implementation. In that case, the constant would change from
     `vector_e1_e2_e3` to `vector_e2_e1_e3`. The result is that code based on the
     old `vector_e1_e2_e3` constant will not compile anymore. That is of course
     preferable to compiling but producing nonsense results at run-time.

- Sometimes we may need to transfer coordinates from one algebra model to another.
  This is not (yet) automated in `Gaigen 2`, so we do it by copying the coordinates one
  by one.
- There are algebraic ways of writing coordinates as components of projections
  involving scalar products or contractions. While those may be useful to see what
  is going on geometrically, they are an inefficient way of retrieving coordinates in a
  program in the rare cases that you really need them and know what you are doing.
  Therefore we permit you to to retrieve them more directly. For instance, the following two lines produce identical results on an orthonormal basis:

```
bivector B = ...;
float e1e2Coordinate_a = B.e1e2();
float e1e2Coordinate_b = _Float(reverse(e1 ^ e2)<<B);
```

Use this capability sparingly, especially when learning geometric algebra, or you
will tend to revert to componentwise linear algebra throughout and not see the
benefits of computing directly with the geometric elements themselves.

The `vector` type

Because we experienced compilation problems with versions of GCC lower than 3.0, we had to explicitly qualify the `vector` type in all examples. For instance, we write `e3ga::vector` instead of just `vector` when we are in the `e3ga` model. This prevents old GCC versions from confusing `e3ga::vector` with `std::vector`, which is an array type provided by the C++ standard template library.

2.13.1 DRAWING BIVECTORS

In this first example, we draw a grid of 2-D bivectors. The code is shown in Figure 2.9 and the output is shown in Figure 2.10. We take two vectors `v1` and `v2`. Vector `v1` is fixed to $\mathbf{e}_1$, and `v2` is rotated 360 degrees in 24 steps of 15 degrees.

The vectors are rendered by the default multivector drawing function `draw()`. We provide two ways to draw the bivectors: as a parallelogram or as a disc. The discs are rendered by `draw()`, but the parallelograms we render ourselves. To switch between the two bivector drawing modes, click anywhere and select the mode from the popup menu.

2.13.2 EXERCISE: HIDDEN SURFACE REMOVAL

In computer graphics, 3-D models are often built from convex polygons. Each polygon is defined by an ordered list of vertices. Most of the time, triangles (3 vertices) or quads (4 vertices) are used. When a solid model is rendered opaque, polygons that face away from the camera are invisible. Because these back-facing polygons are invisible, no time needs to be spent on rasterizing them if they can be singled out early on.

Back-facing polygons can be identified by computing the orientation of the projected (2-D) vertices of the polygon. The convention is that 3-D models are constructed such that the vertices of a polygon have a counterclockwise order when observed from the outside of the model (see Figure 2.11). Back-facing polygons have a clockwise vertex order.

In the example, the surface is triangulated, so we need to find a way to determine the relative orientation of a triangle formed by the endpoints of three vectors $\mathbf{a}$, $\mathbf{b}$, $\mathbf{c}$ in 2-D. It is not $\mathbf{a} \wedge \mathbf{b} \wedge \mathbf{c}$, for that would be zero. Instead, we should consider one of the vertices $\mathbf{a}$ as an anchor, and use the bivector spanned by the difference vectors $(\mathbf{b} - \mathbf{a})$ and $(\mathbf{c} - \mathbf{a})$ relative to the standard bivector $\mathbf{e}_1 \wedge \mathbf{e}_2$. After this hint, implementation should be straightforward.

We have provided the code that renders a 3-D model from 2-D vertices (Figure 2.13). As you can see on the left in Figure 2.12, the code renders the model without back-face culling. The model is rendered as a wireframe so that you can see the back-facing polygons. The right side of Figure 2.12 is the result you should get when you have correctly implemented back-face culling.

```
e3ga::vector v1, v2, v1_plus_v2;
bivector B;

float step = 2 * M_PI / (nbBivectorX * nbBivectorY);
for (float a = 0; a < 2 * M_PI; a += step) {
 // vector 1 is fixed to e1
 v1 = e1;

 // compute vector 2:
 v2 = cos(a) * e1 + sin(a) * e2;

 // compute the bivector:
 B = v1 ^ v2;

 // draw vector 1 (red), vector 2 (green)
 glColor3f(1.0f, 0.0f, 0.0f);
 draw(v1);
 glColor3f(0.0f, 1.0f, 0.0f);
 draw(v2);

 // draw outer product v1^v2:
 glColor3f(0.0f, 0.0f, 1.0f);
 if (!g_drawParallelogram) {
  draw(B);
 }
 else {
  v1_plus_v2 = v1 + v2;
  // draw QUAD with vertices
  // origin -> v1 -> (v1+v2) -> v2
  glBegin(GL_QUADS);
  glVertex2f(0.0f, 0.0f);
  glVertex2f(v1.e1(), v1.e2());
  glVertex2f(v1_plus_v2.e1(), v1_plus_v2.e2());
  glVertex2f(v2.e1(), v2.e2());
  glEnd();
 }

 // ...
}
```

Figure 2.9: Code for drawing bivectors.

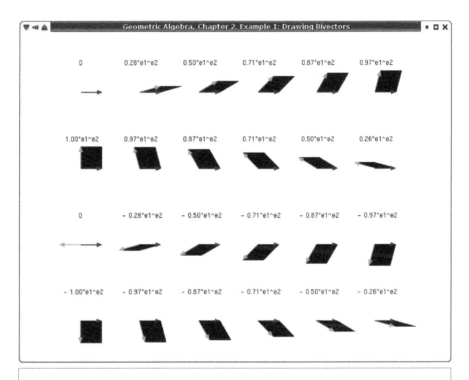

Figure 2.10: Drawing bivectors screenshot (Example 1).

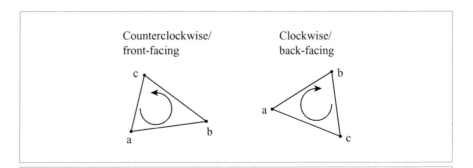

Figure 2.11: The orientation of front- and back-facing polygons.

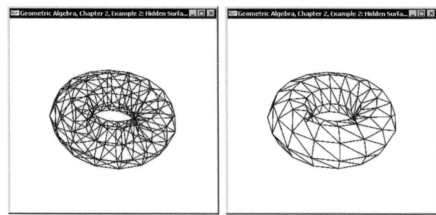

Figure 2.12: On the left, a wireframe torus without back-face culling. On the right, the same torus with back-face culling.

In the sample code, use the left mouse button to rotate the model. The middle and right mouse buttons pop up a menu that allows you to select another model.

If you wonder where the 2-D projected vertices originally came from, we used the OpenGL feedback mechanism to obtain the 2-D viewport coordinates of standard GLUT models. These models can be rendered using a simple call to `glutSolid...()`. GLUT provides functions for rendering teapots, cubes, spheres, cones, tori, dodecahedrons, octahedrons, tetrahedrons, and icosahedrons. See the `getGLUTmodel2D()` function at the bottom of the source file.

2.13.3 SINGULARITIES IN VECTOR FIELDS

As a more advanced application of the outer product, we will show how to use it to locate singularities in a vector field. A vector field is defined by a function V that assigns a vector to every point in space.

For this example, we will work in a 3-D space with a basis $\{e_1, e_2, e_3\}$. So for every point p in space characterized by a position vector $\mathbf{p} = x\,e_1 + y\,e_2 + z\,e_3$, the function V assigns a vector $V(p)$. A simple example of a vector field is the function

$$V(p) = x\,e_1 + 2y\,e_2 + 4z\,e_3. \tag{2.15}$$

A *singularity* in a vector field occurs at any point at which $V(p) = 0$. In the vector field in (2.15), there is a singularity at $\mathbf{p} = 0$, since $V(0) = 0$. In vector field analysis, it is

```
// render model
for (unsigned int i = 0; i < g_polygons2D.size(); i++) {
  // get 2D vertices of the polygon:
  const e3ga::vector &v1 = g_vertices2D[g_polygons2D[i][0]];
  const e3ga::vector &v2 = g_vertices2D[g_polygons2D[i][1]];
  const e3ga::vector &v3 = g_vertices2D[g_polygons2D[i][2]];

  // Exercise:
  // Insert code to remove back-facing polygons here.
  // You can extract the e1^e2 coordinate of a bivector 'B' using:
  // float c = B.e1e2();
  // ...

  // draw polygon
  glBegin(GL_POLYGON);
  for (unsigned int j = 0; j < g_polygons2D[i].size(); j++)
    glVertex2f(
      g_vertices2D[g_polygons2D[i][j]].e1(),
      g_vertices2D[g_polygons2D[i][j]].e2());
  glEnd();
}
```

Figure 2.13: The code that renders a model from its 2-D vertices (Exercise 2).

important to find the locations of singularities. For arbitrary vector fields, that is much more difficult than for the field above.

Consider what happens if we place a box around a singularity and do the following:

- Evaluate the vector field $V(p)$ at all the points p on the surface of this box;
- Normalize each vector to $v(p) = V(p)/\|V(p)\|$ (this requires a metric);
- Place the tail of each vector at the origin.

We now find that the heads of these normalized vectors form a unit sphere at the origin, since they point in all directions. On the other hand, if we place a box around a region of space that does not contain a singularity and repeat the above process, then the tips of the vectors only form part of a sphere. Further, almost all points on this partial sphere have two vectors pointing to it.

While mathematically this process will either give us a sphere or a partial sphere, it requires evaluating the vector field at *all* points on the surface of the box. In practice, we can only sample the vector field at a small number of points and then test the sampled vectors to see if we approximately have a covering of a sphere.

This is where trivectors come in handy. To test if the sampled vector field approximately yields a sphere, we first triangulate the sample points, and then for each triangle of sample points $\triangle p_1 p_2 p_3$, we form the trivector $T_i = \frac{1}{6} \mathsf{V}(p_1) \wedge \mathsf{V}(p_2) \wedge \mathsf{V}(p_3)$ of the normalized vector fields evaluated at those locations. This trivector T_i has the same volume as the tetrahedron formed by the center of the sphere and the three normalized vectors. If we sum the trivectors formed by the normalized vectors of all the triangles of the sample points, the magnitude of the resulting trivector T will approximately be the volume of the sphere if there is a singularity inside the cube that we sampled. If there is no singularity inside the sampling cube, then roughly speaking, each trivector appears twice in the sum, but with opposite sign, and thus T will have a magnitude 0.

Figure 2.14 illustrates the process using a small number of points on the cube. Typically, we normalize T by the volume of the unit sphere, so the magnitude of T should be close to 1 if there is a singularity inside the cube, and close to 0 otherwise.

Figure 2.15 shows code to test for a singularity within a cube. We assume that the vector field has been sampled on a regular grid on each face of the cube and that each vector has been prenormalized. The `SumFace()` procedure computes the sum of the trivectors spanned by two triangles made up of the vertices of one square on the face of the cube. The `TestSingularity()` procedure calls the `SumFace()` procedure for each face of the cube, and sums the value computed for each face. It then normalizes the result by the volume of a unit sphere.

The algorithm can be improved and extended to find higher-order singularities and to find curves and surfaces consisting of singularities. For more details, see [45]. A complex example is shown in Figure 2.16, where the vector field is the gradient of the function $(x - \cos(z))^2 + (y - \sin(z))^2$.

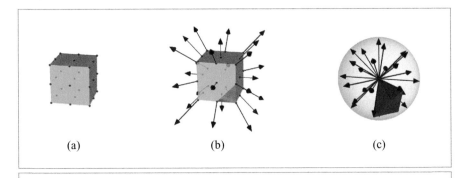

(a)   (b)   (c)

Figure 2.14: Sampling V over a cube and summing the trivectors on the unit sphere.

```
/*
Sum over face
The 'vf' array contains the pre-normalized vector field for the face.
The face is evaluated in a grid-like fashion
at ('gridSize'+1) X ('gridSize'+1) locations.
The resulting (trivector) volume is returned.
*/
trivector sumFace(const vector *vf, int gridSize) {
  trivector sum3d;
  for (int i1 = 0; i1 < gridSize; i1++) {
    for (int i2 = 0; i2 < gridSize; i2++) {
      // cvf = 'current vector field' and points into the vf array
      cvf = vf + i1 * (gridSize + 1) + i2;
      trivector a = _trivector(cvf[0] ^ cvf[gridSize + 2] ^ cvf[gridSize + 1]);
      trivector b = _trivector(cvf[0] ^ cvf[1] ^ cvf[gridSize + 2]);
      sum3d += a+b;
    }
  }
  return sum3d/6.0f;
}


/*
Visits each of the 6 faces of the cube, computes the volume.
Returns true is a singularity is detected
*/
bool testSingularity(const vector *cube[6], int gridSize) {
 // visit all 6 faces
 for (int i = 0; i < 6; i++) {
   sum3d += sumFace(cube[i], gridSize);
 }

 // normalize sum
 sum3d /= 4.0f * 3.14159f / 3.0f;

 // detect point singularity
 return ((norm_e(sum3d) > 0.9) && (norm_e(sum3d) < 1.1));
}
```

Figure 2.15: Code to test for singularity (Example 3). The code was edited for readability. For the unedited source code, see the GA sandbox source code package.

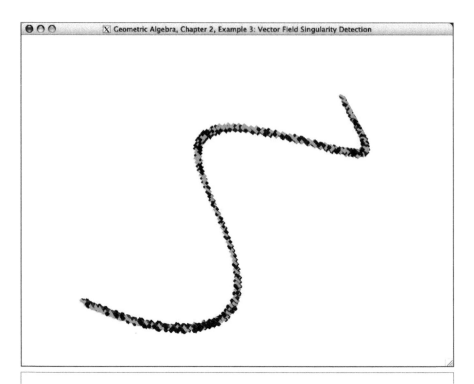

Figure 2.16: A helix-shaped singularity, as detected by Example 3.

3 METRIC PRODUCTS OF SUBSPACES

With the outer product of the previous chapter we can span subspaces. It also enables us to compare lengths on a line, areas in the same plane, and volumes in the same space. We clearly have a need to compare lengths on *different* lines and areas in different planes. The nonmetrical outer product cannot do that, so in this chapter we extend our subspace algebra with a real-valued *scalar product* to serve this (geo)metric need. It generalizes the familiar dot product between vectors to act between blades of the same grade.

Then we carry the algebra further, and investigate how the scalar product and the outer product interact. This automatically leads to an inner product between subspaces of different dimensionality that we call the *contraction*. The contraction of subspace **A** onto subspace **B** computes the part of **B** least like **A**. That also gives us a dual way to characterize subspaces, through blades denoting their orthogonal complement.

With these metric products, we can easily compute geometrically useful operations like the orthogonal projection and determine reciprocal frames for nonorthonormal coordinate systems. We can even use them to embed the 3-D cross product, although we provide strong arguments for using geometric algebra constructs instead.

3.1 SIZING UP SUBSPACES

3.1.1 METRICS, NORMS, AND ANGLES

To establish quantitative measures of subspaces, we need them to be defined in an n-dimensional *metric space* $\mathbb{R}^n$. Such a metric space is just a vector space with a way to compute the norm of an arbitrary vector. That capability can be specified in several ways. The mathematically preferred method is to use a bilinear form Q, which is a scalar-valued function of vectors. That is equivalent to defining an *inner product* $\mathbf{a} \cdot \mathbf{b} = Q[\mathbf{a},\mathbf{b}]$ between two arbitrary vectors $\mathbf{a}$ and $\mathbf{b}$ (also known as the *dot product*). Algebraically, it returns a scalar from two vectors, so it is a mapping $\cdot : \mathbb{R}^n \times \mathbb{R}^n \to \mathbb{R}$, and it is linear and symmetric. It defines a *metric* on the vector space $\mathbb{R}^n$.

For most of this chapter, you can safely develop your intuition by thinking of a *Euclidean metric* in which the dot product is positive definite (the latter meaning that $\mathbf{a} \cdot \mathbf{a}$ is only zero when $\mathbf{a}$ is). Any positive definite metric can be rescaled to a Euclidean metric by choosing one's coordinate axes properly, so this appears quite general enough. These positive definite metrics are called *nondegenerate*. They may seem to be all you need to do Euclidean geometry. But as we already indicated in Chapter 1, there are useful models of Euclidean geometry that use vector spaces with non-Euclidean metrics in their representation of elements in physical space. Such *degenerate metrics* are no longer positive definite, so that for some vector $\mathbf{a}$ the inner product $\mathbf{a} \cdot \mathbf{a}$ can be negative. For some vectors, $\mathbf{a} \cdot \mathbf{a}$ can even be zero without $\mathbf{a}$ being zero; such a vector is called a *null vector*. More detail at this point would distract from our main goal in this chapter, which is to extend the dot product to blades, so we provide Appendix A as additional explanation. As a notation for these metric spaces, we use $\mathbb{R}^{p,q}$ for a space with p positive dimensions and q negative dimensions. A space $\mathbb{R}^{n,0}$ is then an n-dimensional metric space that is effectively a space with a Euclidean metric. We write $\mathbb{R}^n$ if we are not specific on a metric. We will only start using degenerate metrics in Chapter 13.

We commonly use the algebraically defined inner product to compute geometrically useful properties. We compute the length (or *norm*) $\|\mathbf{a}\|$ of a vector $\mathbf{a}$, through

$$\|\mathbf{a}\|^2 = \mathbf{a} \cdot \mathbf{a},$$

and the cosine of the *angle* ϕ between vectors $\mathbf{a}$ and $\mathbf{b}$ through

$$\cos \phi = \frac{\mathbf{a} \cdot \mathbf{b}}{\|\mathbf{a}\| \, \|\mathbf{b}\|}. \tag{3.1}$$

We clearly want such capabilities for general subspaces of the same grade: to be able to assign an absolute measure to their weight (length, area, volume, etc.) and to compare their relative attitudes by an angle measure. We provide for this geometric need algebraically by introducing a real-valued scalar product between blades of the same grade.

3.1.2 DEFINITION OF THE SCALAR PRODUCT *

The *scalar product* is a mapping from a pair of k-blades to the real numbers, and we will denote it by an asterisk ($*$). In mathematical terms, we define a function $* : \bigwedge^k \mathbb{R}^n \times \bigwedge^k \mathbb{R}^n \to \mathbb{R}$. (Do not confuse this terminology with the scalar multiplication in the vector space $\mathbb{R}^n$, which is a mapping $\mathbb{R} \times \mathbb{R}^n \to \mathbb{R}^n$, making a vector $\mathbf{x}$ out of a vector $\mathbf{x}$. As we have seen in the previous chapter, that is essentially the outer product.)

The inner product of vectors is a special case of this scalar product, as applied to vectors. When applied to k-blades, it should at least be backwards compatible with that vector inner product in the case of 1-blades. In fact, the scalar product of two k-blades $\mathbf{A} = \mathbf{a}_1 \wedge \cdots \wedge \mathbf{a}_k$ and $\mathbf{B} = \mathbf{b}_1 \wedge \cdots \wedge \mathbf{b}_k$ can be defined using all combinations of inner products between their vector factors, and this provides that compatibility. It implies that the metric introduced in the original vector space $\mathbb{R}^n$ automatically tells us how to measure the k-blades in $\bigwedge^k \mathbb{R}^n$. The precise combination must borrow the antisymmetric flavor of the spanning product to make it independent of the factorization of the blades $\mathbf{A}$ and $\mathbf{B}$, so that it becomes a quantitative measure that can indeed be interpreted as an absolute area or (hyper)volume.

Let us just define it first, and then show that it works in the next few pages. We conveniently employ the standard notation of a determinant to define the scalar product. It may look intimidating at first, but it is compact and computable.

For k-blades $\mathbf{A} = \mathbf{a}_1 \wedge \ldots \wedge \mathbf{a}_k$ and $\mathbf{B} = \mathbf{b}_1 \wedge \ldots \wedge \mathbf{b}_k$ and scalars α and β, the *scalar product* $* : \bigwedge^k \mathbb{R}^n \times \bigwedge^k \mathbb{R}^n \to \mathbb{R}$ is defined as

$$\alpha * \beta = \alpha\beta$$

$$\mathbf{A} * \mathbf{B} = \begin{vmatrix} \mathbf{a}_1 \cdot \mathbf{b}_k & \mathbf{a}_1 \cdot \mathbf{b}_{k-1} & \ldots & \mathbf{a}_1 \cdot \mathbf{b}_1 \\ \mathbf{a}_2 \cdot \mathbf{b}_k & \mathbf{a}_2 \cdot \mathbf{b}_{k-1} & \ldots & \mathbf{a}_2 \cdot \mathbf{b}_1 \\ \vdots & & \ddots & \vdots \\ \mathbf{a}_k \cdot \mathbf{b}_k & \mathbf{a}_k \cdot \mathbf{b}_{k-1} & \ldots & \mathbf{a}_k \cdot \mathbf{b}_1 \end{vmatrix} \qquad (3.2)$$

$$\mathbf{A} * \mathbf{B} = 0 \quad \text{between blades of unequal grades}$$

Note that the symmetry of the determinant implies some useful symmetries in the scalar product, which we can use in derivations:

$$\mathbf{B} * \mathbf{A} = \mathbf{A} * \mathbf{B} = \widetilde{\mathbf{A}} * \widetilde{\mathbf{B}}. \qquad (3.3)$$

Here the tilde denotes the reversion operation of (2.11) or (2.12).

3.1.3 THE SQUARED NORM OF A SUBSPACE

The (squared) *norm* of a blade can now be defined in terms of its scalar product through

$$\text{squared norm}: \quad \|\mathbf{A}\|^2 = \mathbf{A} * \widetilde{\mathbf{A}}. \qquad (3.4)$$

Let us verify that this algebraic expression indeed gives us a sensible geometric measure of the weight of the subspace represented by the blade, as an area or (hyper)volume.

- **Vectors.** The scalar product of two vectors $\mathbf{a}$ and $\mathbf{b}$ is clearly equal to the standard dot product of the vectors, $\mathbf{a} \cdot \mathbf{b}$. In particular, $\mathbf{a} * \mathbf{a}$ will give us the squared length of the vector $\mathbf{a}$.

- **2-Blades.** For a 2-blade $\mathbf{A}$, factorizable by the outer product as $\mathbf{a}_1 \wedge \mathbf{a}_2$, we obtain

$$
\begin{aligned}
\|\mathbf{A}\|^2 &= (\mathbf{a}_1 \wedge \mathbf{a}_2) * (\mathbf{a}_1 \wedge \mathbf{a}_2)^{\widetilde{}} \\
&= (\mathbf{a}_1 \wedge \mathbf{a}_2) * (\mathbf{a}_2 \wedge \mathbf{a}_1) \\
&= \begin{vmatrix} \mathbf{a}_1 \cdot \mathbf{a}_1 & \mathbf{a}_1 \cdot \mathbf{a}_2 \\ \mathbf{a}_2 \cdot \mathbf{a}_1 & \mathbf{a}_2 \cdot \mathbf{a}_2 \end{vmatrix} \\
&= (\mathbf{a}_1 \cdot \mathbf{a}_1)(\mathbf{a}_2 \cdot \mathbf{a}_2) - (\mathbf{a}_1 \cdot \mathbf{a}_2)^2.
\end{aligned}
$$

This expression is more easily interpreted when we introduce the angle ψ between $\mathbf{a}_1$ and $\mathbf{a}_2$ and use (3.1):

$$
\begin{aligned}
\|\mathbf{A}\|^2 &= (\mathbf{a}_1 \cdot \mathbf{a}_1)(\mathbf{a}_2 \cdot \mathbf{a}_2) - (\mathbf{a}_1 \cdot \mathbf{a}_2)^2 \\
&= \|\mathbf{a}_1\|^2 \|\mathbf{a}_2\|^2 \left(1 - (\cos\psi)^2\right) \\
&= (\|\mathbf{a}_1\| \|\mathbf{a}_2\| \sin\psi)^2
\end{aligned}
$$

We recognize this as the squared area of the parallelogram spanned by $\mathbf{a}_1$ and $\mathbf{a}_2$, precisely as we had hoped when we defined the outer product. Moreover, the properties of the determinant make this independent of the factorization of the blade: factorizing instead as $\mathbf{A} = \mathbf{a}_1 \wedge (\mathbf{a}_2 + \lambda \mathbf{a}_1)$ results in the same value for the scalar product, as you can easily verify.

- **k-Blades.** We know from the previous chapter that the k-volume associated with a k-blade $\mathbf{A} = \mathbf{a}_1 \wedge \cdots \wedge \mathbf{a}_k$ is proportional to the determinant of the matrix $[\![\mathbf{A}]\!] = [\![\mathbf{a}_1 \cdots \mathbf{a}_k]\!]$. Once you realize that the scalar product definition of the squared norm can be written in terms of a matrix product as $\|\mathbf{A}\|^2 = \det([\![\mathbf{A}]\!]^T [\![\mathbf{A}]\!])$, you can use the properties of determinants to simplify this to $\det([\![\mathbf{A}]\!])^2$. So indeed, for k-blades, we do compute the squared k-dimensional hypervolume.

3.1.4 THE ANGLE BETWEEN SUBSPACES

Applying the scalar product to two different blades of the same grade, we also would hope that the scalar product $\mathbf{A} * \mathbf{B}$ has a geometrical meaning that expresses the cosine of the relative angle ϕ between $\mathbf{A}$ and $\mathbf{B}$ in terms of the scalar product (in analogy to the dot product equation (3.1)). This is indeed the case, and the precise definition is

$$
angle: \quad \cos\phi = \frac{\mathbf{A} * \widetilde{\mathbf{B}}}{\|\mathbf{A}\| \|\mathbf{B}\|}. \tag{3.5}
$$

Let us verify that the scalar product indeed yields the correct angle.

- **Vectors.** This formula clearly works for vectors, since it reverts to the well-known vector formula $\cos \phi = (\mathbf{a} \cdot \mathbf{b})/(\|\mathbf{a}\| \|\mathbf{b}\|)$.

- **2-Blades.** For 2-blades $\mathbf{A} = \mathbf{a}_1 \wedge \mathbf{a}_2$ and $\mathbf{B} = \mathbf{b}_1 \wedge \mathbf{b}_2$, we would define their relative angle conceptually by finding out what it takes to turn one onto the other, as illustrated in Figure 3.1(a) and (b). In 3-D, this involves first finding a common factor $\mathbf{c}$ with $\|\mathbf{c}\| = 1$. Next, we reshape the two 2-blades to have $\mathbf{c}$ as one component and a vector perpendicular to $\mathbf{c}$ as the other component:

$$\mathbf{a}_1 \wedge \mathbf{a}_2 = \mathbf{a} \wedge \mathbf{c} \text{ with } \mathbf{a} \cdot \mathbf{c} = 0,$$
$$\mathbf{b}_1 \wedge \mathbf{b}_2 = \mathbf{b} \wedge \mathbf{c} \text{ with } \mathbf{b} \cdot \mathbf{c} = 0.$$

Now evaluating the scalar product shows that the angle between the original 2-blades $\mathbf{A}$ and $\mathbf{B}$ has effectively been reduced to the angle between the vectors $\mathbf{a}$ and $\mathbf{b}$, for which we have a clear definition through the inner product formula:

$$
\begin{aligned}
\mathbf{A} * \widetilde{\mathbf{B}} &= (\mathbf{a}_1 \wedge \mathbf{a}_2) * (\mathbf{b}_1 \wedge \mathbf{b}_2)^{\sim} \\
&= (\mathbf{a} \wedge \mathbf{c}) * (\mathbf{b} \wedge \mathbf{c})^{\sim} \\
&= (\mathbf{a} \wedge \mathbf{c}) * (\mathbf{c} \wedge \mathbf{b}) \\
&= \begin{vmatrix} \mathbf{a} \cdot \mathbf{b} & \mathbf{a} \cdot \mathbf{c} \\ \mathbf{c} \cdot \mathbf{b} & \mathbf{c} \cdot \mathbf{c} \end{vmatrix} \\
&= \begin{vmatrix} \mathbf{a} \cdot \mathbf{b} & 0 \\ 0 & 1 \end{vmatrix} \\
&= \mathbf{a} \cdot \mathbf{b} \\
&= \|\mathbf{a}\| \|\mathbf{b}\| \cos \phi \\
&= \|\mathbf{A}\| \|\mathbf{B}\| \cos \phi,
\end{aligned}
$$

where we used that $\|\mathbf{A}\| = \|\mathbf{a}\|$, easily derived from $\|\mathbf{A}\|^2 = \|\mathbf{a} \wedge \mathbf{c}\|^2 = (\mathbf{a} \cdot \mathbf{a})(\mathbf{c} \cdot \mathbf{c}) - (\mathbf{a} \cdot \mathbf{c})^2 = \mathbf{a} \cdot \mathbf{a} = \|\mathbf{a}\|^2$, and the fact that norms are positive.

So the formula works for 2-blades, giving the angle we expect. Of course, in practice you would evaluate it directly algebraically using the definition, and without the geometric reshaping involving $\mathbf{c}$—the derivation was merely meant to show that the resulting value indeed gives the angle we desire.

- **k-Blades.** The 2-blade example also motivates the result for general k-blades. All common factors between the two blades can be taken out (as $\mathbf{c}$ was in the example above), and they don't affect the cosine value. Of the remainder, there are three possibilities:

  1. **Only scalars are left.** The blades were then multiples of each other, so their angle is 0, the cosine equals 1, and the scalar product is the product of their norms.

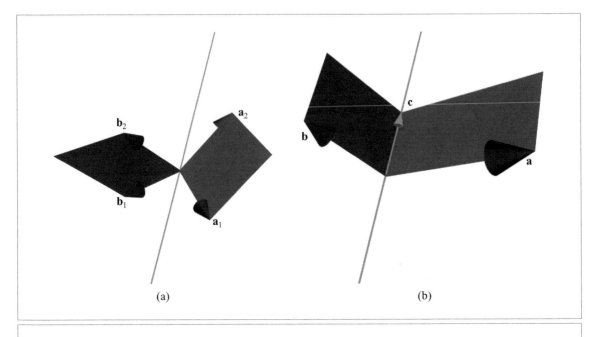

(a)   (b)

Figure 3.1: Computing the scalar product of 2-blades.

2. *We are left with one vector in each term.* We can then rotate one vector onto the other with a well-defined rotation, so their relative angle and its cosine are well defined through the usual dot product formula. We saw that in the 2-blade example above.

3. *We are left with totally disjoint subblades of at least grade 2.* We then need at least two rotations in orthogonal 2-blades to bring the subblades into alignment (see [52]). The cosine computed by (3.5) is now equal to the product of the cosines of these orthogonal rotations, and therefore zero if at least one of them is over a right angle. In that case, the blades should be considered perpendicular. (An example of such a situation in 4-D space involves the 2-blades $\mathbf{A} = \mathbf{e}_1 \wedge \mathbf{e}_3$ and $\mathbf{B} = (\cos \alpha \, \mathbf{e}_1 - \sin \alpha \, \mathbf{e}_2) \wedge (\cos \beta \, \mathbf{e}_3 - \sin \beta \, \mathbf{e}_4)$. Verify that $\mathbf{A} \wedge \mathbf{B} = -\sin \alpha \sin \beta \, \mathbf{e}_1 \wedge \mathbf{e}_2 \wedge \mathbf{e}_3 \wedge \mathbf{e}_4$, so that they are indeed disjoint, while their cosine equals $\mathbf{A} * \widetilde{\mathbf{B}} = \cos \alpha \cos \beta$.)

Reinterpreting a zero cosine within this larger context, it means that two blades are perpendicular if they require at least one right-angle rotation to align them. This sounds like a reasonable enough extension of the concept of perpendicularity to such higher dimensional spaces. In this case, the algebraic imperative should probably inform our geometric intuition, which is ill-developed for more than three dimensions.

As an aside at this point, you may remark that the two useful geometrical properties we can recover from the scalar product (norm and cosine) both involve a reversion of the second argument. If we would have absorbed that reversion into the definition of the scalar product, those formulas would be simpler, and the definition would be tidied up as well (since we would get a matrix, of which element (i, j) would be $\mathbf{a}_i \cdot \mathbf{b}_j$). This is true, and done in some mathematical texts such as [41] (though the reversion then pops up soon after, in the definition of the contraction product). We have chosen (3.2) since that coincides with the scalar product that you will find in geometric algebra literature, though defined in a different manner (and then quite naturally leading to that relatively reversed form, as we will see in Chapter 6).

3.2 FROM SCALAR PRODUCT TO CONTRACTION

In the computation of the relative weight and the attitude of k-dimensional subspaces of the same grade, we sensibly reduced the scalar products to a lower-dimensional situation by factoring out a common blade. This dimension reduction of subspaces is of course very useful geometrically, since it focuses the computation on the geometrically different factors of the subspaces. We now formalize it algebraically, in terms of the outer product and scalar product so that our subspace algebra remains consistent. Fortunately, that formalization can be done relatively simply using the mathematical structure of the subspace algebra. The geometric interpretation then follows the algebraic justification.

3.2.1 IMPLICIT DEFINITION OF CONTRACTION ⌋

We first define a contraction operation implicitly. Our implicit definition is fashioned by using the scalar product on blades of the same grade with a common factor $\mathbf{A}$, and then insisting that the scalar product of the two blades be equal to the scalar product of those blades with the common factor $\mathbf{A}$ removed. That obviously constrains the algebraic properties of the removal operation and in fact makes the removal operation into a product that we identify with the contraction.

Let $\mathbf{B}$ and $\mathbf{Y}$ be blades of the same grade, with a common factor $\mathbf{A}$. Writing $\mathbf{Y} = \mathbf{X} \wedge \mathbf{A}$, let us try to remove this common factor $\mathbf{A}$ from $\mathbf{Y}$. (Figure 3.2 illustrates this for the 2-blades $\mathbf{x} \wedge \mathbf{a}$ and $\mathbf{B}$, with common vector factor $\mathbf{a}$.) We attempt to rewrite the scalar product $\mathbf{Y} * \mathbf{B}$ as a scalar product of $\mathbf{X}$ with "$\mathbf{A}$ removed from $\mathbf{B}$". For now we denote the latter by $\mathbf{A} \rfloor \mathbf{B}$, an obviously asymmetrical notation for an obviously asymmetrical geometric concept. So we should have

$$(\mathbf{X} \wedge \mathbf{A}) * \mathbf{B} = \mathbf{X} * (\mathbf{A} \rfloor \mathbf{B}) \tag{3.6}$$

as a desired property of this new element $\mathbf{A} \rfloor \mathbf{B}$. This somewhat defines the properties of the new blade $\mathbf{A} \rfloor \mathbf{B}$ relative to this chosen $\mathbf{Y}$ (and hence $\mathbf{X}$). We could pick another $\mathbf{Y}$ (still

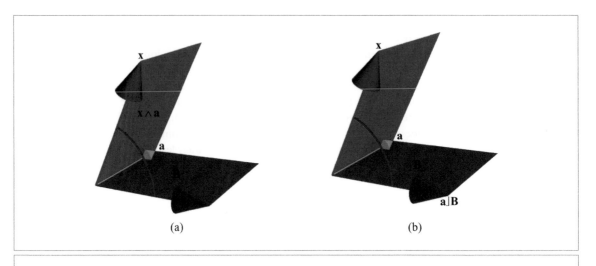

(a)   (b)

Figure 3.2: From scalar product to contraction. The scalar product between a 2-blade $\mathbf{Y} = \mathbf{x} \wedge \mathbf{a}$ and a 2-blade $\mathbf{B}$ can be reduced to a computation on vectors by taking $\mathbf{a}$ out. That results in the evaluation of $\mathbf{x}$ with the vector $\mathbf{a} \rfloor \mathbf{B}$, the contraction of $\mathbf{a}$ on $\mathbf{B}$.

with a common factor of $\mathbf{A}$ with $\mathbf{B}$, so actually we have picked another $\mathbf{X}$) and learn a bit more about the value of $\mathbf{A} \rfloor \mathbf{B}$. If fact, we can try all possibilities for $\mathbf{X}$, of all grades, and if the metric is nondegenerate this actually completely determines what $\mathbf{A} \rfloor \mathbf{B}$ must be (see structural exercise 1 for an example). If the metric is degenerate, vectors can have an inner product of zero without being zero themselves, and this means that the value of $\mathbf{A} \rfloor \mathbf{B}$ cannot be determined completely (see structural exercise 2).

These properties imply that (at least in nondegenerate metrics) this element $\mathbf{A} \rfloor \mathbf{B}$ is not just a weird shorthand notation for a reduced blade, but that we have actually defined a new product in our subspace algebra, making a new blade from its two factors. We call $\mathbf{A} \rfloor \mathbf{B}$ the *contraction of $\mathbf{A}$ onto $\mathbf{B}$*.

We can easily compute the grade of this new element. The left side of (3.6) is nonzero if $\mathrm{grade}(\mathbf{X}) + \mathrm{grade}(\mathbf{A}) = \mathrm{grade}(\mathbf{B})$, and the right side is nonzero if $\mathrm{grade}(\mathbf{A} \rfloor \mathbf{B}) = \mathrm{grade}(\mathbf{X})$. Therefore,

$$\mathrm{grade}(\mathbf{A} \rfloor \mathbf{B}) = \mathrm{grade}(\mathbf{B}) - \mathrm{grade}(\mathbf{A})$$

(with only positive grades allowed), and we see that *the contraction is a grade-reducing product between blades*. As a function mapping defined between blades of different grades $\rfloor : \bigwedge^k \mathbb{R}^n \times \bigwedge^l \mathbb{R}^n \rightarrow \bigwedge^{l-k} \mathbb{R}^n$, it is bilinear since both the outer product and the scalar product are. For the same reason, it is distributive over addition, so that we could lift its definition to the full Grassmann algebra context. But as always, we will mostly use it on blades, and we keep our explanation simple by restricting ourselves to blades in this chapter.

When the grades of $\mathbf{A}$ and $\mathbf{B}$ are the same, $\mathbf{X}$ is a nonzero scalar; call this scalar ξ. Then the left-hand side of (3.6) is $(\xi \wedge \mathbf{A}) * \mathbf{B} = \xi(\mathbf{A} * \mathbf{B})$, while the right-hand side equals $\xi(\mathbf{A}\rfloor\mathbf{B})$. So *for same-grade blades, the contraction is identical to the scalar product*. We can therefore view the contraction as the general grade-reduction product applying to any pair of blades, automatically reducing to the more specific scalar product when it can. This reduces the number of symbols, so it is what one uses in practice—although it somewhat obscures the fact that the scalar product is more fundamental.

3.2.2 COMPUTING THE CONTRACTION EXPLICITLY

The definition of the contraction by (3.6) has two disadvantages: it is implicit, and it only works for nondegenerate metrics. We need a universally valid, explicitly constructive, computational formula for $\mathbf{A}\rfloor\mathbf{B}$.

We give it now, in an axiomatic form that immediately suggests a recursive program to evaluate the contraction. Of course we have to show that this procedure indeed computes the same product as the implicit definition of (3.6) (when that is defined), and that takes a bit of work.

The *contraction* $\rfloor$ is a product producing a $(l - k)$-blade from a k-blade and an l-blade (so it is a bilinear mapping $\rfloor : \bigwedge^k \mathbb{R}^n \times \bigwedge^l \mathbb{R}^n \to \bigwedge^{l-k} \mathbb{R}^n$), with the following defining properties:

$$\alpha\rfloor\mathbf{B} = \alpha\,\mathbf{B} \tag{3.7}$$

$$\mathbf{B}\rfloor\alpha = 0 \quad \text{if grade}(\mathbf{B}) > 0 \tag{3.8}$$

$$\mathbf{a}\rfloor\mathbf{b} = \mathbf{a} \cdot \mathbf{b} \tag{3.9}$$

$$\mathbf{a}\rfloor(\mathbf{B} \wedge \mathbf{C}) = (\mathbf{a}\rfloor\mathbf{B}) \wedge \mathbf{C} + (-1)^{\text{grade}(\mathbf{B})}\mathbf{B} \wedge (\mathbf{a}\rfloor\mathbf{C}) \tag{3.10}$$

$$(\mathbf{A} \wedge \mathbf{B})\rfloor\mathbf{C} = \mathbf{A}\rfloor(\mathbf{B}\rfloor\mathbf{C}), \tag{3.11}$$

where α is a scalar, $\mathbf{a}$ and $\mathbf{b}$ are vectors, and $\mathbf{A}$, $\mathbf{B}$, and $\mathbf{C}$ are blades (which could be scalars or vectors as well as higher-dimensional blades).

It follows that the contraction has useful bilinear and distributive properties:

$$(\mathbf{A} + \mathbf{B})\rfloor\mathbf{C} = \mathbf{A}\rfloor\mathbf{C} + \mathbf{B}\rfloor\mathbf{C} \tag{3.12}$$

$$\mathbf{A}\rfloor(\mathbf{B} + \mathbf{C}) = \mathbf{A}\rfloor\mathbf{B} + \mathbf{A}\rfloor\mathbf{C} \tag{3.13}$$

$$(\alpha\,\mathbf{A})\rfloor\mathbf{B} = \alpha\,(\mathbf{A}\rfloor\mathbf{B}) = \mathbf{A}\rfloor(\alpha\,\mathbf{B}) \tag{3.14}$$

Before we relate this new explicit definition to the implicit definition of (3.6), note that it can indeed be used to compute the contraction product of arbitrary blades by the following procedure. We first split off a vector $\mathbf{a}$ from any first argument k-blade $\mathbf{A}_k$ in the expression $\mathbf{A}_k \rfloor \mathbf{B}_l$. Then, writing $\mathbf{A}_k \equiv \mathbf{A}_{k-1} \wedge \mathbf{a}$, we use (3.11) as

$$\mathbf{A}_k \rfloor \mathbf{B}_l = (\mathbf{A}_{k-1} \wedge \mathbf{a}) \rfloor \mathbf{B}_l = \mathbf{A}_{k-1} \rfloor (\mathbf{a} \rfloor \mathbf{B}_l). \tag{3.15}$$

Here, $\mathbf{a} \rfloor \mathbf{B}_l$ is an $(l-1)$-blade. Therefore we have reduced the contraction of a k-blade onto an l-blade to that of a $(k-1)$-blade onto an $(l-1)$-blade. Proceeding by splitting off additional vectors, we can therefore reduce the expression completely to that of a scalar 0-blade and a $(k-l)$-blade, or an $(l-k)$-blade and a scalar (which gives 0 if $l \neq k$ by (3.8)).

For instance, let us compute the contraction of $\mathbf{A} = \mathbf{e}_1 \wedge \mathbf{e}_2$ onto $\mathbf{B} = \mathbf{e}_1 \wedge \mathbf{e}_3 \wedge \mathbf{e}_2$, with $\{\mathbf{e}_i\}_{i=1}^n$ an orthonormal basis in the Euclidean metric space $\mathbb{R}^{n,0}$, so that $\mathbf{e}_i \cdot \mathbf{e}_j = 0$ for $i \neq j$, and $\mathbf{e}_i \cdot \mathbf{e}_i = 1$. We indicate the rewriting steps:

$$
\begin{aligned}
(\mathbf{e}_1 \wedge \mathbf{e}_2) \rfloor (\mathbf{e}_1 \wedge \mathbf{e}_3 \wedge \mathbf{e}_2) = & \\
= \mathbf{e}_1 \rfloor (\mathbf{e}_2 \rfloor (\mathbf{e}_1 \wedge \mathbf{e}_3 \wedge \mathbf{e}_2)) & \qquad \text{by (3.11)} \\
= \mathbf{e}_1 \rfloor ((\mathbf{e}_2 \rfloor \mathbf{e}_1) \wedge (\mathbf{e}_3 \wedge \mathbf{e}_2) - \mathbf{e}_1 \wedge (\mathbf{e}_2 \rfloor (\mathbf{e}_3 \wedge \mathbf{e}_2))) & \qquad \text{by (3.10)} \\
= \mathbf{e}_1 \rfloor ((\mathbf{e}_2 \rfloor \mathbf{e}_1) \wedge (\mathbf{e}_3 \wedge \mathbf{e}_2) - \mathbf{e}_1 \wedge ((\mathbf{e}_2 \rfloor \mathbf{e}_3) \wedge \mathbf{e}_2 - \mathbf{e}_3 \wedge (\mathbf{e}_2 \rfloor \mathbf{e}_2))) & \qquad \text{by (3.10)} \\
= \mathbf{e}_1 \rfloor (\mathbf{e}_1 \wedge \mathbf{e}_3) & \qquad \text{by (3.9)} \\
= (\mathbf{e}_1 \rfloor \mathbf{e}_1) \wedge \mathbf{e}_3 - \mathbf{e}_1 \wedge (\mathbf{e}_1 \rfloor \mathbf{e}_3) & \qquad \text{by (3.10)} \\
= \mathbf{e}_3. &
\end{aligned}
$$

We showed this computation in all its detail; with (3.16), below, such evaluations become one-liners, done on sight.

If you want to have the contraction defined for general multivectors in the context of a Grassmann algebra, you just use the linear properties of the contraction to write its arguments as a sum of blades, and use distributivity to reduce the result to a sum of contractions of blades.

3.2.3 ALGEBRAIC SUBTLETIES

In Appendix B.2, we give the actual proof that the explicit definition of (3.7)–(3.11) agrees with the implicit definition (3.6). Reading through the proof will familiarize you with the algebra and its techniques, but it is better to first understand the geometrical meaning of the contraction as explained in the next section. So you had better leave the proof for a second pass.

Still, a really useful thing to come out of the detailed proof in the Appendix B is a formula that permits us to do the contraction $\mathbf{x} \rfloor \mathbf{A}$ of a vector $\mathbf{x}$ on a blade $\mathbf{A}$ by passing the operation $\mathbf{x} \rfloor$ through the factored blade. It is

$$\mathbf{x} \rfloor (\mathbf{a}_1 \wedge \mathbf{a}_2 \wedge \cdots \wedge \mathbf{a}_k) = \sum_{i=1}^{k} (-1)^{i-1} \, \mathbf{a}_1 \wedge \mathbf{a}_2 \wedge \cdots \wedge (\mathbf{x} \rfloor \mathbf{a}_i) \wedge \cdots \wedge \mathbf{a}_k. \tag{3.16}$$

Since $\mathbf{x} \rfloor \mathbf{a}_i = \mathbf{x} \cdot \mathbf{a}_i$, this reduces the contraction of a vector onto a blade to a series of inner products. This basically implements the product occurring on the left-hand side of (3.15) as a one-liner, and therefore facilitates the evaluation of arbitrary contractions between blades. The special case for bivectors is

$$\mathbf{x} \rfloor (\mathbf{a}_1 \wedge \mathbf{a}_2) = (\mathbf{x} \cdot \mathbf{a}_1)\, \mathbf{a}_2 - (\mathbf{x} \cdot \mathbf{a}_2)\, \mathbf{a}_1 \tag{3.17}$$

We will use it often in computations where we chose to drop the wedges for the scalar multiples.

3.3 GEOMETRIC INTERPRETATION OF THE CONTRACTION

All the above formulas are the unavoidable algebraic consequences of our simple desire to design a product that could factor the metric scalar product. Now the time has come to investigate the geometric properties of this new product between subspaces. We begin with the following observations to develop our intuition about the contraction of two blades:

1. $\mathbf{A} \rfloor \mathbf{B}$ is a blade when $\mathbf{A}$ and $\mathbf{B}$ are, so $\mathbf{A} \rfloor \mathbf{B}$ represents an oriented subspace with specific attitude, orientation, and weight.

2. The blade $\mathbf{A} \rfloor \mathbf{B}$ represents a subspace that is contained in $\mathbf{B}$. To show this, factor one vector $\mathbf{a}$ out of $\mathbf{A}$, giving $\mathbf{A} = \mathbf{A}' \wedge \mathbf{a}$. Then, by (3.11), we obtain

$$\mathbf{A} \rfloor \mathbf{B} = (\mathbf{A}' \wedge \mathbf{a}) \rfloor \mathbf{B} = \mathbf{A}' \rfloor (\mathbf{a} \rfloor \mathbf{B}).$$

   The term $\mathbf{a} \rfloor \mathbf{B}$ is of the form (3.16), and it is definitely in $\mathbf{B}$ since it only contains vectors of $\mathbf{B}$. Now split another vector off and recurse—the property of remaining in $\mathbf{B}$ inherits. Recursion stops when all that is left of $\mathbf{A}$ is a scalar; then (3.7) shows that the final result is still in $\mathbf{B}$. At any point in this recursion, we may encounter a 0 result, notably when the grade of $\mathbf{A}$ exceeds that of $\mathbf{B}$. But 0 is the empty blade, and as such contained in any blade (of any grade), so it is also in $\mathbf{B}$.

3. For a vector $\mathbf{x}$, having $\mathbf{x} \rfloor \mathbf{A} = 0$ means that $\mathbf{x}$ is perpendicular to *all* vectors in $\mathbf{A}$. This follows immediately from the expansion (3.16): the right-hand side can only be zero if all $\mathbf{x} \rfloor \mathbf{a}_i = \mathbf{x} \cdot \mathbf{a}_i$ are zero; therefore $\mathbf{x}$ is perpendicular to all vectors in a basis of the subspace $\mathbf{A}$; therefore $\mathbf{x}$ is perpendicular to all of $\mathbf{A}$.

4. The outcome of $\mathbf{A} \rfloor \mathbf{B}$ is perpendicular to the subspace $\mathbf{A}$. The proof is simple: take a vector $\mathbf{a}$ of $\mathbf{A}$, then $\mathbf{a} \wedge \mathbf{A} = 0$. Now by (3.11) for the contraction, $\mathbf{a} \rfloor (\mathbf{A} \rfloor \mathbf{B}) = (\mathbf{a} \wedge \mathbf{A}) \rfloor \mathbf{B} = 0 \rfloor \mathbf{B} = 0$, so $\mathbf{a}$ is perpendicular to $\mathbf{A} \rfloor \mathbf{B}$ by item 3 above. But $\mathbf{a}$ was just an arbitrary vector in $\mathbf{A}$. Choosing a set of vectors that forms a basis for $\mathbf{A}$, we can thus show that all of them are perpendicular to $\mathbf{A} \rfloor \mathbf{B}$. Therefore, the whole subspace $\mathbf{A}$ is perpendicular to the subspace $\mathbf{A} \rfloor \mathbf{B}$.

5. The norm of the blade $\mathbf{A} \rfloor \mathbf{B}$ is proportional to the norm of $\mathbf{A}$, the norm of $\mathbf{B}$, and the cosine of the angle between $\mathbf{A}$ and its projection onto $\mathbf{B}$. The derivation of this

is a straightforward application of the norm and angle definitions. It may be found in Appendix B.4, but is probably better appreciated after learning about orthogonal projections in Section 3.6.

6. As we have seen,

$$\text{grade}(\mathbf{A}\rfloor\mathbf{B}) = \text{grade}(\mathbf{B}) - \text{grade}(\mathbf{A}).$$

This confirms the original motivation of $\mathbf{A}\rfloor\mathbf{B}$ as "$\mathbf{A}$ taken out of $\mathbf{B}$": the subspace $\mathbf{B}$ loses the dimension of $\mathbf{A}$ in the contraction. Since blades with negative grades do not exist, the contraction result is zero when $\text{grade}(\mathbf{A}) > \text{grade}(\mathbf{B})$.

We attempt to summarize these results as:

The contraction $\mathbf{A}$ on $\mathbf{B}$ of an a-blade $\mathbf{A}$ and a b-blade $\mathbf{B}$ is a specific subblade of $\mathbf{B}$ of grade $b - a$ perpendicular to $\mathbf{A}$, with a weight proportional to the norm of $\mathbf{B}$ and to the norm of the projection of $\mathbf{A}$ onto $\mathbf{B}$.

When we equate blades with subspaces, we see that this is a compact geometric statement about the meaning of the contraction. It combines the two geometrical concepts of *containment in a subspace* and *perpendicularity* into one product with well-defined algebraic properties. Still, the statement is rather involved. We provide a more geometrically intuitive visualization of the contraction in Section 3.6 (in terms of the projection of $\mathbf{A}$ onto $\mathbf{B}$).

For a 2-blade $\mathbf{B}$ and a vector $\mathbf{x}$, the situation in 3-D is depicted in Figure 3.3: $\mathbf{x}\rfloor\mathbf{B}$ is a vector on a line in the plane determined by $\mathbf{B}$, perpendicular to $\mathbf{x}$. (Obviously, all vectors

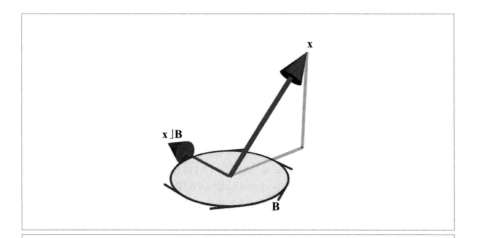

Figure 3.3: The contraction of a vector onto a 2-blade, in a right-handed 3-D Euclidean space.

along the line determined by the $\mathbf{x}\rfloor\mathbf{B}$ would be sensible results, as long as they are linear in the magnitudes of $\mathbf{x}$ and $\mathbf{B}$.) Intuitively, you could think of $\mathbf{x}\rfloor\mathbf{B}$ as *the blade of the largest subspace of* $\mathbf{B}$ *that is most unlike* $\mathbf{x}$.

The geometric meaning of the contraction as the largest subspace most unlike a given space even extends to the extreme cases of scalars interpreted geometrically as weighted points at the origin, as long as you realize that the algebraic blade "zero" should be interpreted geometrically as the empty subspace. We give structural exercises 4–6 to explore that point and convince you of the consistency across the whole range of validity.

We stated that the norm of $\mathbf{A}\rfloor\mathbf{B}$ is in general proportional to the norms of $\mathbf{B}$ and the projection of $\mathbf{A}$ onto $\mathbf{B}$. In the example of Figure 3.3, this is perhaps most easily shown by introducing orthonormal coordinates $\{\mathbf{e}_i\}_{i=1}^{3}$. Choose them such that $\mathbf{B} = \|\mathbf{B}\|\,\mathbf{e}_1 \wedge \mathbf{e}_2$ and $\mathbf{x} = \|\mathbf{x}\|\,(\mathbf{e}_1 \cos\phi + \mathbf{e}_3 \sin\phi)$, then you compute easily that $\mathbf{x}\rfloor\mathbf{B} = \|\mathbf{x}\|\,\|\mathbf{B}\|\,\cos\phi\,\mathbf{e}_2$, confirming all aspects of attitude, orientation, and weight.

When $\mathbf{x}$ is perpendicular to $\mathbf{B}$, the contraction is zero. In the example we should interpret that geometrically as the nonuniqueness of a 1-D subspace of the 2-blade $\mathbf{B}$ that is perpendicular to $\mathbf{x}$ (any homogeneous line in $\mathbf{B}$ would do). The small weight of the contraction blade in an almost perpendicular situation is then an indication of the numerical significance or numerical instability of the result. That interpretation is confirmed by the opposite situation: when $\mathbf{x}$ is contained in $\mathbf{B}$, the cosine is 1, and the contraction is the orthogonal vector within the plane. Nothing is lost in the projection to the plane that determines the norm of $\mathbf{x}\rfloor\mathbf{B}$, and in that sense $\mathbf{x}$ is as stably perpendicular to its projection as it can ever be.

3.4 THE OTHER CONTRACTION ⌊

We motivated the original definition of $\mathbf{A}\rfloor\mathbf{B}$ in (3.6) in terms of geometrical subspaces as "$\mathbf{A}$ taken out of $\mathbf{B}$". This is clearly asymmetrical in $\mathbf{A}$ and $\mathbf{B}$, and we could also have used the same geometrical intuition to define an operation $\mathbf{B}\lfloor\mathbf{A}$, interpreted as "take $\mathbf{B}$ and remove $\mathbf{A}$ from it." The two are so closely related that we really only need one to set up our algebra, but occasionally formulas get simpler when we switch over to this other contraction. Let us briefly study their relationship.

Define the *right contraction* implicitly by taking $\mathbf{A}$ out of the scalar product:

$$\textit{right contraction}:\quad \mathbf{B} * (\mathbf{A} \wedge \mathbf{X}) = (\mathbf{B}\lfloor\mathbf{A}) * \mathbf{X}. \tag{3.18}$$

This is a simple interchange of the order of factors relative to (3.6). The relationship between the two contractions is easily established from the reversion symmetry of the scalar product (3.3) as

$$\mathbf{B}\lfloor\mathbf{A} = (\widetilde{\mathbf{A}}\rfloor\widetilde{\mathbf{B}})^{\sim} = (-1)^{a(b+1)}\mathbf{A}\rfloor\mathbf{B}, \tag{3.19}$$

with $a = \mathrm{grade}(\mathbf{A})$ and $b = \mathrm{grade}(\mathbf{B})$. The first equation follows from a straightforward computation using the reversion signs: $\widetilde{\mathbf{X}} * (\mathbf{B}\lfloor\mathbf{A})^{\sim} = \mathbf{X} * (\mathbf{B}\lfloor\mathbf{A}) = (\mathbf{B}\lfloor\mathbf{A}) * \mathbf{X} = \mathbf{B} * (\mathbf{A}\wedge\mathbf{X}) = \mathbf{B} * (\widetilde{\mathbf{X}}\wedge\widetilde{\mathbf{A}})^{\sim} = (\widetilde{\mathbf{X}}\wedge\widetilde{\mathbf{A}}) * \mathbf{B} = (\widetilde{\mathbf{X}} * \mathbf{A}\rfloor\widetilde{\mathbf{B}})$. So $\mathbf{B}\lfloor\mathbf{A}$ and $\mathbf{A}\rfloor\mathbf{B}$ differ only by a grade-dependent sign, which can be computed by using (2.12) repeatedly.

Although it therefore does not lead to a fundamentally new product, this right contraction is convenient at times for compact notation of relationships. It can be developed axiomatically along the same lines as the regular (left) contraction, and has a similar (but reverse) grade reduction property:

$$\mathrm{grade}(\mathbf{B}\lfloor\mathbf{A}) = \mathrm{grade}(\mathbf{B}) - \mathrm{grade}(\mathbf{A}).$$

For vectors, both contractions reduce to the familiar inner product, so we can see them as generalized inner products now acting on blades (or even on multivectors).

The terms we have used for the contractions correspond to the usage in [39]. It is somewhat unfortunate that the left contraction is denoted by a right parenthesis symbol $\rfloor$, and the right contraction by a left symbol $\lfloor$. The terminology is that something is being contracted from the left or from the right, and the hook on the parenthesis points to the contractor rather than to the contractee. We pronounce the left contraction $\mathbf{A}\rfloor\mathbf{B}$ as "A (contracted) on B," and the right contraction $\mathbf{B}\lfloor\mathbf{A}$ as "B (contracted) by A." We will mostly use the left contraction to express our results.

Most other authors use definitions of inner products that differ from both left and right contraction. The geometry behind those products is essentially the same as for the contractions, but their algebraic implementation generally leads to more involved and conditional expressions for advanced results. We explore these issues in detail in Section B.1 of Appendix B.

3.5 ORTHOGONALITY AND DUALITY

3.5.1 NONASSOCIATIVITY OF THE CONTRACTION

Algebraically, the contraction is nonassociative, for $\mathbf{A}\rfloor(\mathbf{B}\rfloor\mathbf{C})$ is not equal to $(\mathbf{A}\rfloor\mathbf{B})\rfloor\mathbf{C}$. It cannot be, for even the grades of both expressions are unequal, being respectively $\mathrm{grade}(\mathbf{C}) - \mathrm{grade}(\mathbf{B}) - \mathrm{grade}(\mathbf{A})$, and $\mathrm{grade}(\mathbf{C}) - \mathrm{grade}(\mathbf{B}) + \mathrm{grade}(\mathbf{A})$. So what do the two expressions $\mathbf{A}\rfloor(\mathbf{B}\rfloor\mathbf{C})$ and $(\mathbf{A}\rfloor\mathbf{B})\rfloor\mathbf{C}$ represent, and how are they different?

- The geometrical interpretation of the expression $\mathbf{A}\rfloor(\mathbf{B}\rfloor\mathbf{C})$ is: first restrict the outcome to $\mathbf{B}\rfloor\mathbf{C}$, the subspace of $\mathbf{C}$ that is perpendicular to $\mathbf{B}$; then from that subspace pick the subspace perpendicular to $\mathbf{A}$. We can of course construct this as the subspace that is in $\mathbf{C}$ and perpendicular to both $\mathbf{A}$ and to $\mathbf{B}$. The equivalence of both procedures is the geometrical interpretation of the defining property (3.11), which reads

$$(\mathbf{A}\wedge\mathbf{B})\rfloor\mathbf{C} = \mathbf{A}\rfloor(\mathbf{B}\rfloor\mathbf{C}) \quad \text{(universally valid)} \tag{3.20}$$

- The other possibility of composing the contractions $(\mathbf{A}\rfloor\mathbf{B})\rfloor\mathbf{C}$ is not part of our defining properties. It can also be simplified to an expression using the outer product, though the result is not universal but conditional:

$$(\mathbf{A}\rfloor\mathbf{B})\rfloor\mathbf{C} = \mathbf{A} \wedge (\mathbf{B}\rfloor\mathbf{C}) \quad \text{when } \mathbf{A} \subseteq \mathbf{C} \tag{3.21}$$

The proof is nontrivial, and is in Section B.3 of Appendix B.

The geometrical interpretation of this formula is a bit of a tongue-twister. The left-hand side, $(\mathbf{A}\rfloor\mathbf{B})\rfloor\mathbf{C}$, asks us to take the part of a subspace $\mathbf{B}$ that is most unlike $\mathbf{A}$ (in the sense of orthogonal containment), and then remove that from $\mathbf{C}$. The right-hand side, $\mathbf{A} \wedge (\mathbf{B}\rfloor\mathbf{C})$, suggests that we have then actually only taken $\mathbf{B}$ out of $\mathbf{C}$, and have effectively put $\mathbf{A}$ back into the result. That feels more or less correct, but not quite. To make it hold, we need the condition that $\mathbf{A}$ was in $\mathbf{C}$ to begin with— we could not "reconstruct" any other parts of $\mathbf{A}$ by the double complementarity of $(\mathbf{A}\rfloor\mathbf{B})\rfloor\mathbf{C}$.

We will refer to (3.20) and (3.21) together as the *duality formulas* for reasons that become clear below.

3.5.2 THE INVERSE OF A BLADE

There is no unique inverse $\mathbf{A}^{-1}$ of a blade $\mathbf{A}$ that would satisfy the equation $\mathbf{A}\rfloor\mathbf{A}^{-1} = 1$, for we can always add a blade perpendicular to $\mathbf{A}$ to $\mathbf{A}^{-1}$ and still satisfy the equation. Still, we can define a unique blade that works as an inverse of a k-blade $\mathbf{A}_k$ relative to the contraction. We define it as[1]

$$\textit{inverse of a blade: } \mathbf{A}_k^{-1} \equiv \frac{\widetilde{\mathbf{A}}_k}{\|\mathbf{A}_k\|^2} = (-1)^{k(k-1)/2}\frac{\mathbf{A}_k}{\|\mathbf{A}_k\|^2}. \tag{3.22}$$

Note that this is a blade of the same grade as $\mathbf{A}_k$, representing a subspace with the same attitude, and differing from $\mathbf{A}_k$ only by its weight and possibly its orientation. You can easily verify that this is indeed an inverse of $\mathbf{A}$ for the contraction:

$$\mathbf{A}_k\rfloor\mathbf{A}_k^{-1} = \mathbf{A}_k\rfloor\frac{\widetilde{\mathbf{A}}_k}{\|\mathbf{A}_k\|^2} = \frac{\mathbf{A}_k\rfloor\widetilde{\mathbf{A}}_k}{\mathbf{A}_k * \widetilde{\mathbf{A}}_k} = \frac{\mathbf{A}_k\rfloor\widetilde{\mathbf{A}}_k}{\mathbf{A}_k\rfloor\widetilde{\mathbf{A}}_k} = 1, \tag{3.23}$$

using the equivalence of scalar product and contraction for blades of equal grade.

The inverse of a vector $\mathbf{a}$ is thus

$$\mathbf{a}^{-1} = \mathbf{a}/\|\mathbf{a}\|^2,$$

---

1   Later, when we have introduced the associative geometric product in Chapter 6, this will be found to be "the" inverse, relative to that product. Since that inverse can be defined completely in terms of the contraction, we have chosen to define it now and use the same symbol to denote it, to prevent a proliferation of notations.

and as you would expect, a unit vector is its own inverse. That is not true for general blades, as shown by the inverses of the unit blades $\mathbf{e}_1 \wedge \mathbf{e}_2$ and $\mathbf{e}_1 \wedge \mathbf{e}_2 \wedge \mathbf{e}_3$ (defined in the standard orthonormal basis):

$$(\mathbf{e}_1 \wedge \mathbf{e}_2)^{-1} = \mathbf{e}_2 \wedge \mathbf{e}_1 = -\mathbf{e}_1 \wedge \mathbf{e}_2$$

$$(\mathbf{e}_1 \wedge \mathbf{e}_2 \wedge \mathbf{e}_3)^{-1} = \mathbf{e}_3 \wedge \mathbf{e}_2 \wedge \mathbf{e}_1 = -\mathbf{e}_1 \wedge \mathbf{e}_2 \wedge \mathbf{e}_3$$

When we use unit pseudoscalars $\mathbf{I}_n = \mathbf{e}_1 \wedge \mathbf{e}_2 \wedge \cdots \wedge \mathbf{e}_n$ for an n-dimensional Euclidean metric space $\mathbb{R}^{n,0}$, the inverse is simply the reverse:[2]

$$\mathbf{I}_n^{-1} = \widetilde{\mathbf{I}}_n.$$

Inverses of unit pseudoscalars are important in the formulation of duality; for consistency in orientations, you should always remember to include the reverse.

For a null blade, which has norm zero (see Appendix A), the inverse is not defined. In computations involving the contractions of null blades, one can substitute it by the recip-rocal of the blade, which we will meet in Section 3.8. This is a more involved concept, only well defined in "balanced" algebras. We have preferred to keep the formulas in our initial explanations simple, by focusing on nondegenerate metrics and the inverse rather than general metrics and the reciprocal. When we start using the degenerate metrics seriously (in Chapter 13), we will be more careful.

3.5.3 ORTHOGONAL COMPLEMENT AND DUALITY

Given a k-blade $\mathbf{A}_k$ in the space $\mathbb{R}^n$ with unit pseudoscalar $\mathbf{I}_n$, its *dual* is obtained by the dualization mapping $* : \bigwedge^k \mathbb{R}^n \rightarrow \bigwedge^{n-k} \mathbb{R}^n$ defined by

$$\text{dualization: } \mathbf{A}_k^* = \mathbf{A}_k \rfloor \mathbf{I}_n^{-1}.$$

The operation "taking the dual" is linear in $\mathbf{A}_k$, and it results in a blade with the same magnitude as $\mathbf{A}_k$ and a well-defined orientation. The reason for the inverse pseudoscalar is clear when we use it on a (hyper) volume blade such as $\mathbf{a} \wedge \mathbf{b} \wedge \mathbf{c}$. We have seen in the previous chapter how such an n-blade is proportional to the pseudoscalar $\mathbf{I}_n$ by a scalar that is the oriented volume. With the definition of dual, that oriented scalar volume is simply its dual, $(\mathbf{a} \wedge \mathbf{b} \wedge \mathbf{c})^*$, without extraneous signs.

Figure 3.4 shows a 2-D example of dualization: the dual of vector in a space with coun-terclockwise orientation is the clockwise vector perpendicular to it. This is easily proved: choose coordinates such that $\mathbf{a} = \alpha \mathbf{e}_1$ and $\mathbf{I}_2 = \mathbf{e}_1 \wedge \mathbf{e}_2$. Then

$$\mathbf{a}^* = \alpha \mathbf{e}_1 \rfloor (\mathbf{e}_2 \wedge \mathbf{e}_1) = -\alpha \mathbf{e}_2.$$

---

2   In a general metric space $\mathbb{R}^{p,q}$ (see A.1 in Appendix A), this changes to $\mathbf{I}_{p,q}^{-1} = (-1)^q \widetilde{\mathbf{I}}_{p,q}$, as you can easily verify.

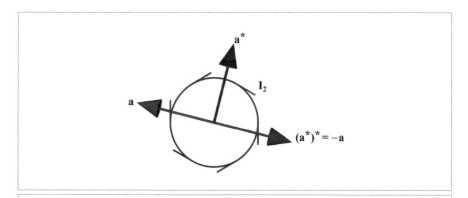

Figure 3.4: Duality of vectors in 2-D, with a counterclockwise-oriented pseudoscalar I_2. The dual of the red vector **a** is the blue vector $\mathbf{a}^*$ obtained by rotating it clockwise over $\pi/2$. The dual of that blue vector is $(\mathbf{a}^*)^*$, which is $-\mathbf{a}$.

This vector is indeed perpendicular to **a**. But note that the expression $\mathbf{a}\rfloor\widetilde{I}_2$ requires no coordinates to denote such a vector perpendicular to **a**. In fact, for a vector **a** in the I_2-plane, $\rfloor\widetilde{I}_2$ *acts like an operator that rotates* **a** *clockwise over* $\frac{1}{2}\pi$ *in the plane* I_2, independent of any coordinate system. We will get back to such operators in Chapter 7.

Taking the dual again does *not* result in the original vector, but in its opposite:

$$(\mathbf{a}^*)^* = -\alpha\,\mathbf{e}_2\rfloor(\mathbf{e}_2 \wedge \mathbf{e}_1) = -\alpha\,\mathbf{e}_1 = -\mathbf{a}.$$

This is a property of other dimensionalities as well. Let us derive the general result:

$$(\mathbf{A}_k{}^*)^* = (\mathbf{A}_k\rfloor\mathbf{I}_n^{-1})\rfloor\mathbf{I}_n^{-1} = \mathbf{A}_k \wedge (\mathbf{I}_n^{-1}\rfloor\mathbf{I}_n^{-1})$$
$$= (-1)^{n(n-1)/2}\,\mathbf{A}_k \wedge (\mathbf{I}_n\rfloor\widetilde{\mathbf{I}}_n) = (-1)^{n(n-1)/2}\,\mathbf{A}_k \wedge 1 = (-1)^{n(n-1)/2}\,\mathbf{A}_k$$

(we used (3.21) in this derivation). There is a dimension-dependent sign, with the pattern $+ + - - + + - - \cdots$, so for 2-D and 3-D, this minus sign in the double reversion occurs. If we need to be careful about signs, we should use an undualization operation to retrieve the proper element of which the dual would be **A**. It is simply defined through:

undualization: $\mathbf{A}^{-*} \equiv \mathbf{A}\rfloor\mathbf{I}_n.$

If there is any ambiguity concerning the pseudoscalar relative to which the duality is taken, then we will write it out in full.

Figure 3.5 illustrates dualization in $\mathbb{R}^{3,0}$ with its Euclidean metric. We define a right-handed pseudoscalar $\mathbf{I}_3 \equiv \mathbf{e}_1 \wedge \mathbf{e}_2 \wedge \mathbf{e}_3$ relative to the standard orthonormal basis $\{\mathbf{e}_1, \mathbf{e}_2, \mathbf{e}_3\}$. A general vector $\mathbf{a} = a_1\mathbf{e}_1 + a_2\mathbf{e}_2 + a_3\mathbf{e}_3$ is dualized to

$$\mathbf{a}^* = \mathbf{a}\rfloor\mathbf{I}_3^{-1}$$

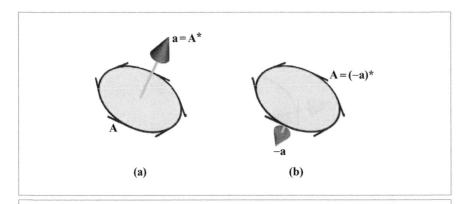

Figure 3.5: Duality of vectors and bivectors in 3-D, with a right-handed pseudoscalar. (a) The dual of a bivector **A** is the vector **a**. Grabbing the vector with your right hand has the fingers moving with the orientation of the bivector. (b) The vector whose dual is the bivector **A** is then −**a**. Now bivector and vector appear to have a left-handed relationship.

$$= (a_1\mathbf{e}_1 + a_2\mathbf{e}_2 + a_3\mathbf{e}_3)\rfloor(\mathbf{e}_3 \wedge \mathbf{e}_2 \wedge \mathbf{e}_1)$$
$$= -a_1\,\mathbf{e}_2 \wedge \mathbf{e}_3 - a_2\,\mathbf{e}_3 \wedge \mathbf{e}_1 - a_3\,\mathbf{e}_1 \wedge \mathbf{e}_2.$$

By our geometric interpretation of the contraction, this 2-blade $\mathbf{A} \equiv \mathbf{a}^*$ denotes a plane that is the *orthogonal complement* to **a**. Note that **A** has the same coefficients as **a** had on its orthonormal basis of vectors, but now on a 2-blade basis that can be associated with the orthonormal vector basis in a natural manner. In this way, a vector is naturally associated with a 2-blade, in 3-D space. (Of course, this is similar to what we would do classically: we use **a** as the normal vector for the plane **A**, but that only works in 3-D space.)

Whenever you have forgotten the signs involved in a desired dualization, it is simplest to make a quick check using a standardized situation of vectors and 2-blades along the axes in an orthonormal basis $\{\mathbf{e}_i\}_{i=1}^n$ (such as $\mathbf{e}_1{}^* = \mathbf{e}_1\rfloor(\mathbf{e}_2 \wedge \mathbf{e}_1) = -\mathbf{e}_2$, so this is a clockwise rotation). But this usage of coordinates should only be a check: with enough practice, you will be able to avoid the extraneous coordinates in the specification of your actual geometrical computations. This will save unnecessary writing and maintain clear geometrical relationships of the elements introduced.

3.5.4 THE DUALITY RELATIONSHIPS

There is a dual relationship between the contraction and the outer product, which we can see explicitly by using the two properties (3.20) and (3.21) when **C** is a unit pseudoscalar $\mathbf{I}_n$ for the space $\mathbb{R}^n$. Since all blades are contained in the pseudoscalar, both properties now become universally valid and can be written using the duality operator:

$$(\mathbf{A} \wedge \mathbf{B})^* = \mathbf{A}\rfloor(\mathbf{B}^*)$$
$$(\mathbf{A}\rfloor\mathbf{B})^* = \mathbf{A} \wedge (\mathbf{B}^*) \quad \text{for } \mathbf{A} \subseteq \mathbf{I}.$$
$$\tag{3.24}$$

These *duality relationships* are very important in simplification of formulas. You can often evaluate an expression a lot more compactly by taking the dual, change a contraction into an outer product, use its properties, undualize, and so on. We will see many examples of this technique in the coming chapters.

3.5.5 DUAL REPRESENTATION OF SUBSPACES

The duality relationships permit us to represent geometrical subspaces in a dual manner.

We have seen in Section 2.8 how a blade $\mathbf{A}$ can represent a subspace directly, checking whether a vector $\mathbf{x}$ is in it by testing whether $\mathbf{x} \wedge \mathbf{A} = 0$. We introduce the *dual representation* of a subspace $\mathcal{A}$ simply by taking the dual of the defining equation $\mathbf{x} \wedge \mathbf{A} = 0$ using (3.24). We obtain

$$\mathbf{D} = \mathbf{A}^* \text{ is the dual representation of } \mathcal{A} \quad : \quad (\mathbf{x} \in \mathcal{A} \iff \mathbf{x} \rfloor \mathbf{D} = 0).$$

The blade $\mathbf{D}$ that dually represents the subspace $\mathcal{A}$ is also a direct representation of the orthogonal complement of the subspace $\mathcal{A}$. You can confirm this by finding all vectors $\mathbf{y}$ for which $\mathbf{y} \wedge \mathbf{D} = 0$ and again using (3.24). This mirrors and generalizes the practice in elementary linear algebra to have a normal vector $\mathbf{n}$ represent a homogeneous hyperplane through the equation $\mathbf{x} \cdot \mathbf{n} = 0$.

Once we really start doing geometry in Part II, we will very flexibly switch between the direct representation and the dual representation, and this will be a powerful way of finding the simplest expressions for our geometrical operations. It is therefore pleasant to have both representations present within our algebra of blades.

Readers who are acquainted with Grassmann-Cayley algebras will note that we have used the contraction to construct the dual representation, and that this therefore involves the metric of the space. Grassmann-Cayley algebra has a seemingly nonmetric way of making dualities, using mathematical constructions called 1-forms. We view these as a disguised form of metric. Since we will be mostly working in $\mathbb{R}^n$ and usually have an obvious metric, the metric road to duality through the contraction is more convenient in our applications. It saves us from having to introduce a lot of mathematical terminology that we do not really need.

3.6 ORTHOGONAL PROJECTION OF SUBSPACES

With the contraction and the inverse, we have the ingredients to construct the orthogonal projection of a subspace represented by a blade $\mathbf{X}$ onto a subspace represented by a blade $\mathbf{B}$. We assume that this blade $\mathbf{B}$ has an inverse relative to the contraction, the blade $\mathbf{B}^{-1}$.

To introduce the construction, consider Figure 3.6, which depicts the projection of a vector on a 2-blade. The vector $\mathbf{x} \rfloor \mathbf{B}$ is a vector in the $\mathbf{B}$-plane perpendicular to $\mathbf{x}$, and that

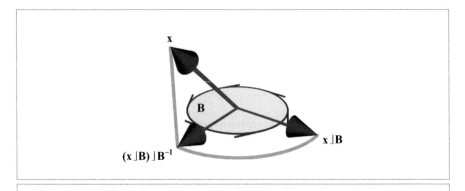

Figure 3.6: Projection of a vector **x** onto a subspace **B**.

means that it is also perpendicular to the projection of **x** on **B**. Therefore we can simply rotate $\mathbf{x} \rfloor \mathbf{B}$ over $\pi/2$ in the **B**-plane to obtain the projection. A rotation with the correct sign is performed by the dual within that plane (i.e., by the operation $\rfloor \mathbf{B}^{-1}$). The string of operations then yields $(\mathbf{x} \rfloor \mathbf{B}) \rfloor \mathbf{B}^{-1}$, as depicted in the figure.

Inspired by this 3-D example, we define the *(orthogonal) projection* $\mathsf{P}_\mathbf{B}[\ ] : \mathbb{R}^n \to \mathbb{R}^n$ as

$$\text{orthogonal projection of vector } \mathbf{x} \text{ onto } \mathbf{B}: \quad \mathsf{P}_\mathbf{B}[\mathbf{x}] \equiv (\mathbf{x} \rfloor \mathbf{B}) \rfloor \mathbf{B}^{-1}. \qquad (3.25)$$

This mapping is linear in **x**, but nonlinear in **B**. In fact, only the attitude of **B** affects the outcome; its weight and orientation are divided out. In this operation, **B** acts as an unoriented, unweighted subspace.

A projection should be idempotent (applying it twice should be the same as applying it once). This is easily verified using the duality properties of Section 3.5.1:

$$\mathsf{P}_\mathbf{B}[\mathsf{P}_\mathbf{B}[\mathbf{x}]] = \left(\left((\mathbf{x} \rfloor \mathbf{B}) \rfloor \mathbf{B}^{-1}\right) \rfloor \mathbf{B}\right) \rfloor \mathbf{B}^{-1} = \left((\mathbf{x} \rfloor \mathbf{B}) \rfloor \mathbf{B}^{-1}\right) \wedge (\mathbf{B} \rfloor \mathbf{B}^{-1})$$
$$= \left((\mathbf{x} \rfloor \mathbf{B}) \rfloor \mathbf{B}^{-1}\right) \wedge 1 = \mathsf{P}_\mathbf{B}[\mathbf{x}].$$

To investigate its properties further, let us write **x** as $\mathbf{x} = \mathbf{x}_\perp + \mathbf{x}_\|$, where $\mathbf{x}_\perp \rfloor \mathbf{B} = 0$, while $\mathbf{x}_\| \rfloor \mathbf{B} \neq 0$. In a space with a Euclidean metric, we would say that $\mathbf{x}_\perp$ is perpendicular to **B**. The projection kills this perpendicular part of **x**,

$$\mathsf{P}_\mathbf{B}[\mathbf{x}] = (\mathbf{x}_\perp \rfloor \mathbf{B}) \rfloor \mathbf{B}^{-1} + (\mathbf{x}_\| \rfloor \mathbf{B}) \rfloor \mathbf{B}^{-1} = 0 + \mathbf{x}_\| \wedge (\mathbf{B} \rfloor \mathbf{B}^{-1}) = \mathbf{x}_\|,$$

leaving the part $\mathbf{x}_\|$ that is contained in $\mathbf{B}^{-1}$ and hence in **B**. This is just what you would expect of a projection.

When you consider the projection of a general blade **X** onto the blade **B**, the principles are the same. The contraction $\mathbf{X} \rfloor \mathbf{B}$ produces a subblade of **B** that is perpendicular to **X** and of grade $(b - x)$, where $b \equiv \text{grade}(\mathbf{B})$ and $x \equiv \text{grade}(\mathbf{X})$. The projection is a subblade of $\mathbf{B}$ of the same grade as $\mathbf{X}$. Such a blade can be made from $\mathbf{X} \rfloor \mathbf{B}$ by dualization of the contraction. The correct sign and magnitude to be in agreement with the formula for the vector projection implies the use of $\mathbf{B}^{-1}$. In total, we obtain for the *orthogonal projection* of a blade $\mathbf{X}$ onto a blade $\mathbf{B}$:

$$\text{projection of } \mathbf{X} \text{ onto } \mathbf{B} : \quad \mathsf{P}_{\mathbf{B}}[\mathbf{X}] \equiv (\mathbf{X} \rfloor \mathbf{B}) \rfloor \mathbf{B}^{-1}. \tag{3.26}$$

Note that if you try to project a subspace of too high a grade onto $\mathbf{B}$, the contraction automatically causes the result to be zero. Even when $\text{grade}(\mathbf{X}) \leq \text{grade}(\mathbf{B})$ this may happen; it all depends on the relative geometric positions of the subspaces, as it should.

The reasoning to achieve the projection formula (3.26) was rather geometrical, but it can also can be derived in a more algebraic manner. Section B.4 in Appendix B gives a proof in terms of the present chapter, but the next chapter gives a perhaps more satisfying proof by naturally extending the projection of a vector to act on a blade, in Section 4.2.2.

Since the projection is more intuitive than the contraction, you may prefer to make (3.26) the formulation of the geometry of the contraction. Through a contraction by $\mathbf{B}$ on both sides, we obtain

$$\mathbf{X} \rfloor \mathbf{B} = \mathsf{P}_{\mathbf{B}}[\mathbf{X}] \rfloor \mathbf{B},$$

and this inspires the following characterization of the contraction:

> The contraction $\mathbf{A} \rfloor \mathbf{B}$ is the subblade of $\mathbf{B}$ of grade $b - a$ that is dual (by $\mathbf{B}^{-1}$) to the projection of $\mathbf{A}$ onto $\mathbf{B}$.

As long as you realize that "dual by $\mathbf{B}$" is shorthand for $\rfloor \mathbf{B}$, the geometrical properties of the contraction listed in Section 3.3 follow easily.

This geometric characterization of $\mathbf{A} \rfloor \mathbf{B}$ probably makes a lot more intuitive sense to you than our earlier description of $\mathbf{A} \rfloor \mathbf{B}$ as the part of $\mathbf{B}$ least like $\mathbf{A}$, for the description in terms of projection and perpendicularity (which is what the dual signifies) better matches the usual primitive operations of linear algebra. Yet algebraically, the contraction is the simpler concept, for *unlike the projection of $\mathbf{A}$ onto $\mathbf{B}$, it is linear in both $\mathbf{A}$ and $\mathbf{B}$.* That makes it a better choice than the projection as a primitive operation on subspaces, algebraically on a par with the outer product $\mathbf{A} \wedge \mathbf{B}$, even though we have to get used to its geometry.

To return to (3.26), it is actually somewhat better to define the projection through

$$\text{projection of } \mathbf{X} \text{ onto } \mathbf{B} : \quad \mathsf{P}_{\mathbf{B}}[\mathbf{X}] \equiv (\mathbf{X} \rfloor \mathbf{B}^{-1}) \rfloor \mathbf{B}. \tag{3.27}$$

Writing it in this manner makes it obviously an element of $\mathbf{B}$ rather than of $\mathbf{B}^{-1}$. For nonnull blades, there is no difference in outcome, since it simply moves the normalization $1/\|\mathbf{B}\|^2$. For the null-blades that may occur in degenerate metrics (see Appendix A), the inverse does not exist and needs to be replaced by the reciprocal relative to the contraction.

The reciprocal of **B** may then differ from **B** by more than scaling, and even have a different attitude. The projection (3.26) is no longer guaranteed to produce a subblade of **B**, as we would want, but (3.27) always will.

3.7 THE 3-D CROSS PRODUCT

In 3-D Euclidean space $\mathbb{R}^{3,0}$, one is used to having the cross product available. In the algebra as we are constructing it now, we have avoided it, for two reasons: we can make it anyway if we need it, and better still, another construction can take its place that generalizes to arbitrary dimensions for all uses of the cross product. We demonstrate these points in this section.

3.7.1 USES OF THE CROSS PRODUCT

First, when do we use a cross product in classical vector computations in 3-D Euclidean space?

- *Normal Vectors*. The cross product is used to determine the vector **a** perpendicular to a plane **A**, called the *normal vector* of the plane (see Figure 3.7(a)). This vector can be obtained from two vectors **x** and **y** in the plane as their *cross product* **x** × **y**. This works in 3-D space only (though it is often used in 2-D space as well, through the cheat of embedding it in a 3-D space). This representation is then used to character-ize the plane, for instance, to perform reflections in it when the plane is the tangent plane to some object that is to be rendered in computer graphics. Unfortunately, this representation of the tangent plane does not transform simply under linear trans-formations as a regular vector, and requires special code to transform the normal vector (you need to use the inverse transpose mapping, scaled by a determinant, as we will show in Section 4.3.6).

- *Rotational Velocities*. We also use the cross product to compute the velocity of a point at location **x** turning around an axis **a** (also indicated by a vector). Then the instantaneous velocity is proportional to **a** × **x** (see Figure 3.7(b)). Yet the indication of a rotation by a rotation axis works only in 3-D space; even in 2-D, the axis points out of the plane of the space, and is therefore not really a part of it. In 4-D, a rota-tion in a plane needs a plane of axes to denote it, since there are two independent directions perpendicular to any plane. Even for computations in 3-D Euclidean geometry, such higher-dimensional rotations are relevant: we need them in the 5-D operational model $\mathbb{R}^{4,1}$ to perform 3-D motions efficiently (in Chapter 13).

- *Intersecting Planes*. A third use is to compute the intersection of two homogeneous planes **A** and **B** in 3-D space: if both are characterized by their normals **a** and **b**, the line of intersection is along the vector **a** × **b** (see Figure 3.7(c)). This construction is a bit of a trick, specific for that precise situation, and it does not generalize in a straightforward manner to the intersection of other homogeneous subspaces such

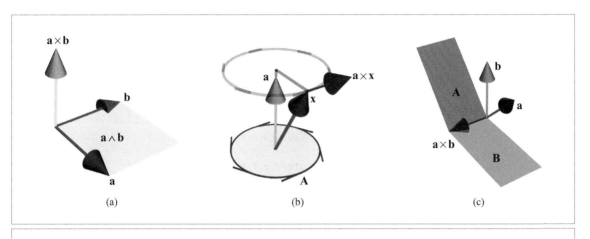

Figure 3.7: Three uses of the cross product.

as lines, or to other dimensions. You can also use it to intersect general lines in 2-D through the embedding in homogeneous coordinates, but that's about it.

All these uses have their limitations, and none extends easily to higher-dimensional spaces. The cross product is, basically, a 3-D trick, and we need to replace it with something more universally applicable.

3.7.2 THE CROSS PRODUCT INCORPORATED

Let us take the characterization of the plane spanned by two vectors **a** and **b** as the defining example to redo the cross product with our subspace products. Using our subspace algebra, we would characterize the plane by the 2-blade $\mathbf{a} \wedge \mathbf{b}$ and the subspace normal to it in the space with pseudoscalar $\mathbf{I}_n$ by the orthogonal complement $(\mathbf{a} \wedge \mathbf{b}) \rfloor \mathbf{I}_n^{-1}$. In 3-D Euclidean space $\mathbb{R}^{3,0}$, the inverse pseudoscalar $\mathbf{I}_3^{-1}$ equals $-\mathbf{I}_3$, and the orthogonal complement is then indeed a vector, computed as $(\mathbf{a} \wedge \mathbf{b})^* = (\mathbf{b} \wedge \mathbf{a}) \rfloor \mathbf{I}_3$.

The classical method computes the normal vector as $\mathbf{a} \times \mathbf{b}$. Both ways of computing the normal vector must be equivalent, so we obtain a definition of the cross product in terms of the outer product and contraction:

$$\mathbf{a} \times \mathbf{b} = (\mathbf{a} \wedge \mathbf{b})^* = (\mathbf{a} \wedge \mathbf{b}) \rfloor \mathbf{I}_3^{-1} \qquad (3.28)$$

Note that this definition indicates explicitly that there are two geometrical concepts involved in the cross product: spanning, and taking the orthogonal complement. The latter is related to the metric of the embedding space (since it ultimately contains the inner product), and this makes the cross product a rather involved construction. In the next chapter, we will see that it also makes its transformation laws complicated.

Let us verify (3.28) by a coordinate-based computation in an orthonormal basis $\{e_1, e_2, e_3\}$ for the 3-D Euclidean space $\mathbb{R}^{3,0}$. Let $\mathbf{a} = a_1\mathbf{e}_1 + a_2\mathbf{e}_2 + a_3\mathbf{e}_3$ and $\mathbf{b} = b_1\mathbf{e}_1 + b_2\mathbf{e}_2 + b_3\mathbf{e}_3$. Then

$$\mathbf{a} \times \mathbf{b} = (a_2b_3 - a_3b_2)\,\mathbf{e}_1 + (a_3b_1 - a_1b_3)\,\mathbf{e}_2 + (a_1b_2 - a_2b_1)\,\mathbf{e}_3. \qquad (3.29)$$

In (2.3), we have a coordinate expression for $\mathbf{a} \wedge \mathbf{b}$:

$$\mathbf{a} \wedge \mathbf{b} = (a_1b_2 - a_2b_1)\,\mathbf{e}_1 \wedge \mathbf{e}_2 + (a_2b_3 - a_3b_2)\,\mathbf{e}_2 \wedge \mathbf{e}_3 + (a_3b_1 - a_1b_3)\,\mathbf{e}_3 \wedge \mathbf{e}_1.$$

It is easy to take the dual of this by using $(\mathbf{e}_1 \wedge \mathbf{e}_2)\rfloor(\mathbf{e}_3 \wedge \mathbf{e}_2 \wedge \mathbf{e}_1) = \mathbf{e}_3$ and the like. The result indeed agrees with the above.

So in terms of coordinates, we are computing very similar quantities whether we use $\mathbf{a} \wedge \mathbf{b}$ or $\mathbf{a} \times \mathbf{b}$. Yet $\mathbf{a} \wedge \mathbf{b}$ is a simpler concept geometrically, because it does not depend on a metric, and it is usable in n-dimensional space (not just 3-D). You used to be forced into using the dual concept $\mathbf{a} \times \mathbf{b}$ since you could only treat vectors in standard linear algebra. Now that we know that it is actually the dual of a bivector $\mathbf{a} \wedge \mathbf{b}$, we had better not dualize it and use $\mathbf{a} \wedge \mathbf{b}$ "as is."

This does not lose any geometry, for all computations with the cross product depicted in Figure 3.7 can be recast into geometric algebra. Let us check them off:

- **Normal Vectors.** We have just seen that a plane can be characterized directly by its bivector rather than by a normal vector constructed from two vectors in it.

- **Velocities.** For the velocity representation involving the cross product $\mathbf{a} \times \mathbf{x}$, we note that our algebra provides a suggestive rewriting through the duality properties:

$$\mathbf{a} \times \mathbf{x} = (\mathbf{a} \wedge \mathbf{x})^* = -(\mathbf{x} \wedge \mathbf{a})^* = -\mathbf{x}\rfloor\mathbf{a}^* = \mathbf{x}\rfloor\mathbf{A}, \qquad (3.30)$$

  where $\mathbf{A} \equiv \mathbf{a}\rfloor I_3$ is the 2-blade whose dual is $\mathbf{a}$. This is depicted in Figure 3.8. So we can replace the computation of the velocity of $\mathbf{x}$ during a rotation around the axis $\mathbf{a}$ by a computation involving the rotation plane $\mathbf{A}$. That actually works in n-dimensional space (even in $n = 2$, where a rotation "axis" does not really exist!). (From Chapter 7 onward, we will even be able to define a rotation in n-dimensional space directly in terms of its rotation plane $\mathbf{A}$, as $\exp(\mathbf{A})$.)

- **Intersecting Planes.** The intersection of the two homogeneous planes of Figure 3.7(c) can be written in terms of the bivectors as

$$\mathbf{a} \times \mathbf{b} = \left((\mathbf{A}\rfloor I_3^{-1}) \wedge (\mathbf{B}\rfloor I_3^{-1})\right)\rfloor I_3^{-1} = (\mathbf{B}\rfloor I_3^{-1})\rfloor \left((\mathbf{A}\rfloor I_3^{-1})\rfloor I_3\right)$$
$$= (\mathbf{B}\rfloor I_3^{-1})\rfloor\mathbf{A} = (\mathbf{B}^*)\rfloor\mathbf{A}.$$

  We shall see (in Chapter 5) that this final expression generalizes to intersections of subspaces of arbitrary dimension as the meet of $\mathbf{B}$ and $\mathbf{A}$. That will be the algebraic representation for the general incidence operator on subspaces.

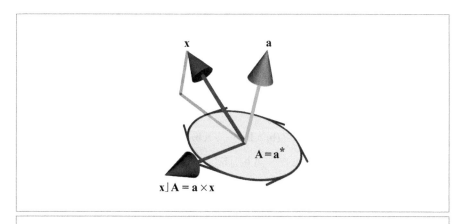

Figure 3.8: The 3-D cross product $\mathbf{a} \times \mathbf{x}$ can be constructed as the contraction $\mathbf{x} \rfloor \mathbf{A}$ on the dual $\mathbf{A}$ of $\mathbf{a}$.

In summary, we can use our other products and the blades to replace the specific and peculiar 3-D cross product on vector representations in a manner that works for all dimensions. Therefore, we will do so. Apart from revisiting it in the context of linear transformations in Chapter 4, where we find more arguments against its use, we will not employ the cross product any more (except occasionally to show the classical form of some of our results). If you have used the cross product a lot, you may wonder what happened to some of the identities that were useful in geometric computations (such as the bac-cab formula). Structural exercises 10–12 show you what to substitute for them.

3.8 APPLICATION: RECIPROCAL FRAMES

Although we avoid using coordinates in our computations, they are often required to present their results. We therefore need a way to retrieve the coefficient x_i of some vector $\mathbf{x}$, expressible as $\mathbf{x} = \sum x_i \mathbf{b}_i$ on some basis $\{\mathbf{b}_i\}_{i=1}^n$. If the basis happens to be orthonormal, then this is simple: $x_i = \mathbf{x} \cdot \mathbf{b}_i$, as is easily verified. However, we would like the flexibility to choose our bases arbitrarily to correspond to whatever the important directions are in any given problem. We then need a more general expression.

In a metric space $\mathbb{R}^n$ with chosen basis $\{\mathbf{b}_i\}_{i=1}^n$ and pseudoscalar $\mathbf{I}_n = \mathbf{b}_1 \wedge \mathbf{b}_2 \wedge \cdots \wedge \mathbf{b}_n$ we can do this as follows. Associate with each basis vector $\mathbf{b}_i$ a *reciprocal basis vector* $\mathbf{b}^i$, defined as

$$\mathbf{b}^i \equiv (-1)^{i-1}(\mathbf{b}_1 \wedge \mathbf{b}_2 \wedge \cdots \wedge \check{\mathbf{b}}_i \wedge \cdots \wedge \mathbf{b}_n)\rfloor \mathbf{I}_n^{-1}. \tag{3.31}$$

Here the inverted arc denotes the removal of the vector $\mathbf{b}_i$, so this vector $\mathbf{b}^i$ is the dual of an $(n-1)$-blade spanned by all vectors except $\mathbf{b}_i$ (as in (3.16)). The reciprocals of the basis vectors form a basis $\{\mathbf{b}^i\}_{i=1}^n$ for the vector space.

The two bases $\{\mathbf{b}_i\}_{i=1}^n$ and $\{\mathbf{b}^i\}_{i=1}^n$ are mutually orthonormal, for

$$
\begin{aligned}
\mathbf{b}_i \cdot \mathbf{b}^j &= (-1)^{j-1} \mathbf{b}_i \rfloor \left((\mathbf{b}_1 \wedge \cdots \wedge \check{\mathbf{b}}_j \wedge \cdots \wedge \mathbf{b}_n) \rfloor \mathbf{I}_n^{-1} \right) \\
&= (-1)^{j-1} (\mathbf{b}_i \wedge \mathbf{b}_1 \wedge \cdots \wedge \check{\mathbf{b}}_j \wedge \cdots \wedge \mathbf{b}_n) \rfloor \mathbf{I}_n^{-1} \\
&= \delta_i^j \, (\mathbf{b}_1 \wedge \cdots \wedge \mathbf{b}_n) \rfloor \mathbf{I}_n^{-1} \\
&= \delta_i^j \, \mathbf{I}_n \rfloor \mathbf{I}_n^{-1} \\
&= \delta_i^j,
\end{aligned}
\tag{3.32}
$$

where the selector symbol δ_i^j is defined to be 1 when $i = j$, and 0 otherwise.

In spaces without an orthonormal basis, it is common to express the coefficients of a vector $\mathbf{x}$ by a superscript, so that $\mathbf{x} = \sum_i x^i \mathbf{b}_i$. (Some authors, like [15], then use a summation convention, in which summation is implied when the same index appears above and below, but we will not employ it.) It is now straightforward to verify that $x^i = \mathbf{x} \cdot \mathbf{b}^i$:

$$
\mathbf{x} \cdot \mathbf{b}^i = \left(\sum_j x^j \mathbf{b}_j \right) \cdot \mathbf{b}^i = \sum_j x^j (\mathbf{b}_j \cdot \mathbf{b}^i) = \sum_j x^j \delta_j^i = x^i.
$$

Therefore:

> *Even on a nonorthonormal basis, the coefficients of a vector representation can be computed by an inner product with appropriately chosen basis vectors.*

It should be noted that orthonormal basis vectors have the same attitude as their reciprocal:

$$
\mathbf{b}_i = \pm \mathbf{b}^i \text{ if } \{\mathbf{b}_i\}_{i=1}^n \text{ is an orthonormal basis,}
$$

with the $+$ sign for positive vectors for which $\mathbf{b}_i \cdot \mathbf{b}_i = +1$, and the $-$ sign for negative vectors for which $\mathbf{b}_i \cdot \mathbf{b}_i = -1$. In a Euclidean metric space $\mathbb{R}^{n,0}$, the reciprocal basis vectors therefore equal the basis vectors, and the distinction is merely notational.

Reciprocal frames are especially useful, as they allow consistent and convenient treatment of nonorthonormal bases. These are known techniques from standard linear algebra. Usually, the reciprocal basis vectors are formulated in terms of minors of certain determinants. It is satisfying to see how easily (3.31) defines the reciprocal basis as a geometrical construction, namely as the orthogonal complement of the span of the other vectors. The geometrically interpretable algebraic formula shows clearly that, for a general basis, the coefficient of $\mathbf{b}_i$ depends on all vectors; for an orthogonal basis, it would only depend on $\mathbf{b}_i$ itself.

3.9 FURTHER READING

When reading other literature in geometric algebra, you will find that most authors use slightly different inner products. These alternatives are spelled out and compared in Section B.1 of Appendix B. We maintain that the contractions are more pure mathematically and geometrically, and they lead to fewer conditional computations in your code. That is why we use them in this book.

We tried to convey their geometrical relevance for computer science in [17] (though with limited success), inspired by [39] and [41]. The latter gives links to the mathematical origins of the construction.

3.10 EXERCISES

3.10.1 DRILLS

1. Let $\mathbf{a} = \mathbf{e}_1 + \mathbf{e}_2$ and $\mathbf{b} = \mathbf{e}_2 + \mathbf{e}_3$ in a 3-D Euclidean space $\mathbb{R}^{3,0}$ with orthonormal basis $\{\mathbf{e}_1, \mathbf{e}_2, \mathbf{e}_3\}$. Compute the following expressions, giving the results relative to the basis $\{1, \mathbf{e}_1, \mathbf{e}_2, \mathbf{e}_3, \mathbf{e}_1 \wedge \mathbf{e}_2, \mathbf{e}_2 \wedge \mathbf{e}_3, \mathbf{e}_3 \wedge \mathbf{e}_1, \mathbf{e}_1 \wedge \mathbf{e}_2 \wedge \mathbf{e}_3\}$. Show your work.

   (a) $\mathbf{e}_1 \rfloor \mathbf{a}$
   (b) $\mathbf{e}_1 \rfloor (\mathbf{a} \wedge \mathbf{b})$
   (c) $(\mathbf{a} \wedge \mathbf{b}) \rfloor \mathbf{e}_1$
   (d) $(2\mathbf{a} + \mathbf{b}) \rfloor (\mathbf{a} + \mathbf{b})$
   (e) $\mathbf{a} \rfloor (\mathbf{e}_1 \wedge \mathbf{e}_2 \wedge \mathbf{e}_3)$
   (f) $\mathbf{a}^*$
   (g) $(\mathbf{a} \wedge \mathbf{b})^*$
   (h) $\mathbf{a} \rfloor \mathbf{b}^*$

2. Compute the cosine of the angle between the following subspaces given on an orthonormal basis of a Euclidean space:

   (a) $\mathbf{e}_1$ and $\alpha \mathbf{e}_1$
   (b) $(\mathbf{e}_1 + \mathbf{e}_2) \wedge \mathbf{e}_3$ and $\mathbf{e}_1 \wedge \mathbf{e}_3$
   (c) $(\cos \phi \, \mathbf{e}_1 + \sin \phi \, \mathbf{e}_2) \wedge \mathbf{e}_3$ and $\mathbf{e}_2 \wedge \mathbf{e}_3$
   (d) $\mathbf{e}_1 \wedge \mathbf{e}_2$ and $\mathbf{e}_3 \wedge \mathbf{e}_4$

3. Set up and draw the reciprocal frame for vectors $\mathbf{b}_1$ and $\mathbf{b}_2$ on an orthogonal basis $\{\mathbf{e}_1, \mathbf{e}_2\}$ represented as $\mathbf{b}_1 = \mathbf{e}_1$ and $\mathbf{b}_2 = \mathbf{e}_1 + \mathbf{e}_2$. Use the reciprocal frame to compute the coordinates of the vector $\mathbf{x} = 3\mathbf{e}_1 + \mathbf{e}_2$ on the $\{\mathbf{b}_1, \mathbf{b}_2\}$-basis.

3.10.2 STRUCTURAL EXERCISES

1. In 2-D Euclidean space $\mathbb{R}^{2,0}$ with orthonormal basis $\{\mathbf{e}_1, \mathbf{e}_2\}$, let us determine the value of the contraction $\mathbf{e}_1 \rfloor (\mathbf{e}_1 \wedge \mathbf{e}_2)$ by means of its implicit definition (3.6) with

$\mathbf{A} = \mathbf{e}_1$ and $\mathbf{B} = \mathbf{e}_1 \wedge \mathbf{e}_2$. Let $\mathbf{X}$ range over the basis of the blades: $\{1, \mathbf{e}_1, \mathbf{e}_2, \mathbf{e}_1 \wedge \mathbf{e}_2\}$. This produces four equations, each of which gives you information on the coefficient of the corresponding basis element in the final result. Show that $\mathbf{e}_1 \rfloor (\mathbf{e}_1 \wedge \mathbf{e}_2) = 0(1) + 0(\mathbf{e}_1) + 1(\mathbf{e}_2) + 0(\mathbf{e}_1 \wedge \mathbf{e}_2)$.

2. (*continued from previous*) Change the metric such that $\mathbf{e}_2 \cdot \mathbf{e}_2 = 0$. This is a nondegenerate metric, of which $\mathbf{e}_2$ is a null vector (see Appendix A). Show that you cannot now determine the coefficient of $\mathbf{e}_2$ in the value of $\mathbf{e}_1 \rfloor (\mathbf{e}_1 \wedge \mathbf{e}_2)$ through the procedure based on (3.6). Then use the explicit definition of the contraction to show that the contraction is still well defined, and equal to $\mathbf{e}_1 \rfloor (\mathbf{e}_1 \wedge \mathbf{e}_2) = \mathbf{e}_2$.

3. Derive the following dualities for the right contraction, corresponding to (3.20) and (3.21) for the usual (left) contraction:

$$\mathbf{C} \lfloor (\mathbf{B} \wedge \mathbf{A}) = (\mathbf{C} \lfloor \mathbf{B}) \lfloor \mathbf{A} \quad \text{universally valid} \qquad (3.33)$$

$$\mathbf{C} \lfloor (\mathbf{B} \lfloor \mathbf{A}) = (\mathbf{C} \lfloor \mathbf{B}) \wedge \mathbf{A} \quad \text{when } \mathbf{A} \subseteq \mathbf{C} \qquad (3.34)$$

Then give the counterpart of (3.24). (Hint: use (3.19).)

4. Verify the geometric interpretation of the usual inner product between vectors, in the light of viewing it as a specific example of the contraction. In agreement with Section 3.3, show that $\mathbf{x} \cdot \mathbf{a}$ can be interpreted as an element of the 0-dimensional subspace of $\mathbf{a}$ perpendicular to the subspace $\mathbf{x}$.

5. The equation $\mathbf{x} \rfloor \alpha = 0$ (in (3.8)) also has a consistent geometric interpretation in the sense of Section 3.3. Since the scalar α denotes the point at the origin, $\mathbf{x} \rfloor \alpha$ has the following semantics: the subspace of vectors perpendicular to $\mathbf{x}$, contained in the 0-blade α. Give a plausible correctness argument of this statement.

6. Similar to the previous two exercises, verify the geometric semantics of (3.7).

7. Duality in 1-D Euclidean space should avoid the extra sign involved in double duality, as specified in (3.24). Show this explicitly, by taking the dual of a vector $\mathbf{a}$ relative to a suitably chosen unit pseudoscalar for the 1-D space and dualizing again.

8. We have seen in Section 2.4 how in 3-D space a trivector $\mathbf{a} \wedge \mathbf{b} \wedge \mathbf{c}$ can be written as

$$\mathbf{a} \wedge \mathbf{b} \wedge \mathbf{c} = \det([\![\mathbf{a}\ \mathbf{b}\ \mathbf{c}]\!])\, \mathbf{e}_1 \wedge \mathbf{e}_2 \wedge \mathbf{e}_3,$$

with $[\![\mathbf{a}\ \mathbf{b}\ \mathbf{c}]\!]$ the 3×3 matrix having the three 3-D vectors $\mathbf{a}, \mathbf{b}, \mathbf{c}$ as columns (a construction borrowed from standard linear algebra). Express this determinant fully in terms of our subspace algebra.

9. In a plane with unit pseudoscalar $\mathbf{I}_2$, we can rotate a vector by a right angle using the fact that contraction $\mathbf{x} \rfloor \mathbf{I}_2$ is a perpendicular to $\mathbf{x}$. Therefore, you can construct an orthogonal basis for the plane from any vector in it. Use this capability to give a coordinate-free specification of a rotation of a vector $\mathbf{x}$ over ϕ radians in that plane. Make sure you get the rotation direction correctly related to the plane's orientation. (We will do rotations properly in Chapter 7.)

10. Using the definition of the cross product (3.28), verify that you can compute the volume spanned by three vectors **a**, **b**, and **c** as $\mathbf{a} \cdot (\mathbf{b} \times \mathbf{c})$. What is the corresponding formula using $\wedge$ and $\rfloor$?

11. Derive the notorious *bac-cab* formula for the cross product (i.e., $\mathbf{a} \times (\mathbf{b} \times \mathbf{c}) = \mathbf{b}(\mathbf{a} \cdot \mathbf{c}) - \mathbf{c}(\mathbf{a} \cdot \mathbf{b})$), directly from its definition (3.28). What is the corresponding formula using $\wedge$ and $\rfloor$, and what is its geometric interpretation?

12. The inner product formula for cross products is $(\mathbf{a} \times \mathbf{b}) \cdot (\mathbf{c} \times \mathbf{d}) = (\mathbf{a} \cdot \mathbf{c})(\mathbf{b} \cdot \mathbf{d}) - (\mathbf{a} \cdot \mathbf{d})(\mathbf{b} \cdot \mathbf{c})$. Derive it from (3.28). What is the corresponding formula using $\wedge$ and $\rfloor$, and what is its geometric interpretation?

13. In a nonorthonormal basis, the outer product $\mathbf{b}^i \wedge \mathbf{b}_i$ of a vector and its corresponding reciprocal is not generally zero. However, when summed over all basis vectors, all those 2-blades cancel out:

$$\sum_i \mathbf{b}^i \wedge \mathbf{b}_i = 0. \tag{3.35}$$

Show this by expressing $\mathbf{b}^i$ on the usual basis $\{\mathbf{b}_j\}$, and using a symmetry argument on the resulting double summation.

3.11 PROGRAMMING EXAMPLES AND EXERCISES

3.11.1 ORTHONORMALIZATION

In this example we use the contraction product and the outer product to orthonormalize a set of three vectors. The code is given in Figure 3.9, and Figure 3.10 shows a screenshot

```
void computeOrthoVectors(const e3ga::vector nonOrtho[3], e3ga::vector ortho[3]) {
  // compute ortho vector 1:
  // unit_e() returns a unit multivector (Euclidean metric)
  ortho[0] = unit_e(nonOrtho[0]);

  // compute ortho vector 2:
  // << is the operator used for the left contraction
  ortho[1] = unit_e(ortho[0] << (ortho[0] ^ nonOrtho[1]));

  // compute ortho vector 3:
  ortho[2] = unit_e((ortho[1] ^ ortho[0]) <<
    (ortho[0] ^ ortho[1] ^ nonOrtho[2]));
}
```

Figure 3.9: Orthonormalization code (Example 1).

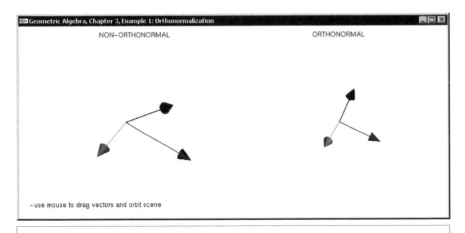

Figure 3.10: Orthonormalization: nonorthonormal vectors on the left, orthonormal vectors on the right (Example 1).

of its output. The first vector is normalized using the function `unit_e()`. This function takes any multivector and returns its unit in the sense that the Euclidean norm (sum of the squares of all coordinates) is 1. `unit_e()` assumes that its input is nonzero.

The second vector is computed in two steps. First, the bivector containing the first two vectors is computed. Then, the first vector is removed from the bivector using the left contraction (the $<<$ operator), resulting in the second vector, orthogonal to the first. This is in fact a computation of the dual within the plane of the two vectors.

The third vector is determined by computing the trivector spanned by all three vectors and removing the bivector spanned by the first two vectors (this is duality in their common space). In 3-D this step is actually redundant, as the third vector is fully determined by the first two vectors. It can be computed using dualization of the bivector or using the cross product (see the next exercise). We will generalize this example to Gram-Schmidt orthogonalization in programming exercise 6.7.2 of Chapter 6.

What happens when the input vectors become dependent?

3.11.2 EXERCISE: IMPLEMENTING THE CROSS PRODUCT

When you download the code for exercise 2 of this Chapter, you will find that it contains a bare-bones `crossProduct()` function:

```
/// returns a x b
e3ga::vector crossProduct(const e3ga::vector &a,
                          const e3ga::vector &b) {
  // exercise: compute the cross product, return it:
  return _vector(0);
}
```

Fill in the function such that it computes the cross product, according to the method of in Section 3.7. The function `dual()` is available to compute the dual of multivectors. If you need the pseudoscalar, use the constant `I3`, or its inverse, `I3i`.

You can check that your implementation works by running the example: drag the input vectors (red and green) around, and verify that the blue vector stays orthogonal to them.

3.11.3 RECIPROCAL FRAMES

In this example, we explore the construction of reciprocal frames as explained in Section 3.8. The example program allows you to manipulate three vectors and see the reciprocal frame of these three vectors. Drag the mouse (using any button) to change the vectors and orbit the scene. When you play around with the example, note the following:

- When you make a vector longer, its reciprocal vector becomes shorter.
- The reciprocal of a vector is always orthogonal to the other two vectors in the original frame. The easiest way to verify this is by "orbiting" the scene, but you may want to take it upon yourself to draw the orthogonal plane as an exercise.

Figure 3.11 lists the code to compute a reciprocal frame.

Note that the example code uses a class called `mv`. In the code shown so far, we always used specialized multivector classes such as `vector`, `bivector`, and `trivector`. The general multivector class `mv` is required here because we have the need for variables that can hold different multivector types. For example, in the following loop, the multivector `P` holds four types of values: first a scalar, then a vector, then a bivector, and finally a trivector.

```
mv P = (i & 1) ? -1.0f : 1.0f; // = pow(-1, i)
for (unsigned int j = 0; j < nbVectors; j++)
    if (j != i) P ^= IF[j];
```

Working with general multivectors is not as efficient as working with specialized multivectors, but sometimes we cannot avoid them.

3.11.4 COLOR SPACE CONVERSION

The reciprocal frame algorithm can be used to do color space conversion. Common conversions (e.g., RGB to YUV) are from one orthogonal frame to another, so that the reciprocal frame is not really necessary. The following is an example when computing the reciprocal frame is required.

Suppose you want to detect the light emitted by red, green, and blue LEDs in a digital image. The colors of the LEDs are unlikely to be pure red, pure green, and pure blue in the RGB color space. But it is possible to transform the color space such that the LEDs do register as such coordinate directions. First, measure the RGB values of the different LEDs in the digital image. Then, compute the reciprocal frame of these three

```
/**
Computes the reciprocal frame 'RF' of input frame 'IF'
Throws std::string when vectors in 'IF' are not independent,
or if one of the IF[i] is null.
*/
void reciprocalFrame(e3ga::vector *IF, e3ga::vector *RF, int nbVectors) {
    // Treat special cases ('nbVectors' equals 0 or 1)
    // ... (not shown here)

    // compute pseudoscalar 'I' of space spanned by input frame:
    mv I = IF[0];
    for (unsigned int i = 1; i < nbVectors; i++) I ^= IF[i];
    if (_Float(norm_r2(I)) == 0.0)
        throw std::string("reciprocalFrame(): vectors are not independent");

    // compute inverse of 'I':
    mv Ii = inverse(I);

    // compute the vectors of the reciprocal framevector
    for (unsigned int i = 0; i < nbVectors; i++) {
        // compute outer product of all vectors except IF[i]
        mv P = (i & 1) ? -1.0f : 1.0f; // = pow(-1, i)
        for (unsigned int j = 0; j < nbVectors; j++)
            if (j != i) P ^= IF[j];

        // compute reciprocal vector 'i':
        RF[i] = _vector(P << Ii);
    }
    return;
}
```

Figure 3.11: Reciprocal frame code (Example 3). Edited for readability: some code was removed at the beginning of the function that dealt with special cases for which (nbVectors < 2).

"color vectors" and use the reciprocal frame to convert the image colors. The code that implements this is shown in Figure 3.12.

The example program lets you play around by sampling different color values and seeing the result of the conversion in real time. Figure 3.13 shows an example.

You can sample colors at any point in the viewport, including the color-bar at the top. The code that draws this bar is also based on the subspace algebra: A unit vector in the "white" direction in the color space is initialized, then the dual of this vector is computed and factorized:

```
                    // get 'white' vector:
                    e3ga::vector white = _vector(unit_e(e1 + e2 + e3));

                    // Get two vectors, orthogonal to white:
                    // factorizeBlade() find two vectors such that
                    // dual(white) == O[1] ^ O[2]
                    e3ga::vector O[2];
                    factorizeBlade(dual(white), O);
```

```
/**
Converts colors in 'source' images to 'dest' image, according
to the input color frame 'IFcolors'. Reciprocal vectors are returned
in 'RFcolors'.
*/
void colorSpaceConvert(
            const unsigned char *source,
            unsigned char *dest,
            unsigned int width, unsigned int height,
            const e3ga::vector *IFcolors,
            e3ga::vector *RFcolors) {
  // compute reciprocal frame
  reciprocalFrame(IFcolors, RFcolors, 3);

  for (unsigned int i = 0; i < (width * height) * 3; i += 3) {
    // convert RGB pixel to vector:
    e3ga::vector c(vector_e1_e2_e3, (float)source[i + 0], (float)source[i + 1],
                                    (float)source[i + 2]);

    // compute colors in in destination image:
    float red = _Float(c << g_RFcolors[0]);
    float green = _Float(c << g_RFcolors[1]);
    float blue = _Float(c << g_RFcolors[2]);

    // clip colors:
    if (red < 0.0f) red = 0.0f;
    else if (red > 255.0f) red = 255.0f;
    if (green < 0.0f) green = 0.0f;
    else if (green > 255.0f) green = 255.0f;
    if (blue < 0.0f) blue = 0.0f;
    else if (blue > 255.0f) blue = 255.0f;

    // set colors in destination image
    dest[i + 0] = (unsigned char)(red + 0.5f); // +0.5f for correct rounding
    dest[i + 1] = (unsigned char)(green + 0.5f);
    dest[i + 2] = (unsigned char)(blue + 0.5f);
  }
}
```

Figure 3.12: Color space conversion code (Example 4).

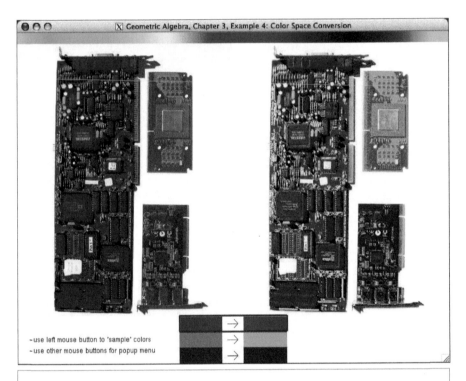

Figure 3.13: Color space conversion screenshot (Example 4). On the left is the original image: a photo of some computer parts that contains red, green, and blue patches. On the right is an example of a converted image: the colors of the parts have been converted to "pure" red, green, and blue.

We now have a frame which spans the RGB color space. We can generate all fully saturated colors by performing a rotation in the O[0] ∧ O[1]-plane:

```
// alpha runs from 0 to 2 PI
for (float angle = 0.0f; angle < PI2; angle += STEP) {
  // generate all fully saturated colors:
  e3ga::vector C = _vector(white + cos(angle) * O[0] +
                                   sin(angle) * O[1]);

  // set current color:
  glColor3fv(C.getC(vector_e1_e2_e3));

  // draw small patch in the current color:
  // ...
}
```

4 LINEAR TRANSFORMATIONS OF SUBSPACES

Linear transformations of a vector space $\mathbb{R}^n$ change its vectors. When this happens, the blades spanned by those vectors change quite naturally to become the spans of the transformed vectors. That defines the extension of a linear transformation to the full subspace algebra. This embedding gives us more powerful tools to apply linear transformations immediately to subspaces, without needing to first decompose those subspaces into vectors.

We study the resulting structure in this chapter. The algebra dictates how we should do the outer products and contractions of transformed blades, and in that way gives us the transformation formulas for the products themselves. Transforming contractions is a lot more involved than transforming outer products (since it involves the metric of the space), but the effort pays off by providing a compact coordinate-free formula for the inverse of a linear transformation.

In this book we will mostly be interested in orthogonal transformations. We can easily derive some of their properties in this chapter and see why they are special (they are the only transformations that are structure-preserving for the contraction). Their real importance and ease of representation will be revealed only in Chapter 7.

At first reading, you can skim through this chapter, taking in only the principle of the *outermorphism*, which takes the structure preservation of the outer product as its tenet and the transformation formulas for the other products. The main facts are summarized in Section 4.6.

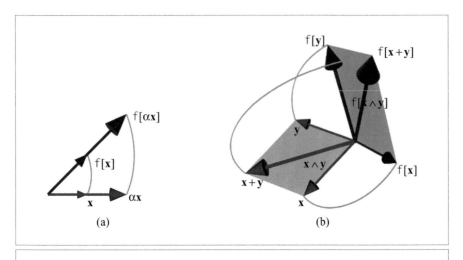

Figure 4.1: The defining properties of a linear transformation.

4.1 LINEAR TRANSFORMATIONS OF VECTORS

We are interested in linear transformations, mapping a vector space $\mathbb{R}^n$ onto itself.[1] Such a *linear transformation* $f : \mathbb{R}^n \to \mathbb{R}^n$ has the defining properties

$$f[\alpha\mathbf{x} + \beta\mathbf{y}] = \alpha\,f[\mathbf{x}] + \beta\,f[\mathbf{y}], \qquad (4.1)$$

where $\alpha, \beta \in \mathbb{R}$ are scalars and $\mathbf{x}, \mathbf{y} \in \mathbb{R}^n$ are vectors. It is convenient to see this as two conditions:

$$\begin{cases} f[\alpha\mathbf{x}] = \alpha\,f[\mathbf{x}] \\ f[\mathbf{x} + \mathbf{y}] = f[\mathbf{x}] + f[\mathbf{y}] \end{cases} \qquad (4.2)$$

The first condition means that a line through the origin remains a straight line through the origin, with a preservation of ratios of vectors along the lines (Figure 4.1(a)). The second condition means that the parallelogram-based addition is preserved (see Figure 4.1(b)).

Examples of such linear transformations on subspaces include scaling, rotation (but only around an axis through origin), and reflection (but only relative to subspace containing the origin), but *not* translation, which tends to produce nonhomogeneous, offset spaces.

---

1   In this chapter, we perform linear transformations within the same space $\mathbb{R}^n$, not from one space to another. Though the same principles apply to both, the additional notation involved in such space-to-space transformations would hide the basic structural simplicity we need to expose here.

Linear transformations therefore do *not* include certain important transformations that we definitely want to include in our treatment of geometry. Yet linear transformations are important, because we will see in Part II how we can construct those desirables using linear transformations in higher-dimensional operational models of affine or Euclidean space. Also, linear mappings provide a local description of a wide class of arbitrary mappings, which is a successful way to study those in differential geometry.

4.2 OUTERMORPHISMS: LINEAR TRANSFORMATIONS OF BLADES

We start with a specific linear transformation f in the vector space $\mathbb{R}^n$, which maps vectors to vectors. We will use sans serif type to denote these linear transformations to distinguish them from the blades (and other elements we introduce later), and denote their action by square brackets to avoid confusion with the grouping brackets of the products, and remind ourselves of their linearity. So $f[\mathbf{x}]$ denotes the action of the linear transformation f on the vector $\mathbf{x}$.

We would like to find a natural extension that makes f act on arbitrary blades, or even arbitrary multivectors. We will argue that this natural extension should be done according to the following simple rules:

$$\begin{aligned} f[\alpha] &= \alpha \text{ for scalar } \alpha \\ f[\mathbf{A} \wedge \mathbf{B}] &= f[\mathbf{A}] \wedge f[\mathbf{B}] \\ f[\mathbf{A} + \mathbf{B}] &= f[\mathbf{A}] + f[\mathbf{B}] \end{aligned} \qquad (4.3)$$

where $\mathbf{A}$ and $\mathbf{B}$ are blades of arbitrary grade (even grade 0), although the results immediately generalize to general multivectors by the imposed linearity. (The third rule is a consequence of the second and (4.2), at least for same-grade blades, but we prefer to have it explicit so that linearity can be easily extended to multivectors.)

An extension of a map of vectors to vectors in this manner to the whole of the Grassmann algebra is called *extension as a (linear) outermorphism*, since the second property shows that we obtain a morphism (i.e., a mapping) that commutes with the outer product. The properties in (4.3) fully define the outermorphism corresponding to the linear transformation f.

Outermorphisms have nice algebraic properties that are essential to their geometrical usage:

- ***Blades Remain Blades***. Geometrically, oriented subspaces are transformed to oriented subspaces.

- ***Grades Are Preserved***. The linear transformation f turns vectors into vectors. Then it follows immediately from the second rule that $grade(f[\mathbf{A}]) = grade(\mathbf{A})$ for blades.

Geometrically, this means that the dimensionality of subspaces does not change under a linear transformation.

- **Preservation of Factorization.** If **A** and **B** have a blade **C** in common (so that they may be written as $\mathbf{A} = \mathbf{A}' \wedge \mathbf{C}$ and $\mathbf{B} = \mathbf{C} \wedge \mathbf{B}'$, for appropriately chosen **A**′ and **B**′), then f[**A**] and f[**B**] have f[**C**] in common. Geometrically, this means that the meet (intersection) of subspaces is preserved.

If you are happy with (4.3) as a definition, you can move on to Section 4.2.2. If you need some motivation to convince yourself of its consistency with the algebra of subspaces as we developed it thus far, read the next section.

4.2.1 MOTIVATION OF THE OUTERMORPHISM

Let us take a step back from the algebraic generalization of a linear transformation in (4.3) and show its geometric plausibility.

In the beginning, we have nothing more than the linear transformation f from vectors to vectors $\mathsf{f} : \mathbb{R}^n \to \mathbb{R}^n$. It obviously satisfies the linearity axioms of (4.2), graphically depicted in Figure 4.1.

We want linear transformations on all k-blades. Starting with 2-blades, we introduce a linear transformation f_2 mapping 2-blades to 2-blades (i.e., $\mathsf{f}_2 : \bigwedge^2 \mathbb{R}^n \to \bigwedge^2 \mathbb{R}^n$). Linearity of f_2 now means linearity for 2-blades, so satisfying $\mathsf{f}_2[\alpha \mathbf{A}] = \alpha \mathsf{f}_2[\mathbf{A}]$ and $\mathsf{f}_2[\mathbf{A} + \mathbf{B}] = \mathsf{f}_2[\mathbf{A}] + \mathsf{f}_2[\mathbf{B}]$ — where **A** and **B** are 2-blades. But this mapping f_2 cannot be totally arbitrary. One way to construct the 2-blades is by using two vectors. If $\mathbf{A} = \mathbf{x} \wedge \mathbf{y}$, how should we relate f (acting on vectors in $\mathbb{R}^n$) to f_2 (acting on 2-blades in $\bigwedge^2 \mathbb{R}^n$), so that we get a consistent structure to our subspace algebra? Figure 4.1 provides the clue: the parallelogram construction is preserved under f by the linearity axioms—and such a construction occurs not only in defining the sum of vectors, but *also in defining the 2-blade through the outer product* (compare Figure 2.2 to Figure 2.3(a)). So we must connect the two linear transformations f and f_2 in a structurally consistent manner by setting

$$\mathsf{f}_2[\mathbf{x} \wedge \mathbf{y}] = \mathsf{f}[\mathbf{x}] \wedge \mathsf{f}[\mathbf{y}].$$

This 2-blade is linear in **x** and **y**, and so are both sides of this equation, guaranteeing that the construction is internally consistent. For instance: $\mathsf{f}_2[\alpha(\mathbf{x} \wedge \mathbf{y})] = \mathsf{f}_2[(\alpha \mathbf{x}) \wedge \mathbf{y}] = \mathsf{f}[\alpha \mathbf{x}] \wedge \mathsf{f}[\mathbf{y}] = \alpha \mathsf{f}[\mathbf{x}] \wedge \mathsf{f}[\mathbf{y}] = \alpha \mathsf{f}_2[\mathbf{x} \wedge \mathbf{y}]$, which is a proof that f_2 thus defined indeed has one of the linearity properties. Since it is so consistent, we can consider f and f_2 as the same linear transformation, just overloaded to apply to arguments of different grade, so we denote them both by f.

The story for 3-blades is similar—the parallelepiped construction can be interpreted as a span (outer product) or as an addition diagram (linearity). Equating the two suggests defining

$$f[\mathbf{x} \wedge \mathbf{y} \wedge \mathbf{z}] = f[\mathbf{x}] \wedge f[\mathbf{y}] \wedge f[\mathbf{z}].$$

Associativity of the outer product gives us associativity for the outermorphism f, and then f naturally extends to all grades.

There is also a strong suggestion of how we should relate a linear transformation among scalars (i.e., 0-blades) to the linear transformation f of vectors in a consistent manner. Remember that by (2.5), the standard product of a vector with a scalar is just the outer product in disguise. As a consequence, the first linearity condition of (4.2) can be read in our exterior algebra as $f[\alpha \wedge \mathbf{x}] = \alpha \wedge f[\mathbf{x}]$. To keep the outermorphism property, it is therefore natural to define

$$f[\alpha] = \alpha$$

as the extension of f to scalars. The geometric semantics of this is that *the point at the origin remains fixed under a linear transformation*, in all its qualities, including weight and sign.

This is how the whole ladder of subspaces is affected naturally by the linear transformation of the underlying vector space, preserving the structure of the spanning product that went into its construction: the span of the transforms is the transform of the span.

4.2.2 EXAMPLES OF OUTERMORPHISMS

Let us look at some simple examples of such extensions of linear transformations.

1. *Uniform Scaling.* This is the linear transformation $S[\mathbf{x}] = \alpha\mathbf{x}$. On an n-blade $\mathbf{A} = \mathbf{a}_1 \wedge \mathbf{a}_2 \wedge \cdots \wedge \mathbf{a}_n$, this gives

    $$S[\mathbf{A}] = S[\mathbf{a}_1] \wedge S[\mathbf{a}_2] \wedge \cdots \wedge S[\mathbf{a}_n] = \alpha^n \mathbf{A}. \tag{4.4}$$

    For 2-blades represented as parallelograms, this contains the well-known result that as each of the sides is multiplied by α, the area is multiplied by α^2; but it is more, since it also contains the statement that the attitude of the 2-blade remains the same, and so does its orientation, even when α is negative. And since 2-blades have no fixed shape, the same applies to *any* area in the plane: as the linear measure gets scaled by α, the area measure scales by α^2.

    For a 3-blade $\mathbf{I}_3 = \mathbf{a}_1 \wedge \mathbf{a}_2 \wedge \mathbf{a}_3$ in 3-D space, we obtain $S[\mathbf{I}_3] = \alpha^3 \mathbf{I}_3$, as expected. When α is negative, there is thus an orientation change of the volume. Again, nothing shatteringly new in its geometric interpretation, but note how in the formulation of such statements, their computation and their proof are all an intrinsic part of the algebra at a very elementary level. That is how we would want it.

2. *Parallel Projection onto a Line.* In the plane with 2-blade $\mathbf{a} \wedge \mathbf{b}$, let the linear transformation P be such that $P[\mathbf{a}] = \mathbf{a}$, while $P[\mathbf{b}] = 0$. This is a projection in the **b**-direction onto the **a**-line (see Figure 4.2). Since any vector $\mathbf{x}$ in this plane can be written as an **a**-component plus a **b**-component, this determines the transformation

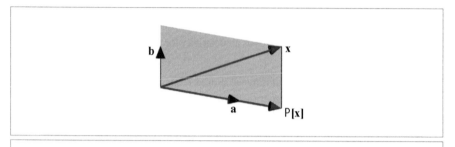

Figure 4.2: Projection onto a line a in the b-direction.

$$P[\mathbf{x}] = P[\alpha\,\mathbf{a} + \beta\,\mathbf{b}] = \alpha\,\mathbf{a} \qquad (4.5)$$

(where α can be computed as $(\mathbf{x}\wedge\mathbf{b})/(\mathbf{a}\wedge\mathbf{b})$ using the reciprocal frame of Section 3.8 or the techniques of Section 2.7.1). Extending this P as an outermorphism, we find that $P[\mathbf{a} \wedge \mathbf{b}] = P[\mathbf{a}] \wedge P[\mathbf{b}] = \mathbf{a} \wedge 0 = 0$. Any 2-blade in the plane $\mathbf{a} \wedge \mathbf{b}$ becomes 0: this transformation makes areas disappear.

You may have expected the answer to be $P[\mathbf{a} \wedge \mathbf{b}] = P[\mathbf{a}]$, because intuitively the plane $\mathbf{a}\wedge\mathbf{b}$ becomes the line $\mathbf{a}$. But this is not what an outermorphism does (it always preserves grade), so we must be careful with such naïve geometrical motivations for the results of algebraic computations. The image of the plane of vectors is indeed the line of vectors, but the plane of vectors is not equivalent to the 2-blade of the plane!

3. **Planar Rotation.** Consider two independent unit vectors $\mathbf{u}$ and $\mathbf{v}$ that span a 2-blade $\mathbf{u}\wedge\mathbf{v}$. This 2-blade determines a plane through the origin in a Euclidean space $\mathbb{R}^{n,0}$. Let R be a rotation around the origin in this plane, a linear transformation. Let the rotation be such that it turns $\mathbf{u}$ to $\mathbf{v}$, so $R[\mathbf{u}] = \mathbf{v}$. Since the whole plane rotates, the vector originally at $\mathbf{v}$ also rotates to a unit vector $\mathbf{w} \equiv R[\mathbf{v}]$. Since the rotation around the origin is linear, the parallelogram spanned by $\mathbf{u}$ and $\mathbf{v}$ transforms to another parallelogram spanned by $\mathbf{v} = R[\mathbf{u}]$ and $\mathbf{w} = R[\mathbf{v}]$ (see Figure 4.3). The sketch shows that $\mathbf{u} \wedge \mathbf{v}$ and $\mathbf{v} \wedge \mathbf{w}$ are identical 2-blades, so $\mathbf{u} \wedge \mathbf{v} = \mathbf{v} \wedge \mathbf{w}$. This permits us to compute the effect of the rotation on a 2-blade:

$$R[\mathbf{u} \wedge \mathbf{v}] = R[\mathbf{u}] \wedge R[\mathbf{v}] = \mathbf{v} \wedge \mathbf{w} = \mathbf{u} \wedge \mathbf{v}. \qquad (4.6)$$

It follows that the 2-blade $\mathbf{u} \wedge \mathbf{v}$ is preserved under the rotation. This corresponds well to our insight that a rotation plane is an invariant of a rotation. But note how specific (4.6) is: it states that *all* properties of the plane—attitude, area measure, and orientation—are preserved. It is remarkable that although the vectors

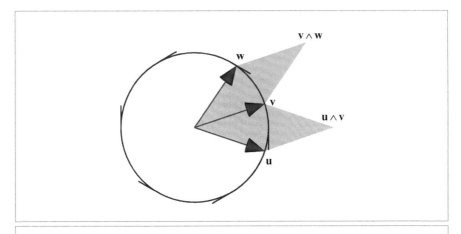

Figure 4.3: A rotation around the origin of unit vectors in the plane of the page, described by 2-blades.

u and **v** themselves are not preserved (they rotate), their 2-blade is. We might express this as *a rotation has no real eigenvectors in its plane, but it has a real eigenblade of grade 2: the plane itself.*

Note that we have not specified any space in which we perform the rotation, assuming only that it has as least 2 dimensions for the 2-blade to be nonzero. So our picture and reasoning apply to any space of more than 1 dimension.

4. ***Point Reflections.*** In a point reflection through the origin, all vectors change sign. So this is a uniform scaling by -1. Then (4.4) shows that an n-blade changes by $(-1)^n$: blades of even grades are unchanged, and blades of odd grades obtain the opposite orientation. Note that this does not depend on the dimensionality of the space in which they are embedded. As an example, point reflection in 3-D space changes the orientation of 3-blades, that is, the handedness of objects: a right-hand glove becomes a left-hand glove (see also structural exercise 1). We will see soon that this cannot be undone by a rotation, using an argument that only involves the outer product.

5. ***Orthogonal Projection.*** In (3.25) we met the orthogonal projection of a vector **x** onto a blade **B** as $P_B[x] = ((x \rfloor B) \rfloor B^{-1})$. Since the mapping is linear, we can extend it as an outermorphism to construct the projection of a higher-order subspace **X** onto **B**. Let us first extend this orthogonal projection from the vector **x** to the bivector **x**∧**y**. By outermorphism, the projection of (**x**∧**y**) onto **B** should be $P_B[x] \wedge P_B[y]$. Now we have a straightforward derivation in which we challenge you to identify the rewriting rules.

$$P_B[\mathbf{x}] \wedge P_B[\mathbf{y}] = ((\mathbf{x}\rfloor\mathbf{B})\rfloor\mathbf{B}^{-1}) \wedge ((\mathbf{y}\rfloor\mathbf{B})\rfloor\mathbf{B}^{-1})$$
$$= (((\mathbf{x}\rfloor\mathbf{B})\rfloor\mathbf{B}^{-1})\rfloor(\mathbf{y}\rfloor\mathbf{B}))\rfloor\mathbf{B}^{-1}$$
$$= ((((\mathbf{x}\rfloor\mathbf{B})\rfloor\mathbf{B}^{-1}) \wedge \mathbf{y})\rfloor\mathbf{B})\rfloor\mathbf{B}^{-1}$$
$$= -((\mathbf{y} \wedge ((\mathbf{x}\rfloor\mathbf{B})\rfloor\mathbf{B}^{-1}))\rfloor\mathbf{B})\rfloor\mathbf{B}^{-1}$$
$$= -(\mathbf{y}\rfloor(((\mathbf{x}\rfloor\mathbf{B})\rfloor\mathbf{B}^{-1})\rfloor\mathbf{B}))\rfloor\mathbf{B}^{-1}$$
$$= -(\mathbf{y}\rfloor((\mathbf{x}\rfloor\mathbf{B}) \wedge (\mathbf{B}^{-1}\rfloor\mathbf{B})))\rfloor\mathbf{B}^{-1}$$
$$= -(\mathbf{y}\rfloor(\mathbf{x}\rfloor\mathbf{B}))\rfloor\mathbf{B}^{-1}$$
$$= -((\mathbf{y} \wedge \mathbf{x})\rfloor\mathbf{B})\rfloor\mathbf{B}^{-1}$$
$$= ((\mathbf{x} \wedge \mathbf{y})\rfloor\mathbf{B})\rfloor\mathbf{B}^{-1}$$
$$= P_B[\mathbf{x} \wedge \mathbf{y}].$$

The final result is, therefore, that we can just apply the projection formula directly to the blade $\mathbf{x} \wedge \mathbf{y}$ to get the outermorphism. Similar steps can be used to provide an inductive proof of the general result for blades of (3.26).

Geometrically, the outermorphism property implies that the projection formula generalizes to higher-order blades in a pleasant way. Our algebra permits the direct projection of subspaces without the necessity of breaking them up into vectors, projecting those, and recomposing the result.

These examples show how merely having the outer product already refines and extends our analysis and application of linear transformations.

4.2.3 THE DETERMINANT OF A LINEAR TRANSFORMATION

We have seen in Chapter 2 how in an n-dimensional space, the blade of highest grade that can be constructed without being identical to 0 is an n-blade, which is a pseudoscalar for the space. The grade-preservation property of a linear transformation f implies that a linear transformation on a pseudoscalar $\mathbf{I}_n$ produces another pseudoscalar. Moreover, all pseudoscalars are scalar multiples of each other, since the space of n-blades $\bigwedge^n \mathbb{R}^n$ is a 1-D linear space. Therefore we find $f[\mathbf{I}_n] = \delta\,\mathbf{I}_n$, with δ a scalar. This defines δ as the change in pseudoscalar magnitude and orientation, as a ratio of the transformed n-dimensional hypervolume (for that is what a pseudoscalar is) to the original hypervolume. It is called the *determinant* of f, denoted det(f). So we have the important implicit definition

$$determinant: \quad f[\mathbf{I}_n] \equiv \det(f)\,\mathbf{I}_n. \tag{4.7}$$

This scalar number det(f) is indeed equivalent in value to that concept in linear algebra, so we are not abusing the name. There, too, the determinant is a ratio of signed hypervolume measures.

The usual way of teaching linear algebra in the applied sciences relies heavily on matrix representations. You might be excused for believing that linear algebra is about matrices rather than about linear transformations, and that the determinant is just a property of a square matrix rather than a fundamental property of a linear transformation. But it is just that, and it can be defined without referring to matrices. We have just done so in (4.7), even managing to avoid coordinates altogether. We briefly show how we can use this geometrical approach to retrieve the determinants of some common transformations.

- *Determinant of a Rotation.* The example of the rotation in the Euclidean plane indicated by the blade $\mathbf{u} \wedge \mathbf{v}$ demonstrated that $\mathsf{R}[\mathbf{u} \wedge \mathbf{v}] = \mathbf{u} \wedge \mathbf{v}$. Since $\mathbf{u} \wedge \mathbf{v}$ is proportional to the pseudoscalar $\mathbf{I}_2$ of the plane, this implies

$$\mathsf{R}[\mathbf{I}_2] = \mathbf{I}_2.$$

  Therefore the determinant of a 2-D rotation equals 1.

  If the rotation plane is embedded within an n-dimensional Euclidean metric space $\mathbb{R}^{n,0}$, then we can span a pseudoscalar for the n-dimensional embedding space using $\mathbf{I}_2$ combined with $(n - 2)$ vectors perpendicular to the plane. Each of those vectors is not affected by the rotation, so for the part of the space they span we have $\mathsf{R}[\mathbf{I}_{n-2}] = \mathbf{I}_{n-2}$ (where $\mathbf{I}_{n-2}$ is a pseudoscalar for the $(n - 2)$-dimensional space), by the outermorphism property. We thus find: $\mathsf{R}[\mathbf{I}_n] = \mathsf{R}[\mathbf{I}_2 \wedge \mathbf{I}_{n-2}] = \mathsf{R}[\mathbf{I}_2] \wedge \mathsf{R}[\mathbf{I}_{n-2}] = \mathbf{I}_2 \wedge \mathbf{I}_{n-2} = \mathbf{I}_n$. So the determinant of a rotation still equals 1, even in an n-dimensional space $(n \geq 2)$.

- *Determinant of a Point Reflection.* We have seen that a point reflection satisfies $\mathsf{f}[\mathbf{I}_n] = (-1)^n \mathbf{I}_n$. Thus its determinant equals 1 in even dimensions and -1 in odd dimensions. This suggests that in even dimensions, a reflection can be performed as a rotation, and indeed it can.

- *Determinant of a Projection onto a Line.* The projection P onto a line has a determinant that varies with the dimensionality of the space $\mathbb{R}^n$. We have seen how any blade with grade exceeding 1 becomes zero. Therefore any pseudoscalar of $\mathbb{R}^n$ with $n > 1$ is projected to 0, and $\det(\mathsf{P}) = 0$. However, for $n = 1$ the line must necessarily be the whole space $\mathbb{R}^1$. Now the projection is the identity, so $\det(\mathsf{P}) = 1$.

We can continue the theme of determinants. Applying two linear transformations $\mathsf{f} : \mathbb{R}^n \to \mathbb{R}^n$ and $\mathsf{g} : \mathbb{R}^n \to \mathbb{R}^n$, first f then g, we obtain a composite transformation that is again linear (as you can easily show) and that can therefore also be extended as an outermorphism. We denote this composite transformation by $(\mathsf{g} \circ \mathsf{f})$. We compute its determinant:

$$\det(\mathsf{g} \circ \mathsf{f}) \, \mathbf{I}_n = (\mathsf{g} \circ \mathsf{f})[\mathbf{I}_n] = \mathsf{g}[\mathsf{f}[\mathbf{I}_n]] = \det(\mathsf{f}) \, \mathsf{g}[\mathbf{I}_n] = \det(\mathsf{g}) \det(\mathsf{f}) \, \mathbf{I}_n.$$

Therefore, we get the composition rule of determinants:

$$\det(\mathsf{g} \circ \mathsf{f}) = \det(\mathsf{g}) \det(\mathsf{f}). \tag{4.8}$$

This is a well-known result, derived within this context of the outer product in a straightforward algebraic manner with satisfying geometrical semantics.

4.3 LINEAR TRANSFORMATION OF THE METRIC PRODUCTS

The linear transformation f has been extended to an outermorpishm to transform an outer product completely naturally as

$$f[A \wedge B] = f[A] \wedge f[B].$$

Of course, we should also understand how the scalar product and the contraction transform; those three combined will then enable us to transform arbitrary geometric compositions and will give us the full extent of linear transformations in our subspace algebra.

The scalar product is easily transformed, but the contraction takes a bit more work and requires us to introduce an important concept: the *adjoint* of a linear transformation.

4.3.1 LINEAR TRANSFORMATION OF THE SCALAR PRODUCT

The scalar product returns a scalar, so it transforms under a linear transformation as

$$f[A * B] = A * B. \tag{4.9}$$

This looks straightforward, but when you remember that $A * \widetilde{A}$ is the squared norm of the blade A, does this now mean that no linear transformation can change the norm? Or, a related question since $A * \widetilde{B}$ is proportional to the cosine of the angle between A and B, does this mean that no linear transformation can change the angle? That would imply that all linear transformations are orthogonal transformations.

Resolve these problems for yourself—or take a hint from structural exercise 6.

4.3.2 THE ADJOINT OF A LINEAR TRANSFORMATION

The transformation $f[A \rfloor B]$ for the contraction $A \rfloor B$ will follow from our definitions so far, but can only be formulated compactly when we introduce an additional construction: the *adjoint* of f. We do this first.

The *adjoint* $\overline{f}$ of the linear transformation $f : \bigwedge \mathbb{R}^n \to \bigwedge \mathbb{R}^n$ is a linear transformation $\overline{f} : \bigwedge \mathbb{R}^n \to \bigwedge \mathbb{R}^n$, defined implicitly for vectors by the equation

$$\text{adjoint transformation}: \quad \overline{f}[a] * b = a * f[b] \tag{4.10}$$

for all a and b of $\mathbb{R}^n$. In nondegenerate metrics, this defines it fully. In degenerate metrics, we have similar incompleteness issues as for (3.6)—you can then use an explicit, coordinate-based formula. This is explored in structural exercise 9.

It is strange to have the definition so implicit. To show that the adjoint is actually a familiar concept from linear algebra, we momentarily convert the equation to matrix notation.

Let $[\![f]\!]$ be the matrix of the mapping f and $[\![\,\overline{f}\,]\!]$ be the matrix of $\overline{f}$, and convert the inner product to a matrix product. Vectors like $\mathbf{a}$ are represented as a column matrix $[\![\mathbf{a}]\!]$. By transferring (4.10) to matrix notation in the case of a Euclidean orthonormal basis and by using the matrix transpose, we obtain

$$[\![\mathbf{a}]\!]^{T}\,[\![f]\!][\![\mathbf{b}]\!] = ([\![\,\overline{f}\,]\!][\![\mathbf{a}]\!])^{T}\,[\![\mathbf{b}]\!] = [\![\mathbf{a}]\!]^{T}\,[\![\,\overline{f}\,]\!]^{T}\,[\![\mathbf{b}]\!],$$

so $[\![\,\overline{f}\,]\!] = [\![f]\!]^{T}$. This implies that for vectors in a Euclidean orthonormal basis, the adjoint of f is the *transpose* mapping, specified in a coordinate-free manner.

For blades (and even for multivectors), we extend $\overline{f}$ as an outermorphism. This leads to

$$\text{adjoint transformation}: \quad \overline{f}[\mathbf{A}] * \mathbf{B} = \mathbf{A} * f[\mathbf{B}], \tag{4.11}$$

which we could have taken as the definition of the adjoint for blades. For general f, it follows easily from the symmetry of the scalar product that

$$\overline{\overline{f}} = f,$$

since $\overline{\overline{f}}[\mathbf{A}] * \mathbf{B} = \mathbf{A} * \overline{f}[\mathbf{B}] = f[\mathbf{A}] * \mathbf{B}$, for all $\mathbf{b}$. Another useful property is

$$\overline{f^{-1}} = \overline{f}^{\,-1},$$

which is easily shown from the definition (4.10).

Some examples:

- If f is the uniform scaling defined by $f[\mathbf{x}] = \alpha\,\mathbf{x}$, then $\overline{f}[\mathbf{x}] \cdot \mathbf{a} = \mathbf{x} \cdot (\alpha\,\mathbf{a}) = (\alpha\,\mathbf{x}) \cdot \mathbf{a}$ for all $\mathbf{x}$ and $\mathbf{a}$. This yields $\overline{f}[\mathbf{x}] = \alpha\,\mathbf{x}$, so in this case $\overline{f} = f$, the adjoint equals the original transformation. As an outermorphism on blades, the adjoint is also $\overline{f}[\mathbf{X}] = f[\mathbf{X}]$.

- A special case of the uniform scaling is the point-reflection into the origin, which has $\alpha = -1$. Again, $\overline{f} = f$, but now also $f^{-1} = f$.

- In Figure 4.2, we met the line projection $P[\mathbf{x}] = \mathbf{a}\,(\mathbf{x}\wedge\mathbf{b})/(\mathbf{a}\wedge\mathbf{b}) = \mathbf{a}\,(\mathbf{x}\wedge\mathbf{b})\rfloor(\mathbf{a}\wedge\mathbf{b})^{-1}$. Its adjoint is $\overline{P}[\mathbf{x}] = (\mathbf{x}\cdot\mathbf{a})\,\mathbf{b}\rfloor(\mathbf{a}\wedge\mathbf{b})^{-1}$. This, therefore, is proportional to the dual of $\mathbf{b}$ in the $(\mathbf{a}\wedge\mathbf{b})$ plane.

4.3.3 LINEAR TRANSFORMATION OF THE CONTRACTION

To demonstrate how the contraction product transforms under linear transformations, we first show

$$f[\,\overline{f}[\mathbf{A}]\rfloor\mathbf{B}\,] = \mathbf{A}\rfloor f[\mathbf{B}], \tag{4.12}$$

which we derive simply using the scalar product definition:

$$\mathbf{X} * (\mathbf{A}\rfloor f[\mathbf{B}]) = (\mathbf{X}\wedge\mathbf{A}) * f[\mathbf{B}]$$
$$= \overline{f}[\mathbf{X}\wedge\mathbf{A}] * \mathbf{B}$$

$$= (\bar{f}[\mathbf{X}] \wedge \bar{f}[\mathbf{A}]) * \mathbf{B}$$
$$= \bar{f}[\mathbf{X}] * (\bar{f}[\mathbf{A}] \rfloor \mathbf{B})$$
$$= \mathbf{X} * f[\bar{f}[\mathbf{A}] \rfloor \mathbf{B}].$$

If $\bar{f}$ is invertible (which happens precisely when f is invertible), we can define $\mathbf{A}' = \bar{f}^{-1}[\mathbf{A}]$. Substituting that in (4.12) and dropping the prime gives the promised transformation law for the contraction product:

$$\text{contraction transformation}: \quad f[\mathbf{A} \rfloor \mathbf{B}] = \bar{f}^{-1}[\mathbf{A}] \rfloor f[\mathbf{B}]. \qquad (4.13)$$

By the linearity of the functions and operators involved, this is of course immediately extendable to arbitrary multivectors as $f[A \rfloor B] = \bar{f}^{-1}[A] \rfloor f[B]$.

The result of (4.13) is very powerful, but unfortunately rather abstract. We have not found an easy geometric picture that will convince you of its necessary truth. But the interpretation of the contraction $\mathbf{A} \rfloor \mathbf{B}$ as "the part of $\mathbf{B}$ that remains when $\mathbf{A}$ is taken out in a perpendicular manner" (to paraphrase) helps remember the placement of f and its derived mappings. Obviously, the transformed result is a part of $\mathbf{B}$, so it should transform as $f[\mathbf{B}]$; and taking $\mathbf{A}$ out could explain the inverse f^{-1}; doing so in an orthogonal manner justifies the adjoint $\bar{f}$.

4.3.4 ORTHOGONAL TRANSFORMATIONS

If a linear transformation of vectors preserves their inner product, we call it an *orthogonal transformation*. It then satisfies

$$\text{orthogonal transformation}: \quad f[\mathbf{a}] \cdot f[\mathbf{b}] = \mathbf{a} \cdot \mathbf{b}, \text{ for all } \mathbf{a}, \mathbf{b} \in \mathbb{R}^n.$$

Since orthogonal transformations are invertible, we may set $\mathbf{a} = f^{-1}[\mathbf{x}]$, to obtain $\mathbf{x} \cdot f[\mathbf{b}] = f^{-1}[\mathbf{x}] \cdot \mathbf{b}$ for all $\mathbf{x}, \mathbf{b}$. It follows that

$$\bar{f} = f^{-1},$$

so for an orthogonal transformation, the adjoint equals the inverse transformation. You probably knew this fact from linear algebra, in terms of *the inverse of an orthogonal matrix is its transpose*. In the present context, "inverse equals adjoint" is a statement about the mappings rather than their matrices, and by outermorphism, it also holds for the outermorphism extension of the mapping to blades.

With this, the transformation formula of the contraction is much simpler when f is an orthogonal transformation:

$$f[\mathbf{A} \rfloor \mathbf{B}] = f[\mathbf{A}] \rfloor f[\mathbf{B}].$$

Therefore, for orthogonal transformations, the contraction transforms in a structure-preserving manner: the contraction of the transformed blades is the transformation of the contraction. Orthogonal transformations are thus "innermorphisms" as well as outermorphisms (since the contraction is actually an inner product for blades).

Familiar examples are reflections and rotations. In fact, they are the prototypical orthogonal transformations, and we will discover in Chapter 7 that the general case can always be written as a rotation followed by a reflection (or vice versa).

4.3.5 TRANSFORMING A DUAL REPRESENTATION

The dual of a blade $\mathbf{X}$ is $\mathbf{X}^* \equiv \mathbf{X}\rfloor\mathbf{I}_n^{-1}$. When $\mathbf{X}$ undergoes a linear transformation f, the dual is transformed as well. We should define the transformation f^* of the dual by demanding that it preserve the duality relationship: the f^*-transform of the dual should be the dual of the f-transformed $\mathbf{X}$. In formula:

$$\textit{dual transformation}: \quad \mathsf{f}^*[\mathbf{X}^*] \equiv (\mathsf{f}[\mathbf{X}])^*.$$

That specifies it. We introduce $\mathbf{D} = \mathbf{X}^*$, and find

$$
\begin{aligned}
\mathsf{f}^*[\mathbf{D}] &= (\mathsf{f}[\mathbf{D}^{-*}])^* \\
&= \mathsf{f}[\mathbf{D}\rfloor\mathbf{I}_n]\rfloor\mathbf{I}_n^{-1} \\
&= (\overline{\mathsf{f}}^{-1}[\mathbf{D}] \rfloor \mathsf{f}[\mathbf{I}_n]) \rfloor \mathbf{I}_n^{-1} \\
&= \overline{\mathsf{f}}^{-1}[\mathbf{D}] \wedge (\mathsf{f}[\mathbf{I}_n] \rfloor \mathbf{I}_n^{-1}) \\
&= \det(\mathsf{f})\,\overline{\mathsf{f}}^{-1}[\mathbf{D}], \tag{4.14}
\end{aligned}
$$

as the necessary definition of the linear transformation on duals. Since this is not generally equal to $\mathsf{f}[\mathbf{D}]$, it implies that blades that are intended as dual representations do not transform in the same way as blades that are intended as direct representations. In a proper representation of geometry, we therefore need to indicate how a blade is to be interpreted before we can act on it appropriately with a linear transformation.

Note that the outermorphism transformation law of direct blades is nonmetric, since it only involves the outer product. By contrast, $\mathsf{f}^* = \det(\mathsf{f})\,\overline{\mathsf{f}}^{-1}$ is metric, since it is expressible in terms of the adjoint $\overline{\mathsf{f}}$ whose definition involves the scalar product. This makes sense, since in our algebra of subspaces dualization is a metric concept.

An orthogonal transformation has a determinant equal to ± 1 (see structural exercise 10), and has $\overline{\mathsf{f}}^{-1} = \mathsf{f}$. For such transformations, the dual representation therefore transforms rather nicely: $\mathsf{f}^*[\mathbf{X}^*] = \pm\mathsf{f}[\mathbf{X}^*]$. For rotations, which have determinant $+1$, this is completely structure-preserving: the dual of the transform is the transform of the dual. For rotations it is therefore not necessary to know whether a blade is a dual or direct representation before you can transform it. For orthogonal transformations containing a reflection, which have determinant -1, the extra minus sign in the dual is caused by the reflection of the pseudoscalar of the space. It now makes a difference whether you desire to take the dual of the transform relative to the original pseudoscalar (then you need the -1), or relative to the transformed pseudoscalar (then there is no sign).

4.3.6 APPLICATION: LINEAR TRANSFORMATION OF THE CROSS PRODUCT

As we discussed in Section 3.7, the cross product is an inherently 3-D construct, and we can replace all its uses by elements of our subspace algebra that generalize to n-dimensional space:

$$\mathbf{a} \times \mathbf{b} = (\mathbf{a} \wedge \mathbf{b})^* = (\mathbf{a} \wedge \mathbf{b}) \rfloor \mathbf{I}_3^{-1}.$$

It is interesting to derive an additional argument as to why the cross product should not be used: the normal vector $\mathbf{a} \times \mathbf{b}$ of $\mathbf{a}$ and $\mathbf{b}$, used as a characterization of the homogeneous plane spanned by $\mathbf{a}$ and $\mathbf{b}$, behaves rather awkwardly under a linear transformation f. For its definition clearly shows that it is a dual representation of a 2-blade, and therefore it should transform according to (4.14):

$$\mathbf{a} \times \mathbf{b} \;\; \mapsto \;\; \det(\mathsf{f}) \, \overline{\mathsf{f}}^{-1}[\mathbf{a} \times \mathbf{b}]. \tag{4.15}$$

This is not equal to $\mathsf{f}[\mathbf{a}] \times \mathsf{f}[\mathbf{b}]$, so the normal vector of the transforms does not equal the transform of the normal vector.

In computer graphics and engineering applications, it is customary to use "vector" as a synonym for a tuple of coordinates as well as for a geometrical 1-D direction. The three coefficients of $\mathbf{n} = \mathbf{a} \times \mathbf{b}$ are a vector in the sense of a 3-tuple, but the geometry of $\mathbf{n}$ makes this vector transform unlike a geometrical direction, so it is called a *normal vector*. Even though it is given on the same basis as a regular vector, it requires its own methods to be transformed. The transformation of such a normal vector under the linear transformation with matrix $[\![\mathsf{f}]\!]$ on regular vectors must explicitly be implemented as the matrix $\det(\mathsf{f}) \, [\![\overline{\mathsf{f}}^{-1}]\!] = \det(\mathsf{f}) \, [\![\mathsf{f}]\!]^{-T}$, where $[\![\mathsf{f}]\!]^{-T}$ is the inverse of the transposed matrix. Within the code, such a different transformation for seemingly similar 3-tuples can only be invoked if their data type is kept explicitly. Even then, novices easily get confused; this is a common source of programming error. We provide a programming exercise at the end of this chapter to explore the differences.

In our subspace algebra, we have two choices. We can introduce normal vectors, which then of course need to transform according to $\mathsf{f}^* = \det(\mathsf{f}) \, \overline{\mathsf{f}}^{-1}$. They do so automatically when we define them explicitly in the code. Or we can just characterize the same quantity using the original 2-blade $\mathbf{a} \wedge \mathbf{b}$, which transforms according to the outermorphism f as $\mathsf{f}[\mathbf{a} \wedge \mathbf{b}] = \mathsf{f}[\mathbf{a}] \wedge \mathsf{f}[\mathbf{b}]$. As a characterization of either a plane or a rotation (through its rotation plane), the 2-blade is just as admissible as the classical normal vector, and now you don't need to remember how it transforms. This transformational simplicity reinforces the practical considerations of Section 3.7, and reaffirms our decision to drop the cross product from sound practice in geometrical representation.

4.4 INVERSES OF OUTERMORPHISMS

With the transformation formula for the contraction, we can derive a closed form, coordinate-free formula for the inverse of an outermorphism. First, we compute

$$\overline{f}[\mathbf{A} \rfloor \mathbf{I}_n^{-1}] = f^{-1}[\mathbf{A}] \rfloor \overline{f}[\mathbf{I}_n^{-1}] = \det(\overline{f}) \, f^{-1}[\mathbf{A}] \rfloor \mathbf{I}_n^{-1}.$$

Since $\det(\overline{f}) = \det(\overline{f}) \, \mathbf{I}_n^{-1} * \mathbf{I}_n = \overline{f}[\mathbf{I}_n^{-1}] * \mathbf{I}_n = \mathbf{I}_n^{-1} * f[\mathbf{I}_n] = \det(f) \, \mathbf{I}_n^{-1} * \mathbf{I}_n = \det(f)$, we can substitute $\det(\overline{f})$ by $\det(f)$. Now do the contraction of both sides on $\mathbf{I}_n$, giving[2]

$$f^{-1}[\mathbf{A}] = \frac{\overline{f}[\mathbf{A} \rfloor \mathbf{I}_n^{-1}] \rfloor \mathbf{I}_n}{\det f}. \tag{4.16}$$

In terms of duals, (4.16) reads

$$f^{-1}[\mathbf{A}]^* = \frac{\overline{f}[\mathbf{A}^*]}{\det f}.$$

Mathematicians may object to the use of duality in this formula: taking the inverse of a mapping does not necessarily require a metric space, whereas duality is of course a metric concept. It turns out that the two dualities in (4.16) cancel each other, in the sense that the result does not depend on the precise form of the metric any more, and is therefore actually nonmetric. We feel that if you have a metric (as you often do), you should use it. If you don't have one, you can temporarily introduce a convenient one (e.g., Euclidean) and compute on.

Here are some examples of evaluation of the inverse on various blades:

1. *Pseudoscalar.*

$$f^{-1}[\mathbf{I}_n] = \frac{\overline{f}[1] \rfloor \mathbf{I}_n}{\det f} = \frac{1}{\det f} \mathbf{I}_n,$$

    showing that $\det(f^{-1}) = (\det f)^{-1}$.

2. *Scalar.*

$$f^{-1}[\alpha] = \frac{\overline{f}[\alpha \mathbf{I}_n^{-1}] \rfloor \mathbf{I}_n}{\det f} = \frac{\alpha \det \overline{f} \, (\mathbf{I}_n^{-1} \rfloor \mathbf{I}_n)}{\det f} = \frac{\alpha \det \overline{f}}{\det f} = \alpha,$$

    as expected since the inverse is also an outermorphism.

3. *Vectors.*

$$f^{-1}[\mathbf{a}] = \frac{\overline{f}[\mathbf{a} \rfloor \mathbf{I}_n^{-1}] \rfloor \mathbf{I}_n}{\det f}.$$

---

2   When rereading this book, you may want to replace the contractions in formula (4.16) by the geometric product of Chapter 6.

For vectors there is no simplification of the basic equation, but when you follow this computation for the matrix of f^{-1} on some basis, you find that you have essentially the minor-based construction of the inverse from classical linear algebra (see structural exercise 15). But remember, that construction only applies to f acting on vectors, whereas (4.16) is much more powerful because it also applies to the outermorphism extension on arbitrary blades and multivectors.

4.5 MATRIX REPRESENTATIONS

A matrix is a convenient way of representing a linear transformation relative to a given basis. It is specific to that basis, and therein lies its strength (it is efficient) and its weakness (you cannot tell what it does directly from its form). Matrices are used in implementations of geometric algebra, though usually hidden from the user. You do not need them to specify your linear transformations, but they are an efficient implementation of the specified transformation at the very lowest level.

We introduce our notation for matrices, and show how they can be written completely in terms of the products from subspace algebra to convey their geometric specificity. Then we construct the matrix of the outermorphism of a linear transformation f acting on all multivectors of $\bigwedge \mathbb{R}^n$.

4.5.1 MATRICES FOR VECTOR TRANSFORMATIONS

If you have a basis $\{\mathbf{b}_i\}_{i=1}^n$ (not necessarily orthonormal) for the vector space $\mathbb{R}^n$, you can use this to define a *matrix representation* of a linear transformation f of that space. This can then be used to transform arbitrary vectors, since they can be written as a weighted sum of basis vectors, over which f distributes.

We define a column representation of the vector $\mathbf{x}$, in which the element in the jth row of the column is labeled by a superscript as $[\![\mathbf{x}]\!]^j$. Then

$$[\![\mathbf{x}]\!]^j = x^j = \mathbf{x} \cdot \mathbf{b}^j,$$

using the reciprocal frame from Section 3.8. We define the matrix through specification of its element (j, i), which is the j-coordinate of the transformed $\mathbf{b}_i$:

$$[\![f]\!]_i^j \equiv f[\mathbf{b}_i] \cdot \mathbf{b}^j. \qquad (4.17)$$

For this matrix, the superscript index j labels the row of matrix, and the subscript index i labels the column (so it is consistent with viewing a vector as a matrix with 1 column and n rows). Equation (4.17) implies that the ith column of the matrix is the image of the ith basis vector $\mathbf{b}_i$ under the transformation f.

The transformation of an arbitrary vector $\mathbf{x}$ can be composed from the matrix contributions as

$$f[\mathbf{x}] = \sum_{i=1}^{n} f[(\mathbf{x} \cdot \mathbf{b}^i)\,\mathbf{b}_j] = \sum_{j=1}^{n}\sum_{i=1}^{n} (\mathbf{x} \cdot \mathbf{b}^i)(f[\mathbf{b}_i] \cdot \mathbf{b}^j)\,\mathbf{b}_j = \sum_{j=1}^{n}\sum_{i=1}^{n} [\![f]\!]_i^{\,j}\,[\![\mathbf{x}]\!]^i\,\mathbf{b}_j .$$

Taking the inner product with $\mathbf{b}^j$ and using $\mathbf{b}_i \cdot \mathbf{b}^j = \delta_i^{\,j}$ (according to (3.32)), the $\mathbf{b}_j$-coefficient of the transformed vector is

$$[\![f[\mathbf{x}]]\!]^{\,j} = f[\mathbf{x}] \cdot \mathbf{b}^j = \sum_{i=1}^{n} [\![f]\!]_i^{\,j}\,[\![\mathbf{x}]\!]^i .$$

We have thus retrieved the familiar multiplication of a vector by a matrix, which may be denoted in shorthand over all rows as

$$[\![f[\mathbf{x}]]\!] = [\![f]\!]\,[\![\mathbf{x}]\!] .$$

Writing out (4.17) using the techniques of (3.32), we find the matrix element fully expressed in geometric algebra (remember that $\mathbf{I}_n = \mathbf{b}_1 \wedge \cdots \wedge \mathbf{b}_n$, and $\check{\mathbf{b}}_j$ means that $\mathbf{b}_j$ is omitted):

$$
\begin{aligned}
[\![f]\!]_i^{\,j} &= (-1)^{j-1}\,f[\mathbf{b}_i]\rfloor\big(\mathbf{b}_1 \wedge \cdots \wedge \check{\mathbf{b}}_j \wedge \cdots \wedge \mathbf{b}_n)\rfloor \mathbf{I}_n^{-1}\big) \\
&= (-1)^{j-1}\,(f[\mathbf{b}_i] \wedge \mathbf{b}_1 \wedge \cdots \wedge \check{\mathbf{b}}_j \wedge \cdots \wedge \mathbf{b}_n)\rfloor \mathbf{I}_n^{-1} \\
&= (\mathbf{b}_1 \wedge \cdots \wedge \mathbf{b}_{j-1} \wedge f[\mathbf{b}_i] \wedge \mathbf{b}_{j+1} \wedge \cdots \wedge \mathbf{b}_n)\rfloor \mathbf{I}_n^{-1} . \quad (4.18)
\end{aligned}
$$

So the recipe to construct the matrix element is that $f[\mathbf{b}_i]$ takes the place of $\mathbf{b}_j$ in the first factor of the expression $\mathbf{I}_n \rfloor \mathbf{I}_n^{-1}$.

The general expression of (4.18) denotes explicitly which geometrical concepts are involved in the definition of the matrix of a transformation: a particular basis (not necessarily orthonormal) and a particular choice of pseudoscalar for the space. The latter is automatically determined once one specifies the basis vectors in the desired order, as $\mathbf{I}_n = \mathbf{b}_1 \wedge \cdots \wedge \mathbf{b}_n$. But (4.18) demonstrates that there is a rather involved object at the basis of the common way of doing linear algebra. This has advantages in its compactness; but taking it fully as the basis of expression, one loses the power to express simpler concepts when those would suffice.

4.5.2 MATRICES FOR OUTERMORPHISMS

A linear mapping can be represented as a matrix, and this holds not only for the original mapping on vectors, but also on the outermorphism mappings on each of the k-blades (or k-vectors). This follows the same procedure as for vectors, but now on a basis for the $\binom{n}{k}$-dimensional space $\bigwedge^k \mathbb{R}^n$ of k-blades (or k-vectors) in $\mathbb{R}^n$.

We first need to establish a basis and a reciprocal basis for that space. Let us illustrate the principle for bivectors; the generalization is straightforward. If the basis of the vector space is $\{\mathbf{b}_i\}_{i=1}^n$, then we can take as our basis 2-blades:

$$\mathbf{b}_{ij} \equiv \mathbf{b}_i \wedge \mathbf{b}_j.$$

There is no natural order among them (except perhaps in 3-D, where you can take them in order of the missing index: $\{\mathbf{b}_{23}, \mathbf{b}_{31}, \mathbf{b}_{12}\}$, with cyclically varying indices). You should see ij as a single index running through its $\binom{n}{2}$ possible values. We denote that by a capital index, so a general bivector would be expressible as $\mathbf{X} = \sum_I X^I \mathbf{b}_I$, with appropriately chosen coefficients X^I.

In a metric space $\mathbb{R}^n$, the reciprocal basis corresponding to the basis $\{\mathbf{b}_I\}$ is then constructed using the reciprocal basis vectors:

$$\mathbf{b}^{ij} = \left(\mathbf{b}^i \wedge \mathbf{b}^j\right)^{\sim}.$$

The reversion makes the scalar product between the basis vectors behave as if it were an orthonormal basis under the scalar product (or, if you prefer, the contraction):

$$\mathbf{b}_I * \mathbf{b}^I = \mathbf{b}_{ij} * \mathbf{b}^{ij} = (\mathbf{b}_i \wedge \mathbf{b}_j) * (\mathbf{b}^j \wedge \mathbf{b}^i) = (\mathbf{b}_i * \mathbf{b}^i)(\mathbf{b}_j * \mathbf{b}^j) = 1,$$

and the scalar product is zero for unequal indices. Therefore

$$\mathbf{b}_I * \mathbf{b}^J = \delta_I^J.$$

The coefficient X^I of a bivector $\mathbf{X}$ on the basis $\{\mathbf{b}_I\}$ is then $\mathbf{X} * \mathbf{b}^I$, so that we can write

$$\mathbf{X} = \sum_I (\mathbf{X} * \mathbf{b}^I)\, \mathbf{b}_I.$$

With this preparation, the matrix of the outermorphism of f can be defined by complete analogy with the vector case as

$$[\![\mathsf{f}]\!]_I^J = \mathsf{f}[\mathbf{b}_I] * \mathbf{b}^J.$$

This is enough to compute the matrix. If the linear transformation f is expensive to evaluate, you may prefer a form that expresses this matrix directly in terms of the vectors $\mathsf{f}[\mathbf{b}_i]$. Let us expand the general term, reverting to the specific vector indices:

$$\begin{aligned}
[\![\mathsf{f}]\!]_{i_1 i_2}^{j_1 j_2} &= \mathsf{f}[\mathbf{b}_{i_1 i_2}] * \mathbf{b}^{j_1 j_2} \\
&= (\mathsf{f}[\mathbf{b}_{i_1}] \wedge \mathsf{f}[\mathbf{b}_{i_2}]) * (\mathbf{b}^{j_2} \wedge \mathbf{b}^{j_1}) \\
&= (\mathsf{f}[\mathbf{b}_{i_2}] * \mathbf{b}^{j_2})(\mathsf{f}[\mathbf{b}_{i_1}] * \mathbf{b}^{j_1}) - (\mathsf{f}[\mathbf{b}_{i_2}] * \mathbf{b}^{j_1})(\mathsf{f}[\mathbf{b}_{i_1}] * \mathbf{b}^{j_2}) \\
&= [\![\mathsf{f}]\!]_{i_2}^{j_2} [\![\mathsf{f}]\!]_{i_1}^{j_1} - [\![\mathsf{f}]\!]_{i_2}^{j_1} [\![\mathsf{f}]\!]_{i_1}^{j_2}.
\end{aligned}$$

By either method, the outermorphism matrix can be constructed from the matrix acting on the basis vectors. In any implementation, this trivial computation of the outermorphism should be hidden from the user, who should simply be allowed to apply the linear transformation to any element of the algebra. The programming exercise in Section 4.9.2 lets you enjoy this functionality, and compares the efficiency of the matrix approach to the regular subspace algebra implementation of the outermorphism.

4.6 SUMMARY

We summarize the most important results of this chapter:

- A linear transformation $f : \mathbb{R}^n \to \mathbb{R}^n$ can always be extended as an *outermorphism* to a linear transformation working on blades in all of $\bigwedge \mathbb{R}^n$, also denoted by f.

- Under a linear transformation f (extended as an outermorphism), the products of the subspace algebra transform as follows:

$$f[\mathbf{A} \wedge \mathbf{B}] = f[\mathbf{A}] \wedge f[\mathbf{B}]$$
$$f[\mathbf{A} * \mathbf{B}] = \mathbf{A} * \mathbf{B}$$
$$f[\mathbf{A} \rfloor \mathbf{B}] = \overline{f}^{-1}[\mathbf{A}] \rfloor f[\mathbf{B}],$$

where $\overline{f}$ is the *adjoint* transformation (basically, the transpose of matrix algebra). The structure of the outer product is therefore preserved by any linear transformation.

- There is a coordinate-free formula for the inverse of a linear transformation $f : \mathbb{R}^n \to \mathbb{R}^n$, which reads

$$f^{-1}[\mathbf{A}] = \frac{\overline{f}[\mathbf{A} \rfloor \mathbf{I}_n^{-1}] \rfloor \mathbf{I}_n}{\det f}.$$

- For orthogonal transformations, $\overline{f}^{-1} = f$, so the structure of the contraction is preserved by any orthogonal transformations.

4.7 SUGGESTIONS FOR FURTHER READING

In this single chapter, we have explained all you need to know for this book about linear transformations in general. We will home in on the orthogonal transformations in particular, and the powerful ways geometric algebra offers to represent them, in Chapter 7. But of course a lot more can be said about linear transformations and how they can be analyzed using the tools of the algebra of subspaces.

Some of the following literature is better studied when you have also learned about the geometric product (after Chapter 6), but it seems most appropriate to give you this list of material on linear transformations in the eponymous chapter.

- A very accessible article by Hestenes [27] gives a good entry to linear transformations as viewed by geometric algebra, and uses its tools to expose the structure of linear algebra.

- Working with blades as computational elements is very similar to developing a multilinear algebra (though in a metric manner). Classically, that quickly gets into tensor representations of multilinear mappings. Doran and Lasenby ([15], Chapter 4) relate this clearly to the geometric algebra representation.

- Numerical techniques in linear algebra rely heavily on techniques like the singular value decomposition (SVD) and eigenvector analysis. The extension of such tools to geometric algebra should be straightforward and would give us the general eigenspaces. As yet, little numerical work exists that uses it directly, though [1] gives some initial results.

- We will have more to say about the representation of orthogonal transformations in the "Further Reading" list of Chapter 7.

4.8 STRUCTURAL EXERCISES

1. Point mirroring in 3-D space leads to a change of orientation of the volume 3-blades. We know this spatial inversion better from reflection in a mirror. Show that this has indeed the same effect. (Hint: Let the mirror plane be characterized by a 2-blade $\mathbf{B}$, and let $\mathbf{a}$ be a vector perpendicular to $\mathbf{B}$ (for example, $\mathbf{a} = \mathbf{B}^*$). Then define the linear transformation performing the mirror reflection, and apply it to a sensibly chosen 3-blade in this setup. Why does your result generalize to arbitrary 3-blades?)

2. Let us compute the determinant according to (4.7) in a 2-D space and compare it to the classical determinant. Take a basis $\{\mathbf{b}_1, \mathbf{b}_2\}$, not necessarily orthonormal. Let the linear mapping f be such that $\mathsf{f}[\mathbf{b}_1] = \mathbf{x}$ and $\mathsf{f}[\mathbf{b}_2] = \mathbf{y}$. Develop $\mathbf{x}$ and $\mathbf{y}$ onto the basis $\mathbf{x} = x_1\mathbf{b}_1 + x_2\mathbf{b}_2$ and $\mathbf{y} = y_1\mathbf{b}_1 + y_2\mathbf{b}_2$. Use $\mathbf{I}_2 = \mathbf{x} \wedge \mathbf{y}$ and compute the determinant according to (4.7). Now compute the matrix of f on the given basis, and compute its classical determinant. The results should match.

3. You may want to apply a linear mapping f to a k-dimensional subspace. You could then be tempted to use (4.7) with its pseudoscalar $\mathbf{I}_k$ substituted for $\mathbf{I}_n$ to define what the determinant of f is on this subspace. Why doesn't this work?

4. Consider the linear transformation of vectors in the $\mathbf{a} \wedge \mathbf{b}$ plane determined by what happens to the vectors $\mathbf{a}$ and $\mathbf{b}$: $\mathsf{f}[\mathbf{a}] = 5\mathbf{a}-3\mathbf{b}$ and $\mathsf{f}[\mathbf{b}] = 3\mathbf{a}-5\mathbf{b}$. Use classical linear algebra methods to find eigenvectors and their eigenvalues. Now use our algebra to determine the determinant, and an eigen-2-blade with its corresponding eigenvalue, and then interpret the geometry of the transformation.

5. Design a nontrivial linear map $f : \mathbb{R}^2 \rightarrow \mathbb{R}^2$ that has an eigenvector and an eigen-2-blade, both with eigenvalue 1.

6. When deriving the linear transformation of the scalar product $f[A * B] = A * B$ in Section 4.3.1, we raised the issue that this appears to mean that every linear transformation leaves the squared norm $A * \widetilde{A}$ invariant. Show that this is of course not true. (Hint: What is the formula for the squared norm of the transformed A actually?)

7. To continue with the previous problem after you know about the adjoint in Section 4.3.2, rewrite the correct expression for the squared norm of $f[A]$ in the form $A * g[\widetilde{A}]$ and determine g in terms of f. This is the *metric mapping* corresponding to the transformation f, and it shows that the transformed space can be treated as a space with a new inner product $a \cdot b \equiv a * g[b]$.

8. Continuing from the previous problem, show that the metric mapping corresponding to an orthogonal transformation is the identity. Therefore, orthogonal transformations preserve norms (and cosines of angles).

9. Show that in a non-degenerate metric space $\mathbb{R}^n$ with arbitrary basis $\{b_i\}_{i=1}^n$, the adjoint of a linear transformation f can be constructed as

$$\overline{f}[x] = \sum_{i=1}^{n} (x * f[b_i])\, b^i. \tag{4.19}$$

10. Show that an orthogonal transformation has a determinant of ± 1.

11. Give an expression for $\overline{f}[A \rfloor B]$. Hint: Consider the symmetry of (4.10).

12. Give an example of a linear transformation for which the transformed cross product $f[a \times b]$ is not parallel to the cross product of the transforms $f[a] \times f[b]$. That of course implies it is not perpendicular to $f[a]$ and $f[b]$, so it has ceased to be a normal vector. This theme is further explored in the programming exercise of Section 4.9.3 below.

13. For the shear $x \mapsto f_s[x] \equiv x + s\,(x \cdot e_1)\,e_2$ (on the standard orthonormal basis of $\mathbb{R}^{n,0}$), compute the transformation matrix $[\![f_s]\!]$ (to act on vectors). Also compute the matrix $[\![f_s^*]\!]$. Verify the results in a picture of the shear of a planar line and its normal vector.

14. Verify that (4.18) indeed gives the identity matrix for the identity mapping.

15. The classical closed-form formula for the inverse of a matrix $[\![A]\!]$ is

$$[\![A]\!]^{-1} = \frac{\mathrm{adj}([\![A]\!])}{\det([\![A]\!])}, \tag{4.20}$$

where $\mathrm{adj}([\![A]\!])$ is the classical adjoint matrix, of which the $(i, j)^{\mathrm{th}}$ element equals $(-1)^{i+j} \det([\![A_{ji}]\!])$, with $[\![A_{ji}]\!]$ a minor matrix obtained from $[\![A]\!]$ by omitting the jth row and the ith column. Show that this terrific coordinate-based construction is identical to the coordinate-free formula (4.16). Equation (4.20) is very hard to compute with algebraically, though we will say that it is easy to implement. (We should mention that in practice, one implements matrix inversion by Gaussian elimination, so that (4.20) is usually treated as little more than a mathematical curiosity, neither good for derivation nor for implementation.)

16. Continuing the previous exercise, give an expression in matrix form of the dual mapping $f^* = \det(f)\,\overline{f}^{-1}$. This should endow the involved algebraic concept of a matrix of minors with a clear geometrical meaning.

17. In standard linear algebra, one way to encode a subspace is as the image of a matrix. The subspace spanned by the basis $\{\mathbf{b}_1,\dots,\mathbf{b}_k\}$ is then the image of the matrix $[\![\mathbf{B}]\!] = [\![\mathbf{b}_1 \cdots \mathbf{b}_k]\!]$. The orthogonal projection of a vector $\mathbf{x}$ onto this subspace in $[\![\mathbf{B}]\!]$ is computed using the *projection matrix* as the vector

$$[\![\mathbf{B}]\!]([\![\mathbf{B}]\!]^T[\![\mathbf{B}]\!])^{-1}[\![\mathbf{B}]\!]^T[\![\mathbf{x}]\!].$$

Show that this is, in fact, the same mapping as our $(\mathbf{x}\rfloor\mathbf{B})\rfloor\mathbf{B}^{-1}$ of (3.25). How would you describe the extension to an outermorphism in standard linear algebra?

4.9 PROGRAMMING EXAMPLES AND EXERCISES

4.9.1 ORTHOGONAL PROJECTION

This example lets you manipulate three vectors. One of the vectors gets projected onto the 2-blade spanned by the other two. The code is very simple:

```
// g_vectors[] is a global array of 3 vectors.

// compute bivector (*4 to make it a bit larger):
bivector B = _bivector(4.0f * g_vectors[0] ^ g_vectors[1]);

// project g_vectors[2] onto the bivector
// The symbol '<<' is the left contraction
e3ga::vector P = _vector((g_vectors[2] << inverse(B)) << B);
```

The output of this example is shown in Figure 4.4. In the next example, we use the outer-morphism matrix representation of the projection to do the same thing.

4.9.2 ORTHOGONAL PROJECTION, MATRIX REPRESENTATION

Outermorphisms are great because they can be summarized into their respective matrix representation, one matrix for each grade of blades. These outermorphism matrices can then be applied to any blade instead of the original outermorphism defined explicitly in terms of subspace products. That matrix approach is usually faster.

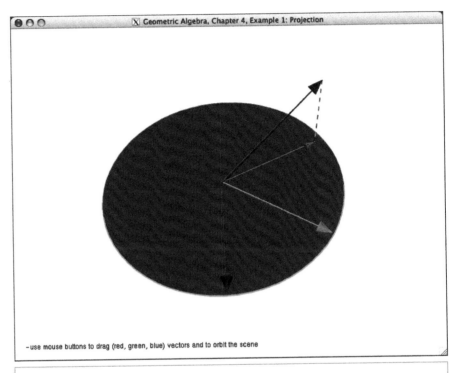

Figure 4.4: Projection. The blue vector is projected onto the bivector spanned by the red and green vector.

In this example, we redo the previous example, this time using the outermorphism matrices to apply the transformation. See the code in Figure 4.5. First, we compute the images of all basis vectors under the linear transformation—in this case, orthogonal projection. Those images are used to initialize the matrix representation M of the transformation on vectors:

```
om M(imageOfE1, imageOfE2, imageOfE3);
```

The om class (for outermorphism matrix) contains a matrix for each grade part $1, 2, \cdots n$ of elements in the n-dimensional space $\mathbb{R}^n$. Given the images of the basis vectors, it initializes all outermorphism matrices using the method described in Section 4.5.2. Once the om class is initialized, it can be applied to any blade, for example,

```
e3ga::vector P = _vector(apply_om(M, g_vectors[2]));
```

The example program may take some time to start up because it also contains a little benchmark that is called at the start of the main() function. The benchmark times

```
// g_vectors[] is a global array of 3 vectors.

// compute bivector (*4 to make it a bit larger):
bivector B = _bivector(4.0f * g_vectors[0] ^ g_vectors[1]);

// we need the images of the 3 basis vectors under the
// projection:
e3ga::vector imageOfE1 = _vector((e1 << inverse(B)) << B);
e3ga::vector imageOfE2 = _vector((e2 << inverse(B)) << B);
e3ga::vector imageOfE3 = _vector((e3 << inverse(B)) << B);
// initialize the matrix representation
om M(imageOfE1, imageOfE2, imageOfE3);

// apply the matrix to the vector:
e3ga::vector P = _vector(apply_om(M, g_vectors[2]));
```

Figure 4.5: Matrix representation of projection code.

how many seconds it takes to do 1,000,000 projections, either using the regular method of computing the projection as $(\mathbf{X} \rfloor \mathbf{B}) \rfloor \mathbf{B}^{-1}$ or using the precomputed outermorphism matrix representation. The results are printed to the console. On one of our machines, the result was

```
10000000 projections using matrix representation: 0.128367 secs
10000000 projections using regular GA: 0.255311 secs
```

So using the outermorphism matrix representation was about twice as fast as using the explicit product method. The result on your machine depends on your CPU architecture as well as your compiler, but in general, the outermorphism matrix representation is faster.

4.9.3 TRANSFORMING NORMAL VECTORS

As we explained in Section 4.3.6, normal vectors transform differently from regular vectors under a linear transformation. We use non-uniform scaling as an example. The dramatically different results are illustrated in Figure 4.6.

The code for initializing the non-uniform scaling outermorphism matrix is:

```
// initialize the outermorphism
// g_scale is a global array of floats
om M(
  _vector(g_scale[0] * e1),
```

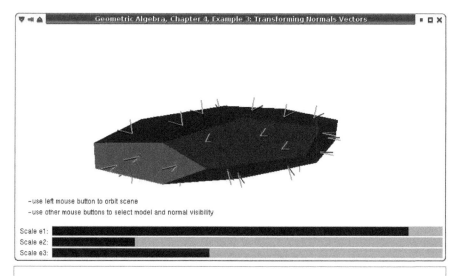

Figure 4.6: Transforming normal vectors. The screenshot shows a squashed dodecahedron. The correct normals—computed with 2-blades—are shown in green, the bad normals in red. It is clear that the red normals are not orthogonal to the surface, especially for the two top polygons.

```
_vector(g_scale[1] * e2),
_vector(g_scale[2] * e3));
```

The good and bad normals are then computed as

```
// compute the normals
// g_normals3D is a global array of vectors
// g_attitude3D is a global array of bivectors
e3ga::vector badNormal, goodNormal;

badNormal = unit_e(apply_om(M, g_normals3D[i]));
goodNormal = unit_e(dual(apply_om(M, g_attitude3D[i])));
```

As in Section 2.13.2, the 3-D models are extracted from GLUT, this time using two orthogonal projections. See the getGLUTmodel3D() function.

5 INTERSECTION AND UNION OF SUBSPACES

Geometric algebra contains operations to determine the *union* and *intersection* of subspaces, the `join` and `meet` products.

These products are of course important in geometry, and it is therefore disappointing to learn that they are not very tidy algebraically. In particular, they are not (bi-)linear: a small disturbance in their arguments may lead to major changes in their outcome as geometric degeneracies occur. This will give their treatment a different flavor than the products we introduced so far.

But `meet` and `join` are still very useful. Even when applied to the subspaces at the origin, `meet` and `join` generalize some specific formulas from 3-D linear algebra into a more unified framework and extend them to subspaces intersecting in n-dimensional space. Their full power will be unleashed later, in Part II, when we can use them to intersect offset subspaces and even spheres, circles, and the like. Yet it is good to understand their algebraic structure first, and we now have all the tools to do so.

5.1 THE PHENOMENOLOGY OF INTERSECTION

When we intersect two planes through the origin in 3-D, the outcome is usually a line. In terms of subspaces as blades, two grade-2 elements produce a grade-1 element.

However, if the two planes happen to coincide, we would want the geometric outcome of their intersection to be that plane of coincidence, which is of grade 2. None of the products we have seen so far can do this grade switching in the result as a consequence of the geometric relationship of their arguments, so there must be something new going on algebraically. In fact, an incidence product encoding this geometry cannot be linear, since even a small disturbance of one of the input planes can lead to this discontinuity in the result.

That nonlinearity prohibits extending the intersection and union easily from blades (which represent subspaces) to general multivectors (which do not). That we cannot do this makes sense for geometrical reasons as well, because any definition of geometrical union and intersection should be based on containment of the result in the arguments, or vice versa, and this is only well defined for subspaces. Algebraically, we will therefore have to *limit intersection and union to blades*. These are the first products that must be constrained to operate within the limited algebra of subspaces, not in the full Grassmann algebra of multivectors.

But even then, an algebraic problem that we can foresee geometrically is that the desired outcome does not have all the properties of a blade, because it is not meaningful to assign a unique magnitude and orientation (i.e., a sign) to the blade representing the subspace of intersection. This is illustrated in Figure 5.1(a) and (b) for two planes represented by 2-blades. Since those 2-blades may be reshaped without changing their value as blades, both depictions are permitted—but they each suggest a different intersection magnitude. It is equally easy to change the orientation of the possible intersection line. Therefore, the outcome of the intersection of two blades is a subspace, but one of which only the attitude matters.

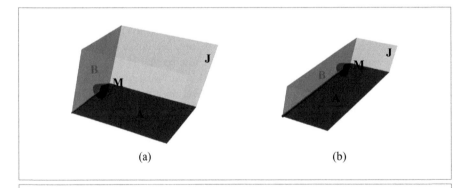

(a)   (b)

Figure 5.1: The ambiguity of the magnitude of the intersection of two planes, **A** and **B**. Both figures are acceptable solutions to the problem of finding blades representing the union **J** and intersection **M** of the subspaces represented by the same 2-blades **A** and **B**.

We are going to design two products between blades to compute with intersections. They will be called `meet` and `join`, and denoted by ∩ and ∪ to signify that they are meant to represent the *geometric intersection and union* of two blades. The setlike notation will not be confusing (we hardly use sets in this book), and in fact is a helpful reminder that the resulting elements are not fully quantified blades and that the products are nonlinear.[1]

5.2 INTERSECTION THROUGH OUTER FACTORIZATION

Consider two blades **A** and **B**, which happen to contain some common blade. To be precise, let **M** be the largest common divisor of **A** and **B** in the sense of the outer product. This is the algebraic formalization of their *geometric intersection*; we will call it their `meet` and denote it by **A** ∩ **B**.

Algebraically, we should be able to factor out **M** from both **A** and **B**, since it is contained in both. We do this in a particular order, writing

$$\mathbf{A} = \mathbf{A}' \wedge \mathbf{M} \text{ and } \mathbf{B} = \mathbf{M} \wedge \mathbf{B}'. \tag{5.1}$$

If **A** and **B** are disjoint, then **M** is a scalar (a 0-blade).

A and **B** together reside within a blade **J**, their smallest common multiple in terms of the outer product. This is a pseudoscalar of the subspace in which this meet intersection problem actually resides. We will call it their `join` and denote it by **A** ∪ **B**, for it is the *geometric union* of the subspaces. It is clear that `join` and `meet` are related through the factorization, for we can write

$$\mathbf{A} \cup \mathbf{B} = \mathbf{A}' \wedge \mathbf{M} \wedge \mathbf{B}' \text{ and } \mathbf{A} \cap \mathbf{B} = \mathbf{M}. \tag{5.2}$$

We already observed, when we discussed the geometry of Figure 5.1, that we should expect this factorization by **M** not to be unique. Indeed, in (5.1) we may multiply **M** by a scalar γ. Then **A**' must be multiplied by $1/\gamma$ to preserve **A**, and similar for **B**. As a consequence, this would multiply the `join` result of (5.2) by $1/\gamma$. So we can always trade off a scalar factor between the `meet` and the `join`, of any weight or sign. This ambiguity need not be a problem in geometrical usage of the outcome. For instance, a projection of a vector **x** to the `meet` subspace **M** is given by $(\mathbf{x} \rfloor \mathbf{M}^{-1}) \rfloor \mathbf{M}$, and this is invariant to the scalar ambiguity since it involves both **M** and $\mathbf{M}^{-1}$.

---

1   The reader should be warned that the terminology of "`join`" and "`meet`" is used in some literature in a different sense, directly corresponding to our outer product, and our operation of contraction with a dual, respectively. Those are then truly linear products, though they do not always compute the geometric union and intersection (they return zero in degenerate situations). To add to the confusion, that literature uses the notations ∨ for their "`join`" and ∧ for their "`meet`."

The first use of meet and join in nonmetric projective geometry seems to have induced people to neglect the magnitude and sign completely. Yet this is a pity, for there are situations in which consistent use of *relative magnitudes* conveys useful geometric information (for instance, on the sines of intersection angles). To enable this, we will develop consistent formulas for meet and join based on the *same* factorization. We can then guarantee that meet and join of the same subspaces can be used consistently, and we will demonstrate how that can be applied. Of course you can always ignore this quantitative precision in any application where you do not need it—in that case, the order of the factors in (5.1) and (5.2) can be chosen arbitrarily.

5.3 RELATIONSHIPS BETWEEN MEET AND JOIN

For practical use, we have to make the computational relationships between meet and join more explicit than merely relating them through their factorization. To do so, we first need formulas for $\mathbf{A}'$ and $\mathbf{B}'$. Neither contain any factors also in $\mathbf{M}$, so we can use the contraction to define them as the part of $\mathbf{A}$ not in $\mathbf{M}$ and the part of $\mathbf{B}$ not in $\mathbf{M}$. But to be quantitative, we have to be careful about the order of the arguments and about their normalization:

$$\mathbf{B}' = \mathbf{M}^{-1} \rfloor \mathbf{B}, \text{ and } \mathbf{A}' = \mathbf{A} \lfloor \mathbf{M}^{-1}, \tag{5.3}$$

where we employed both contractions for simplicity of expression. If you are uncomfortable with using the contractions in this direct manner, you may derive the former more formally from the identity $\mathbf{B}' = \mathbf{M}^{-1} \rfloor (\mathbf{M} \wedge \mathbf{B}')$, which holds since a basis for $\mathbf{B}'$ can be chosen that is orthogonal to all of the factors of $\mathbf{M}^{-1}$ and hence of $\mathbf{M}$. The expression for $\mathbf{A}'$ can be derived in a similar manner. This shows why we need to use the inverse of $\mathbf{M}$ to achieve proper normalization.

But this means we can only proceed if $\mathbf{M}$ has an inverse. This may seem to restrict the kind of spaces in which we can do intersections, excluding those with null vectors, and that would be a serious limitation in practice. However, we will lift that apparent restriction in Section 5.7 (when we show that both join and meet are independent of the particular metric, as you may already suspect, since after all they are based on factorization by the nonmetric outer product). For now, assume that all blades are in the algebra of a Euclidean vector space.

Denoting the grades of the elements by the corresponding lowercase letters (where j is the grade of the join $\mathbf{J}$), we have various simple relationships between them

$$a' = a - m, \quad b' = b - m, \quad j = a + b - m, \quad m + j = a + b, \tag{5.4}$$

and these help in keeping track of the various quantitative relationships we are going to derive. Together with consideration of order and normalization, all can then be remembered easily.

With (5.3), the join in terms of the meet can be written in two ways:

$$\mathbf{J} = \mathbf{A} \cup \mathbf{B} = \mathbf{A} \wedge (\mathbf{M}^{-1} \rfloor \mathbf{B}) = (\mathbf{A} \lfloor \mathbf{M}^{-1}) \wedge \mathbf{B}. \tag{5.5}$$

We can also solve for the meet in terms of the join. We first establish

$$1 = \mathbf{J} * \mathbf{J}^{-1} = (\mathbf{A} \wedge (\mathbf{M}^{-1} \rfloor \mathbf{B})) * \mathbf{J}^{-1} = \mathbf{A} * ((\mathbf{M}^{-1} \rfloor \mathbf{B}) \rfloor \mathbf{J}^{-1})$$
$$= (\mathbf{A}' \wedge \mathbf{M}) * (\mathbf{M}^{-1} \wedge (\mathbf{B} \rfloor \mathbf{J}^{-1})) = \mathbf{A}' * (\mathbf{B} \rfloor \mathbf{J}^{-1}).$$

Then $1 = (\mathbf{B} \rfloor \mathbf{J}^{-1}) * (\mathbf{A} \lfloor \mathbf{M}^{-1}) = (\mathbf{M}^{-1} \wedge (\mathbf{B} \rfloor \mathbf{J}^{-1})) * \mathbf{A} = (\mathbf{M}^{-1} \wedge (\mathbf{B} \rfloor \mathbf{J}^{-1})) \rfloor \mathbf{A}$. Now $\mathbf{M} = \mathbf{M} \wedge 1 = \mathbf{M} \wedge ((\mathbf{M}^{-1} \wedge (\mathbf{B} \rfloor \mathbf{J}^{-1})) \rfloor \mathbf{A}) = (\mathbf{M} \rfloor (\mathbf{M}^{-1} \wedge (\mathbf{B} \rfloor \mathbf{J}^{-1}))) \rfloor \mathbf{A} = (\mathbf{B} \rfloor \mathbf{J}^{-1}) \rfloor \mathbf{A}$, so that we obtain:

$$\mathbf{M} = \mathbf{A} \cap \mathbf{B} = (\mathbf{B} \rfloor \mathbf{J}^{-1}) \rfloor \mathbf{A}. \tag{5.6}$$

This formula to compute $\mathbf{M}$ from $\mathbf{J}$ (given $\mathbf{A}$ and $\mathbf{B}$) is often used in applications, since when subspaces $\mathbf{A}$ and $\mathbf{B}$ are in general position it is easy to specify a blade $\mathbf{J}$ for their join.

The dual of this relationship shows the structure of the meet more clearly: taking the inner product with $\mathbf{J}^{-1}$ on both sides of (5.6), we obtain $\mathbf{M} \rfloor \mathbf{J}^{-1} = ((\mathbf{B} \rfloor \mathbf{J}^{-1}) \rfloor \mathbf{A}) \rfloor \mathbf{J}^{-1} = (\mathbf{B} \rfloor \mathbf{J}^{-1}) \wedge (\mathbf{A} \rfloor \mathbf{J}^{-1})$. So relative to the join, *the dual* meet *is the outer product of the duals*:

$$\mathbf{M} \rfloor \mathbf{J}^{-1} = (\mathbf{B} \rfloor \mathbf{J}^{-1}) \wedge (\mathbf{A} \rfloor \mathbf{J}^{-1}). \tag{5.7}$$

This is often compactly denoted as

$$dual\ meet: \quad (\mathbf{A} \cap \mathbf{B})^* = \mathbf{B}^* \wedge \mathbf{A}^*, \tag{5.8}$$

but then you have to remember that this is not the dual relative to the pseudoscalar $\mathbf{I}_n$ of the total space, but only of the pseudoscalar of the subspace within which the intersection problem resides (i.e., of the join $\mathbf{J} = \mathbf{A} \cup \mathbf{B}$).

Some more expressions relating the four quantities $\mathbf{A}$, $\mathbf{B}$, $\mathbf{M}$, and $\mathbf{J}$ are given in the structural exercises. It should be noted that such relationships between meet and join do not give us a formula or algorithm to compute either. In higher-dimensional subspaces, the search for a join of arbitrary blades requires care, for it can easily lead to an exponential algorithm. An $O(n)$ algorithm will be given in Section 21.7, but we cannot explain that at this point since it uses the geometric product of Chapter 6.

5.4 USING MEET AND JOIN

In practice, the join is often more easily determined than the meet, since the most interesting intersections and unions of subspaces tend to occur when they are in general position within some subspace with a known pseudoscalar (two planes in space, a line and a plane in space, etc.). Then the join is just the pseudoscalar of that common subspace, and (5.6) gives the meet. A numerical example conveys this most directly.

We intersect two planes represented by the 2-blades $\mathbf{A} = \frac{1}{2}(\mathbf{e}_1 + \mathbf{e}_2) \wedge (\mathbf{e}_2 + \mathbf{e}_3)$ and $\mathbf{B} = \mathbf{e}_1 \wedge \mathbf{e}_2$. Note that we have normalized them to facilitate interpreting the relative quantitative aspects. These are homogeneous planes in general position in 3-D space, so their $\mathtt{join}$ is proportional to $\mathbf{I}_3 \equiv \mathbf{e}_1 \wedge \mathbf{e}_2 \wedge \mathbf{e}_3$. It makes sense to orient $\mathbf{J}$ with $\mathbf{I}_3$ so that we simply take $\mathbf{J} = \mathbf{I}_3$. This gives for the $\mathtt{meet}$:

$$
\begin{aligned}
\mathbf{A} \cap \mathbf{B} &= \tfrac{1}{\sqrt{3}} \left((\mathbf{e}_1 \wedge \mathbf{e}_2) \rfloor (\mathbf{e}_3 \wedge \mathbf{e}_2 \wedge \mathbf{e}_1) \right) \rfloor \left((\mathbf{e}_1 + \mathbf{e}_2) \wedge (\mathbf{e}_2 + \mathbf{e}_3) \right) \\
&= \tfrac{1}{\sqrt{3}} \, \mathbf{e}_3 \rfloor \left((\mathbf{e}_1 + \mathbf{e}_2) \wedge (\mathbf{e}_2 + \mathbf{e}_3) \right) \\
&= -\tfrac{1}{\sqrt{3}} (\mathbf{e}_1 + \mathbf{e}_2) = -\sqrt{\tfrac{2}{3}} \, (\frac{\mathbf{e}_1 + \mathbf{e}_2}{\sqrt{2}})
\end{aligned}
\tag{5.9}
$$

(the last step expresses the result in normalized form). Figure 5.2 shows the answer; the sign of $\mathbf{A} \cap \mathbf{B}$ is the right-hand rule applied to the turn required to make $\mathbf{A}$ coincide with $\mathbf{B}$, in the correct orientation. We will show that the magnitude of the $\mathtt{meet}$ equals the sine of the smallest angle between them, so that in this example their angle is $\mathrm{asin}(-\sqrt{2/3})$, measured from $\mathbf{A}$ to $\mathbf{B}$.

Classically, one would compute the intersection of two homogeneous planes in 3-space by first converting them to normal vectors and then taking the cross product. We can see that this gives the same answer in this nondegenerate case in 3-space, using the definition of the cross product (3.28) and our duality equations (3.20), (3.21), and remembering that the dual 2-blades are vectors, using the undualization of pg. 81:

$$
\mathbf{A}^* \times \mathbf{B}^* = (\mathbf{A}^* \wedge \mathbf{B}^*)^* = -(\mathbf{B}^* \wedge \mathbf{A}^*)^* = (\mathbf{B}^* \wedge \mathbf{A}^*)^{-*} = \mathbf{B}^* \rfloor \mathbf{A} = \mathbf{A} \cap \mathbf{B}
\tag{5.10}
$$

So the classical result is a special case of (5.6) or (5.8). But formulas (5.6) and (5.8) are much more general: they apply to the intersection of subspaces of *any* grade, within a space of *any* dimension.

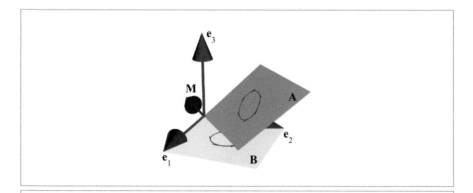

Figure 5.2: The $\mathtt{meet}$ of two oriented planes.

5.5 JOIN AND MEET ARE MOSTLY LINEAR

Once the join has been selected, the formula of (5.6) for the meet shows that the meet is linear in **A** and **B** since it can be expressed by contraction products, which are clearly linear. If we change **A** and/or **B** such that the join does not change, this remains true. In this sense, the meet is mostly linear. However, as soon as some degeneracy occurs or is resolved, the join changes in a nonlinear manner and the meet formula enters a new domain (within which it is again linear). You can tell that this happens when the meet with your selected join returns zero. That signals degeneracy and the need to pick another join.

As a geometric example, assume that in 3-D we have a homogeneous line **a** (a vector) and a homogeneous plane **B** (a 2-blade), as in Figure 5.3. As long as the line is not contained in the plane (so that they are in general position), the pseudoscalar I_3 can be used as the join **J**, and the meet varies nicely with both arguments.

$$\mathbf{M} = \mathbf{a} \cap \mathbf{B} = (\mathbf{B} \rfloor I_3^{-1}) \rfloor \mathbf{a} = \mathbf{B}^* \cdot \mathbf{a} = \mathbf{b} \cdot \mathbf{a}.$$

This is a scalar, geometrically denoting the common point at the origin, with a magnitude proportional to the cosine of angle between the line **a** and the normal vector $\mathbf{b} \equiv \mathbf{B} \rfloor I_3^{-1}$ of the plane; that is, proportional to the sine of the angle of line and plane (and weighted by their magnitudes). It changes sign as the line enters the plane from below rather than

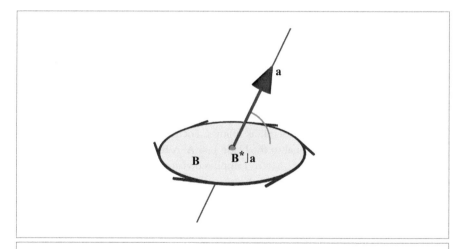

Figure 5.3: A line meeting a plane in the origin in a point. If the join is taken to be the right-handed pseudoscalar, the intersection point is positive when the line pierces the oriented plane as shown. Other normal coordinates can be chosen to bear this out: Let $\mathbf{B} = \mathbf{e}_1 \wedge \mathbf{e}_2$, $\mathbf{a} = a_1\mathbf{e}_1 + a_2\mathbf{e}_2 + a_3\mathbf{e}_3$, then with $\mathbf{J} = \mathbf{e}_1 \wedge \mathbf{e}_2 \wedge \mathbf{e}_3$ you find $\mathbf{M} = \mathbf{a} \cap \mathbf{B} = \mathbf{B}^* \rfloor \mathbf{a} = a_3 = \|\mathbf{a}\| \sin \phi$, which is positive in the situation shown.

above, with above and below determined by the orientation of the plane **B** relative to the pseudoscalar chosen for the `join` (i.e., the orientation of the common space). This shows the use of the sign of the `meet`; it gives the sense of intersection and allows us to eliminate surface intersections of rays coming from inside an object (if we orient its boundary properly and consistently). It also shows why the sign of a scalar (i.e., the orientation of a point at the origin) can be important in the algebra of subspaces.

Precisely when the line becomes coincident with the plane, this expression for the `meet` becomes zero. This is the signal that it is actually no longer the proper `meet`, for the `join` must be changed to a normalized version $\mathbf{I}_2$ of the plane **B**, which is now the smallest common subspace. The problem has essentially become 2-D. We find then find that the `meet` is the line **a**, weighted:

$$\mathbf{M} = (\mathbf{B}\rfloor\mathbf{I}_2^{-1})\rfloor\mathbf{a} = \beta\mathbf{a},$$

with $\beta \equiv \mathbf{B}\rfloor\mathbf{I}_2^{-1}$ the signed magnitude of the **B**-plane. This expression is also linear in both arguments **a** and **B**, as long as we vary them so that the `join` does not change (so we may only rotate and scale the line within the plane $\mathbf{I}_2$, and only change weight or orientation of the plane **B**, *not* its attitude).

This example generalizes to k-spaces in n-dimensional space: the `meet` is linear as long as the `join` does not change, and it degrades gracefully to zero to denote that such a change of `join` becomes necessary.

5.6 QUANTITATIVE PROPERTIES OF THE `MEET`

If you normalize the `join`, you can interpret the value of the `meet` as proportional to the sine between **A** and its projection on **B** (or vice versa, depending on the relative grades). We encountered a particular instance of this in the example of Figure 5.2.

We can see that this holds in general, as follows. Focus on $\mathbf{A}'$ relative to the space **B**. The `join` should be proportional to the blade $\mathbf{J} = \mathbf{A}' \wedge \mathbf{B}$. Let the pseudoscalar of this space be **I**, then normalizing the `join` to **I** implies division of **J** by the scalar $\mathbf{J}\rfloor\mathbf{I}^{-1} = \mathbf{J}^*$. This rescaling of the `join` implies that the `meet` should be rescaled to become $\mathbf{M}\mathbf{J}^*$, so proportional to the scalar $\mathbf{J}^*$. Now inspect $\mathbf{J}^* = (\mathbf{A}' \wedge \mathbf{B})^*$. This is proportional to the volume spanned by $\mathbf{A}'$ and **B**—and we know from the previous chapters that the magnitude of a spanned volume involves the sine of the relative angle between the arguments. Alternatively, we can rewrite $\mathbf{J}^* = \mathbf{A}'\rfloor(\mathbf{B}^*) = \mathbf{A}' * \mathbf{B}^*$. This scalar product is proportional to the cosine of the angle between $\mathbf{A}'$ and the orthogonal complement of **B**. That can be converted to the sine of the complementary angle, retrieving the same interpretation:

> The magnitude of the `meet` $\mathbf{A} \cap \mathbf{B}$ of normalized blades **A** and **B** within a normalized `join` denotes the sine of the angle from **A** to **B**.

The sine measure is quite natural as an indication of the relative attitudes of homogeneous spaces. In classical numerical analysis, the absolute value of the sine of the angle between subspaces is a well-known measure for the distance between subspaces in terms of their orthogonality: it is 1 if the spaces are orthogonal, and decays gracefully to 0 as the spaces get more parallel.

The sign of the sine is worth studying in more detail, for it indicates from which direction **A** approaches **B**. However, we have to be careful with this interpretation: there may be a sign change depending on whether we compute **A** ∩ **B** or **B** ∩ **A**. One should study this sign only relative to the choice of sign for the pseudoscalar for the space spanned by the join during normalization. Let us therefore compare **B** ∩ **A** with **A** ∩ **B** relative to the same join, by means of the dualization formula (5.8):

$$
\begin{aligned}
\mathbf{B} \cap \mathbf{A} &= (\mathbf{A}^* \wedge \mathbf{B}^*)^{-*} \\
&= (-1)^{(j-a)(j-b)} \, (\mathbf{B}^* \wedge \mathbf{A}^*)^{-*} \\
&= (-1)^{(j-a)(j-b)} \, \mathbf{A} \cap \mathbf{B},
\end{aligned}
$$

using (2.13) to swap the arguments of the outer product. Therefore, it depends on the grades of the elements whether the meet is symmetric or antisymmetric. Two lines through the origin in a plane ($a = b = 1, j = 2$) meet in antisymmetric fashion: **A** ∩ **B** = −**B** ∩ **A**. This makes sense, since if we swap the lines then we are measuring the sine of an opposite angle, and this is of opposite sign. On the other hand, a line and a plane through the origin in space ($a = 1, b = 2, j = 3$) meet symmetrically: **A**∩**B** = **B**∩**A**. There is still a sine involved, which changes sign as the relative orientation changes so that we can tell whether the line passes from the front or the back of the plane. But in the computation, it apparently does not matter whether the line meets the plane or vice versa.

This subtle interplay of signs of orientation of the join, the relative attitudes in space, and the order of arguments in the meet requires some experience to interpret properly. We give some examples of the ordering sign for common situations in Table 5.1.

5.7 LINEAR TRANSFORMATION OF MEET AND JOIN

Even though the meet and join are not completely linear in their arguments, they do transform tidily under invertible linear transformations in a structure-preserving manner (by which we mean that the transform of the meet equals the meet of the transforms). This paradoxical result holds because such transformations cannot change the relative attitudes of the blades involved in any real way: if **A** was not contained in **B** before a linear transformation f, then f[**A**] will also not be contained in f[**B**] after the transformation. In that sense, the preservation of meet and join is a structural property of linear transformations. The proof of this fundamental property is not hard, since we know how the outer product and the contraction transform.

Table 5.1: The order of the arguments for a computed meet may affect the sign of the result. This table shows the signs for some common geometrical situations. A plus denotes no sign change, a minus a change. The vector space model in which all elements pass through the origin is denoted as *orig*. This is the algebra of the homogeneous subspaces in Part I.

For convenient referencing, we have also listed some results for the 4D homogeneous model (*hom*) and the 5D conformal model of 3-dimensional Euclidean space (*conf*), which will only be treated in Part II. In the bottom block, 'line' and 'plane' can always substituted for 'circle' and 'sphere'. The order sensitivity does not depend on the model used, since only the 'co-dimensions' $(j - a)$ and $(j - b)$ matter.

Elements in meet	join Space	a,b,j *orig*	a,b,j *hom*	a,b,j *conf*	Sign
Two origin points	Point	0,0,0	1,1,1	2,2,2	+
Origin point and origin line	Line	0,1,1	1,2,2	2,3,3	+
Two origin lines	Plane	1,1,2	2,2,3	3,3,4	−
Two origin lines	Line	1,1,1	2,2,2	3,3,3	+
Two origin planes	Space	2,2,3	3,3,4	4,4,5	−
Origin line and origin plane	Space	1,2,3	2,3,4	3,4,5	+
Origin line and origin plane	Plane	1,2,2	2,3,3	3,4,4	+
Two parallel lines	Plane		2,2,3	3,3,4	−
Two intersecting lines	Plane		2,2,3	3,3,4	−
Two skew lines	Space		2,2,4	3,3,5	+
Two intersecting planes	Space		3,3,4	4,4,5	−
Two parallel planes	Space		3,3,4	4,4,5	−
Line and plane	Space		2,3,4	3,4,5	+
Line and plane	Plane		2,3,3	3,4,4	+
Point and line	Plane		1,2,3	2,3,4	+
Point and plane	Space		1,3,4	2,4,5	−
Point and circle	Sphere			1,3,4	+
Point and sphere	Space			1,4,5	+
Point pair and circle	Sphere			2,3,4	+
Point pair and sphere	Sphere			2,4,4	+
Circle and sphere	Space			3,4,5	+
Circle and sphere	Sphere			3,4,4	+
Circle and circle	Space			3,3,5	+
Circle and circle	Sphere			3,3,4	−
Tangent vector and sphere	Sphere			2,4,4	+
Sphere and sphere	Space			4,4,5	−

First, the `join` is made by a factorization in terms of the outer product. Since a linear transformation is an outermorphism, the linear mapping f preserves the outer product factorization, and we obtain trivially that

$$f[\mathbf{A} \cup \mathbf{B}] = f[\mathbf{A}] \cup f[\mathbf{B}].$$

The `meet` also transforms in a structure preserving manner:

$$f[\mathbf{A} \cap \mathbf{B}] = f[\mathbf{A}] \cap f[\mathbf{B}].$$

The reason is simply that the defining relationships of (5.1) and (5.2) between $\mathbf{A}, \mathbf{B}, \mathbf{J} = \mathbf{A} \cup \mathbf{B}$ and $\mathbf{M} = \mathbf{A} \cap \mathbf{B}$ only involve the outer product; therefore a linear transformation f, acting as an outermorphism, preserves all these relationships between the transformed entities.

When converting the expression $f[\mathbf{A}] \cap f[\mathbf{B}]$ to a computational form involving the contraction analogous to (5.6), these outermorphic correspondences imply that one should use *duality relative to the transformed* `join` $f[\mathbf{J}]$, not the original `join` $\mathbf{J}$. So the transformation of (5.6) reads explicitly:

$$f[\mathbf{A} \cap \mathbf{B}] = \left(f[\mathbf{B}] \rfloor f[\mathbf{J}]^{-1} \right) \rfloor f[\mathbf{A}]$$

This is really different from $\left(f[\mathbf{B}] \rfloor \mathbf{J}^{-1} \right) \rfloor f[\mathbf{A}]$, since $f[\mathbf{J}]$ is in general not even proportional to $\mathbf{J}$. This dependence on the `join` dualization is a good reason to use the explicit (5.7) rather than the overly compact (5.8).

Since a linear transformation usually changes the metric measures of elements (except when it is an orthogonal transformation), the preservation of `meet` and `join` under general linear transformations shows that *these are actually nonmetric products*. For the `join`, that is perhaps not too surprising (since it is like an outer product), but the occurrence of the two contractions in the computation of the `meet` makes it look decidedly metric. The nonmetric nature of the result must mean that these two contractions effectively cancel in their metric properties. In that sense, we have merely used the contraction to write things compactly. Mathematicians encoding union and intersection for the nonmetric projective spaces have devised a special and rather cumbersome notation for the nonmetric duality that is actually involved here. (It is called the Hodge dual, and its proper definition requires the introduction of the n-dimensional space of 1-forms.) We will not employ it, and just use our metric contraction instead.

But it is relevant to note that the precise form of the metric does not matter. If we ever need to compute `meet` and `join` in spaces with unusual metrics, we can always assume that we are in a Euclidean metric if that simplifies our computations. This is why we had no compunction about using the inverses $\mathbf{M}^{-1}$ and $\mathbf{J}^{-1}$ in our derivations; they can always be made to exist by embedding the whole computation temporarily in a Euclidean metric. We will apply this principle in the algorithm to compute `meet` and `join` in Chapter 21.

5.8 OFFSET SUBSPACES

So far we have only treated subspaces containing the origin, and although we have been able to do that case in general, it is of course not the most relevant case in applications. We postpone the treatment of the intersection of subspaces offset from the origin to their proper formalization as elements of the homogeneous model in Chapter 11. There we will show that parallel lines have a finite `meet` and that skew lines in space meet in a scalar proportional to their orthogonal Euclidean distance.

More surprisingly still, in Chapter 13 we will introduce an operational model of Euclidean geometry in which the `meet` of spheres, circles, lines, and planes can be computed by straightforward application of the subspace intersection formulas of the present chapter.

5.9 FURTHER READING

The `meet` and `join` are strangely positioned within the literature on algebras for subspaces: they are either neglected (presumably because they are algebraically not very tidy), or an attempt is made to take them as axiomatic products replacing the outer product and contraction.

- The `meet` and `join` are treated seriously and extensively in Stolfi's classical book on oriented projective geometry [60]. It is richly illustrated, and sharpens the intuition of working with oriented subspaces. It also gives an algorithm for `meet` and `join` in terms of the matrix representation of subspaces. Unfortunately, it does not treat metric geometry.

- When `meet` and `join` are taken as basic products, they are linearized: the `join` is redefined as the outer product of its arguments, and the `meet` is defined through duality (our (5.8)). It is then called the *regressive product*. This alternative algebra of subspaces tends to be nonmetric, with nonmetric duality, and is not easily extended to geometric algebra. Still, the work is mathematically interesting; a rather complete account is [3].

- We noted that the outcomes of `meet` and `join` are not fully qualified subspaces, since there is an ambiguity about their absolute weight and orientation. Within the context of geometric algebra, they are perhaps more properly represented as projection operators than as blades. This has been explored in [8]. An interesting subalgebra results, which forms the basis of algorithms to compute `meet` and `join`.

- The fundamental nature of `meet` and `join` for the treatment of linear algebra is displayed in [27]. When reading that and other literature founded in geometric algebra, beware that the use of an inner product that is not the contraction (see Appendix B) tends to create seemingly exceptional outcomes for `meet` and join when scalars or pseudoscalars are involved. The contractions avoids those exceptions; this issue is explained in [17] as one of the reasons to prefer them.

5.10 EXERCISES

5.10.1 DRILLS

Compute join $\mathbf{A} \cup \mathbf{B}$ and meet $\mathbf{A} \cap \mathbf{B}$ for the following blades:

1. $\mathbf{A} = \mathbf{e}_1$ and $\mathbf{B} = \mathbf{e}_2$.
2. $\mathbf{A} = \mathbf{e}_2$ and $\mathbf{B} = \mathbf{e}_1$.
3. $\mathbf{A} = \mathbf{e}_1$ and $\mathbf{B} = 2\mathbf{e}_1$.
4. $\mathbf{A} = \mathbf{e}_1$ and $\mathbf{B} = (\mathbf{e}_1 + \mathbf{e}_2)/\sqrt{2}$.
5. $\mathbf{A} = \mathbf{e}_1$ and $\mathbf{B} = \cos\phi\,\mathbf{e}_1 + \sin\phi\,\mathbf{e}_2$.
6. $\mathbf{A} = \mathbf{e}_1 \wedge \mathbf{e}_2$ and $\mathbf{B} = \cos\phi\,\mathbf{e}_1 + \sin\phi\,\mathbf{e}_2$.
7. $\mathbf{A} = \mathbf{e}_1 \wedge \mathbf{e}_2$ and $\mathbf{B} = \mathbf{e}_2$.
8. $\mathbf{A} = \mathbf{e}_1 \wedge \mathbf{e}_2$ and $\mathbf{B} = \mathbf{e}_2 + 0.00001\,\mathbf{e}_3$.

5.10.2 STRUCTURAL EXERCISES

1. There is an interesting reciprocal relationship between $\mathbf{A}, \mathbf{B}, \mathbf{J}$, and $\mathbf{M}$.

$$(\mathbf{B}\rfloor\mathbf{J}^{-1}) * (\mathbf{A}\lfloor\mathbf{M}^{-1}) = 1$$

   Verify the steps in the following proof: $1 = \mathbf{M}^{-1} * \mathbf{M} = \mathbf{M}^{-1} * ((\mathbf{B}\rfloor\mathbf{J}^{-1})\rfloor\mathbf{A}) = (\mathbf{M}^{-1} \wedge (\mathbf{B}\rfloor\mathbf{J}^{-1})) * \mathbf{A} = (\mathbf{B}\rfloor\mathbf{J}^{-1}) * (\mathbf{A}\lfloor\mathbf{M}^{-1})$. Then prove in a similar manner:

$$(\mathbf{M}^{-1}\rfloor\mathbf{B}) * (\mathbf{J}^{-1}\lfloor\mathbf{A}) = 1$$

2. Find the error in this part of a 'proof' of the meet transformation formula of page 135 (from the first printing of this book):

   Let us first establish how the inverse of the join transforms by transforming the join normalization equation $\mathbf{J}^{-1} * \mathbf{J} = 1$:

$$1 = \mathsf{f}[1] = \mathsf{f}[\mathbf{J}^{-1} * \mathbf{J}] = \overline{\mathsf{f}}^{\,-1}[\mathbf{J}^{-1}] * \mathsf{f}[\mathbf{J}],$$

   so that $\overline{\mathsf{f}}^{\,-1}[\mathbf{J}^{-1}] = \mathsf{f}[\mathbf{J}]^{-1}$.

3. Compute meet and join of two vectors $\mathbf{a}$ and $\mathbf{b}$ in general position, and show that the magnitude of their meet (relative to their join) is the sine of their angle. Relate the sign of the sine to the order of intersection. In this case, the meet should be antisymmetric.

4. Compute the `meet` and `join` of two parallel vectors **u** and **v**. The `meet` should now be symmetric. (Hint: Use one of them as the `join`.)

5. As an exercise in symbolic manipulation of the products so far, let us consider the `meet` of $\mathbf{a} \wedge \mathbf{B}$ and $\mathbf{a} \wedge \mathbf{C}$, where **a** is a vector and the blades **B** and **C** have no common factor. The answer should obviously be proportional to **a**, but what precisely is the proportionality factor? (Hint: If you get stuck, the next exercise derives the answer as $(\mathbf{a} \wedge \mathbf{B} \wedge \mathbf{C})^*$.)

6. Verify the steps in the following computation of the answer to the previous exercise. They are rather ingenious; note the third step especially, and the conversion to a scalar product (check the grades involved!). The `join` for dualization should be a blade proportional to $\mathbf{a} \wedge \mathbf{B} \wedge \mathbf{C}$ (if it is zero our suppositions are wrong, and vice versa). Here goes:

$$
\begin{aligned}
(\mathbf{a} \wedge \mathbf{B}) \cap (\mathbf{a} \wedge \mathbf{C}) &= \left((\mathbf{a} \wedge \mathbf{C})^* \wedge (\mathbf{a} \wedge \mathbf{B})^*\right)^{-*} = \left((\mathbf{a}\rfloor \mathbf{C}^*) \wedge (\mathbf{a}\rfloor \mathbf{B}^*)\right)^{-*} \\
&= \left(\mathbf{a}\rfloor \left(\mathbf{C}^* \wedge (\mathbf{a}\rfloor \mathbf{B}^*)\right)\right)^{-*} = \left(\mathbf{a}\rfloor \left(\mathbf{C}^* \wedge (\mathbf{a} \wedge \mathbf{B})^*\right)\right)^{-*} \\
&= \mathbf{a} \wedge \left(\mathbf{C}^* \rfloor (\mathbf{a} \wedge \mathbf{B})\right) = \mathbf{a} \wedge \left(\mathbf{C}^* * (\mathbf{a} \wedge \mathbf{B})\right) \\
&= \mathbf{a} \wedge \left((\mathbf{a} \wedge \mathbf{B}) * \mathbf{C}^*\right) = \mathbf{a} \wedge \left((\mathbf{a} \wedge \mathbf{B})\rfloor \mathbf{C}^*\right) \\
&= \mathbf{a} \wedge (\mathbf{a} \wedge \mathbf{B} \wedge \mathbf{C})^*.
\end{aligned}
$$

7. Use the previous derivation to derive the general *factorization of the* `meet`:

$$(\mathbf{A} \wedge \mathbf{B}) \cap (\mathbf{A} \wedge \mathbf{C}) = \mathbf{A}\,(\mathbf{A} \wedge \mathbf{B} \wedge \mathbf{C})^*, \tag{5.11}$$

where **A**, **B**, and **C** have no common factors.

5.11 PROGRAMMING EXAMPLES AND EXERCISES

5.11.1 THE `MEET` AND `JOIN`

This example allows you to interactively select and manipulate two multivectors. The multivectors can be vector-valued or 2-blade-valued. Either the `meet` or the `join` of the multivectors are drawn:

```
// M1 and M2 are the two multivectors
mv X;
if (g_draw == DRAW_MEET) X = meet(M1, M2);
else X = join(M1, M2);
// ... (set color, scale)
draw(X);
```

Note that we use multivectors (class mv) here because neither the input nor the output has a fixed multivector type. As demonstrated in the next example, working with the mv class in general and the meet() and join() functions in particular is much slower than the specialized classes and the ordinary products.

To make it easier to produce degenerate cases—such as two parallel vectors—we round the coordinates of multivectors M1 and M2 to multiples of 0.2. This causes them to move in a stepwise fashion.

5.11.2 EFFICIENCY

In Gaigen 2, the implementation of the meet and join is very slow compared to the other products. This example performs a benchmark to demonstrate this. It creates 1,000,000 pairs of random vectors and bivectors. It then times how long it takes to compute the outer product of these pairs, and how long it takes to compute the join of these pairs. In our benchmark the join is about 100 times slower than outer product. There are several reasons for this:

- To compute the meet and join, a specialized (factorization) algorithm is used, whereas computing the outer product is as simple as multiplying and summing coordinates in the right way. See Section 21.7 for a description of the meet and join algorithm used in this example.
- The algorithm uses the mv class instead of the specialized types such as vector and bivector. The mv class uses coordinate compression, which is slow.
- The ordinary subspace products are just very efficient in Gaigen 2.

It may be possible to optimize the meet and join to a level where they are about 10 times slower than the regular products. But in general, you should try to avoid the meet and join in your programs if you care about efficiency. If you know the relative position of elements involved, just use the formula $\mathbf{B}^* \rfloor \mathbf{A}$ in the appropriate subspace instead of $\mathbf{A} \cap \mathbf{B}$.

5.11.3 FLOATING POINT ISSUES

As stated above, the meet and join are computed by a factorization algorithm. Before the factorization starts, the algorithm computes what the grade of the output should be. This involves comparing a condition number (similar to that of a matrix) to a small threshold value. Hence, the algorithm will flip-flop between grades in degenerate cases (e.g., near-parallel vectors).

This example (see Figure 5.4) searches for the point where the join of two (near-)parallel vectors changes from a vector to a 2-blade. It starts with a very small probe epsilon of 10^{-10}, and tests if $\mathbf{e}_1 \cup (\mathbf{e}_1 + 10^{-10}\, \mathbf{e}_2)$ is a 2-blade. If not, it grows the probe epsilon, and loops. In the example, the flip-flop occurs when $\mathbf{b} = \mathbf{e}_1 + 1.007748 - \times 10^{-7}\, \mathbf{e}_2$, which is to be expected, because the meet-join algorithm uses an epsilon of 10^{-7}.

```
// get two vectors, initialize 'a' to 'e1'
e3ga::vector a, b;
a = e1;

float probeEpsilon = 1e-10f;

while (true) {// the loop will be broken when the join is a bivector:
  // add a tiny bit of 'e2' to b:
  b = e1 + probeEpsilon * e2;

  // compute the join
  mv X = join(a, b);

  // get analysis of 'X'
  mvAnalysis AX(X);

  // check if blade, and if a blade, then is it a 2-blade or a vector?
  if (!AX.isBlade()) {
    // this should never happen
    printf("Error: the join of a and b is not a blade!\n");
    return -1;
  }
  else {
    // compute string "join(..., ...)"
    std::string str = "join(" + a.toString_e() + ", " + b.toString_e() + ")";

    if (AX.bladeSubclass() == mvAnalysis::BIVECTOR) {
      printf("%s is a 2-blade\n", str.c_str());
      return 0; // terminate
    }
    else printf("%s is a vector\n", str.c_str());
  }

  // Grow 'probeEpsilon' a little such that it won't take forever to reach
  // the point where join(a, b) is a 2-blade:
  probeEpsilon *= 1.01f;
}
```

Figure 5.4: Searching for the point at which the join of two (near-)parallel vectors becomes a 2-blade (Example 3).

6 THE FUNDAMENTAL PRODUCT OF GEOMETRIC ALGEBRA

We have seen how the outer product and the contraction characterize rather different properties of subspaces: qualitative spanning and quantitative measurements. Together, they have given us an enriched view of the linear algebra of subspaces. This much has been known for some time, and is part of the branch of applied mathematics that is called Grassmann-Cayley algebra.

In this chapter we will start afresh and introduce the basics of Clifford algebra to develop a powerful *geometric algebra*. This geometric algebra will incorporate *operators on subspaces* into our framework, and permit us to displace the constructions of the subspace algebra in a structure-preserving manner. The crucial construction is to unify the qualitative and quantitative subspace products into a single *geometric product*, more fundamental than either. The geometric product is invertible, and it allows us to manipulate and solve equations about geometrical quantities almost as if they were regular arithmetical expressions. The true power of this geometric product will become clear in the next chapter, when we use it to define the versor product construction for operators. This chapter defines the geometric product, first for vectors and then for general multivectors. Subsequently, we show how the geometric product indeed subsumes the earlier products (which is a bit of tedious but necessary algebra), and we end with the use of its invertibility to define general projection and rejection operations through geometric division.

6.1 THE GEOMETRIC PRODUCT FOR VECTORS

6.1.1 AN INVERTIBLE PRODUCT FOR GEOMETRY

Consider a fixed and known vector $\mathbf{a}$ and an unknown vector $\mathbf{x}$, both in a Euclidean vector space $\mathbb{R}^n$. Let us assume that all we know about $\mathbf{x}$ is the scalar value α of its inner product with $\mathbf{a}$. Then $\mathbf{x}$ must satisfy $\mathbf{x} \cdot \mathbf{a} = \alpha$. This implies that the endpoint of $\mathbf{x}$ lies on a hyperplane perpendicular to the direction of $\mathbf{a}$. In Figure 6.1, this is sketched in $\mathbb{R}^{3,0}$ as the yellow plane. Geometrically, it is clear that we cannot retrieve $\mathbf{x}$ from α and $\mathbf{a}$. Algebraically, this means that there is no unique "inner division". If there were, we could invert the inner product and retrieve $\mathbf{x}$ from its product with $\mathbf{a}$ by means of some formula like $(\mathbf{x} \cdot \mathbf{a})/\mathbf{a} = \mathbf{x}$.

The outer product is not much better in this respect. Suppose that we were told the value of the outer product of $\mathbf{a}$ and $\mathbf{x}$ is to be the bivector $\mathbf{B}$. That bivector must of course be an area element of the plane shared by $\mathbf{a}$ and $\mathbf{x}$. The equation $\mathbf{x} \wedge \mathbf{a} = \mathbf{B}$ defines a line in space, offset from the origin. (You can see this as follows: let $\mathbf{p}$ be a solution of $\mathbf{p} \wedge \mathbf{a} = \mathbf{B}$. Then $\mathbf{x}$ satisfies $\mathbf{x} \wedge \mathbf{a} = \mathbf{p} \wedge \mathbf{a}$, so that $(\mathbf{x} - \mathbf{p}) \wedge \mathbf{a} = 0$. We saw in Section 2.8.1 that this implies $(\mathbf{x} - \mathbf{p}) = \lambda \mathbf{a}$. Therefore $\mathbf{x} = \mathbf{p} + \lambda \mathbf{a}$, a general point on the line through $\mathbf{p}$ in the direction $\mathbf{a}$.) This line has been sketched in blue in Figure 6.1. The endpoint of $\mathbf{x}$ must be on this line; but knowing the line does not specify $\mathbf{x}$. This is the geometrical reason why we cannot algebraically retrieve $\mathbf{x}$ from knowing the outer product $\mathbf{B}$ and $\mathbf{a}$; there is no "outer division", and no formula such that $(\mathbf{x} \wedge \mathbf{a})/\mathbf{a} = \mathbf{x}$ for all $\mathbf{x}$.

We thus see that, when taken separately, the two products with $\mathbf{a}$ are insufficient to retrieve $\mathbf{x}$; yet they are somehow complementary. Indeed, combining the two pieces

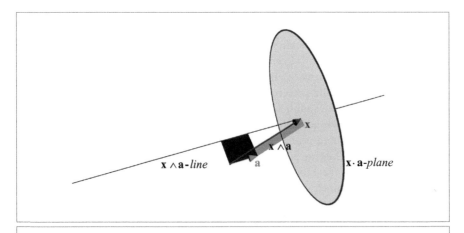

Figure 6.1: Combination of the noninvertible subspace products leads to the invertible geometric product (see text).

of information is obviously enough to fully determine $\mathbf{x}$, with its endpoint at the intersection of the hyperplane and the line, as illustrated in Figure 6.1 for 3-D space. Therefore a product of $\mathbf{x}$ and $\mathbf{a}$ that contains both the inner product and outer product information should be invertible.

6.1.2 SYMMETRY AND ANTISYMMETRY

There is a clean way to construct a composite product from the inner product $\mathbf{x} \cdot \mathbf{a}$ and the outer product $\mathbf{x} \wedge \mathbf{a}$. It is based on their symmetries. The inner product $\mathbf{x} \cdot \mathbf{a}$ is symmetric in $\mathbf{x}$ and $\mathbf{a}$, for it retains its value when $\mathbf{x}$ and $\mathbf{a}$ are interchanged. The outer product $\mathbf{x} \wedge \mathbf{a}$ is antisymmetric; it changes sign under exchange of $\mathbf{x}$ and $\mathbf{a}$.

We can now make a new product between $\mathbf{x}$ and $\mathbf{a}$ such that the inner product is its symmetric part and the outer product its antisymmetric part. That defines it uniquely. This product is called the *geometric product* (though some call it the *Clifford product* after its 1872 inventor William Kingdon Clifford). It is so central that we use the empty symbol ' ' to denote it, writing $\mathbf{x}\,\mathbf{a}$ for the geometric product of $\mathbf{x}$ and $\mathbf{a}$.

The demands on its symmetric and antisymmetric parts give the equations

$$\mathbf{x} \cdot \mathbf{a} = \tfrac{1}{2}(\mathbf{x}\,\mathbf{a} + \mathbf{a}\,\mathbf{x}) \tag{6.1}$$

and

$$\mathbf{x} \wedge \mathbf{a} = \tfrac{1}{2}(\mathbf{x}\,\mathbf{a} - \mathbf{a}\,\mathbf{x}). \tag{6.2}$$

By adding these equations, we find that the geometric product of the vectors $\mathbf{x}$ and $\mathbf{a}$ must be

$$\textit{geometric product for vectors}: \quad \mathbf{x}\,\mathbf{a} \equiv \mathbf{x} \cdot \mathbf{a} + \mathbf{x} \wedge \mathbf{a}. \tag{6.3}$$

This product of two vectors produces a multivector that consists of a scalar part and a 2-blade part. That is unusual; our previous products always produced outcomes of a single grade. But it is precisely because the parts of different grades do not mix in an addition that they can be retrieved separately. That makes the geometric product invertible.

6.1.3 PROPERTIES OF THE GEOMETRIC PRODUCT

Let us check the algebraic properties of this new product between vectors:

- *Commutativity*. The geometric product of two general vectors is not commutative, for the equation $\mathbf{x}\,\mathbf{a} = \mathbf{a}\,\mathbf{x}$ would imply that $\mathbf{x} \wedge \mathbf{a} = 0$. This means that commutativity only happens when $\mathbf{x}$ and $\mathbf{a}$ are parallel. On the other hand, the product is also not anticommutative, for that would imply $\mathbf{x} \cdot \mathbf{a} = 0$, which is also a special relationship of $\mathbf{x}$ and $\mathbf{a}$. As a consequence of this lack of general commutativity, we should be very careful about the order of the factors in the product $\mathbf{x}\,\mathbf{a}$.

- *Linearity and Distributivity*. The geometric product is linear and distributive, since both the inner and the outer product are, and these properties inherit under addition.

- *Associativity*. Definition 6.3 does not specify how to compute the geometric product of more than two vector factors. We have motivated our definition because we wanted an invertible product so that $(\mathbf{x}\,\mathbf{a})/\mathbf{a}$ would be equal to $\mathbf{x}$ (with division defined in terms of the geometric product). This suggests that we should define the product to be *associative*. The desired equation then holds, since we could rewrite it to $(\mathbf{x}\,\mathbf{a})/\mathbf{a} = \mathbf{x}\,(\mathbf{a}/\mathbf{a}) = \mathbf{x}$. Moreover, in an associative algebra, each invertible element has a unique inverse, so the division would be uniquely defined. (We clarify that point later, in Section 6.1.5.)

We give the fully general algebraic definition of the geometric product in Section 6.2.1. But first, we would like to familiarize you with the use of having such a product for vectors, to aid your intuition and acceptance of this new construction.

6.1.4 THE GEOMETRIC PRODUCT FOR VECTORS ON A BASIS

When asked to evaluate the geometric product of two vectors $\mathbf{a}$ and $\mathbf{b}$ given in a coordinate basis, we can simply evaluate their inner and outer products and add them. However, it is more direct to expand the geometric product in terms of a sum of geometric products of the coordinate vectors. We then need to establish what those basis products are.

Let us take an orthonormal basis $\{\mathbf{e}_i\}$ in a metric space $\mathbb{R}^n$. The geometric product of a basis vector with itself evaluates to a scalar derived from the metric:

$$\mathbf{e}_i\,\mathbf{e}_i = \mathbf{e}_i \cdot \mathbf{e}_i + \mathbf{e}_i \wedge \mathbf{e}_i = \mathbf{e}_i \cdot \mathbf{e}_i = \mathsf{Q}[\mathbf{e}_i, \mathbf{e}_i],$$

where we used either the inner product or the bilinear form Q of the metric space (see Appendix A). In a Euclidean space $\mathbb{R}^{n,0}$, this is of course simply equal to 1 for all orthonormal basis vectors.

For two different vectors from the orthonormal basis, we get

$$\mathbf{e}_i\,\mathbf{e}_j = \mathbf{e}_i \cdot \mathbf{e}_j + \mathbf{e}_i \wedge \mathbf{e}_j = \mathbf{e}_i \wedge \mathbf{e}_j.$$

This does not simplify further, but it does show that $\mathbf{e}_i\,\mathbf{e}_j = -\mathbf{e}_j\,\mathbf{e}_i$ when $i \neq j$. We sometimes denote $\mathbf{e}_i\,\mathbf{e}_j$ as $\mathbf{e}_{ij}$ to show more clearly that it is a single element of the algebra. This element has an unusual property:

$$\mathbf{e}_{ij}^2 = (\mathbf{e}_i\,\mathbf{e}_j)\,(\mathbf{e}_i\,\mathbf{e}_j) = \mathbf{e}_i\,(\mathbf{e}_j\,\mathbf{e}_i)\,\mathbf{e}_j = -\mathbf{e}_i\,(\mathbf{e}_i\,\mathbf{e}_j)\,\mathbf{e}_j = -(\mathbf{e}_i\,\mathbf{e}_i)\,(\mathbf{e}_j\,\mathbf{e}_j) = -1. \qquad (6.4)$$

Therefore the algebra of the real Euclidean vector space $\mathbb{R}^{2,0}$ contains an element $\mathbf{e}_{12}$ that squares to -1 under the geometric product! It is not the imaginary unit from complex numbers, but a real 2-blade, representing a unit plane.

In a 2-D vector space, the element $\mathbf{e}_{12}$ completes the algebra. Multiplying $\mathbf{e}_{12}$ with $\mathbf{e}_1$ reverts to something we already had:

$$\mathbf{e}_{12}\,\mathbf{e}_1 = \mathbf{e}_1\,\mathbf{e}_2\,\mathbf{e}_1 = -\mathbf{e}_2\,\mathbf{e}_1\,\mathbf{e}_1 = -\mathbf{e}_2\,(\mathbf{e}_1 \cdot \mathbf{e}_1),$$

so this is simply a multiple of $\mathbf{e}_2$. For the 2-D Euclidean space $\mathbb{R}^{2,0}$ with orthonormal basis, the full multiplication table is:

	1	$\mathbf{e}_1$	$\mathbf{e}_2$	$\mathbf{e}_{12}$
1	1	$\mathbf{e}_1$	$\mathbf{e}_2$	$\mathbf{e}_{12}$
$\mathbf{e}_1$	$\mathbf{e}_1$	1	$\mathbf{e}_{12}$	$\mathbf{e}_2$
$\mathbf{e}_2$	$\mathbf{e}_2$	$-\mathbf{e}_{12}$	1	$-\mathbf{e}_1$
$\mathbf{e}_{12}$	$\mathbf{e}_{12}$	$-\mathbf{e}_2$	$\mathbf{e}_1$	-1

Now we can use the linearity and distributivity to compute the geometric product of any two vectors. For $\mathbf{a} = a_1\mathbf{e}_1 + a_2\mathbf{e}_2$ and $\mathbf{b} = b_1\mathbf{e}_1 + b_2\mathbf{e}_2$:

$$\begin{aligned}
\mathbf{a}\,\mathbf{b} &= (a_1\mathbf{e}_1 + a_2\mathbf{e}_2)\,(b_1\mathbf{e}_1 + b_2\mathbf{e}_2) \\
&= a_1 b_1(\mathbf{e}_1\,\mathbf{e}_1) + a_2 b_2(\mathbf{e}_2\,\mathbf{e}_2) + a_1 b_2(\mathbf{e}_1\,\mathbf{e}_2) + a_2 b_1(\mathbf{e}_2\,\mathbf{e}_1) \\
&= (a_1 b_1 + a_2 b_2) + (a_1 b_2 - a_2 b_1)\,\mathbf{e}_{12}
\end{aligned}$$

By the extension of these techniques, the geometric product can be computed for vectors in n-dimensional space.

6.1.5 DIVIDING BY A VECTOR

Since the geometric product on vectors is invertible, a vector $\mathbf{a}$ should have an inverse. This inverse $\mathbf{a}^{-1}$ is easy to find:

$$\textit{inverse of a vector}: \quad \mathbf{a}^{-1} = \frac{\mathbf{a}}{\mathbf{a} \cdot \mathbf{a}} = \frac{\mathbf{a}}{\|\mathbf{a}\|^2}. \tag{6.5}$$

This indeed works, for

$$\mathbf{a}^{-1}\,\mathbf{a} = \frac{1}{\mathbf{a} \cdot \mathbf{a}}\,\mathbf{a}\,\mathbf{a} = \frac{1}{\mathbf{a} \cdot \mathbf{a}}\left(\mathbf{a} \cdot \mathbf{a} + \mathbf{a} \wedge \mathbf{a}\right) = \frac{1}{\mathbf{a} \cdot \mathbf{a}}\left(\mathbf{a} \cdot \mathbf{a} + 0\right) = 1.$$

Vectors with zero norms (the *null vectors* of Appendix A) do not have inverses.

The associativity of the geometric product makes the inverse of a vector *unique*. If $\mathbf{a}'$ would also be an inverse of $\mathbf{a}$, then $\mathbf{a}'\,\mathbf{a} = 1$. Now right-multiply both sides by $\mathbf{a}^{-1}$, regroup by associativity, and you get $\mathbf{a}' = \mathbf{a}^{-1}$. Therefore, there is only one inverse.[1]

---

1   Note that the inverse for the geometric product is the same we used as "an" inverse for the inner product in Section 3.5.2; it was then not unique, but we conveniently picked one that would be useful in the wider context that we have now reached.

Having the inverse allows us to divide by vectors, so that we indeed can retrieve $\mathbf{x}$ from knowing the value of $(\mathbf{x}\mathbf{a})$ and $\mathbf{a}$, as was our goal:

$$(\mathbf{x}\mathbf{a})\mathbf{a}^{-1} = \mathbf{x}(\mathbf{a}\mathbf{a}^{-1}) = \mathbf{x}.$$

This shows the necessity of the associativity property. We often prefer to denote this by a division sign as $(\mathbf{x}\mathbf{a})/\mathbf{a}$, but note that the noncommutativity of the geometric product implies that division is also noncommutative. So using the notation $/\mathbf{a}$ is permitted as long as we remember that this means "division *on the right* by $\mathbf{a}$."

Geometrically, the inverse of a vector $\mathbf{a}$ is a vector in the same direction as $\mathbf{a}$, and properly rescaled.

6.1.6 RATIOS OF VECTORS AS OPERATORS

Having an algebraic definition of the division of vectors already helps to solve geometric problems. In a 2-D Euclidean space $\mathbb{R}^{2,0}$, we can pose the similarity problem illustrated in Figure 6.2:

> Given two vectors $\mathbf{a}$ and $\mathbf{b}$, and a third vector $\mathbf{c}$ (in the plane of $\mathbf{a}$ and $\mathbf{b}$), determine $\mathbf{x}$ so that $\mathbf{x}$ is to $\mathbf{c}$ as $\mathbf{b}$ is to $\mathbf{a}$ (i.e., solve $\mathbf{x} : \mathbf{c} = \mathbf{b} : \mathbf{a}$).

It is geometrically intuitive what we would want: a proportionality involving both the relative length and angle of $\mathbf{b}$ and $\mathbf{a}$ should be transferred to $\mathbf{x}$ and $\mathbf{c}$.

We take a leap of faith, and read this ratio in terms of the geometric product. So we guess that the solution to this might be the solution to the equation

$$\mathbf{x}\mathbf{c}^{-1} = \mathbf{b}\mathbf{a}^{-1}.$$

The solution is immediate through right-multiplication of both sides by $\mathbf{c}$:

$$\mathbf{x} = (\mathbf{b}\mathbf{a}^{-1})\mathbf{c}. \qquad (6.6)$$

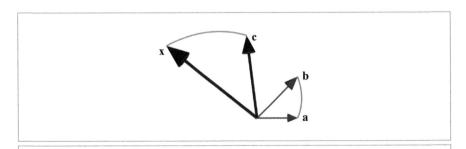

Figure 6.2: Ratios of vectors.

This is a fully computable expression. For instance, with $\mathbf{a} = \mathbf{e}_1$ (so that $\mathbf{a}^{-1} = \mathbf{e}_1$), $\mathbf{b} = \mathbf{e}_1 + \mathbf{e}_2$, and $\mathbf{c} = 2\mathbf{e}_2$ in the standard orthonormal basis, we obtain $\mathbf{x} = ((\mathbf{e}_1 + \mathbf{e}_2)\mathbf{e}_1^{-1})\,2\mathbf{e}_2 = 2(1 - \mathbf{e}_1\,\mathbf{e}_2)\,\mathbf{e}_2 = 2(\mathbf{e}_2 - \mathbf{e}_1)$. Draw a picture like Figure 6.2 to convince yourself of the correctness of this outcome.

In fact, we might see $(\mathbf{b}\,\mathbf{a}^{-1})$ in (6.6) as an *operator* that acts on $\mathbf{c}$ to produce $\mathbf{x}$. The operator is parametrized by $\mathbf{a}$ and $\mathbf{b}$, and it should be capable of both rotation and scaling to produce $\mathbf{x}$ from $\mathbf{c}$. From the construction above, we would suspect that it only depends on the relative angle and size of the vectors $\mathbf{a}$ and $\mathbf{b}$. If that is so, we may as well take $\mathbf{a}$ to be the unit vector $\mathbf{e}_1$ and $\mathbf{b} = \rho\,(\cos\phi\,\mathbf{e}_1 + \sin\phi\,\mathbf{e}_2)$, with ρ the relative length and ϕ the relative angle (from $\mathbf{a}$ to $\mathbf{b}$). Then we compute that the rotation/scaling operator is

$$\mathbf{b}\,\mathbf{a}^{-1} = \rho\,(\cos\phi\,\mathbf{e}_1 + \sin\phi\,\mathbf{e}_2)\,\mathbf{e}_1 = \rho\,(\cos\phi - \sin\phi\,\mathbf{e}_{12}).$$

You may verify that $\mathbf{b}\,\mathbf{a}^{-1}$ acts on the basis vector $\mathbf{e}_1$ to produce $\rho\,(\cos\phi\,\mathbf{e}_1 + \sin\phi\,\mathbf{e}_2)$, and on $\mathbf{e}_2$ it yields $\rho\,(\cos\phi\,\mathbf{e}_2 - \sin\phi\,\mathbf{e}_1)$. Moreover, since the geometric product is linear, these results can be used to produce the result on a general vector $\mathbf{c} = c_1\,\mathbf{e}_1 + c_2\,\mathbf{e}_2$, which yields the $\mathbf{x}$ of our problem:

$$\mathbf{x} = \rho\,(c_1\cos\phi - c_2\sin\phi)\,\mathbf{e}_1 + \rho\,(c_1\sin\phi + c_2\cos\phi)\,\mathbf{e}_2. \tag{6.7}$$

This is precisely the solution we would expect to the original problem, if we would have expanded it in coordinates. It is clearly a rotation combined with a scaling. You would represent it in terms of a matrix operation as

$$\left[\!\!\left[\begin{array}{c} x_1 \\ x_2 \end{array} \right]\!\!\right] = \left[\!\!\left[\begin{array}{cc} \rho\cos\phi & -\rho\sin\phi \\ \rho\sin\phi & \rho\cos\phi \end{array} \right]\!\!\right] \left[\!\!\left[\begin{array}{c} c_1 \\ c_2 \end{array} \right]\!\!\right], \tag{6.8}$$

when expressed on the $\{\mathbf{e}_1, \mathbf{e}_2\}$ basis.

It is highly satisfactory that our geometric product not only produces this result, but that it does so in the form (6.6): $\mathbf{x} = (\mathbf{b}/\mathbf{a})\,\mathbf{c}$. That expression is immediately derivable from the original problem statement and completely formulated in terms of the elements of the problem, rather than using an extraneous coordinate system. If you have to write code, that is how you would want to specify it in a high-level programming language for geometry; in comparison, (6.7) and especially (6.8) feel like assembly code, with their use of coordinates reminiscent of registers.

The coordinate-free operator $\mathbf{b}/\mathbf{a}$ is a good example of the kind of operational power that the geometric product gives us. We will have much more to say about such operators in Chapter 7.

6.2 THE GEOMETRIC PRODUCT OF MULTIVECTORS

In the definition of the geometric product for vectors, we followed a geometric motivation and defined it in terms of inner product and outer product, loosely following its historic process of invention by Clifford. We then claimed it was actually more fundamental than either. If that is indeed the case, we should be able to start with the geometric product and define it algebraically without reference to the other products. We do that now, for it allows us to extend the geometric product properly beyond vector terms and use it as the foundation of our geometric algebra.

6.2.1 ALGEBRAIC DEFINITION OF THE GEOMETRIC PRODUCT

We start with a metric vector space $\mathbb{R}^n$ and its linear space of subspaces $\bigwedge \mathbb{R}^n$. Its metric is characterized by a bilinear form Q (see Appendix A), or, equivalently, by a inner product of vectors.

We define the *geometric product* from $\bigwedge \mathbb{R}^n \times \bigwedge \mathbb{R}^n$ to $\bigwedge \mathbb{R}^n$ by the following properties:

- **Scalars**. The geometric product is an extension of the usual scalar multiplication in $\mathbb{R}^n$, so the expressions

$$\alpha \beta \text{ and } \alpha \mathbf{x}, \quad \alpha, \beta \in \mathbb{R}$$

can from now on be read as involving the geometric product. We will explicitly define this multiplication by a scalar to be *commutative* with any element A:

$$\alpha A = A \alpha \text{ for all } \alpha \in \mathbb{R}, A \in \bigwedge \mathbb{R}^n.$$

- **Scalar Squares**. The geometric product $\mathbf{x}^2 \equiv \mathbf{x}\mathbf{x}$ of a vector $\mathbf{x}$ with itself is defined to be a scalar, equivalent to the metric quantity $\mathbf{x} \cdot \mathbf{x} = Q[\mathbf{x},\mathbf{x}]$. This ties the geometric product to the metric of the vector space $\mathbb{R}^n$.
- **Distributivity and Linearity**. The geometric product is defined to be distributive over the addition of elements:

$$A (B + C) = A B + A C \text{ and } (A + B) C = A C + B C$$

This also defines the general linearity of the geometric product (since A can be a scalar).

- **Associativity**. The geometric product is defined to be associative:

$$A (B C) = (A B) C$$

so that we may write $A B C$ without confusion about the result.

- **Commutativity Not Required!**. The geometric product is *not* defined to be either purely commutative or purely anticommutative (although it may be either, for suitably chosen factors). This is essential, as it permits the geometric product to unite the commutative properties of metric computations with the anticommutative properties of spanning to produce a product that is complete in its geometric properties.

Note that our original definition (6.3) of the geometric product as a sum of inner and outer product is *not* part of these algebraic defining properties. We will actually rederive (6.3), and with it will define the outer product and contraction in terms of the geometric product. Such a procedure demonstrates that it is indeed the more fundamental of the three, our order of treatment in this book notwithstanding.

The geometric product makes our algebra of $\bigwedge \mathbb{R}^n$ into a true *geometric algebra*, and we will use that term from now on. It transcends the subspace algebra we have used so far, and has a much more rich, powerful, and consistent structure. While subspace algebra was similar to Grassmann-Cayley algebra, geometric algebra closely resembles a nongeometrical mathematical construction called *Clifford algebra*. The terms are often used interchangeably by others, but we will make a distinction. For the moment, you can think of geometric algebra as the geometrically significant part of Clifford algebra. We will be able to make this distinction more precise in Section 7.7.2.

6.2.2 EVALUATING THE GEOMETRIC PRODUCT

Since the above are all the properties of the geometric product, they should enable the expansion of arbitrary expressions of the geometric product of multivectors. Let us do some of this before we proceed—we will develop faster techniques to compute with the geometric product, but it is important to realize that these definitions are indeed complete.

The completeness is most easily shown when we have an orthonormal basis for the metric vector space $\mathbb{R}^n$. We should demonstrate that we can compute the geometric product of any two basis vectors. This should of course agree with our special case for vectors, computed in Section 6.1.4, but we are not allowed to use those. We should work fully from the algebraic definition just given.

The geometric product of a basis vector with itself is easy, since by definition it can be expressed in terms of the bilinear form or the inner product that is part of the definition of the metric vector space:

$$\mathbf{e}_i \, \mathbf{e}_i = \mathbf{e}_i \cdot \mathbf{e}_i = \mathsf{Q}[\mathbf{e}_i, \mathbf{e}_i]$$

But the geometric product of two different basis vectors $\mathbf{e}_i \, \mathbf{e}_j$ does not appear to allow simplification by the axioms:

$$\mathbf{e}_i \, \mathbf{e}_j = ?$$

We know from the vector computations that it cannot be simplified, but we would at least like to be able to show that $\mathbf{e}_j \, \mathbf{e}_i = -\mathbf{e}_i \, \mathbf{e}_j$ to be in correspondence with the geometric product as we defined it earlier for vectors.

There is a neat trick called *polarization* that comes to the rescue. The bilinear form or inner product of the metric vector space can be evaluated on any two vectors. The bilinear nature gives an identity for $Q[\mathbf{x}, \mathbf{y}]$ or $\mathbf{x} \cdot \mathbf{y}$ that can be manipulated into a symmetric shape:

$$
\begin{aligned}
\mathbf{x} \cdot \mathbf{y} &= \tfrac{1}{2}\big((\mathbf{x}+\mathbf{y}) \cdot (\mathbf{x}+\mathbf{y}) - (\mathbf{x} \cdot \mathbf{x}) - (\mathbf{y} \cdot \mathbf{y})\big) \\
&= \tfrac{1}{2}\big((\mathbf{x}+\mathbf{y})(\mathbf{x}+\mathbf{y}) - (\mathbf{x}\mathbf{x}) - (\mathbf{y}\mathbf{y})\big) \\
&= \tfrac{1}{2}(\mathbf{x}\mathbf{y}+\mathbf{y}\mathbf{x})
\end{aligned}
\tag{6.9}
$$

Therefore the inner product $\mathbf{x} \cdot \mathbf{y}$ of two general vectors is the *symmetric part of their geometric product*. We are thus able to derive part of our motivating definitions of Section 6.1.2 from the algebraic definition above.

This symmetry property gives us the idea to manipulate our different basis vectors by splitting the product in its symmetric and antisymmetric parts:

$$
\begin{aligned}
\mathbf{e}_i\,\mathbf{e}_j &= \tfrac{1}{2}(\mathbf{e}_i\,\mathbf{e}_j + \mathbf{e}_j\,\mathbf{e}_i) + \tfrac{1}{2}(\mathbf{e}_i\,\mathbf{e}_j - \mathbf{e}_j\,\mathbf{e}_i) \\
&= \mathbf{e}_i \cdot \mathbf{e}_j + \tfrac{1}{2}(\mathbf{e}_i\,\mathbf{e}_j - \mathbf{e}_j\,\mathbf{e}_i) \\
&= 0 + \tfrac{1}{2}(\mathbf{e}_i\,\mathbf{e}_j - \mathbf{e}_j\,\mathbf{e}_i),
\end{aligned}
$$

since in the orthonormal basis $\mathbf{e}_i \cdot \mathbf{e}_j = 0$ for $i \neq j$. It follows that

$$
\mathbf{e}_i\,\mathbf{e}_j = -\mathbf{e}_j\,\mathbf{e}_i.
$$

So the new definition indeed permits us to derive this important property of the multiplication of basis vectors.

With the multiplication of the basis elements established, we can use associativity, linearity, and distributivity to compute the geometric product of any two elements in the linear space $\bigwedge \mathbb{R}^n$.

6.2.3 GRADES AND THE GEOMETRIC PRODUCT

At this point you may object that we have not really shown that the geometric product, constructing elements starting from a metric vector space $\mathbb{R}^n$, really generates the same linear structure of elements of different grades that we had before. We are therefore formally not allowed to write $\bigwedge \mathbb{R}^n$ for the space in which geometric algebra lives.

That is a correct objection; but if the spaces are not the same, they are certainly algebraically isomorphic, and that is good enough to identify them geometrically. The reason is that the orthonormal basis of the vector space $\mathbb{R}^n$ leads to a basis for the product structure that satisfies $\mathbf{e}_i\,\mathbf{e}_j = -\mathbf{e}_j\,\mathbf{e}_i$. That is the essence in generating the higher-grade elements. This property is identical to the outer product antisymmetric property $\mathbf{e}_i \wedge \mathbf{e}_j = -\mathbf{e}_j \wedge \mathbf{e}_i$. The identity of these properties means that we can use the geometric product to at least faithfully reconstruct the basis of the ladder of subspaces $\bigwedge \mathbb{R}^n$ that we originally made using the outer product.

And even though the geometric product has richer properties than the outer product, we cannot make other elements beyond the ladder of subspaces. Consider for example $\mathbf{e}_1 (\mathbf{e}_1 \mathbf{e}_2)$ as an attempt to make something new. Had we used the outer product, the construction $\mathbf{e}_1 \wedge (\mathbf{e}_1 \mathbf{e}_2) = \mathbf{e}_1 \wedge (\mathbf{e}_1 \wedge \mathbf{e}_2)$ would have been zero. For the geometric product, the result is not zero, but it reverts to something we already have:

$$\mathbf{e}_1 (\mathbf{e}_1 \mathbf{e}_2) = \mathbf{e}_1 \, \mathbf{e}_1 \, \mathbf{e}_2 = (\mathbf{e}_1^2) \, \mathbf{e}_2,$$

so this is $\pm \mathbf{e}_2$, depending on the metric. You can generalize this argument to show that nothing beyond the elements of $\bigwedge \mathbb{R}^n$ can be made; the scalar squares foil any such attempt. Therefore, the geometric product of a metric vector space $\mathbb{R}^n$ "lives" in precisely the same structure $\bigwedge \mathbb{R}^n$ as the outer product of the same space $\mathbb{R}^n$.

However, this analysis brings out an important difference between the geometric product and the outer product. When multiplying the extended basis elements of grade k and grade l by the outer product, we are left with a single element of grade $k + l$ (or zero). With the geometric product, the product of two basis elements of grade k and l may have any of the grades

$$|k - l|, |k - l| + 2, \ldots, (k + l - 2), (k + l).$$

The highest grade $(k + l)$ occurs when all basis vectors in the elements are different. (The geometric product is then essentially the same as the outer product of those elements.) But each vector in common between the two basis elements reduces the grade by two as it combines to produce a scalar. The extreme case is when all the vectors in one are contained in the other, leaving only $|k - l|$ factors as a result. (The geometric product is then the left or right contraction of one argument onto the other.)

If we now have arbitrary elements A_k and B_l of grade k and l, respectively, these can be decomposed on the bases of $\bigwedge^k \mathbb{R}^n$ and $\bigwedge^l \mathbb{R}^n$. When we multiply them using the geometric product, any or all of the possible grades between $|k - l|$ and $(k + l)$ may occur. Therefore *the geometric product produces multivectors of mixed grade*. The grade() operation no longer has a single integer value in geometric algebra.

The algebraic invertibility of the geometric product can now be understood in principle. The series of terms in the geometric product of the two elements A_k and B_l apparently give us a complete inventory of their relative geometric relationship, allowing full reconstruction of one when we are given the other.

6.3 THE SUBSPACE PRODUCTS RETRIEVED

The geometric product is the fundamental product in geometric algebra—you will not need any other product, since it contains all geometric relationships between its arguments. Yet we have seen that the subspace products (by which we mean the outer product, scalar product, and contraction) are also useful geometrically. In fact, the whole geometrical concept of subspace requires the outer product to be encoded in our algebra.

Since we want to have those products to "do geometry," we should show that they are included in our geometric algebra based on the geometric product. There are two routes:

- We could use the symmetries of the geometric product to retrieve outer product and contraction (basically reversing the construction that motivated the geometric product in Section 6.1.2). This is actually only partly successful. It does not define the subspace products fully, but it does show that they are consistent with the symmetry structure of the geometric product. When we perform the analysis in Section 6.3.1 below, we obtain many useful relationships between the various products.

- We can identify the subspace products of blades as certain well-defined grades of their geometric product. This indeed defines them fully, though it gives us less algebraic insight in their relationships. We do this in Section 6.3.2.

In the practice of applying the subspace products, both approaches are useful. Depending on the geometrical problem that one tries to solve computationally, either may feel like the more direct route. That is why we present both.

6.3.1 THE SUBSPACE PRODUCTS FROM SYMMETRY

The familiar outer product of a vector $\mathbf{a}$ with a blade $\mathbf{B}$ can be related to the geometric product by the following two expressions:

$$\mathbf{a} \wedge \mathbf{B} = \tfrac{1}{2}(\mathbf{a}\,\mathbf{B} + \widehat{\mathbf{B}}\,\mathbf{a}), \qquad (6.10)$$

$$\mathbf{B} \wedge \mathbf{a} = \tfrac{1}{2}(\mathbf{B}\,\mathbf{a} + \mathbf{a}\,\widehat{\mathbf{B}}), \qquad (6.11)$$

where $\widehat{\mathbf{B}} = (-1)^{\mathrm{grade}(\mathbf{B})}\,\mathbf{B}$ is the grade involution of $\mathbf{B}$ introduced in Section 2.9.5. (Writing the equations in this form makes it easier to lift them to general multivectors B.)

The proof of these statements may be found in Section C.1 of Appendix C. It demonstrates that the two equations above indeed identify the same outer product structure that we had before, at least when one of the factors is a vector, and it proves the associativity $(\mathbf{x} \wedge \mathbf{y}) \wedge \mathbf{z} = \mathbf{x} \wedge (\mathbf{y} \wedge \mathbf{z})$ of the product thus defined. Because of that associativity, this outer product in terms of the geometric product can be extended to general blades, and by linearity to general multivectors. Only the case of two scalars is formally not included, but other than that this outer product is isomorphic to the outer product we had before.

The contractions can be related to the geometric product in similar fashion when they involve a vector factor $\mathbf{a}$:

$$\mathbf{a}\rfloor\mathbf{B} = \tfrac{1}{2}(\mathbf{a}\,\mathbf{B} - \widehat{\mathbf{B}}\,\mathbf{a}), \qquad (6.12)$$

$$\mathbf{B}\lfloor\mathbf{a} = \tfrac{1}{2}(\mathbf{B}\,\mathbf{a} - \mathbf{a}\,\widehat{\mathbf{B}}). \qquad (6.13)$$

The proof of these statements may be found in Section C.2 of Appendix C. Unfortunately, because of lack of associativity, we cannot prove the defining equation (3.11) $\mathbf{A}\rfloor(\mathbf{B}\rfloor\mathbf{C}) = (\mathbf{A} \wedge \mathbf{B})\rfloor\mathbf{C}$ (nor its counterpart for the right contraction). Neither can we define the contraction result on two scalar arguments in this manner. So although the products defined by (6.12) and (6.13) are consistent with the earlier contractions, the contractions based on the symmetries of the geometric product are not pinned down very precisely. This algebraic freedom partly explains the variation in inner products in the geometric algebra literature exposed in Appendix B.

But at least for a vector factor, the constructions above define the subspace products uniquely. Conversely, this means that the geometric product of a vector with an arbitrary multivector can be decomposed using the contraction and the outer product.

$$\mathbf{a}\,\mathbf{B} = \mathbf{a}\rfloor\mathbf{B} + \mathbf{a} \wedge \mathbf{B}, \tag{6.14}$$

$$\widehat{\mathbf{B}}\,\mathbf{a} = \widehat{\mathbf{B}}\lfloor\mathbf{a} + \widehat{\mathbf{B}} \wedge \mathbf{a} = -\mathbf{a}\rfloor\mathbf{B} + \mathbf{a} \wedge \mathbf{B}, \tag{6.15}$$

where we used (3.19) to convert the right contraction to a left contraction. These equations subsume and generalize (6.3).

The subspace product definitions permit us to change the order of multiplications in a geometric product, which is often convenient in evaluating expressions:

$$\widehat{\mathbf{B}}\,\mathbf{a} = \mathbf{a}\,\mathbf{B} - 2\mathbf{a}\rfloor\mathbf{B} = -\mathbf{a}\,\mathbf{B} + 2\mathbf{a} \wedge \mathbf{B} \tag{6.16}$$

and

$$\mathbf{a}\,\mathbf{B} = \widehat{\mathbf{B}}\,\mathbf{a} + 2\,\mathbf{a}\rfloor\mathbf{B} = -\widehat{\mathbf{B}}\,\mathbf{a} + 2\,\mathbf{a} \wedge \mathbf{B}. \tag{6.17}$$

In all these equations, you may note that right-multiplication of $\mathbf{B}$ by $\mathbf{a}$ is always accompanied by a grade involution the formulas become simplest or most symmetrical when you define them in terms of $\mathbf{a}\,\mathbf{B}$, $\mathbf{a}\rfloor\mathbf{B}$, and $\mathbf{a} \wedge \mathbf{B}$, combined with $\widehat{\mathbf{B}}\,\mathbf{a}$, $\widehat{\mathbf{B}}\lfloor\mathbf{a}$ ($= -\mathbf{a}\rfloor\mathbf{B}$), and $\widehat{\mathbf{B}} \wedge \mathbf{a}$ ($= \mathbf{a} \wedge \mathbf{B}$). This grade involution is apparently the natural geometric sign when moving a vector to the right of a blade. We will see this phenomenon reappear throughout the equations of geometric algebra.

You may be puzzled by a paradox in the associativity of the various products. According to (6.14), the geometric product is the sum of the outer product and contraction. Both the geometric product and the outer product are associative, whereas the contraction is not. How can the sum of something associative and something nonassociative ever be associative itself? The solution is: don't look at it that way. Instead, start with the geometric product, which is defined to be associative. Then derive the outer product as (6.10) (i.e., as half the sum of a geometric product and its swapped version). This is associative since addition is associative. Then derive the contraction as (6.12) (i.e., as half the difference of a geometric product and its swapped version). This is nonassociative, since subtraction is nonassociative. Now there is no paradox.

By linearity, all equations in this section are easily extended from a blade $\mathbf{B}$ to a general multivector B.

6.3.2 THE SUBSPACE PRODUCTS AS SELECTED GRADES

An alternative way of obtaining the subspace products from the geometric product is as parts of well-chosen grades using the k-grade selection operator $\langle\ \rangle_k$. For the geometric product on vectors, this is simple:

$$\mathbf{a} \cdot \mathbf{b} = \langle \mathbf{a}\,\mathbf{b} \rangle_0 \ \text{ and } \ \mathbf{a} \wedge \mathbf{b} = \langle \mathbf{a}\,\mathbf{b} \rangle_2. \tag{6.18}$$

This generalizes as follows:

$$\mathbf{A}_k \wedge \mathbf{B}_l \equiv \langle \mathbf{A}_k\,\mathbf{B}_l \rangle_{k+l} \tag{6.19}$$

$$\mathbf{A}_k \rfloor \mathbf{B}_l \equiv \langle \mathbf{A}_k\,\mathbf{B}_l \rangle_{l-k} \tag{6.20}$$

$$\mathbf{A}_k \lfloor \mathbf{B}_l \equiv \langle \mathbf{A}_k\,\mathbf{B}_l \rangle_{k-l} \tag{6.21}$$

$$\mathbf{A}_k * \mathbf{B}_l \equiv \langle \mathbf{A}_k\,\mathbf{B}_l \rangle_0 \tag{6.22}$$

Blades of negative grade are zero—so the left contraction $\rfloor$ is zero when $k > l$, and the right contraction $\lfloor$ is zero when $k < l$. By linearity of the geometric product, all these definitions can be lifted from blades to k-vectors and then to multivectors as a sum over the appropriate grades.

In contrast to the symmetry-based approach, these equations are complete definitions for all arguments. The proofs that these equations give the same products we had in the earlier chapters may be found in Section C.3 of Appendix C. It involves some new manipulation techniques that are useful to study.

A surprising property of these definitions is that the selection of certain grades of the geometric product of blades apparently produces another blade. Beware that this does not generalize to the selection of other grade parts!

Once these correspondences with the old subspace products have been established, some of the properties of the subspace products can be used to simplify grade-based expressions and vice versa. For instance, the symmetry property $\mathbf{A} * \mathbf{B} = \mathbf{B} * \mathbf{A}$ of the scalar product can be easily lifted to multivectors as $A * B = B * A$. This implies

$$\langle A\,B \rangle_0 = \langle B\,A \rangle_0 \tag{6.23}$$

which is a useful cyclic reordering property of the grade-0 symbol.

The grade approach to the subspace products is a feasible way of implementing all products in geometric algebra based on a single implemented product (the geometric product). This is explored in programming exercise 6.7.1.

6.4 GEOMETRIC DIVISION

With the integration of subspace products with the geometric product, we have a more powerful algebra to analyze subspaces. We now combine our new capability of division by a subspace with the earlier techniques. This not only generalizes the projection, but it also produces the compact representation of a new construction: subspace reflection.

6.4.1 INVERSES OF BLADES

The geometric product is invertible, so dividing by a multivector has a unique meaning, equivalent to multiplication by the inverse of the multivector. However, not all multivectors have inverses. Fortunately, we are mostly interested in two kinds: blades, and multivectors that can be written as a product of invertible vectors. The latter are called versors, and we treat them in Chapter 7; they are obviously invertible (their inverse is formed by the inverses of the vector factors, in reverse order).

Blades are also invertible, if they have a nonzero norm (i.e., if they are not null-blades; see Appendix A). The inverse of a blade $\mathbf{A}$ is then

$$\text{inverse of a blade } \mathbf{A}: \quad \mathbf{A}^{-1} = \frac{\mathbf{A}}{\mathbf{A} * \mathbf{A}} = \frac{\widetilde{\mathbf{A}}}{\mathbf{A} * \widetilde{\mathbf{A}}} = \frac{\widetilde{\mathbf{A}}}{\|\mathbf{A}\|^2} \tag{6.24}$$

where $\widetilde{\mathbf{A}}$ is the reverse of Section 2.9.5. This formula is based on the property that the squared norm of a blade is a scalar, which makes the division well defined and unambiguous (since a scalar commutes with the geometric product, its right-division and left-division coincide). Its validity is most easily verified using an orthogonal factorization of the blade $\mathbf{A}$ as a product of orthogonal vectors: $\mathbf{A} = \mathbf{a}_1 \mathbf{a}_2 \cdots \mathbf{a}_k$. Such a factorization can be made for invertible blades by the Gram-Schmidt procedure of Section 6.7.2. By computing the geometric product $\mathbf{A}\widetilde{\mathbf{A}}$ vector by vector, you find that it is equal to $\mathbf{A} * \widetilde{\mathbf{A}}$, so that the inverse formula is indeed correct.

This inverse of a blade is unique, by the associativity argument also used in Section 6.1.5.

6.4.2 DECOMPOSITION: PROJECTION TO SUBSPACES

We can express a vector $\mathbf{x}$ trivially relative to an invertible blade $\mathbf{A}$ as $\mathbf{x} = \mathbf{x}(\mathbf{A}\mathbf{A}^{-1})$. Moving the brackets by associativity and invoking (6.14), we get a pleasantly suggestive rewriting. Let us first explore this for a 1-blade $\mathbf{A}$, the vector $\mathbf{a}$.

$$\begin{aligned} \mathbf{x} &= (\mathbf{x}\,\mathbf{a})\,\mathbf{a}^{-1} \\ &= (\mathbf{x}\cdot\mathbf{a} + \mathbf{x}\wedge\mathbf{a})\,\mathbf{a}^{-1} \\ &= (\mathbf{x}\cdot\mathbf{a})\,\mathbf{a}^{-1} + (\mathbf{x}\wedge\mathbf{a})\,\mathbf{a}^{-1} \end{aligned} \tag{6.25}$$

The first term in (6.25) is a vector (since it is a scalar times a vector). We recognize it as $(\mathbf{x}\cdot\mathbf{a})\,\mathbf{a}^{-1} = (\mathbf{x}\cdot\mathbf{a}^{-1})\,\mathbf{a} = (\mathbf{x}\rfloor\mathbf{a}^{-1})\rfloor\mathbf{a}$; that is, as the *orthogonal projection* of $\mathbf{x}$ onto $\mathbf{a}$ (see Section 3.6). We can now write it as a division:

projection of $\mathbf{x}$ *onto* $\mathbf{a}$: $(\mathbf{x}\rfloor\mathbf{a})/\mathbf{a}$.

It is the component of **x** in the **a**-direction. The second term in (6.25) must then be the component of **x** that contains no **a**-components at all (since the two terms must add to produce **x**). We follow Hestenes [33] in calling this the *rejection* of **x** by **a**:

rejection of **x** *by* **a**: $(\mathbf{x} \wedge \mathbf{a})/\mathbf{a}$

We can imagine its construction visually as in Figure 6.3: span the bivector **x** ∧ **a**. This is a reshapeable area element, and it is equivalent to a rectangle perpendicular to **a** spanned by some vector **r** perpendicular to **a**. That rectangular element can be written as the outer product **r** ∧ **a**, but because **r** is perpendicular to **a** (implying that **r** · **a** = 0), we can even write it as a geometric product:

$$\mathbf{x} \wedge \mathbf{a} = \mathbf{r} \wedge \mathbf{a} = \mathbf{r} \cdot \mathbf{a} + \mathbf{r} \wedge \mathbf{a} = \mathbf{r}\mathbf{a},$$

This rewriting is helpful because the geometric product is invertible; that makes this equation for **r** solvable. Through right-division by **a** on **r a** = **x** ∧ **a**, we obtain the solution **r** = (**x** ∧ **a**)/**a**, which is indeed the rejection of **x** by **a**.

We thus see that the identity **x** = (**x a**)/**a**, when written out in terms of the inner product and the outer product, is actually a *decomposition* of the vector **x** relative to **a**, providing its **a**-component and non-**a**-component. It offers us the possibility of describing a vector relative to another vector, but does so in a satisfyingly coordinate-free manner.

Returning to the general blade **A** and again invoking (6.14), we find the decomposition

$$\mathbf{x} = (\mathbf{x}\,\mathbf{A})\,\mathbf{A}^{-1} = (\mathbf{x}\rfloor\mathbf{A})\,\mathbf{A}^{-1} + (\mathbf{x} \wedge \mathbf{A})\,\mathbf{A}^{-1}. \qquad (6.26)$$

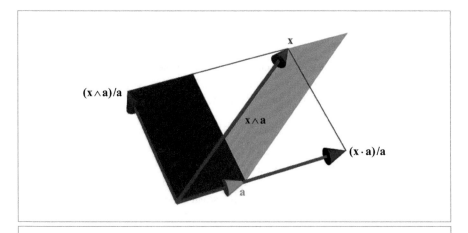

Figure 6.3: Projection and rejection of **x** relative to **a**.

You may expect that the first term is a 1-blade fully contained in $\mathbf{A}$, and that it should be equal to the projection of $\mathbf{x}$ onto $\mathbf{A}$.

$$\text{projection of } \mathbf{x} \text{ onto } \mathbf{A}: \quad (\mathbf{x} \rfloor \mathbf{A})/\mathbf{A}.$$

We have encountered this before (in Section 3.6) as $(\mathbf{x} \rfloor \mathbf{A}) \rfloor \mathbf{A}^{-1}$, which commutes with all products. But structural exercise 7 explores why we can replace the contraction by a geometric division in this formula.

The second term is again called the rejection of $\mathbf{x}$ by $\mathbf{A}$,

$$\text{rejection of } \mathbf{x} \text{ by } \mathbf{A}: \quad (\mathbf{x} \wedge \mathbf{A})/\mathbf{A},$$

since it is a vector perpendicular to $\mathbf{A}$. To prove that fact compactly, we combine the subspace products and the grade selection of the geometric product (to be frank, it took us about an hour to make it this simple). If $(\mathbf{x} \wedge \mathbf{A}) \mathbf{A}^{-1}$ is perpendicular to $\mathbf{A}$, it should be perpendicular to any vector in $\mathbf{A}$. Let us pick one, $\mathbf{a}$, and compute the inner product:

$$
\begin{aligned}
\mathbf{a} \cdot \left((\mathbf{x} \wedge \mathbf{A}) \mathbf{A}^{-1} \right) &= \langle \mathbf{a}(\mathbf{x} \wedge \mathbf{A}) \mathbf{A}^{-1} \rangle_0 \\
&= \frac{1}{2} \langle \mathbf{a} \mathbf{x} \mathbf{A} \mathbf{A}^{-1} + \mathbf{a} \widehat{\mathbf{A}} \mathbf{x} \mathbf{A}^{-1} \rangle_0 \\
&= \frac{1}{2} \langle \mathbf{a} \mathbf{x} - \mathbf{A} \mathbf{a} \mathbf{x} \mathbf{A}^{-1} + 2(\mathbf{a} \wedge \widehat{\mathbf{A}}) \mathbf{x} \mathbf{A}^{-1} \rangle_0 \\
&= \frac{1}{2} \langle \mathbf{a} \mathbf{x} - \mathbf{a} \mathbf{x} \mathbf{A}^{-1} \mathbf{A} - 0 \rangle_0 \\
&= \frac{1}{2} \langle \mathbf{a} \mathbf{x} - \mathbf{a} \mathbf{x} \rangle_0 = 0.
\end{aligned}
$$

Identify the steps we took—they are all based on formulas in this chapter and you will see an instructive example of how the grade approach and the symmetry approach to the subspace products can be combined. In the rejection, we can substitute the geometric product in $(\mathbf{x} \wedge \mathbf{A}) \mathbf{A}^{-1}$ by a right contraction (see structural exercise 8).

Although we can replace the geometric products by contractions in both projection and rejection, there is not necessarily an advantage in doing so. The geometric product is invertible, and this often helps to simplify expressions, so that would plead in favor of leaving it. On the other hand, the contraction helps remind us of the containment relationships (subspaces taken out of other subspaces), and makes it easier to apply duality relationships to convert the subspace products.

Since projection and rejection are linear transformations, we can extend them from vectors to general blades as outermorphisms (and even to multivectors, by linearity). For the projection, we have done this before in Section 4.2.2, and we derived that it boils down to substituting the general blade $\mathbf{X}$ for the vector $\mathbf{x}$, to obtain

$$\text{projection of } \mathbf{X} \text{ onto } \mathbf{A}: \quad \mathbf{X} \mapsto (\mathbf{X} \rfloor \mathbf{A}) \mathbf{A}^{-1}. \tag{6.27}$$

However, the outermorphic extension of the rejection quickly disappoints, since it becomes a rather trivial operation (although indeed linear). For $(\mathbf{X} \wedge \mathbf{A}) \mathbf{A}^{-1}$ is zero as soon as $\mathbf{X}$ contains at least one common vector with $\mathbf{A}$ (and if both $\mathbf{X}$ and $\mathbf{A}$ were bivectors in 3-D, this would always be the case). The easiest way to express the concept of the rejection of a general blade $\mathbf{X}$ by a subspace $\mathbf{A}$ is simply as the difference of $\mathbf{X}$ and its projection: $\mathbf{X} \mapsto \mathbf{X} - (\mathbf{X}\rfloor\mathbf{A})\,\mathbf{A}^{-1}$. However, this is not a proper subspace operation; it does not necessarily produce a blade (see structural exercise 9), so it should be used with care. The rejection is not as tidy as it appears at first sight, when we introduced it for vectors.

6.4.3 THE OTHER DIVISION: REFLECTION

We have seen that the geometric product is noncommutative. This implies that geometric division (which is just geometric multiplication by the inverse) is not commutative either. We have also seen that division of $(\mathbf{x}\,\mathbf{a})$ by $\mathbf{a}$ on the right (i.e., right division) produces $\mathbf{x}$, as you would hope. Let us investigate the result of left division:

$$
\begin{aligned}
\mathbf{a}^{-1}\,\mathbf{x}\,\mathbf{a} &= \mathbf{a}^{-1}\,(\mathbf{x}\,\mathbf{a}) \\
&= \frac{1}{\mathbf{a}\,\mathbf{a}}\,\mathbf{a}\,(\mathbf{x}\,\mathbf{a}) \\
&= \mathbf{a}\,\mathbf{x}\,\mathbf{a}\,\frac{1}{\mathbf{a}\,\mathbf{a}} \quad \text{[since scalars commute]} \\
&= (\mathbf{a}\,\mathbf{x})\,\mathbf{a}^{-1} \\
&= (\mathbf{a}\cdot\mathbf{x})\,\mathbf{a}^{-1} + (\mathbf{a}\wedge\mathbf{x})\,\mathbf{a}^{-1} \\
&= (\mathbf{x}\cdot\mathbf{a})\,\mathbf{a}^{-1} - (\mathbf{x}\wedge\mathbf{a})\,\mathbf{a}^{-1} \quad\quad (6.28)
\end{aligned}
$$

Compare this to the decomposition of (6.26) (which was made with right division):

$$
(\mathbf{x}\,\mathbf{a})\,\mathbf{a}^{-1} = (\mathbf{x}\cdot\mathbf{a})\,\mathbf{a}^{-1} + (\mathbf{x}\wedge\mathbf{a})\,\mathbf{a}^{-1}.
$$

We observe that in (6.28) the non-$\mathbf{a}$-component of $\mathbf{x}$ (which we called the rejection of $\mathbf{a}$) is *subtracted* from the projection of $\mathbf{x}$ onto $\mathbf{a}$, rather than added. Figure 6.4 shows

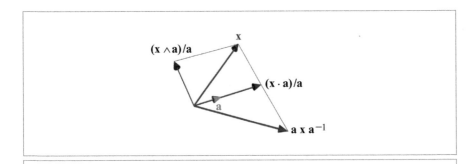

Figure 6.4: Reflection of $\mathbf{x}$ in $\mathbf{a}$.

the effect: the vector **x** is *reflected* in the **a**-line. Only when **x** and **a** have the same direction is there no difference between the two types of division (but they then trivially both result in **x**).

The bad news is that we have to be careful about the order of division, but the good news is that we have found a simple way to make line reflections: we can reflect **x** through **a** by *sandwiching* **x** between **a** and $\mathbf{a}^{-1}$ as $\mathbf{a}^{-1}\mathbf{x}\mathbf{a}$ or equivalently $\mathbf{a}\mathbf{x}\mathbf{a}^{-1}$. This is actually one of the basic constructions in geometric algebra, so common that it could be considered as a product in its own right, the "versor product" of **x** by **a**. It can be extended to blades, and is then a powerful way to represent orthogonal transformations.

The next chapter is fully devoted to this important operation.

6.5 FURTHER READING

With the geometric product, you are almost ready to read the literature on geometric algebra. However, since that typically involves the special representations of operators by rotors and versors, we recommend that you wait for one more chapter.

But if you are interested in the historical roots, an inspirational piece (without rotors) that focuses on the development of number systems for geometry is David Hestenes' *Origins of Geometric Algebra*, Chapter 1 in [29]. It traces the developments from Euclid via Descartes to Grassmann, and, implicitly, Clifford. Leo can recommend it as the piece that got him hooked, back in 1997.

6.6 EXERCISES

6.6.1 DRILLS

1. Let $\mathbf{a} = \mathbf{e}_1 + \mathbf{e}_2$ and $\mathbf{b} = \mathbf{e}_2 + \mathbf{e}_3$ in a 3-D Euclidean space with orthonormal basis $\{\mathbf{e}_1, \mathbf{e}_2, \mathbf{e}_3\}$. Compute the following expressions, giving the results relative to the basis $\{1, \mathbf{e}_1, \mathbf{e}_2, \mathbf{e}_3, \mathbf{e}_2 \wedge \mathbf{e}_3, \mathbf{e}_3 \wedge \mathbf{e}_1, \mathbf{e}_1 \wedge \mathbf{e}_2, \mathbf{e}_1 \wedge \mathbf{e}_2 \wedge \mathbf{e}_3\}$. Show your work.

   (a) $\mathbf{a}\,\mathbf{a}$
   (b) $\mathbf{a}\,\mathbf{b}$
   (c) $\mathbf{b}\,\mathbf{a}$
   (d) $(\mathbf{e}_1 \wedge \mathbf{e}_2)\,\mathbf{a}$
   (e) $\mathbf{a}\,(\mathbf{e}_1 \wedge \mathbf{e}_2)$
   (f) $(\mathbf{e}_1 \wedge \mathbf{e}_2 \wedge \mathbf{e}_3)\,\mathbf{a}$
   (g) $\mathbf{a}^{-1}$
   (h) $\mathbf{b}\,\mathbf{a}^{-1}$
   (i) $(\mathbf{e}_1 \wedge \mathbf{e}_2)^{-1}$

2. Make a full geometric product multiplication table for the 8 basis elements $\{1, e_1, e_2, e_3, e_1 \wedge e_2, e_2 \wedge e_3, e_3 \wedge e_1, e_1 \wedge e_2 \wedge e_3\}$; (a) in a Euclidean metric $\mathbb{R}^{3,0}$ and (b) in a metric $\mathbb{R}^{2,1}$ with $e_1 \cdot e_1 = -1$.

6.6.2 STRUCTURAL EXERCISES

1. Section 6.1.1 demonstrated the noninvertibility of contraction and outer product. Show by a geometrical example that the cross product of two vectors is not invertible either. Also give an algebraic argument based on its (invertible) relationship to the outer product.

2. The pseudoscalar is the highest-order blade in the algebra of $\bigwedge \mathbb{R}^n$. It receives its name because in many dimensions it is like a scalar in its commutation properties with vectors under the geometric product. In which dimensions does it commute with all vectors?

3. The outer product can be defined as the completely antisymmetric summed average of all permutations of geometric products of its factors, with a sign for each term depending on oddness or evenness of the permutation. For the 3-blade, this means:

$$\mathbf{x} \wedge \mathbf{y} \wedge \mathbf{z} = \frac{1}{3!} (\mathbf{x} \mathbf{y} \mathbf{z} - \mathbf{y} \mathbf{x} \mathbf{z} + \mathbf{y} \mathbf{z} \mathbf{x} - \mathbf{z} \mathbf{y} \mathbf{x} + \mathbf{z} \mathbf{x} \mathbf{y} - \mathbf{x} \mathbf{z} \mathbf{y})$$

Derive this formula.

4. The parts of a certain grade of a geometric product of blades are not necessarily blades. Show that in a 4-D space with orthonormal basis $\{e_i\}_{i=1}^4$, a counterexample is $\langle e_1 (e_1 + e_2) (e_2 + e_3) (e_1 + e_4) \rangle_2$. (You may want to use software for this. If you find a simpler counterexample, let us know...)

5. Show that the definition of the scalar product as $\mathbf{A} * \mathbf{B} = \langle \mathbf{A} \mathbf{B} \rangle_0$ is equivalent to the determinant definition of (3.2). You will then also understand why the matrix in the latter definition has the apparently reversed $\mathbf{a}_i \cdot \mathbf{b}_{k-j}$ as element (i,j) for k-blades.

6. Originally, we motivated the contraction as the counterpart of an outer product relative to the scalar product, which led to the implicit definition (3.6):

$$(\mathbf{X} \wedge \mathbf{A}) * \mathbf{B} = \mathbf{X} * (\mathbf{A} \rfloor \mathbf{B}).$$

Prove this part of the definition using the grade-based definitions of $\wedge$, $*$, and $\rfloor$ in Section 6.3.2.

7. In the formula $(\mathbf{x} \rfloor \mathbf{A}^{-1}) \mathbf{A}$, we can replace the geometric product by a contraction, so that it is in fact the projection $(\mathbf{x} \rfloor \mathbf{A}^{-1}) \rfloor \mathbf{A}$. Show this, using the suggestion that $\mathbf{x} \rfloor \mathbf{A}^{-1}$ might be a subblade of $\mathbf{A}$—which you first need to demonstrate. After that, decompose $\mathbf{x} \rfloor \mathbf{A}^{-1}$ as a product of orthogonal vectors, and evaluate the two formulas to show their equivalence.

8. As a counterpart of the previous exercise, show that $(\mathbf{x} \wedge \mathbf{A}^{-1}) \mathbf{A} = (\mathbf{x} \wedge \mathbf{A}^{-1}) \rfloor \mathbf{A}$. (Hint: Write the second $\mathbf{A}$ as a wedge product of orthogonal vectors, and peel them off one by one).

9. In a 4-D space with orthonormal basis $\{\mathbf{e}_i\}_{i=1}^4$, project the 2-blade $\mathbf{X} = (\mathbf{e}_1 + \mathbf{e}_2) \wedge (\mathbf{e}_3 + \mathbf{e}_4)$ onto the 2-blade $\mathbf{A} = (\mathbf{e}_1 \wedge \mathbf{e}_3)$. Then determine the rejection as the difference of $\mathbf{X}$ and its projection. Show that this is not a blade. (See also structural exercise 5 of Chapter 2.)

10. Let an orthonormal coordinate system $\{\mathbf{e}_i\}_{i=1}^3$ be given in 3-D Euclidean space $\mathbb{R}^{3,0}$. Compute the support vector (i.e., the vector of the point on the line closest to the origin) of the line with direction $\mathbf{u} = \mathbf{e}_1 + 2\mathbf{e}_2 - \mathbf{e}_3$, through the point $\mathbf{p} = \mathbf{e}_1 - 3\mathbf{e}_2$. What is the distance of the line to the origin?

6.7 PROGRAMMING EXAMPLES AND EXERCISES

6.7.1 EXERCISE: SUBSPACE PRODUCTS RETRIEVED

The geometric product is the fundamental product of geometric algebra. Other products are derived from it. In these exercises, we follow Section 6.3 and implement two different ways of retrieving the left contraction and the outer product from the geometric product.

Exercise 1a: The Symmetry Approach (for Vectors Only)

Implement the outer product of a vector and any multivector using (6.10):

$$\mathbf{a} \wedge \mathbf{B} = \tfrac{1}{2}(\mathbf{a}\,\mathbf{B} + \widehat{\mathbf{B}}\,\mathbf{a}).$$

Implement the left contraction of a vector and any multivector using (6.12):

$$\mathbf{a}\rfloor\mathbf{B} = \tfrac{1}{2}(\mathbf{a}\,\mathbf{B} - \widehat{\mathbf{B}}\,\mathbf{a}).$$

The downloadable example code provides a bare-bones framework for doing this. You should complete the following functions:

```
// exercise 1a: complete in this function
mv outerProduct_1a(const e3ga::vector &a, const mv &B) {
  printf("Warning: outerProduct_1a() not implemented yet!\n");
  return 0.0f;
}

// exercise 1a: complete in this function
mv leftContraction_1a(const e3ga::vector &a, const mv &B) {
  printf("Warning: leftContraction_1a() not implemented yet!\n");
  return 0.0f;
}
```

After you have completed the functions, compile and run the example. The testing code will complain if you made a mistake in the implementation. You may need the following functions:

- `gradeInvolution(const mv &X)` computes the grade involution of a multivector.

- `gp(const mv &X, const mv &Y)` computes the geometric product of two multivectors. The * operator is bound to it, see Table 2.4.

Exercise 1b: The Grade Approach

Equations (6.19) and (6.20) provide another way to obtain the outer product and the left contraction, respectively:

$$\mathbf{A}_k \wedge \mathbf{B}_l \equiv \langle \mathbf{A}_k \mathbf{B}_l \rangle_{k+l}$$

$$\mathbf{A}_k \rfloor \mathbf{B}_l \equiv \langle \mathbf{A}_k \mathbf{B}_l \rangle_{l-k}$$

Implement this by filling in `outerProduct_1b()` and `leftContraction_1b()` in the example code.

```
// exercise 1b: complete in this function
mv outerProduct_1b(const mv &A, const mv &B) {
  printf("Warning: outerProduct_1b() not implemented yet!\n");
  return 0.0f;
}

// exercise 1b: complete in this function
mv leftContraction_1b(const mv &A, const mv &B) {
  printf("Warning: leftContraction_1b() not implemented yet!\n");
  return 0.0f;
}
```

You may need the following functions:

- `takeGrade(const mv &X, int gradeUsageBitmap)` extracts grade parts from multivector. The `gradeUsageBitmap` is a *bitwise or* of the constants `GRADE_0`, `GRADE_1`, `GRADE_2`, and `GRADE_3`, which have values 1, 2, 4, 8, respectively. So, to extract grade k, you can also use `takeGrade(X, 1 << k)`. In the context of integers, the `<<` operator means *bitwise shift left*, of course.
- If you want to know whether a grade part is present in a multivector variable X, you can use `((X.gu() & GRADE_k) != 0)`, where k is the grade part index. For example `((X.gu() & GRADE_2) != 0)` is `true` when the bivector grade part is present in X.

6.7.2 GRAM-SCHMIDT ORTHOGONALIZATION

Geometric algebra does *not* require the representation of its elements in terms of a particular basis of vectors. Therefore, the specific treatment of issues like orthogonalization are much less necessary. Yet it is sometimes convenient to have an orthogonal basis, and such a basis is simple to construct using our products. We saw a first glimpse of this in the example of Section 3.11.1, using the contraction. Now that we have the geometric product we can give a more general and more complete treatment of orthogonalization.

Suppose we have a set of three vectors $\mathbf{v}_1$, $\mathbf{v}_2$, $\mathbf{v}_3$ in a Euclidean space, as in Figure 6.5(a), and would like to form them into an orthogonal basis. The perpendicularized frame will have its vectors denoted as $\mathbf{b}_1$, $\mathbf{b}_2$, $\mathbf{b}_3$; we arbitrarily keep $\mathbf{v}_1$ as the first of those (Figure 6.5(b)):

$$\mathbf{b}_1 \equiv \mathbf{v}_1.$$

Then we form the rejection of $\mathbf{v}_2$ by $\mathbf{v}_1$, which is automatically perpendicular to $\mathbf{v}_1$, by forming $\mathbf{v}_2 \wedge \mathbf{b}_1$ (Figure 6.5(c)) and dividing out $\mathbf{b}_1$ to orthogonalize it (Figure 6.5(d)):

$$\mathbf{b}_2 \equiv (\mathbf{v}_2 \wedge \mathbf{b}_1)/\mathbf{b}_1$$

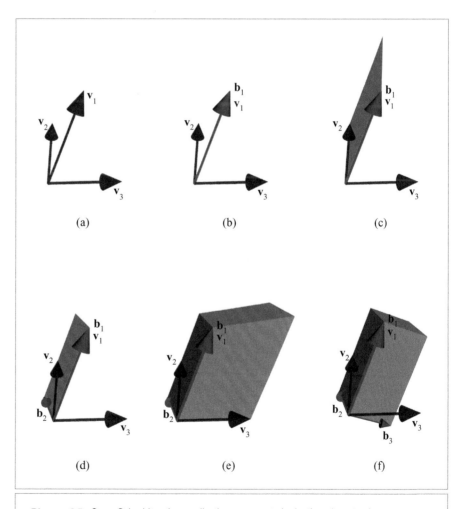

Figure 6.5: Gram-Schmidt orthogonalization as repeated rejections (see text).

That is our second vector of the frame. Now we take the rejection of $\mathbf{v}_3$ by $\mathbf{b}_1 \wedge \mathbf{b}_2$, which is perpendicular to both $\mathbf{b}_1$ and $\mathbf{b}_2$. Graphically, this is done by forming the trivector $\mathbf{v}_3 \wedge \mathbf{b}_1 \wedge \mathbf{b}_2$ (Figure 6.5(e)), and straightening it by dividing it by $\mathbf{b}_1 \wedge \mathbf{b}_2$ (Figure 6.5(f)). Algebraically, $\mathbf{b}_3$ is:

$$\mathbf{b}_3 \equiv (\mathbf{v}_3 \wedge \mathbf{b}_1 \wedge \mathbf{b}_2)/(\mathbf{b}_1 \wedge \mathbf{b}_2),$$

and we are done. This is the Gram-Schmidt orthogonalization procedure, rewritten in geometric algebra.

Figure 6.6 gives code listing for orthogonalizing an n-dimensional basis. Note that we view the selection of the first vector as a (rather trivial) rejection to produce clean code. Also note that the function throws a `std::string` when it detects a null blade. The rest of the example is identical to that of Section 3.11.1.

The result of the Gram-Schmidt orthogonalization implies that vectors spanning a subspace can be orthogonalized if they are invertible. This has consequences for the blade representing that subspace, for using the new basis we can write it as a geometric product of vectors $\mathbf{b}_1 \, \mathbf{b}_2 \cdots \mathbf{b}_k$ rather than as an outer product of vectors $\mathbf{v}_1 \wedge \mathbf{v}_2 \wedge \cdots \wedge \mathbf{v}_k$. This is often useful in algebraic manipulation inside proofs, since the geometric product has

```
/**
Uses GA to perform Gram-Schmidt orthogonalization.
Throws std::string when input vectors 'vIn' are dependent.
Results are returned in 'vOut'.
*/
void GramSchmidtGA(const e3ga::vector vIn[], e3ga::vector vOut[], int nbVectors) {
  mv B = 1;

  for (int i = 0; i < nbVectors; i++) {
    mv newB = vIn[i] ^ B;

    // check for dependence of input vectors:
    if (_Float(norm_r2(newB)) == 0.0f)
      throw std::string("input vectors are dependent");

    // compute orthogonal vector 'i':
    vOut[i] = _vector(newB * inverse(B));

    B = newB;
  }
}
```

Figure 6.6: Gram-Schmidt orthogonalization code (Example 2).

richer algebraic properties; for instance, it is invertible, whereas the outer product is not. Since orthogonal vectors anticommute, we have: *an invertible blade can be written as a geometric product of anticommuting vectors.*

In non-Euclidean metrics, null vectors and null blades occur and those are noninvertible. This implies we cannot use the division the orthogonalization algorithm requires. Yet even in such a space, a blade can be written as a geometric product of anticommuting vectors; we just have to compute them in a different manner. We recommend the method described in Section 19.4 as the numerically stable way of finding these anticommuting vectors. The method amounts to computing the metric matrix of the blade and computing its eigenvalue decomposition; the eigenvectors are then used to compute the anticommuting vectors that span the blade.

7 ORTHOGONAL TRANSFORMATIONS AS VERSORS

Reflection in a line is represented by a sandwiching construction involving the geometric product. Though that may have seemed a curiosity in the previous chapter, we will show that it is crucial to the representation of operators in geometric algebra. Geometrically, all orthogonal transformations can be considered as multiple reflections. Algebraically, this leads to their representation as a geometric product of unit vectors.

An even number of reflections gives a rotation, represented as a rotor—the geometric product of an even number of unit vectors. We show that rotors encompass and extend complex numbers and quaternions, and present a real 3-D visualization of the quaternion product. Rotors transcend quaternions in that they can be applied to elements of any grade, in a space of any dimension.

The distinction between subspaces and operators fades when we realize that any subspace generates a reflection operator, which can act on any element. The concept of a versor (a product of vectors to be used as an operator in a sandwiching product) combines all these representations of orthogonal transformations. We show that versors preserve the structure of geometric constructions and can be universally applied to any geometrical element. This is a unique feature of geometric algebra, and it can simplify code considerably.

The chapter ends with a discussion of the difference between geometric algebra and Clifford algebra, and a preliminary consideration of issues in efficient implementation to convince you of the practical usability of the versor techniques in writing efficient code for geometry.

7.1 REFLECTIONS OF SUBSPACES

We have seen in Section 6.4 how we can construct the reflection of a vector $\mathbf{x}$ in a line through the origin characterized by a vector $\mathbf{a}$ as

$$\textit{reflection of } \mathbf{x} \textit{ in a-line: } \mathbf{x} \mapsto 2(\mathbf{x} \cdot \mathbf{a})\,\mathbf{a}^{-1} - \mathbf{x} = \mathbf{a}\,\mathbf{x}\,\mathbf{a}^{-1}.$$

The magnitude or orientation of $\mathbf{a}$ are irrelevant to the outcome, since the inversion removes any scalar factor. Only the attitude of the line matters.

Since line reflection is a linear transformation on $\mathbf{x}$, we should be able to extend it from vectors to general blades $\mathbf{X}$ as an outermorphism. The result is

$$\textit{reflection of } \mathbf{X} \textit{ in a-line: } \mathbf{X} \mapsto \mathbf{a}\,\mathbf{X}\,\mathbf{a}^{-1}. \tag{7.1}$$

This is indeed relatively straightforward to prove by induction using $\mathbf{X}_k = \mathbf{x}_k \wedge \mathbf{X}_{k-1}$ and assuming that it holds for $\mathbf{X}_{k-1}$. Then

$$
\begin{aligned}
(\mathbf{a}\,\mathbf{x}_k\,\mathbf{a}^{-1}) \wedge (\mathbf{a}\,\mathbf{X}_{k-1}\,\mathbf{a}^{-1}) &= \tfrac{1}{2}(\mathbf{a}\,\mathbf{x}_k\,\mathbf{a}^{-1}\,\mathbf{a}\,\mathbf{X}_{k-1}\,\mathbf{a}^{-1} + \mathbf{a}\,\widehat{\mathbf{X}}_{k-1}\,\mathbf{a}^{-1}\,\mathbf{a}\,\mathbf{x}_k\,\mathbf{a}^{-1}) \\
&= \tfrac{1}{2}(\mathbf{a}\,(\mathbf{x}_k\,\mathbf{X}_{k-1})\,\mathbf{a}^{-1} + \mathbf{a}\,(\widehat{\mathbf{X}}_{k-1}\,\mathbf{x}_k)\,\mathbf{a}^{-1}) \\
&= \mathbf{a}\left(\tfrac{1}{2}(\mathbf{x}_k\,\mathbf{X}_{k-1} + \widehat{\mathbf{X}}_{k-1}\,\mathbf{x}_k)\right)\mathbf{a}^{-1} \\
&= \mathbf{a}\,(\mathbf{x}_k \wedge \mathbf{X}_{k-1})\,\mathbf{a}^{-1} \\
&= \mathbf{a}\,\mathbf{X}_k\,\mathbf{a}^{-1}
\end{aligned}
$$

and the induction basis is of course the trivial statement for scalars $\mathbf{a}\,\xi\,\mathbf{a}^{-1} = \xi$. See structural exercise 1 for another way of *not* proving this. As always, we can extend the linear transformation from blades $\mathbf{X}$ to general multivectors X by linearity.

By the simple trick of swapping the sign in (7.1), we can modify the line reflection formula into a hyperplane reflection formula. For the reflection of a vector $\mathbf{x}$ in the plane $\mathbf{A}$ is equivalent to swapping the sign of the rejection of $\mathbf{x}$ by that plane (i.e., it is $\mathbf{x} - 2(\mathbf{x} \wedge \mathbf{A})\,/\,\mathbf{A}$). By inserting a pseudoscalar and its inverse and using duality, we can rewrite this as $\mathbf{x} - 2(\mathbf{x}\rfloor\mathbf{a})\,/\,\mathbf{a}$, with $\mathbf{a} = \mathbf{A}^*$ the normal vector of the hyperplane. This can be rewritten as before in terms of a geometric product, yielding

$$\textit{reflection of } \mathbf{x} \textit{ in dual hyperplane a: } \mathbf{x} \mapsto -\mathbf{a}\,\mathbf{x}\,\mathbf{a}^{-1}.$$

Note that the precise sign of the dualization of the hyperplane $\mathbf{A}$ to produce $\mathbf{a}$ is not important, due to the absorption of any scalar factors by the subsequent inverse. Figure 7.1 compares the two kinds of reflections.

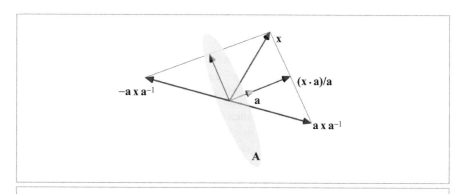

Figure 7.1: Line and plane reflection of a vector **x** in a vector **a**, used as line direction or as normal vector for the plane **A**.

This hyperplane reflection is actually what we mean by a reflection—in 3-D, it reflects in a plane, which is like looking into a mirror. It extends to blades by outermorphism as the linear transformation $\mathsf{a}: \bigwedge^k \mathbb{R}^n \to \bigwedge^k \mathbb{R}^n$ defined by

$$\text{reflection of } X \text{ in dual hyperplane } \boldsymbol{a}: \ X \mapsto \mathsf{a}[X] \equiv \mathbf{a}\,\widehat{X}\,\mathbf{a}^{-1}, \tag{7.2}$$

where $\widehat{X} = (-1)^{\text{grade}(X)} X$ is the grade involution.

To be a reflection, its determinant should be -1 in a space of any dimensionality. With the pseudoscalar of n-dimensional space denoted as $\mathbf{I}_n$, we can check this easily using definition (4.7) of a determinant in geometric algebra:

$$\det(\mathsf{a}) = (\mathbf{a}\,\widehat{\mathbf{I}}_n\,\mathbf{a}^{-1})\,\mathbf{I}_n^{-1} = \mathbf{a}\,(\widehat{\mathbf{I}}_n\,\mathbf{a}^{-1})\,\mathbf{I}_n^{-1} = \mathbf{a}\,(-\mathbf{a}^{-1}\mathbf{I}_n)\,\mathbf{I}_n^{-1} = -1.$$

Here we used the fact that **a** is contained in the space $\mathbf{I}_n$, and therefore $\mathbf{a} \wedge \mathbf{I}_n = 0$, so that $\mathbf{a}\,\mathbf{I}_n = -\widehat{\mathbf{I}}_n\,\mathbf{a}$. So the determinant is indeed -1. If you would repeat the determinant computation for the line reflection, you would find $(-1)^{n+1}$ for the determinant in n-dimensional space. This shows that in 3-D, a line reflection is actually not a reflection but a rotation (see structural exercise 3). By contrast, hyperplane reflections are indeed proper reflections in any dimension.

7.2 ROTATIONS OF SUBSPACES

Having reflections as a sandwiching product leads naturally to the representation of rotations. For by a well-known theorem, any rotation can be represented as an even number of reflections. In geometric algebra, this statement can be converted immediately into a computational form.

7.2.1 3-D ROTORS AS DOUBLE REFLECTORS

Two reflections make a rotation, even in $\mathbb{R}^3$ (see Figure 7.2). Since an even number of reflections absorbs any sign, we may make these reflections either both line reflections $\mathbf{a}\,\mathbf{x}\,\mathbf{a}^{-1}$ or both (dual) hyperplane reflections $-\mathbf{a}\,\mathbf{x}\,\mathbf{a}^{-1}$, whichever feels most natural or is easiest to visualize. The figure uses line reflections.

As the figure shows, first reflecting in $\mathbf{a}$, then in $\mathbf{b}$, gives a rotation over an axis perpendicular to the $\mathbf{a}\wedge\mathbf{b}$-plane, over an angle that is *twice* the angle from $\mathbf{a}$ to $\mathbf{b}$ (and this angle and plane also give the sense of rotation, clockwise or anticlockwise in the plane). In this construction, only the plane and relative angle of the vectors $\mathbf{a}$ and $\mathbf{b}$ matter. GAViewer or the programming example in Section 7.10.2 each provide an interactive version. We strongly recommend playing with either if you need to tune your intuition.

It is simple to convert this geometrical idea into algebra. The operation

$$\mathbf{x} \mapsto \mathbf{b}\,(\mathbf{a}\,\mathbf{x}\,\mathbf{a}^{-1})\,\mathbf{b}^{-1} = \mathbf{b}\,\mathbf{a}^{-1}\,\mathbf{x}\,\mathbf{a}\,\mathbf{b}^{-1} = (\mathbf{b}/\mathbf{a})\,\mathbf{x}\,(\mathbf{b}/\mathbf{a})^{-1}$$

is the double reflection, and therefore produces the rotation in the $\mathbf{a}\wedge\mathbf{b}$ plane (we moved some scalar squared norms around to get a pleasant expression). We observe that this rotation is generated by an element $R = \mathbf{b}/\mathbf{a} = \mathbf{b}\,\mathbf{a}^{-1}$, as applied to a vector by the recipe

$$\textit{rotation of } \mathbf{x}\colon\ \mathbf{x}\ \mapsto\ R\,\mathbf{x}\,R^{-1}.$$

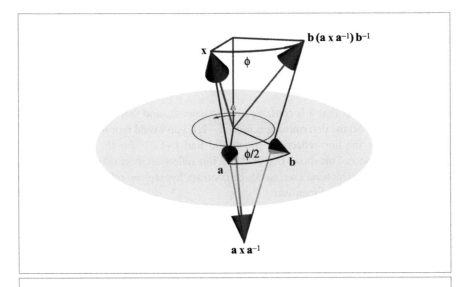

Figure 7.2: A rotation in a plane is identical to two reflections in vectors in that plane, separated by half the rotation angle.

Note that this element R is not necessarily a blade, since it is a geometric product. That is why we do not denote it by a bold symbol. It defines a linear transformation that we denote by $\mathsf{R}[\ ]$.

Since rotation is a linear transformation, it can be extended as an outermorphism. You can easily show by mimicking the proof of (7.1) that a blade $\mathbf{X}$ rotates to $\mathsf{R}[\mathbf{X}]$ defined by

$$\textit{rotation of } \mathbf{X}: \ \mathbf{X} \ \mapsto \ \mathsf{R}[\mathbf{X}] \equiv R\,\mathbf{X}\,R^{-1}. \tag{7.3}$$

There are no extra signs in this transformation formula, unlike the reflection formula of (7.2); the double reflection has canceled them all. As a consequence, the determinant of a rotation is $+1$ in a space of any dimension:

$$\det(\mathsf{R}) = (R\,\mathsf{I}_n\,R^{-1})/\mathsf{I}_n = R\,R^{-1}\,\mathsf{I}_n\,\mathsf{I}_n^{-1} = 1.$$

Geometrically, this means that there is no orientation change of the pseudoscalar.

Equation (7.3) can even be extended to arbitrary multivectors, for the rotation is linear. That means that we are now capable of rotating *any* multivector, not just a vector, with the same formula. In 3-D, this is already helpful: we can rotate a plane (represented as a 2-blade) directly without having to dualize it first to a normal vector (see Figure 7.3). On the other hand, if it had been given dually, we could rotate it in that form as well. These capabilities extend to higher-dimensional spaces and in Part II will permit us to rotate Euclidean circles, spheres, and other elements, all using the same operation.

It is common practice to take out the scaling factor in $R = \mathbf{b}/\mathbf{a}$, reducing it to a unit element called a *rotor*. That obviously makes no difference to the application to a blade,

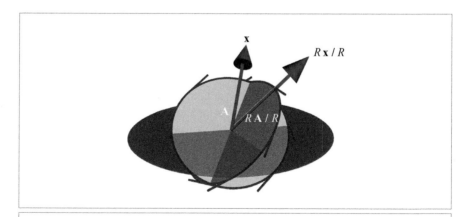

Figure 7.3: The same operation RX/R rotates a vector, a 2-blade, or any element of the algebra.

since any scaling factor in R is canceled by the reciprocal factor in R^{-1}. To compute the normalization of a rotor, we need to compute the scalar product of a mixed-grade multivector, which we have not done before. Using (6.22) and (6.23), it is straightforward:

$$\|R\|^2 = R * \widetilde{R} = \langle \mathbf{b}\,\mathbf{a}^{-1}\,\mathbf{a}^{-1}\,\mathbf{b}\rangle_0 = \langle \mathbf{b}^2\,(\mathbf{a}^{-1})^2\rangle_0 = \mathbf{b}^2/\mathbf{a}^2.$$

Therefore $\|R\| = \pm\|\mathbf{b}\|/\|\mathbf{a}\|$, and dividing it out produces a properly normalized rotor.

To construct a rotor to represent the rotation from $\mathbf{a}$ to $\mathbf{b}$, we can either do what we just did (start with general $\mathbf{a}$ and $\mathbf{b}$ and taking out the norm ratio), or just define it as the ratio of two *unit* vectors $\mathbf{b}$ and $\mathbf{a}$ from the start. When we do the latter, $R = \mathbf{b}\,\mathbf{a}^{-1} = \mathbf{b}\,\mathbf{a}$, and $R^{-1} = \mathbf{a}\,\mathbf{b} = \widetilde{R}$. It follows that the inverse of a rotor is its reverse:

$$R\,\widetilde{R} = 1,$$

so that the rotation of $\mathbf{X}$ can be performed as $\mathsf{R}[\mathbf{X}] = R\,\mathbf{X}\,\widetilde{R}$. Performing the normalization once is often better in practice than having to compute its inverse with each application, so we will use these normalized rotors constructed from unit vectors as the representation of rotations in this chapter.

7.2.2 ROTORS PERFORM ROTATIONS

It is natural to relate the rotor $\mathbf{b}/\mathbf{a}$ to the geometrical relationships of the two vectors: their common plane $\mathbf{a} \wedge \mathbf{b}$ and their relative angle. We can use those geometric elements to encode it algebraically, by developing the geometric product of the unit vectors in terms of their inner and outer product, and those in terms of angle and plane. Since $\mathbf{b}$ and $\mathbf{a}$ were assumed to be unit vectors, we have $\mathbf{b}/\mathbf{a} = \mathbf{b}\,\mathbf{a}$, and compute

$$R = \mathbf{b}\,\mathbf{a} = \mathbf{b} \cdot \mathbf{a} + \mathbf{b} \wedge \mathbf{a} = \cos(\phi/2) - \mathbf{I}\,\sin(\phi/2), \tag{7.4}$$

where $\phi/2$ is the angle from $\mathbf{a}$ to $\mathbf{b}$, and $\mathbf{I}$ is the unit 2-blade for the $(\mathbf{a} \wedge \mathbf{b})$-plane. This rotor involving the angle $\phi/2$ actually rotates over ϕ (as Figure 7.2 suggests, and as we will show below).

The action of a rotor may appear a bit magical at first. It is good to see in detail how the sandwiching works to produce a rotation of a vector $\mathbf{x}$ in a Euclidean space. To do so, we introduce notations for the various components of $\mathbf{x}$ relative to the rotation plane. What we would hope when we apply the rotor to $\mathbf{x}$ is that

- The component of $\mathbf{x}$ perpendicular to the rotation plane (i.e., the rejection $\mathbf{x}_\uparrow$ defined by $\mathbf{x}_\uparrow \equiv (\mathbf{x} \wedge \mathbf{I})/\mathbf{I}$) remains unchanged;
- The component of $\mathbf{x}$ within the plane (i.e., the projection $\mathbf{x}_\| \equiv (\mathbf{x}\rfloor\mathbf{I})/\mathbf{I}$) gets shortened by $\cos\phi$;
- A component of $\mathbf{x}$ perpendicular to the projection in the plane (i.e., $\mathbf{x}_\perp \equiv \mathbf{x}_\|\rfloor$ $\mathbf{I} = \mathbf{x}_\|\,\mathbf{I}$) gets added, with a scaling factor $\sin\phi$.

It seems a lot to ask of the simple formula $R\mathbf{x}\widetilde{R}$, but we can derive that this is indeed precisely what it does. The structure of the derivation is simplified when we denote $c \equiv \cos(\phi/2)$, $s \equiv \sin(\phi/2)$, and note beforehand that the rejection and projection satisfy the commutation relations $\mathbf{x}_\uparrow \mathbf{I} = \mathbf{I}\mathbf{x}_\uparrow$ and $\mathbf{x}_\| \mathbf{I} = -\mathbf{I}\mathbf{x}_\|$ (these relations actually define them fully, by the relationships of Section 6.3.1). Also, we have seen in (6.4) that in a Euclidean space $\mathbf{I}^2 = -1$, which is essential to make the whole thing work. Then

$$\begin{aligned}
R\mathbf{x}\widetilde{R} &= R(\mathbf{x}_\uparrow + \mathbf{x}_\|)\widetilde{R} \\
&= (c - s\mathbf{I})(\mathbf{x}_\uparrow + \mathbf{x}_\|)(c + s\mathbf{I}) \\
&= c^2\mathbf{x}_\uparrow - s^2(\mathbf{I}\mathbf{x}_\uparrow\mathbf{I}) + cs(\mathbf{x}_\uparrow\mathbf{I} - \mathbf{I}\mathbf{x}_\uparrow) + cs(\mathbf{x}_\|\mathbf{I} - \mathbf{I}\mathbf{x}_\|) + c^2\mathbf{x}_\| - s^2(\mathbf{I}\mathbf{x}_\|\mathbf{I}) \\
&= (c^2 + s^2)\mathbf{x}_\uparrow + (c^2 - s^2)\mathbf{x}_\| + 2cs\,\mathbf{x}_\|\mathbf{I} \\
&= \mathbf{x}_\uparrow + \cos\phi\,\mathbf{x}_\| + \sin\phi\,\mathbf{x}_\perp,
\end{aligned}$$

which is the desired result. Note especially how the vector $\mathbf{x}_\perp$, which was not originally present, is generated automatically. It is very satisfying that the whole process is driven by the algebraic commutation rules that encode the various geometrical perpendicularity and containment relationships. This shows that we truly have an algebra capable of mimicking geometry.

The unit vectors in the directions $\mathbf{x}_\uparrow$, $\mathbf{x}_\|$, and $\mathbf{x}_\perp$ form an orthonormal basis for the relevant subspace of the vector space involved in this rotation. The rotor application has constructed this frame automatically from the vector $\mathbf{x}$ that needs to be rotated. This is in contrast to rotation matrices, which use a fixed frame for the total space that is unrelated to the elements to be processed. Such a fixed frame then necessitates a lot of coordinate coefficients to represent an arbitrary rotation. Even when the frame has been well chosen (so that for instance $\mathbf{e}_1 \wedge \mathbf{e}_2$ is the rotation plane), the sine and cosine of the angle occur twice in the rotation matrix:

$$[\![R]\!] = \begin{bmatrix} \cos\phi & -\sin\phi & 0 & \cdots & 0 \\ \sin\phi & \cos\phi & 0 & \cdots & 0 \\ 0 & 0 & 1 & \cdots & 0 \\ \vdots & \vdots & \vdots & \ddots & \vdots \\ 0 & 0 & 0 & \cdots & 1 \end{bmatrix}.$$

When multiplying rotations, this double occurrence causes needless double work that rotors avoid (like quaternions; we show the relationship in Section 7.3.5). So although it would seem like a waste to construct a new frame for each vector, the rotation representation we have shown can actually be more efficient than a rotation matrix implementation. After all, we never actually construct the frame; we merely perform $R\mathbf{x}\widetilde{R}$.

There is a classical way of generating a 3-D rotation matrix that is also based on the construction $\mathbf{x}_\uparrow + \cos\phi\,\mathbf{x}_\| + \sin\phi\,\mathbf{x}_\perp$. It is *Rodrigues' formula*, which uses a unit vector $\mathbf{a}$

in the direction of the rotation axis to construct $\mathbf{x}_\dagger = \mathbf{a}\,(\mathbf{a}\cdot\mathbf{x})$, $\mathbf{x}_\parallel = \mathbf{x} - \mathbf{a}\,(\mathbf{a}\cdot\mathbf{x})$, $\mathbf{x}_\perp = \mathbf{a}\times\mathbf{x}$, resulting in the rotation matrix

Rodrigues' formula: $\;\;[\![R]\!] = [\![\mathbf{a}]\!][\![\mathbf{a}]\!]^T + \cos\phi\,([\![1]\!] - [\![\mathbf{a}]\!][\![\mathbf{a}]\!]^T) + \sin\phi\,[\![\mathbf{a}^\times]\!]$,

where $[\![\mathbf{a}^\times]\!]$ is the matrix corresponding to the cross product operation. This is a coordinate-free specification of an operator based on geometric principles. The geometric principle may be the same as before, but note that this formula is an explicit construction rather than an automatic consequence. Unfortunately it only works in 3-D (as the use of the cross product betrays). Moreover, it constructs a matrix that only applies to vectors rather than a universal rotation operation.

We emphasize that for a rotation, the *bivector angle* $\mathbf{I}\phi$ contains all information: both the angle and the plane in which it should be measured. From this bivector angle, one can immediately construct the rotor performing the corresponding rotation. We will see a straightforward method for that in Section 7.4, and may write $R_{\mathbf{I}\phi}$ to foreshadow this.

7.2.3 A SENSE OF ROTATION

Using the transformation formula $\mathbf{x} \mapsto R\mathbf{x}\widetilde{R}$, we see that a rotor R and "minus that rotor" $(-R)$ give the same resulting rotation. This does not necessarily mean that the representation of rotations by rotors is two-valued: these rotors can be distinguished when doing relative rotations of connected objects. Such relative rotations can be achieved in two ways: by going clockwise or counterclockwise. You may think that you cannot tell from the result which it was, but it is useful to discriminate them in some applications (it can prevent you from curling up the wires on your robot). Let us call this property the *sense* of a rotation. It comes for free with the rotor representation.

We derive the rotation angle for the negative rotor $-R_{\mathbf{I}\phi}$ by rewriting it into standard form:

$$
\begin{aligned}
-R_{\mathbf{I}\phi} &= -\cos(\phi/2) + \mathbf{I}\,\sin(\phi/2) \\
&= \cos\big((2\pi + \phi)/2\big) - \mathbf{I}\,\sin\big((2\pi + \phi)/2\big) \\
&= R_{\mathbf{I}(2\pi+\phi)}.
\end{aligned}
\tag{7.5}
$$

It is now obvious that $R_{\mathbf{I}\phi}$ and $-R_{\mathbf{I}\phi}$ lead to the same result *on a vector* since a rotation over $2\pi + \phi$ is the same as a rotation over ϕ, see Figure 7.4. Yet the following real-life experiment called the *plate trick* shows that this is actually not true for connected objects.

> Hold out your hand in front of your shoulder, a hand-length away, palm upwards and carrying a plate. Now make a motion with your arm that rotates the plate horizontally in its plane over 2π. After completion, you will have your elbow sticking up awkwardly in the air. Continue the plate rotation over another 2π (you may have to wriggle your body a little to keep the plate turning in its plane). Perhaps

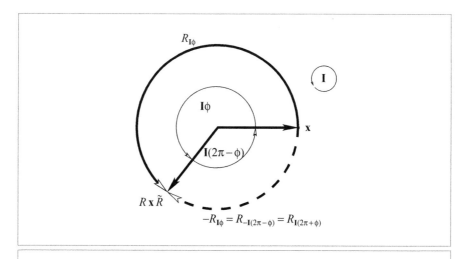

Figure 7.4: Sense of rotation.

surprisingly, both you and the plate are now back in their original position: a 4π rotation equals the identity on coupled bodies. This is a more subtle result than the usual statement: a 2π rotation equals the identity for isolated elements (like a plate by itself).

The shortest way to achieve an angle of $2\pi + \phi$ for the plate (with the same position of the elbow) is to turn the other way over $4\pi - (2\pi + \phi) = 2\pi - \phi$. Therefore it makes sense to say that $-R_{\mathbf{I}\phi}$ is a rotation over that effective angle, but in the opposite direction. That geometrical insight is confirmed by evaluating the negated rotor by a different algebraic route:

$$
\begin{aligned}
-R_{\mathbf{I}\phi} &= -\cos(\phi/2) + \mathbf{I}\ \sin(\phi/2) \\
&= \cos((2\pi - \phi)/2) + \mathbf{I}\ \sin((2\pi - \phi)/2) \\
&= R_{-\mathbf{I}(2\pi-\phi)}.
\end{aligned}
$$

This is indeed the rotation over the complementary angle $2\pi - \phi$, in the plane $-\mathbf{I}$ with opposite orientation, see Figure 7.4. Therefore we can uniquely assign the rotor's angles in the range $[0, 4\pi)$ to actual rotations of different magnitudes and senses, as in Figure 7.5.

Comparing this figure with the sign changes of the sine and cosine of half the rotation angle gives a clear test for which the rotation exactly is encoded by a given rotor R.

- The cosine $\cos(\phi/2) = \langle R \rangle_0$ changes sign as $|\phi/2|$ exceeds $\pi/2$, so exactly when the absolute value of the effective rotation angle ϕ exceeds π. A positive value of the

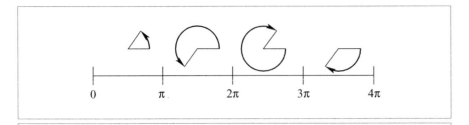

Figure 7.5: The unique rotor-based rotations in the range $\phi = [0, 4\pi)$.

scalar part of the rotor tells you that this is a rotation over the smallest angle (whether clockwise or counterclockwise).

- The sign of $\sin(\phi/2) = \langle R \rangle_2 \, \mathbf{I}$ gives the sense of rotation: positive indicates a rotation following the orientation of $\mathbf{I}$, negative follows the orientation of $-\mathbf{I}$.

The occurrence of $\mathbf{I}$ in the expression for the sense of rotation is necessary: since $\mathbf{I}$ defines what orientation we mean by (counter)clockwise, the sense of rotation should also change when we change the sign of $\mathbf{I}$.

Mathematically, it is often said that the rotors constitute a double covering of the rotation group—one physical rotation is being represented in two distinct ways (R and $-R$). We now see how this sign actually conveys geometrically significant information about the rotation *process* rather than the rotation *result*. Hestenes [29] calls the rotors *oriented rotations*, which is a good term to have. It is the half-angle representation that enables us to distinguish the various orientations.

7.3 COMPOSITION OF ROTATIONS

The composition of rotations follows automatically from their representation as a geometric product:

The rotor of successive rotations, first R_1 then R_2, is their geometric product $R_2 R_1$.

This is easily shown by associativity of the geometric product, since $R_2 (R_1 \, \mathbf{x} \, \widetilde{R}_1) \, \widetilde{R}_2 = (R_2 R_1) \, \mathbf{x} \, (R_2 R_1)\widetilde{}$. That the result is indeed a rotor follows from $(R_2 R_1)(R_2 R_1)\widetilde{} = R_2 R_1 \widetilde{R}_1 \widetilde{R}_2 = 1$.

We expand this composition in detail in this section to sharpen our intuition, both algebraically and geometrically, and to relate it to other rotation representations such as complex numbers and quaternions.

7.3.1 MULTIPLE ROTATIONS IN 2-D

If we rotate in a single plane with pseudoscalar $\mathbf{I}$, we are effectively dealing with rotations in a 2-D Euclidean subspace $\mathbb{R}^{2,0}$. Performing the 2-D rotation $R_{\mathbf{I}\phi_2}$ after $R_{\mathbf{I}\phi_1}$ results in the total rotation $R_{\mathbf{I}(\phi_2+\phi_1)}$, as you would expect. This also shows that planar rotations commute.

The algebraic demonstration is straightforward:

$$
\begin{aligned}
R_{\mathbf{I}\phi_2}\, R_{\mathbf{I}\phi_1} &= \\
&= \left(\cos(\phi_2/2) - \mathbf{I}\,\sin(\phi_2/2) \right) \left(\cos(\phi_1/2) - \mathbf{I}\,\sin(\phi_1/2) \right) \\
&= \left(\cos(\phi_2/2)\cos(\phi_1/2) - \sin(\phi_2/2)\sin(\phi_1/2) \right) \\
&\quad - \mathbf{I}\left(\cos(\phi_2/2)\sin(\phi_1/2) + \sin(\phi_2/2)\cos(\phi_1/2) \right) \\
&= \cos((\phi_2+\phi_1)/2) - \mathbf{I}\,\sin((\phi_2+\phi_1)/2) \\
&= R_{\mathbf{I}(\phi_2+\phi_1)}.
\end{aligned}
$$

Algebraically, this looks like the standard computation using a product of complex numbers, since $\mathbf{I}^2 = -1$. We prefer to view it as a calculation in the real geometric algebra of coplanar elements.

In 2-D, rotations do not actually require the rotor sandwiching product $R\mathbf{x}\widetilde{R}$ to be applied. Since any vector $\mathbf{x}$ in the $\mathbf{I}$-plane satisfies the anticommutation relationship $\mathbf{x}\mathbf{I} = -\mathbf{I}\mathbf{x}$, we can bring a rotor to the other side:

$$
\begin{aligned}
R_{\mathbf{I}\phi}\,\mathbf{x}\,R_{-\mathbf{I}\phi} &= \left(\cos(\phi/2) - \mathbf{I}\,\sin(\phi/2) \right)\mathbf{x}\left(\cos(\phi/2) + \mathbf{I}\,\sin(\phi/2) \right) \\
&= \mathbf{x}\left(\cos(\phi/2) + \mathbf{I}\,\sin(\phi/2) \right)\left(\cos(\phi/2) + \mathbf{I}\,\sin(\phi/2) \right) \\
&= \mathbf{x}\left(\cos\phi + \mathbf{I}\,\sin\phi \right) \\
&= \left(\cos\phi - \mathbf{I}\,\sin\phi \right)\mathbf{x}.
\end{aligned}
\tag{7.6}
$$

The two final lines show alternative forms for the one-sided planar rotation. We have met the final form, using left multiplication, when we did the motivating problem in Section 6.1.6.

In summary, in a plane (and in a plane only!) the half-angle rotors in the sandwiching product can be converted to whole-angle, one-sided products using either left or right multiplication.

7.3.2 REAL 2-D ROTORS SUBSUME COMPLEX NUMBERS

We have just shown how the rotation of a vector $\mathbf{x}$ in a plane $\mathbf{I}$ containing it can be simplified from the two-sided sandwiching form to a postmultiplication:

$$
\mathbf{x} \;\mapsto\; \mathbf{x}\,(\cos\phi + \mathbf{I}\,\sin\phi).
\tag{7.7}
$$

Because $\mathbf{I}^2 = -1$, this is reminiscent of complex numbers, a well-known tool to perform rotations in the complex plane. Yet our approach must be subtly different, for the vector

$\mathbf{x}$ anticommutes with the 2-blade $\mathbf{I}$ (i.e., $\mathbf{x}\mathbf{I} = -\mathbf{I}\mathbf{x}$), whereas if $\mathbf{x}$ and $\mathbf{I}$ had been complex numbers they should have commuted with each other. Also, we are in a real plane, not in a complex plane at all. What is going on here? How can such different algebras lead to the same (or at least isomorphic) results?

The answer lies in the special role of the real axis in the complex plane. The selection of such a special reference direction destroys the geometrical symmetries of the plane and changes the algebra of the symbols. Let $\mathbf{e}$ be the unit vector in the direction of the real axis, then a complex number X corresponding to the vector $\mathbf{x}$ in the plane denotes how to rotate and scale $\mathbf{e}$ to get to $\mathbf{x}$. In terms of geometric algebra, this is the ratio $X = \mathbf{x}/\mathbf{e}$ (see Section 6.1.6). So

   a complex number is a geometric ratio of a vector to a fixed vector.

To be specific, the complex number corresponding to a vector $\mathbf{a} = a_1\mathbf{e}_1 + a_2\mathbf{e}_2$ in the $\mathbf{I} \equiv \mathbf{e}_1 \wedge \mathbf{e}_2$-plane relative to the 'real axis' $\mathbf{e}_1$ is

$$A = \mathbf{a}/\mathbf{e}_1 = a_1 - a_2\,\mathbf{I}. \tag{7.8}$$

The original vector addition can of course be lifted to these complex numbers by linearity of the geometric product:

$$A + B = (\mathbf{a}/\mathbf{e}_1 + \mathbf{b}/\mathbf{e}_1) = (\mathbf{a} + \mathbf{b})/\mathbf{e}_1 = (a_1 + b_1) - (a_2 + b_2)\,\mathbf{I}.$$

Two such complex numbers multiply according to the geometric product:

$$A\,B = (a_1 - a_2\,\mathbf{I})\,(b_1 - b_2\,\mathbf{I}) = (a_1b_1 - a_2b_2) - (a_1b_2 + a_2b_1)\,\mathbf{I}.$$

This product is obviously commutative. With sum and product thus defined, our complex numbers are clearly isomorphic to the usual complex numbers and their multiplication, if you set $i \equiv -\mathbf{I}$.

Yet we will not use complex numbers to do geometry in the plane, for they lose the distinction between vectors and operators; the vectors have effectively become represented as rotation/scaling operators. The capability to describe such operators compactly was part of the attraction of complex numbers when they were first introduced. But we really want both vectors and operators in our geometry, so we want *all* elements that can be made in the basis:

$$\{1, \mathbf{e}_1, \mathbf{e}_2, \mathbf{e}_1 \wedge \mathbf{e}_2 = \mathbf{I}\}.$$

That is precisely what the geometric algebra of the plane provides, in an integrated manner that also contains the algebraic and geometrical relationships between vectors and operators. Complex numbers only use the basis $\{1,\mathbf{I}\}$. They are only half of what is required to do all of Euclidean planar geometry.

To show the power of this way of looking at complex numbers, programming exercise 7.10.5 computes the fractal Julia sets using only real vectors. That formulation makes their extension to n-dimensional space straightforward.

7.3.3 MULTIPLE ROTATIONS IN 3-D

Let us investigate what happens in Euclidean 3-D space when we perform the rotor $R_{I_2 \phi_2}$ after $R_{I_1 \phi_1}$, with different planes I_2 and I_1. It is convenient to have a shorthand for the trigonometric functions involved; let us use $c_i' = \cos(\phi_i/2)$ and $s_i' = \sin(\phi_i/2)$, with the prime to remind us of the halving of the angle. The total rotor after multiplication has only grade 0 and grade 2 terms, since grade 4 cannot exist in 3-D space. The grade-2 term, which is in general a bivector, can be written as a 2-blade, since in the geometric algebra of a 3-D space, all bivectors are 2-blades. Thus in 3-D space, we compute for the rotor composition

$$c_t' - I_t s_t' = (c_2' - I_2 s_2')(c_1' - I_1 s_1')$$
$$= c_1' c_2' + s_1' s_2' \langle I_2 I_1 \rangle_0 - c_2' s_1' I_1 - c_1' s_2' I_2 + s_1' s_2' \langle I_2 I_1 \rangle_2.$$

We have split the result in the scalar part (i.e., 0-blade) and 2-blade parts. Note how the geometric product generates five terms out of the product of two factors of two terms, since $I_2 I_1$ has both a 0-grade part and a 2-grade part (and in more than 3-D, there would even be a 4-grade part).

I_1 and I_2 are standard rotations with rotor angles of $\pi/2$ in the planes of the rotations we want to compose (they correspond to 180 degree rotations). The scalar $\langle I_2 I_1 \rangle_0$ is the cosine $c_\perp$ of the angle $\phi_\perp$ between those planes, and $\langle I_2 I_1 \rangle_2$ is the oriented plane $I_\perp$ perpendicular to both, weighted by the sine $s_\perp$ of the angle $\phi_\perp$ from I_1 to I_2. Substituting this, using nonprimed $c_\perp$ and $s_\perp$ for cosine and sine of a nonhalved angle, we get

$$c_t' - I_t s_t' = (c_1' c_2' + s_1' s_2' c_\perp) - (c_2' s_1' I_1 + c_1' s_2' I_2 - s_1' s_2' s_\perp I_\perp). \tag{7.9}$$

That gives the total rotor; if you need its plane and angle separately you should take the normalized grade-2 part and use an arc tangent function on the scaling factors of the two parts to retrieve the angle. (We will encapsulate this later in the logarithm function for 3-D rotation, in Section 10.3.3.)

We emphasize that these computations do not need to be written out explicitly. In a program, the product of two rotors is just $R_2 R_1$. We spelled them out in coordinates chiefly to convince you that this simple multiplication indeed implements all the correct details of the composition of rotations.

7.3.4 VISUALIZING 3-D ROTATIONS

Consider the equation given by the scalar part of (7.9), and write it out in full detail:

$$\cos(\phi_t/2) = \cos(\phi_1/2)\cos(\phi_2/2) + \sin(\phi_1/2)\sin(\phi_2/2)\cos(\phi_\perp).$$

This is precisely the cosine law for sides from spherical trigonometry, depicted in Figure 7.6(a). It means that we can imagine the *multiplicative* composition of rotors in 3-D as the *addition* of half-angle spherical arcs, as in Figure 7.6(b).

That is also confirmed by remembering that a rotation is a double reflection in a "vee" formed by two unit vectors in the rotation plane through the origin, separated by half the rotation angle as in Figure 7.2. The actual absolute orientation of these vectors in the plane is immaterial (as you may check; it must be rotationally invariant, since any vector out of the plane can be rotated by the construction!). Now, composing two rotations (in possibly different planes) is identical to composing two double reflections; it is natural to rotate the two vees of vectors so that the first and last vectors of both vees coincide. Then it is obvious that those two reflections cancel each other in the composition (algebraically, they are divided by themselves), while the other two remain to give the vee for the resulting rotation. To complete the visualization, surround these unit vectors by a sphere, and you see the characteristics of the rotation sphere representation: each vee of vectors determines an arc of half the rotation angle, and their composition is the completion of a spherical triangle.

The addition of freely sliding spherical arcs on great circles is such a simple means to compose rotations that it deserves to be better known, whether you use geometric algebra or not. Structural exercise 8 gives some practice in its geometry and the accompanying algebra.

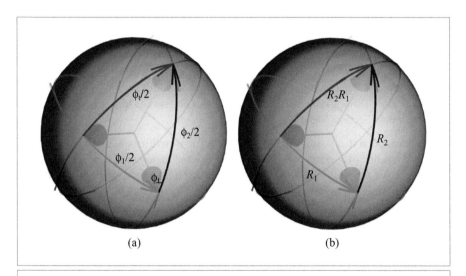

(a)   (b)

Figure 7.6: (a) A spherical triangle. (b) Composition of rotations through concatenation of rotor arcs. $R_2 R_1$ is the composite rotor of doing first R_1, then R_2, and is the arc completing the spherical triangle.

7.3.5 UNIT QUATERNIONS SUBSUMED

The rotors in 3-D space are closely related to quaternions. In our view, unit quaternions are rotors separated from their natural context in the geometric algebra of real 3-D Euclidean space $\mathbb{R}^{3,0}$. Because of their mathematical origin, people view them as imaginary, and that makes them unfortunately much more mysterious than they need to be. Identifying them with rotors helps, since those are real operators in a real vector space, with (in 2-D and 3-D) a scalar part (related to the cosine of the angle) and a 2-blade part (containing sine and rotation plane). The 2-blades have a negative square, but that does not make them imaginary. The spherical arc visualization of rotors renders them completely real, in both the English and the mathematical senses of the word. The same real visualization works for unit quaternions.

Let us spell out the correspondence between rotors and unit quaternions precisely. A *quaternion* consists of two parts, a scalar part and a complex vector part:

$$quaternion:\quad q = q_0 + \vec{\mathbf{q}}.$$

We will consider only *unit quaternions*, characterized by $q_0^2 + \|\vec{\mathbf{q}}\|^2 = 1$. The nonscalar part of a unit quaternion is often seen as a kind of vector that denotes the rotation axis, but expressed on a strange basis of complex vector quantities i, j, k that square to -1 and anticommute. For us, $\vec{\mathbf{q}}$ and its basis elements are not vectors but basis 2-blades of the coordinate planes:

$$i = -\mathbf{I}_3\,\mathbf{e}_1 = \mathbf{e}_3\mathbf{e}_2,\ j = -\mathbf{I}_3\,\mathbf{e}_2 = \mathbf{e}_1\mathbf{e}_3,\ k = -\mathbf{I}_3\,\mathbf{e}_3 = \mathbf{e}_2\mathbf{e}_1,$$

Note that $ij = k$ and cyclic, and $ijk = -1$. The three components of an element on this 2-blade basis represent not the rotation unit axis vector $\mathbf{e}$, but the rotation *plane* $\mathbf{I}$. The two are related simply by geometric duality (i.e., quantitative orthogonal complement) as

$$axis\ \mathbf{e}\ \text{to 2-blade}\ \mathbf{I}:\quad \mathbf{e} \equiv \mathbf{I}^* = \mathbf{I}\,\mathbf{I}_3^{-1},\ \text{so}\ \mathbf{e}\,\mathbf{I}_3 = \mathbf{I},$$

and their coefficients are similar, though on totally different bases (basis vectors for the axis $\mathbf{e}$, basis 2-blades for the rotation plane $\mathbf{I}$).

The standard notation for a unit quaternion $q = q_0 + \vec{\mathbf{q}}$ separates it into a scalar part and a supposedly complex vector part $\vec{\mathbf{q}}$ denoting the axis. This naturally corresponds to a rotor $R = q_0 - \vec{\mathbf{q}}\mathbf{I}_3$ having a scalar part and a 2-blade part:

$$unit\ quaternion\ q_0 + \vec{\mathbf{q}}\quad \leftrightarrow\quad rotor\ q_0 - \mathbf{q}\,\mathbf{I}_3. \tag{7.10}$$

(The minus sign derives from the rotor definition (7.4).) In the latter, $\mathbf{q}$ is now a *real* vector denoting a rotation axis. When combining these quantities, the common geometric product naturally takes over the role of the rather *ad hoc* quaternion product. We embed unit quaternions as rotors, perform the multiplication, and transfer back:

$$q\,p\ =\ (q_0 + \vec{\mathbf{q}})\,(p_0 + \vec{\mathbf{p}})\quad \text{(quaternion product!)}$$

$$\leftrightarrow (q_0 - \mathbf{q}\mathbf{I}_3)(p_0 - \mathbf{p}\mathbf{I}_3) \quad \text{(geometric product!)}$$
$$= q_0 p_0 + \langle \mathbf{I}_3 \mathbf{q} \mathbf{I}_3 \mathbf{p} \rangle_0 - (\mathbf{q}\, p_0 + q_0 \mathbf{p} - \langle \mathbf{I}_3 \mathbf{q} \mathbf{I}_3 \mathbf{p} \rangle_2 \, \mathbf{I}_3^{-1}) \, \mathbf{I}_3$$
$$= q_0 p_0 - \langle \mathbf{q}\mathbf{p} \rangle_0 - (\mathbf{q}\, p_0 + q_0 \mathbf{p} + \langle \mathbf{q}\mathbf{p} \rangle_2 \, \mathbf{I}_3^{-1}) \, \mathbf{I}_3$$
$$= q_0 p_0 - \mathbf{q} \cdot \mathbf{p} - (\mathbf{q}\, p_0 + q_0 \mathbf{p} + \mathbf{q} \times \mathbf{p}) \, \mathbf{I}_3$$
$$\leftrightarrow (q_0 p_0 - \vec{\mathbf{q}} \cdot \vec{\mathbf{p}}) + (p_0 \vec{\mathbf{q}} + q_0 \vec{\mathbf{p}} + \vec{\mathbf{q}} \times \vec{\mathbf{p}}). \tag{7.11}$$

There is one conversion step in there that may require some extra explanation: $\langle \mathbf{q}\mathbf{p} \rangle_2 \mathbf{I}_3^{-1} = (\mathbf{q} \wedge \mathbf{p})\mathbf{I}_3^{-1} = \mathbf{q} \times \mathbf{p}$, by (3.28). The inner product and cross product in (7.11) are just defined as the usual combinations of the coefficients of the complex vectors.

With the above, we have retrieved the usual multiplication formula from quaternion literature, but using only quantities from the real geometric algebra of the 3-D Euclidean space $\mathbb{R}^{3,0}$. This shows that the unit quaternion product is really just the geometric product on rotors. The quaternion product formula betrays its three-dimensional origin clearly in its use of the cross product, whereas the geometric product formula is universal and works for rotors in n-dimensional space.

The geometric algebra method gives us a more natural context to use the quaternions. In fact, we don't use them, for like complex numbers in 2-D they are only half of what is needed to do Euclidean geometry in 3-D. We really need both rotation operators and vectors, separately, in a clear algebraic relationship. The rotation operators are rotors that obey the same multiplication rule as unit quaternions; the structural similarity between (7.9) and (7.11) should be obvious. This is explored in structural exercise 10. Of course, the rule *must* be the same, since both rotors and quaternions can effectively encode the composition of 3-D rotations.

We summarize the advantages of rotors: In contrast to unit quaternions, rotors can rotate k-dimensional subspaces, not only in 3-D but even in n-dimensional space. Geometrically, they provide us with a clear and real visualization of unit quaternions, exposed in the previous section as half-angle arcs on a rotation sphere, which can be composed by sliding and addition. It is a pity that the mere occurrence of some elements that square to -1 appears to have stifled all sensible attempts to visualization in the usual approach to quaternions, making them appear unnecessarily complex. Keep using them if you already did, but at least do so with a real understanding of what they are. This is explored in Structural Exercise 10.

7.4 THE EXPONENTIAL REPRESENTATION OF ROTORS

In Section 7.3.1, we made a basic rotor as the ratio of two unit vectors, which is effectively their geometric product. Multiple applications then lead to:

> In a Euclidean space $\mathbb{R}^{n,0}$, a rotor is the geometric product of an even number of unit vectors.

The inverse of a rotor composed of such unit vectors is simply its reverse. This is not guaranteed in general metrics, which have unit vectors that square to -1. But if $R\widetilde{R}$ would be -1, R would not even produce a linear transformation, for it would reverse the sign of scalars. Therefore, we should prevent this and have as a definition for the more general spaces:

A rotor R is the geometric product of an even number of unit vectors, such that $R\widetilde{R} = 1$.

Even within those more sharply defined rotors, mathematicians such as Riesz [52] make a further important distinction between rotors that are "continuously connected to the identity" and those that are not. This property implies that some rotors can be performed gradually in small amounts (such as rotations), but that in the more general metrics there are also rotors that are like reflections and generate a discontinuous motion. Only the former are candidates for the proper orthogonal transformations that we hope to represent by rotors.

You can always attempt to construct rotors as products of vectors, but checking whether you have actually made a proper rotor becomes cumbersome. Fortunately, there is an alternative representation in which this is trivial, and moreover, it often corresponds more directly to the givens in a geometric problem. It is the exponential representation, which computes a Euclidean rotor immediately from its intended rotation plane and angle. That construction generalizes unchanged to other metrics.

7.4.1 PURE ROTORS AS EXPONENTIALS OF 2-BLADES

We have seen how in Euclidean 3-D space, a rotor $R_{\mathbf{I}\phi}$ can be written as the sum of a scalar and a 2-blade, involving a cosine and a sine of the scalar angle ϕ. We can also express the rotor in terms of its bivector angle using the *exponential form* of the rotor:

$$R_{\mathbf{I}\phi} = \cos(\phi/2) - \mathbf{I}\sin(\phi/2) = e^{-\mathbf{I}\phi/2}. \tag{7.12}$$

The exponential on the right-hand side is defined by the usual power series. The correctness of this exponential rewriting can be demonstrated by collecting the terms in this series with and without a net factor of $\mathbf{I}$. Because $\mathbf{I}^2 = -1$ in the Euclidean metric, that leaves the familiar scalar power series of sine and cosine. To show the structure of this derivation more clearly, we define $\psi = -\phi/2$.

$$\begin{aligned}
e^{\mathbf{I}\psi} &= 1 + \frac{\mathbf{I}\psi}{1!} + \frac{(\mathbf{I}\psi)^2}{2!} + \frac{(\mathbf{I}\psi)^3}{3!} + \cdots, \\
&= (1 - \frac{\psi^2}{2!} + \frac{\psi^4}{4!} - \cdots) + \mathbf{I}\,(\frac{\psi}{1!} - \frac{\psi^3}{3!} + \frac{\psi^5}{5!} - \cdots) \\
&= \cos\psi + \mathbf{I}\sin\psi.
\end{aligned} \tag{7.13}$$

After some practice, you will no longer need to use the scalar plus bivector form of the rotor $R_{\mathbf{I}\phi}$ to perform derivations but will be able to use its exponential form instead.

For instance, if you have the component $\mathbf{x}_\parallel$ of a vector $\mathbf{x}$ that is contained in $\mathbf{I}$, then $\mathbf{x}_\parallel \mathbf{I} = -\mathbf{I}\mathbf{x}_\parallel$. From this you should dare to state the commutation rule for the versor $R_{\mathbf{I}\phi}$ immediately as $\mathbf{x}_\parallel R_{\mathbf{I}\phi} = R_{-\mathbf{I}\phi}\mathbf{x}_\parallel$ and use it to show directly that

$$R_{\mathbf{I}\phi}\,\mathbf{x}_\parallel\,\widetilde{R}_{\mathbf{I}\phi} = e^{-\mathbf{I}\phi/2}\,\mathbf{x}_\parallel\,e^{\mathbf{I}\phi/2} = (\mathbf{x}_\parallel\,e^{\mathbf{I}\phi/2})\,e^{\mathbf{I}\phi/2} = \mathbf{x}_\parallel\,e^{\mathbf{I}\phi},$$

or, if you would rather, $e^{-\mathbf{I}\phi}\,\mathbf{x}_\parallel$. So within the $\mathbf{I}$-plane, the formula $\mathbf{x} \mapsto \mathbf{x}\,e^{\mathbf{I}\phi}$ performs a rotation. This result is (7.6), now with a more compact computational derivation.

Clearly, the exponential representation in (7.13) is algebraically isomorphic to the exponential representation of a unit complex number by the correspondence exposed in Section 7.3.2. The result,

$$e^{i\pi} + 1 = 0,$$

famously involving "all" relevant computational elements of elementary calculus, is obtained from (7.12) by setting $i = -\mathbf{I}$ and $\phi = 2\pi$, as $e^{-\mathbf{I}\pi} = -1$. Its geometric meaning is that a rotation over 2π in any plane $\mathbf{I}$ has the rotor -1 (not $+1$; remember the plate trick!).

7.4.2 TRIGONOMETRIC AND HYPERBOLIC FUNCTIONS

Though the motivation of the exponential form of a rotor was through Euclidean rotations, the Taylor series definition can be used in arbitrary metric spaces $\mathbb{R}^n$. When we write out the exponential $\exp(\mathbf{A})$ for a pure rotor (with $\mathbf{A}$ a 2-blade from $\bigwedge \mathbb{R}^n$) the even powers all become scalar, because a 2-blade $\mathbf{A}$ squares to a scalar in any metric (as all blades do). The odd powers become a multiple of $\mathbf{A}$ for the same reason.

In a Euclidean metric, a basic 2-blade squares to -1, and this generates the trigonometric functions sine and cosine we saw appear above. In general metrics, a 2-blade may have a positive square, or even a zero square (for a null 2-blade). Therefore, the computation will essentially reduce to some scalar power series out of the familiar list:

$$\exp x \equiv 1 + \frac{x}{1!} + \frac{x^2}{2!} + \cdots,$$

$$\sinh x \equiv x + \frac{x^3}{3!} + \frac{x^5}{5!} + \cdots,$$

$$\cosh x \equiv 1 + \frac{x^2}{2!} + \frac{x^4}{4!} + \cdots,$$

$$\sin x \equiv x - \frac{x^3}{3!} + \frac{x^5}{5!} - \cdots,$$

$$\cos x \equiv 1 - \frac{x^2}{2!} + \frac{x^4}{4!} - \cdots.$$

With this preparation we obtain, for any blade $\mathbf{A} \in \bigwedge \mathbb{R}^n$:

$$\exp(\mathbf{A}) = \begin{cases} \cos \alpha + \mathbf{A} \frac{\sin \alpha}{\alpha} = \cos \alpha + \mathbf{U} \sin \alpha & \text{if } \mathbf{A}^2 = -\alpha^2 \\ 1 + \mathbf{A} = 1 + \alpha \mathbf{U} & \text{if } \mathbf{A}^2 = 0 \\ \cosh \alpha + \mathbf{A} \frac{\sinh \alpha}{\alpha} = \cosh \alpha + \mathbf{U} \sinh \alpha & \text{if } \mathbf{A}^2 = \alpha^2 \end{cases} \tag{7.14}$$

The alternative forms pull out the unit-blade $\mathbf{U}$ in the $\mathbf{A}$ direction (so that $\mathbf{A} = \mathbf{U}\alpha$, for positive α). Note the particularly simple form for null-blades; hardly any term survives in the expansion.

We will need all of these expressions when we model Euclidean geometry in Part II. The trigonometry (for $\mathbf{A}^2 < 0$) describes the composition of Euclidean rotations, the null case ($\mathbf{A}^2 = 0$) will represent Euclidean translations, and the hyperbolic case ($\mathbf{A}^2 > 0$) will perform scalings.

7.4.3 ROTORS AS EXPONENTIALS OF BIVECTORS

Pure rotors are exponentials of 2-blades, and we have just defined them for all metrics. As the exponential representation of the 3-D Euclidean rotation is such a convenient parameterization of the rotor, the question arises whether all rotors in all spaces can be written in such a form. Since 2-blades coincide with bivectors only in 2-D and 3-D, we will at least need to admit exponentials of bivectors (rather than just 2-blades) as the general form of rotors. Could this be the most general form?

Detailed investigation shows that matters are mathematically more complicated. It is unfortunately not true that any rotor in any space can be written as the exponential of a bivector. However, Riesz [52] shows that in the Euclidean spaces $\mathbb{R}^{n,0}$ and $\mathbb{R}^{0,n}$ and in the Minkowski spaces $\mathbb{R}^{n-1,1}$ and $\mathbb{R}^{1,n-1}$ there exists a bivector B such that every orthogonal transformation $\mathsf{L}[\mathbf{x}]$ continuously connected to the identity can be written as:

$$\mathsf{L}[\mathbf{x}] = e^{-B/2} \, \mathbf{x} \, e^{B/2}. \tag{7.15}$$

So for those spaces, a rotor that is continuously connected to the identity can be expressed as the exponential of a bivector. We are fortunate that our main interests in this book are precisely these Euclidean and Minkowski spaces, for in no other spaces does this statement hold for all orthogonal transformations continuously connected to the identity, or their rotors.

Moreover, in these n-dimensional Euclidean and Minkowski spaces (and again only in those), an arbitrary bivector B can be written as the sum of commuting 2-blades. This allows us to (de)compose the bivector exponential as

$$e^{-B/2} = e^{-(\mathbf{B}_k + \cdots + \mathbf{B}_1)/2} = e^{-\mathbf{B}_k/2} \cdots e^{-\mathbf{B}_1/2}, \tag{7.16}$$

where the 2-blades $\mathbf{B}_i$ are orthogonal in the sense that they all commute. In effect, any rotor can then be made from pure rotors.

Even when you use noncommuting bivectors in the construction (7.16) of a rotor from pure rotors, the result will be a rotor (since rotors connected to the identity form a group under the geometric product). However, in general you are not allowed to add the exponents of successive exponentials in geometric algebra:

$$e^{\mathbf{B}} e^{\mathbf{A}} \neq e^{\mathbf{B}+\mathbf{A}},$$

for the series expansion of the left-hand side (which may change when $\mathbf{A}$ and $\mathbf{B}$ are swapped) is simply different than the expansion on the right (which is symmetric in $\mathbf{A}$ and $\mathbf{B}$). The terms to second order already show this:

$$
\begin{aligned}
e^{\mathbf{B}} e^{\mathbf{A}} &= (1 + \mathbf{B} + \tfrac{1}{2}\mathbf{B}^2 + \cdots)(1 + \mathbf{A} + \tfrac{1}{2}\mathbf{A}^2 + \cdots) \\
&= 1 + \mathbf{B} + \mathbf{A} + \tfrac{1}{2}(\mathbf{B}^2 + 2\,\mathbf{B}\,\mathbf{A} + \mathbf{A}^2) + \cdots \\
&\neq 1 + (\mathbf{B} + \mathbf{A}) + \tfrac{1}{2}(\mathbf{B}^2 + \mathbf{B}\,\mathbf{A} + \mathbf{A}\,\mathbf{B} + \mathbf{A}^2) + \cdots \\
&= 1 + (\mathbf{B} + \mathbf{A}) + \tfrac{1}{2}(\mathbf{B} + \mathbf{A})^2 + \cdots \\
&= e^{\mathbf{B}+\mathbf{A}}.
\end{aligned}
$$

So even to second order, $e^{\mathbf{B}+\mathbf{A}}$ would only equal $e^{\mathbf{B}}e^{\mathbf{A}}$ if $\mathbf{A}\mathbf{B} = \mathbf{B}\mathbf{A}$ (i.e., if $\mathbf{A}$ and $\mathbf{B}$ commute). However, when this condition holds, it can be shown that addition of the exponents is indeed permitted. Therefore,

$$e^{\mathbf{B}+\mathbf{A}} = e^{\mathbf{B}} e^{\mathbf{A}} \text{ if } \mathbf{A}\mathbf{B} = \mathbf{B}\mathbf{A}.$$

It is not "only if" because of some accidental exceptions involving rotations over multiples of π in properly chosen rotation planes (e.g., take $\mathbf{A} = 3\pi\,\mathbf{e}_2 \wedge \mathbf{e}_3$ and $\mathbf{B} = 4\pi\,\mathbf{e}_3 \wedge \mathbf{e}_1$; the 2-blades do not commute, but since $\exp(\mathbf{A}) = -1$ is scalar, the exponentials do).

An alternative form to (7.16) is to write the exponential as a product of vectors:

$$e^{-\mathbf{B}/2} = (\mathbf{b}_{2k}\,\mathbf{b}_{2k-1}) \cdots (\mathbf{b}_2\,\mathbf{b}_1),$$

in which the $\mathbf{b}_i$ are unit vectors (so that $\mathbf{b}_i^2 = \pm 1$), related to the 2-blades $\mathbf{B}_i$ of (7.16) by $\mathbf{B}_i = \mathbf{b}_{2i-1} \wedge \mathbf{b}_{2i}$. Still, there are subtle issues: if you use this to *construct* the exponential as a product of vectors, you may accompany an odd number of the $\mathbf{b}_i$ by a minus sign, resulting in the rotor $-e^{-\mathbf{B}/2}$ rather than $e^{-\mathbf{B}/2}$. This would still work well to represent the orthogonal transformation, since the sandwiching product in (7.15) leads to the same result. Indeed, in most spaces another bivector C can be found so that $e^{-C/2} = -e^{-\mathbf{B}/2}$ for those spaces, so this is in fact an identical construction. The only exceptions are the Minkowski spaces up to dimension 4. In $\mathbb{R}^{1,1}$, one can find no C for any B; in $\mathbb{R}^{2,1}$ and $R^{1,2}$, one can find no C for B such that $B^2 \geq 0$; and in $R^{3,1}$ and $\mathbb{R}^{1,3}$, only for $B^2 = 0$ can no C be found. The final case has some geometrical relevance in this book; it occurs as the conformal model of a 2-D Euclidean space.

In these small Minkowski spaces there apparently exist rotors that are not continuously connected to the identity. Generally, it is true that:

> *In Euclidean and Minkowski spaces, rotors connected to the identity are exponentials of bivectors.*

We have also shown the reverse statement, that in these spaces any exponential of a bivector is a rotor.

7.4.4 LOGARITHMS

Since we have the exponential expression $R = e^{-B/2}$ to make a rotor from a bivector B, we also would like the inverse: given a rotor, extract the bivector that could generate it. This would be a *logarithm* function for bivector exponentials.

Having such a logarithm is very relevant for interpolation, for it would allow us to define the Nth root of a rotor R as

$$R^{1/N} = \exp(\log(R)/N).$$

The result is a rotor that performs the rotation from X to $R X \widetilde{R}$ as N smaller rotations, which can be drawn as interpolation results:

$$R X \widetilde{R} = \left(R^{1/N} \left(R^{1/N} \cdots \left(R^{1/N} X \widetilde{R}^{1/N} \right) \cdots \widetilde{R}^{1/N} \right) \widetilde{R}^{1/N} \right) \quad (N \text{ factors in total}).$$

For 3-D rotations, we do this in Section 10.3.3. When rotors are used to represent general 3-D rigid body motions in Chapter 13, the rotor logarithm will allow us to interpolate such motions in closed form.

But in geometric algebra, logarithms are somewhat involved. One problem is that the logarithm does not have a unique value. For instance, even with a simple rotation in a single 2-blade, we have seen how $R_{\mathbf{I}\phi} = R_{\mathbf{I}(\phi+4\pi k)}$, so that we can always add a multiple of 4π to the outcome. One usually takes one value (for instance the one with the smallest norm) as the principal value of the logarithm. We will do so implicitly (some denote that principal value as $\text{Log}(R)$, with a capital L, as a reminder, but we will just use the $\log R$ notation).

A second problem is finding a closed form formula. If the bivector is a 2-blade, its exponential expansion involves standard trigonometric or hyperbolic functions, and its principal logarithm can be found using the inverse functions atan or atanh (we do this for rotors in $\mathbb{R}^{3,0}$ in Section 10.3.3). However, the general rotor is the exponent of a bivector, not a 2-blade. Since a bivector does not usually square to a scalar, there are now no simple expansions of the exponential, and many mixed terms result. If we want to get back to the basic trigonometric or hyperbolic functions (to get geometrically significant parameters like bivector angles, translation vectors, and scalings), we then need to factorize the total expression. That would effectively split the bivector into mutually commuting 2-blades with sensible geometric meaning, and would make the logarithm extractable in closed form. Unfortunately this factorization is hard to do in general. In this book, we will derive specific formulas for specific transformations we encounter (Euclidean rotations, Euclidean rigid body motions, rigid body motions with positive scaling) in the appropriate chapters of Part II.

7.5 SUBSPACES AS OPERATORS

The rotations we have just treated so extensively are generated as an even number of hyperplane reflections. We now study the reflections in general subspaces. Like rotors, they also employ a sandwiching product, effectively using subspaces as operators on other subspaces. The analysis reveals that we need to keep track of how a blade represents a subspace (dually or directly) to process it correctly. Our understanding of general reflections then allows us to specify the conditions for containment and perpendicularity of subspaces in n-dimensional space as compact commutation relationships.

More patterns appear: projections to subspaces can also be written as sandwiching, but now use the contraction. And operators may be transformed like objects: the reflection of a rotation operator in the motivating example of Chapter 1 now finds its justification.

7.5.1 REFLECTION BY SUBSPACES

If we have a blade $\mathbf{A}$ representing a subspace, the reflection in it should invert the rejection of a vector by that blade, so

$$\text{reflection of } \mathbf{x} \text{ in subspace } \mathbf{A}: \ \mathbf{x} \mapsto \mathbf{x} - 2(\mathbf{x} \wedge \mathbf{A})\,\mathbf{A}^{-1} = -\hat{\mathbf{A}}\,\mathbf{x}\,\mathbf{A}^{-1}.$$

Extending this as an outermorphism, each grade in $\mathbf{X}$ contributes a factor $(-1)^{a+1}$, for a total formula that reads

$$\text{reflection of } \mathbf{X} \text{ in subspace } \mathbf{A}: \ \mathbf{X} \mapsto (-1)^{x(a+1)}\,\mathbf{A}\,\mathbf{X}\,\mathbf{A}^{-1},$$

with $x = \text{grade}(\mathbf{X})$ and $a = \text{grade}(\mathbf{A})$. We can use a subspace in this manner as a reflector.

The resulting equation does not match our earlier formula for the hyperplane in (7.2), since we characterized that by its dual $\mathbf{a} = \mathbf{A}^*$. So let us derive the formula for such a reflection in a dually represented blade as well, setting $\mathbf{D} = \mathbf{A}\,\mathbf{I}_n^{-1}$.

$$(-1)^{x(a+1)}\,\mathbf{A}\,\mathbf{X}\,\mathbf{A}^{-1} = (-1)^{x(a+1)}\,\mathbf{D}\,\mathbf{I}_n\,\mathbf{X}\,\mathbf{I}_n^{-1}\,\mathbf{D}^{-1}$$
$$= (-1)^{x(n+a+2)}\,\mathbf{D}\,\mathbf{X}\,\mathbf{D}^{-1}$$
$$= (-1)^{xd}\,\mathbf{D}\,\mathbf{X}\,\mathbf{D}^{-1},$$

since $d = \text{grade}(\mathbf{D}) = \text{grade}(\mathbf{A}\rfloor\mathbf{I}_n^{-1}) = n - \text{grade}(\mathbf{A}) = n - a$. We have thus found

$$\text{reflection of } \mathbf{X} \text{ in dual subspace } \mathbf{D}: \ \mathbf{X} \mapsto (-1)^{xd}\,\mathbf{D}\,\mathbf{X}\,\mathbf{D}^{-1}.$$

This matches (7.2) when $\mathbf{D}$ is a vector. It is an instance of the general sandwiching formula involving geometric products, since we could have written the dual blade as a geometric product of orthogonal factors using the Gram-Schmidt orthogonalization procedure. Incidentally, this also shows geometrically why a blade should square to a scalar: double reflection must be the identity, and that is represented by a scalar rotor.

These reflections of oriented subspaces are illustrated in Figure 7.7. The difference in the formulas for the different characterizations (direct or dual) of the mirrors imply that our software will apparently need to realize whether a mirroring blade is given in its direct representation or in its dual representation when we perform a reflection. The same is true for the blade $\mathbf{X}$ that gets reflected, if you want duality relative to the original pseudoscalar rather than to the reflected pseudoscalar. We ask you to derive the proper expressions yourself in structural exercise 11. The set of equations that results for the reflection operator is collected in Table 7.1.

That is how involved the geometry of reflection is when you want to keep track of the orientation of the spaces. However, realize the power inherent in these formulas: we can now *reflect any oriented subspace into any subspace within a space of any dimensionality*, and get a result of the correct attitude, magnitude, and orientation. That is worth a bit of precision in the administration. Of course, if you don't want to keep track of the orientation, the formulas all become identical to $\mathbf{A}\mathbf{X}\mathbf{A}^{-1}$.

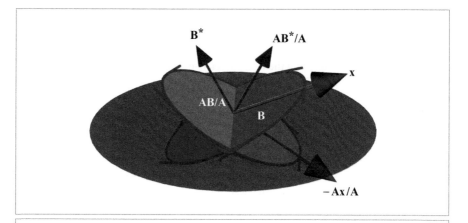

Figure 7.7: A plane acting as a reflector of oriented subspaces. The reflection of a direct blade $\mathbf{X}$ in a subspace $\mathbf{A}$ is $(-1)^{x(a+1)}\mathbf{A}\mathbf{X}/\mathbf{A}$. This formula gives a different sign for the red vector $\mathbf{x}$ and the yellow bivector $\mathbf{B}$. The reflection of a dual element $\mathbf{Y}$ in $\mathbf{A}$ is $\mathbf{A}$ is $(-1)^{(y+1)(a+1)}\mathbf{A}\mathbf{X}/\mathbf{A}$. This implies that the blue normal vector $\mathbf{b} = \mathbf{B}^*$ reflects differently from the regular vector $\mathbf{x}$ to correctly remain the normal vector of the reflected $\mathbf{B}$.

Table 7.1: Reflection of an oriented subspace $\mathbf{X}$ in a subspace $\mathbf{A}$. When either is represented dually rather than directly, the signs change as indicated. When the dual representation $\mathbf{Y} = \mathbf{X}^*$ is the input, one usually desires to have the outcome also in dual form relative to the same original unreflected pseudoscalar. That result has been indicated, where $a = \text{grade}(\mathbf{A})$, $d = \text{grade}(\mathbf{D})$, $x = \text{grade}(\mathbf{X})$. $y = \text{grade}(\mathbf{Y})$. (Duality with respect to the reflected pseudoscalar is formula-preserving and would obey the first column with $\mathbf{X} \to \mathbf{Y}$ and $x \to y$.)

	$\mathbf{X}$ direct	$\mathbf{Y} = \mathbf{X}^*$ dual
$\mathbf{A}$ direct	$(-1)^{x(a+1)}\,\mathbf{A}\,\mathbf{X}\,\mathbf{A}^{-1}$	$(-1)^{(y+1)(a+1)+(n-1)}\,\mathbf{A}\,\mathbf{Y}\,\mathbf{A}^{-1}$
$\mathbf{D} = \mathbf{A}^*$ dual	$(-1)^{xd}\,\mathbf{D}\,\mathbf{X}\,\mathbf{D}^{-1}$	$(-1)^{(y+1)d}\,\mathbf{D}\,\mathbf{Y}\,\mathbf{D}^{-1}$

7.5.2 SUBSPACE PROJECTION AS SANDWICHING

In the reflection formula

$$\mathbf{X} \;\mapsto\; (-1)^{x(a+1)}\mathbf{A}\,\mathbf{X}\,\mathbf{A}^{-1},$$

the subspace $\mathbf{A}$ acts on the subspace $\mathbf{X}$ as a reflector. The subspace $\mathbf{A}$ is then effectively used as an operator, using the geometric product in its sandwiching.

In the same abstract sense, a subspace can act as an orthogonal projector. We have seen the formula for that in Section 3.6 as $(\mathbf{X}\rfloor\mathbf{A})\rfloor\mathbf{A}^{-1}$. That formula can actually also be written in sandwich form, but now using the (nonassociative) contractions instead of the (associative) geometric product. This even includes the necessary orientation sign above, by courtesy of (3.19):

$$\mathbf{X} \;\mapsto\; (\mathbf{X}\rfloor\mathbf{A})\rfloor\mathbf{A}^{-1} = (-1)^{x(a+1)}(\mathbf{A}\lfloor\mathbf{X})\rfloor\mathbf{A}^{-1} = (-1)^{x(a+1)}\mathbf{A}\lfloor(\mathbf{X}\rfloor\mathbf{A}^{-1}).$$

In this sense, we can switch from a subspace used as a reflector to a subspace used as a projector by replacing the geometric products with the contractions. Both sandwiching operators are grade-preserving. The algebraic properties of geometric product and contraction lead to different properties on repetition, for the reflector is an involution (doing it twice is the identity), while the projector is idempotent (doing it twice is like doing it once).

7.5.3 TRANSFORMATIONS AS OBJECTS

We just saw how subspaces can be operators acting through the sandwiching product. Conversely, operators transform as subspaces.

For instance, we may want to rotate the plane I of a rotation $R_1 = R_{I\phi} = \cos(\phi/2) - I \sin(\phi/2)$ to become a different rotation plane—such nested rotations are common in robotics and hierarchical modeling, where the shoulder rotates the elbow rotation, which in turn rotates the wrist rotation. What is the rotor of this new rotation of R_2 applied to R_1? It is not $(R_2 R_1)$, for that would merely apply the rotation R_2 after we have applied R_1.

The clue is that the rotation R_2 should rotate the I plane of R_1 to become $R_2 I \widetilde{R}_2$. That makes the new rotor

$$R'_1 = \cos(\phi/2) - (R_2\, I\, \widetilde{R}_2)\, \sin(\phi/2) = R_2\big(\cos(\phi/2) - I \sin(\phi/2)\big)\, \widetilde{R}_2 = R_2\, R_1 \widetilde{R}_2,$$

where we used the commutativity of scalars to absorb all terms under the application of R_2. The result is that the rotor R_1 is rotated by R_2, precisely as the phrasing of the problem suggested. (Structural exercise 9 should illustrate this on the spherical image.) Comparing to (7.3), we see that *rotors can be rotated just like subspaces* or any other element of the algebra. We could actually have derived that immediately, by using the linearity of the outermorphism, but is good to have a geometrical argument for this algebraic result.

The same reasoning and derivation holds for reflections. We can even reflect rotation operators, as the example in Chapter 1 showed. The result of the reflection of a rotor $R_{I\phi}$ in a hyperplane dually characterized by $\mathbf{n}$ is a rotation in the reflected plane:

$$\mathbf{n}\, R_{I\phi}\mathbf{n}^{-1} = \mathbf{n}\left(\cos(\phi/2) - I \sin(\phi/2)\right)\mathbf{n}^{-1}$$
$$= \cos(\phi/2) - (\mathbf{n}\, I\, \mathbf{n}^{-1})\, \sin(\phi/2)$$
$$= R_{\mathbf{n}I\mathbf{n}^{-1}\phi}.$$

We may summarize these principles as

> *concatenated transformations use the geometric product, but nested transformations use the sandwiching product.*

If you compare this to linear algebra, you know that concatenated rotations would be done by a total rotation matrix that is the product of the successive rotation matrices: $[\![R_2]\!]\, [\![R_1]\!]$, whereas a nested transformation requires a sandwich: $[\![R_2]\!]\, [\![R_1]\!]\, [\![R_2]\!]^{-1}$, so that is similar. One would then apply this to a vector $\mathbf{x}$ as $[\![R]\!]\, [\![\mathbf{x}]\!]$, whereas in geometric algebra the application to a general element X (vector, operator, etc.) would be $R X R^{-1}$. So in linear algebra, the application to an object obeys the same rule as concatenation of operators, whereas in geometric algebra it is like their nesting.

7.6 VERSORS GENERATE ORTHOGONAL TRANSFORMATIONS

We have seen a single reflection in a hyperplane, and how an even number of successive reflections generates a rotation. An odd number of reflections is a rotation-plus-reflection, sometimes called an antirotation. All are of the form $X \mapsto \pm V X V^{-1}$. We call such a sandwiching a versor product, and the element V a versor. Since these operations are so powerful, it pays to analyze them in more detail. We especially need to be careful about their signs (as the analysis of reflections already showed), and we are interested in their structural properties.

7.6.1 THE VERSOR PRODUCT

The subsequent application of sandwiching products $\mathbf{x} \mapsto -\mathbf{v}\,\mathbf{x}\,\mathbf{v}^{-1}$ using the vectors $\mathbf{v}_1, \mathbf{v}_2, \cdots, \mathbf{v}_k$ leads to an overall operation that is

$$\mathbf{x} \;\mapsto\; (-1)^k\, \mathbf{v}_k \cdots \mathbf{v}_2\, \mathbf{v}_1\, \mathbf{x}\, \mathbf{v}_1^{-1}\, \mathbf{v}_2^{-1} \cdots \mathbf{v}_k^{-1}.$$

Let us define a *k-versor* as an element of the geometric algebra $\bigwedge \mathbb{R}^n$ that can be obtained by multiplying k vectors using the geometric product:

$$versor: \quad V = \mathbf{v}_k \cdots \mathbf{v}_2\, \mathbf{v}_1.$$

Then we can write the total sandwiching product on the vector $\mathbf{x}$ as

$$\mathbf{x} \mapsto \widehat{V}\,\mathbf{x}\,V^{-1}, \tag{7.17}$$

where $\widehat{V}$ is the grade involution of V, equal to $+V$ if k is even and $-V$ when k is odd. The inverse of V is of course simply obtained by the inverse vector factors in opposite order:

$$V = \mathbf{v}_k \cdots \mathbf{v}_2\, \mathbf{v}_1, \quad \text{then} \quad V^{-1} = \mathbf{v}_1^{-1}\, \mathbf{v}_2^{-1} \cdots \mathbf{v}_k^{-1}.$$

For a unit V, the inverse is the reverse. Null vectors (which square to zero; see Section A.4 in Appendix A) cannot be used as factors in this operation, for they do not have inverses.

The operation of (7.17) is a linear transformation on $\mathbf{x}$, and it can be extended as an outermorphism to produce a general form that we call the *versor product* $V : \bigwedge^k \mathbb{R}^n \to \bigwedge^k \mathbb{R}^n$ on a blade $\mathbf{X}$:

$$versor\ product\ of\ V\ on\ \mathbf{X}: \quad \mathbf{X} \mapsto V[\mathbf{X}] \equiv (-1)^{xv}\, V\,\mathbf{X}\,V^{-1}. \tag{7.18}$$

Here $x \equiv \mathrm{grade}(\mathbf{X})$ and $v \equiv \mathrm{grade}(V)$. The latter is a slight abuse of notation, since V does not have a unique grade; it is allowed since only the parity of v matters, and this is all odd or all even for a versor.

Linearity permits us to extend this definition beyond blades to a general multivector X, where we have to take the sum over the grade parts of X. If the grades are mixed, this cannot be simplified, but if the grades of X are all odd or even (and this is typically the case), this is a simple generalization of (7.18): substitute the general X for the blade $\mathbf{X}$.

Versors clearly multiply by the geometric product, since they are themselves constructed as the geometric product of vectors, and their corresponding versor transformations compose naturally:

$$\widehat{V}_2 \, (\widehat{V}_1 \, \mathbf{x} \, V_1^{-1}) \, V_2^{-1} = (\widehat{V}_2 \, \widehat{V}_1) \, \mathbf{x} \, (V_2 \, V_1)^{-1} = (V_2 \, V_1)\widehat{} \, \mathbf{x} \, (V_2 \, V_1)^{-1}.$$

And vice versa: *the versor of a composition of operators is the geometric product of their versors.* Thus versors reveal the true geometrical meaning of the algebraically introduced geometric product: it multiplies geometrical operators.

7.6.2 EVEN AND ODD VERSORS

We have seen that some versors represent reflections and that two reflections make a rotation. The different geometrical feeling between these kinds of operations is how they treat handedness; rotations preserve it, while reflections manage to turn a right hand into a left hand. This is well represented in the signs of the determinants of their transformations if we view them as linear transformations, both in geometric algebra and classically. But the versor representation makes distinguishing them even simpler; the important difference is between *odd and even versors* (i.e., versors made up of an odd or even number of reflections). This is precisely the difference between an odd and even number of vector factors in the versor. And since the geometric product contains only all even or all odd terms, this corresponds to the oddness or evenness of their grades. All odd-grade versors are reflections, and all even-grade versors are rotations in their vector space $\mathbb{R}^n$. This is easily proved using (7.18)

$$\det(\mathsf{V}) = \mathsf{V}[\mathbf{I}_n]/\mathbf{I}_n = (-1)^{n\nu} \, V \mathbf{I}_n \, V^{-1} \, \mathbf{I}_n^{-1} = (-1)^{n\nu+(n+1)\nu} \, V V^{-1} \, \mathbf{I}_n \, \mathbf{I}_n^{-1} = (-1)^\nu.$$

This result is independent of the metric of the space $\mathbb{R}^n$.

Because of this difference in properties, it makes sense to have the versor product for even and odd versors listed separately:

$$even \; versors : \mathbf{X} \mapsto V \mathbf{X} \, V^{-1} \tag{7.19}$$

$$odd \; versors : \mathbf{X} \mapsto V \widehat{\mathbf{X}} \, V^{-1} \tag{7.20}$$

As before, these formulas even apply to arbitrary multivectors X. For even versors, this is the substitution of $\mathbf{X}$ by X; for odd versors, one should sum over the results for odd and even grades of X separately—the grade involution takes care of the proper signs of the various grade parts of X.

7.6.3 ORTHOGONAL TRANSFORMATIONS ARE
VERSOR PRODUCTS

Not every linear transformation can be written in the form of a versor product. In fact, the ones that can are precisely the *orthogonal transformations*. You may already suspect this from their determinants, but we can also show this more directly. The crucial property is to verify what happens to an inner product of vectors after transformation:

$$\mathsf{V}[\mathbf{x}] \cdot \mathsf{V}[\mathbf{y}] = (-\widehat{V}\,\mathbf{x}\,V^{-1}) \cdot (-\widehat{V}\,\mathbf{y}\,V^{-1})$$
$$= \langle V\mathbf{x}\,V^{-1}\,V\mathbf{y}\,V^{-1}\rangle_0 = \langle V\mathbf{x}\mathbf{y}\,V^{-1}\rangle_0 = \langle \mathbf{x}\mathbf{y}\,V^{-1}\,V\rangle_0 = \langle \mathbf{x}\mathbf{y}\rangle_0$$
$$= \mathbf{x} \cdot \mathbf{y} = \mathsf{V}[\mathbf{x} \cdot \mathbf{y}].$$

The inner product is preserved, so this is an orthogonal transformation. The derivation is easily reversed to show that any such linear transformation can be written as a versor product.

Because the versor product is an orthogonal transformation, it transforms the contraction in a structure-preserving manner. (The adjoint equals the inverse, which is represented by the reverse of the versor, modulo an irrelevant scalar factor; see structural exercise 5.) And because the versor product is a linear transformation, it can be extended as an outermorphism—therefore it also preserves the outer product. In fact, *the versor product preserves the geometric product*. For even versors, this is a one-liner:

$$\mathsf{V}[\mathbf{A}]\,\mathsf{V}[\mathbf{B}] = V\mathbf{A}\,V^{-1}\,V\mathbf{B}\,V^{-1} = V\mathbf{A}\mathbf{B}\,V^{-1} = \mathsf{V}[\mathbf{A}\mathbf{B}],$$

and for odd versors it is not much harder to prove once you realize that $\widehat{\mathbf{A}}\,\widehat{\mathbf{B}} = (\mathbf{A}\mathbf{B})\widehat{}$.

Since a versor product is also grade-preserving, all constructions that are made as grade selections of a geometric product are obviously preserved by the versor product. This includes all subspace products:

$$\mathsf{V}[\mathbf{A}\mathbf{B}] = \mathsf{V}[\mathbf{A}]\,\mathsf{V}[\mathbf{B}]$$
$$\mathsf{V}[\mathbf{A} \wedge \mathbf{B}] = \mathsf{V}[\mathbf{A}] \wedge \mathsf{V}[\mathbf{B}]$$
$$\mathsf{V}[\mathbf{A} * \mathbf{B}] = \mathsf{V}[\mathbf{A}] * \mathsf{V}[\mathbf{B}] \quad (= \mathbf{A} * \mathbf{B})$$
$$\mathsf{V}[\mathbf{A}\rfloor\mathbf{B}] = \mathsf{V}[\mathbf{A}] \rfloor \mathsf{V}[\mathbf{B}]$$
$$\mathsf{V}[\mathbf{A}\lfloor\mathbf{B}] = \mathsf{V}[\mathbf{A}] \lfloor \mathsf{V}[\mathbf{B}]$$

Such structure-preservation properties easily extend to functions of multivectors, notably the exponential of bivectors:

$$\mathsf{V}[\exp(B)] = \exp(\mathsf{V}[B]). \tag{7.21}$$

So, *the transformation of a rotor can be found by transforming its bivector*. We will use this frequently, so it is good to convince you explicitly why this holds. Since B is a bivector, we have no bothersome signs:

$$\exp(\mathsf{V}[B]) = 1 + VB\,V^{-1} + \tfrac{1}{2!}(VB\,V^{-1})^2 + \cdots$$
$$= 1 + VB\,V^{-1} + \tfrac{1}{2!}VB\,V^{-1}\,VB\,V^{-1} + \cdots$$

$$\begin{aligned}
&= 1 + V B V^{-1} + \tfrac{1}{2!}(V B^2 \, V^{-1}) + \cdots \\
&= V (1 + B + \tfrac{1}{2!} B^2 + \cdots) \, V^{-1} \\
&= V [\exp(B)],
\end{aligned}$$

so (7.21) is simply a consequence of the structure preservation of the geometric product by the versor product.

This property of structure preservation makes versor-based transformations easy to work with. In fact, we are going to make it the basis of all geometrical computations in geometric algebra by choosing a proper space to represent geometries in. We call these *operational models* of the geometries, and will especially develop the operational models $\mathbb{R}^{n+1,1}$ of a Euclidean space $\mathbb{E}^n$ in Chapters 13 and 16. In that model, all Euclidean transformations (including translations) are orthogonal transformations encoded by versors.

7.6.4 VERSORS, BLADES, ROTORS, AND SPINORS

We now have many elements that can be used in the versor-type sandwiching products.

- **Versor**. A versor is a geometric product of invertible vectors.
- **Rotor**. A rotor R is a geometric product of an even number of unit vectors such that $R^{-1} = \widetilde{R}$. It can be written as the exponential of a bivector in most spaces of interest (see Section 7.4.3 for the fine print).
- **Blade**. A blade is an outer product of vectors. If it is to be used in a reflection operation, it uses a sandwiching product, and therefore it should be invertible. Invertible blades can always be written as a geometric product of mutually orthogonal vectors (by the Gram-Schmidt procedure).

Therefore we have the following relationships:

All invertible blades are versors (but few versors are blades).
Rotors are even unit versors (whose inverse is their reverse), and vice versa.
All even unit blades whose inverse is their reverse are rotors, but few rotors are blades.

A prototypical case of a blade acting in a versor product is the 2-blade $\mathbf{I} = \mathbf{e}_1 \wedge \mathbf{e}_2$ in $\mathbb{R}^{n,0}$. If it is the special case of a rotor (for rotor angle $-\pi$), it generates the rotation $\mathbf{x} \mapsto \mathbf{I} \mathbf{x} \widetilde{\mathbf{I}}$ over $-\pi$ in the $\mathbf{I}$-plane. The same blade could also be used as a reflector; then it would generate $-\mathbf{I} \mathbf{x} \widetilde{\mathbf{I}}$.

In the literature of mathematical physics there are elements called *spinors*, traditionally associated with the description of rotations in quantum mechanics. These are closely related to rotors. It is useful to understand this link, since some of the spinor literature is relevant to geometry.

Spinors are not introduced as geometric products of vectors, but as elements that preserve grade under a sandwiching product in a Clifford algebra. Consider the set of elements S

that can transform a vector $\mathbf{x}$ into a vector by the operation $S\mathbf{x}S^{-1}$. (This is called the *Clifford group*.) When such elements are normalized to $S\widetilde{S} = \pm 1$ and of even grades, they are called spinors, making up a *spin group* (though some authors appear to permit odd spinors as well [51]).

The *special spin group* is the subgroup of the spin group consisting of the elements for which $S\widetilde{S} = +1$. Its elements are most closely related to the rotors, but careful study shows (see e.g., [33], pg. 106) that there are some special spinors that are not rotors. They consist of the weighted sum of a rotor and its dual, but they are rare (they only occur in spaces whose dimensionality mod 4 equals 0). So it is almost true that "special spinor" and "rotor" are equivalent terms. In summary:

All rotors are special spinors; almost all special spinors are rotors.

This way of looking at rotors is interesting, for it casts a slightly different light on their exponential representation. Can we show that the exponential of a general bivector, when used in a versor product, is indeed a transformation from vectors to vectors? Let us expand the exponentials:[1]

$$
\begin{aligned}
e^{-B/2}\,\mathbf{x}\,e^{B/2} &= (1 - \tfrac{1}{2}B + \tfrac{1}{2}(B/2)^2 + \cdots)\,\mathbf{x}\,(1 + \tfrac{1}{2}B + \tfrac{1}{2}(B/2)^2 + \cdots) \\
&= \mathbf{x} - \tfrac{1}{2}B\mathbf{x} + \tfrac{1}{2}\mathbf{x}B + \tfrac{1}{8}\left(B^2\mathbf{x} - 2B\mathbf{x}B + \mathbf{x}B^2\right) + \cdots \\
&= \mathbf{x} + (\mathbf{x}\rfloor B) + \tfrac{1}{2!}((\mathbf{x}\rfloor B)\rfloor B) + \tfrac{1}{3!}(((\mathbf{x}\rfloor B)\rfloor B)\rfloor B) + \cdots .
\end{aligned}
\tag{7.22}
$$

The result clearly produces a vector: each contraction of a vector with a bivector gives a vector, so the successive nestings keep producing vector terms. For a simple rotation represented by a Euclidean 2-blade $\mathbf{B} = \mathbf{I}\phi$, we depict the series of terms in Figure 7.8; it generates increasingly accurate approximations for the rotation result, correct in magnitude and geometry. Each subsequent contraction by $\mathbf{I}\phi$ rotates and scales the previous contribution. Note that only the first term $\mathbf{x}$ may contain a component $(\mathbf{x}\wedge\mathbf{B})/\mathbf{B}$ that is not contained in $\mathbf{B}$.

7.7 THE PRODUCT STRUCTURE OF GEOMETRIC ALGEBRA

7.7.1 THE PRODUCTS SUMMARIZED

We now have a number of products with geometrical meanings for subspaces and their operators. They are all based on the geometric product. We have the *outer product* to span subspaces. We have the *scalar product* to compute norms and angles between subspaces

---

1　If the grouping of the elements into compact contractions seems inspired, there will be a more structural way of deriving this equation when you have learned to differentiate in Chapter 8.

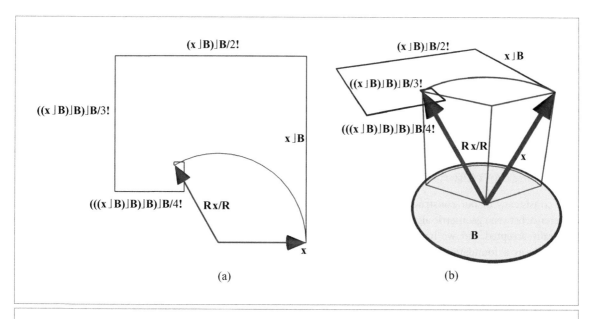

Figure 7.8: The rotor product in Euclidean spaces as a Taylor series (a) in 2-D, and (b) in 3-D. The subsequent terms are denoted by the blue lines, converging to the rotation result.

of the same grade. It is subsumed by the *contractions*, which extends this capability to different grades. Then there is the *versor product*, which can apply subspaces as operators acting on other subspaces to produce reflections and rotations. Sandwiching using the contractions produces projection operators. Finally, there is the *geometric product*, which acts as a multiplication of versor operators and as the foundation of the whole system. The basic principles by which these varied operators are constructed is always the geometric product and (anti-)commutation combined with addition or grade selection.

All these products are bilinear and distributive over addition. We have met two more products that are of geometrical significance, but which only have these properties in a piecewise manner: meet and join. To retain their meaning of "intersection" and "union", these can only be applied to blades, and should adapt themselves in a nonlinear manner to the geometric degeneracy of their arguments. This makes them algebraically less tidy than the basic products above.

With this collection of products, the foundation of geometric algebra is virtually complete (only one more operation and product will be introduced in the next chapter: differentiation and the associated commutator product with a bivector). Any element or operation from linear algebra can now be substituted by a corresponding element and coordinate-free operator from geometric algebra. For simplicity of structure and universality of code, this is always advantageous, though it may come at a computational price—we treat that issue briefly below (Section 7.7.3) and extensively in Part III.

The algebraic foundation by itself cannot be applied immediately to geometric problems in applications: a modeling step is required to identify the proper algebraic concepts to encode features of the situation. For Euclidean, affine, and projective geometry, there are standard recommended ways of modeling. These are explained in Part II, which is essential reading if you want to use geometric algebra effectively.

7.7.2 GEOMETRIC ALGEBRA VERSUS CLIFFORD ALGEBRA

The consistency of our constructions so far allows us to express our opinion on the difference between geometric algebra and Clifford algebra. The following is by no means generally accepted, but we have found it a useful distinction for practical purposes, especially as a foundation for developing efficient implementations for the various admissible operations in Part III.

- *Clifford algebra* is defined in the same multivector space $\bigwedge \mathbb{R}^n$ of a metric space $\mathbb{R}^n$ as geometric algebra. It has the same definition of the geometric product to construct elements from other elements. It moreover permits us to construct elements by a *universal addition*, also defined between any two elements.

- By contrast, in our view of *geometric algebra* we only permit *exclusively multiplicative* constructions and combinations of elements. The obvious exceptions to this are the two basis elements in the whole construction: the vector space and its field $\mathbb{R}$, which were linear from the start, and their duals (since duality is an isomorphic construction). Thus the only elements in the geometric algebra $\bigwedge \mathbb{R}^n$ that we allow to be added constructively are of grade 0 (scalars), grade 1 (vectors), grade $(n-1)$ (covectors), and grade n (pseudoscalars).

Of course, many of the products in geometric algebra are bilinear and allow generalization over addition through their distributivity. But we view that additive structure only as convenient for the *decomposition* of those products, never as a *construction* of new elements. The distributivity property is convenient in implementations, since it allows the representation of an arbitrary element on a basis. We then store the coefficients it has on that basis, and are allowed to reconstruct the element by recomposing the terms, but never should we play the game of making new elements by adding arbitrarily weighted basis elements, as in Clifford algebra. The reason is simply that we have no geometric interpretation for such elements.

By contrast, all elements produced by multiplication using any of our products do have a geometrical interpretation. The *blades* among them, from the subalgebra involving only the inner and outer products (and of course including duality, `meet`, and `join`) are clearly subspaces. They can even be drawn. The elements involving the geometric product are *versors* representing orthogonal transformations, and they act on the subspace elements through the versor product to again provide drawable elements. There is therefore never a doubt about the geometrical nature of any of the multiplicatively constructed elements.

A similar contrast exists between Grassmann algebra, which permits arbitrary addition, and what we called *subspace algebra* in our early chapters, permitting only the multiplicative constructions. Unfortunately, mathematics has developed the additive Grassmann and Clifford algebras to a much greater extent than their multiplicative parts. Much of that work is irrelevant to their geometrical usage. It may even be incomplete, for what would have been good and useful geometrical theorems may not be stated because they are not generally valid when addition is allowed, and therefore are considered less pure. When consulting the mathematical literature, be on the lookout for results on "simple" multivectors, which require factorizability by the outer product (and are therefore about blades) or the geometric product (and are thus about versors).

The sole exception we have been forced to make so far to our multiplicative principle involves exponentiation. When one multiplies two rotors, a rotor results. Starting from rotors that can be represented as the exponentials of 2-blades, we can construct the exponentials of general bivectors as geometrically significant operators. Permitting their logarithm as an operation in our algebra, we should therefore *permit general addition of bivectors* as a constructive operation—but the resulting elements may then subsequently only be used for exponentiation. As such, this is still our multiplicative principle, merely expressed in logarithmic form.

In the grade approach to geometric algebra, the multiplicative principle is much less easily formulated. The direct translation from the subspace product motivation would permit us only to use the grade parts that define those products; that is the limited set exposed in Section 6.3.2. Beyond that, there may be more geometrically relevant grades (such the minimum and maximum grade of a geometric product that we will meet in Section 21.7), but the general issue of admissible grades in the multiplicative principle has not yet been thoroughly explored.

The *multiplicative principle* is beginning to be acknowledged by mathematicians. It will be interesting to see whether this will unearth hitherto dormant results in Clifford algebra with patently geometrical applications. For now, we have made the multiplicative principle the basis of our implementation (and it is part of the reason why it is among the fastest known). We have not yet encountered geometrical situations that we cannot represent and process.

7.7.3 BUT—IS IT EFFICIENT?

The use of blades as elements of computation and of rotors as operators to perform orthogonal transformations on them permits us to encode a lot of geometry in a compact, coordinate-free, and universal manner. As a consequence, we need to distinguish fewer data types in geometric constructions. That in turn simplifies the flow of algorithms. The resulting code looks a lot more compact and readable, since all operators can be encoded in terms of geometrical elements of the application, rather than in unrelated coordinate systems. But is that code also more efficient?

This question is not easily answered. There are many facets to the issue because there are many different kinds of operations in a geometry, and the balance may differ per application. When we do a practical comparison for a ray tracer in Part III (see Table 22.1), the fastest implementation of geometric algebra is 25 percent slower than an optimized, explicitly written-out classical implementation. That cost of geometric algebra is about the same as the performance that can be achieved by the currently commonly used homogeneous coordinates and quaternions (which provide much less universality than geometric algebra). We believe a 5 to 10 percent overhead for the use of geometric algebra should be achievable in such applications. This would be an acceptable price to pay for the much cleaner structure of the code, with a much reduced number of data types and an elimination of the corresponding special operations that need to be explicitly defined on them.

Let us briefly discuss some of the relevant issues in making an efficient implementation of geometric algebra, with special emphasis on the orthogonal transformation of blades, which will be the structural backbone of geometric modeling in Part II.

- *For the composition of orthogonal transformations, rotors are superior in up to 10 dimensions*. The geometric algebra of an n-dimensional space has a general basis of 2^n elements. Rotors, which are only even-dimensional, in principle require 2^{n-1} parameters for their specification (though typical rotors use only a part of this). Linear transformations specified by matrices need $n \times n$ matrices with n^2 parameters (and typically need them all). Rotors are therefore more efficient for storage of transformations in less than 7 dimensions (and for the practical dimensionalities of 3, 4, 5 about twice as efficient). Composing transformations as rotors takes 2^n operations, and composing them as matrices requires n^3 operations. Therefore in fewer than 10 dimensions, rotors are more efficient than matrices (in the practical dimensions 3, 4, 5 about four times more). The reason for this gain by rotors in composition is partly that they do not represent general linear transformations, just the orthogonal transformations, and that they can really exploit that algebraic limitation, whereas matrices cannot. Unit quaternions are rotors in 3-D, and of course well known to be efficient for the composition of rotations.

- *For the linear transformation of vectors, matrices are always superior*. To perform an $n \times n$ matrix on an n-dimensional vector (orthogonal or not) takes n^2 operations. A general rotor could require as much as $2^{n-1} \times n \times 2^{n-1} = n\,2^{2(n-1)}$ in a straightforward implementation of its two-sided product. This is always more, in the practical dimensions about four times more. Part of the computations can be saved by realizing that grade-preservation of the rotor operation must mean that some terms cancel (so that they do not need to be computed). Other techniques may reduce the computation further, but not enough to make the direct rotor approach competitive. The conversion of a rotor to a matrix may therefore be an advantageous way to apply it to a vector. This is what one typically does for the unit quaternions, which are 3-D rotors; we treat their conversion matrix in Sections 7.10.3 and 7.10.4.

- **For the linear transformation of general blades, outermorphism matrices beat vector matrices and rotors.** A general k-blade is specifiable on a $\binom{n}{k}$-dimensional basis. When we want to transform a k-blade by an orthogonal transformation, we have three possibilities: rotors, outermorphism matrices (see Section 4.5), or matrices on the constituent vectors followed by recomposition. These take, respectively, $\binom{n}{k} 2^n/2$ operations (using grade preservation-based reduction), $\binom{n}{k}^2$ operations, and $kn^2 + k\binom{n}{k}$ operations (the final term is an estimate of the complexity of the outer product construction). Of those three, the outermorphism matrix is cheapest (even for $k = 1$, when it reverts to the vector matrix of the previous item). Thus the generalized linear algebra that geometric algebra offers pays off in a different form.

- **For optimal orthogonal transformation, use rotors in a code generator.** Although outermorphism matrices are relatively cheap to use on k-blades, they are not the optimum. They do not employ all the structure of the computation, for they use neither the fact that the element to be transformed is a k-versor or a k-blade (rather than a general k-vector), nor can they use our knowledge that the transformation is orthogonal. Therefore they can still be improved by explicitly spelling out the products involved for each specific type of multivector. If you can predict the grade of the elements beforehand, such grade-based accelerations can be built in at compile time. But even at run time it may be worth testing the multivector type of an element and jumping to some specialized part of the code. This engenders no overhead at all if you can predict and specify the multivector type for each variable in your code. In this approach, the rotor multiplication formula is crucial, since it can be used by the symbolic code generator to derive all required formulas in a unified manner.

The bottom line is: geometric algebra works, the structural simplicity it brings can be used directly in high-level programming, and the computational overhead can be kept low (in the order of 5-10 percent). But the actual low implementational level on which the computations take place needs to be carefully designed, for a literal implementation of the geometric algebra products can rapidly become too expensive. We devote Part III of this book completely to these implementational issues.

7.8 FURTHER READING

In this chapter, most of the literature on the foundations of geometric algebra has become accessible, and you might even read some of the mathematical literature on Clifford algebra.

The obviously geometrically relevant literature on rotors (and spinors) has been absorbed into geometric algebra so you can read about it in the language of this book. The basic sources are [33] and [15], who also relate it to Lie groups in much more detail than we do in this book. Other references about the actual use of rotors in geometric applications will be supplied with the appropriate chapters in Part II.

For an entry to the more mathematical literature, we can recommend *Clifford Algebras and Spinors* [41]. Together with [52] and [51], it gives the precise mathematics. Though all are short on actual geometry, they can be used to answer questions about the validity of perceived patterns that one may be tempted to use in implementations.

7.9 EXERCISES

7.9.1 DRILLS

1. Compute $R_1 \equiv R_{\mathbf{e}_1 \wedge \mathbf{e}_2\, \pi/2}$ and apply to $\mathbf{e}_1$.
2. Compute $R_2 \equiv \exp(\mathbf{e}_3 \wedge \mathbf{e}_1\, \pi/4)$ and apply to $\mathbf{e}_2 \wedge \mathbf{e}_4$.
3. Compute $R_2\, R_1$ and apply to $\mathbf{e}_1 \wedge \mathbf{e}_2$.
4. Compute the axis and angle of $R_2\, R_1$.
5. Compute the product of the rotors $R_{\mathbf{e}_{14}\pi/2}$ and $R_{\mathbf{e}_{23}\pi/2}$ and apply to $\mathbf{e}_{12}$.
6. Reflect $(\mathbf{e}_1 + \mathbf{e}_2) \wedge \mathbf{e}_3$ in the plane $\mathbf{e}_1 \wedge \mathbf{e}_4$.
7. Reflect the dual plane reflector $\mathbf{e}_1$ in the plane $\mathbf{e}_1 \wedge \mathbf{e}_3$.

7.9.2 STRUCTURAL EXERCISES

1. The generalization of the line reflection from $\mathbf{a}\,\mathbf{x}\,\mathbf{a}^{-1}$ to $\mathbf{a}\,\mathbf{X}\,\mathbf{a}^{-1}$ seems straightforward when we remember that a k-blade can be written as the geometric product of k mutually orthogonal vectors: $\mathbf{X} = \mathbf{x}_1\,\mathbf{x}_2\cdots\mathbf{x}_k$, and then simply compute the outermorphism as $(\mathbf{a}\,\mathbf{x}_1\,\mathbf{a}^{-1})\,(\mathbf{a}\,\mathbf{x}_2\,\mathbf{a}^{-1})\cdots(\mathbf{a}\,\mathbf{x}_k\,\mathbf{a}^{-1}) = \mathbf{a}\,\mathbf{X}\,\mathbf{a}^{-1}$. The result is correct but the proof is wrong as it stands. Why? (Hint: Can you guarantee the factorization after reflection?)
2. We have seen that for a Euclidean unit 2-blade, $\mathbf{I}^2 = -1$. Interpret this geometrically in terms of versors.
3. Verify that a line reflection in 3-D can be performed as a rotation. Which rotation? Give the axis and angle. Verify that this reflection can be applied to any blade.
4. Show that the fact that the geometric product transforms naturally under application of a versor, together with linearity, implies that the contraction is preserved. (Hint: An intermediate step uses linearity to show that the outer product is preserved.)
5. Show from the definition of the adjoint (in Section 4.3.2) that the adjoint of a transformation that can be written as a versor product with a versor V is a versor product with the versor V^{-1}. Relate this to the orthogonality of a versor-based transformation.

6. We can reflect mirrors into mirrors to compute the effective mirror of a total reflection. Why can you ignore all signs in that computation and therefore universally use $\mathbf{M}_2\,\mathbf{M}_1\,\mathbf{M}_2$ for the reflection of mirror 1 in mirror 2 regardless of whether they have been represented directly or dually?

7. Match the computation of the composition of 2-D rotations in Section 7.3.1 to that of the 3-D rotations in Section 7.3.3, both algebraically and in the geometric visualization.

8. To study the spherical image of rotation composition, take a rotation in the $\mathbf{e}_1\mathbf{e}_3$ plane over $\pi/2$, followed by a rotation in the $\mathbf{e}_3\mathbf{e}_2$ plane over $\pi/2$. As rotors, these are $(1 - \mathbf{e}_1\mathbf{e}_3)/\sqrt{2}$ and $(1 - \mathbf{e}_3\mathbf{e}_2)/\sqrt{2}$. Draw two great circles, with poses corresponding to the rotation planes $\mathbf{e}_1\mathbf{e}_3$ and $\mathbf{e}_3\mathbf{e}_2$. On these great circles, the rotations over $\pi/2$ are represented as oriented arcs of length $\pi/4$ (the corresponding rotor angle). These arcs are freely movable along their great circles. To compose the rotations, make them meet so that you can perform R_1 and then R_2. This is a depiction as in Figure 7.6. The arc completing the spherical triangle is in a skew plane with a length that looks like it might be $\pi/3$. Do an actual computation to confirm this value for the rotor angle and a plane of $(-\mathbf{e}_3\mathbf{e}_2 - \mathbf{e}_1\mathbf{e}_3 + \mathbf{e}_2\mathbf{e}_1)/\sqrt{3}$. The resulting rotation is over $-2\pi/3$ in this plane. Rewrite this using (7.10), and show that the rotation axis is $(-\mathbf{e}_1 - \mathbf{e}_2 + \mathbf{e}_3)$.

9. Draw the rotated rotor $R_2\,R_1\,\widetilde{R}_2$ as an arc in the spherical image. (Hint: What would you expect it to be based on its geometric meaning? Warning: It is not simply the R_1 arc rotated over the R_2-arc!)

10. Establish the precise correspondence between the quantities in the rotor composition (7.9) and the quaternion product of (7.11). (Warning: This is a painful exercise in keeping things straight, and not very rewarding.)

11. Derive the formulas for the reflection of a dual blade $\mathbf{Y} = \mathbf{X}^*$ from the formulas for reflection of a directly represented blade $\mathbf{X}$. Derive the last column of Table 7.1 from the column before. Make sure you take the dual of both input and output relative to the same unreflected pseudoscalar $\mathbf{I}_n$.

12. A special case of reflection is when $\mathbf{A}$ is the scalar 1. Derive the algebraic outcome and interpret geometrically. Another special case is $\mathbf{A} = \mathbf{I}_n$; compute that and interpret. Why is the latter outcome not the dual of the former?

13. You can project onto a rotor and get a geometrically meaningful result. Give the geometric interpretation of the projection $\mathsf{P}_R[\mathbf{x}] \equiv (\mathbf{x}\rfloor R)\,R^{-1}$. (Hint: Think "chord.") For rotors, it matters whether you put the inverse on the first or the last factor: what is $(\mathbf{x}\rfloor R^{-1})\,R$?

14. In $\mathbb{R}^{4,0}$ with the usual basis, perform a rotation in the $\mathbf{e}_1 \wedge \mathbf{e}_2$ plane followed by a rotation in the $\mathbf{e}_3 \wedge \mathbf{e}_4$ plane. Compute the rotor of the composition, and show that this is the exponent of a bivector, not of a 2-blade. (Hint: See structural exercise 5 of Chapter 2.) Note also that the rotor is not of the simple form "scalar plus 2-blade" of Section 7.4 (or even "scalar plus bivector").

7.10 PROGRAMMING EXAMPLES AND EXERCISES

7.10.1 REFLECTING IN VECTORS

The code of the first example is a straightforward implementation of the line reflection equation:

```
e3ga::vector reflectVector(const e3ga::vector &a,
        const e3ga::vector &x) {
  return _vector(a * x * inverse(a));
}
```

The program allows you to interactively manipulate both a (red) and x (green). You can use the popup menu to switch to a mode that shows you that this also works for bivectors.

7.10.2 TWO REFLECTIONS EQUAL ONE ROTATION

This example displays an interactive version of Figure 7.2. The input vector (green) is successively reflected in two different (red) vectors. The end result is that the input vector is rotated, as expected. To reflect the input vector we invoke reflectVector() twice:

```
// update the reflected/ rotated vectors
g_reflectedVector = reflectVector(g_reflectionVector1,
        g_inputVector);
g_rotatedVector = reflectVector(g_reflectionVector2,
        g_reflectedVector);
```

Figure 7.9 shows a screenshot.

7.10.3 MATRIX-ROTOR CONVERSION 1

To connect to programs and libraries not based on geometric algebra (such as OpenGL), you may need to convert back and forth between rotors and matrices. This example provides the code for the 3-D case. The algorithms are based on geometric intuition—see the next exercise for more efficient solutions.

Rotor To Matrix Conversion

The columns of a (rotation) matrix are the images of the basis vectors under the transformation. To convert from rotor to matrix, we transform e_1, e_2, and e_3 and copy them into the matrix. The implementation is straightforward:

```
void rotorToMatrix(const rotor &R, float M[9]) {
  // compute images of the basis vectors:
  rotor Ri = _rotor(inverse(R));
  e3ga::vector image[3] = {
   _vector(R * e1 * Ri), // image of e1
   _vector(R * e2 * Ri), // image of e2
```

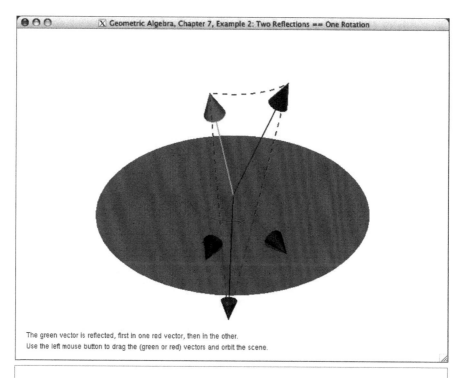

The green vector is reflected, first in one red vector, then in the other.
Use the left mouse button to drag the (green or red) vectors and orbit the scene.

Figure 7.9: Interactive version of Figure 7.2.

```
    _vector(R * e3 * Ri) // image of e3
  };

  // copy coordinates to matrix:
  for (int i = 0; i < 3; i++)
    for (int j = 0; j < 3; j++)
      M[j * 3 + i] = image[i].getC(vector_e1_e2_e3)[j];
}
```

Rotation Matrix To Rotor Conversion

Conversion from matrix to rotor is more complicated. Again we start with the fact that the columns of the matrix are the images of the basis vectors. We should remember that rotors are ambiguous: we can always increase the angle by 4π to get another rotor that is equivalent. And since a rotation matrix does not specify the sense of rotation, R and $-R$ are both acceptable solutions. We compute the smallest rotor (i.e., with the smallest angle) as our solution in three steps:

- First, compute the smallest rotor R_1 that rotates $\mathbf{e}_1$ to the image of $\mathbf{e}_1$ under the matrix transform.
- Then, compute the smallest rotor R_2 that rotates $R_1 \mathbf{e}_2/R_1$ to its image of $\mathbf{e}_2$ under the matrix transform. Because of orthogonality, this rotor will leave $R_1 \mathbf{e}_1/R_1$ unchanged.
- Finally, compute the full rotor: $R = R_2 R_1$.

Because of orthogonality, $\mathbf{e}_3$ automatically transforms correctly to $R \mathbf{e}_3/R$. The code for this algorithm:

```
// note: very imprecise in some situations; do NOT use this
// function in practice

rotor matrixToRotor(const float M[9]) {
  e3ga::vector imageOfE1(vector_e1_e2_e3,
        M[0 * 3 + 0], M[1 * 3 + 0], M[2 * 3 + 0]);
  e3ga::vector imageOfE2(vector_e1_e2_e3,
        M[0 * 3 + 1], M[1 * 3 + 1], M[2 * 3 + 1]);

  rotor R1 = rotorFromVectorToVector(_vector(e1), imageOfE1);
  rotor R2 = rotorFromVectorToVector(_vector(R1 * e2 * inverse(R1)),
        imageOfE2);

  return _rotor(R2 * R1);
}
```

There is a compact formula that computes the smallest rotor that rotates a unit vector $\mathbf{a}$ to another unit vector $\mathbf{b}$, in 3-D. It is

$$R = \frac{1 + \mathbf{b}\,\mathbf{a}}{\sqrt{2\,(1 + \mathbf{b} \cdot \mathbf{a})}}. \tag{7.23}$$

(We discuss it in more context in Section 10.3.2, but we use it now.) It is implemented in the function `rotorFromVectorToVector()`. The rotor formula is unstable when $\mathbf{a} \cdot \mathbf{b} \simeq -1$, which happens near a rotation over 180 degrees (the rotation plane is then not accurately determined, neither geometrically nor algebraically). This also makes the code listed above unstable. We work around this limitation in two ways in the stable version of the function, shown in Figure 7.10. This first is to pick the first basis vector such that a 180-degree rotation is not required. This is tested using `if (M[0 * 3 + 0] > -0.9f) {/*...*/}`. The second is to provide a default rotation plane (2-blade) to be used by `rotorFromVectorToVector()`—this plane must be orthogonal to the image of the first basis vector. `rotorFromVectorToVector()` uses this plane in situations where the rotation is near 180 degrees to come up with a solution for this geometrically degenerate case.

7.10.4 EXERCISE: MATRIX-ROTOR CONVERSION 2

The conversion functions we presented above are (geometrically) intuitive, but they are not the most efficient solutions. A much better way is to perform the rotation on the

```
rotor matrixToRotorStable(const float M[9]) {
  e3ga::vector imageOfE1(vector_e1_e2_e3,
    M[0 * 3 + 0], M[1 * 3 + 0], M[2 * 3 + 0]);
  e3ga::vector imageOfE2(vector_e1_e2_e3,
    M[0 * 3 + 1], M[1 * 3 + 1], M[2 * 3 + 1]);
  e3ga::vector imageOfE3(vector_e1_e2_e3,
    M [0 * 3 + 2], M[1 * 3 + 2], M[2 * 3 + 2]);

  if (M[0 * 3 + 0] > - 0.9f)
  {
    rotor R1 = rotorFromVectorToVector(_vector(e1), imageOfE1);
    rotor R2 = rotorFromVectorToVector(_vector(R1 * e2 * inverse(R1)),
      imageOfE2,_bivector(dual(imageOfE1)));
    return _rotor(unit_e(R2 * R1));
  }
  else if (M[1 * 3 + 1] > - 0.9f)
  {
    rotor R1 = rotorFromVectorToVector(_vector(e2), imageOfE2);
    rotor R2 = rotorFromVectorToVector(_vector(R1 * e3 * inverse(R1)),
      imageOfE3,_bivector(dual(imageOfE2)));
    return _rotor(unit_e(R2 * R1));
  }
  else
  {
    rotor R1 = rotorFromVectorToVector(_vector(e3), imageOfE3);
    rotor R2 = rotorFromVectorToVector(_vector(R1 * e1 * inverse(R1)),
      imageOfE1,_bivector(dual(imageOfE3)));
    return _rotor(unit_e(R2 * R1));
  }
}
```

Figure 7.10: Rotation matrix to rotor conversion.

unit basis vectors symbolically and encode the results. This is straightforward. On an orthonormal basis $\{\mathbf{e}_i\}$, with associated bivector basis $\mathbf{e}_{ij} \equiv \mathbf{e}_i \wedge \mathbf{e}_j$, let the rotor to be converted be

$$R = w + x\,\mathbf{e}_{23} + y\,\mathbf{e}_{31} + z\,\mathbf{e}_{12}.$$

The normalization of the rotor implies that $w^2 + x^2 + y^2 + z^2 = 1$. Then one computes

$$
\begin{aligned}
R\,\mathbf{e}_1\widetilde{R} &= (w + x\,\mathbf{e}_{23} + y\,\mathbf{e}_{31} + z\,\mathbf{e}_{12})\,\mathbf{e}_1\,(w - x\,\mathbf{e}_{23} - y\,\mathbf{e}_{31} - z\,\mathbf{e}_{12}) \\
&= (w^2 + x^2 - y^2 - z^2)\,\mathbf{e}_1 + 2\,(-wz + xy)\,\mathbf{e}_2 + 2\,(wy + xz)\,\mathbf{e}_3 \\
&= \left(1 - 2(y^2 + z^2)\right)\mathbf{e}_1 + 2\,(-wz + xy)\,\mathbf{e}_2 + 2\,(wy + xz)\,\mathbf{e}_3.
\end{aligned}
$$

The transformation of the other basis vectors is obtained by cyclicity (resubstituting the indices $1 \rightarrow 2 \rightarrow 3 \rightarrow 1$ and values $x \rightarrow y \rightarrow z \rightarrow x$). The result is the matrix that implements the linear transformation of applying the rotor R to a vector:

$$[\![R]\!] = \begin{bmatrix} 1 - 2y^2 - 2z^2 & 2yx + 2wz & 2zx - 2wy \\ 2xy - 2wz & 1 - 2z^2 - 2x^2 & 2zy + 2wx \\ 2xz + 2wy & 2yz - 2wx & 1 - 2x^2 - 2y^2 \end{bmatrix}$$

This basically is also how one converts a quaternion into a matrix. If you already have software for that, you can use it, though you may need to to initialize a rotor from its (quaternion) coordinates, which is the implementation of the correspondence of (7.10):

```
float w, x, y, z;
rotor R(rotor_scalar_e1e2_e2e3_e3e1, w, -z, -x, -y);
```

Here w, x, y, and z are the coordinates of a classic quaternion (which are not always defined in the same way, so beware what correspondence between the quaternion units i, j, k, and the basis bivectors should be used!).

Implement this rotor to matrix conversion and test its speed when applied to vectors. The example code provides the basic framework for testing and timing. In our solution, this classic version was about four times faster than the geometric version.

The converse function, to convert a rotation matrix into a rotor, can also be sped up. Here we can use the standard conversion of a matrix to a unit quaternion. Consider the form of the matrix above and use combinations of the elements to retrieve the four parameters. Addition of selected off-diagonal elements gives products of two of the parameters; the diagonal elements can be added with appropriate signs to give the square of only one variable, which is enough to compute the rest. Any trusted site on quaternion computations gives the details.

7.10.5 JULIA FRACTALS

Fractals are usually introduced using complex numbers. But with the subsumption of complex numbers into geometric algebra, as explained in Section 7.3.2, they are just as easily generated using vectors in a real geometric algebra. This has the additional advantage that they can be extended to more than two dimensions without changing the algorithm. We explore this for *Julia fractals*, based on [37].

In the classic computation of 2-D fractal images, the image space is considered to be a complex plane. Each pixel (indicated by its complex coordinates) is inside the fractal set if, under repeated application of some mathematical function, the result does not tend to infinity.

For the Julia set, the complex function is an iterative computation, computing the complex number X_{k+1} as the pth power of the previous number and some additive constant complex number C:

$$X_{k+1} = X_k^p + C. \tag{7.24}$$

The initial X_0 to which the function is applied is defined as

$$X_0 = x + i\,y, \tag{7.25}$$

where x and y are the coordinates of the pixel in the image. The fractal picture is obtained by coloring the pixel according to the value of X_{k+1} after a fixed number of iterations. By varying C, different images are obtained. The constant integer p is commonly 2, but other values are possible.

The function above can be converted into geometric algebra and then just becomes an operation on the vectors of a real plane. We have seen in (7.8) that each complex number X is associated with a vector as $X = \mathbf{x}/\mathbf{e}$, with $\mathbf{e}$ denoting the unit direction vector of the real axis.

Taking in particular the fractal with $p = 2$ involves using the complex square of X_k. With the substitution $X = \mathbf{x}/\mathbf{e}$, this can be computed by a geometric product involving only real vectors:

$$\mathbf{x}_{k+1}\,\mathbf{e}^{-1} = \mathbf{x}_k\,\mathbf{e}^{-1}\,\mathbf{x}_k\,\mathbf{e}^{-1} + \mathbf{c}\,\mathbf{e}^{-1},$$

which is equivalent to

$$\mathbf{x}_{k+1} = \mathbf{x}_k\,\mathbf{e}\,\mathbf{x}_k + \mathbf{c}.$$

This is clearly a vector, proportional (by $\mathbf{x}_k^2$) to the reflection of the constant unit vector $\mathbf{e}$ in the previous vector $\mathbf{x}_k$, with $\mathbf{c}$ added. The initial complex number X_0 is replaced by the vector $\mathbf{x}$ in the image plane for which we want to compute the value of the fractal function.

With these substitutions, the fractals are computed in a real geometric algebra. Nothing in this new formulation refers to the plane of vectors, so fractals are easily extended to n-dimensional Euclidean space by taking the initial $\mathbf{x}$ and $\mathbf{c}$ as vectors in that space. In 3-D, this leads to what are known as *quaternionic fractals*, though without actually using quaternions. An example of a 3-D fractal is shown in Figure 7.13.

We have implemented the basic algorithm; see the code listing in Figure 7.11. An example of the output is shown in Figure 7.12. Note that we terminate the evaluation of the value in the inner loop after a maximum of maxIter iterations. This is done to make the example more responsive—you can zoom, translate, and change the value of $\mathbf{c}$ using the mouse buttons. By default maxIter = 10, but you can modify this value by pressing 1 to 9 on the keyboard.

Exercise 5a

Experiment with changing the power p in the fractal algorithm. You first need to derive the corresponding vector update equation!

```
void computeFractal(const e2ga::vector &translation, const e2ga::vector &c,
      mv::Float zoom, int maxIter,
      std::vector<unsigned char> &rgbBuffer, int width, int height)
{
  int idx = 0;

  // we use e = e1 ('__e1_ct__' stands for 'e1 constant type')
  __e1_ct__ e;

  // for each pixel in the image, evaluate fractal function:
  for (int imageY = 0; imageY < height; imageY++) {
    for (int imageX = 0; imageX < width; imageX++) {
      float imageXf = (float)(imageX - width/2);
      float imageYf = (float)(imageY - height/2);
      e2ga::vector p(vector_e1_e2, imageXf, imageYf);
      e2ga::vector x = _vector(zoom * p - translation);

      for (int i = 0; i < maxIter; i++) {
        x = _vector(x * e * x + c); // p = 2
        if (_Float(norm_e2(x)) > 1e4f) break; //1e4 = 'infinity'
      }

      // convert to grey-scale value:
      float valF = _Float(norm_e(x)) /10.0f;
      unsigned char val = (valF > 255) ? 255 : (unsigned char)(valF + 0.5f);

      rgbBuffer[idx + 0] = rgbBuffer[idx + 1] = rgbBuffer[idx + 2] = val;

      idx += 3;
    }
  }
}
```

Figure 7.11: 2-D Julia fractal code.

Exercise 5b

If you are feeling adventurous, try implementing the n-D version.

7.10.6 EXTRA EXAMPLE: ROTATIONS USED IN OUR USER INTERFACE

The following code is used to orbit the scene in a lot of different examples:

```
// Called by GLUT when mouse is dragged:
void MouseMotion(int x, int y) {
```

```
e3ga::vector mousePos = mousePosToVector(x, y);
e3ga::vector motion = _vector(mousePos - g_prevMousePos);

// update rotor
if (g_rotateModelOutOfPlane)
  g_modelRotor = _rotor(e3ga::exp(0.005f * (motion ^ e3ga::e3))
      * g_modelRotor);
else g_modelRotor = _rotor(e3ga::exp(0.00001f * (motion ^ mousePos))
      * g_modelRotor);

// remember mouse pos for next motion:
g_prevMousePos = mousePos;
}
```

The function starts with determining the mouse motion relative to the previous mouse event. Since it ends with storing the current mouse position for the next mouse event, the interesting part must be in the middle.

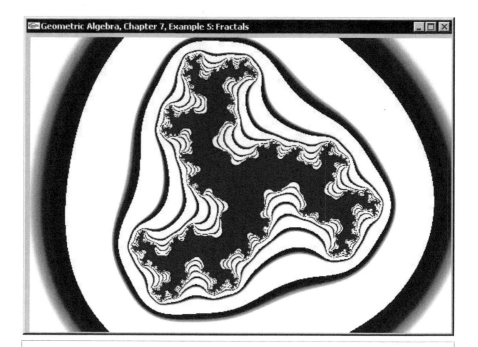

Figure 7.12: A 2-D Julia fractal, computed using the geometric product of real vectors.

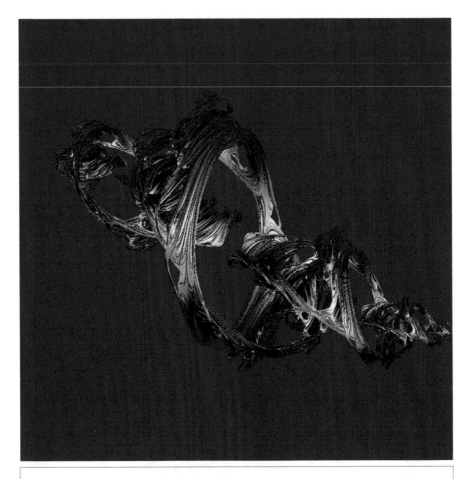

Figure 7.13: A 3-D Julia fractal. Image from [37], by courtesy of the Lasenby family.

The middle section of the function updates the g_modelRotor in one of two different ways, depending on the value of the boolean g_rotateModelOutOfPlane:

1. If g_rotateModelOutOfPlane is false, the rotation is in the screen plane. The updating rotor is formed by computing the exponent of the 2-blade spanned by the motion and the mousePos.

2. If g_rotateModelOutOfPlane is true, the rotation is outside the screen plane. The updating rotor is formed by computing the exponent of the 2-blade spanned by the motion and the vector e3, which is orthogonal to the screen plane.

8 GEOMETRIC DIFFERENTIATION

Differentiation is the process of computing with changes in quantities. When the changes are small, those computations can be linear to a good approximation, and it is not too hard to develop a calculus for geometry by analogy to classical analysis.

When formulated with geometric algebra, it becomes possible to differentiate not only with respect to a scalar (as in real calculus) or a vector (as in vector calculus), but also with respect to general multivectors and k-blades. The differentiation operators follow the rules of geometric algebra: they are themselves elements that must use the noncommutative geometric product in their multiplication when applied to other elements. As you might expect, this has precisely the right geometric consequences for the differentiation process to give geometrically significant results.

This chapter is a bit of a sideline to the main flow of thought in this book. Although the later chapters occasionally use differentiation in their examples, it is not essential. You can easily skip it at first reading, and move on to Part II on the modeling of geometries. We include the subject because it is important for geometric optimization and differential geometry. These techniques are beginning to appear in practical applications of geometric algebra.

8.1 GEOMETRICAL CHANGES BY ORTHOGONAL TRANSFORMATIONS

The geometrical elements we have constructed are of various types, and within the context of the geometry they can change in different ways. Each of these kinds of changes should find their place in a suitably defined calculus for geometric elements.

- *Orthogonal Transformations.* Elements of a geometry change when they are transformed, and the class of transformations that is permitted determines the kind of geometry one has. We are especially interested in Euclidean geometry and the accompanying transformations of rotation, reflection, and translation (and, by a stretch of the term Euclidean, scaling). We have already seen that rotations and reflections can be represented by versors, since they are orthogonal transformations. In Part II, we will show that it is possible to set up a model of Euclidean geometry so that translations and scaling are also representable by versors, which will unify the whole structure of operators.

    Orthogonal transformations represented by versors thus become central to doing geometric algebra. Among these, we are especially interested in rotors, since they cause the smooth continuous changes that are typical of motions. In their representation as exponents of bivectors, the calculus of rotors is surprisingly easy to treat: all differentiation reduces to computing commutators with the bivectors of the transformations. This has a natural connection with the Lie algebras that are used classically to compute the calculus of continuous transformation groups.

- *Parameterizations.* An element of the geometry is often parametrized in terms of other elements. A specific case is location-dependence, which is parameterization by the positional vector $\mathbf{x}$, or time-dependence on a scalar time parameter τ. A more involved instance of parameterization is explicit geometric relational dependence, such as, for example, when an element X is reflected using a plane mirror $\mathbf{a}$ to make $\mathbf{a}\,\widehat{X}\,\mathbf{a}^{-1}$. As the parameter element changes (for instance because it is transformed, such as when the mirror $\mathbf{a}$ rotates), the parametrized element changes as well. Geometric algebra provides a calculus to compute with such changes.

    This calculus consists of a scalar operator called the *directional derivative* to measure how the parametrized element reacts to a known change in the parameter (and the result is of the same type as the original), and of a *total geometric derivative* that specifies the change relative to any change in the parameter (and that returns an operator of a different type than its argument). The latter is more general (the directional derivative merely describes its components), and particularly useful in geometric integration theory (not treated in this book; see Section 8.8 for pointers).

In all of this, we have to be a bit careful about just copying the classical linear techniques, such as Taylor series definitions. Simply adding linear approximations of perturbations to a blade may not add up to a perturbed blade (but instead result in some nonfactorizable multivector), so we need to develop things in a structure-preserving manner. That is why we start with the calculus of versors, and develop the more classical derivatives in the remainder of the chapter.

8.2 TRANSFORMATIONAL CHANGES

First, let us consider an element X that has been changed by a rotor R. In the Euclidean and Minkowski spaces that interest us, the rotor can be written as the exponential of a bivector $R = \exp(-B/2)$, and when we develop this in a power series in B, we get

$$e^{-B/2} X e^{B/2} = X + \tfrac{1}{2}(XB - BX) + \cdots \tag{8.1}$$

The first-order term involves a combination that we will encounter a lot in our considerations, so it pays to define it as a new and useful product in geometric algebra. We briefly introduce it and its properties in Section 8.2.1. Then we play around with variations of changes to this basic transformation equation.

- We study what kind of changes small rotors can effect in an element X in Section 8.2.2. Once we have encoded motions as rotors (in Part II), those will be what we mean by "moving X slightly." Those motions together form a Lie algebra, which we connect to geometric algebra in Section 8.2.3.

- Those small changes in X can be propagated simply to other motions that X may undergo, as we show in Section 8.2.4.

- The most involved change is when the parameters of a motion themselves get moved—for instance, when a rotation plane translates or a mirror starts rotating. We study that in Section 8.2.5.

Each of these cases can be described by a well-chosen commutator product, some exactly, some to first order in the magnitude of the change.

8.2.1 THE COMMUTATOR PRODUCT

The *commutator product* of two general elements of geometric algebra is defined as the product $\times : \bigwedge \mathbb{R}^n \times \bigwedge \mathbb{R}^n \to \bigwedge \mathbb{R}^n$ defined by

$$X \times B \equiv \tfrac{1}{2}(XB - BX).$$

It is clearly bilinear and distributive, since it consists of a sum of geometric products, bilinear in the arguments. We have purposely not used the bold blade notation for its arguments, since its typical use involves more general multivectors.

This product is not associative. Instead of the identity $(A \times B) \times C = A \times (B \times C)$, so that $(A \times B) \times C - A \times (B \times C)$ would be zero, we have

$$(A \times B) \times C - A \times (B \times C) = B \times (C \times A),$$

which is more symmetrically expressed as the *Jacobi identity*:

$$(A \times B) \times C + (C \times A) \times B + (B \times C) \times A = 0. \tag{8.2}$$

You can prove this easily yourself in structural exercise 1.

Even though the commutator product can be defined for general multivectors, we will not need it in completely general form: in our calculus of rotors, one of the two arguments (say the second argument B) is always a bivector. This has a property of grade-preservation (as we soon show):

$$\mathrm{grade}(X \times B) = \mathrm{grade}(X) \quad \text{when} \ \ \mathrm{grade}(B) = 2.$$

When used in this way, the commutator product is a grade-preserving product $\times : \bigwedge^k \mathbb{R}^n \times \bigwedge^2 \mathbb{R}^n \to \bigwedge^k \mathbb{R}^n$, extended to the whole space $\bigwedge \mathbb{R}^n$. This property of grade-preservation is important geometrically, for clearly we want all terms in a Taylor series like (8.1), showing the perturbation of X, to be of the same grade as X.

We prove this grade-preserving property in a slightly roundabout way. We first note that the terms XB and BX contain the grades $x - 2$, x, and $x + 2$ (where $x = \mathrm{grade}(X)$), since they are geometric products. The subtraction in the commutator product can kill some terms, so the whole range of grades may not be there. To investigate this, we take the reverse of the commutator to find

$$(X \times B)^{\widetilde{\ }} = \tfrac{1}{2}(\widetilde{B}\,\widetilde{X} - \widetilde{X}\,\widetilde{B}) = \tfrac{1}{2}(\widetilde{X}\,B - B\,\widetilde{X}) = \widetilde{X} \times B.$$

We observe that the commutator product gets the same overall sign under reversion as X (namely $(-1)^{x(x-1)/2}$). Among the potential terms of grade $x - 2$, x, and $x + 2$, only the grade of x has precisely that same grade-dependent sign for all grades. (This is due to the sign pattern of the reversion over the grades, which is $+ + - - + + - - + + \cdots$, so that two grades up or two grades down may have opposite signs from grade x, for general x.) Therefore $X \times B$ must be of grade x, and the commutator product with a bivector is grade-preserving.

As an aside, having the commutator product permits listing a pleasing series of equations, expressing the geometric product in terms of other products when one of the arguments is a scalar α, a vector $\mathbf{a}$, or a bivector A (not necessarily a 2-blade):

$$\alpha X = \alpha \wedge X$$
$$\mathbf{a} X = \mathbf{a} \wedge X + \mathbf{a} \rfloor X$$
$$A X = A \wedge X + A \rfloor X + A \times X$$

All equations hold for any multivector X.

You may have failed to notice the rather subtle difference between the commutator symbol $\times$ and the cross-product symbol $\times$. Fortunately, there is little danger of confusing them in formulas, since we will use the commutator product only when one of the arguments is a bivector (which is an uppercase symbol, only bold if we know that it is a 2-blade), and the cross product only when both arguments are 3-D vectors (which are always lowercase bold).

8.2.2 ROTOR-INDUCED CHANGES

After this introduction of the commutator product, we resume our treatment of the geometrical changes. Using the commutator notation, the transformation of X by the rotor $\exp(-B/2)$ can be developed in a Taylor series as

$$e^{-B/2} X e^{B/2} = X + X \times B + \tfrac{1}{2}(X \times B) \times B + \tfrac{1}{3!}((X \times B) \times B) \times B + \cdots \quad (8.3)$$

You can prove this yourself, guided by structural exercise 2. The series continues the pattern as the generalization of the earlier (7.22) for vectors only. Since the commutator product with the bivector is grade-preserving, X remains of the same grade under this transformation (as it should, since the versor product is fully structure-preserving for all products).

Now suppose that the rotor is close to the identity. It is then the exponential of a small bivector $-\delta B/2$, with $\delta \approx 0$. We can write, in orders of δ:

$$e^{-\delta B/2} X e^{\delta B/2} = X + X \times \delta B + O(\delta^2). \quad (8.4)$$

We read this as specifying the small change in an element X under a small orthogonal transformation. Such a small transformation must be represented by an even versor, which we can normalize to a rotor. The transformation caused by an odd versor cannot be continuously connected to the identity (i.e., done in small steps); you can perform a small amount of rotation, but not a small amount of reflection. We call small changes caused by small transformations *perturbations*.

To preserve the geometric meaning of X, we must demand that any small change δX to it must be writable as the application of a small rotor to it. These are the only kinds of small changes we should consider in our calculus. They are the proper generalization in geometric algebra of the additive change δX in a quantity X, beyond the scalars and vectors of the classical framework. Any small changes that cannot be written in this form may disrupt the algebraic structure of X, and with that its geometric interpretation.

Elements of geometric algebra should only be perturbed by rotors.

We found in (8.4) that to first order, such a change can be written as $X \times (\delta B)$, with δB a small bivector.

Remembering that rotors represent orthogonal transformations, you can see how even for a Euclidean vector $\mathbf{x}$ a general additive change $\delta \mathbf{x}$ is not permitted. Orthogonal transformations must preserve the norm, and this can only happen if the change $\delta\mathbf{x}$ is perpendicular to $\mathbf{x}$. This must mean that $\mathbf{x} \cdot (\delta\mathbf{x}) = 0$. The general element of grade 1 and linear in $\mathbf{x}$ with that property is $\delta\mathbf{x} = \mathbf{x} \rfloor \delta B$, with δB a small general bivector (for indeed $\mathbf{x} \cdot (\mathbf{x} \rfloor \delta B) = \mathbf{x} \rfloor (\mathbf{x} \rfloor \delta B) = (\mathbf{x} \wedge \mathbf{x}) \rfloor \delta B = 0$). And for vectors,

$$\delta\mathbf{x} = \mathbf{x} \rfloor \delta B = \tfrac{1}{2}(\mathbf{x}\,\delta B - \delta B\,\mathbf{x}) = \mathbf{x} \times \delta B,$$

so that it indeed has the desired form of a commutator product.

This limitation of the changes may appear unnecessarily restrictive, since it even forbids a simple translational change $\delta\mathbf{x}$ to a vector $\mathbf{x}$. Indeed it does, for $\mathbf{x}$ denotes a 1-D direction, and that should only be turned by a rotor. But fortunately this limitation to rotors does not automatically mean that we cannot translate geometrical points in any direction. It merely necessitates us to find a way to represent that geometrical point in geometric algebra such that its translation is a rotor. In such a representation, any small translation would be permitted. We will present such a representation in Chapter 13. For now, please accept that the principle of allowing only rotor-type changes is not a geometrical limitation, but merely an algebraic structuring of the treatment of such changes.

8.2.3 MULTIPLE ROTOR-INDUCED CHANGES

When two small changes occur successively, by $\exp(-\delta_1 A/2)$ and $\exp(-\delta_2 B/2)$, respectively, the resulting total change is

$$e^{-\delta_2 B/2}\, e^{-\delta_1 A/2}\, X\, e^{\delta_1 A/2}\, e^{\delta_2 B/2} =$$
$$= X + X \times (\delta_1 A + \delta_2 B) +$$
$$+ \tfrac{1}{2}\left((X \times \delta_1 A) \times \delta_1 A + 2(X \times \delta_1 A) \times \delta_2 B + (X \times \delta_2 B) \times \delta_2 B\right) + O(\delta^3).$$

To first order in the δs, the changes act independently and additively, but there is an interesting and asymmetrical structure in the second-order changes. This is most clearly seen when we attempt to undo the changes in opposite order. Many terms cancel (obviously those of first grade), and the Jacobi identity can be used to merge two terms, giving the result

$$e^{\delta_2 B/2}\, e^{\delta_1 A/2}\, e^{-\delta_2 B/2}\, e^{-\delta_1 A/2}\, X\, e^{\delta_1 A/2}\, e^{\delta_2 B/2}\, e^{-\delta_1 A/2}\, e^{-\delta_2 B/2} =$$
$$= X + X \times (\delta_1 A \times \delta_2 B) + O(\delta^3).$$

To the first relevant order, this changes X by an additive commutator with a bivector. Therefore, the commutator combination of two changes together acts like a new versor-type change, according to the bivector $(\delta_1 A \times \delta_2 B)$:

$$e^{\delta_2 B/2}\, e^{\delta_1 A/2}\, e^{-\delta_2 B/2}\, e^{-\delta_1 A/2}\, X\, e^{\delta_1 A/2}\, e^{\delta_2 B/2}\, e^{-\delta_1 A/2}\, e^{-\delta_2 B/2} =$$

$$\approx e^{-\delta_1 \delta_2 A \times B / 2} \, X \, e^{\delta_1 \delta_2 A \times B / 2}$$

The new versor is of a smaller order than the two original changes (δ^2 rather than δ). Studying this combination of changes in transformations gets us into Lie algebra, classically used to analyze small continuous transformations. It can for instance be employed (in control theory) to prove that a few standard transformations suffice to achieve any transformation. In geometric algebra, the Lie algebra computations reduce to making a bivector basis for the space of transformations. That amounts to choosing a few bivectors as basic and trying to make the others by commutator products, commutators of commutators, and so on. This is possible because the algebra of bivectors is closed under the commutator product. If you can make a basis for the whole bivector space, this proves that any motion can be achieved by doing commutators of motions.

As an example, let us consider the combination of two rotations in Euclidean 3-space, in the $A = \mathbf{e}_1 \wedge \mathbf{e}_2$ plane and the $B = \mathbf{e}_2 \wedge \mathbf{e}_3$ plane, and investigate if we can make any rotation by a combination of these basic rotations. The commutator of the bivectors is $A \times B = -\mathbf{e}_3 \wedge \mathbf{e}_1$, so that performing a small rotation over angle ϕ in the A plane followed by a small rotation ψ in the B plane, and then reversing them, leads to a small rotation $\phi\psi$ in the $\mathbf{e}_1 \wedge \mathbf{e}_3$ plane. That rotation was not among our basic transformations, but it clearly completes the set of bivectors for rotors. It shows that with the two rotations we can make the third independent rotation. Directions in 3-D space are controllable with only two basic rotations.

By contrast, translations in 3-D really need three independent components to reach an arbitrary position. The reason is that translations commute, so that any commutator is zero. Geometrically, this implies that no independent translation can be created from two translations in a plane. (We will meet the bivectors of translations only later, in Chapter 13, but the argument is simple enough not to require precise representation.)

As a third example, consider the maneuvering of a car. You can only steer and drive (forward or backward), yet you can reach any position in any orientation. The car is obviously controllable. The basic parallel parking maneuver that allows a car to move sideways is actually a (simplified) sequence of two commutators of the steer and drive actions. For more details, see [16].

8.2.4 TRANSFORMATION OF A CHANGE

In Section 8.2.2, we showed the nature of small changes in elements like X caused by small rotors. Such changes can propagate through additional versors. For instance, if we have the transformation $X \mapsto V X / V$, and X is perturbed by a versor with characterizing bivector A, we can rewrite the result in terms of a perturbation of the original result:

$$V(e^{-\delta A/2} X e^{\delta A/2}) V^{-1} = (V e^{-\delta A/2} V^{-1})(V X V^{-1})(V e^{\delta A/2} V^{-1}).$$

Therefore, the result of the mapping gets perturbed by the mapped perturbation

$$Ve^{-\delta A/2} V^{-1} = e^{-V\delta AV^{-1}/2},$$

by (7.21). You can simply substitute the error bivector δA by $V\delta A/V$ to get the total error bivector of the perturbation on VX/V. No need for first-order approximations, the result is exact, and also holds for odd V.

8.2.5 CHANGE OF A TRANSFORMATION

Things are somewhat more subtle when it is not X that is perturbed in the mapping $X \mapsto VX/V$, but the versor V (which may be odd or even, though we temporarily drop the sign to show the structure of the argument more clearly). This happens, for instance, when you reflect in a plane that has some uncertainty in its parameters. When the versor V becomes $e^{-\delta A/2} V e^{\delta A/2}$, the total perturbation is

$$(e^{-\delta A/2} V e^{\delta A/2}) X (e^{-\delta A/2} V^{-1} e^{\delta A/2}).$$

We need to express this in terms of a versor operation on the transformation result VX/V to find out how that is perturbed. When we do so, the transformation versor on VX/V can be rewritten to first order as

$$e^{-\delta A/2} V e^{\delta A/2} V^{-1} \approx (V + V \times \delta A) V^{-1} = 1 + (V \times \delta A)/V. \tag{8.5}$$

These are the first few terms of the Taylor series of the exponential $\exp((V \times \delta A)/V)$, considered as a function of δA. So we find for the versor operator computing the perturbed result to first order:

$$e^{-\delta A/2} V e^{\delta A/2} V^{-1} = e^{(V \times \delta A)/V}.$$

This should be written as the versor $e^{-\delta B/2}$, and that demand defines the bivector of the local perturbation δB as

$$\delta B = -2(V \times \delta A)/V = \delta A - V\delta A/V. \tag{8.6}$$

This method of computation of a versor using only a first-order Taylor series is fine, as long as you remember that this is only valid to first order. The resulting versor is *not* the exact result valid for a big change to V. We can imagine the local validity of this technique when we rotate a mirror around a general axis. To a good approximation, the reflection rotates around the projection of the rotation axis onto the mirror, and that is described by the first-order rotor so that the reflection describes a circular arc. However, as the mirror rotates more, this projected axis changes and higher-order effects kick in; the circular arc was just a local second-order approximation to what is actually a caustic. (We will treat this application in Section 8.5.2.)

8.3 PARAMETRIC DIFFERENTIATION

After this treatment of the transformational changes of an element, we study the second type of change we mentioned in the introduction.

Parametric differentiation is concerned with changes in elements in their dependence on their defining constituents. As such, it generalizes both the usual scalar differentiation and the derivative from vector calculus. All differentiation is based on functional dependence of scalar functions. In the usual approach, when these scalar functions are coordinate functions of a parameterized spatial curve or a vector field, the derivatives themselves can be reassembled into a geometric quantity such as the tangent vector to the curve or the divergence of the vector field. Such elements are truly geometric in that they do not depend on the coordinate functions that were introduced, but this is not always clear from either their derivation, their form, or their use.

Geometric algebra offers a way of computing with derivatives without using coordinates in the first place, by developing a calculus to apply them to its elements constructed using its products. However, proper coordinate-free definitions of the geometrical derivatives along these lines would require us to view them as a ratio of integrals. This would lead us a bit too far astray—you are referred to [26] for such a treatment. Here we will follow a more direct coordinate-based route, starting from scalar differentiation, but we quickly rise above that to attain truly geometric differentiation, expressed in coordinate-free formulas and techniques.

We construct our differentiation operators from specific to general, in the order of scalar differentiation, directional vector differentiation, total vector differentiation, and multivector differentiation. The final concept is the most general and contains the others, but we prefer to build up to it gradually.

8.4 SCALAR DIFFERENTIATION

Scalar differentiation of a multivector-valued function $F(\tau)$ relative to its scalar parameter τ is defined in the usual manner:

$$\frac{d}{d\tau} F(\tau) \equiv \lim_{\epsilon \to 0} \frac{F(\tau + \epsilon) - F(\tau)}{\epsilon}.$$

Geometric algebra has little to add to this form of differentiation, even though the function can now take values in the algebra. This type of differentiation is simply a scalar operator that commutes with all elements of the algebra. Therefore, it can be freely moved in a geometric product of multivector-valued functions, and obeys the product rule:

$$\frac{d}{d\tau}[F(\tau)\,G(\tau)] = \frac{d}{d\tau}[F(\tau)]\,G(\tau) + F(\tau)\,\frac{d}{d\tau}[G(\tau)]. \tag{8.7}$$

Yet we will see later that scalar differentiation is a particular instance of a more general multivector differentiation, and in preparation for that we denote it as ∂_τ.

Since the function F is typically defined using geometric algebra products, the differentiation result may also allow compact formulation using those products, so it is worth carrying out these differentiations symbolically. The following gives a simple example:

Let a vector $\mathbf{x}$ follow a curve on an orbit parameterized as $\mathbf{x}(\tau)$ by the time parameter τ. If we want to differentiate the scalar-valued vector function $\mathbf{x} \mapsto \mathbf{x}^2$ (involving the geometric product) along the curve, this is done in careful detail as follows:

$$
\begin{aligned}
\partial_\tau \mathbf{x}(\tau)^2 &= \partial_\tau[\mathbf{x}(\tau)\,\mathbf{x}(\tau)] \\
&= \partial_\tau[\mathbf{x}(\tau)]\,\mathbf{x}(\tau) + \mathbf{x}(\tau)\,\partial_\tau[\mathbf{x}(\tau)] \quad \text{(product rule)} \\
&= 2\,\partial_\tau[\mathbf{x}(\tau)] \cdot \mathbf{x}(\tau) \qquad\qquad\qquad \text{(inner product definition)} \\
&= 2\,\dot{\mathbf{x}}(\tau) \cdot \mathbf{x}(\tau) \qquad\qquad\qquad\quad \text{(dot for time derivative)}
\end{aligned}
$$

The result of the scalar operator ∂_τ applied to the scalar-valued function $\mathbf{x}(\tau)^2$ is therefore a scalar, as you would expect. We will often leave the parameterization understood, and would then denote this in shorthand as $\partial_\tau \mathbf{x}^2 = 2\,\dot{\mathbf{x}} \cdot \mathbf{x}$.

The scalar differentiation can easily be applied to the constructions of geometric algebra. As an example, we show the scalar differentiation of a time-dependent rotor equation. Let the rotor be $R = e^{-\mathbf{I}\phi/2}$, where the bivector angle $\mathbf{I}\phi$ is a function of τ so that both rotation plane and rotation angle may vary. We use the rotor to produce a time-dependent, rotated version $\mathbf{X}(\tau) = R(\tau)\,\mathbf{X}_0\,R(\tau)^{-1}$ of some constant blade $\mathbf{X}_0$. For constant $\mathbf{I}$, scalar differentiation with respect to time gives (using chain rule and commutation rules)

$$
\begin{aligned}
\partial_\tau \mathbf{X}(\tau) &= \partial_\tau[e^{-\mathbf{I}\phi/2}\mathbf{X}_0 e^{\mathbf{I}\phi/2}] \\
&= -\tfrac{1}{2}\partial_\tau[\mathbf{I}\phi](e^{-\mathbf{I}\phi/2}\mathbf{X}_0 e^{\mathbf{I}\phi/2}) + \tfrac{1}{2}(e^{-\mathbf{I}\phi/2}\mathbf{X}_0 e^{\mathbf{I}\phi/2})\partial_\tau[\mathbf{I}\phi] \\
&= \tfrac{1}{2}(\mathbf{X}\,\partial_\tau[\mathbf{I}\phi] - \partial_\tau[\mathbf{I}\phi]\,\mathbf{X}) \\
&= \mathbf{X} \times \partial_\tau[\mathbf{I}\phi].
\end{aligned}
\qquad (8.8)
$$

This retrieves the commutator form of the change in a rotor transformation: $\mathbf{X}$ changes in first order by its commutator with the derivative of the bivector of the change. This agrees with our analysis of changes in a rotor-based transformation as a commutator in (8.4). For the more general case of a variable $\mathbf{I}$, see Structural Exercise 6.

The simple expression that results assumes a more familiar form when $\mathbf{X}$ is a vector $\mathbf{x}$ in 3-space, the attitude of the rotation plane is fixed so that $\frac{d}{dt}\mathbf{I} = 0$, and we introduce a scalar angular velocity $\omega \equiv \frac{d}{dt}\phi$. It is then common practice to introduce the vector dual to the plane as the angular velocity vector $\boldsymbol{\omega}$, so $\boldsymbol{\omega} \equiv (\omega\mathbf{I})^* = \omega\mathbf{I}/\mathbf{I}_3$. We obtain

$$
\tfrac{d}{dt}\mathbf{x} = \mathbf{x} \times \tfrac{d}{dt}(\mathbf{I}\phi) = \mathbf{x} \times (\omega\,\mathbf{I}_3) = \mathbf{x}\rfloor(\omega\,\mathbf{I}_3) = (\mathbf{x}\wedge\boldsymbol{\omega})\,\mathbf{I}_3 = \boldsymbol{\omega}\times\mathbf{x},
$$

where the final symbol $\times$ is the 3-D vector cross product. This shows the correspondence of our scalar differentiation with the classical way of expressing the change.

As before when we treated other operations, we find that an equally simple geometric algebra expression is much more general than the classical expression; here (8.8) describes the differential rotation of k-dimensional subspaces in n-dimensional space rather than merely of vectors in 3-D.

8.4.1 APPLICATION: RADIUS OF CURVATURE OF A PLANAR CURVE

In the differential geometry of planar curves in the Euclidean plane $\mathbb{R}^{2,0}$, you often want a local description of a parameterized curve $\mathbf{r}(\tau)$ in terms of its local tangent circle. That characterizes the curve well to second order; the local curvature is the reciprocal of the radius of this tangent circle. The following derivation is a good example of a proper classical coordinate-free treatment, borrowed from [50], which we are then able to complete to a closed-form solution using geometric algebra.

Let the local tangent circle be characterized by its center $\mathbf{c}$ and its radius ρ. Then a point $\mathbf{r}$ lies on it if it satisfies $(\mathbf{c} - \mathbf{r})^2 = \rho^2$. Now we let $\mathbf{r}$ be the parameterized curve point $\mathbf{r}(\tau)$, which relative to its parameter τ has first derivative $\dot{\mathbf{r}}(\tau)$ and second derivative $\ddot{\mathbf{r}}(\tau)$. (This handy overdot notation of these "fluxions" is common in physics texts.) Taking derivatives of the defining equation, we get the following list of requirements on $\mathbf{c}$ and ρ:

$$(\mathbf{c} - \mathbf{r})^2 = \rho^2$$

$$2\,(\mathbf{c} - \mathbf{r}) \cdot \dot{\mathbf{r}} = 0$$

$$-2\,\dot{\mathbf{r}} \cdot \dot{\mathbf{r}} + 2\,(\mathbf{c} - \mathbf{r}) \cdot \ddot{\mathbf{r}} = 0$$

Our source [50] stops here, but we can continue because we have geometric algebra. The occurrence of $(\mathbf{c} - \mathbf{r})$ in an inner product with both $\dot{\mathbf{r}}$ and $\ddot{\mathbf{r}}$ makes us wonder what $(\mathbf{c} - \mathbf{r}) \rfloor (\dot{\mathbf{r}} \wedge \ddot{\mathbf{r}})$ might be, since that contains both terms by (3.17). Because of the equations above, it is fortunately independent of both $\mathbf{c}$ and ρ:

$$(\mathbf{c} - \mathbf{r}) \rfloor (\dot{\mathbf{r}} \wedge \ddot{\mathbf{r}}) = ((\mathbf{c} - \mathbf{r}) \cdot \dot{\mathbf{r}})\,\ddot{\mathbf{r}} - ((\mathbf{c} - \mathbf{r}) \cdot \ddot{\mathbf{r}})\,\dot{\mathbf{r}} = -(\dot{\mathbf{r}} \cdot \dot{\mathbf{r}})\,\dot{\mathbf{r}} = -\dot{\mathbf{r}}^3 \qquad (8.9)$$

Moreover, since in 2-D any trivector is zero, so is $(\mathbf{c} - \mathbf{r}) \wedge (\dot{\mathbf{r}} \wedge \ddot{\mathbf{r}})$. Therefore, the contraction in (8.9) can be replaced by a geometric product. Since that is invertible, we can perform right division and obtain

$$\mathbf{c} = \mathbf{r} - \frac{\dot{\mathbf{r}}^3}{\dot{\mathbf{r}} \wedge \ddot{\mathbf{r}}}$$

as the closed-form solution for $\mathbf{c}$, and by substitution we obtain ρ^2:

$$\rho^2 = \left(\frac{\dot{\mathbf{r}}^3}{\dot{\mathbf{r}} \wedge \ddot{\mathbf{r}}} \right)^2.$$

In both expressions, we recognize the occurrence of the reciprocal of the rejection of $\ddot{\mathbf{r}}$ by $\dot{\mathbf{r}}$—so only the component of $\ddot{\mathbf{r}}$ orthogonal to $\dot{\mathbf{r}}$ contributes to these geometric quantities (the other part is related to reparameterization, and is geometrically less interesting). The center of the tangent circle is clearly in the direction orthogonal to $\dot{\mathbf{r}}$.

The ensuing expression for the curvature requires a square root of a square of a vector; its sign should be related to choosing a positive direction for vectors orthogonal to $\dot{\mathbf{r}}$. Using the pseudoscalar $\mathbf{I}_2$ of the plane for dualization, we use $\dot{\mathbf{r}}^*$ as the positive direction relative to $\dot{\mathbf{r}}$. Then the curvature is

$$\kappa = 1/\rho = \frac{(\dot{\mathbf{r}} \wedge \ddot{\mathbf{r}})^*}{\|\dot{\mathbf{r}}\|^3}.$$

This is easily converted to the familiar coordinate form by setting $\mathbf{r}(\tau) = x(\tau)\,\mathbf{e}_1 + y(\tau)\,\mathbf{e}_2$, with the parameter derivatives of the functions x and y denoted by overdots:

$$\kappa = \frac{\dot{x}\ddot{y} - \dot{y}\ddot{x}}{(\dot{x}^2 + \dot{y}^2)^{3/2}}.$$

This expression takes considerably more work to derive when using coordinates from the start.

8.5 DIRECTIONAL DIFFERENTIATION

Let $F(\mathbf{x})$ be an element of geometric algebra dependent on a vector $\mathbf{x}$. (If $\mathbf{x}$ is the position vector, this would be a position-dependent quantity, such as a vector field or a bivector field in the space $\mathbb{R}^n$.) We may want to know how $F(\mathbf{x})$ changes at a particular value of $\mathbf{x}$ if we would move in the direction $\mathbf{a}$. It will clearly vary by an amount of the same grade type as F itself, so such a directional differentiation is a scalar operator on $F(\mathbf{x})$. It is denoted by $(\mathbf{a} * \partial_{\mathbf{x}})$—we will explain why soon—and defined as

$$(\mathbf{a} * \partial_{\mathbf{x}})\, F(\mathbf{x}) \equiv \lim_{\epsilon \to 0} \frac{F(\mathbf{x} + \epsilon\,\mathbf{a}) - F(\mathbf{x})}{\epsilon}.$$

Since it is a scalar operator, it commutes with all elements. You might expect that this implies that it acts very much like differentiation in real calculus, but that is incorrect: the geometric products in the functions it acts on make for rather different (but geometrically correct) results.

As an example, the function $\mathbf{x} \mapsto \mathbf{x}^2$ is defined everywhere, and gives a scalar field on the vector space $\mathbb{R}^n$. Its directional derivative is

$$(\mathbf{a} * \partial_{\mathbf{x}})[\mathbf{x}^2] = \lim_{\epsilon \to 0} \frac{(\mathbf{x} + \epsilon\,\mathbf{a})^2 - \mathbf{x}^2}{\epsilon} \qquad \text{(definition)}$$

$$= \lim_{\epsilon \to 0} \frac{\epsilon\,\mathbf{x}\,\mathbf{a} + \epsilon\,\mathbf{a}\,\mathbf{x} + \epsilon^2\mathbf{a}^2}{\epsilon} \qquad \text{(definition)}$$

$$= \mathbf{x}\,\mathbf{a} + \mathbf{a}\,\mathbf{x} \qquad \text{(limit process)}$$

$$= 2\,\mathbf{a} \cdot \mathbf{x} \qquad \text{(inner product definition)}$$

You see the familiar result: there is no variation when $\mathbf{a}$ is perpendicular to $\mathbf{x}$, and maximum variation in the $\mathbf{x}$-direction.

Since the differentiation is a scalar operator, it can be moved freely through expressions, and obeys a product rule like (8.7).

8.5.1 TABLE OF ELEMENTARY RESULTS

We do some basic derivations and collect them in Table 8.1, which contains other results that follow the same pattern. (In our derivations here, we assume that all vectors reside in the same space, but the table is slightly more general and requires projection of the parameter vectors to the space of $\mathbf{x}$, hence the occurrence of $\mathsf{P}[\mathbf{a}]$. We explain this in Section 8.6.)

- The *identity function* $F(\mathbf{x}) = \mathbf{x}$ has the derivative you would expect:

$$(\mathbf{a} * \partial_{\mathbf{x}})\,\mathbf{x} = \lim_{\epsilon \to 0} \frac{(\mathbf{x} + \epsilon\,\mathbf{a}) - \mathbf{x}}{\epsilon} = \mathbf{a}.$$

- Scalar differentiation of the *inner product* leads to a substitution of $\mathbf{x}$ by its change $\mathbf{a}$:

$$(\mathbf{a} * \partial_{\mathbf{x}})\,(\mathbf{x} \cdot \mathbf{b}) = \mathbf{a} \cdot \mathbf{b}.$$

- The *inner product with a vector-valued linear function* f unexpectedly pulls in the adjoint function $\overline{\mathsf{f}}$ of Section 4.3.2:

$$(\mathbf{a} * \partial_{\mathbf{x}})\,(\mathsf{f}[\mathbf{x}] \cdot \mathbf{b}) = (\mathbf{a} * \partial_{\mathbf{x}})\,(\mathbf{x} \cdot \overline{\mathsf{f}}[\mathbf{b}]) = \mathbf{a} \cdot \overline{\mathsf{f}}[\mathbf{b}].$$

- The *scalar derivative of the inverse* $1/\mathbf{x} = \mathbf{x}^{-1}$ gives a surprising result:

$$(\mathbf{a} * \partial_{\mathbf{x}})\,\mathbf{x}^{-1} = \lim_{\epsilon \to 0} \frac{1}{\epsilon}\left(\frac{1}{\mathbf{x} + \epsilon\,\mathbf{a}} - \frac{1}{\mathbf{x}}\right)$$

$$= \lim_{\epsilon \to 0} \frac{1}{\epsilon}\left(\frac{\mathbf{x} + \epsilon\,\mathbf{a}}{\mathbf{x}^2 + 2\epsilon\,\mathbf{a} \cdot \mathbf{x}} - \frac{1}{\mathbf{x}}\right)$$

$$= \lim_{\epsilon \to 0} \frac{1}{\epsilon}\left(\frac{\mathbf{x}\,(1 + \epsilon\,\mathbf{x}^{-1}\,\mathbf{a})}{\mathbf{x}^2\,(1 + 2\epsilon\,\mathbf{a} \cdot \mathbf{x}^{-1})} - \frac{1}{\mathbf{x}}\right)$$

$$= \lim_{\epsilon \to 0} \frac{1}{\epsilon}\left(\frac{\mathbf{x}}{\mathbf{x}^2}\,(1 + \epsilon\,\mathbf{x}^{-1}\,\mathbf{a})\,(1 - 2\epsilon\,\mathbf{a} \cdot \mathbf{x}^{-1}) - \frac{1}{\mathbf{x}}\right)$$

Table 8.1: Directional derivatives and vector derivatives of common functions in an m-dimensional vector manifold $\mathbb{R}^m$ within a larger vector manifold $\mathbb{R}^n$. Here $\mathbf{x}$, $\mathbf{a}$ are vectors, $\mathbf{A}$ is a blade, $\mathsf{P}[\,]$ is shorthand for the projection $\mathsf{P}_{\mathsf{I}_m}[\,] : \mathbb{R}^n \to \mathbb{R}^m$ locally mapping vectors onto the lower-dimensional manifold.

$$(\mathbf{a} * \partial_{\mathbf{x}})\,\mathbf{x} = \mathsf{P}[\mathbf{a}]$$

$$(\mathbf{a} * \partial_{\mathbf{x}})\,(\mathbf{x} \cdot \mathbf{b}) = \mathsf{P}[\mathbf{a}] \cdot \mathbf{b}$$

$$(\mathbf{a} * \partial_{\mathbf{x}})\,\mathbf{x}^{-1} = -\mathbf{x}^{-1}\,\mathsf{P}[\mathbf{a}]\,\mathbf{x}^{-1}$$

$$(\mathbf{a} * \partial_{\mathbf{x}})\,\|\mathbf{x}\|^k = k\,(\mathsf{P}[\mathbf{a}] \cdot \mathbf{x})\,\|\mathbf{x}\|^{k-2}$$

$$(\mathbf{a} * \partial_{\mathbf{x}})\,\frac{\mathbf{x}}{\|\mathbf{x}\|^k} = \frac{\mathsf{P}[\mathbf{a}] - k\,(\mathsf{P}[\mathbf{a}] \cdot \mathbf{x})/\mathbf{x}}{\|\mathbf{x}\|^k}$$

$$\partial_{\mathbf{x}}\,\mathbf{x} = m$$

$$\partial_{\mathbf{x}} \cdot \mathbf{x} = m$$

$$\partial_{\mathbf{x}} \wedge \mathbf{x} = 0$$

$$\partial_{\mathbf{x}}(\mathbf{x} \cdot \mathbf{a}) = \mathsf{P}[\mathbf{a}]$$

$$\partial_{\mathbf{x}}(\mathbf{x} \wedge \mathbf{a}) = (m-1)\,\mathsf{P}[\mathbf{a}]$$

$$\partial_{\mathbf{x}}(\mathbf{x}\mathbf{A}) = m\,\mathsf{P}[\mathbf{A}]$$

$$\partial_{\mathbf{x}}(\mathbf{x} \rfloor \mathbf{A}) = \mathrm{grade}(\mathbf{A})\,\mathsf{P}[\mathbf{A}]$$

$$\partial_{\mathbf{x}}(\mathbf{x} \wedge \mathbf{A}) = (m - \mathrm{grade}(\mathbf{A}))\,\mathsf{P}[\mathbf{A}]$$

$$\partial_{\mathbf{x}}(\widehat{\mathbf{A}}\mathbf{x}) = (m - 2\,\mathrm{grade}(\mathbf{A}))\,\mathsf{P}[\mathbf{A}]$$

$$\partial_{\mathbf{x}}\|\mathbf{x}\| = \frac{\mathbf{x}}{\|\mathbf{x}\|}$$

$$\partial_{\mathbf{x}}\|\mathbf{x}\|^k = k\,\|\mathbf{x}\|^{k-2}\mathbf{x}$$

$$\partial_{\mathbf{x}}\frac{\mathbf{x}}{\|\mathbf{x}\|^k} = \frac{m-k}{\|\mathbf{x}\|^k}$$

$$\partial_{\mathbf{x}}\,(\mathsf{f}[\mathbf{x}] \cdot \mathbf{y}) = \mathsf{P}[\overline{\mathsf{f}}[\mathbf{y}]]$$

$$= \mathbf{x}^{-1}\,(\mathbf{x}^{-1}\,\mathbf{a} - 2\mathbf{a}\cdot\mathbf{x}^{-1})$$
$$= -\mathbf{x}^{-1}\,\mathbf{a}\,\mathbf{x}^{-1}.$$

This clearly differs from the classical result in real analysis (ignoring all commutation restrictions, we would get the familiar $-\mathbf{a}/\mathbf{x}^2$). The construction can be immediately interpreted geometrically, as in Figure 8.1. When you realize that $\mathbf{x}^{-1}$ is the inversion of $\mathbf{x}$ in the unit sphere, you see that a change $\mathbf{a}$ in $\mathbf{x}$ is a $1/\mathbf{x}^2$-scaled version of the reflection of $\mathbf{a}$ relative to the plane perpendicular to $\mathbf{x}$, which is exactly what the differentiation result signifies.

- For *powers of the norm*, which are scalar functions, we retrieve a semblance of the usual calculus result.

$$(\mathbf{a} * \partial_{\mathbf{x}})\,\|\mathbf{x}\|^{k} = \lim_{\epsilon\to 0}\frac{1}{\epsilon}\left(\|\mathbf{x}+\epsilon\mathbf{a}\|^{k} - \|\mathbf{x}\|^{k}\right)$$

$$= \lim_{\epsilon\to 0}\frac{1}{\epsilon}\left(\|\mathbf{x}\|^{k}(\sqrt{1+2\epsilon\mathbf{x}^{-1}\cdot\mathbf{a}})^{k} - \|\mathbf{x}\|^{k}\right)$$

$$= \lim_{\epsilon\to 0}\frac{1}{\epsilon}\left(\|\mathbf{x}\|^{k}(1+k\epsilon\mathbf{x}^{-1}\cdot\mathbf{a}) - \|\mathbf{x}\|^{k}\right)$$

$$= k\,\mathbf{x}\cdot\mathbf{a}\,\|\mathbf{x}\|^{k-2}.$$

The other entries of Table 8.1 can be demonstrated using similar techniques.

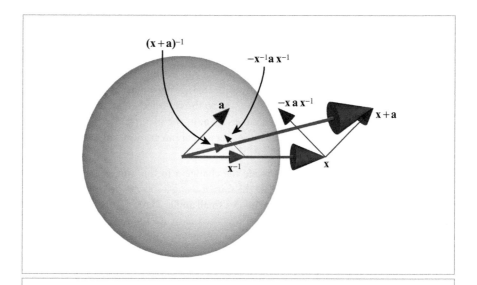

Figure 8.1: Directional differentiation of a vector inversion. The small additive perturbation vector $\mathbf{a}$ is reflected in the plane with normal $\mathbf{x}$ to make $-\mathbf{x}\,\mathbf{a}\,\mathbf{x}^{-1}$, and the result scaled by $1/\mathbf{x}^2$ to produce $-\mathbf{x}^{-1}\,\mathbf{a}\,\mathbf{x}^{-1}$ as the correct difference (to first order) between $(\mathbf{x}+\mathbf{a})^{-1}$ and $\mathbf{x}^{-1}$.

8.5.2 APPLICATION: TILTING A MIRROR

Consider the situation where we have a planar mirror in the origin with normal vector $\mathbf{n}$ (not necessarily a unit normal). The reflection of an element X in this mirror is given by (see Section 7.1):

$$X \;\mapsto\; \mathsf{n}[X] \equiv \mathbf{n}\,\widehat{X}\,\mathbf{n}^{-1}.$$

We now perturb the mirror, for instance by a small rotation, and want to know what happens to the reflection result. Let us do this in two steps: first we see how any change in $\mathbf{n}$ affects the reflection result; then we relate the change in $\mathbf{n}$ to the parameters of the perturbing rotational action on the mirror.

- For the first step, we apply the directional derivative for an $\mathbf{a}$-change in $\mathbf{n}$:

$$\begin{aligned}
(\mathbf{a} * \partial_{\mathbf{n}})[\mathbf{n}\,\widehat{X}\,\mathbf{n}^{-1}] &= \mathbf{a}\,\widehat{X}\,\mathbf{n}^{-1} + \mathbf{n}\,\widehat{X}\,(-\mathbf{n}^{-1}\,\mathbf{a}\,\mathbf{n}^{-1}) \\
&= (\mathbf{a}\,\mathbf{n}^{-1})\,(\mathbf{n}\,\widehat{X}\,\mathbf{n}^{-1}) - (\mathbf{n}\,\widehat{X}\,\mathbf{n}^{-1})\,(\mathbf{a}\,\mathbf{n}^{-1}) \\
&= 2(\mathbf{a}\,\mathbf{n}^{-1}) \times (\mathbf{n}\,\widehat{X}\,\mathbf{n}^{-1}) \\
&= (\mathbf{n}\,\widehat{X}\,\mathbf{n}^{-1}) \times (2\,\mathbf{n}^{-1} \wedge \mathbf{a})
\end{aligned}$$

The final simplification holds because the scalar part $\mathbf{n}^{-1} \cdot \mathbf{a}$ of $\mathbf{n}^{-1}\mathbf{a}$ does not contribute to the commutator product result.

The result shows that it is the part of $\mathbf{a}$ perpendicular to $\mathbf{n}$ that causes changes to the reflection. This is, of course, just what we would have expected, since the magnitude of $\mathbf{n}$ does not affect the reflection $\mathsf{n}[X]$ at all. A small orthogonal change to a vector is effectively a rotation, so the directional derivative is eminently suited to process the rotational change. But there is more: to first order, the change in the reflection $\mathsf{n}[X]$ can be written as a commutator. Therefore, it can be represented (at least locally) as a rotor transformation. Comparing with (8.4), we see that the bivector $\mathbf{B}$ of the transforming rotor equals $\mathbf{B} = 2\mathbf{n}^{-1} \wedge \mathbf{a}$. So the reflected element $\mathsf{n}[X]$ describes a rotation as the mirror normal changes by $\mathbf{a}$, in the plane $\mathbf{n}^{-1} \wedge \mathbf{a}$, by a rotation angle $\|\mathbf{n}^{-1} \wedge \mathbf{a}\|$. Recognizing this is in fact a local integration, since it reverses the differentiation process.

- In the second step, we need to relate the change $\mathbf{a}$ in the mirror normal $\mathbf{n}$ to an actual transformation. Let us rotate the mirror using a rotor $\exp(-\mathbf{I}\phi/2)$, with $\mathbf{I}$ the unit 2-blade of the rotation plane and ϕ a small angle. Then, according to (8.3), the normal vector $\mathbf{n}$ changes to first order by the vector $\mathbf{n} \times \mathbf{I}\phi$. This is therefore what we should use as our $\mathbf{a}$.

- Combining the two results, the bivector of the total transformation in the reflected X is

$$\mathbf{B} = 2\mathbf{n}^{-1} \wedge \mathbf{a} = 2\,\phi\,\mathbf{n}^{-1} \wedge (\mathbf{n} \times \mathbf{I}) = 2\,\phi\,\mathbf{n}^{-1} \wedge (\mathbf{n} \rfloor \mathbf{I}). \qquad (8.10)$$

That result is valid in any number of dimensions. It gives the bivector of the resulting rotation of $\mathsf{n}[X]$, which specifies both the rotation plane and its angle.

To get a better feeling for the geometry of (8.10) in 3-D, introduce the unit rotational axis of the mirror motion $\mathbf{m} = \mathbf{I}^*$, normalize $\mathbf{n}$ to unity, and express the result as a rotational axis $\mathbf{b} = \mathbf{B}^*$. Some manipulation gives

$$\mathbf{b} = 2\,\phi\,\mathbf{n}\rfloor(\mathbf{n} \wedge \mathbf{m}) = 2\,\phi(\mathbf{m} \wedge \mathbf{n})/\mathbf{n}.$$

This axis is the rejection of $\mathbf{m}$ by $\mathbf{n}$, or (if you prefer) the projection of the axis $\mathbf{m}$ onto the plane with normal vector $\mathbf{n}$. That projection obtains a factor $\sin \psi$ of the angle ψ between $\mathbf{n}$ and $\mathbf{m}$. The rotation angle β for the reflection $\mathsf{n}[X]$ of X under the rotation of ϕ around the $\mathbf{m}$ axis is the norm of $\mathbf{b}$, which evaluates as

$$\beta = 2\phi\,\sin(\psi). \tag{8.11}$$

This is a rather powerful result acquired with fairly little effort, only at the very last moment requiring some trivial trigonometry. Figure 8.2 sketches the situation. Two

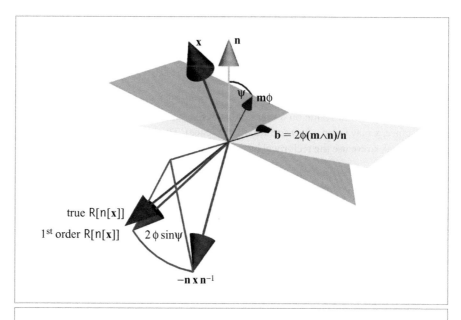

Figure 8.2: Changes in reflection of a rotating mirror. The yellow mirror with normal $\mathbf{n}$ rotates around the $\mathbf{m}$ axis over an angle ϕ, producing the green mirror plane. This changes the reflection $-\mathbf{n}\mathbf{x}\mathbf{n}^{-1}$ of a vector $\mathbf{x}$ to the gray vector. That change is to first order described as the rotation of $-\mathbf{n}\mathbf{x}\mathbf{n}^{-1}$ around an axis that is the projection of $\mathbf{m}$ on the $\mathbf{n}$ plane, over an angle $2\phi\sin\psi$, where ψ is the angle between $\mathbf{n}$ and $\mathbf{m}$. This involved and geometrically quantitative figure is the result of only a few lines of coordinate-free computation in geometric algebra.

special cases make perfect sense: if $\psi = 0$, then **n** and **m** are aligned, and indeed no rotation over **m** changes the reflection of X; and if $\psi = \pi/2$, then **n** and **m** are perpendicular, and any rotation ϕ of the rotation plane becomes a 2ϕ rotation of the reflection n[X].

We will get back to this rotated reflection in its full generality in Section 13.7.

8.6 VECTOR DIFFERENTIATION

In scalar differentiation, we consider a vector function as a changing in time (or some such scalar parameter). We may also want to consider $F(\mathbf{x})$ as a function of position as encoded by the vector variable **x**, and differentiate directly relative to that variable. This is most easily defined by developing it on a basis, doing a directional differentiation with respect to each of the components, and reassembling the result in one quantity. It is the ∇-operator of vector analysis, but we will denote it as $\partial_{\mathbf{x}}$. This explicitly specifies the variable relative to which we differentiate and prepares for a generalization beyond vectors and toward differential geometry. On a basis $\{\mathbf{e}_i\}_{i=1}^m$ for the space $\mathbb{R}^m$ in which **x** resides, let x^i denote the coordinate functions of the vector **x** so that it can be written as

$$\mathbf{x} = \sum_{i+1}^m x^i \mathbf{e}_i.$$

We will be setting up this vector differentiation in a very general framework, in which the space $\mathbb{R}^m$ of **x** may reside on a manifold (curved subspace) within a larger space $\mathbb{R}^n$ (for instance, **x** may lie on a 2-D surface in 3-D space). The basis for $\mathbb{R}^m$ may then not be orthonormal, so we use the reciprocal basis of Section 3.8, and compute x^i as $x^i = \mathbf{e}^i \cdot \mathbf{x}$.

The directional derivative in the coordinate direction of $\mathbf{e}_i$ is simply the scalar derivative of the coordinate function:

$$(\mathbf{e}_i * \partial_{\mathbf{x}}) = \tfrac{\partial}{\partial x^i} = \partial_{x^i}.$$

As their notation suggests, we can assemble the results of each of these directional operators and consider them as the components of a more general *vector derivative operation* defined on this basis as

$$\partial_{\mathbf{x}} \equiv \sum_{i=1}^m \mathbf{e}^i \, (\mathbf{e}_i * \partial_{\mathbf{x}}) = \sum_i \mathbf{e}^i \, \tfrac{\partial}{\partial x^i}. \tag{8.12}$$

(When you study reciprocal frames, expressions like these are actually coordinate-free when they contain the upper and lower indices that cancel; in physics, lower-index vectors are called covariant and upper-index vectors contravariant, but we will not follow that terminology here.)

The operator $\partial_{\mathbf{x}}$ computes the total change in its argument when **x** changes in all possible ways, but it keeps track of those changes in a geometrical manner, registering the $\mathbf{e}_i$-related scalar change in the magnitude of the $\mathbf{e}^i$ component of the total change. Preserving this geometrical information is surprisingly powerful, and in advanced geometric calculus it is shown that this operator can be inverted by integration (see [26]).

You should interpret the grade of the operator $\partial_{\mathbf{x}}$ as a vector (i.e., as the grade of its subscript). As a geometrical vector operator, it should conform to the commutation rules for geometric products. We will not use the square application brackets here, for it is more productive to see this as a geometric element rather than as a linear operator, and to move it to other places in the sequence of symbols for computational purposes. The subscript $\mathbf{x}$ in $\partial_{\mathbf{x}}$ denotes which vector variable is being differentiated (and this is necessary when there is more than one).

As an example, we apply the vector differentiation to the function $F(\mathbf{x}) = \mathbf{x}^2$, relative to its vector parameter $\mathbf{x}$:

$$
\begin{aligned}
\partial_{\mathbf{x}}\,\mathbf{x}^2 &= \sum_i \mathbf{e}^i \partial_{x^i}\Big(\sum_{j,k} x^j x^k\,\mathbf{e}_j \cdot \mathbf{e}_k\Big) && \text{(coordinate definition)} \\[2mm]
&= \sum_i \mathbf{e}^i\Big(\sum_k x^k\,\mathbf{e}_i \cdot \mathbf{e}_k + \sum_j x^j\,\mathbf{e}_j \cdot \mathbf{e}_i\Big) && \text{(coordinate independence)} \\[2mm]
&= 2\sum_i \mathbf{e}^i\,(\mathbf{e}_i \cdot \mathbf{x}) && \text{(linearity)} \\[2mm]
&= 2\,\mathbf{x}.
\end{aligned}
$$

(8.13)

We obtain the result $2\mathbf{x}$, which you might have expected from pattern matching with scalar differentiation (though that is a dangerous principle to apply). The result is not a vector, but a vector field that has the value $2\mathbf{x}$ at a location $\mathbf{x}$. This vector field is in fact the *gradient* of the scalar function $\mathbf{x}^2$ (i.e., the direction in which it varies most, with a magnitude that indicates the amount of variation).

The recognition of the multiplication in $\partial_{\mathbf{x}}\,F(\mathbf{x})$ as the geometric product makes it quite natural to expand this in terms of the inner and outer product, simply applying (6.14):

$$
\partial_{\mathbf{x}}\,F(\mathbf{x}) = \partial_{\mathbf{x}} \rfloor F(\mathbf{x}) + \partial_{\mathbf{x}} \wedge F(\mathbf{x}).
$$

For a vector-valued function F, the first term corresponds to the usual *divergence* operator $\mathrm{div}[F(\mathbf{x})] \equiv \nabla \cdot F(\mathbf{x})$, and the second term is related to the *curl* operator $\mathrm{rot}[F(\mathbf{x})] \equiv \nabla \times F(\mathbf{x})$, written in terms of the 3-D cross product; it is actually its dual. As with the other uses of the cross product, replacing the curl by an outer-product-based construction ensures validity in arbitrary dimensionality. If F is scalar-valued, then only the $\partial_{\mathbf{x}} \wedge F(\mathbf{x})$ term remains, and is identical to the *gradient* operator $\mathrm{grad}[F(\mathbf{x})] = \nabla F(\mathbf{x})$. For a symmetric vector function F^+ (equal to its adjoint), the part $\partial_{\mathbf{x}} \wedge F^+[\mathbf{x}]$ equals zero, for a skew-symmetric vector function F^- (opposite to its adjoint), the part $\partial_{\mathbf{x}} \cdot F^-[\mathbf{x}]$ equals zero.

8.6.1 ELEMENTARY RESULTS OF VECTOR DIFFERENTIATION

We have introduced the vector differentiation as the geometric algebra equivalent of the ∇-operator from vector analysis. Although the definition as we have given it uses coordinates, the vector differentiation is a proper geometrical operation that is not dependent on any chosen coordinate system. When you apply it, you should avoid coordinates, and instead use results from a table of standard functions (combined with product rule and chain rule of differentiation). We give such a collection of useful elementary results in Table 8.1, and derive some of its more educational entries below.

- **Identity Function x.** The identity function $F(\mathbf{x}) = \mathbf{x}$ has a derivative that depends on the dimensionality of the space $\mathbb{R}^m$ in which $\mathbf{x}$ resides.

$$\partial_{\mathbf{x}} \mathbf{x} = \sum_i \mathbf{e}^i \frac{\partial}{\partial x^i} [\sum_j x^j \mathbf{e}_j]$$

$$= \sum_{i,j} \delta^j_i \, \mathbf{e}^i \, \mathbf{e}_j = \sum_i \mathbf{e}^i \, \mathbf{e}_i$$

$$= \sum_i \mathbf{e}^i \cdot \mathbf{e}_i + \sum_i \mathbf{e}^i \wedge \mathbf{e}_i = \sum_i 1 + 0 = m.$$

(Here we used $\sum_i \mathbf{e}^i \wedge \mathbf{e}_i = 0$, given as (3.35).) This algebraic derivation gives a clue for the correct geometrical way to look at this: all changes in all directions are to be taken into account. In m-dimensional space, there are m directions, and each of these provide a unit change in coordinates with each unit step, for a total of m.

Since the vector differentiation applies as a geometric product, you can split the result in an inner and outer product part that the computation above has shown to obey $\partial_{\mathbf{x}} \cdot \mathbf{x} = m$ and $\partial_{\mathbf{x}} \wedge \mathbf{x} = 0$. The outer product result $\partial_{\mathbf{x}} \wedge \mathbf{x} = 0$ shows that you can think of $\partial_{\mathbf{x}}$ as being like a vector in the $\mathbf{x}$ direction, and the inner product result then shows that it is like $m/\mathbf{x}$ (but view these as no more than mnemonics; $\partial_{\mathbf{x}}$ is of course not a vector but an operator).

- **Inner Product x · a.** When we study the change in the scalar quantity $F(\mathbf{x}) = \mathbf{x} \cdot \mathbf{a}$ (geometrically the projected component of $\mathbf{x}$ onto a vector $\mathbf{a}^{-1}$), we should in general allow for the variations of $\mathbf{x}$ to be in its m-dimensional manifold (curved subspace), whereas $\mathbf{a}$ may be a vector of the encompassing space $\mathbb{R}^n$ (for instance, $\mathbf{x}$ on a sphere, $\mathbf{a}$ a general vector in 3-D space; $\mathbf{x} \cdot \mathbf{a}$ is well defined everywhere on the sphere, so it has a derivative).

Two things happen to the measured changes caused by variations in $\mathbf{x}$. First, even when $\mathbf{x}$ and $\mathbf{a}$ are in the same m-dimensional space, the quantity $\mathbf{x} \cdot \mathbf{a}$ can only pick up the changes in the direction $\mathbf{a}$, so summing over all directions only this 1-D variation remains. Second, $\mathbf{x}$ cannot really vary in the $\mathbf{a}$-direction, since it has to remain in its

m-dimensional manifold, or more accurately, in the tangent space at $\mathbf{x}$ isomorphic to $\mathbb{R}^m$, for which $\{\mathbf{e}_i\}_{i=1}^m$ is the basis. It is the projection of the $\mathbf{a}$-direction onto this tangent space that must be the actual gradient.

The algebraic computation confirms this, with indices i and j ranging over coordinates for the space in which $\mathbf{x}$ resides, and k over the space of $\mathbf{a}$, using a local coordinate basis for the total n-dimensional space in which the problem is defined:

$$
\begin{aligned}
\partial_{\mathbf{x}}\,(\mathbf{x}\cdot\mathbf{a}) &= \sum_i \mathbf{e}^i\,\tfrac{\partial}{\partial x^i}[\sum_j\sum_k x^j a^k \mathbf{e}_j\cdot\mathbf{e}_k] \\
&= \sum_i\sum_k a^k\,\mathbf{e}^i\,(\mathbf{e}_i\cdot\mathbf{e}_k) \\
&= \sum_i a^i\,\mathbf{e}_i = \mathsf{P}_{\mathbf{I}_m}[\mathbf{a}],
\end{aligned}
$$

since the summation of the $\mathbf{a}$ components is only done for the elements in the basis of the tangent space at $\mathbf{x}$ with pseudoscalar $\mathbf{I}_m$. In tables, we will use $\mathsf{P}[\mathbf{a}]$ as shorthand.

- **Outer Product** $\mathbf{x}\wedge\mathbf{a}$. When we compute the variation of the bivector $\mathbf{x}\wedge\mathbf{a}$, this can be rewritten as the variation of $\mathbf{x}\mathbf{a}-\mathbf{x}\cdot\mathbf{a}$. The variation over $\mathbf{x}$ in the first term causes a factor m (the dimensionality of the space that $\mathbf{x}$ resides in), but of course it picks up only the part $\mathsf{P}[\mathbf{a}]$ of $\mathbf{a}$. The second term we have seen above, and the total variation is now $\partial_{\mathbf{x}}\,(\mathbf{x}\wedge\mathbf{a}) = (m-1)\,\mathsf{P}[\mathbf{a}]$.

- **Norm** $\|\mathbf{x}\|$. Geometrically, what would you expect the derivative of the norm to be? Since it is a scalar function, the vector derivative will be the gradient of the norm, i.e., the direction in which it increases most steeply weighted by the weight of increase. So the answer should be $\mathbf{x}/\|\mathbf{x}\|$, the unit vector in the $\mathbf{x}$ direction. The algebraic computation confirms that it is:

$$
\begin{aligned}
\partial_{\mathbf{x}}\,\|\mathbf{x}\| &= \sum_i \mathbf{e}^i\partial_{x^i}\Big(\sum_{j,k} x^j x^k\,\mathbf{e}_j\cdot\mathbf{e}_k\Big)^{1/2} \\
&= \sum_i \mathbf{e}^i\,\Big(\sum_k x^k\,\mathbf{e}_i\cdot\mathbf{e}_k + \sum_j x^j\,\mathbf{e}_j\cdot\mathbf{e}_i\Big)/(2\|\mathbf{x}\|) \\
&= \sum_i \mathbf{e}^i\,(\mathbf{e}_i\cdot\mathbf{x})/\|\mathbf{x}\| \\
&= \mathbf{x}/\|\mathbf{x}\|.
\end{aligned}
$$

This result depends on the metric through the norm $\|\mathbf{x}\|$.

- **Adjoint as Derivative**. When we introduced the adjoint $\overline{\mathsf{f}}$ of a function f in Section 4.3.2, we only had an implicit definition through $\mathbf{x}*\mathsf{f}[\mathbf{y}] = \overline{\mathsf{f}}[\mathbf{x}]*\mathbf{y}$. Using the vector derivative, we can define the adjoint explicitly as

$$\overline{\mathsf{f}}[\mathbf{x}] \equiv \partial_\mathbf{y} \left(\mathsf{f}[\mathbf{y}] * \mathbf{x} \right), \tag{8.14}$$

where both $\mathbf{x}$ and $\mathbf{y}$ are in the same space $\mathbb{R}^m$ (to avoid the need for a projection). You can prove it immediately by rewriting the argument of the differentiation using the earlier definition. This definition can also be applied to nonlinear functions, and it then computes a local adjoint, which may be different at every location $\mathbf{x}$.

8.6.2 PROPERTIES OF VECTOR DIFFERENTIATION

The vector differentiation operator is clearly linear. It also obeys a *product rule*, though we need to take care of its noncommutativity. Therefore, it becomes inconvenient to denote its application by square brackets; we need a more specific notation. Dropping the reference to $\mathbf{x}$ for readability, we express the product rule as

$$\partial\,(FG) = \partial \grave{F} G + \partial F \grave{G},$$

where in each term the accent denotes on what factor the scalar differentiation part of the ∂ should should act—the geometric vector part is not allowed to roam, so we cannot simply say that the operator acts on the element just to the right of it. To give an example:

$$\partial_\mathbf{x}\,(\mathbf{x}\,\mathbf{x}) = \partial_\mathbf{x}\,\grave{\mathbf{x}}\,\mathbf{x} + \partial_\mathbf{x}\,\mathbf{x}\,\grave{\mathbf{x}} = \partial_\mathbf{x}\,\grave{\mathbf{x}}\,\mathbf{x} + \partial_\mathbf{x}\,(2(\grave{\mathbf{x}}\cdot\mathbf{x}) - \grave{\mathbf{x}}\,\mathbf{x}) = 2\partial_\mathbf{x}\,(\grave{\mathbf{x}}\cdot\mathbf{x}) = 2\mathbf{x}.$$

Note that the subtle swap to get the elements into standard order precisely kills the term $\partial_\mathbf{x}\,\grave{\mathbf{x}}\,\mathbf{x} = m\,\mathbf{x}$.

Because of the noncommutativity, there are other product rules, such as

$$F\partial G = \grave{F}\partial G + F\partial \grave{G},$$

with the accents again denoting how to match each differentiation with its argument.

There is also a *chain rule*, which looks a bit complicated. Let the coordinate $\mathbf{x}$ be hidden by a vector-valued function $\mathbf{y}$, so that the dependence of F on $\mathbf{x}$ is $F(\mathbf{y}(\mathbf{x}))$. Then the chain rule of vector differentiation is

$$\partial_\mathbf{x}\,F(\mathbf{y}(\mathbf{x})) = \partial_\mathbf{x}\,\left(\mathbf{y}(\grave{\mathbf{x}}) * \partial_\mathbf{y}\right) F(\mathbf{y}).$$

The two geometric products in this equation can be executed in either order due to associativity. If we start from the right, this states that we should first consider F as a function of $\mathbf{y}$ and do a directional differentiation in the $\mathbf{y}(\mathbf{x})$-direction; that typically gives something involving both $\mathbf{y}(\mathbf{x})$ and $\mathbf{y}$. We should *not* substitute $\mathbf{x}$ in the latter, but differentiate the $\mathbf{x}$-dependence in the former. This can be confusing, so let us do an example.

Let $G(\mathbf{y}) = \mathbf{y}^2$, and $\mathbf{y}(\mathbf{x}) = (\mathbf{x}\cdot\mathbf{a})\mathbf{b}$. If we would just evaluate G as a function of $\mathbf{x}$ by substitution, we would get $G(\mathbf{x}) = (\mathbf{x}\cdot\mathbf{a})^2\,\mathbf{b}^2$, so that $\partial_\mathbf{x}\,G(\mathbf{x}) = 2(\mathbf{x}\cdot\mathbf{a})\,\mathbf{a}\,\mathbf{b}^2$. The chain rule application should produce the same answer.

We first evaluate from the right, so we start with the directional differentiation of $G(\mathbf{y}) = \mathbf{y}^2$. For a general vector $\mathbf{z}$, the directional derivative $(\mathbf{z} * \partial_{\mathbf{y}}) \mathbf{y}^2 = 2\mathbf{z} \cdot \mathbf{y}$, so with $\mathbf{z} = \mathbf{y}(\mathbf{x})$ the result is $2\mathbf{y}(\mathbf{x}) \cdot \mathbf{y} = 2(\mathbf{x} \cdot \mathbf{a})(\mathbf{b} \cdot \mathbf{y})$. Note that we kept $\mathbf{y}$. In the second step, this expression needs to be differentiated to $\mathbf{x}$, giving $\partial_{\mathbf{x}}(2(\mathbf{x} \cdot \mathbf{a})(\mathbf{b} \cdot \mathbf{y})) = 2\mathbf{a}(\mathbf{b} \cdot \mathbf{y})$. That is the answer, but we prefer it in terms of $\mathbf{x}$, so we should substitute the expression for $\mathbf{y}$ in terms of $\mathbf{x}$, giving the same result as before.

If instead we had evaluated from the left, we would first need to evaluate $\partial_{\mathbf{x}}(\mathbf{y}(\dot{\mathbf{x}}) * \partial_{\mathbf{y}}) = \partial_{\mathbf{x}}((\mathbf{x} \cdot \mathbf{a})\mathbf{b}) * \partial_{\mathbf{y}} = \mathbf{a}(\mathbf{b} * \partial_{\mathbf{y}})$. Do not be bothered by the presence of $\partial_{\mathbf{y}}$ in this derivation; since it is not differentiating anything, it behaves just like a vector. Now we apply the resulting operator to $G(\mathbf{y}) = \mathbf{y}^2$, giving $2\mathbf{a}(\mathbf{b} \cdot \mathbf{y})$ as in the other evaluation order. Here, too, you would need to substitute the expression $\mathbf{y}(\mathbf{x})$ to get the result in terms of $\mathbf{x}$.

The operator we just evaluated can be rewritten using the definition of the adjoint of the function $\mathbf{y}(\mathbf{x}) = (\mathbf{x} \cdot \mathbf{a})\mathbf{b}$, which is $\overline{\mathbf{y}}(\mathbf{x}) = (\mathbf{x} \cdot \mathbf{b})\mathbf{a}$. We then recognize $\mathbf{a}(\mathbf{b} * \partial_{\mathbf{y}})$ as the adjoint of the $\mathbf{y}$-function applied on $\partial_{\mathbf{y}}$, i.e., $\overline{\mathbf{y}}[\partial_{\mathbf{y}}]$. We can also use the adjoint to write the actual answer for our differentiation of the squaring function G as $2\overline{\mathbf{y}}(\mathbf{y}(\mathbf{x}))$, which actually holds for any function $\mathbf{y}$ used to wrap the argument $\mathbf{x}$.

The implicit understanding of how to deal with the substitutions in the equation is a bit cumbersome. A more proper notation for the process may be to keep the $\mathbf{x}$ in there at all steps:

$$\partial_{\mathbf{x}} G(\mathbf{y}(\mathbf{x})) = \dot{\partial}_{\mathbf{x}} (\mathbf{y}(\dot{\mathbf{x}}) * \partial_{\mathbf{y}(\mathbf{x})}) G(\mathbf{y}(\mathbf{x})) = \overline{\mathbf{y}}[\partial_{\mathbf{y}(\mathbf{x})}] G(\mathbf{y}(\mathbf{x})). \tag{8.15}$$

The final rewriting uses the differential definition of the adjoint of (8.14) (which also holds for nonlinear vector functions $\mathbf{y}$). This usage was motivated in the example. It means that we treat the differentiation operator $\partial_{\mathbf{y}(\mathbf{x})}$ just as the vector it essentially is. Then the differentiation with respect to $\mathbf{y}(\mathbf{x})$ should be understood as above, but the lack of an accent denotes that that particular $\mathbf{x}$-dependence should not be differentiated by $\partial_{\mathbf{x}}$.

So in the end, the chain rule is essentially a transformation of the differentiation operator: *when an argument gets wrapped into a function, the differentiation with respect to that argument gets wrapped into the adjoint of that function.*

8.7 MULTIVECTOR DIFFERENTIATION

We can extend these forms of differentiation beyond vectors to general multivectors, though for geometric algebra, the extension to differentiation with respect to blades and versors is most useful. Another extension is the differentiation with respect to a linear function of multivectors, which finds uses in optimization. We will not treat that here, but refer to Chapter 11 in [15].

8.7.1 DEFINITION

The definition of directional multivector differentiation is a straightforward extension of the idea behind the directional vector differentiation. You simply vary the argument X of a function additively in its A-component, so that A should at least be of the same grade as X (as for instance when X is perturbed by a transformation, to first order). The definition reflects this grade-matching in its use of the scalar product:

$$(A * \partial_X) F(X) \equiv \lim_{\epsilon \to 0} \frac{F(X + \epsilon A) - F(X)}{\epsilon}.$$

We emphasize that this is a scalar operator, since the grade of the result is the same as that of the original function.

As in the case of the vector derivative, we can see the directional multivector derivative as merely one component of a more general multivector derivative. We introduce coordinates now for the total 2^m-dimensional space of multivectors in the tangent space $\mathbb{R}^m$ at X. To distinguish it clearly from the m-dimensional vector basis, let us denote this multivector basis by a running capital index: $\{\mathbf{e}_I\}_{I=1}^{2^m}$. As with the vector basis in the vector derivative, this may not be orthonormal, so we also employ a reciprocal basis $\{\mathbf{e}^I\}_{I=1}^{2^m}$; see also Section 3.8. Then the multivector derivative is defined as

$$\partial_X = \sum_I \mathbf{e}^I (\mathbf{e}_I * \partial_X),$$

where $\mathbf{e}_I$ in principle runs over all 2^m elements 1, $\mathbf{e}_i$, $\mathbf{e}_i \wedge \mathbf{e}_j$, and so on, and the scalar product selects only the basis elements that are components of X.

This clearly contains vector differentiation as a special case. But also scalar differentiation is included: if we let X be a scalar $X = \tau$, only the basis element $\mathbf{e} = 1$ is selected, so $\partial_\tau = 1 (1 * \partial_\tau) = (1 * \partial_\tau) = \frac{d}{d\tau}$, conforming to our earlier definition of this symbol. For scalars, directional differentiation and multivector differentiation coincide.

As with the vector derivative, the coordinate-based definition should be used to derive elementary coordinate-free results, which should then be the basis of all actual computations. We have collected some in Table 8.2, including results on scalar functions that often occur in optimization problems. The pattern of derivation of these equations is completely analogous to that for vector differentiation.

8.7.2 APPLICATION: ESTIMATING ROTORS OPTIMALLY

This example is taken from [36]. We are given k labeled vectors $\mathbf{u}_i$, which have been rotated to become k correspondingly labeled vectors $\mathbf{v}_i$. We want to try and retrieve that rotor from this data. If both sets of vectors are measured with some noise (as they usually are), we cannot exactly reconstruct the rotor R, but we have to estimate it. Let us use as our criterion for fit the minimization of the total squared distance between our estimated

Table 8.2: Elementary results of multivector differentiation. The multivector varies in the space $\bigwedge \mathbb{R}^m$, contained in the larger space $\bigwedge \mathbb{R}^n$. The map $\mathsf{P}[]$ projects from the latter to the former.

$$(A * \partial_X)\, X \;\; = \;\; \mathsf{P}[A]$$

$$(A * \partial_X)\widetilde{X} \;\; = \;\; \mathsf{P}[\widetilde{A}]$$

$$(A * \partial_X)\, X^k \;\; = \;\; \mathsf{P}[A]\, X^{k-1} + X\,\mathsf{P}[A]\, X^{k-2} + \cdots + X^{k-1}\,\mathsf{P}[A]$$

$$\partial_X X \;\; = \;\; \binom{m}{\mathrm{grade}(X)}$$

$$\partial_X \|X\|^2 \;\; = \;\; 2\widetilde{X}$$

$$\partial_X (X * A) \;\; = \;\; \mathsf{P}[A]$$

$$\partial_X (\widetilde{X} * A) \;\; = \;\; \mathsf{P}[\widetilde{A}]$$

$$\partial_X (X^{-1} * A) \;\; = \;\; \mathsf{P}[-X^{-1} A X^{-1}]$$

$$\partial_X \|X\|^k \;\; = \;\; k\, \|X\|^{k-2}\widetilde{X}$$

rotation vectors compared to where we measured them. This is an old problem, known in biometrics literature as the Procrustes problem and in astronautics as Wahba's problem.

So we need to find the rotor R that minimizes

$$\Gamma(R) = \sum_{i=1}^{k} (\mathbf{v}_i - R\,\mathbf{u}_i\,\widetilde{R})^2 = \sum_{i=1}^{k} (\mathbf{v}_i^2 + \mathbf{u}_i^2 - 2\langle \mathbf{v}_i\, R\, \mathbf{u}_i\, \widetilde{R}\rangle_0). \tag{8.16}$$

Preferably, we would like to differentiate this with respect to R and set the resulting derivative to zero to find the optimal solution. However, the rotor normalization condition $R\widetilde{R} = 1$ makes this mathematically somewhat involved. It is easier to temporarily replace the rotor R by a versor V and consequently to replace $\widetilde{R}$ by V^{-1}, and then to differentiate relative to the unconstrained V to compute the optimum V_*. Clearly the terms without R (or V) do not affect the optimum, so

$$V_* = \text{argmax}_V \left(\sum_{i=1}^{k} \langle \mathbf{v}_i \, V \mathbf{u}_i \, V^{-1} \rangle_0 \right).$$

Now we differentiate by ∂_V and use the product rule. We can use some of the results from Table 8.2 once we realize that this is differentiation of a scalar product and use its symmetry and reordering properties (as (6.23)):

$$\partial_V \Gamma(V) = \sum_{i=1}^{k} \partial_V \langle \mathbf{v}_i \, V \mathbf{u}_i \, V^{-1} \rangle_0$$

$$= \sum_{i=1}^{k} \left(\partial_V [\dot{V} * (\mathbf{u}_i V^{-1} \mathbf{v}_i)] + \partial_V [(\dot{V}^{-1}) * (\mathbf{v}_i V \mathbf{u}_i)] \right)$$

$$= \sum_{i=1}^{k} \left(\mathbf{u}_i V^{-1} \mathbf{v}_i - V^{-1} (\mathbf{v}_i V \mathbf{u}_i) V^{-1} \right)$$

$$= 2 V^{-1} \sum_{i=1}^{k} (V \mathbf{u}_i V^{-1}) \wedge \mathbf{v}_i.$$

Therefore the rotor R_* that minimizes $\Gamma(R)$ must be the one that satisfies

$$\sum_{i=1}^{k} (R_* \mathbf{u}_i \widetilde{R}_*) \wedge \mathbf{v}_i = 0. \tag{8.17}$$

This algebraic result makes geometric sense. For each $\mathbf{v}_i$, it ignores the components that are just scalings of the corresponding rotated $\mathbf{u}_i$; the rotation cannot affect those parts anyway. Only the perpendicular components matter, and those should cancel overall if the rotation is to be optimal—if not, a small extra twist could align the vectors better.

The result so far does not give us the optimal rotor R_* explicitly; it has merely restated the optimization condition in a manner that shows what the essential components of the data are that determine the solution. Our reference [36] now cleverly uses vector differentiation to manipulate the equation to a form that can be solved by standard linear algebra. First, they observe that if we introduce the linear function

$$\mathsf{f}[\mathbf{x}] = \sum_{i=1}^{k} \mathbf{u}_i \, (\mathbf{v}_i \cdot \mathbf{x}),$$

the condition (8.17) can be written as

$$\partial_\mathbf{x} \wedge (R_* \, \mathsf{f}[\mathbf{x}] \, \widetilde{R}_*) = \frac{1}{2} \sum_{i-1}^{k} \left(\partial_\mathbf{x} (\mathbf{v}_i \cdot \mathbf{x}) \, (R_* \, \mathbf{u}_i \, \widetilde{R}_*) - (R_* \, \mathbf{u}_i \, \widetilde{R}_*)(\mathbf{v}_i \cdot \mathbf{x}) \partial_\mathbf{x} \right)$$

$$= - \sum_{i=1}^{k} (R_* \mathbf{u}_i \widetilde{R}_*) \wedge \mathbf{v}_i = 0.$$

This kind of pulling out a differentiation operator is a good trick to remember. The resulting equation expresses the fact that $R_* \mathsf{f}[\mathbf{x}] \widetilde{R}_*$ is a symmetric function of $\mathbf{x}$. The function f itself is therefore the $\widetilde{R}_*$-rotated version of a symmetrical function.

We could proceed symbolically with geometric algebra to find the symmetric part of $\mathsf{R}_*[\mathsf{f}]$ (by adding the adjoint $\overline{\mathsf{f}}[\overline{\mathsf{R}}_*]$ and dividing by two), and its inverse (using (4.16)), and taking that out of the function; what remains is then the rotation by the desired rotor. However, [36] at this point switches over to using numerical linear algebra. In linear algebra, any linear function f has a *polar decomposition* in a symmetric function followed by an orthogonal transformation, and this can be computed using the singular value decomposition (SVD) of its matrix $[\![\mathsf{f}]\!]$ as $\mathsf{f} \equiv [\![U]\!] [\![S]\!] [\![V]\!]^T = ([\![U]\!] [\![V]\!]^T) ([\![V]\!] [\![S]\!] [\![V]\!]^T)$. Using this result, the matrix of the optimal rotation R_* is $[\![V]\!] [\![U]\!]^T$, where $[\![U]\!]$ and $[\![V]\!]$ are derived from the SVD of $[\![\mathsf{f}]\!] \equiv \sum_{i=1}^{k} [\![\mathbf{u}_i]\!] [\![\mathbf{v}_i]\!]^T$. This rotation matrix is easily converted back into the optimal rotor R_*, see Section 7.10.4.

This simple matrix computation algorithm is indeed the standard solution to the Procrustes problem. In the usual manner of its derivation, formulated in terms of involved matrix manipulations, one may have some doubts as to whether the SVD (with its inherent use of the Frobenius metric on matrices) is indeed the optimal solution to original optimization problem (which involved the Euclidean distance). In the formulation above, the intermediate result (8.17) shows that this is indeed correct, and that the SVD is merely used to compute the decomposition rather than to perform the actual optimization.

From a purist point of view, it is of course a pity that the last part of the solution had to revert temporarily to a matrix formulation to compute a rotor. We expect that appropriate numerical techniques will be developed soon completely within the framework of geometric algebra.

8.8 FURTHER READING

The main reference for further reading on geometric calculus is the classic book by Hestenes [33], which introduced much of it. It contains a wealth of material, including an indication of how geometric calculus could be used to rephrase differential geometry. His web site contains more material, including some new introductions such as [26].

The approach in Doran and Lasenby's book [15] is tailored towards physicists, and it has good and practical introductions to the techniques of geometric calculus. Read them for directed integration theory. They are practitioners who use it daily, and they give just the right amount of math to get applicable results.

8.9 EXERCISES

8.9.1 DRILLS

1. Compute the radius of the tangent circle for the circular motion $\mathbf{r}(\tau) = \exp(-I\tau)\,\mathbf{e}_1$ in the plane $\mathbf{I} = \mathbf{e}_1 \wedge \mathbf{e}_2$, at the general location $\mathbf{r}(\tau)$.

2. Compute the following derivatives:

    1. $(\mathbf{a} * \partial_{\mathbf{x}})\,\mathbf{x}^3$
    2. $\partial_{\mathbf{x}}\,\mathbf{x}^3$
    3. $(\mathbf{a} * \partial_{\mathbf{x}})\,(\mathbf{x}\,\mathbf{b}/\mathbf{x})$
    4. $\partial_{\mathbf{x}}\,(\mathbf{x}\,\mathbf{b}/\mathbf{x})$
    5. $\dot{\mathbf{x}}\,\dot{\partial}_{\mathbf{x}}$
    6. $\dot{\mathbf{x}} \wedge \dot{\partial}_{\mathbf{x}}$
    7. $\dot{\mathbf{x}} \cdot \dot{\partial}_{\mathbf{x}}$.

3. Show that the coordinate vectors are related to differentiation through $\mathbf{e}_k = \frac{\partial}{\partial x^k}\mathbf{x}$.

4. Show that the reciprocal frame vectors are the gradients of coordinate functions: $\mathbf{e}^k = \partial_{\mathbf{x}}\,x^k$.

8.9.2 STRUCTURAL EXERCISES

1. Prove the Jacobi identity (8.2) and relate it to nonassociativity of the bivector algebra.

2. Derive the Taylor expansion of a rotor transformation:

$$e^{-B/2}\,X\,e^{B/2} = X + X \times B + \tfrac{1}{2}\,((X \times B) \times B) + \cdots.$$

Do this by assuming that the first-order term is correct for small bivectors, it is easily derived by setting $\exp(-B/2) \approx 1 - B/2$. Now write a versor involving a finite B as versors involving $B/2$, $B/4$, $B/8$, and so on and build up the total form through repeated application of the smallest bivector forms. That should give the full expansion.

3. The *Baker-Campbell-Hausdorff formula* writes the product of two exponentials as a third, and gives a series expansion of its value:

$$e^C = e^A\,e^B$$

with

$$C = A + B + A \times B + \tfrac{1}{3}\,(A \times (A \times B) + B \times (B \times A)) + \cdots.$$

Show that these first terms of the series are correct. This formula again shows the importance of the commutator $A \times B$ in quantifying the difference with fully commuting variables. We should warn you that the general terms of the series are more complicated than the first few suggest.

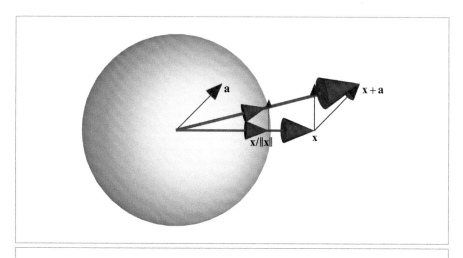

Figure 8.3: The directional derivative of the spherical projection.

4. *Directional differentiation of spherical projection.*
   Suppose that we project a vector $\mathbf{x}$ on the unit sphere by the function $\mathbf{x} \mapsto \mathsf{P}[\mathbf{x}] = \mathbf{x}/\|\mathbf{x}\|$. Compute its directional derivative in the $\mathbf{a}$ direction, as a standard differential quotient using Taylor series expansion. Use geometric algebra to write the result compactly, and give its geometric meaning. (Hint: See Figure 8.3.)

5. Justify the following form of Taylor's expansion formula of a function F around the location $\mathbf{x}$:

$$F(\mathbf{x} + \mathbf{a}) = e^{\mathbf{a} * \partial_{\mathbf{x}}} F(\mathbf{x}),$$

   where you can interpret the exponent in a natural manner as a symbolic expansion instruction.

6. For variable $\mathbf{I}(\tau)$, the resulting $\partial_\tau \mathbf{X}(\tau)$ of (8.8) can still be written as a commutator $\mathbf{X} \times B$ with a bivector B. Derive the explicit expression for B:

$$B = \mathbf{I}\, \partial_\tau[\phi] + \partial_\tau[\mathbf{I}]\, (e^{\mathbf{I}\phi} - 1)/\mathbf{I}.$$

   Hint: One way is to use the result $B = -2\, \partial_\tau[R]\, \widetilde{R}$ from [15].

PART II
MODELS OF GEOMETRIES

9 MODELING GEOMETRIES

So far we have been treating only homogeneous subspaces of the vector spaces (i.e., subspaces containing the origin). We have spanned them, projected them, and rotated them, but we have not moved them off the origin to make more interesting geometrical structures such as lines floating in space. You might fear that we need to extend our framework considerably to incorporate such new geometrical elements and their algebra, introducing offset blades as algebraic primitives.

However, significant extensions turn out to be unnecessary, by using a simple trick that is a lot more generic than it appears at first: these offset elements of geometry in a vector space $\mathbb{R}^n$ can also be represented by blades, but in a representational space $\mathbb{R}^{n+1}$ with one extra dimension. The geometric algebra of that space then gives us most of what we need to compute in $\mathbb{R}^n$. Since the blades of that higher-dimensional space $\mathbb{R}^{n+1}$ are its homogeneous subspaces, this is called the *homogeneous model* of geometry (though "homogenized" might be more accurate, for we have made things homogeneous in $\mathbb{R}^{n+1}$ that were not in $\mathbb{R}^n$).

In this view, more complicated geometrical objects do not require new operations or techniques in geometric algebra, merely the standard computations in a higher-dimensional space, followed by an interpretation step. The geometric algebra approach considerably extends the classical techniques of *homogeneous coordinates*, so it pays to redevelop this fairly well-known material. The homogeneous model permits us to represent offset subspaces as blades, and transformations on them as linear transformations and their outermorphisms. As we develop the details in Chapter 11, we find that the geometric algebra approach exposes some weaknesses in the homogeneous model. It turns out that we cannot really define a useful inner product in the representation space $\mathbb{R}^{n+1}$ that represents the metric aspects of the original space $\mathbb{R}^n$ well; we can only revert to the inner product of $\mathbb{R}^n$. As a consequence, we also have no compelling geometric product, and our geometric algebra of $\mathbb{R}^{n+1}$ is impoverished, being reduced to outer product and nonmetric uses of duality (such as `meet` and `join`). This restricts the natural use of the homogeneous model to applications in which the metric is less important than the aspects of spanning and intersection. The standard example is the projective geometry involved in imaging by multiple cameras, and we treat that application in detail in Chapter 12. Still, the quantitative capabilities of geometric algebra do help in assigning some useful relative metrical meaning to ratios of computed elements.

The better model to treat the metric aspects of Euclidean geometry is a representation that can make full use of the power of geometric algebra. That is the *conformal model* of Chapter 13, which requires *two* extra dimensions. It provides an isometric model of Euclidean geometry. In this representation, all Euclidean transformations become representable as versors, and are therefore manifestly structure-preserving. This gives a satisfyingly transparent structure for dealing with objects and operators, far transcending the classical homogeneous coordinate techniques. We initially show how this indeed extends the homogeneous model with metric capabilities, such as the smooth interpolation of rigid body motions in Chapter 13. Then in Chapter 14 we find that there are other elements of Euclidean geometry naturally represented as blades in this model: spheres, circles, point pairs, and tangents. These begin to suggest applications and algorithms that transcend the usual methods. To develop the tools for those, we look at the new constructions in detail in Chapter 15. In the last chapter on the conformal model, we find the reason behind its name: all conformal (angle-preserving) transformations are versors, and this now also gives us the possibility to smoothly interpolate rigid body motions with scaling. In all of these chapters, the use of the interactive software is important to convey how natural and intuitive these new tools can become.

But first, we should make more explicit how the regular n-dimensional geometric algebra, used as a *vector space model*, gives us tools to treat the directional aspects of an n-dimensional space. This capability of computing with an algebra of directions will transfer to the more powerful models as a directional submodel at every location in space.

10 THE VECTOR SPACE MODEL: THE ALGEBRA OF DIRECTIONS

When we developed geometric algebra in the first part of this book, we illustrated the principles with pictures in which vectors are represented as arrows at the origin, bivectors as area elements at the origin, and so on. This is the purest way to show the geometric properties corresponding to the algebra.

The examples showed that you can already use this algebra of the mathematical vector space $\mathbb{R}^n$ to model useful aspects of Euclidean geometry, for it is the *algebra of directions* of n-dimensional Euclidean space. We explore some more properties of this model in this chapter, with special emphasis on computations with directions in 2-D and 3-D. Most topics are illustrated with programming exercises at the end of the chapter.

First we show how the vector space model can be used to derive fundamental results in the mathematics of angular relationships. We give the basic laws of trigonometry in the plane and in space, and show how rotors can be used to label and classify the crystallographic point groups.

Then we compute with 3-D rotations in their rotor representation, establishing some straightforward techniques to construct a rotor from a given geometrical situation, either deterministically or in an optimal estimation procedure. The logarithm of a 3-D rotor enables us to interpolate rotations.

Finally, we give an application to external camera calibration, to show how the vector space model can mix directional and locational aspects.

10.1 THE NATURAL MODEL FOR DIRECTIONS

There are n independent 1-D directions in an n-dimensional physical space, and they can conveniently be drawn as vectors at the origin. Mathematically, they form a vector space $\mathbb{R}^n$, for they can be added and scaled with real numbers to produce other legitimate 1-D directions. The metric of the directions in the physical space (typically Euclidean) can be used to induce a metric in this mathematical representation. That gives a model of the directions in physical space in terms of the geometric algebra of a metric vector space $\mathbb{R}^n$.

The *vector space model* thus constructed is indeed a good computational representation of spatial directions at the origin. We have used it in all our illustrations of the geometrical properties of geometric algebra in Part I. This already gave a list of powerful abilities, directly applicable to computations with directions. We list the main results.

- The k-dimensional directions in n-dimensional space can be composed as outer products of k 1-D directions (represented as vectors). These k-blades can be decomposed on an $\binom{n}{k}$-dimensional basis. Only for grades 0, 1, $(n-1)$, and n can k-blades be constructed by arbitrary addition of basis elements.

- Directions have an attitude, a weight, and an orientation.

- Relative angles between k-dimensional directions can be computed using the contraction, even when they are of different grades. A k-direction can be represented by its dual, the direction of its orthogonal complement.

- Intersection and union of k-directions is defined by `meet` and `join`, which are a specific combination of outer product and duality. The orthogonal projection of k-directions is also well defined by the contraction.

- Directions can be used to reflect other directions using sandwiching products (where some care is required to process their orientation properly).

- Directions can be rotated using rotors and multiplied by the geometric product to produce such rotors.

Beyond these structural properties, true for the abstract directions of a general vector space $\mathbb{R}^n$, we need specific techniques to use the blades of the vector space model to solve particular geometrical problems, notably in the Euclidean spaces $\mathbb{R}^2$ and $\mathbb{R}^3$.

10.2 ANGULAR RELATIONSHIPS

The vector space model is the natural model to treat angular relationships at a single location. To show this, we derive the elementary laws of sines and cosines in a planar triangle; some similar relationships in a spherical triangle; and the point groups of crystallography. The results are not new, but intended as examples of how you can now think about such problems in a purely directional manner using geometric algebra. Especially the ability to divide and multiply vectors will simplify both the computations and the definition of the oriented angular parameters in the configurations.

10.2.1 THE GEOMETRY OF PLANAR TRIANGLES

The combination of rotors in the same plane is sufficient to derive the various relationships between sides and angles in triangles. We repeat the derivation of these laws as given in [29], pp. 68–70, since this application shows the simplicity and power of geometric algebra. Quantities that are required to characterize the properties are completely definable in terms of the original elements of the problem.

In Figure 10.1(a) we have indicated a triangle in the 2-D Euclidean **I**-plane, composed of three vectors **a**, **b**, **c** that have the relationship

$$\mathbf{a} + \mathbf{b} + \mathbf{c} = 0. \tag{10.1}$$

These vectors indicate weighted directions, and their weights can be drawn as their lengths. Although they have been drawn offset from the origin, there are no actual positional aspects to this triangle and its relationships. This is shown by redrawing all vectors involved as emanating from the origin, in a purist version of the triangle, as Figure 10.1(b). The relevant geometric algebra of both figures is the same.

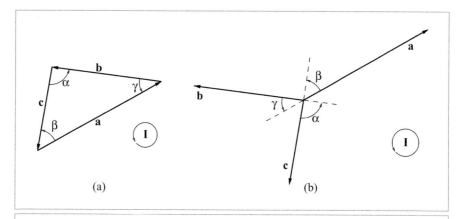

(a)   (b)

Figure 10.1: A triangle $\mathbf{a} + \mathbf{b} + \mathbf{c} = 0$ in a directed plane **I**, and an equivalent configuration for treatment with the vector space model.

Solving this equation $\mathbf{a} + \mathbf{b} + \mathbf{c} = 0$ for $\mathbf{c}$, and squaring, we get

$$c^2 = (\mathbf{a} + \mathbf{b})^2 = \mathbf{a}^2 + \mathbf{b}^2 + \mathbf{ab} + \mathbf{ba} = \mathbf{a}^2 + \mathbf{b}^2 + 2\mathbf{a} \cdot \mathbf{b}. \qquad (10.2)$$

We may introduce the angle α between $\mathbf{c}$ and $-\mathbf{b}$ (in that order), and similarly β and γ (see Figure 10.1), and we can introduce the lengths of the sides $\|\mathbf{a}\|$, $\|\mathbf{b}\|$, and $\|\mathbf{c}\|$. The picture defines what is meant, but in geometric algebra we would rather define those elements unambiguously as properties of the geometric ratios of the original vectors. Section 6.1.6 gives the principle. Carefully observing the required angles and signs leads to the exact definitions:

$$-\mathbf{b}/\mathbf{a} \equiv \|\mathbf{b}\|/\|\mathbf{a}\| \, e^{\mathrm{I}\gamma}, \quad -\mathbf{c}/\mathbf{b} \equiv \|\mathbf{c}\|/\|\mathbf{b}\| \, e^{\mathrm{I}\alpha}, \quad -\mathbf{a}/\mathbf{c} \equiv \|\mathbf{a}\|/\|\mathbf{c}\| \, e^{\mathrm{I}\beta}. \qquad (10.3)$$

Daring to make such definitions is a skill that you should master, for it is the transition from the classical methods of thinking about angles (with the associated headaches on the choice of signs) to the automated computations of geometric algebra. Make sure you understand the precise relationship between these definitions and the figure they represent!

When combined with the basic property (10.1), the angle definitions (10.3) fully define all relationships in the triangle. It just takes geometrically inspired algebraic manipulation to bring them out. For instance, we can multiply these equations. Remembering that exponentials of commuting arguments are additive, we obtain by (10.3)

$$e^{\mathrm{I}(\alpha+\beta+\gamma)} = e^{\mathrm{I}\alpha} e^{\mathrm{I}\gamma} e^{\mathrm{I}\beta} = (-\mathbf{c}/\mathbf{b})(-\mathbf{b}/\mathbf{a})(-\mathbf{a}/\mathbf{c}) = -1 = e^{\mathrm{I}\pi} \qquad (10.4)$$

This implies that

$$\alpha + \beta + \gamma = \pi \bmod (2\pi), \qquad (10.5)$$

which is a rather familiar result.

To obtain other classics, we split the geometric product in a contraction and an outer product, thereby separating the equations into their scalar and bivector parts. This automatically introduces the trigonometric functions as components of the rotors.

We multiply both sides of (10.3) by $\|\mathbf{a}\|^2$, and so on, and obtain six equations:

$$-\mathbf{b} \cdot \mathbf{a} = \|\mathbf{b}\| \, \|\mathbf{a}\| \, \cos\gamma, \quad -\mathbf{b} \wedge \mathbf{a} = \|\mathbf{b}\| \, \|\mathbf{a}\| \, \mathrm{I} \sin\gamma,$$
$$-\mathbf{c} \cdot \mathbf{b} = \|\mathbf{c}\| \, \|\mathbf{b}\| \, \cos\alpha, \quad -\mathbf{c} \wedge \mathbf{b} = \|\mathbf{c}\| \, \|\mathbf{b}\| \, \mathrm{I} \sin\alpha,$$
$$-\mathbf{a} \cdot \mathbf{c} = \|\mathbf{a}\| \, \|\mathbf{c}\| \, \cos\beta, \quad -\mathbf{a} \wedge \mathbf{c} = \|\mathbf{a}\| \, \|\mathbf{c}\| \, \mathrm{I} \sin\beta.$$

Then the earlier results can be put into the classical form. Equation (10.2) is the law of cosines:

$$\|\mathbf{c}\|^2 = \|\mathbf{a}\|^2 + \|\mathbf{b}\|^2 - 2\|\mathbf{a}\| \, \|\mathbf{b}\| \, \cos\gamma. \qquad (10.6)$$

Taking the outer product of (10.1) with $\mathbf{a}$, $\mathbf{b}$, and $\mathbf{c}$, we obtain

$$\mathbf{a} \wedge \mathbf{b} = \mathbf{b} \wedge \mathbf{c} = \mathbf{c} \wedge \mathbf{a}, \tag{10.7}$$

which leads to the law of sines in the $\mathbf{I}$-plane:

$$\frac{\sin \alpha}{\|\mathbf{a}\|} = \frac{\sin \beta}{\|\mathbf{b}\|} = \frac{\sin \gamma}{\|\mathbf{c}\|}. \tag{10.8}$$

We have divided out the plane $\mathbf{I}$ in which this holds to achieve this classical form. But in fact, (10.7) is more specific and is valid in any plane $\mathbf{I}$ in n-dimensional space.

In the classical formulation, the area of the triangle is $\frac{1}{2}\|\mathbf{a}\| \, \|\mathbf{b}\| \sin \gamma$ (or a similar expression). We see that in the directed plane $\mathbf{I}$, we can define the *oriented area* Δ of the triangle naturally by the equivalent ratios

$$\Delta = \frac{\mathbf{a} \wedge \mathbf{b}}{2\,\mathbf{I}} = \frac{\mathbf{b} \wedge \mathbf{c}}{2\,\mathbf{I}} = \frac{\mathbf{c} \wedge \mathbf{a}}{2\,\mathbf{I}}. \tag{10.9}$$

This is a proper geometric quantity that relates the area to the orientation of the plane it is measured in.

10.2.2 ANGULAR RELATIONSHIPS IN 3-D

In a 3-D Euclidean space, geometrical directions can be indicated by vectors or bivectors (which are always 2-blades). The scalars and trivectors have trivial directional aspects and are mostly used for their orientations and magnitudes.

Relative angles between the directional elements are fully represented by their geometric ratios. Let us consider only unit elements, so that we can fully focus on the angles. Between two unit directional elements, there are three possibilities:

- *Two Vectors*. The geometric ratio of two unit vectors $\mathbf{u}$ and $\mathbf{v}$ is a rotor $R = \mathbf{v}/\mathbf{u}$. It contains in its components both the rotation plane $\mathbf{I}$ and the relative angle ϕ of the two vectors. These can be retrieved from the rotor as the bivector angle $\mathbf{I}\phi$, using the logarithm function defined below in Section 10.3.3. Note that only the product of plane and angle is a well-defined geometric quantity, since each separately has an ambiguity of magnitude and orientation. In that sense, scalar angles are ungeometrical and should be avoided in computations, since they necessitate the nongeometrical choice of standard orientation for the $\mathbf{I}$-plane. Since you probably only need the angles to use them in a rotation operator anyway, you may as well keep their bivector with their magnitude as a single *bivector angle*.

- *Two Bivectors (2-Blades)*. The geometric ratio of two unit 2-blades $\mathbf{U}$ and $\mathbf{V}$ also defines a rotor $R = \mathbf{V}/\mathbf{U}$. This is most easily seen by introducing their normal vectors $\mathbf{u} \equiv \mathbf{U}^* = \mathbf{U}/I_3$ and $\mathbf{v} \equiv \mathbf{V}^* = \mathbf{V}/I_3$. Substituting gives $R = \mathbf{V}/\mathbf{U} = \mathbf{v}/\mathbf{u}$. Therefore, this reduces to the previous case. The bivector angle between two bivectors is automatically measured in a plane perpendicular to the common line of the two planes, and this plane and the angle are found as the bivector angle $\log(\mathbf{V}/\mathbf{U})$.

- **Vector and Bivector (2-Blade).** When we have a unit bivector $\mathbf{U}$ and a unit vector $\mathbf{v}$, we previously defined the cosine of their angle through the contraction, as in Figure 3.2(b). With the geometric product, we can proceed slightly differently, defining the full bivector angle. We try to determine a unit vector $\mathbf{w}$ in the plane $\mathbf{U}$ and perpendicular to $\mathbf{v}$, such that it can rotate $\mathbf{v}$ into the plane over an angle α as $\mathbf{v}e^{\mathbf{w}\mathbf{I}_3\alpha}$, after which that rotated version of $\mathbf{v}$ and $\mathbf{w}$ together span $\mathbf{U}$. The sketch of Figure 10.2 shows that this mouthful is the algebraic demand

$$\mathbf{U} = \mathbf{v}\,e^{\mathbf{w}\mathbf{I}_3\alpha}\,\mathbf{w}.$$

This fully determines both $\mathbf{w}$ and α as aspects of the geometric product $\mathbf{v}\mathbf{U}$. It is simplest to make a rotor out of the element $\mathbf{v}\mathbf{U}$ by undualization to show its bivector:

$$\mathbf{v}\,\mathbf{U}\,\mathbf{I}_3 = e^{\mathbf{w}\mathbf{I}_3\alpha}\,\mathbf{w}\mathbf{I}_3 = e^{\mathbf{w}\mathbf{I}_3(\pi/2+\alpha)}.$$

Taking the logarithm retrieves this bivector, which gives all parameters in one operation (though it is a pity not to get the actual bivector angle $\alpha\,\mathbf{w}\,\mathbf{I}_3$).

Angular relationships between three directions can also be defined. They are much more involved, because there are various standard ways of characterizing the parameters of the spherical triangle, as depicted in Figure 10.3. These are its vertices (represented

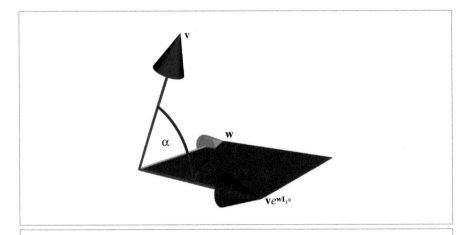

Figure 10.2: The angle between a vector and a bivector (see text).

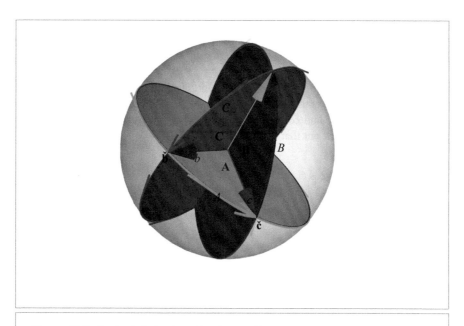

Figure 10.3: A spherical triangle and its characterizing parameters.

by three vectors), its sides (the angles between the vectors), its angles (the angles between the bivectors containing its sides), and its altitudes (angle between vector and plane bivector). The various combinations of these quantities provide the laws of spherical geometry. Geometric algebra again permits compact and computationally complete specification of the relationships.

The bivectors containing the vertices can be defined through

$$\check{\mathbf{a}}/\check{\mathbf{b}} \equiv e^{\mathbf{C}}, \quad \check{\mathbf{b}}/\check{\mathbf{c}} \equiv e^{\mathbf{A}}, \quad \check{\mathbf{c}}/\check{\mathbf{a}} \equiv e^{\mathbf{B}}, \tag{10.10}$$

where $\check{\mathbf{a}}$, $\check{\mathbf{b}}$, $\check{\mathbf{c}}$ are the unit vectors pointing at the vertices of the spherical triangle. The weights A, B, C of the bivectors $\mathbf{A}, \mathbf{B}, \mathbf{C}$ are the edge lengths, measured as angles on the unit sphere. The bivectors can be used to define the angles of the spherical triangle, through

$$\check{\mathbf{B}}/\check{\mathbf{A}} \equiv -e^{\mathbf{I}_3\mathbf{c}}, \quad \check{\mathbf{C}}/\check{\mathbf{B}} \equiv -e^{\mathbf{I}_3\mathbf{a}}, \quad \check{\mathbf{A}}/\check{\mathbf{C}} \equiv -e^{\mathbf{I}_3\mathbf{b}}, \tag{10.11}$$

with $\check{\mathbf{A}}$, $\check{\mathbf{B}}$, $\check{\mathbf{C}}$ now denoting the unit bivectors and $\mathbf{I}_3$ the unit pseudoscalar of $\mathbb{R}^{3,0}$. The weights of the vectors $\mathbf{a}$, $\mathbf{b}$, $\mathbf{c}$ thus defined are the internal angles a, b, c of the spherical triangle in Figure 10.3.

Multiplication of the unit vector expressions leads to

$$e^{\mathbf{A}} e^{\mathbf{B}} e^{\mathbf{C}} = 1,$$

and multiplication of the unit bivector expressions gives

$$e^{I_3 c} \, e^{I_3 b} \, e^{I_3 a} = -1.$$

We have met the former in the guise of the multiplication of rotors in Section 7.3.4. That should give you the confidence that these expressions can indeed lead to the correct expressions for sine and cosine formulas, when split in their different grades. To be specific, you can work out that the scalar part of the unit vector expression gives the identity

$$\cos C = \cos A \, \cos B + \sin A \sin B \cos c$$

known as the cosine law for sides. The minus sign in the bivector expression leads to subtle differences between the laws for the sides and for the angles. The scalar part is now

$$\cos c = -\cos a \, \cos b + \sin a \sin b \cos C.$$

which is known as the cosine law for angles.

For the full details on these and other laws of spherical trigonometry, we refer to Appendix A of [29], highly recommended reading for anybody using spherical trigonometry. When we get more versed in geometric algebra, we should probably not reduce the geometry of these spherical triangles to the set of classical scalar relationships, learning instead to compute directly with the rotors, vectors, and bivectors involved.

10.2.3 ROTATION GROUPS AND CRYSTALLOGRAPHY

A famous old puzzle in the mathematics of physics has been the classification of crystallographic groups: how many ways are there to make crystals in space? This is clearly a geometric problem, and one would hope that geometric algebra could help out. The vector space model, which provides the algebra of directions, can be used to solve a subproblem: how can one select the local symmetry of reflections and rotations to combine well, in the sense that multiple reflections and rotations generate only a finite number of elements? Such a set of operators is called a *point group*.

We use a unit vector $\mathbf{a}$ (with $\mathbf{a}^2 = 1$) to denote a local plane of reflection. When used as the normal vector of this plane, it generates the reflection of a vector $\mathbf{x}$ as $-\mathbf{a} \mathbf{x} \mathbf{a}$. This transformation is insensitive to the sign of $\mathbf{a}$. However, we can in principle distinguish the planes $\mathbf{a}$ and $-\mathbf{a}$ by their orientation, and for accurate classification of the point groups we should use such oriented planes.

In a 3-D crystal, there are several symmetry planes for reflection. In the algebra, they are each denoted by their vector versors. As the reflections combine as operators, new symmetries appear as the products of these versors. To form a crystal, the combined set of operators should form a finite set, so that a single point (atom) transformed by all operators of the crystal is only equivalent to a finite number of other atoms around the same point.

An even number of reflections generates a rotation, which is represented by the geometric product of the corresponding normal vectors to give an operator like $R \equiv \mathbf{a}\,\mathbf{b}$. A six-fold symmetry at the local point implies that applying the rotation operator $\mathbf{x} \mapsto R\,\mathbf{x}\,\tilde{R}$ six times gives back the original $\mathbf{x}$. Two possible conditions would satisfy this demand: $R^6 = 1$ and $R^6 = -1$. The latter is more specific, and uses the possibility of geometric algebra to represent *oriented rotations* (see also Section 7.2.3). Since we need to reflect other operators (not just points), we use the -1. That gives the most accurate classification of the symmetries of the crystal. In general, $R^{2p} = 1$ for a p-fold symmetry.

To use a particular example in 2-D, suppose we have two generating vectors $\mathbf{a}$ and $\mathbf{b}$ representing reflecting planes. These induce symmetries in the plane. To be a four-fold symmetry, we have the three demands:

$$\mathbf{a}^4 = 1, \quad \mathbf{b}^4 = 1, \quad (\mathbf{ab})^8 = 1.$$

Clearly the generating vectors of this point group should be unit vectors with a relative angle of $\pi/4$ to satisfy these demands. The rotation and reflection operators that are possible by the combinations of these vectors are listed in Table 10.1. There are 16 symmetry operators, which together form the group $2H_4$ of [30]. Each of these elements is a local symmetry V_i of the crystal. If you start with a single location vector $\mathbf{x}$ and generate all elements $\mathbf{x}_i = \hat{V}_i\,\mathbf{x}\,V_i{}^{-1}$, you find 8 equivalent locations for atoms in the symmetry group of this crystal. This is demonstrated in the programming example in Section 10.7.2.

The defining relations on two vectors is sufficient to determine all points groups in 2-D. In 3-D, three generating vectors are required. If we have the demands in the form

$$(\mathbf{ab})^{2p} = (\mathbf{bc})^{2q} = (\mathbf{ca})^{2r} = 1,$$

Table 10.1: The operators of the point group $2H_4$, with defining relations $\mathbf{a}^4 = \mathbf{b}^4 = (\mathbf{ab})^8 = 1$. The terms positive and negative are relative (but consistent), and based on the ordering of $\mathbf{a}$ and $\mathbf{b}$. These generate the symmetries of the crystal.

Positive Rotations	Negative Rotations	Positive Reflections	Negative Reflections
1	-1	$\mathbf{a}$	$-\mathbf{a}$
$\mathbf{ab}$	$-\mathbf{ab}$	$\mathbf{aba}$	$-\mathbf{aba}$
$(\mathbf{ab})^2$	$(\mathbf{ba})^2$	$-\mathbf{bab}$	$\mathbf{bab}$
$-\mathbf{ba}$	$\mathbf{ba}$	$-\mathbf{b}$	$\mathbf{b}$

then the rotors generated by this can be written as

$$(\mathbf{ab}) = e^{\mathbf{I}\mathbf{c}'\pi/p}, \quad (\mathbf{bc}) = e^{\mathbf{I}\mathbf{a}'\pi/q}, \quad (\mathbf{ca}) = e^{\mathbf{I}\mathbf{b}'\pi/r},$$

in which $\mathbf{I}$ is the pseudoscalar of the 3-D space, and $\mathbf{a}'$ and so on are the poles (unit axes) of the rotations. Careful analysis performed by Coxeter (in a different, non-GA representation) reveals that a necessary condition for finitely representable solutions is:

$$\tfrac{1}{p} + \tfrac{1}{q} + \tfrac{1}{r} > 1.$$

The integer solutions to this equation then generate the point groups in 3-D space. The complete classification may be found in [30].

Geometric algebra thus classifies the groups by sets of vectors in relative positions; these not only represent the reflection planes (which is how more classical solutions also characterize the symmetry), but they can immediately be used to generate the operators that perform all the operations in the group simply by the geometric product.

The extension to crystallographic groups imposes the additional condition that the point groups should remain closed under translation. We cannot treat those with the same simplicity within the vector space model. Our reference [30] develops the crystallographic groups further, using the conformal model described in Chapter 13 to include translations with the same ease.

10.3 COMPUTING WITH 3-D ROTORS

The Euclidean space $\mathbb{R}^{3,0}$ is a structural model of orientations in 3-D. In this model, rotations are represented by rotors. Those are structure-preserving and easily interpolated. We have already treated many of their structural properties when we introduced them in the general setting of n-dimensional algebras. Even though they did not then necessarily represent 3-D rotations, those provided a good and immediate illustration. We revisit and extend that material to provide the practical tools for 3-D rotors: how to find them from frames, how to determine their bivectors, and how to interpolate them.

10.3.1 DETERMINING A ROTOR FROM A ROTATION PLANE AND AN ANGLE

The most natural way to find a rotor in any geometrical problem is to know the rotation plane represented by the 2-blade $\mathbf{I}$ and the desired rotation angle ϕ (measured to be positive in the orientation of $\mathbf{I}$). Their product gives the geometric quantity $\mathbf{I}\phi$, which does not depend on the choice of the orientation of the plane. We called this the bivector angle in Section 7.2.2 (that is actually a misnomer in more than three dimensions, where it should then more properly be called a 2-blade angle, but that is rather a mouthful). The rotor in terms of the bivector angle is

$$R = e^{-\mathbf{I}\phi/2}.$$

We have seen how this is essentially a quaternion in Section 7.3.5, and in the programming exercise of Sections 7.10.3 and 7.10.4 showed how to convert between rotation matrices, quaternions, and rotors. In Section 7.3.4, we gave a pleasant visualization of the composition of rotations as the addition of half-angle arcs on a unit sphere.

If you have been able to determine the cosine c of the rotation angle in some way (for instance, through a scalar product), you can use this fairly directly to determine the rotor as

$$R = \sqrt{(1 + c)/2} - \sqrt{(1 - c)/2}\, \mathbf{I}. \tag{10.12}$$

The square roots are expressions of the cosine and sine of the half angle in terms of the cosine. When you take the standard positive values of the square roots, the orientation of **I** denotes the direction of rotation.

10.3.2 DETERMINING A ROTOR FROM A FRAME ROTATION IN 3-D

A rotor can be found when you know how enough vectors rotate. In any number of dimensions, there is a cheap way to make the unique rotor turning unit vector **a** to unit vector **b** in their common plane. This of course involves the half angle between them, which may seem expensive to compute. However, a unit vector in that half-angle direction is easily found by adding the vectors **a** and **b** and normalizing. The rotor is then the geometric ratio of this vector with **a**. That gives

$$R = \frac{1 + \mathbf{b}\,\mathbf{a}}{\sqrt{2\,(1 + \mathbf{a} \cdot \mathbf{b})}}. \tag{10.13}$$

You should realize that this is one of many rotations that can turn **a** into **b**. It is the simplest in the sense that it only involves their common plane. The formula is unstable when **a** and **b** are close to opposite; when they are truly opposite, there is no unique rotation plane to turn one into the other.

In general, you would need to know the image of several vectors before you can determine the exact relative rotor. In 3-D Euclidean space, there is a compact formula to retrieve the rotor from the three vectors of a frame $\{\mathbf{e}_i\}$ (which does not even have to be orthogonal) and their images $\{\mathbf{f}_i\}$. It is

$$R \sim 1 + \sum_{i=1}^{3} \mathbf{f}^i\, \mathbf{e}_i,$$

which needs to be properly scaled to become a rotor (for which of course $R\widetilde{R} = 1$). Note that it uses the reciprocal frame of the vectors $\mathbf{f}_i$ (see Section 3.8). It does not give the correct result for rotations over π (returning zero instead), and is unstable near rotations close to π. These consequences were already explored in the programming exercise of Section 7.10.3.

We give a rather advanced derivation of this formula, invoking vector derivatives. The only reason for doing this is to show the general n-dimensional pattern, should you ever need it, before we home in on the 3-D-only case.

$$\begin{aligned}
\sum \mathbf{f}^i\, \mathbf{e}_i &= \sum R\, \mathbf{e}^i\, \widetilde{R}\, \mathbf{e}_i \\
&= R\, \partial_{\mathbf{a}}\, (\widetilde{R}\, \mathbf{a}) \\
&= n - 2R\left(0\langle \widetilde{R}\rangle_0 + 2\langle \widetilde{R}\rangle_2 + 4\langle \widetilde{R}\rangle_4 + \cdots \right) \\
&\overset{3D}{=} 3 - 4\, R\langle \widetilde{R}\rangle_2 \\
&= 3 - 4\, R\, (\widetilde{R} - \langle R\rangle_0) \\
&= 4\,\langle R\rangle_0\, R - 1
\end{aligned}$$

and the result follows, since this shows that R is proportional to $1 + \sum_i \mathbf{f}^i \mathbf{e}_i$ in 3-D.

10.3.3 THE LOGARITHM OF A 3-D ROTOR

As we saw in Section 7.4.3, general rotors are the exponentials of bivectors. In the 3-D vector space model of $\mathbb{R}^{3,0}$, it is fairly straightforward to retrieve that bivector from the rotor. In such a 3-D space, the rotors are exponentials of 2-blades, since all bivectors are 2-blades in 3-D.

An explicit principal logarithm (as explained in Section 7.4.4) is easy to give for a rotor $R = \exp(-\mathbf{I}\phi/2)$. It can be determined by writing out the expression for the rotor in its grades and reassembling those using standard trigonometric functions on the grade parts:

$$\begin{aligned}
-\mathbf{I}\phi/2 &= \log(R) \\
&= \log\!\left(\exp(-\mathbf{I}\phi/2) \right) \\
&= \log\!\left(\cos(\phi/2) - \mathbf{I}\, \sin(\phi/2) \right) \\
&= \frac{\langle R\rangle_2}{\|\langle R\rangle_2\|}\, \mathrm{atan}\left(\frac{\|\langle R\rangle_2\|}{\|\langle R\rangle_0\|} \right).
\end{aligned} \tag{10.14}$$

There are some special cases that should be borne in mind. Obviously, when the scalar $\langle R\rangle_0$ equals zero, the division in the atan is ill-defined, even though there is still a well-defined rotation plane, and therefore a well-defined logarithm. In an implementation, you should use the atan2 function to provide numerical stability and additionally make (10.14) valid for the second and third quadrant. In two other cases, namely when the rotor equals 1 or -1, the rotation plane is ambiguous. For the identity rotation, the small angle makes the behavior around $R = 1$ still numerically stable, and the logarithm is virtually zero. But the ambiguity at $R = -1$ cannot be resolved without making arbitrary choices. These cases are recognizable in the pseudocode of Figure 13.5 (where the rotation logarithm is presented as part of the logarithm of the general rigid body motion).

This formula (10.14) can be extended from a rotor to a nonunit versor; there is then an additional term of $\log(\|R\|)$.

10.3.4 ROTATION INTERPOLATION

When you want to interpolate between two known orientations, you can do this by dividing the rotor between the extreme poses in equal amounts. This requires being able to take the nth root of a rotor through using its logarithm. For the Euclidean rotors in the 3-D vector space model of $\mathbb{R}^{3,0}$, this can be done explicitly.

Let us suppose we know the initial and final rotor in the interpolation sequence as R_1 and R_2, respectively. Then the total rotation that needs to be performed is characterized by the rotor $R \equiv R_2/R_1$. To perform this total rotation in n steps, one needs to apply the rotor $r \equiv R^{1/n}$, n times. In the geometric algebra formulation, this is simply

$$R = R_2/R_1 = e^{-\mathbf{I}\phi/2} \quad \Rightarrow \quad r = e^{-\mathbf{I}\phi/(2n)} = e^{\log(R)/n}.$$

This formula requires the bivector corresponding to the rotor, which we derived in (10.14) as its principal logarithm. The rotor that needs to be applied to the original element after k applications is $r^k R_1$, and of course we should have $r^n R_1 = R_2$. That gives the simple program of programming exercise 10.7.1. See also Figure 10.4.

An alternative is to give the resulting rotor not in multiplicative form, but to use the trigonometric functions to give a closed expression of the interpolation going on between

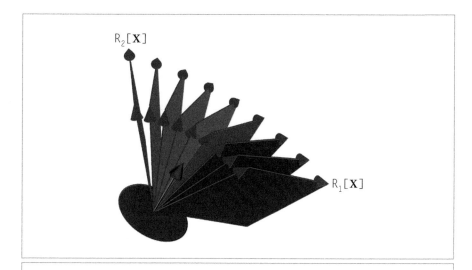

Figure 10.4: The interpolation of rotations illustrated on a bivector **X**. The poses $R_1[\mathbf{X}]$ and $R_2[\mathbf{X}]$ are interpolated by performing the rotor R_2/R_1 in eight equal steps.

R_1 and R_2. This is the way it is found in the quaternion literature, where this is known as *slerp interpolation* (for spherical-linear interpolation). To derive the formula, note that the requirement $R_2/R_1 = \cos(\phi/2) - \mathbf{I}\sin(\phi/2)$ gives $\mathbf{I}R_1 = (R_2 - R_1\cos(\phi/2))/\sin(\phi/2)$. Then the linear interpolation is achieved by rotation over a fraction $\lambda\phi/2$ of the angle, from R_1 towards R_2. This is the rotor $e^{-\lambda\mathbf{I}\phi/2}R_1$, which may be expressed as

$$
\begin{aligned}
R_\lambda &= \left(\cos(\lambda\phi/2) - \mathbf{I}\sin(\lambda\phi/2)\right)R_1 \\
&= \frac{R_1\sin(\phi/2)\cos(\lambda\phi/2) - R_1\cos(\phi/2)\sin(\lambda\phi/2) + R_2\sin(\lambda\phi/2)}{\sin(\phi/2)} \\
&= \frac{\sin((1-\lambda)\phi/2)}{\sin(\phi/2)}R_1 + \frac{\sin(\lambda\phi/2)}{\sin(\phi/2)}R_2
\end{aligned}
\tag{10.15}
$$

This is the linear interpolation formula for rotations (see [56]). It is valid in n-dimensional space.

In our software, we prefer not to use this explicit form, but instead the structurally simpler formulation in terms of the incremental rotor r. In the bivector formulation, one can easily design more sophisticated interpolation algorithms that interpolate between several rotors, such as bivector splines.

10.4 APPLICATION: ESTIMATION IN THE VECTOR SPACE MODEL

For applications in synthetically generated computer graphics, the generative techniques for rotations suffice, but in the analytical fields of computer vision and robotics you need to determine rotations based on noisy data.

10.4.1 NOISY ROTOR ESTIMATION

If your data is noisy, you should of course not establish the rotation based on three frame vectors as in Section 10.3.2, but instead use a rotor estimation technique. You could try to use the frame estimation on triplets of rotors and average their bivectors, but it is better to do a proper estimation minimizing a well-defined cost function.

We encountered such a technique in Section 8.7.2, when we used the rotation estimation problem to give an example of multivector differentiation. Its outcome is a useful result that you can easily implement even without understanding the details of its derivation.

10.4.2 EXTERNAL CAMERA CALIBRATION

When you have a set of cameras of unknown relative positions and attitudes observing one scene, you can only integrate their views if you have *calibrated* the setup (i.e., estimated the parameters of their relative geometry). You can collect the data for such a calibration by moving a spherical marker around in the scene and synchronizing the cameras to record their various views of it at the corresponding times, as illustrated in Figure 10.7. The observed data is of course inherently noisy, and you need to do some processing to determine the "best" estimate for their relative poses. We describe an algorithm for this, taken from [38], culminating in the programming exercise of Section 10.7.3. Our source assumes that the cameras have been calibrated internally. This involves a determination of the parameters of their optics and internal geometry, so that we can interpret a pixel in an image as coming from a well-determined spatial direction relative to its optical axis. Geometrically, it turns the camera into a measurement instrument for spatial directions.

Let us consider $M + 1$ cameras. We arbitrarily take one of them as our reference camera number 0, and characterize the location of the optical center of camera j relative to the center of camera 0 by the translation vector $\mathbf{t}_j$ and its orientation by the rotor R_j. We mostly follow the notation of [38] for easy reference, which uses bold capitals for vectors in the world, and lowercase for vectors within the cameras. We will simplify the situation by assuming that the marker is visible in all cameras at all times (our reference deals with occlusions; this is not hard but leads to extra administration).

The marker is shown N times at different locations $\mathbf{X}_i$ in the real world. Relative to camera j, it is seen at the location $\mathbf{X}_{ij}$ given implicitly by

$$\mathbf{X}_i = \mathbf{t}_j + R_j \mathbf{X}_{ij} \widetilde{R}_j.$$

However, all that camera j can do is register that it sees the center of the marker in its image, and (using the internal calibration) know that it should be somewhere along the ray in direction $\mathbf{x}_{ij}$ from its optical center. The scaling factor along this ray is σ_{ij}; if we would know its value, the camera would be a 3-D sensor that would measure $\sigma_{ij} \mathbf{x}_{ij} = \mathbf{X}_{ij}$, and then R_j and $\mathbf{t}_j$ could be used to compute the true location of the measured points. But the only data we have are the $\mathbf{x}_{ij}$ for all the cameras. All other parameters must be estimated. This is the *external calibration* problem.

The estimation of all parameters is done well when the reconstructed 3-D points are not too different from their actual locations. When we measure this deviation as the sum of the squared differences, it implies that we want to minimize the scalar quantity

$$\Gamma = \sum_{j=1}^{M} \sum_{i=1}^{N} \left(\mathbf{X}_i - \mathbf{t}_j - R_j \, \sigma_{ij} \, \mathbf{x}_{ij} \, \widetilde{R}_j \right)^2.$$

Now partial differentiation with respect to the various parameters can be used to derive partial solutions, assuming other quantities are known. This employs the geometric differentiation techniques from Chapter 8. The results are simple to interpret geometrically and are the estimators you would probably have proposed even without being able to show that they are the optimal solutions.

- **Optimal translation t_j given R_j, σ_{ij}, X_i, and the data x_{ij}.** This involves differentiation of the cost function relative to t_j. The zero derivative is attained at

$$t_j = \frac{1}{N} \sum_{i=1}^{N} \left(X_i - R_j \, \sigma_{ij} \, x_{ij} \, \widetilde{R}_j \right). \qquad (10.16)$$

This is simply the average difference of where camera j would reconstruct the points based on its presumed rotations and their true average location.

- **Optimal rotation R_j given X_i and the data x_{ij}.** The differentiation with respect to a rotor was treated in Section 8.7.2 when optimizing (8.16). Here the result corresponding to (8.17) is that the optimal R_j must satisfy

$$\sum_{i=1}^{N} \left((X_i - t_j) \wedge (R_j \, \sigma_{ij} \, x_{ij} \, \widetilde{R}_j) \right) = 0.$$

The geometrical interpretation is that the optimal rotor rotates to minimize the transverse components of all X_i when reconstructed by camera j. Substituting the expression for the optimal t_j leads to

$$\sum_{i=1}^{N} \left((X_i - \underline{X}) \wedge (R_j \, \sigma_{ij} \, x_{ij} \, \widetilde{R}_j) \right) = 0, \qquad (10.17)$$

where $\underline{X} \equiv \frac{1}{N} \sum_{k=1}^{N} X_k$ is the centroid of the world points. As in Section 8.7.2, this optimal R can be found by a singular value decomposition of a linear function defined in terms of the other parameters.

- **Optimal scaling σ_{ij} given R_j, t_j, X_i, and the data x_{ij}.** This requires scalar differentiation of Γ, and results in

$$\sigma_{ij} = (X_i - t_j) \cdot (R_j \, x_{ij}^{-1} \, \widetilde{R}_j). \qquad (10.18)$$

This is almost a division of the estimated X_{ij} by the rotated x_{ij}, except that the inner product makes only the parts along the ray contribute (so that a scalar results).

- **Optimal world points given t_j, R_j, σ_{ij}, and the data x_{ij}.** Setting the vector derivative of Γ with respect to X_i to zero yields

$$X_i = \frac{1}{M} \sum_{j=1}^{M} (t_j + R_j \, \sigma_{ij} \, x_{ij} \, \widetilde{R}_j). \qquad (10.19)$$

This is simply the average of the estimations of the location of world point i by each camera.

- *Optimal world points X_i given $\mathbf{t}_j$, R_j, and the data $\mathbf{x}_{ij}$.* The optimum for σ_{ij} of (10.18) can be substituted in (10.19) to find a system of linear equations for $\mathbf{X}_i$ using R_j, $\mathbf{t}_j$, and the data $\mathbf{x}_{ij}$. That system can be solved optimally by a least squares technique.

- *Optimal translations $\mathbf{t}_j$ given R_j and the data $\mathbf{x}_{ij}$.* Another substitution of (10.18) and combination with the result for the $\mathbf{X}_i$ leads to a system of homogeneous linear equations for the $\mathbf{t}_j$ using an estimate of the R_j and the data $\mathbf{x}_{ij}$. That system can be solved optimally using an SVD.

With these composite results, the following iterative scheme can be formulated:

1. Make an initial guess of the R_j based on the data using a standard stereo vision algorithm (the geometrical basis for such algorithms will be explained in Section 12.2).
2. Estimate the $\mathbf{t}_j$ using the estimate of R_j and the data $\mathbf{x}_{ij}$.
3. Estimate the world points $\mathbf{X}_i$ using R_j, $\mathbf{t}_j$, and the data $\mathbf{x}_{ij}$.
4. Estimate the scaling σ_{ij} using (10.18).
5. Obtain a new estimation for R_j using (10.17) and iterate from step 2.

The authors of [38] report that a dozen or so iterations are required for convergence (depending on the number of cameras). The resulting calibration algorithm is fully linear. In modern calibration practice, it is customary to use the outcome of such linear algorithms as an initial estimate for a few steps of subsequent nonlinear optimization to improve the estimation.

This algorithm is the basis of the programming example of Section 10.7.3. The compact and direct derivation of the partial solution formulas (10.16) through (10.19) are an exemplary usage of geometric algebra for these kinds of geometrical optimization problems in the vector space model.

10.5 CONVENIENT ABUSE: LOCATIONS AS DIRECTIONS

The vector space model is the natural model to compute with directions and the ultimate tool for this purpose. It will remain clearly recognizable as a submodel providing the algebra of Euclidean directions, even as we move on to more sophisticated models in the next chapters.

In a purist point of view, the vector space model would not be used for other tasks. Yet we can, of course, model location in the vector space model, in the same way as it has always been done in elementary linear algebra. We just view a location as obtained by traveling in a certain direction denoted by a direction vector $\mathbf{p}$, over a distance given by its norm. This treats a direction vector as a *position vector*, and particularly for problems only involving point objects, it is not bad practice. The calibration example of the previous section shows that it can be very effective, and since that is a problem in which locations are actually observed as directions, the vector space model is in fact its natural setting.

When you also have geometrical elements other than pure directions (such as line or plane offset from the origin), you run into the familiar problems that the use of classical linear algebra also entailed: translations of such elements require administration of object types and corresponding data structures. For instance, you can characterize a line by a position vector and a direction vector, but you should keep them clearly separate, for under a translation over a vector $\mathbf{t}$, the position needs to change but the direction should not. Uniformity is only obtained by having a single algebraic element representing the line, with a representation of translation that can operate on it directly. The vector space model does not provide that in a structural manner. You need to encode this structure explicitly or use at least the homogeneous model.

Examples of the "convenient abuse" of directions as locations abound in all graphics and robotics literature, as well as in typical physics textbooks. Hestenes [29] shows how geometric algebra can be used effectively in the vector space model to do all of classical physics. The vector space model does not lack computational power, and its rotors help considerably in simplifying classical problems like the orbits of planets in a gravitational field (which involve locations, but viewed from the sun so that their directional treatment becomes natural). But this computational power can only be wielded by manually keeping track of what geometry is represented by each element and which operations are permitted to be performed on it. That is less a problem for physics (which tends to connect its equations by natural language anyway), but it is a major source of programming errors in computer graphics (as reported in [23, 44]). The models of the next chapters will provide alternatives in which a more extended algebra is used to perform simultaneously both the computations and the bookkeeping of geometrical elements.

10.6 FURTHER READING

The vector space model may seem prevalent in almost all linear algebra texts, since it is the most basic way to treat geometry with vectors. However, in geometric algebra, the full vector space model naturally includes blades and rotors. Not many texts incorporate those in their treatment of basic geometry.

Your best background material for advanced use of the vector space model of geometric algebra are texts in introductory physics (such as [29] and [15]). For current use in practical problems, the papers using geometric algebra in professional journals on computer vision and robotics are your best source, though these increasingly use the more powerful conformal model to address problems involving direction and location.

10.7 PROGRAMMING EXAMPLES AND EXERCISES

10.7.1 INTERPOLATING ROTATIONS

Interpolating rotations is an important problem with many applications. It is straightforward to implement once you are able to compute the logarithm of a rotor. This can be implemented as follows (see Section 10.3.3):

```
bivector log(const rotor &R) {
  // get the bivector/2-blade part of R
  bivector B = _bivector(R);

  // compute the 'reverse norm' of the bivector part of R:
  mv::Float R2 = _Float(norm_r(B));

  // check to avoid divide-by-zero
  //    (and also below zero due to FP roundoff):
  if (R2 <= 0.0){
    if (_Float(R) < 0) {
      // the user asks for log(-1)
      // we return a 360 degree rotation in an arbitrary plane:
      return _bivector((float)M_PI * (e1 ^ e2));
    }
    else
      return bivector();   // return log(1) = 0
  }

  // return the log:
  return _bivector(B * ((float)atan2(R2, _Float(R)) / R2));
}
```

When you look for this function in the GA sandbox source code package, note that it resides in the file e3ga_util.cpp in the libgasandbox directory and not in the main source file of this example. The same is true for the exp() function listed below. This exp() is a specialization of the generic exponentiation algorithm, as described in Sections 7.4 and 21.3.

```
rotor exp(const bivector &x) {
  // compute the square
  mv::Float x2 = _Float(x << x);

  // x2 must always be <= 0, but round off error can make it
  // positive:
  if (x2 > 0.0f) x2 = 0.0f;

  // compute half angle:
  mv::Float ha = sqrt(- x2);
```

```
    if (ha == (mv::Float)0.0) return _rotor((mv::Float)1.0);
    // return rotor:
    return _rotor((mv::Float)cos(ha) +
        ((mv::Float)sin(ha) / ha) * x);
}
```

Now that we can go back and forth between rotor and bivector representations of rotations, interpolating rotations becomes straightforward: given two rotors src and dst, we first compute their difference as the ratio inverse(src) * dst. Then we compute the log() of the difference, and scale by the interpolation parameter alpha (which will run from 0 to 1). The final step is reconstructing the rotor using exp(), and putting it back in the original frame by multiplying with src:

```
rotor interpolateRotor(const rotor &src, const
    rotor &dst, mv::Float alpha) {
    // return src * exp(alpha * log(inverse(src) * dst));
    return _rotor(src * exp(_bivector(alpha *
        log(_rotor(inverse(src) * dst)))));
}
```

The example uses this code to display a rotating/translating frame with a trail following behind it, as seen in Figure 10.5. Interpolation of translations is done the classical way:

```
e3ga::vector interpolateVector(
    const e3ga::vector &src, const e3ga::vector &dst,
        mv::Float alpha) {
    return _vector((1.0f − alpha) * src + alpha * dst);
}
```

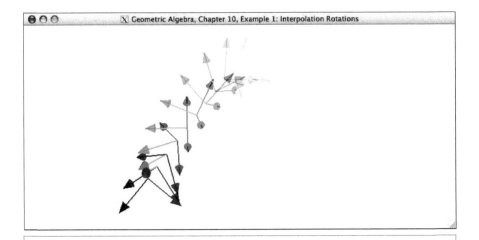

Figure 10.5: Interpolation of rotations (Example 1).

In later examples—when we gain the additional ability to represent translation and scalings as rotors—we will redo this example. Only the `log()` and `exp()` functions will change; the interpolation function remains essentially the same (see Sections 13.10.4 and 16.10.3).

10.7.2 CRYSTALLOGRAPHY IMPLEMENTATION

We have made an implementation to play with the vectors generating symmetries of crystal lattices. An example of the output is Figure 10.6. The application works for the point groups in 3-D. It shows three input vectors **a**, **b**, **c** drawn as black line segments, which are the basic generators of the symmetry operators from Section 10.2.3. You can drag those around to investigate possible nearby symmetries.

The operators are computed from the initial generators in a brute force manner, by repeatedly multiplying them with each other to make new versors until no new symmetry operators are found. To get a true point group, **a**, **b**, and **c** should be chosen in particular ways,

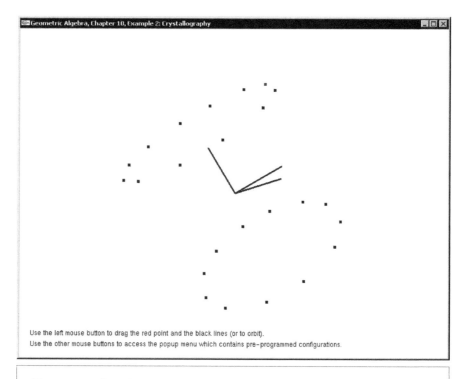

Use the left mouse button to drag the red point and the black lines (or to orbit).
Use the other mouse buttons to access the popup menu which contains pre-programmed configurations.

Figure 10.6: Crystallography (Example 2). This example shows the 24 symmetries of a hexagonal crystal generated from a single red point.

and for some should be considered as composite operators like (**ac**). If the generators are not in the proper configuration to generate a point group, their combination may lead to an infinite number of versors, but we arbitrarily cut off the generation at 200 different versors. The largest true point group of cubic symmetry has 48 operators.

To draw the point groups, the code applies the operators to a single (draggable) input point. A popup menu can be used to initialize the vectors to preprogrammed configurations (cubic, hexagonal, tetragonal, orthorhombic, triclinic, monoclinic, and trigonal), corresponding to the seven main point groups. For more details on how to generate all the subgroups (32 in total), consult [30].

10.7.3 EXTERNAL CAMERA CALIBRATION

This example implements the external calibration of [38], as summarized in Section 10.4.2. It assumes that the cameras have already been calibrated internally (we used the method of Zhang [63]), and that we have an initial external calibration (we used the 8-point algorithm, see e.g., [25]).

The data provided with the example is actual calibration data from a geometric-algebra-based motion capture system built by the authors. The data contains the initial 8-point calibration estimation, but we have deliberately decreased the quality of the initial estimation to make the effect of the refinement more pronounced. Figure 10.7 shows the reconstruction of the markers seen by the cameras after completion of the calibration.

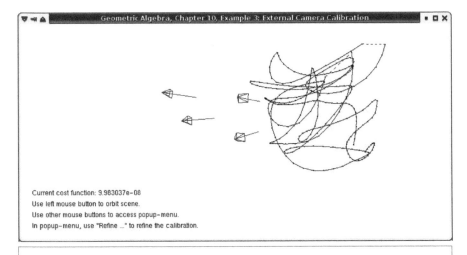

Figure 10.7: External camera calibration (Example 3). The cameras are drawn as red wire-frame pyramids (view volume) with a line indicating their viewing direction. The marker points used for calibration are drawn as black dots. A line connects the marker points to show how the single marker was waved through the viewing volume of the cameras.

To implement (10.16) through (10.19), we first need some context. The data is kept in a class State. This state contains cameras in array m_cam, and 3-D world points in array m_pt:

```
class State {
public:
  // ... (constructors, etc)

  // the cameras
  std::vector<Camera> m_cam;

  // the reconstructed markers, for each frame
  std::vector<e3ga::vector> m_pt;

  // is reconstruction of markers valid?
  std::vector<bool> m_ptValid;
};
```

The reconstruction of a marker is invalid when it is visible to only one camera.

Each camera carries its own 2-D marker measurements (m_pt), rotation (m_R and m_Rom), translation (m_t), scaling σ_{ij} (m_sigma), and per-marker visibility (m_visible). (We ignored visibility issues when we explained the algorithm, but it is included in [38] and implemented in the code.)

```
class Camera {
public:
  // ... (constructors, etc)

  // rotation
  rotor m_R;
  // rotation matrix
  om m_Rom;

  // translation
  e3ga::vector m_t;

  // for each frame, is a marker visible?
  std::vector<bool> m_visible;

  // for each frame, the '2D' point in the image plane
  // (normalized image coordinates, i.e., the e3 coordinate = -1)
  std::vector<e3ga::vector> m_pt;

  // for each frame the estimated multiplication factor of m_pt
  std::vector<mv::Float> m_sigma;

};
```

Now each of the equations in Section 10.4.2 can be implemented and those functions combined in the total algorithm. As an example, (10.16) leads to the code:

```
void State::updateTranslation() {
  // Iterate over all cameras
  // (start at '1' because translation of first camera is always 0)
  for (unsigned int c = 1; c < m_cam.size(); c++) {
    Camera &C = m_cam[c];

    vector sum;
    int nbVisible = 0;
    for (unsigned int i = 0; i < m_pt.size(); i++) {
      if ((C.m_visible[i]) && m_ptValid[i]) {
        sum += _vector(m_pt[i] - C.m_sigma[i] *
          (apply_om(C.m_Rom, C.m_pt[i])));
        nbVisible++;
      }
    }
    C.m_t = sum * (1.0f / (mv::Float)nbVisible);
  }
}
```

Note that for efficiency we use the outermorphism matrix representation of the rotor to apply it to vectors by means of the function apply_om(). You can consult the rest of the code in the GA sandbox source code package.

We will encounter the motion capture system again in Section 12.5.3, when we reconstruct 3-D markers from the raw 2-D data.

11 THE HOMOGENEOUS MODEL

While the 3-D vector space model can nicely model directions, it is usually considered to be inadequate for use in 3-D computer graphics, primarily because of a desire to treat points and vectors as different elements that are transformed differently by translations. Instead, people commonly use an extension of linear algebra known as *homogeneous coordinates*. This is often described as augmenting a 3-D vector $\mathbf{v}$ with coordinates $(v_1, v_2, v_3)^T$ to a 4-vector $(v_1, v_2, v_3, 1)^T$. This extension makes nonlinear operations such as translations implementable as linear mappings.

For the homogeneous model in geometric algebra, the modeling principle is the same: we embed our n-dimensional base space $\mathbb{R}^n$ (which you may think of as Euclidean $\mathbb{R}^{n,0}$, to be specific) in an $(n+1)$-dimensional representational vector space $\mathbb{R}^{n+1}$, of which we then use the inherent algebra. That produces a complete algebraic framework, which is well suited to compute with oriented flats, subspaces offset from the origin in $\mathbb{R}^n$ represented as blades in $\mathbb{R}^{n+1}$.

The algebra of $\mathbb{R}^{n+1}$ provides generally applicable formulas for translation, rotation, and even affine and projective transformations in the base space $\mathbb{R}^n$. The operations of meet and join always return sensible results for incidences of flats. For instance, for two parallel lines the meet returns the common direction of the lines as a point at infinity weighted by their mutual distance. We derive useful covariant and invariant measures, including the projectively invariant cross ratio, in terms of geometric algebra.

We treat these subjects in the following order: after defining the embedding space in Section 11.1, we introduce the blades of $\mathbb{R}^{n+1}$ representing the flats of the base space $\mathbb{R}^n$ in Section 11.2 through Section 11.6. This will enable you to represent offset points, lines, and planes, either directly or dually. We combine the flats using the incidence operations of meet and join in Section 11.7. The resulting unified generic intersections and unions can simplify the data flow of programs considerably. Moreover, some of the incidence constructions are cross ratios and can be interpreted as defining motion-invariant measures. Then we study motions and transformations in Section 11.8 and show that all direct flats are moved by the same linear transformation, and all dual flats by another. This will simplify your code even more. We provide some initial guidance in constructions using the homogeneous model in Section 11.9, but end on a more somber note: the homogeneous model has some deficiencies in dealing with Euclidean metric properties. They will be fixed—but not in this model.

11.1 HOMOGENEOUS REPRESENTATION SPACE

As in all chapters in Part II on modeling, we are interested in representing geometric elements and operations in the space $\mathbb{R}^n$, which typically has a Euclidean metric. The homogeneous model differs from the vector space model of the previous chapter in its roundabout representation method: it embeds $\mathbb{R}^n$ in a space $\mathbb{R}^{n+1}$ with one more dimension and then uses the algebra of $\mathbb{R}^{n+1}$ to represent those elements of $\mathbb{R}^n$ in a structured manner. That structure is the gain, for when we program the geometry, the homogeneous model provides an automatic and consistent framework of data structures for the elements we represent, as well as a more universal set of geometrical operators.

In this chapter, we will therefore continually have to switch between a geometric element and its representation, much more explicitly than we did before. We will denote those transitions explicitly in the beginning, but as we grow familiar with the model it becomes convenient to let representations coincide with the elements they represent, to avoid getting too pedantic.

Let us call $\mathbb{R}^n$ the *base space* and the higher space $\mathbb{R}^{n+1}$ the *(homogeneous) representation space*. We view this construction as an embedding of $\mathbb{R}^n$ *inside* $\mathbb{R}^{n+1}$ in a straightforward, linear manner; we just add an extra dimension. We denote the unit vector in this "new" dimension by e_0.

We are free to define the metric of the homogeneous representation space as long as we make sure that it coincides with the metric of $\mathbb{R}^n$ whenever both are defined. A metric is completely determined by the inner product, so given the value of $\mathbf{x} \cdot \mathbf{y}$ when both $\mathbf{x}$ and $\mathbf{y}$ are in $\mathbb{R}^n$, we use that same value for the same vectors when viewed as part of the larger space $\mathbb{R}^{n+1}$. Then we can use the same dot notation for the inner product in $\mathbb{R}^{n+1}$; where they coincide, they agree anyway.

We now conveniently choose the metric of $\mathbb{R}^{n+1}$ such that $e_0 \cdot \mathbf{x} = 0$ for any $\mathbf{x} \in \mathbb{R}^n$. That just implies that e_0 is perpendicular to the subspace $\mathbb{R}^n$. This splits the representation space $\mathbb{R}^{n+1}$ nicely; a general vector x in $\mathbb{R}^{n+1}$ can always be written as $x = \xi_0 e_0 + \mathbf{x}$, an e_0-component and a $\mathbb{R}^n$-component (either or both of which may be zero). For convenience, we will denote quantities completely residing in the subspace $\mathbb{R}^n$ by bold font (such as $\mathbf{x}$ or $\mathbf{e}_1$) and those that do not by a regular math font (such as x or e_0).

Our definition of the inner product in $\mathbb{R}^{n+1}$ is not yet complete, for we can only compute the inner product between two arbitrary vectors if we also know how to evaluate $e_0 \cdot e_0$. Of course this quantity is not part of the real geometry we want to describe—that resides completely in the base space $\mathbb{R}^n$. Since we have no geometrical reasons to choose a particular value of $e_0 \cdot e_0$, we should choose it for reasons of computational convenience. The choice $e_0 \cdot e_0 = 0$ would make e_0 noninvertible, which is inconvenient to many computations (such as taking a dual). The other natural choices are $e_0 \cdot e_0 = +1$ and $e_0 \cdot e_0 = -1$; both occur in the literature. We support both in this book, or rather, we do not choose between them simply by conscientiously writing e_0^{-1} whenever we need the inverse of e_0. If the metric would have $e_0 \cdot e_0 = +1$, you can substitute $e_0^{-1} = e_0$; if you prefer $e_0 \cdot e_0 = -1$, you can substitute $e_0^{-1} = -e_0$. Keeping the inverse notation explicit in this manner has the additional advantage that it enables easy checking of the dimensional correctness of expressions at a glance. We reiterate that the actual choice for this part of the metric will *not* affect real geometric quantities computed from the model; the extra e_0-dimension is imaginary (in the literal sense of residing only in the imagination of ourselves or of our computer). No computable geometric property should depend on it, for it is merely a mathematical device to make computations more homogeneous.

The multiplication table for the inner product of the various types of vectors is summarized in Table 11.1 for a 3-D Euclidean space represented by a 4-D homogeneous space. When we do not focus on a metric, we will denote the homogeneous space by $\mathbb{R}^{n+1}$. Should we need to differentiate, then we denote by $\mathbb{R}^{n+1,0}$ the all-Euclidean space in which $e_0^2 = +1$, and by $\mathbb{R}^{n,1}$ the space in which $e_0^2 = -1$.

Table 11.1: Specification of the geometric algebra of the homogeneous model of a 3-D Euclidean space—the inner product table for the canonical basis $\{e_0, \mathbf{e}_1, \mathbf{e}_2, \mathbf{e}_3\}$. For e_0^2, we allow a choice of +1 or −1. We have overloaded the dot notation for the inner product in both the homogeneous space and the Euclidean space without possible confusion, since they coincide for all arguments on which they are both defined.

$\cdot$	e_0	$\mathbf{e}_1$	$\mathbf{e}_2$	$\mathbf{e}_3$
e_0	± 1	0	0	0
$\mathbf{e}_1$	0	1	0	0
$\mathbf{e}_2$	0	0	1	0
$\mathbf{e}_3$	0	0	0	1

11.2 ALL POINTS ARE VECTORS

One of the motivations for a higher dimensional space for our geometry is to make a distinction between points and vectors. A point in the base space $\mathbb{R}^n$ is a location marker, whereas a vector in $\mathbb{R}^n$ is a direction. In the vector space model of the previous chapter, there was hardly a distinction, for we had to use the vectors to mark locations (they were all we had). In the homogeneous model, location and directions are represented differently. This is the first instance of an interface between an element of the base space and its representation, and it is the foundation of the model.

11.2.1 FINITE POINTS

To give the algebra of the homogeneous representation space $\mathbb{R}^{n+1}$ a geometric meaning, we interpret e_0 as the point at the origin. A point at any other location $\mathbf{p}$ is made through translation of the point at the origin over the Euclidean vector $\mathbf{p}$. This is done by adding $\mathbf{p}$ to e_0. This construction therefore gives the representation of the point $\mathcal{P}$ in the base space $\mathbb{R}^n$ at location $\mathbf{p}$ as the vector p in the homogeneous representation space $\mathbb{R}^{n+1}$:

$$point\ to\ vector:\ \ p = e_0 + \mathbf{p}.$$

Algebraically, p is just a regular *vector in* $\mathbb{R}^{n+1}$; geometrically, we interpret it as a *point of* $\mathbb{R}^n$ (see Figure 11.1(a).) We call such a point representation p with unit coefficient for e_0 a *unit point*. You can visualize this construction as in Figure 11.1(a) (necessarily drawn for $n = 2$).

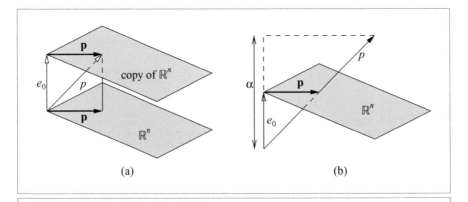

(a)   (b)

Figure 11.1: Elements of the base space $\mathbb{R}^n$ are represented by embedding it in a homogeneous representation space of one more dimension. The extra vector is e_0, and it is perpendicular to all vectors $\mathbf{p}$ of $\mathbb{R}^n$. (a) It is convenient to draw $\mathbb{R}^n$ at the tip of e_0 to denote the location of vectors $p = e_0 + \mathbf{p}$ of $\mathbb{R}^{n+1}$ representing unit points of $\mathbb{R}^n$. (b) A general point $\alpha(e_0 + \mathbf{p})$ with a weight α is still interpreted as having location $\mathbf{p}$.

A more general vector in $\mathbb{R}^{n+1}$ would be a multiple of this: $\alpha\,(e_0 + \mathbf{p})$. We interpret that as a point of $\mathbb{R}^n$ at the same location $\mathbf{p}$, but with a different *weight*. We can retrieve these real geometric quantities *location* (or position vector) and *weight* easily from a point representative $p \in \mathbb{R}^{n+1}$, as illustrated in Figure 11.1(b):

$$\text{weight}: e_0^{-1} \cdot p; \quad \text{location}: \frac{p}{e_0^{-1} \cdot p} - e_0 = \frac{e_0^{-1}\rfloor(e_0 \wedge p)}{e_0^{-1}\rfloor p}. \tag{11.1}$$

The fancy rewriting of the location anticipates a general pattern later. You should view the computational patterns as *selection operations*: in the expression for $p = \alpha\,e_0 + \alpha\,\mathbf{p}$, the computation $e_0^{-1} \cdot p$ takes e_0 out of the terms in the expression containing it. That leaves α. The expression $e_0^{-1}\rfloor(e_0 \wedge p)$ is more involved. The part between brackets kills terms containing e_0, and then the subsequent inner product with e_0^{-1} restores the rest by removing the extra e_0, so it selects the part of p that does not contain e_0 (it is actually a rejection). Such expressions will occur throughout this chapter.

These algebraic formulas are easy enough to implement in any program, but not very efficient. In a coordinate representation of p as $p = (p_0, p_1, p_2, p_3)^T$ on the basis $\{e_0, \mathbf{e}_1, \mathbf{e}_2, \mathbf{e}_3\}$, you would implement these selection operations more directly. Simply take the coordinate p_0 as the weight, and $(p_1, p_2, p_3)^T/p_0$ as the position vector of the location. That is the same as in the usual homogeneous coordinate-based representation. The algebraic formulation of (11.1) is the coordinate-free way to show the essential structure. This structure is useful in symbolic manipulations of equations, and it will generalize to lines and planes.

We actually use the general formulas in our program specifications in Gaigen2. Our code generator recognizes at compile time that they can be implemented as selection operations, so we suffer no performance loss for this structural generality.

11.2.2 INFINITE POINTS AND ATTITUDES

It is clear that the interpretation (11.1) of a homogeneous vector as a point at a specific location only works when $e_0^{-1} \cdot p \neq 0$. If not, the vector has no e_0-component and it is therefore of the form $\alpha\,\mathbf{u}$, a vector completely in $\mathbb{R}^n$. Such vectors have a different geometric interpretation: they are directions in the base space, just as they were in the vector space model:

$$\text{weighted 1-direction}: \quad \alpha\,\mathbf{u} \in \mathbb{R}^n.$$

In fact, in the homogeneous model there are two ways of thinking about such a vector: as a *direction* in $\mathbb{R}^n$ determining a weighted, oriented attitude in space, or as a *point at infinity* in $\mathbb{R}^n$. The latter is also called an *improper point*, and is consistent with it being the limiting case of the finite point $p = e_0 + \mathbf{p}$ when $\mathbf{p}$ gets so large that it dwarfs e_0. Despite this associated interpretation in the base space $\mathbb{R}^n$, $\alpha\,\mathbf{u}$ is a finite element of the representational space $\mathbb{R}^{n+1}$. On the other hand, it is consistent to view $\mathbf{p}$ as a direction because it is in the purely directional subalgebra $\mathbb{R}^n$ of the total space $\mathbb{R}^{n+1}$, which we have used as the algebra of directions before in the vector space model of Chapter 10.

It may be a personal choice to decide which explanation suits your intuition better. At the first encounter, you probably prefer the "direction" interpretation, since it so naturally builds on the familiar geometry of the vector space model. However, it was a major insight in (projective) geometry that many theorems get more universal by incorporating well-defined "improper" points at infinity on par with the "proper" finite points. Two lines in a plane then always intersect in a point, be it finite or infinite (and in the latter case we tend to say that they have the same direction). That is an argument in favor of using the improper point at infinity conceptualization. The semantics make no difference to the algebra or the geometry, of course, but some flexibility here may make you more dextrous in converting geometrical situations into algebraic formulas. The unification in this point of view will also extend to flats of higher grade.

For a direction (or improper point) $\mathbf{p}$, a unit normalization such that $\mathbf{p} \cdot \mathbf{p} = 1$ is natural, but a nonunit weight has a natural geometric meaning as a velocity or as a density.

11.2.3 ADDITION OF POINTS

Since points (finite or infinite) in the base space $\mathbb{R}^n$ are represented as vectors in the representation space $\mathbb{R}^{n+1}$, we may be tempted to add them. Of course, that addition is purely an operation in the representation space, of which a sensible geometric interpretation in the base space needs to be found if it is to be permitted in the model.

Adding a finite point represented by the vector p to itself gives the vector $2\,p = 2e_0 + 2\mathbf{p}$, which we interpret as a point with double the weight at the same location. We can clearly take any multiple, and it is a bonus feature of the homogeneous model that points have such a natural weight associated with them. The weighted points will appear in incidences, for instance as the result of the meet of a line and a plane, where the weight denotes the significance of the intersection (weight ± 1 for a perpendicular intersection, weight close to zero for a grazing intersection). We saw this at the origin in Section 5.5. Having points being more than mere location markers is a useful extension to our quantitative geometry. Such weights for locations were not available in the vector space model.

Adding an infinite point (or 1-direction) to itself has the interpretation we saw in the vector space model; it represents the same attitude, with double the weight. Depending on its use, you may interpret this as a direction with twice the velocity or half the density, both eminently sensible geometric notions. Addition of different direction vectors also makes sense; it is like adding velocities, an elementary operation in the physical interpretation of geometry (and historically actually the motivation for vector addition).

Adding two different points is less clearly interpretable, unless we also view it as a physical operation, and $p+q$ is then the representation of the center of mass of p and q. For denoting their weights by m_p and m_q, we find that their sum

$$p + q = \left(m_p \, (e_0 + \mathbf{p}) \right) + \left(m_q \, (e_0 + \mathbf{q}) \right) = (m_p + m_q)\, e_0 + (m_p\, \mathbf{p} + m_q\, \mathbf{q})$$

is interpreted as a point with weight $m_p + m_q$ at the location $\frac{m_p\,\mathbf{p} + m_q\,\mathbf{q}}{m_p + m_q}$, precisely the correct result for the physical interpretation. It is nice to get this for free, and it suggests that the homogeneous model contains more than just mathematical geometry (as practiced in the classical geometry courses), or alternatively that some things we think of as fundamental descriptive elements of physics are actually geometrical.

In many computer graphics applications focusing on geometry, one prefers to compute with unit points, which only have a location. We believe (with [24]) that this is unnecessarily limiting and that it pays to think of weighted points (especially as one ultimately wants to include physics in rendered worlds anyway!). However, some essential algorithms (such as de Casteljau's curve evaluation algorithm) appear to depend on unit points. If you only want unit points as input and output, you need to introduce a special weighted addition called the *affine combination*, in which the weights add to unity:

$$p = \sum_{i=1}^{k} \alpha_i\, p_i \ \ \text{with} \ \ \sum_{i=1}^{k} \alpha_i = 1 \tag{11.2}$$

It is easy to verify that $e_0^{-1} \cdot p = 1$ if $e_0^{-1} \cdot p_i = 1$ for all p_i, so unit points as input lead to a unit point as output. For two unit points p and q, addition with equal weight gives

$$\frac{1}{2}\,(p + q),$$

which clearly computes the unit point in the middle. For k points p_i, one computes the unit point at the *centroid* by equal weighting:

$$\text{centroid} = \tfrac{1}{k} \sum_{i=1}^{k} p_i.$$

This is a clearly purely geometric quantity, though actually it is a special case of the center of mass for weighted points. Later, in Section 11.8.8, we will see that affine transformations preserve weights, so that the affine combination is an affinely invariant construction—hence its name.

The final addition we should consider is that of an infinite point and a finite point, which in $\mathbb{R}^{n+1}$ looks innocent enough (it is again just vector addition). If the finite point is a unit point p and the infinite point represented by the vector $\mathbf{t}$, this gives the point $p + \mathbf{t}$ at location $\mathbf{p} + \mathbf{t}$. So adding an infinite point (or 1-direction) to a unit point translates it. If the point p is not a unit point but has a weight m, the resulting point is still $p + \mathbf{t}$, which is now interpreted as a point of weight m at the location $\mathbf{p} + \mathbf{t}/m$. An image from physics suggests itself: $\mathbf{t}$ is not a translation (i.e., not a displacement) but an *impulse*, and a heavy point responds to that with a smaller movement (per time unit). But actually, we prefer to reparameterize and define point translations properly scaled to the point so that they can be linear transformations. A translation over $\mathbf{t}$ of a point p with mass m is then $p + m\mathbf{t} = p + (e_0^{-1} \cdot p)\,\mathbf{t}$, which is linear in p. More about translations in Section 11.8.2.

In summary, addition of points in the base space (finite or not) makes geometric sense, so addition of vectors in the representational space is permissible. Since scalar multiplication also has a meaning in terms of weights, we can indeed consider $\mathbb{R}^{n+1}$ as a vector space without getting uninterpretable operations.

11.2.4 TERMINOLOGY: FROM PRECISE TO CONVENIENT

We have seen how in the vector space model of Chapter 10 the vector $\mathbf{p}$ can be used indirectly to represent an offset point of $\mathbb{R}^n$ at the location represented by a direction vector $\mathbf{p}$. That common practice mixes the concepts of directions, locations, and points. In the homogeneous model, we consider the vector $p = e_0 + \mathbf{p}$ (which is really a direction in the representational space $\mathbb{R}^{n+1}$) as a representation of a point in the space $\mathbb{R}^n$, clearly distinct from a position vector $\mathbf{p}$ in that space. Our distinct geometric concepts of point and position are now represented by different algebraic quantities in the homogeneous model, and this helps the structural clarity of our geometrical computations. After a bit of habituation, you will conveniently think of p as a point of $\mathbb{R}^n$, even though it is a vector of $\mathbb{R}^{n+1}$.

When we need precision in our statements, we will denote points of the n-dimensional Euclidean space in script ("The point $\mathcal{P}$..."), the vectors and blades in the corresponding vector space in **bold**, ("...at location $\mathbf{p}$...") and vectors and blades in the $(n + 1)$-dimensional homogeneous space in *italic* ("...represented by the vector p."). But mostly we will simply talk about the (unit) point p, in a convenient abuse of notation and terminology.

11.3 ALL LINES ARE 2-BLADES

Using the addition of points, we can generate the points on the infinite line connecting p and q as $\alpha p + \beta q$, letting α and β vary over all reals. However, this does not represent the line as a single element of computation—it is just a parameterized set of points. We would prefer to represent the line connecting p to q in $\mathbb{R}^n$ as a blade in our algebra of $\mathbb{R}^{n+1}$. As we would hope, the algebraic structure provides just that.

11.3.1 FINITE LINES

Since we have a full geometric algebra for the homogeneous representation space $\mathbb{R}^{n+1}$, we also have the outer product available. From vectors p and q (representing points $\mathcal{P}$ and $\mathcal{Q}$) we can therefore form $p \wedge q$, and investigate its properties. Already at first glance, it looks promising as the representation of the line through $\mathcal{P}$ and $\mathcal{Q}$: p and q are contained in it, in the sense that a vector can be contained in a blade, because $p \wedge (p \wedge q) = 0$ and $q \wedge (p \wedge q) = 0$. Those are statements in $\mathbb{R}^{n+1}$, which suggest an interpretation in $\mathbb{R}^n$ as

the point represented by x is on the line represented by $p \wedge q$ if and only if $x \wedge (p \wedge q) = 0$, giving an algebraic procedure in $\mathbb{R}^{n+1}$ corresponding to the concept of "lying on" in $\mathbb{R}^n$. Immediately the linearity of the outer product shows that any vector of the form $\alpha p + \beta q$ is also in $p \wedge q$, and no other vectors of $\mathbb{R}^{n+1}$ are. So we can indeed use $p \wedge q$ as our line representation. Note that this is what we called the direct representation in Section 2.8.2, since we use an outer product test for containment of a point. The geometry of this representation principle is illustrated in Figure 11.2(a).

Let us look at $p \wedge q$ in more detail. Substituting the quantities $\mathbf{p}$ and $\mathbf{q}$ from the base space $\mathbb{R}^n$ (we use unit points, since any multiples define the same line in the sense of the outer product; we discuss line weight later) gives us

$$p \wedge q = (e_0 + \mathbf{p}) \wedge (e_0 + \mathbf{q}) = e_0 \wedge \mathbf{q} + \mathbf{p} \wedge e_0 + \mathbf{p} \wedge \mathbf{q} = e_0 \wedge (\mathbf{q} - \mathbf{p}) + \mathbf{p} \wedge \mathbf{q}. \quad (11.3)$$

We recognize $\mathbf{q} - \mathbf{p}$ as the vector from $\mathbf{p}$ to $\mathbf{q}$, which is the vector of $\mathbb{R}^n$ denoting the direction of the line from the point at $\mathbf{p}$ to the point at $\mathbf{q}$. It has a direction (the carrier of $\mathbf{q} - \mathbf{p}$), an orientation (from p to q), and a weight (the distance from p to q).

The other term, the 2-blade $\mathbf{p} \wedge \mathbf{q}$, we call the *moment* of the line (although that term is classically used for a similar concept, which is scalar). The moment encodes the distance of the line to the origin, as we will derive below.

But lines can be specified in other ways. When dealing with lines as rays, we would prefer to encode a line by a point p on it (the source of the ray), and its direction vector $\mathbf{a}$, rather than by two points. It should still give a line, so the two representations should be related. And indeed, they can be converted into each other through the algebra of the outer product. When we know two points p and q on the line, we set $\mathbf{a} = q - p = \mathbf{q} - \mathbf{p}$, and then the antisymmetry allows us to write

$$p \wedge q = p \wedge (q - p) = p \wedge (\mathbf{q} - \mathbf{p}) = p \wedge \mathbf{a}.$$

Therefore, exactly the same 2-blade can be made by the point p and the direction vector $\mathbf{a}$ as from two points p and q, *even using the same operator to combine the data!* Using the terminology we introduced for points, a direction like $\mathbf{a}$ is an improper point, and the equation $p \wedge q = p \wedge \mathbf{a}$ shows that *a finite line can always be represented by two points,* one of which may be improper. It is the same line in $\mathbb{R}^n$, represented by the same 2-blade $\mathbb{R}^{n+1}$, as Figure 11.2(b) shows for the representation of a line in $\mathbb{R}^2$.

The reshapability of the 2-blade that represents the line of course permits many more representations. For instance, if we shift both points along the line by the same amount $\lambda \mathbf{a}$ to become $p + \lambda \mathbf{a}$ and $q + \lambda \mathbf{a}$, the new points still span the same line in moment, direction, and even in weight. Just take their outer product to prove this equivalence: $(p + \lambda \mathbf{a}) \wedge (q + \lambda \mathbf{a}) = p \wedge q + (p - q) \wedge \lambda \mathbf{a} + \lambda^2 \mathbf{a} \wedge \mathbf{a} = p \wedge q$, computationally indistinguishable from the line spanned by p and q. A particularly symmetrical form of line representation is by its affine midpoint (i.e., the centroid) and its direction,

$$p \wedge q = \frac{p + q}{2} \wedge (q - p),$$

as you can easily verify.

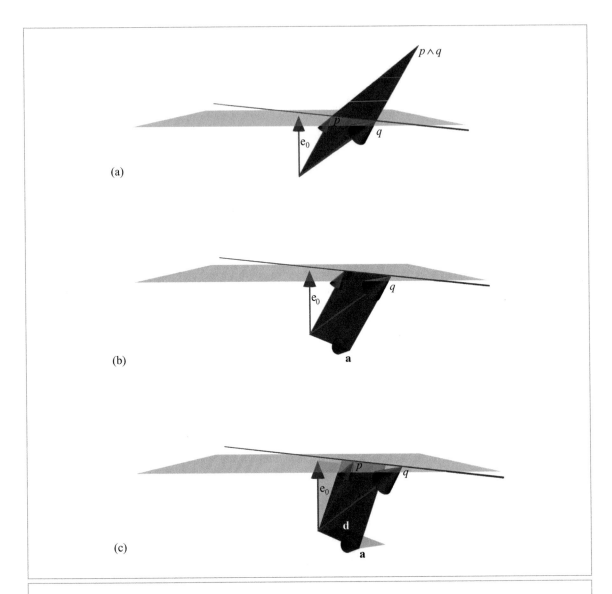

Figure 11.2: Representing offset subspaces in $\mathbb{R}^{n+1}$: an offset line in $\mathbb{R}^2$ is represented as a 2-blade of the homogeneous representation space $\mathbb{R}^3$. In (a) the 2-blade of the line is defined as $p \wedge q$ for two unit points p and q on the line. In (b) the 2-blade has been reshaped to show the direction vector $\mathbf{a} \equiv \mathbf{q} - \mathbf{p} = q - p$ as one of its factors. In (c) the 2-blade has been reshaped to show that the support point $d = e_0 + \mathbf{d}$, with $\mathbf{d} \cdot \mathbf{a} = 0$, is another factor. The moment of the line is $\mathbf{d}\,\mathbf{a}$.

The geometrical equivalence of all these lines is a good feature to have, since it permits the comparison of elements created in different ways. But you will need to get used to the geometrically pure nature of the resulting line element: it is the line, and just the line, though with the properties of weight and orientation. The 2-blade no longer explicitly contains any points. So the 2-blade line representation $p \wedge q$ in the homogeneous model is *not* the localized line segment between p and q, and $p \wedge \mathbf{a}$ is *not* the half line at p in the $\mathbf{a}$-direction. The positional information on the constructors p and q is mostly lost in the outer product; only the distance to the origin remains as positional aspect, and it denotes where the line is located, not where it starts. If you need the line segment, you should keep the points (or at least one of them; the other can be reconstructed from p and $p \wedge q$ as $q = p + e_0^{-1} \rfloor (p \wedge q)$). If you want the ray starting from a point p, you should keep the point p as well as the directed line.

To show that the distance to the origin is contained in $p \wedge q$, we rewrite the 2-blade by using the *(perpendicular) support vector* $\mathbf{d}$ of the line. It is the position vector of a special point d, the *support point*, which is the point on the line closest to the origin. We find it as the rejection of the location of an arbitrary point $\mathbf{p}$ by the line direction $\mathbf{a}$:

$$\text{support vector}: \quad \mathbf{d} = \frac{\mathbf{p} \wedge \mathbf{a}}{\mathbf{a}} = \frac{\mathbf{p} \wedge \mathbf{q}}{\mathbf{q} - \mathbf{p}},$$

where the division is right division. Deriving this is simple: we must have $\mathbf{p} \wedge \mathbf{a} = \mathbf{d} \wedge \mathbf{a}$ (since d is on the line) and $\mathbf{d} \cdot \mathbf{a} = 0$, so that $\mathbf{d}\,\mathbf{a} = \mathbf{p} \wedge \mathbf{a}$. This is illustrated in Figure 11.2(c). Now we can denote the line L as the geometric product of its support point and its direction: $L = d\,\mathbf{a}$.

Of course, after we have used the outer product we no longer have the factors available that constructed it, so the above computation cannot be performed as written. All we have is a 2-blade L, and we should rewrite the expressions for its parameters in terms of operations that can be applied to this 2-blade. When you realize that direction and support are actually defined by the unique rewriting of L as the geometric product $L = (e_0 + \mathbf{d})\,\mathbf{a}$, such expressions are easy to find using selection operations. The quantity $\mathbf{d}\,\mathbf{a}$ is the moment $\mathbf{M}$ of the line, often more convenient to use than its support. The two are related by $\mathbf{d} = \mathbf{M}/\mathbf{a}$.

With this preparation, we can retrieve the relevant geometrical elements from the line 2-blade L by the following algebraic manipulations:

$$\text{direction}: \quad \mathbf{a} = e_0^{-1} \rfloor L,$$

$$\text{moment}: \quad \mathbf{M} = e_0^{-1} \rfloor (e_0 \wedge L),$$

$$\text{support vector}: \quad \mathbf{d} = \frac{e_0^{-1} \rfloor (e_0 \wedge L)}{e_0^{-1} \rfloor L}.$$

We will show in Section 12.1 that in an implementation, these algebraic operations are done by selection of the appropriate coordinate tuples. Despite their apparent algebraic complexity, they can be reduced to simple addressing. This can be done automatically during compilation, and therefore the operations do not cost any real computation time (except the support vector, which involves a division; this is the reason to prefer the use of moments instead, whenever we can).

11.3.2 LINES AT INFINITY

We have composed two finite points, as well as a finite point and an infinite point, and both are interpretable as lines in $\mathbb{R}^n$. The algebra of $\mathbb{R}^{n+1}$ also permits the composition of *two improper points* using the outer product, and you may wonder how we should interpret such an element $\mathbf{p} \wedge \mathbf{q}$ (which is in $\mathbb{R}^{n+1}$, but completely in its subspace $\mathbb{R}^n$). It contains no finite points (the equation $x \wedge (\mathbf{p} \wedge \mathbf{q}) = 0$ has no solution), but it does contain all 1-D directions that can be constructed as weighted sums of $\mathbf{p}$ and $\mathbf{q}$. We can call this either a 2-D direction (for it is a purely 2-blade of $\mathbb{R}^n$, the space of directions of Chapter 10) or an oriented line at infinity, or simply an improper line. When you think of the improper points $\mathbf{p}$ and $\mathbf{q}$ as two stars (either because you view them as points at infinity, or as the directions in which those stars lie), this improper line is the oriented circle on the heavenly sphere passing through both of them.

With this interpretation, the 2-blades in the algebra of the homogeneous representation space are all accounted for as finite or infinite lines. It is satisfying that the composition with the outer product constructs the correct element "line" whatever name we might prefer to give to the constituents (although we had to slightly adapt and refine our geometrical concept of a line to match the algebraic properties). This consistency cleans up our geometry programs considerably:

- We will not need separate data structures for geometrically seemingly different, but algebraically equivalent, elements of the model (such as a line made of two points p and q versus a line constructed of a point p and an improper point $\mathbf{a}$).
- There is a universal constructor (the outer product) to make lines out of such points (improper or not), which is moreover an integral part of the algebra.

You can see how this begins to extend the geometric algebra version of the homogeneous model beyond the algebraically naïve homogeneous coordinate formulation.

11.3.3 DON'T ADD LINES

The linear structure of the bivectors in $\mathbb{R}^{n+1}$ may tempt us to add two lines, hoping to produce another line interestingly related to them. But we will *not* allow this operation! Algebraically, it seems reasonable, but geometrically it is not: the result is in general not interpretable as a line, for it may not be a 2-blade. Here Clifford algebra (which permits addition) and geometric algebra (which should focus on blades and versors and therefore prefers multiplicative constructions) part ways. This is a specific instance of the multiplicative principle we discussed in Section 7.7.2.

The smallest dimension in which the sum of two 2-blades may not be a 2-blade is 4, so that the addition of lines needs to be forbidden for base spaces of three dimensions or more. In a base space of two dimensions, it could be permitted, but universality of the code suggests forbidding it there as well. There must be something special going on in $\mathbb{R}^2$ that we may be able to generalize and that just happens to look like addition. And indeed, what is special about 2-D is that any two lines have a point in common (which may be an improper point at infinity).

- If the common point is finite, the two lines L and M pass through a common point p and can therefore be rewritten as $L = p \wedge \mathbf{u}$ and $M = p \wedge \mathbf{v}$; then $\alpha L + \beta M = \alpha (p \wedge \mathbf{u}) + \beta (p \wedge \mathbf{v}) = p \wedge (\alpha \mathbf{u} + \beta \mathbf{v})$ can generate any line through the point p from just these two. It gives us the idea to generate a *pencil of lines* in the point p from some given lines, and that indeed works in any dimensionality. With a local basis of n lines through one point in n-dimensional space, you can describe all the lines through that point as linear combinations.

- If the common point is the infinite point $\mathbf{u}$, then the lines L and M have a direction in common and can be written as $L = p \wedge \mathbf{u}$ and $M = q \wedge \mathbf{u}$. Now linear combinations produce general lines of the form $(\alpha p + \beta q) \wedge \mathbf{u} = r \wedge \mathbf{u}$, all translated parallel versions of the original lines. This is called a *parallel pencil of lines*.

In a 2-D base space (with its 3-D homogeneous representation space), one of these two cases is guaranteed for two lines, so we can add lines blindly.

In n-dimensional space you can translate in n directions (though one of them produces coincident lines, so it is less interesting). If the lines have no point in common (finite or infinite), they cannot be added in any useful geometric sense. Simplicity of the algebra suggests that we forbid adding of lines in all cases so that we have universally applicable operations. If you really want to make another line through the same point from a given line, you should rotate it around that point, since that gives much better properties (for instance, it preserves the weight, and you of course know the plane and angle of their relative directions since you did the rotation yourself). If you want to make a line parallel to another line, you should just translate it (rather than adding another parallel line to it); that gives a more sensible description of where it goes. We will meet rotation and translation operators in Section 11.8. However, such operations do assume a certain geometry of the space: rotations are Euclidean, translations are at least affine. If the base space merely has a projective geometry, you have no choice but to resort to a pencil-like construction; but then you should remember that you cannot apply it universally.

11.4 ALL PLANES ARE 3-BLADES

A plane Π is determined by three points $\mathcal{P}$, $\mathcal{Q}$, $\mathcal{R}$. By complete analogy to the line, the 3-blade $p \wedge q \wedge r$ represents the plane, and it can be written in several equivalent forms:

$$p \wedge q \wedge r = p \wedge (q - p) \wedge (r - p) = p \wedge (\mathbf{q} - \mathbf{p}) \wedge (\mathbf{r} - \mathbf{p}) = p \wedge \mathbf{A}.$$

The final form shows that it has a location (here determined by p) and a pure 2-blade $\mathbf{A}$ of $\mathbb{R}^n$ as its 2-D direction. In one of many consistent interpretations, this is therefore the connection of the finite point p to a specific line at infinity $\mathbf{A}$, but we can also construct it as the point r connected to the finite line $p \wedge (\mathbf{q} - \mathbf{p})$, and many intermediate forms. As with the line, all these constructors of the plane are equivalent and lead to the same data element.

Planes at infinity, composed of only improper points, also exist in base spaces of sufficiently high dimension. In a general base space $\mathbb{R}^3$ or the Euclidean 3-space $\mathbb{R}^{3,0}$, we would identify the plane at infinity with the heavenly sphere at infinity: it contains all infinite points. It can also be considered as a 3-D direction; that is, an oriented volume (which gives it a distinctly different geometrical feeling).

The antisymmetry of the outer product permits us to replace the point p characterizing the location of the plane in the blade $p \wedge \mathbf{A}$ by any affine combination of points on the plane and still represent the same plane in all its aspects of location, direction, orientation, and weight. A particularly symmetrical way of writing the plane is as

$$p \wedge q \wedge r = \frac{p+q+r}{3} \wedge (p \wedge q + q \wedge r + r \wedge p),$$

in which we recognize the centroid of the triangle formed by the points p, q, and r, as well as the sum of the oriented line carriers of its three sides. Unfortunately, many of the properties of that triangle disappear in the antisymmetric outer product that constructs the oriented and weighted plane of its carrier, though the weight of the 3-blade representing the plane is twice the area of the triangle (if one uses unit points, see structural exercise 4). As for lines, this gives the capability to compare planar elements generated in different ways: coplanar elements differ by a scalar, oppositely oriented elements by a sign, and equal area elements have the same norm. But of course $p \wedge q \wedge r$ is not the triangle through p, q, and r, so any specific information on vertices or edges needs to be kept separately if needed.

We can also define the support point d as the point that allows us to express the plane as a geometric product $\Pi = d \mathbf{A}$, which defines it. When you do the computation, you find a nicely symmetric expression for the support vector $\mathbf{d}$ of the plane in the base space, as

$$\mathbf{d} = \frac{\mathbf{p} \wedge \mathbf{q} \wedge \mathbf{r}}{\mathbf{p} \wedge \mathbf{q} + \mathbf{q} \wedge \mathbf{r} + \mathbf{r} \wedge \mathbf{p}}. \tag{11.4}$$

This equation is completely expressed in terms of quantities of the vector space model, but such expressions about offset subspaces are more easily derived in the homogeneous model. Equation (11.4) has an interesting geometric interpretation in terms of the reciprocal frame of the basis $\{\mathbf{p,q,r}\}$ (see structural exercise 6).

As with the lines, planes should not be added to produce new planes, because they usually will not. In a 3-D base space, however, the planes are represented as trivectors in a 4-D representational space. In that space, all 3-vectors are 3-blades, so adding planes in 3-D is permitted (as is the addition of $(n-1)$-dimensional offset hyperplanes in a general n-dimensional base space). Also, all planes containing a common line form a pencil of planes—factoring out the line shows that adding such planes is just like vector addition, and therefore allowed. Yet it is better practice to produce new planes in the pencil by rotation or translation, if the geometry of the space permits.

11.5 *k*-FLATS AS $(k + 1)$-BLADES

11.5.1 FINITE *k*-FLATS

The pattern of constructing flats continues: taking the outer product of $(k+1)$ points gives a finite $(k+1)$-blade in the homogeneous representation space $\mathbb{R}^{n+1}$, directly representing a k-dimensional offset subspace in the n-dimensional base space $\mathbb{R}^n$. In the specification by some point p at $\mathbf{p}$ and a base space direction $\mathbf{A}$, the offset k-space X is represented by the blade

$$X = p \wedge p_1 \wedge p_2 \wedge \cdots \wedge p_k = p \wedge (\mathbf{p}_1 - \mathbf{p}) \wedge (\mathbf{p}_2 - \mathbf{p}) \wedge \cdots \wedge (\mathbf{p}_k - \mathbf{p}) = p \wedge \mathbf{A} = (e_0 + \mathbf{p}) \wedge \mathbf{A}.$$

We will call such general subspaces *flats* (following [60]), or k-flat for a flat of rank k (the rank is the grade of its direction $\mathbf{A}$, the dimensionality of the offset subspace in the base space).

11.5.2 INFINITE *k*-FLATS

In complete analogy to the infinite lines and planes, an improper k-flat is made up from $(k + 1)$ points at infinity and is therefore a $(k + 1)$-blade in the base space. There is a potential confusion here between the term k-flat and its grade $(k+1)$, but remember that the $(k + 1)$-blade resides in the representational space $\mathbb{R}^{n+1}$ and the k-flat is its interpretation in the base space $\mathbb{R}^n$. The confusion here is that the $(k + 1)$-blade happens to be completely within the copy of $\mathbb{R}^n$ that is in $\mathbb{R}^{n+1}$, but it is of lesser dimensionality than if it had been in the vector space model of the base space. As two examples, the vector (1-blade) $\mathbf{u}$ is a point at infinity (0-blade), and the heavenly sphere in 3-D is 2-dimensional, but represented by a 3-blade in the representational space.

With both finite and infinite flats accounted for in the homogeneous representation space $\mathbb{R}^{n+1}$, we get a highly satisfying semantics for the outer product, not merely referring to the spanning of subspaces by weighted directions (as we had in all previous chapters), but now also including the localizing positions (with the proper localization ambiguity).

11.5.3 PARAMETERS OF *k*-FLATS

The parameters of these general flats are similar to what they were for lines and planes (and, in hindsight, points): for finite flats there is a direction $\mathbf{A}$, a moment $\mathbf{M}$, and a support vector $\mathbf{d}$ (or equivalently a support point d); for infinite flats, only a direction.

The parameters for the finite k-flat are simply retrieved from the representation above, in the manner we have seen before for lines and planes:

$$direction: \quad \mathbf{A} = e_0^{-1} \rfloor X,$$

$$moment: \quad \mathbf{M} = e_0^{-1} \rfloor (e_0 \wedge X),$$

$$support\ vector: \quad \mathbf{d} = \mathbf{M}/\mathbf{A} = \frac{e_0^{-1} \rfloor (e_0 \wedge X)}{e_0^{-1} \rfloor X},$$

$$unit\ support\ point: \quad d = X/\mathbf{A} = \frac{X}{e_0^{-1} \rfloor X}.$$

Having the parameters permits flexible rewriting of X to suit particular computations. Use of the support point d allows us to rewrite the flat not merely as an outer product $d \wedge \mathbf{A}$, but as the geometric product $d\,\mathbf{A}$:

$$(e_0 + position\ vector) \wedge direction = (e_0 + support\ vector)\, direction.$$

That very demand actually defines the support point.

11.5.4 THE NUMBER OF PARAMETERS OF AN OFFSET FLAT

In an n-dimensional space, you need a lot of parameters to determine an offset k-space in all its aspects of direction, orientation, location, and weight. The representation $p \wedge \mathbf{A}$, with $\mathbf{A}$ a k-blade permits us to count them: there are $\binom{n}{k}$ required to determine the direction $\mathbf{A}$ as a k-blade (including its weight), and $(n - k)$ independent degrees of freedom that remain of the $(k + 1)$-blade that determines the moment $\mathbf{M} = \mathbf{p} \wedge \mathbf{A}$ (this is the freedom of the rejection of an n-dimensional position vector by a k-dimensional direction). This gives a total of $\binom{n}{k} + n - k$, and its dependence on n and k is tabulated in Table 11.2 for some low-dimensional cases.

When the points spanning the $(k+1)$-blade are all improper points (or points at infinity), we obtain a pure base space $(k+1)$-blade representing a k-dimensional direction element. There are $\binom{n}{k+1}$ such elements (including their weights). The number of parameters in those blades is also indicated in Table 11.2.

11.6 DIRECT AND DUAL REPRESENTATIONS OF FLATS

As with the proper subspaces in the vector space model of Chapter 10, the offset subspaces of the homogeneous model can be represented in two related ways: directly and dually. We need both representations to compute effectively.

11.6.1 DIRECT REPRESENTATION

In Part I, we got used to visualizing blades as being subspaces through the origin. Of course the blades representing the offset flats are precisely that, though in the $(n+1)$-dimensional

Table 11.2: (a) The number of parameters of offset and weighted k-space elements in n-space. (b) The number of parameters in k-dimensional directions (the improper blades).

(a)	Subspace Grade k (one less than blade grade!)						(b)	Direction Grade k (one less than blade grade!)				
n	0	1	2	3	4	5	n	0	1	2	3	4
0	1						0	0				
1	2	1					1	1				
2	3	3	1				2	2	1			
3	4	5	4	1			3	3	3	1		
4	5	7	8	5	1		4	4	6	4	1	
5	6	9	13	12	6	1	5	5	10	10	5	1
$\vdots$		$\ddots$					$\vdots$		$\ddots$			

homogeneous representation space $\mathbb{R}^{n+1}$. Figure 11.2(a) gave an example for the 2-blade representing a line in the base space $\mathbb{R}^2$ as an origin plane in the representation space $\mathbb{R}^3$. It is good to make explicit how the containment of a vector x in such a blade of $\mathbb{R}^{n+1}$ precisely retrieves the containment relationship of the vector $\mathbf{x}$ (and the associated point $\mathcal{P}$) in the base space of geometrical interest $\mathbb{R}^n$.

The direct interpretation of such a $(k+1)$-blade in the homogeneous representation is done by testing the membership of a general point x using the outer product as in Section 2.8.2. So we test whether a unit point x (at $\mathbf{x}$) is on the flat through unit point p with direction element $\mathbf{A}$ by forming the outer product of x with $p \wedge \mathbf{A}$ and requiring it to vanish:

$$0 = x \wedge p \wedge \mathbf{A} = x \wedge (p - x) \wedge \mathbf{A}$$
$$= x \wedge (\mathbf{p} - \mathbf{x}) \wedge \mathbf{A} = e_0 \wedge (\mathbf{p} - \mathbf{x}) \wedge \mathbf{A} + \mathbf{x} \wedge \mathbf{p} \wedge \mathbf{A}.$$

Since e_0 is orthogonal to the bold base space elements, this leads to the two equations

$$(\mathbf{x} - \mathbf{p}) \wedge \mathbf{A} = 0; \text{ and } \mathbf{x} \wedge \mathbf{p} \wedge \mathbf{A} = 0.$$

Taking the outer product of the former with $\mathbf{x}$ shows that the latter is automatically satisfied when the former is. Therefore the condition $x \wedge (p \wedge \mathbf{A}) = 0$ in the homogeneous representation space is equivalent to $(\mathbf{x} - \mathbf{p}) \wedge \mathbf{A} = 0$ in the base space.

From the vector space model, we know that this is precisely the condition for the vector $(\mathbf{x} - \mathbf{p})$ to lie on the subspace with direction $\mathbf{A}$ passing through the origin in the base space. This of course implies that the position vector $\mathbf{x}$ reaches to the offset subspace at location

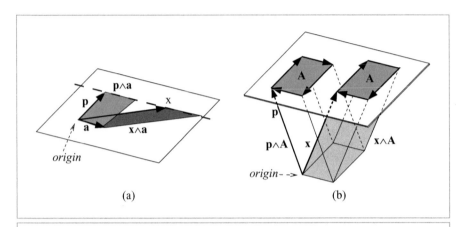

Figure 11.3: Defining offset subspaces fully in the base space by requiring the difference vector $\mathbf{x} - \mathbf{p}$ to be contained in the direction blade $\mathbf{A}$ (or $\mathbf{a}$), as required by the condition $(\mathbf{x}-\mathbf{p})\wedge\mathbf{A} = 0$. Also shown is the alternative interpretation of this equation: $\mathbf{x}\wedge\mathbf{A} = \mathbf{p}\wedge\mathbf{A}$, which holds by the reshapeable nature of the blades involved.

$\mathbf{p}$. For a line in 2-D and a plane in 3-D, these conditions are sketched in Figure 11.3(a) and (b).

We can convert this condition to the familiar parametric equation for a point on the offset $\mathbf{A}$-space at $\mathbf{p}$. Let $\mathbf{A} = \mathbf{a}_1 \wedge \mathbf{a}_2 \wedge \cdots \wedge \mathbf{a}_k$, then the general solution to the equation

$$(\mathbf{x} - \mathbf{p}) \wedge \mathbf{a}_1 \wedge \mathbf{a}_2 \wedge \cdots \wedge \mathbf{a}_k = 0$$

is given by a linear combination of the direction factors:

$$\mathbf{x} - \mathbf{p} = \lambda_1\mathbf{a}_1 + \lambda_2\mathbf{a}_2 + \cdots + \lambda_k\mathbf{a}_k. \tag{11.5}$$

You recognize the various cases for $k = 1$ (a line), $k = 2$ (a plane), and even $k = 0$ (a point). The parameters λ_i for a given vector $\mathbf{x}$ can be computed easily (see structural exercise 7).

For an improper flat $\mathbf{A}$, the equation $x \wedge \mathbf{A} = 0$ has no solution (for we can never make the part $e_0 \wedge \mathbf{A}$ be equal to 0). So an improper flat contains no proper (finite) points. Of course, an improper k-blade does contain k independent directions (i.e., it does contain k independent improper points).

11.6.2 DUAL REPRESENTATION

When we treated subspaces in Section 3.5.5, we found that an equivalent representation is by their orthogonal complement. Algebraically, that is computed through duality relative to the pseudoscalar of the space in which the blade resides.

We can do this here as well, but we must of course compute relative to the representational space $\mathbb{R}^{n+1}$. Let us take as the pseudoscalar for that the $(n + 1)$-blade

$$\mathbf{I}_{n+1} \equiv e_0 \wedge \mathbf{I}_n = e_0 \, \mathbf{I}_n,$$

with $\mathbf{I}_n$ the pseudoscalar for the base space $\mathbb{R}^n$ (which is why we write it as a bold blade). Because of the orthogonality of the representational dimension e_0 to the base space, we can choose to use the outer product or the geometric product, whichever is more convenient for the computation at hand.

Duality with this pseudoscalar requires its inverse:

$$X^* = X \rfloor I_{n+1}^{-1} = X \rfloor (\mathbf{I}_n^{-1} e_0^{-1}) = X^\star e_0^{-1}.$$

We introduced two shorthands for duals here, the six-pointed star for the representational space $\mathbb{R}^{n+1}$ and the five-pointed star for the dual in the base space $\mathbb{R}^n$ (the mnemonic is that six is one more than five). The base space dual should really only be used on elements of that base space, and then should provide the link to the vector space model (which is after all the algebra of the base space). We have to use the proper inverse of e_0 to absorb the ambiguity in choice of sign for the metric of the representational space we mentioned in Section 11.1. It has the additional advantage of showing at a glance whether we have a blade or a dual blade. Unfortunately this is only on paper, for in an implementation e_0^{-1} is substituted by $+e_0$ or $-e_0$, so there is no obvious qualitative distinction.

With the pseudoscalar thus defined, the dual of the general flat $X = p \wedge \mathbf{A}$ can be expressed in various equivalent forms. Each has its own use, so we give them all:

$$
\begin{aligned}
X^* = (p \wedge \mathbf{A})^* &= p \rfloor \mathbf{A}^* \\
&= p \rfloor (\mathbf{A}^\star e_0^{-1}) && \text{(multiplicative form)} \\
&= \widehat{\mathbf{A}^\star} - e_0^{-1} \widehat{(p \rfloor \mathbf{A}^\star)} && \text{(additive form)} \\
&= \widehat{\mathbf{A}^\star} + e_0^{-1} \widehat{\mathbf{M}^\star} && \text{(direction and moment)} \\
&= (e_0^{-1} - \mathbf{d}^{-1}) \, \widehat{\mathbf{M}^\star} && \text{(support and moment)}
\end{aligned}
\tag{11.6}
$$

The grade involution emerges to properly keep track of the orientation. Note that the grade involution extends over the dual, so that $\widehat{\mathbf{A}^\star} = (-1)^n \widehat{\mathbf{A}}^\star$. These signs are partly caused by our choice of pseudoscalar.

The simplest dual element of this form is a hyperplane with a direction characterized by a unit normal vector $\mathbf{n} \equiv \mathbf{A}^\star$ at an oriented distance δ from the origin (positive when in the $\mathbf{n}$-direction). Then we can take the point at location $\mathbf{d} = \delta \, \mathbf{n}$ to localize it, so that we obtain as the dual of the hyperplane:

$$\pi \equiv \Pi^* = -\mathbf{n} + \delta \, e_0^{-1}.$$

This vector is indicated in Figure 11.4 for a hyperplane in $\mathbb{R}^2$, which is a line, and a hyperplane in $\mathbb{R}^1$, which is a point. The latter figure is a cross section of the former.

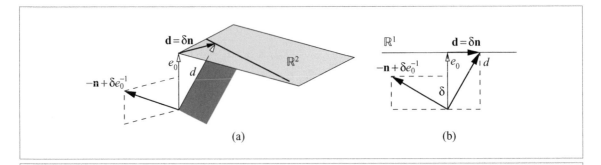

Figure 11.4: The dual hyperplane representation in $\mathbb{R}^2$ and $\mathbb{R}^1$ for a hyperplane with support vector $\mathbf{d} = \delta\mathbf{n}$, with $\mathbf{n}$ a unit vector. It is the vector $-\mathbf{n} + \delta\, e_0^{-1}$; the figures are drawn for $e_0^{-1} = +e_0$.

In the dual representation, testing whether a point x lies on a dual flat X^* is now done by demanding the contraction $x \rfloor X^*$ to be zero. We can write this out to check that it leads to the correct condition on base space elements (after some rewriting):

$$0 = x\rfloor X^* = x\rfloor\left(\widehat{\mathbf{A}^\star} - e_0^{-1}(\mathbf{p}\rfloor\widehat{\mathbf{A}^\star})\right) = (\mathbf{x} - \mathbf{p})\rfloor\widehat{\mathbf{A}^\star} + e_0^{-1}\left((\mathbf{x}\wedge\mathbf{p})\rfloor\widehat{\mathbf{A}^\star}\right).$$

Both terms should be independently zero because of the independence of e_0^{-1} with the bold base space elements. Moreover, when the first term is zero,

$$0 = (\mathbf{x} - \mathbf{p})\rfloor\widehat{\mathbf{A}^\star},$$

the second is as well (for that can be written as $-e_0^{-1}\left((\mathbf{x}\rfloor((\mathbf{x} - \mathbf{p})\rfloor\widehat{\mathbf{A}^\star})\right)$). Therefore, the condition $0 = x\rfloor X^*$ is identical to $(\mathbf{x} - \mathbf{p})$ being perpendicular to the dual of the direction in the base space. This is simply the dual of the direct representation condition $(\mathbf{x}-\mathbf{p})\wedge\mathbf{A} = 0$, so all is consistent and the dual approach leads to the same offset subspace as the direct approach.

For the dual plane representation above, the dual condition yields

$$0 = x\rfloor\pi = (e_0 + \mathbf{x})\cdot(-\mathbf{n} + \delta\, e_0^{-1}) = \delta - \mathbf{x}\cdot\mathbf{n},$$

which indeed retrieves the familiar, purely Euclidean base space condition

$$\mathbf{x}\cdot\mathbf{n} = \delta \tag{11.7}$$

for a point to lie on a hyperplane characterized by a unit normal vector $\mathbf{n}$ at distance δ from the origin. That is the *normal equation* of the hyperplane, also known as its Hesse normal form.

This is all in accordance with how one treats such hyperplanes and equations in the usual homogeneous coordinates representation, where a hyperplane is represented as a covector

$[\![\mathbf{n}; -\delta]\!]$ and the probe point as the vector $[\![\mathbf{x}; 1]\!]^T$. Their matrix product then is $\mathbf{x}\cdot\mathbf{n}-\delta$, and requiring this to be zero gives the same normal equation. We discuss this correspondence in more detail in Section 12.1, where we also show that we have considerably generalized the principle beyond planes that may be represented by vectors, to arbitrary offset flats of any grade such as lines in their dual form.

The same parameters we derived from directly represented blades in Section 11.5.3 can of course be derived from their dual representations. This can be done by dualizing the formula for a parameter derived from X and manipulating it until it contains an expression on terms of X^*, or more directly by using familiar techniques on some of the equivalent expressions in (11.6). The results are listed in Table 11.3 for easy reference.

Table 11.3: All nonzero blades in the homogeneous model of Euclidean geometry, and their parameters. A blade X represents either a finite flat, a dual finite flat, or a flat at infinity (direct or dual), which we denote as direction. These have the generic forms denoted in the top row, where $\mathbf{A}$ denotes a purely Euclidean blade of appropriate grade. Dual directions transform in the same manner as directions. The operations on the blades may look complicated, but in an implementation they can be implemented at compile time as coordinate selections. Rotation and translation are treated in Section 11.8.

	Finite Flat $X = e_0\mathbf{A} + \mathbf{d}\wedge\mathbf{A}$	Dual Finite Flat $X^* = \widehat{\mathbf{A}^{\star}} - e_0^{-1}(\mathbf{d}\rfloor\widehat{\mathbf{A}^{\star}})$	Direction $\mathbf{A}$
(Dual) direction	$\mathbf{A} = e_0^{-1}\rfloor X$	$\widehat{\mathbf{A}^{\star}} = e_0\rfloor(e_0^{-1}\wedge X^*)$	X
Support	$\mathbf{d} = \dfrac{e_0^{-1}\rfloor(e_0\wedge X)}{e_0^{-1}\rfloor X}$	$\mathbf{d} = \dfrac{-e_0\rfloor X^*}{e_0\rfloor(e_0^{-1}\wedge X^*)}$	
Moment	$\mathbf{M} = e_0^{-1}\rfloor(e_0\wedge X)$	$\widehat{\mathbf{M}^{\star}} = e_0\rfloor X^*$	
Rotation	RXR^{-1}	RX^*R^{-1}	RXR^{-1}
Translation	$X + \mathbf{t}\wedge(e_0^{-1}\rfloor X)$	$X^* - e_0^{-1}\wedge(\mathbf{t}\rfloor X^*)$	X

11.7 INCIDENCE RELATIONSHIPS

The flats we have introduced can be combined. The outcome of spanning operations, duality, and incidence operations on flats are again flats. They form a complete system of computing with such relationships between offset subspaces. This could of course only be achieved by encoding both the proper flats and the improper flats. Two planes now always have a line in common, even when they are parallel (in which case it is an improper line at infinity).

These geometrical properties are reflected in our algebra by closure under $\wedge$, *, and $\cap$ (since the meet is the dual of the outer product of the duals, the third follows automatically from the others). In particular, whatever computation we perform, we will always obtain an element of the algebra. There is therefore no need for exceptional treatment of lines, points, directions, and so on; we can always compute on, regardless of the type of the outcome. Only at the very end of our computations, when we need to render the results, we may need to test for the type of element that is the final outcome, and compute its parameters.

11.7.1 EXAMPLES OF INCIDENCE COMPUTATIONS

We give some examples of incidences in the homogeneous model, expanding the natural form $A \cap B = B^* \rfloor A$ of the meet in Chapter 5 down to the consequences for the Euclidean parameters of the flats in the base space. The purpose of this is not to show you how to do such computations, for you would never do them in real life. Your geometric algebra software will automatically take care of computation and interpretation in a structural manner. But to learn to rely on that, you need to see the correspondence to the classical approach at least once. This permits us to point out the differences, and demonstrate how the unification of the structure can simplify the flow of your code.

Two Lines in a Plane

When we have two lines in the same plane $\mathbf{I}$ through the origin in $\mathbb{R}^n$, say $L = e_0\,\mathbf{u} + \mathbf{U}$ and $M = e_0\,\mathbf{v} + \mathbf{V}$, their meet at a point can be computed. For this we need their join. Assuming first that the lines are in general position in the plane $\mathbf{I}$, the join is the common plane representative $e_0\,\mathbf{I}$ in the homogeneous representation space $\mathbb{R}^{n+1}$. Since both arguments of the meet are of the same type, it is convenient to use the dual computation $L \cap M = (M^* \wedge L^*)^{-*}$. We rewrite a dual in terms of the Euclidean dual:

$$L^* = (e_0\,\mathbf{u} + \mathbf{U})\,\mathbf{I}^{-1}\,e_0^{-1} = -\mathbf{u}^\star + \mathbf{U}^\star\,e_0^{-1}.$$

You should realize that in the plane considered, $\mathbf{u}^\star$ is a vector and $\mathbf{U}^\star$ is a scalar. Their commutative properties then enable a fairly quick simplification of results:

$$L \cap M = \left((-\mathbf{v}^\star + \mathbf{V}^\star e_0^{-1}) \wedge (-\mathbf{u}^\star + \mathbf{U}^\star e_0^{-1}) \right)^{-*}$$
$$= (\mathbf{v}^\star \wedge \mathbf{u}^\star)^{-*} + (\mathbf{V}^\star \mathbf{u}^\star e_0^{-1})^{-*} - (\mathbf{U}^\star \mathbf{v}^\star e_0^{-1})^{-*}$$
$$= e_0 (\mathbf{v}^\star \wedge \mathbf{u}^\star)^{-\star} + \mathbf{V}^\star \mathbf{u} - \mathbf{U}^\star \mathbf{v}$$
$$= e_0 (\mathbf{u} \wedge \mathbf{v})^\star + \mathbf{V}^\star \mathbf{u} - \mathbf{U}^\star \mathbf{v}. \tag{11.8}$$

(For the final step, we used $(\mathbf{v}^\star \wedge \mathbf{u}^\star)^{-\star} = \mathbf{v}^\star \cdot \mathbf{u} = \mathbf{u} \cdot \mathbf{v}^\star = (\mathbf{u} \wedge \mathbf{v})^\star$.) That is the result of the meet with this join as the plane of the lines. Depending on whether $(\mathbf{u} \wedge \mathbf{v})^\star$ is zero or not, we have two interpretations: the lines intersect in a finite point or in an infinite point. Therefore, a single computation in the homogeneous model $\mathbb{R}^{n+1}$ captures several cases that are different in $\mathbb{R}^n$. We spell them out to show that they correspond to some familiar expressions.

- **Finite Intersection Point.** If $(\mathbf{u} \wedge \mathbf{v})^\star \neq 0$, we can write the result of (11.8) as a weighted point in its homogeneous representation:

$$L \cap M = (\mathbf{u} \wedge \mathbf{v})^\star \left(e_0 + \frac{\mathbf{V}}{\mathbf{u} \wedge \mathbf{v}} \mathbf{u} + \frac{\mathbf{U}}{\mathbf{v} \wedge \mathbf{u}} \mathbf{v} \right)$$

This shows that the result is indeed a point, at the location $\frac{\mathbf{V}}{\mathbf{u} \wedge \mathbf{v}} \mathbf{u} + \frac{\mathbf{U}}{\mathbf{v} \wedge \mathbf{u}} \mathbf{v}$, and with a weight $(\mathbf{u} \wedge \mathbf{v})^\star$. The correctness of the location is illustrated in Figure 11.5. We already met the geometry of this construction in Section 2.7.2, when we first encountered Cramer's rule as a ratio of planar elements. The solution then was rather ad hoc; we now see that it is simply a special case of the meet operation.

The weight of the meet, $(\mathbf{u} \wedge \mathbf{v})^\star$, is an interesting feature of the computation. If we take $\mathbf{u}$ and $\mathbf{v}$ to be unit vectors (so that the lines L and M are normalized

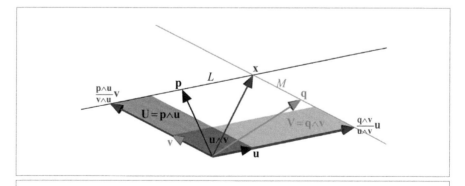

Figure 11.5: The intersection of two offset lines L and M to produce a point x. Their meet tells us that the point x can be reached by translating e_0 over a multiple of $\mathbf{u}$ and a multiple of $\mathbf{v}$. We show graphically why the coefficient for $\mathbf{v}$ is the ratio of the moment $\mathbf{U}$ of the line L and the bivector of directions $\mathbf{u} \wedge \mathbf{v}$: the stretch factor of the vector $\mathbf{v}$ is the same as that of the bivector $\mathbf{u} \wedge \mathbf{v}$ to make $\mathbf{U}$ (taking reshapability into account).

in this sense), this weight attributes a sign to the meet enabling us to know how L and M intersect: positive when $\mathbf{v}$ points counterclockwise of $\mathbf{u}$ (in a plane with a counterclockwise orientation), negative in the other direction. Moreover, the weight is an intersection strength that can be interpreted as a numerical stability of the intersection; when the lines intersect perpendicularly it is strongest, and when they are almost parallel the weight becomes very small (and if the lines would be noisy data, their intersection would indeed get rather unlocalized).

- **Two Parallel Lines.** When the weight $(\mathbf{u} \wedge \mathbf{v})^\star$ equals zero, the lines are parallel. Now the result of the meet in (11.8) can be simplified to

$$L \cap M = \mathbf{V}^\star \mathbf{u} - \mathbf{U}^\star \mathbf{v}.$$

Since $\mathbf{V}^\star$ and $\mathbf{U}^\star$ are scalars, this is a multiple of the common direction $\mathbf{u}$ (or $\mathbf{v}$), so the lines now meet in a weighted improper point, although you may prefer to say that they have a common direction. It is accompanied by a weight, obviously proportional to the weight of the lines L and M, but also to their relative position. If we take the lines to be unit lines (in the sense that $\mathbf{u}$ and $\mathbf{v}$ are unit vectors), the dual moments are simply their oriented distances to the origin, and the result simplifies to $(\delta_M - \delta_L)\,\mathbf{u}$. So *two parallel unit lines intersect in a point at infinity weighted by their distance.* This is an interestingly compact quantitative combination of both their commonality (their direction) and their difference (their distance).

- **Two Coincident Lines.** When the lines are not only parallel, but have distance zero, the result of the meet computation above is zero. As we noted in Chapter 5, this indicates that we have reached a case in which we have the wrong join to compute the actual meet (which should never be zero). Indeed, when the lines coincide, the situation is rather trivial: their join is the line, and so is their meet.

If the meet computations above intimidated you, please realize that you don't have to do such computations by hand; the meet operation will just give the right answer. It switches cases automatically depending on the relative geometry of its input arguments.

But it is good to realize that the difference between the first two configurations (general or parallel) is actually not even a case switch for the meet, since the join does not change. The configurations merely appear as different cases if we need to interpret finite points differently from infinite points (for instance, to draw them using their parameters). That happens only at the end of a chain of computations; if (11.8) was an intermediate result, you could just compute on without splitting the data flow of the program.

So in practice, you do not need to do such computations before you can write code. We have only performed them here in detail to comment on the type of result, which is typical: coordinate-free, weighted, and merging separate geometrical cases. In your code, use $M^* \rfloor L$ if you know that M and L are guaranteed not to be degenerate; if they may be, use an algorithmic implementation of the meet operation $L \cap M$. The latter is much more expensive, as we will find in Section 21.7, and therefore to be avoided.

Two Skew Lines in Space

When two lines $L = p \wedge \mathbf{u} = e_0\,\mathbf{u} + \mathbf{U}$ and $M = q \wedge \mathbf{v} = e_0\,\mathbf{v} + \mathbf{V}$ are in general position in a 3-D space with pseudoscalar $\mathbf{I}_3$, their join in the homogeneous representation space is of course the pseudoscalar $e_0\,\mathbf{I}_3$. The technique for the computation of their meet is similar to the simpler case above (alternatively, you can use the derivation of structural exercise 11). The result now is

$$L \cap M = \left(\mathbf{U} \wedge \mathbf{v} + \mathbf{u} \wedge \mathbf{V}\right)^{\star} = \left((\mathbf{p} - \mathbf{q}) \wedge \mathbf{u} \wedge \mathbf{v}\right)^{\star} = \left(\mathbf{d}\,(\mathbf{u} \wedge \mathbf{v})\right)^{\star}, \qquad (11.9)$$

which is a scalar containing both the directional dissimilarity (in terms of the sine of the angle between $\mathbf{u}$ and $\mathbf{v}$) and the locational dissimilarity (since it is proportional to their weighted distance, which is the oriented length of their separation vector $\mathbf{d} \equiv \left((\mathbf{p} - \mathbf{q}) \wedge \mathbf{u} \wedge \mathbf{v}\right)/(\mathbf{u} \wedge \mathbf{v})$). This is illustrated in Figure 11.6. In this case, you can only retrieve the distance between the lines from the meet when you know the meet of their directions $(\mathbf{u} \wedge \mathbf{v})^{*}$.

When the meet computed in this manner becomes zero, it denotes the need for a new join to compute the true meet. The algebraic form of the meet shows that this can happen in one of two ways: the factor $(\mathbf{u} \wedge \mathbf{v})$ can become zero, which means that the lines are

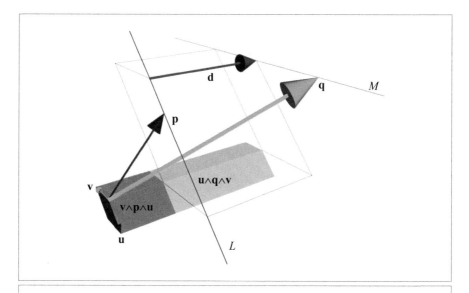

Figure 11.6: The meet of two skew lines $L = (e_0 + \mathbf{p}) \wedge \mathbf{u}$ and $M = (e_0 + \mathbf{q}) \wedge \mathbf{v}$ in their common space equals $((\mathbf{p} - \mathbf{q}) \wedge \mathbf{u} \wedge \mathbf{v})^{\star}$, which can be interpreted as the difference of two oriented volumes. You can perform a rejection to display the separation vector $\mathbf{d}$, perpendicular to $\mathbf{u} \wedge \mathbf{v}$, reexpressing the meet as $(\mathbf{d}\,(\mathbf{u} \wedge \mathbf{v}))^{\star}$.

parallel, or $\mathbf{d}$ can become zero, which means that the lines intersect. Geometrically, either case implies that the lines have become coplanar; then the join is no longer the space $e_0\, I_3$, but their common plane, and the meet reduces essentially to the previous case.

11.7.2 RELATIVE ORIENTATION

When a meet of two elements $\mathbf{A}$ and $\mathbf{B}$ is scalar, it can be interpreted as a relative orientation with numerical significance. We announced this when treating the meet of subspaces through the origin in Chapter 5. In terms of modeling, we were then using the vector space model. The examples given there can now be generalized to offset flats by employing the homogeneous model.

If the meet of $\mathbf{A}$ and $\mathbf{B}$ is scalar, then the elements $\mathbf{A}$ and $\mathbf{B}$ should be complementary within the join, since (5.4) states that the grades are related as $m = a + b - j$, so that $m = 0$ implies $a = j - b$. We can then replace the contraction in the computation of the meet by a scalar product, and use its symmetry to rewrite it to a proper dual relative to the join:

$$\mathbf{A} \cap \mathbf{B} = (\mathbf{B}^*)\rfloor\mathbf{A} = (\mathbf{B}^*) * \mathbf{A} = \mathbf{A} * (\mathbf{B}^*) = \mathbf{A}\rfloor(\mathbf{B}^*) = (\mathbf{A} \wedge \mathbf{B})^*.$$

Therefore the meet in this scalar case is just the oriented volume measure of $\mathbf{A}$ and $\mathbf{B}$ (in that order, and relative to the chosen orientation of the pseudoscalar for the subspace $\mathbf{J}$ that determines the duality). We treat the most common case in 3-D Euclidean geometry with a right-handed pseudoscalar. These are illustrated in Figure 11.7.

- **Two Points on a Line.** In the homogeneous model, a line L is represented by a 2-blade. Introduce the pseudoscalar I_2 of this 2-blade. Now we can compute the meet of two points p and q on this line L as

$$p \cap q = (p \wedge q)^*.$$

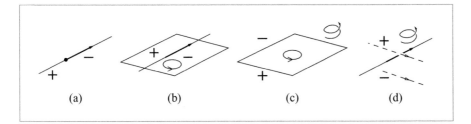

(a)   (b)   (c)   (d)

Figure 11.7: The relative orientation of oriented flats: (a) the sides of a (unit) point on an oriented line, (b) the sides of an oriented line in an oriented plane, (c) the sides of an oriented plane in an oriented space, and (d) the sides of an oriented line in an oriented space. In (d), the probes are oriented lines; in the other examples they are positively weighted points. The right-handed corkscrews in (c) and (d) denote the orientation of the 3-D pseudoscalar.

This is indeed a scalar and it changes sign as p and q change their order on the line L. It does so continuously, being zero when p coincides with q (see Figure 11.7(a)). If we use identical weighted points (for which $e_0^{-1} \cdot p = e_0^{-1} \cdot q$), the weight of the meet is proportional to the *oriented length* of the line segment from p to q; in particular, relative to a join of weight 1, and using its orientation as standard, one has $p \cap q = (p \wedge q)^* = \pm(q - p)$.

That is useful for comparing ratios along the same line. Unfortunately, the actual proportionality constant is $\|\mathbf{u}\| \sqrt{(e_0^2 + \mathbf{d}^2)}$ (compute this in structural exercise 10), which is different for almost all lines. Therefore, lengths of segments on different lines cannot directly be compared in this manner; you would need to revert to the vector space submodel or upgrade to the conformal model. The homogeneous model is poorly suited for metrical computations.

- **A Point and a Line in Their Plane.** A point p and a line $L = q \wedge r$ also have complementary grades in their plane, so we can determine their relative orientation as

$$p \cap L = (p \wedge L)^* = (p \wedge q \wedge r)^*.$$

Taking the sign of this as the relative orientation, the *positive side* of the line L is such that the orientation of p, q, and r is the same as that imposed on the plane by the join. For a plane oriented in the customary counterclockwise fashion, the positive side of L is therefore *left of $q \wedge r$* (see Figure 11.7(b)). In a plane with counterclockwise orientation (what mathematicians think of as positive), the positive side of the line is on your left when you look along its direction (as you can show by rewriting the line to $L = q \wedge (r - q) = q \wedge \mathbf{v}$).

The weight of the meet is proportional to the area of the triangle formed by p, q, and r (by a factor that depends on q and r). As p varies from the left of the line to the right, this changes smoothly from positive to negative, with zero when p is on the line (which signals that we should change the join to compute the actual meet). If we keep the line fixed in its parameters (i.e., keep q and r fixed), then the value of the meet is proportional to the (oriented) perpendicular distance of the point p to the line L. The proportionality factor depends on the location of the plane relative to the origin, but within the plane the measure is usable as a relative oriented distance of points to the line.

- **A Point and a Plane in Space.** Similarly, a point p can be on different sides of a plane $\Pi = q \wedge r \wedge s$ in space. The positive side of Π is where $\text{sign}((p \wedge \Pi)^*)$ is positive, which is the side where the volume $p \wedge q \wedge r \wedge s$ has the same orientation as the volume measure assigned to the space. Usually one would take the right-handed pseudoscalar $e_0 \wedge \mathbf{e}_1 \wedge \mathbf{e}_2 \wedge \mathbf{e}_3$ for that, which is equivalent to the orientation of the tetrahedron formed by the points e_0, $e_0 + \mathbf{e}_1$, $e_0 + \mathbf{e}_2$, $e_0 + \mathbf{e}_3$ (in that order!). The volume measure of $p \wedge \Pi$ determines the quantitative aspects in a smoothly varying manner, with zero occurring only if p is on the plane Π (see Figure 11.7(c)). For a right-handed space, orient the fingers of your right hand with the plane, then your thumb points away from the positive side. If we had defined the relative orientation as the sign of $(\Pi \wedge p)^*$, we would get the opposite answer, for this `meet` is not symmetric in its arguments; see Section 5.6.)

Numerically, the value of the `meet` is proportional to the (oriented) perpendicular distance of the point p to the plane Π.

- **Two Skew Lines in Space**. It is somewhat harder to imagine that two oriented lines L and M in space also have a relative orientation. This implies that there is a positive side and a negative side to line L in space, though not for points as probes (so these sides are hard to color), but for lines like M.

  Once you realize it can be done, this is of course simple. If the lines are given though points on them such as $L = p \wedge q$ and $M = r \wedge s$, then the positive side is the one where $L \wedge M = p \wedge q \wedge r \wedge s$ has the same orientation as the pseudoscalar of the space. This is illustrated in Figure 11.7(d), which shows some of the oriented line probes.

  The weight of the `meet` is now proportional to the (oriented) perpendicular distance between the two lines, weighted by the sine of the angle between their relative directions.

  A standard example would be the line $L = e_0 \wedge \mathbf{e}_3$, with $M = (e_0 + \mathbf{e}_1) \wedge \mathbf{e}_2$ being positive relative to it in right-handed space. In this case, $L \cap M = (e_0 \wedge \mathbf{e}_3 \wedge \mathbf{e}_1 \wedge \mathbf{e}_2)^*$ $= +1$, so this is positive. As an easy way to see the sign, view L as a rotation axis; grab it with your right hand (!) and if M points similar to your fingers, it is positive relative to L. Note that the order does not matter: when L is positive relative to M, then M is positive relative to L, for this `meet` is symmetric in its arguments.

  This relative orientation measure on lines can be used to determine whether a ray M passes inside a triangle described by three consistently oriented boundary carrier lines L_1, L_2, L_3: all signs of relative orientations with M should be equal.

11.7.3 RELATIVE LENGTHS: DISTANCE RATIO AND CROSS RATIO

We have seen several computable measures that are proportional to Euclidean quantities in which we are interested. They are accompanied by extra factors that make them uninterpretable by themselves. The giveaway is often the occurrence of e_0 or e_0^{-1}—they should never be present in a geometrical result, for they are only accidental elements of the representation, not of reality. True, e_0 represents the point at the origin, but even that is really arbitrary; since we allow translations in our geometry, we can move it anywhere. Yet it is possible to perform computations that have an objective meaning in the base space. We show what kind of length measurements are independent of the representation, and under what conditions.

Consider three colinear points $\mathcal{P}, \mathcal{Q}, \mathcal{R}$. We like to keep the representation as general as possible, and do not even demand normalization of their representations. We denote

$p = p_0(e_0 + \mathbf{p})$, and so on. A two-point combination (as occurs for instance in a `meet`) depends on the scaling factors of the representation:

$$p \wedge q = p_0\, q_0 \left(e_0 \wedge (\mathbf{p} - \mathbf{q}) + \mathbf{p} \wedge \mathbf{q}\right) = p_0\, q_0\, d\, (\mathbf{p} - \mathbf{q}),$$

where $d = e_0 + (\mathbf{p} \wedge \mathbf{q})/(\mathbf{p} - \mathbf{q})$ is the support point of the common line. Because of the colinearity of the points, that support point can be factored out of other two-point combinations as well, so, for instance,

$$q \wedge r = q_0\, r_0\, d\, (\mathbf{q} - \mathbf{r}).$$

When we take the ratio of these quantities, d cancels out, and the ratio of two bivectors from $\mathbb{R}^{n+1}$ becomes the ratio of two vectors of $\mathbb{R}^n$:

$$\textit{affine distance ratio}: \quad \frac{p \wedge q}{q \wedge r} = \frac{p_0}{r_0}\, \frac{\mathbf{p} - \mathbf{q}}{\mathbf{q} - \mathbf{r}}. \tag{11.10}$$

This is a scalar, since $(\mathbf{p} - \mathbf{q})$ and $(\mathbf{q} - \mathbf{r})$ are proportional vectors for the colinear points p, q, and r.

For identically normalized points (so that $p_0 = r_0$), this simplifies to being the ratio of oriented distances along a line. Only for transformations that do not change the normalization of points is this a meaningful, invariant quantity that does not change with the transformation. We are therefore interested in this class of transformations. In Section 11.8.8 we show that they are precisely the *affine transformations* of $\mathbb{R}^n$. These encompass all Euclidean transformations, notably the rigid body motions in the base space $\mathbb{R}^n$. We call the quantity $(p \wedge q)/(q \wedge r)$ the *affine distance ratio* to remind us that it is an affine invariant of the points p, q, and r.

The distance ratio would be invariant under an arbitrary linear transformations of $\mathbb{R}^{n+1}$ if it had not involved the normalization constants p_0 and r_0 at all. But we can simply eliminate those by taking a similar ratio of p and r with another point s (also on the same line), and then cancel all normalization factors by taking the ratio of two ratios. This is called the *(projective) cross ratio*:

$$\left(\frac{p \wedge q}{q \wedge r}\right) \left(\frac{r \wedge s}{s \wedge p}\right) = \left(\frac{\mathbf{p} - \mathbf{q}}{\mathbf{q} - \mathbf{r}}\right) \left(\frac{\mathbf{r} - \mathbf{s}}{\mathbf{s} - \mathbf{p}}\right).$$

See Figure 11.8. This combination of four oriented distances of points along the same line is not changed by *any* linear transformation of the homogeneous representation space. Such arbitrary transformations in the homogeneous representation space $\mathbb{R}^{n+1}$ represent precisely the *projective transformations* of the base space $\mathbb{R}^n$ (often called colineations or homographies). Projective transformations can affect distances and even the order of points along a line severely, so it was quite a discovery (independently by Chasles and by Möbius in the 1850s) that an invariant quantity could nevertheless be found. We discuss projective transformations in Section 11.8.9.

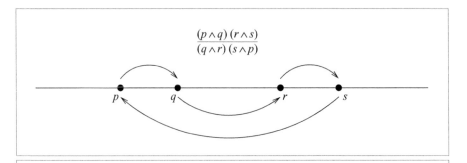

Figure 11.8: The combinations of four points taken in the cross ratio. The directed arcs above the line go into the numerator, the directed arcs below the line into the denominator.

Note the particularly simple form the cross ratio takes in the embedding of the homogeneous model in geometric algebra, where the ratio of bivectors of $\mathbb{R}^{n+1}$ or vectors of $\mathbb{R}^n$ is well-defined so that there is no need to first reduce the factors to scalars (which is necessarily the classical approach). Because of the commutation properties of all elements involved, we can also rewrite the cross ratio to a single fraction:

$$projective\ cross\ ratio\ :\quad \frac{(p \wedge q)\,(r \wedge s)}{(q \wedge r)\,(s \wedge p)} = \frac{(\mathbf{p} - \mathbf{q})\,(\mathbf{r} - \mathbf{s})}{(\mathbf{q} - \mathbf{r})\,(\mathbf{s} - \mathbf{p})}. \qquad (11.11)$$

Symmetry properties of the cross ratio follow from the antisymmetry of the cross product of points (left of the equality), or the subtraction of vectors (right of the equality); see also structural exercise 18.

This cross ratio of points on a common line can be extended simply to a cross ratio for lines passing though a common point and lying in a common plane. One usually does so by creating four points from those lines by intersecting the lines with an arbitrary fifth line, not through the common point, and using the resulting four points in a cross ratio on that line. We can prove the correctness of this simply, as soon as we realize that we are actually taking a cross ratio of *areas* rather than *lines* (just as the construction above is a cross ratio of distances rather than points).

As depicted in Figure 11.9, we take a reference point x and four arbitrary points p, q, r, s, and the four triangles determined by x and two of the other points. The area of the triangle formed by x, p, q is proportional to the weight of $x \wedge p \wedge q$. To show the invariance of the cross ratios, we rewrite them to the cross ratios of points on the arbitrary line L. This requires defining points p' and q' on that line, lying on the same rays from x as p and q do. It changes each area measure, for

$$x \wedge p \wedge q = x \wedge (p - x) \wedge (q - x)$$
$$= x \wedge (\mathbf{p} - \mathbf{x}) \wedge (\mathbf{q} - \mathbf{x})$$

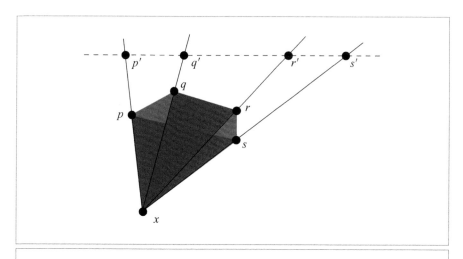

Figure 11.9: The combinations of four lines (actually, areas) taken in the cross ratio.

$$= x \wedge \big(p_0(\mathbf{p} - \mathbf{x})\big) \wedge \big(q_0(\mathbf{q} - \mathbf{x})\big)/(p_0\, q_0)$$
$$= x \wedge (\mathbf{p}' - \mathbf{x}) \wedge (\mathbf{q}' - \mathbf{x})/(p_0\, q_0)$$
$$= x \wedge p' \wedge q'/(p_0\, q_0).$$

When we take the cross ratio of the four areas, these radial rescaling constants will cancel. We have one more trick to play to get back to the cross ratio for lengths: recognizing that all area elements can be written as multiples of $x \wedge L$, we define the support point d of the line L relative to x through the demand $d\, L = x \wedge L$. Then all areas are expressed in terms of geometric products, and the common point support cancels in the ratio:

$$\frac{(x \wedge p \wedge q)\,(x \wedge r \wedge s)}{(x \wedge q \wedge r)\,(x \wedge s \wedge p)} = \frac{(x \wedge p' \wedge q')\,(x \wedge r' \wedge s')}{(x \wedge q' \wedge r')\,(x \wedge s' \wedge p')}$$
$$= \frac{d\,(p' \wedge q')\,d\,(r' \wedge s')}{d\,(q' \wedge r')\,d\,(s' \wedge p')}$$
$$= \frac{(p' \wedge q')\,(r' \wedge s')}{(q' \wedge r')\,(s' \wedge p')}.$$

Since the final result is invariant under projective transformations, the original area cross ratio is invariant as well. Using the same principles, you can define an invariant cross ratio of volumes containing a common line in 3-D, and corresponding quantities in higher-dimensional spaces.

11.8 LINEAR TRANSFORMATIONS: MOTIONS AND MORE

We have made flats by explicit construction using points and directions, and incidence operations among them. Once made, they can of course be moved around by transformations of the base space. If that space is Euclidean, we need to know how to perform the Euclidean operations of translation and rotation to the flats. In fact, it is often a convenient way to construct an offset flat by defining it first at the origin and then rotating and moving it to the proper location. That is, for instance, how we will interpret camera images in Section 12.2, first observing them in the camera frame, and then placing the camera in its actual location to interpret the consequences in the world. Therefore we need to know how to move flats around, both in direct and in dual form.

We can also perform linear transformations on the homogeneous representation space $\mathbb{R}^{n+1}$ and interpret the result they have on the elements in the base space $\mathbb{R}^n$. This is a representational trick, taking the representation space perhaps somewhat more seriously than we should have. Still, we find that in this way we can easily generate affine and projective transformations of the base space, and since we get them virtually for free in this manner, we may as well make use of that.

All this is an extension of common practice in the point-based homogeneous coordinates approach. As before, the algebra dictates how to extend the definitions from points to flats as outermorphisms, and gives the general structural framework for these kinds of transformations on general flats.

11.8.1 LINEAR TRANSFORMATIONS ON BLADES

We repeat from Chapter 4 that linear transformations on blades can be extended as outermorphisms, and therefore they follow simply from specifying what they do to a vector. If $\mathbf{x} \mapsto \mathsf{f}[\mathbf{x}]$ for a vector $\mathbf{x}$, then for a blade $\mathbf{X}$ we have

$$\mathbf{X} \mapsto \mathsf{f}[\mathbf{X}].$$

This makes transformations of offset subspaces very simple, at least when we have them in their direct representation. If the offset subspaces are characterized as duals, the transformation law is somewhat more involved, and we have derived that in Section 4.3.5. If $\mathbf{x} \mapsto \mathsf{f}[\mathbf{x}]$ for a vector $\mathbf{x}$, then a blade $\mathbf{X}^*$ (dually representing a subspace blade $\mathbf{X}$) transforms by the transformation f^* defined as

$$\mathbf{X}^* \mapsto \mathsf{f}^*[\mathbf{X}^*] \equiv \det(\mathsf{f})\,\overline{\mathsf{f}}^{-1}[\mathbf{X}^*]. \tag{11.12}$$

We remind you that for an orthogonal transformation, $\mathsf{f}^* = \det(\mathsf{f})\,\mathsf{f} = \pm\mathsf{f}$ (see Section 4.3.4).

11.8.2 TRANSLATIONS

Translations act on locations, not on directions. The vector space model had no clear way to distinguish those except by giving them separate data structures, and that makes translations in that model clumsy. A vector used as a position vector $\mathbf{p}$ translates to $\mathbf{p} + \mathbf{t}$, whereas a vector used as direction translates to $\mathbf{a}$. A line or plane could be characterized by a support vector $\mathbf{d}$, which is yet another type of vector. This type should translate only along itself to remain the support vector of the translated result: $\mathbf{d}$ translates to $\mathbf{d} + (\mathbf{t} \cdot \mathbf{d})/\mathbf{d}$. So in the vector space model, there are three types of vectors, each with their own translation law. The homogeneous model cleans all this up and makes the translations of flats automatic and universal.

The translation of a flat $X = p \wedge \mathbf{A}$ should of course simply move it to $(p + \mathbf{t}) \wedge \mathbf{A}$, a flat with the same orientation but passing through the translated point $p + \mathbf{t}$ rather than p. Since that adds $\mathbf{t} \wedge \mathbf{A}$ to the original flat X, we need to retrieve the direction $\mathbf{A}$ from X. This is achieved as $\mathbf{A} = e_0^{-1} \rfloor X$, and the *translation formula* becomes

$$translation : \quad X \; \mapsto \; \mathsf{T}_{\mathbf{t}}[X] = X + \mathbf{t} \wedge (e_0^{-1} \rfloor X). \tag{11.13}$$

This is clearly a linear transformation of X, since it consists of linear operations. Structural exercise 15 has you prove that its determinant equals 1. This implies that translations do not affect areas and volumes in the homogeneous representation space $\mathbb{R}^{n+1}$, and since they also do not affect normalization (for $e_0^{-1} \rfloor X = e_0^{-1} \rfloor \mathsf{T}_{\mathbf{t}}[X]$), they preserve areas and volumes in the base space $\mathbb{R}^n$ as well.

Note that (11.13) works on flats of any grade, even for points with their scalar directions. And it is consistent with our general representation: when we have made a flat at the origin with direction $\mathbf{A}$ (represented by the blade $e_0 \mathbf{A}$), then we can get the general form of an offset flat by translating it over $\mathbf{p}$ to the point p:

$$\mathsf{T}_{\mathbf{p}}[e_0 \mathbf{A}] = e_0 \mathbf{A} + \mathbf{p} \wedge \mathbf{A} = p \wedge \mathbf{A}.$$

Equation (11.13) is defined for all blades, even though it was designed with finite subspaces in mind. When we apply it to an improper subspace represented by the base space blade $\mathbf{X}$, we find:

$$\mathsf{T}_{\mathbf{t}}[\mathbf{X}] = \mathbf{X},$$

so that *directions are translation-invariant.*

We also need to know the consequence of a translation on a dually represented offset flat. As we mentioned above, this is not the same as applying the original linear mapping $\mathsf{T}_{\mathbf{t}}$. We denote the dual translation by $\mathsf{T}_{\mathbf{t}}^{*}$, and compute it explicitly, by undualization of its argument, translation, and then redualization to get back to the dual representation. That gives the *dual translation formula*:

$$\mathsf{T}_\mathbf{t}^*[X^*] \equiv (\mathsf{T}_\mathbf{t}[X])^*$$
$$= \left(X + \mathbf{t} \wedge (e_0^{-1} \rfloor X)\right)^* = X^* + \mathbf{t} \rfloor (e_0^{-1} \wedge X^*)$$
$$= X^* - e_0^{-1} \wedge (\mathbf{t} \rfloor X^*). \tag{11.14}$$

You may verify that $\mathsf{T}_\mathbf{t}^* = \overline{\mathsf{T}}_\mathbf{t}^{-1}$, in accordance with (11.12) for general linear transformations, since $\det(\mathsf{T}_\mathbf{t}) = 1$.

We emphasize that the translation formulas apply to any blade (or dual blade), whatever the grade. So there is no need for a separate translation operation for vectors, planes, and so on. Yet there is a distinction between direct and dual representations, for $\mathsf{T}_\mathbf{t}$ and $\mathsf{T}_\mathbf{t}^*$ are different functions of their arguments.

11.8.3 ROTATION AROUND THE ORIGIN

A Euclidean rotation around an axis through the origin characterized by a versor R should affect the Euclidean positional and direction elements in the usual way: both should turn. You might expect that we therefore need to decompose the flat into these components, do the rotation, and recompose. However, there is a nice surprise: since R commutes with e_0, we can make it act on the whole blade using the outermorphism property:

$$X = (e_0 + \mathbf{p}) \wedge \mathbf{A} \mapsto$$
$$\mathsf{R}[X] = (e_0 + R\,\mathbf{p}\,R^{-1}) \wedge (R\,\mathbf{A}\,R^{-1})$$
$$= (R\,(e_0 + \mathbf{p})\,R^{-1}) \wedge (R\,\mathbf{A}\,R^{-1})$$
$$= R\,((e_0 + \mathbf{p}) \wedge \mathbf{A})\,R^{-1}$$
$$= R\,X\,R^{-1}$$

This is a very convenient form, much nicer than the translation formula (11.13).

Dual representations of flats can also be rotated. We can apply the procedure of undualization of the argument, rotation, and redualization, but this results in the same rotation formula, since the even versor R commutes with I_{n+1}:

$$X^* \mapsto (R\,(X^*\,I_{n+1})\,R^{-1})\,I_{n+1}^{-1} = (R\,(X^*)\,R^{-1})\,I_{n+1}\,I_{n+1}^{-1} = R\,(X^*)\,R^{-1}.$$

(Alternatively, you could have realized that for the orthogonal transformation R, (11.12) gives that $\mathsf{R}^* = \det(\mathsf{R})\,\overline{\mathsf{R}}^{-1} = \mathsf{R}$.) So a dually represented flat can be rotated using the same formula as for a directly represented flat.

Here we see an advantage of operations that can be represented by even versors: they transform elements universally, whether they are direct or dual. We will get back to this issue; it is going to be an important feature of the conformal model.

11.8.4 GENERAL ROTATION

A general rotation around an axis at location $\mathbf{t}$ can be constructed in the usual manner by translation over $-\mathbf{t}$, rotating by R, and then putting the result back to $\mathbf{t}$. This gives, after some algebraic simplification,

$$ X \;\mapsto\; RXR^{-1} + (\mathbf{t} - R\mathbf{t}R^{-1}) \wedge \left(e_0^{-1} \rfloor (RXR^{-1}) \right). $$

It is clear that only the component of $\mathbf{t}$ in the rotation plane contributes, since $\mathbf{t} - R\mathbf{t}R^{-1} = 2(\mathbf{t}\rfloor R)R^{-1}$. You could see this formula as the translation of the rotation operator R, and in that light, rotors transform differently from either blades or dual blades.

If X is a dual representative, the general rotation formula is different:

$$ X^* \;\mapsto\; RX^*R^{-1} - e_0^{-1} \wedge \left((\mathbf{t} - R\mathbf{t}R^{-1}) \rfloor (RX^*R^{-1}) \right). $$

This difference is caused by the asymmetry of the translation operator.

11.8.5 RIGID BODY MOTION

Finally, the most common way of characterizing a rigid motion is as a rotation R around the origin, followed by a translation over $\mathbf{t}$. On a direct blade this obviously leads to

$$ X \;\mapsto\; RXR^{-1} + \mathbf{t} \wedge \left(e_0^{-1} \rfloor (RXR^{-1}) \right), $$

and on a dual blade to

$$ X^* \;\mapsto\; RX^*R^{-1} - e_0^{-1} \wedge \left(\mathbf{t} \rfloor (RX^*R^{-1}) \right). $$

Those are all the formulas you will need for rigid body motions on general objects. As you apply them to the specific factors, you will see that the purely Euclidean directions $\mathbf{A}$ are translation-invariant:

$$ \mathbf{A} \;\mapsto\; R\mathbf{A}R^{-1} + \mathbf{t} \wedge \left(e_0^{-1} \rfloor (R\mathbf{A}R^{-1}) \right) = R\mathbf{A}R^{-1}. $$

This is consistent with their interpretation as improper flats at infinity: they cannot be translated to the finite part of space.

11.8.6 CONSTRUCTING ELEMENTS THROUGH MOTIONS

If in an application you need a specific flat and its representation, it is often convenient to construct it first at the origin, and then orient it and move it to the proper location. For instance, to make a line through p, we start with some standard line in the direction $\mathbf{e}_1$ through the origin. This is represented by the blade $e_0 \wedge \mathbf{e}_1$. Now apply a rotor R to orient the line to the desired orientation $\mathbf{u}$. Since $R\, e_0\, \widetilde{R} = e_0$ and $R\, \mathbf{e}_1\, \widetilde{R} = \mathbf{u}$, the result on

$e_0 \wedge \mathbf{e}_1$ is $e_0 \wedge \mathbf{u}$: only the direction has changed. Now apply a translation $\mathsf{T_p}$ to bring the line to the proper location so that $\mathsf{T_p}[e_0] = p$. The translation does not affect the direction element, so that the result is the line $p \wedge \mathbf{u}$. Following these steps in terms of the operators we just defined shows that the general form of a line is indeed $e_0 \wedge \mathbf{u} + p \wedge \mathbf{u}$. The same reasoning to move a flat at the origin with the correct orientation $\mathbf{A}$ to a desired location gives $e_0 \wedge \mathbf{A} + \mathbf{p} \wedge \mathbf{A}$ as its general form.

In this way of looking at the flats, you see that the first operation of orienting the flat correctly acts only on the direction element, and is therefore actually defined in the vector space model. The second step, translation, is the only one invoking the homogeneous model, and it attaches itself only to the locational element e_0, the point at the origin. In this sense, the homogeneous model splits nicely:

$$\mathsf{T_p}[\mathsf{R}[e_0 \wedge \mathbf{A}]] = \mathsf{T_p}[e_0] \wedge \mathsf{R}[\mathbf{A}],$$

and this clearly shows the backward compatibility of the attitudinal elements with the vector space model. We have not lost any of our capabilities to compute with directions, and added the locational aspects, in a neat algebraic extension.

11.8.7 RIGID BODY MOTION OUTERMORPHISMS AS MATRICES

The linearity of the rigid body motions in the homogeneous model permits their representation as matrices. The form of the matrices differs for the various elements they act on, because the basis on which the points, lines, and planes are represented vary widely. We will give these matrices explicitly for the 3-D case, in terms of the specific homogeneous coordinates of the points, lines, and planes, in Section 12.1.

11.8.8 AFFINE TRANSFORMATIONS

An *affine transformation* is a special transformation of the points in the base space. It has the property that it preserves colinearity of points on a line and the ratios of their distances along that line. Any linear transformation of the vector space $\mathbb{R}^n$ has this property, and so do translations.

That is the definition of affine transformations in $\mathbb{R}^n$, and we need to find out what it implies for their representation as mappings of $\mathbb{R}^{n+1}$. As we have shown in Section 11.7.3, the demand on meaningful ratios of distances of points along the same line implies that their e_0-coefficient is preserved under affine transformations. In formula, this means that a point x transforms such that $e_0^{-1} \cdot x$ is the same before the affine transformation A as after. Therefore,

$$e_0^{-1} \cdot x = e_0^{-1} \cdot \mathsf{A}[x] = \overline{\mathsf{A}}[e_0^{-1}] \cdot x.$$

Since this should hold for any vector x, an affine transformation must satisfy

$$\overline{\mathsf{A}}[e_0^{-1}] = e_0^{-1}.$$

You may also read this as

$$\overline{A}[e_0] = e_0,\tag{11.15}$$

because e_0^{-1} and e_0 only differ by a constant after you have chosen the metric.

It is easy to show that this implies that A preserves the $\mathbb{R}^n$-subspace as being the part of $\mathbb{R}^{n+1}$ that is perpendicular to e_0. We denote the pseudoscalar of $\mathbb{R}^n$ as $\mathbf{I}_n$ so that we can write the statement that no part of $\mathbb{R}^n$ has an e_0 component as

$$0 = e_0^{-1} \rfloor \mathbf{I}_n.$$

Now we apply A to this statement and use the transformation property of the contraction (4.13) to obtain

$$0 = A[e_0^{-1} \rfloor \mathbf{I}_n] = \overline{A}^{-1}[e_0^{-1}] \rfloor A[\mathbf{I}_n] = e_0^{-1} \rfloor A[\mathbf{I}_n],$$

so that the transformed pseudoscalar indeed satisfies the same condition. Therefore the affinely transformed elements are still in $\mathbb{R}^n$. In geometrical terms, the preservation of $\mathbb{R}^n$ means that *an affine mapping changes directions into directions*. More poetically, since the directions are the elements at infinity, affine mappings preserve the heavenly sphere (not point by point, but as a set).

We need to find a standard form for the most general linear transformation A on a vector x of $\mathbb{R}^{n+1}$ whose adjoint preserves e_0. It uses the most general linear transformation f in the subspace $\mathbb{R}^n$, and it can add an arbitrary vector $\mathbf{t}$ to translate the result. On a unit vector $\mathbf{x}$ it should therefore have the effect $A[x] = e_0 + f[\mathbf{x}] + \mathbf{t}$. If we extend f to $\mathbb{R}^{n+1}$ by defining $f[e_0] = e_0$, we can write this as $A[x] = f[x] + \mathbf{t}$, but this is not yet a linear transformation since it is not linear in the scale of x. The proper form must be

$$A[x] = f[x] + \mathbf{t} \wedge (e_0^{-1} \cdot x),$$

where the factor on the $\mathbf{t}$ is the weight of the point x. When you try to extend this to arbitrary blades as an outermorphism, you find that this is awkward (some mixed terms appear), until you realize that $e_0^{-1} \cdot x = e_0^{-1} \cdot f[x]$ (do you realize why?). Then the affine transformation of a general multivector is

$$A[X] = f[X] + \mathbf{t} \wedge (e_0^{-1} \rfloor f[X]).\tag{11.16}$$

Clearly the Euclidean motions are affine transformations (rotations only have the f-part, for translations that is the identity). But more transformations can be defined; just taking out the translation component leaves the linear transformations in $\mathbb{R}^n$, which can be decomposed as rotations, translation, and nonuniform scaling.

The affine combination of points in (11.2) is now clearly an affine covariant, in the sense that the affine combination of transformed points equals the transform of the affine combination, as long as we use an affine transformation. Affine combinations of points can be combined using the construction operations such as outer product, `meet`, and `join` to produce other affinely covariant constructions. This is the algebraic and geometrical basis of curve interpolation methods such as de Casteljau's algorithm.

11.8.9 PROJECTIVE TRANSFORMATIONS

Projective transformations are the general linear transformations in the space $\mathbb{R}^{n+1}$, interpreted in the space $\mathbb{R}^n$. Such general transformations can transform finite points to infinite points (e.g., the rotation in $\mathbb{R}^{n+1}$ that turns e_0 into $\mathbf{e}_1$ is among them).

Within the homogeneous model of geometric algebra, the projective transformations can be extended as outermorphisms to act on all blades of $\mathbb{R}^{n+1}$. In the base space $\mathbb{R}^n$, this implies that we can not only transform points, but also offset subspaces such as lines and planes. Therefore the *point conic* classically defined through as the locus of all points x satisfying the equation

$$x \cdot \mathsf{A}[x] = 0,$$

can easily be extended to a *general conic*

$$X \rfloor \mathsf{A}[X] = 0.$$

Figure 11.10 shows a line conic associated with a point conic in this manner.

Beyond this extended capability and its consequences, the geometric algebra of the homogeneous model so far has added little to the classical treatment of projective geometry. Moreover, these capabilities are already present in the weaker Grassmann-Cayley algebra, which is quite sufficient to do nonmetric projective geometry in practice (as [21] shows). Geometric algebra makes the incorporation of Euclidean measures in those techniques somewhat easier, but that is about the extent of its contribution to date.

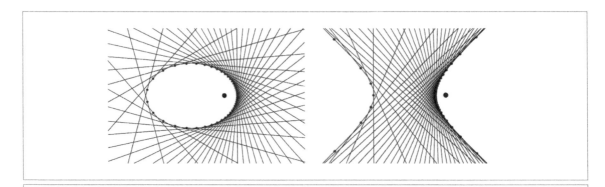

Figure 11.10: In the homogeneous model, the equation $X \rfloor \mathsf{A}[X] = 0$ can be solved for flats of different grades. For points, the solutions determine the familiar conics (in red); for lines, they give line conics (in blue).

On the other hand, geometric algebra changes our perspective on some aspects of projective geometry. The principle of duality between points and lines in the plane, which is classically presented as a unique feature of projective geometry, is just a special case of our usual dualization operation of division by the pseudoscalar. Not only is it applicable in n-dimensions to elements of arbitrary grade, but it is a quantitative computational principle rather than the merely qualitative substitutional symmetry of geometrical theorems it often is in the classical treatments. The full quantitative use of this principle is admittedly a bit confused in the homogeneous model (as we show in Section 11.10), and that has perhaps led to an underestimation of its metric qualities. But dualization will come into its own in the conformal model of Euclidean geometry. In that model, we will also see how regular points in Euclidean geometry are actually point pairs (with one member residing at infinity), and that the identification of opposite points in a spherical model of projective geometry is again not uniquely projective, just a general fundamental principle applied in a slightly different context (see Section 16.7.2).

Related to these issues is the suggestion that the homogeneous model may not be the most natural representation of projective geometry. A geometric algebra representation may be found in which conics can be represented as blades, and in which projective transformations are rotors. Such a projective model would then be the natural way to embed projective geometry within geometric algebra. It would be powerful: one should be able to intersect two conics X and Y by their meet $Y^* \rfloor X$, for instance. The practical desirability of similar computational techniques will be made apparent by our treatment of the conformal model that provides this structure for Euclidean transformations. Having such different geometries expressed in similar algebras will convey the common nature of their fundamental structure.

To return to the homogeneous model, within the context of projective transformations an affine transformation amounts to choosing a particular pseudoscalar subspace $\mathbf{I}_n$ and demanding that it be an eigenblade that transforms to become a multiple of itself; the complement e_0 is then preserved by the adjoint of the transformation. So affine geometry is projective geometry in a selected invariant base space of the homogeneous representation space.

11.9 COORDINATE-FREE PARAMETERIZED CONSTRUCTIONS

The extension to blades and their operators that we achieved by considering the geometric algebra of the homogeneous model can effectively be used in your programming. With a bit of practice, you can design coordinate-free expressions for elements parameterized by other elements. For instance, in 3-D you might have a line L and a point p not on it, and wish to find the line M through p that intersects L perpendicularly, as illustrated in Figure 11.11(a).

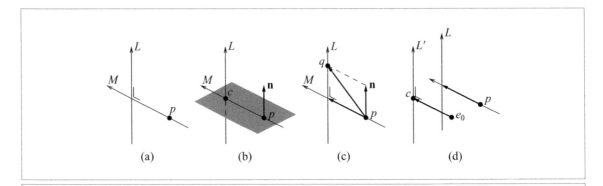

Figure 11.11: (a) The problem: to find the line M through a point p perpendicular to another line L. (b) A solution involving the plane through p orthogonal to L to find the support point c. (c) A solution using the rejection of the connection vector from p to an arbitrary point q on the line. (d) Translating the situation back to the origin and solving it as a standard problem.

There are various routes to finding solution for such problems, some easier than others. In the end you usually find an expression that you could have written down immediately had you looked at the problem in a better way from the start. Practice helps, and these are clearly skills that you should develop if you have to use the homogeneous model a lot. However, we will soon replace the homogeneous model by the conformal model as an operational model of Euclidean geometry, which requires similar skills within a richer set of metrical tools, so check Chapter 13 first to see whether that model might not serve your needs better.

Back to the problem of finding the line through p perpendicular to L. Let us denote the direction of L by $\mathbf{u}$ (it can be found as $\mathbf{u} = e_0^{-1} \rfloor L$), and the moment of L by $\mathbf{U}$ (it is $\mathbf{U} = e_0^{-1} \rfloor (e_0 \wedge L)$).

1. A number of thoughts arise about the relative geometrical situation of p, L, and M, and it is interesting how they immediately invoke algebraic expressions in the homogeneous model.

    - If we can find the unit point c on L that is closest to p, then $M = p \wedge c$.
    - Whatever M is, it has to be perpendicular to L, and therefore it should be in the plane Π though p that is perpendicular to L; that plane is, dually, $\Pi^* = p \rfloor (\mathbf{u} \wedge e_0^{-1}) = -\mathbf{u} + (\mathbf{p} \cdot \mathbf{u}) e_0$, using (11.6).
    - Having the plane Π, the point c is its meet with L, so $c = \Pi^* \rfloor L$.

    The combination of these observations provides the solution for the closest point c as $c = \left(-\mathbf{u} + (\mathbf{p} \cdot \mathbf{u}) e_0^{-1} \right) \rfloor \left(e_0 \mathbf{u} + \mathbf{U} \right) = e_0 - \mathbf{u} \rfloor \mathbf{U} + (\mathbf{p} \cdot \mathbf{u}) \mathbf{u}$. It gives M as $p \wedge c$. Done.

But we can learn from rewriting things. First, $p \wedge c$ can be rewritten as $p \wedge (c - p) = p \wedge (\mathbf{c} - \mathbf{p})$, and now $\mathbf{c} - \mathbf{p}$ can be simplified to $\mathbf{c} - \mathbf{p} = -\mathbf{u} \rfloor \mathbf{U} - (\mathbf{p} \wedge \mathbf{u}) \mathbf{u} = (\mathbf{U} - \mathbf{p} \wedge \mathbf{u}) \mathbf{u}$. You should begin to recognize that final expression. The moment $\mathbf{U}$ of L can be constructed from a general point q on it as $\mathbf{q} \wedge \mathbf{u}$, and the 2-direction of M can then apparently be written as $((\mathbf{q} - \mathbf{p}) \wedge \mathbf{u}) \mathbf{u}$, which is proportional to the rejection $((\mathbf{q} - \mathbf{p}) \wedge \mathbf{u}) / \mathbf{u}$ of the difference vector of any point on L with p, by the direction of L. That is of course the support vector from p to L; now that we see it we recognize it in the problem (see Figure 11.11(c)).

2. So in hindsight, our solution might have been: compute the support vector between the unit point p and and arbitrary unit point q of L as the rejection $((\mathbf{q} - \mathbf{p}) \wedge \mathbf{u}) / \mathbf{u}$, and take the outer product with p to produce M. This is illustrated in Figure 11.11(b). Using the original points q and p here saves extracting their Euclidean parts, and makes it easier to express the result purely in terms of p and L. We just use $L = q \wedge \mathbf{u}$, and with $\mathbf{u} = e_0^{-1} \rfloor L$ obtain

$$M = p \wedge \frac{L - p \wedge (e_0^{-1} \rfloor L)}{e_0^{-1} \rfloor L}.$$

Our computation in this way is even properly normalized: since p is a unit point, M is weighted by the directed distance from p to L. If p is not a unit point, the computation of the direction is incorrect, so a more robust expression would replace p with $p / (e_0^{-1} \cdot p)$ (at least in the direction part). Note that the normalization of L cancels out, so L need not be normalized.

3. Another way to find the location $\mathbf{c}$ intersection point c is to realize that $c \wedge \mathbf{u} = L$ since it lies on L, and that it lies on the same plane with normal $\mathbf{u}$ as p does, so that $c \cdot \mathbf{u} = p \cdot \mathbf{u}$ (alternatively, you could remark that $c - p$ should be perpendicular to $\mathbf{u}$, giving the same equation). Then we add those to obtain $c \mathbf{u} = L + p \cdot \mathbf{u}$, and right division by $\mathbf{u}$ produces the result $c = (L + (p \cdot \mathbf{u})) / \mathbf{u}$. The line M is then $M = p \wedge c = p \wedge (c - p) = p \wedge (L - (p \wedge \mathbf{u})) / \mathbf{u}$, giving the same result as above. Note that this computation involves the geometric product in a direct algebraic manner rather than invoking the meet. It is the purest solution, algebraically speaking.

4. We may suddenly realize that the original question was merely to ask for the line along the support vector of L when seen from p. So we should get the answer if we translate L and p back to the origin (by a translation over $-\mathbf{p}$), compute the support of that translated line L' (which is a standard formula from Table 11.3), and translate back to p, as illustrated in Figure 11.11(d).

   Translating L over $-\mathbf{p}$ means adding $-\mathbf{p} \wedge \mathbf{u}$ to it so that the moment of the translated line is $\mathbf{U} - \mathbf{p} \wedge \mathbf{u}$, with $\mathbf{U}$ the moment of L. The support vector of L' is then is $\mathbf{d} = (\mathbf{U} - \mathbf{p} \wedge \mathbf{u}) / \mathbf{u}$, which is the direction of M with proper sign and weight, so $M = p \wedge \mathbf{d}$. Or, to finish the computation properly, we must translate the support line $e_0 \wedge \mathbf{d}$ of L' back to the point p; then e_0 becomes p and the base space element $\mathbf{d}$ is translation-invariant, so it remains $\mathbf{d}$.

Whenever you solve a problem that is important to a computation, try to derive and interpret the result in several ways. Some of those may integrate better with other aspects of the larger problem and allow for a more efficient combined total solution. Make sure you use the opportunities that the homogeneous model offers to give your results quantitative properties with useful signs and weights; related to that, check that the formulas you derive are robust to normalization assumptions.

11.10 METRIC PRODUCTS IN THE HOMOGENEOUS MODEL

You may have noticed that we have been rather careful in our choice of algebraic operations. In fact, there is hardly more than the outer product, duality, the meet, and outermorphisms involved on the elements of the representation space $\mathbb{R}^{n+1}$. When you realize that mathematicians can introduce duality in a nonmetric way (by employing k-forms), you see that these are all nonmetric operations. The meet is simply the dual of the outer product of duals, and we have seen in Chapter 5 that its outcome does not depend on a metric.

When we did use the purely metric products of contraction or the geometric product, it was either on the subspace $\mathbb{R}^n$ or on elements with the same flat in $\mathbb{R}^{n+1}$, such as when computing the cross ratio. (The probing of a dual representation by the inner product with a point can also be written nonmetrically using a 1-form, so it does not count as metric usage).

Our nonmetrical use of the homogeneous model $\mathbb{R}^{n+1}$ is in fact most clearly demonstrated by our refusal to even choose its metric: we purposely left the option of choosing $e_0^2 = 1$ or $e_0^2 = -1$. That did not return to haunt us in our results so far, which must therefore have been nonmetrical.

The main problem with using the metric of $\mathbb{R}^{n+1}$ is that you cannot use it directly to do Euclidean geometry, for it has no clear Euclidean interpretation. We address this fundamental shortcoming because it helps refine what the properties of a model of a geometry should be. That understanding prepares us for the conformal model that fixes the awkwardness of the homogeneous model in this respect.

11.10.1 NON-EUCLIDEAN RESULTS

We would demand of a representation for Euclidean geometry that it is structure-preserving under translations (i.e., translation-covariant). This means that a result that holds in one location can be translated to another location and still hold. (We also demand rotation covariance, but the homogeneous model inherits that property from the vector space model, so we need not discuss it.) Translation covariance should imply that we can take another point $e_0' = e_0 + \mathbf{t}$ as our new origin instead of e_0, and all geometrical results relative to this new origin should be similar. Yet this change of origin from e_0 to e_0' changes the metric of $\mathbb{R}^{n+1}$, for

$$(e_0')^2 = e_0^2 + \mathbf{t}^2,$$

and whether we choose $e_0^2 = +1$ or $e_0^2 = -1$, the square of e_0' is going to be different.

This deficiency also shows itself in the geometric product, and propagates to versors. Things get structurally awkward with inverses, for *the inverse of the representation is not the representation of the inverse*. This statement is

$$x^{-1} = (e_0 + \mathbf{x})^{-1} = \frac{e_0 + \mathbf{x}}{e_0^2 + \mathbf{x}^2} \neq e_0 + \mathbf{x}/\mathbf{x}^2 = e_0 + \mathbf{x}^{-1},$$

and you see how different the results are. (An element that does frequently occur when computing with inverses is $e_0^{-1} - \mathbf{x}^{-1}$. It has the property $(e_0^{-1} - \mathbf{x}^{-1}) \cdot (e_0 + \mathbf{x}) = 0$ and is therefore some standard perpendicular direction to the vector x, useful in duality computations.)

The lack of well-behaved inverses haunts the use of the geometric product in versor constructions. For instance, the reflection of one point into another seems proper enough at the origin:

$$e_0 \, p \, e_0^{-1} = e_0 \, (e_0 + \mathbf{p}) \, e_0^{-1} = e_0 - \mathbf{p}.$$

That is a reasonable geometric outcome, a point reflection has clearly taken place. But when we use another point as reflector, something strange happens. Let us study that around the origin as well:

$$q \, e_0 \, q^{-1} = (e_0 + \mathbf{q}) \, e_0 \, (e_0 + \mathbf{q})/(e_0^2 + \mathbf{q}^2)$$
$$= \left(e_0 \, (e_0^2 - \mathbf{q}^2) + 2 \, e_0^2 \, \mathbf{q}\right)/(e_0^2 + \mathbf{q}^2).$$

This is a weighted point at location

$$2 \, \frac{e_0^2}{e_0^2 - \mathbf{q}^2} \, \mathbf{q}.$$

Not only is this strangely nonlinear, but it depends on the metric of the representation space since it contains e_0^2. We saw above that this metric depends on where the origin is. This well-defined reflection construction in $\mathbb{R}^{n+1}$ may be useful for something (it is related to stereographic projection), but it is not meaningful for the Euclidean geometry of $\mathbb{R}^n$.

These insights are not new. In the classical use of homogeneous coordinates, you should have learned to be careful with the inner product. The inner product of two vector representatives x and y is

$$x \cdot y = e_0^2 + \mathbf{x} \cdot \mathbf{y},$$

and only the second term has objective meaning for the computation of angles and lengths in the Euclidean space you are modeling. But that quantity might as well have been computed using only the vector space model. On the other hand, the inner product does have a sensible meaning when defining a plane in a dual manner by $\pi = \mathbf{n} - e_0^{-1}\delta$, as we have seen:

$$x \cdot \pi = 0 \quad \Longleftrightarrow \quad \mathbf{x} \cdot \mathbf{n} = \delta.$$

This is essentially a nonmetric use of the inner product, since it is done between a direct element and a dual element: $0 = x \cdot \pi = x \cdot \Pi^* = (x \wedge \Pi)^*$. It is all a bit confusing and less tidy than you might have hoped from geometric algebra. We emphasize that the problem is not geometric algebra itself, but the homogeneous model and our desire to use it for Euclidean geometry. It will be replaced by a much better model for that purpose in Chapter 13.

11.10.2 NONMETRIC ORTHOGONAL PROJECTION

Despite its metric shortcomings, we can learn interesting geometry from the structure we have in the homogeneous model. As an example, let us look at orthogonal projection. In Part I we derived that the projection of a blade X onto a blade A is represented as

$$P_A[X] = (X \rfloor A) \rfloor A^{-1}.$$

The result is in A in a manner that in a Euclidean space is "orthogonally below X." You would expect this to be a metric construction.

The algebraic operation can be performed in the homogeneous model as well, for inverses are well-defined. What can this orthogonal projection mean in a space $\mathbb{R}^{n+1}$ without a clear metric? The homogeneous projection of a point x onto a line L is depicted in Figure 11.12. The result is a point on the line L with an unexpected relationship to its dual: the line $x \wedge P_L[x]$ intersects the dual line L^{-*}.

It turns out that we can rewrite the projection as a `meet`, not only in the homogeneous model, but in general. Leaving out the inverse of A to show the structure of the derivation more clearly, and assuming X and A in general position, we simply rewrite:

$$A \cap (X \wedge A^{-*}) = (X \wedge A^{-*})^* \rfloor A = (X \rfloor A) \rfloor A. \tag{11.17}$$

The orthogonal projection is therefore less metric than we might have assumed, though it does involve an (inverse) dual.

To explain Figure 11.12, the projection in 3-D of a point x on a unit line L is obtained by making the dual L^{-*}. This is also a line, and writing L as $L = (e_0 + \mathbf{d})\,\mathbf{u} = e_0\,\mathbf{u} + \mathbf{M}$, we find $L^{-*} = -e_0^2\,(e_0^{-1}\,\mathbf{M}^\star - \mathbf{u}^\star)$. This has $1/\mathbf{d}$ as its support vector, and runs perpendicular to the plane through L and the origin; both properties are of course very much dependent on the location of the origin. Then the projection is the point of L obtained by intersection with the plane $x \wedge L^{-*}$. That also implies that the line through x and its projection cuts

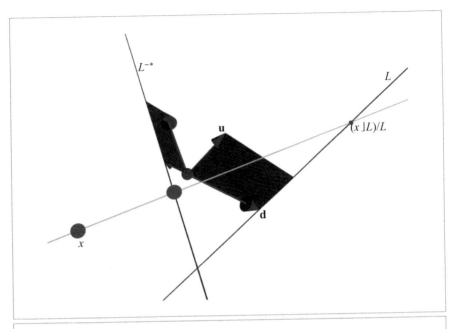

Figure 11.12: The orthogonal projection in the homogeneous model (see text).

L^{-*}. This situation is depicted in Figure 11.12. Euclidean geometry it ain't, but it could be useful in projective geometry.

11.11 FURTHER READING

A good reference for getting an intuition for oriented subspaces in the homogeneous model is Stolfi's classical book [60]. His hand-drawn visualizations are justly famous, and they give several ways of looking at this "raylike" representation of points in one more dimension.

He uses a kind of Grassmann algebra to encode the basic operations between flats, but grounded in `meet` and `join` rather than the outer product and the contraction. That is more limiting for use in Euclidean situations, so that his treatment is mostly nonmetric, not even reverting to the vector space model where this would have been possible. This is typical of most uses of the homogeneous model in projective imaging as well. Some references for that field are given at the end of the next chapter.

11.12 EXERCISES

11.12.1 DRILLS

Compute the 2-blades corresponding to the lines given by the data below. Which of the lines are the same, considered as weighted oriented elements of geometry, and which are the same as offset subspaces?

1. Two points at locations e_1 and e_2.

2. A point at location e_1 and a direction $(e_2 - e_1)$.

3. A point at location e_2 and a direction $(e_2 - e_1)$.

4. Two points with locations $2(e_2 - e_1)$ and $3(e_2 - e_1)$.

5. A point at location e_1 and a direction $2(e_2 - e_1)$.

6. A unit point at location e_1 and a point with weight 2 at location e_2.

11.12.2 STRUCTURAL EXERCISES

1. Let an orthonormal coordinate system $\{e_i\}_{i=1}^3$ be given in 3-D Euclidean space. Compute the support vector of the line with direction $u = e_1 + 2e_2 - e_3$, through the point $p = e_1 - 3e_2$. What is the distance of the line to the origin?

2. Convert the line of the previous exercise into a parametric equation $x = p + \lambda u$; express λ as a function of x for a point x on the line. Interpret geometrically.

3. Show that the support vector d of a k-flat is the rejection of the position vector of an arbitrary point p on it by the k-direction A.

4. Show that the weight of the plane $p \wedge q \wedge r$ is twice the area of the triangle spanned by the normalized points p, q, and r. Give the corresponding statement for a general $(k - 1)$-dimensional flat constructed from k normalized points (see also Section 2.8.3).

5. Three points a, b, c form a plane, and these points can be used to address any other point x in that plane as a linear combination:

$$x = \alpha a + \beta b + \gamma c.$$

Using normalized points, one can do this with an affine combination. The resulting scalars α, β, γ are called *barycentric coordinates* (literally, "weight-based"). Compute α, β, γ in terms of the points a, b, c, and express the result using the relative vectors $a = a - c, b = b - c$ and $x = x - c$. This should give you:

$$\alpha = \frac{x \wedge b}{a \wedge b}, \quad \beta = \frac{x \wedge a}{b \wedge a}, \quad \gamma = 1 - \frac{x \wedge (b - a)}{a \wedge b}. \qquad (11.18)$$

Interpret the result geometrically in terms of areas in the plane (most easily seen when x is inside the triangle formed by a, b, and c). What are the barycentric coordinates of the center of gravity?

These barycentric coordinates can be used to interpolate any scalar property ϕ given at each of the vertices of a triangle to an intermediate ϕ_x value at x, through:

$$\phi_x = \alpha\,\phi_a + \beta\,\phi_b + \gamma\,\phi_c \tag{11.19}$$

This equation will be used in the ray tracer of Chapter 23.

6. The inverse of the support vector of a plane spanned by three points at locations **p**, **q**, and **r** is equal to the sum of the reciprocal vectors corresponding to these three vectors in their own frame $\{\mathbf{p},\mathbf{q},\mathbf{r}\}$. Show this using (11.4) and (3.31).

7. In the parametric equation for an offset flat (11.5), the vector **x** determines the values of the λ_i uniquely. Compute a formula for λ_i. (Hint: Eliminate the other λ_j, with $j \neq i$, by suitably chosen outer products with $\mathbf{a}_j$ vectors. Alternatively, use the idea of a reciprocal basis from Section 3.8.)

8. The base space $\mathbb{R}^n$ is a subspace of $\mathbb{R}^{n+1}$, and it has a subalgebra of the geometric algebra of that full space. Hestenes [27] has devised the *projective split* relative to a fixed element to express a subalgebra of a space in standard form. Using e_0 and e_0^{-1} (the elements of $\mathbb{R}^{n+1}$ that are not also in $\mathbb{R}^n$), a direct blade of $\mathbb{R}^{n+1}$ can be split by his method. This leads to a natural factorization of the blade. The following does this, and results in the familiar parts that we recognize as support and direction (see also Table 11.3).

$$
\begin{aligned}
X &= e_0\,e_0^{-1}\,X \\
&= e_0 \wedge (e_0^{-1}\rfloor X) + e_0\rfloor(e_0^{-1}\wedge X) \\
&= e_0\,(e_0^{-1}\rfloor X) + \frac{e_0^{-1}\rfloor(e_0\wedge X)}{e_0^{-1}\rfloor X}\,(e_0^{-1}\rfloor X) \quad \text{if } e_0^{-1}\rfloor X \neq 0 \\
&= \left(e_0 + \frac{e_0^{-1}\rfloor(e_0\wedge X)}{e_0^{-1}\rfloor X}\right)(e_0^{-1}\rfloor X) \quad \text{if } e_0^{-1}\rfloor X \neq 0
\end{aligned}
$$

Perform a similar split for the dual blade to derive the corresponding column of Table 11.3.

9. Construct the dual representation of the midplane between two points p and q.

10. We suggested in Section 11.7 that the meet of two points p and q on a line $L = d\mathbf{u}$ is proportional to their distance. Explore precisely how, so give $p\cap q$ quantitatively.

11. The meet of two skew lines $p \wedge \mathbf{u}$ and $q \wedge \mathbf{v}$ can be computed as $M^*\rfloor L$. Verify the steps in the following derivation of (11.9) using this formula.

$$
\begin{aligned}
(p \wedge \mathbf{u})\cap(q \wedge \mathbf{v}) &= \left((q\wedge \mathbf{v})\rfloor(\mathrm{I}_3^{-1}\wedge e_0^{-1})\right)\rfloor(p\wedge \mathbf{u}) \\
&= \left(q\rfloor(\mathbf{v}\rfloor\mathrm{I}_3^{-1}\wedge e_0^{-1})\right)\rfloor(p\wedge \mathbf{u}) \\
&= \left(q\rfloor(\mathbf{v}\rfloor\mathrm{I}_3^{-1})\wedge e_0^{-1} + (\mathbf{v}\rfloor\mathrm{I}_3^{-1})\right)\rfloor(p\wedge \mathbf{u}) \\
&= \left(q\rfloor(\mathbf{v}\rfloor\mathrm{I}_3^{-1})\right)\rfloor\mathbf{u} + (\mathbf{v}\rfloor\mathrm{I}_3^{-1})\rfloor(p\wedge \mathbf{u})
\end{aligned}
$$

$$= \mathbf{u} \rfloor \left(\mathbf{q} \rfloor (\mathbf{v} \rfloor \mathbf{I}_3^{-1}) \right) + (\mathbf{p} \wedge \mathbf{u}) \rfloor (\mathbf{v} \rfloor \mathbf{I}_3^{-1})$$
$$= (\mathbf{u} \wedge \mathbf{q} \wedge \mathbf{v}) \rfloor \mathbf{I}_3^{-1} + (\mathbf{p} \wedge \mathbf{u} \wedge \mathbf{v}) \rfloor \mathbf{I}_3^{-1}$$
$$= \left((\mathbf{p} - \mathbf{q}) \wedge \mathbf{u} \wedge \mathbf{v} \right) \rfloor \mathbf{I}_3^{-1}$$

12. Take two lines $L_1 = p_1 \wedge \mathbf{a}_1$ and $L_2 = p_2 \wedge \mathbf{a}_2$ that satisfy $\mathbf{a}_1 \wedge \mathbf{a}_2 \neq 0$ but $(p_1 \wedge \mathbf{a}_1) \wedge$ $(p_2 \wedge \mathbf{a}_2) = 0$. These lines are not skew, but intersect in some plane in space. Their join is now proportional to the common plane $\mathbf{a}_2 \wedge p_2 \wedge \mathbf{a}_1 = p_2 \wedge \mathbf{a}_1 \wedge \mathbf{a}_2$. Let the unit blade representing this be I. Show that their meet is now:

$$\mathbf{J} = I \quad \text{(proportional to } p_2 \wedge \mathbf{a}_1 \wedge \mathbf{a}_2) \qquad (11.20)$$
$$\mathbf{M} = e_0 (\mathbf{a}_1 \wedge \mathbf{a}_2)^\star + \mathbf{a}_1 (p_2 \wedge \mathbf{a}_2)^\star - \mathbf{a}_2 (p_1 \wedge \mathbf{a}_1)^\star + \tfrac{1}{2}((p_1 + p_2) \wedge \mathbf{a}_1 \wedge \mathbf{a}_2)/I_2,$$

with duality relative to I_2, the unit blade of $\mathbf{a}_1 \wedge \mathbf{a}_2$ (i.e., as $\mathbf{X} \rfloor \mathbf{I}_2^{-1}$ whether or not $\mathbf{X}$ is in I_2). Interpret the result, and show its geometric meaning by relating it to the special case of (11.8) (where the common plane passed through the origin). In that case, the inner products could be replaced by geometric products, and you may recognize Cramer's rule (since the ratios of areas in the same plane are ratios of 2-D determinants). This $\{\mathbf{a}_1, \mathbf{a}_2\}$-basis is a natural coordinate system for this problem in the Euclidean plane. In 3-D, this is extended by the last term, which is perpendicular to the common plane (for it is in fact a rejection).

The proportionality factor $\alpha = (\mathbf{a}_1 \wedge \mathbf{a}_2)^\star$ of e_0 in the expression for $\mathbf{M}$ is the weight of the point at $\mathbf{d}$. It indicates the numerical stability of the intersection, and may be used as a measure of the significance of the interpretation as an intersection point.

13. In Section 11.7.2, we stated, "In a plane with counterclockwise orientation, the positive side of the line is on your left when you look along its direction." Convince yourself that this statement gives the same positive side independent of whether you look at the plane from above or below, so that it is a truly geometrically invariant definition. That is good, for it would be useless otherwise.

14. Translate the representational space $e_0 I_n$, both in its direct representation and in its dual representation. Do the algebraic results match your geometric expectation?

15. Show explicitly that the determinants of the translation formulas (11.13) and (11.14) for a flat and for a dual flat both equal 1.

16. Show explicitly that translation is an outermorphism; that is, $T_t[X] \wedge T_t[Y] = T_t[X \wedge Y]$. (Hint: You will find that this holds because of the sign changes in the distributive formula for the contraction (3.10).)

17. There is a way to patch up the homogeneous model so that translation becomes representable in a versor-like form, and you may find this used in the somewhat older literature (such as [4]). It uses a different metric in which $e_0 \cdot e_0 = 0$, and represents a point at location $\mathbf{x}$ as $x = 1 + \mathbf{x} I_4$, where $I_4 = e_0 I_3$ is the pseudoscalar of the homogeneous representation space. Show that in this approach, the element $T = (1 + \mathbf{t} I_4/2)$ acts more or less as a translation versor on points, in the sense that $T \mathbf{x} T$ is the translated point (note the absence of the reversion!).

However, the representation of the higher-grade objects (such as lines and planes) in such a model is *ad hoc*, in that various objects are not related to each other by an outer product-like spanning operation, or a meet-like product for intersection. Work out some cases and show that the versor form of the translation is not perfect: some objects should be translated as $TX\widetilde{T}$, others as TXT, whereas one would have hoped that such fundamental operations would be independent of their argument.

Since the invention of the conformal model of Chapter 13 (which fixes all these defects by using null vectors in a different manner), this *motor algebra* has fallen into disuse, and we mention it here only for completeness.

18. Setting the cross ratio defined in (11.11) to λ, show that by permutations of the points p, q, r, s you can obtain any of λ, $1/\lambda$, $(1 - \lambda)$, $(1 - 1/\lambda)$, $1/(1 - \lambda)$ or $1/(1 - 1/\lambda))$. These are permutation symmetries that you should take into account when checking the similar situation of four unlabeled points (or lines) before and after projective transformations. You can make a symmetrical function of them all as a test for equality of situations: $(\lambda^2 - \lambda + 1)^3/(\lambda(\lambda - 1))^2$; see [55].

19. You are to draw a sequence of equidistant telegraph poles along a straight road in a picture showing the landscape seen in a bird's eye view, with the horizon 6 cm from the first pole and the separation between first and second pole 1 cm (see Figure 11.13). Compute where the third pole should be. Extend this to computing the location of the kth pole. (Hint: Compute the cross ratio of the first two poles to the point at infinity in a "straight" photograph. Then realize that the cross ratio is a projective invariant.)

20. In the homogeneous coordinate approach, an affine transformation A is represented by a matrix acting on the homogeneous coordinate representation of a vector

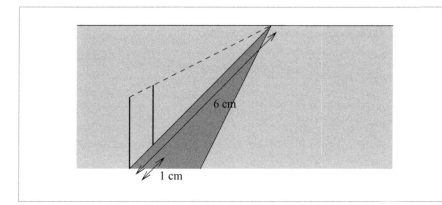

Figure 11.13: The beginning of a line of equidistant telegraph poles. Structural exercise 19 asks you to draw the rest.

$[\![x_1, x_2, x_3, 1]\!]^T$. Interpret the defining condition $\overline{\mathsf{A}}[e_0] = e_0$ of (11.15) in terms of a property of such an affine matrix. Confirm your answer in Table 12.2.

21. Redo some of the orthogonal projection examples in the vector space model of a 3-D Euclidean space using the meet interpretation of (11.17).

11.13 PROGRAMMING EXAMPLES AND EXERCISES

In the GA sandbox source code package, the implementation of the homogeneous model of 3-D Euclidean space is called h3ga. The basis vectors are e1, e2, e3, and e0. The metric is Euclidean (i.e., $\mathbf{e}_1 \cdot \mathbf{e}_1 = \mathbf{e}_2 \cdot \mathbf{e}_2 = \mathbf{e}_3 \cdot \mathbf{e}_3 = e_0 \cdot e_0 = 1$). Various specialized multivector types of blades are available, as dictated by our interpretation of the algebra (see Table 11.4).

Note in this table that bivector and lineAtInfinity are basically the same types; only their names differ. Gaigen 2 has no problem with this.

An outermorphism class (om) is also available. The specialized outermorphism class omPoint can be applied to points only; it is the equivalent of a classic 4 × 4 homogeneous matrix. This is demonstrated in Section 12.5.1.

11.13.1 WORKING WITH POINTS

Our first foray into programming with the homogeneous model is an experiment with drawing and translating points. Two types of points are provided in h3ga: the point class and the normalizedPoint class. The e0 coordinate of a normalizedPoint is always 1, so only normalized points can be stored in it. normalizedPoint variables do not store their

Table 11.4: Specialized multivector types in the h3ga.

Name	Sum of Basis Blades
vector	e1, e2, e3
point	e1, e2, e3, e0
normalizedPoint	e1, e2, e3, e0=1
line	e1∧e2, e2∧e3, e3∧e1, e1∧e0, e2∧e0, e3∧e0
lineAtInfinity	e1∧e2, e2∧e3, e3∧e1
bivector	e1∧e2, e2∧e3, e3∧e1
plane	e1∧e2∧e3, e1∧e2∧e0, e2∧e3∧e0, e3∧e1∧e0
planeAtInfinity	e1∧e2∧e3
rotor	scalar, e1∧e2, e2∧e3, e3∧e1

e0 coordinate; this saves one coordinate and is also more efficient during computations. So a point has 4 coordinates, a normalizedPoint only 3.

Since OpenGL is internally based on homogeneous coordinates, the homogeneous model is in many aspects a perfect match for it. Accordingly, both types of points can easily be used as vertices in OpenGL, as shown below.

The example code draws two triangles. You can manipulate the vertices of the triangles with the mouse. One triangle is made up from points, the other from normalized-Points. The code to draw the triangles with points is

```
glBegin(GL_LINE_LOOP);
for (int i = 0; i < NB_POINTS; i++) {
  const point &P = g_points[i];
  glVertex4fv(P.getC(point_e1_e2_e3_e0));
  // or:
  // glVertex4f(P.e1(), P.e2(), P.e3(), P.e0());
}
glEnd();
```

Note that we use glVertex4fv() to pass an array containing all 4 coordinates in OpenGL. These are the e_1, e_2, e_3, and e_0 coordinates, or—in OpenGL terminology—the x, y, z, and w coordinates. Alternatively, we could use glVertex4f() and pass each coordinate separately, as shown in the line that is commented out.

To load normalizedPoints in OpenGL, glVertex3fv() is used:

```
glBegin(GL_LINE_LOOP);
for (int i = 0; i < NB_NORMALIZED_POINTS; i++) {
  const normalizedPoint &P = g_normalizedPoints[i];
  glVertex3fv(P.getC(normalizedPoint_e1_e2_e3_e0f1_0));
  //note e0 'fixed' at 1.0
  // or:
  // glVertex3f(P.e1(), P.e2(), P.e3());
}
glEnd();
```

Since both the normalizedPoint type and OpenGL assume that the e_0 coordinate is 1, this works as expected.

Translation of the points is implemented using the generic translation formula of (11.13).

```
g_points[idx]=_point(g_points[idx]+(T ^ (e0 << g_points[idx])));
```

where T is the translation vector. The possible weight of the points is taken into account automatically. In the case of normalizedPoints, the left contraction (e0 << g_points[g_dragPoint]) is evaluated at compile time, and has no run-time cost. That is, the following code

```
g_normalizedPoints[idx] =
  _normalizedPoint(g_normalizedPoints[idx] +
  (T ^ (e0 << g_normalizedPoints[idx])));
```

is equivalent in performance to:

```
g_normalizedPoints[idx] = g_normalizedPoints[g_dragPoint] + T;
```

but it is more generic since it conforms to the general pattern. Geometric algebra has enough structure to permit automated generation of implementations. Having such a code generator permits you to introduce small optimizations like the normalized points very easily. We discuss these techniques in Part III.

11.13.2 INTERSECTING PRIMITIVES

This example allows you

1. To drag points;
2. To create new points;
3. To combine points to form lines and planes.

The intersections of the lines and planes are computed and drawn. Figure 11.14 shows a screenshot.

The points are kept in a global array `g_points`. Give two indices (`pointIdx1` and `pointIdx2`), a line is computed from two points as

```
line L = _line(g_points[pointIdx1] ^ g_points[pointIdx2]);
```

The code for computing a plane through three points is very similar:

```
plane P = _plane(g_points[pointIdx1] ^ g_points[pointIdx2] ^
    g_points[pointIdx3]);
```

The intersections of the lines and planes are computed in a very generic fashion. In principle, the code for computing the intersections spells out the meet:

```
// compute 'pseudoscalar' of the space spanned by P1 and P2
mv I = unit_e(join(P1, P2));

// compute P1* . P2
mv intersection = (P1 << inverse(I)) << P2;
```

Here, P1 and P2 are two primitives (lines or planes). P1 and P2 have been initialized such that the grade of P1 is lower than or equal to the grade of P2.

However, this basic code would make it very hard for the user to create degenerate situations such as two identical lines, a plane and a line contained in it, or a plane contained in a plane. Mouse manipulation will not succeed to bring such situations about. That is why we use a trick to make the primitives snap to each other. We project P1 onto P2. If the projection of P1 is close enough to P1 itself, we set P1 equal to its projection:

```
mv projection = (P1 << inverse(P2)) << P2;

// check if projection of P1 onto P2 is 'close' to P1
const float CLOSE = 0.02f;
if (_Float(norm_e(projection - P1)) < CLOSE) {
```

```
    // yes:
    P1 = projection;
}
```

Note that this construction will (unfortunately) not snap skew lines to make them intersect.

Creating Points

The example also allows the user to create points by clicking at any point in the viewport. The user expects a point to be created under the mouse cursor, but the distance from the viewport is in principle unspecified; to keep the user interface simple, we assume that using the distance of the previously dragged point is a good choice.

To implement this, we first create a new point at the required distance from the camera:

```
// create a new point at 'distance' from camera
normalizedPoint pt = _normalizedPoint(vectorAtDepth(distance,
    mousePos) - e3 * distance);
```

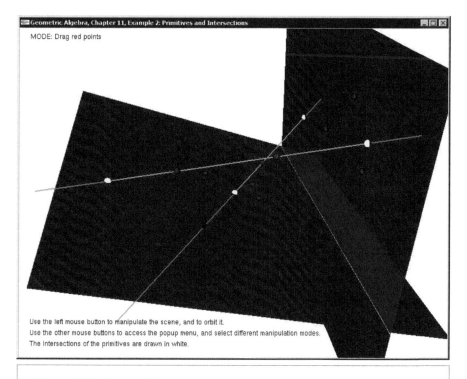

Figure 11.14: Example 2 in action.

The function `vectorAtDepth()` returns a the 2-D vector `mousePos` scaled such that the projection (at `distance`) is equal `mousePos`. Subtracting `e3 * distance` then gives the required point. But this point is still in the camera (modelview) frame.

To transform the point to the global frame, we retrieve the OpenGL modelview matrix, invert it, and apply it to the point as an outermorphism:

```
// retrieve modelview matrix from OpenGL:
float modelviewMatrix[16];
glGetFloatv(GL_MODELVIEW_MATRIX, modelviewMatrix);

// invert matrix
float inverseModelviewMatrix[16];
invert4x4Matrix(modelviewMatrix, inverseModelviewMatrix);

// use matrix to initialize an outermorphism; apply it to the point
omPoint M(inverseModelviewMatrix);
pt = apply_om(M, pt);
```

The point is then added to the global list of points. We will make more use of matrices and outermorphisms in Sections 12.5.1 and 12.5.2.

11.13.3 DON'T ADD LINES

As described in Section 11.3.3, the sum of two homogeneous lines is another line only when they have a factor in common. It then is the weighted average of the two lines. Otherwise, their sum is not a blade, and hence not geometrically interpretable. The following example demonstrates this.

You can drag around the four points that span the lines. The example computes the two lines and draws them. It also adds the lines, and tries to draw the result:

```
// compute the lines
line L1 = _line(pt[0] ^ pt[1]);
line L2 = _line(pt[2] ^ pt[3]);
line L1plusL2 = _line(L1 + L2);
// ...
draw(L1);
draw(L2);
// ...
draw(L1plusL2);
```

`L1plusL2` will show up only when

- `L1` and `L2` share a point;
- `L1` and `L2` have the same direction.

To facilitate creating such situations, the example makes the points snap to each other when they get within close range. The example also adjust the directions when the lines are near-parallel, such that the lines become truly parallel.

11.13.4 PERSPECTIVE PROJECTION

This example projects 3-D models onto an image plane, as shown in Figure 11.15. The camera and the image plane are determined by four points:

```
normalizedPoint cameraPoint = g_points[CAMERA_PT_IDX];

plane imagePlane = _plane(
  g_points[IMAGE_PLANE_PT_IDX + 0] ^
  g_points[IMAGE_PLANE_PT_IDX + 1] ^
  g_points[IMAGE_PLANE_PT_IDX + 2]
);
```

The points are stored in a global array g_points. They can be dragged around using the mouse. The projection is performed by constructing the line vertex ∧ cameraPoint and intersecting it with the imagePlane:

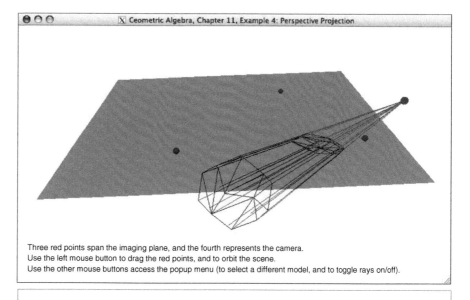

Three red points span the imaging plane, and the fourth represents the camera.
Use the left mouse button to drag the red points, and to orbit the scene.
Use the other mouse buttons access the popup menu (to select a different model, and to toggle rays on/off).

Figure 11.15: Perspective projection (Example 4). The dodecahedron gets projected onto the blue image plane. This is done by constructing rays (lines) between the vertices of the dodecahedron and the camera (the red point on the right). The image plane is determined by the other three points. The meet of rays and plane produces the intersection points.

```
const normalizedPoint &vertex = g_vertices3D[g_polygons3D[i][j]];
// compute projected vertex
point pv = _point(dual(vertex ^ cameraPoint) << imagePlane);
```

We perform back-face culling (see also Section 2.13.2), by comparing the orientation of the image plane with the orientation of the plane spanned by the projected vertices:

```
// array PV contains the projected vertices
mv::Float ori = _Float((PV[0] ^ PV[1] ^ PV[2]) *
  inverse (imagePlane));
if (ori > 0.0f) {
  // draw the polygon
  // ...
}
```

Note that when you reverse orientation of the image plane (by reversing the order of the points that define it), you get the opposite definition of which faces are the back-faces. In that case, you will see the backside of the model projected onto the plane.

(Disclaimer: It is of course faster to use a matrix representation of the projection outer-morphism, but this example is intended as an enlightening exercise on an alternative geo-metrical manner to produce the same result.)

Exercise 4a: Orthogonal Projection

Adjust the example to use orthogonal projection rather than perspective projection. Note that such an orthogonal projection is fully specified by the image plane, so that the camera position is irrelevant. The solution to this exercise is not simply to use the orthogonal pro-jection formula $(\mathbf{A}\rfloor\mathbf{B}^{-1})\rfloor\mathbf{B}$, due to the metric imperfections of the homogeneous model discussed in Section 11.10.

12 APPLICATIONS OF THE HOMOGENEOUS MODEL

The homogeneous model is well suited to applications in which incidences of offset flat subspaces are central, but less so when metric properties are also important. In this, it plays a role similar to Grassmann-Cayley algebra. This chapter gives details on its direct use in applications.

In the first part, we discuss the coordinate representations of the homogeneous elements. This naturally embeds the powerful Plücker coordinates for line computations, which are seen to be are a natural extension of homogeneous point coordinates (using the outer product). We show how you can master them and derive new application formulas simply. Those coordinates for lines, as well as points and planes, permit compact formulation of the affine transformations as matrices.

In the second part, we give an advanced application of the homogeneous model. It is well suited to encode the projective geometry involved in the imaging of the world by one or more pinhole cameras. The homogeneous model permits easy specification of how observations in the various cameras are connected, which permits using observations by one camera to guide the search for corresponding features in another camera. We treat the fairly advanced subjects of stereo-vision based on point and line matches in 2 or 3 cameras.

Both subjects of this chapter are incidental to the main flow of the book, though the section on Plücker coordinates prepares you for the implementation of geometric algebra in Part III.

12.1 HOMOGENEOUS PLÜCKER COORDINATES IN 3-D

In the previous chapter, we saw how the homogeneous model permits direct and dual representation of flats. This extends the commonly used homogeneous coordinates for points and hyperplanes used throughout computer graphics and robotics literature. In some more specialized texts, you may also find a representation of lines by *Plücker coordinates*. These are coordinates tailored to the description of lines, and they permit direct computation with lines as basic elements. That provides for faster and simpler code than if you described lines by a direction vector and a position vector, or as two plane equations, or by two points. Repeated attempts have been made to introduce them into mainstream computer graphics, where line computations are obviously important, but with limited success.

It is clear why Plücker coordinates are less well known than they deserve to be. In the usual texts, they are presented as a strange mathematical trick, not clearly related to the homogeneous coordinates. They require 6-D vectors and corresponding matrices, which appear extraneous to the usual 4-D data structures in homogeneous coordinate software. So most people continue to encode lines as composite elements in a data structure consisting of two vectors. This denies lines the status of being convenient elements of computation, and that in turn affects solutions to practical problems; if there is a natural way of describing and solving some problem using lines, it is less likely to be found. An example is visual self-localization by a robot in an office environment; a typical solution looks to match point features, neglecting the numerous straight lines usually present in such an environment. Reducing everything to point-based or plane-based computations may be suboptimal and neglects the perfectly useful and stably measurable straight edges. But to actually use them in your estimation processes, you need the ability to compute with them.

In this chapter, we show how to look at the Plücker coordinates in a manner that makes them natural and convenient to use. There are no new concepts here; this is an application of the structure of the previous chapter, but is good to relate the algebra to the specific coordinate techniques that the proponents of Plücker coordinates (such as [57]) suggest.

12.1.1 LINE REPRESENTATION

Let us again look at the representation of lines, and focus specifically on a 3-D base space. The line representation $p \wedge q$ of (11.3) as

$$p \wedge q = e_0 \wedge (\mathbf{q} - \mathbf{p}) + \mathbf{p} \wedge \mathbf{q} \qquad (12.1)$$

involves six coordinates: three for the 2-blade $e_0 \wedge (\mathbf{q} - \mathbf{p})$ and three for the 2-blade $\mathbf{p} \wedge \mathbf{q}$. There is one dependency relationship (since $e_0 \wedge (\mathbf{q} - \mathbf{p}) \wedge (\mathbf{p} \wedge \mathbf{q}) = 0$), and that reduces the degrees of freedom of the line to five. Geometrically, you can interpret the first term of (12.1) as the direction vector and the second term as the moment. We have denoted those in Figure 12.1. This figure cannot show the fourth dimension e_0; with that extra

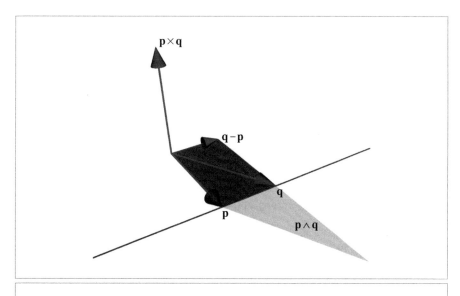

Figure 12.1: Plücker coordinates of a line in 3-D.

dimension it should look more like Figure 11.2(a), which is its lower-dimensional version for lines in 2-D.

We can group the six coordinates of a line in 3-D as two vectors of three components, by employing the cross product:

$$p \wedge q = e_0 \wedge (\mathbf{q} - \mathbf{p}) + \mathbf{p} \wedge \mathbf{q} = (\mathbf{p} - \mathbf{q})\,e_0 + (\mathbf{p} \times \mathbf{q})\,\mathsf{I}_3. \tag{12.2}$$

We recognize (minus) the direction vector and the dual of the moment 2-blade (which is the properly weighted normal vector of the plane through the line and the origin). Together, these form the *Plücker coordinates* of the 3-D line, denoted by

$$-\{\mathbf{p} - \mathbf{q}; \mathbf{p} \times \mathbf{q}\} \tag{12.3}$$

(see [57]). Classically, these are often treated as six slots for storing numbers, without much algebraic structure, and manipulations with these numbers may seem rather arbitrary, especially when combined with the homogeneous coordinate representations of points and planes. Table 12.1 gives some examples. The operations are simple and fast and that explains the interest of the Plücker coordinates, but the structure between these "magic formulas" is lost, so that implementations are hard to extend to cases not covered in such tables (or to higher dimensions). The usual nonalgebraic encoding of the type of element represented by different kinds of brackets around the coordinates does little to clarify the structure of the table.

Table 12.1: Common Plücker coordinate computations with a line $\{\mathbf{a}, \mathbf{m}\}$, a plane $[\mathbf{n}, \delta]$, and a point $(\mathbf{p} : 1)$. The type of bracket denotes the type of object nonalgebraically. (Expanded from [57].) Note that $\mathbf{a}$ is minus the usual direction vector.

$(\mathbf{p} : 1)$	Point at location $\mathbf{p}$
$[\mathbf{n}, -\delta]$	Plane with normal $\mathbf{n}$ and offset δ
$\mathbf{n} \cdot \mathbf{p} - \delta$	Distance from point to plane
$\{\mathbf{a}, \mathbf{m}\} = \{\mathbf{p} - \mathbf{q}, \mathbf{p} \times \mathbf{q}\}$	Line through two points, from Q to P
$(\mathbf{m} \cdot \mathbf{m})/(\mathbf{a} \cdot \mathbf{a})$	Squared distance of line to origin
$(\mathbf{m} \times \mathbf{a} : \mathbf{a} \cdot \mathbf{a})$	Point on line closest to origin
$[\mathbf{a} \times \mathbf{m} : \mathbf{m} \cdot \mathbf{m}]$	Plane through line, perpendicular to origin plane
$(\mathbf{m} \times \mathbf{n} + \delta \mathbf{a} : \mathbf{a} \cdot \mathbf{n})$	Point in line and plane
$[\mathbf{a} \times \mathbf{p} - \mathbf{m} : \mathbf{m} \cdot \mathbf{p}]$	Plane through line and point
$[\mathbf{a} \times \mathbf{n} : \mathbf{m} \cdot \mathbf{n}]$	Plane through line, additional direction $\mathbf{n}$

In the geometric algebra point of view, we see that the six numbers are coefficients on a specific basis, in (12.2) demonstrated by the extra symbols e_0 and $\mathbf{I}_3$. This gives the full algebraic meaning to the Plücker coordinates, and endows them automatically with relationships to the other elements through their membership in geometric algebra. Their actual construction in (12.2) is a clear instance of this: by taking the outer product of two homogeneous points, we get the line representation, providing the straightforward connection between point coordinates and line coordinates through the standard operation of outer product. The same standard operation should connect the table entries for point, line, and plane through line and point. This obvious geometrical fact is not at all clear from the corresponding entries in Table 12.1. Only by translating the bracket notation back into its actual geometric basis and applying the standard constructions of geometric algebra can we understand and extend such tables.

12.1.2 THE ELEMENTS IN COORDINATE FORM

In the homogeneous model of 3-D space, the representation of the quantities in terms of coordinates is easily computed. We give a complete inventory of the repertoire of geometric elements in the homogeneous model. We study the coordinate representation of the offset flat subspaces of various grades (points, lines, and planes) in both their direct and in their dual representations.

1. **Point**. We have seen that the 1-D homogeneous subspace representing a 0-D point in the homogeneous model is of the form $e_0 + \mathbf{p}$ (or a multiple). In a homogeneous model of 3-space with basis $\{\mathbf{e}_1, \mathbf{e}_2, \mathbf{e}_3, e_0\}$, it therefore has the coordinates

$(p_1, p_2, p_3, 1)$ (or a multiple), with (p_1, p_2, p_3) the Euclidean coordinates of its position vector. In Table 12.1 this is written as $(\mathbf{p} : 1)$, following [57].

2. **Hyperplane.** If we have a plane characterized as all points $\mathbf{x}$ such that $\mathbf{x} \cdot \mathbf{n} = \delta$, this can be written as $(e_0 + \mathbf{x}) \cdot (-\mathbf{n} + \delta e_0^{-1}) = 0$, so as $x \cdot (-\mathbf{n} + \delta e_0^{-1}) = 0$. Therefore $-\mathbf{n} + \delta e_0^{-1}$ (or any multiple of it) is the dual of the blade representing the hyperplane. It is represented on a basis $\{\mathbf{e}_1, \mathbf{e}_2, \mathbf{e}_3, e_0^{-1}\}$, which appears different from the basis for points until you realize that $e_0^{-1} = \pm e_0$, depending on the metric chosen. Therefore the bases are the same, and you cannot tell these elements apart as vectors.

If $-\mathbf{n} + \delta e_0^{-1}$ is the dual of a blade, then the blade itself must be

$$(-\mathbf{n} + \delta e_0^{-1})\,(e_0\,I_3) = -\mathbf{n}e_0 I_3 + \delta I_3$$

(where we take the undualization in the $(n + 1)$-dimensional homogeneous model, so relative to the pseudoscalar $e_0\,I_3$). This has the Plücker coefficients $[n_1, n_2, n_3 : -\delta]$ used in Table 12.1, but on the trivector basis

$$\{\mathbf{e}_2\,\mathbf{e}_3\,e_0,\ \mathbf{e}_3\,\mathbf{e}_1\,e_0,\ \mathbf{e}_1\,\mathbf{e}_2\,e_0,\ -\,\mathbf{e}_1\,\mathbf{e}_2\,\mathbf{e}_3\},$$

which is clearly different from the vector basis for points. In standard literature on Plücker coordinates, this difference in basis is denoted nonalgebraically by square brackets: $[\mathbf{n} : -\delta]$. Others consider planes as row vectors, and points as column vectors, so then their brackets could be interpreted as specifying this difference.

If you want a consistent representation of the direct and dual-oriented planes, you should be very explicit in your choice of basis. The dual plane also has a representation $[\mathbf{n} : -\delta]$, but the basis is $\{-\mathbf{e}_1, -\mathbf{e}_2, -\mathbf{e}_3, -e_0^{-1} = \mp e_0\}$ (the sign of the last element related to the sign in $e_0^{-1} = \pm e_0$). Such headaches are the consequences of splitting up a single geometric element (the dual plane) in its coordinates and basis, each of which separately have no objective geometric meaning.

3. **Lines.** We have seen above in (12.2) that a line is represented by minus the usual direction vector $\mathbf{a} = \mathbf{p} - \mathbf{q}$ and a moment vector $\mathbf{m} = \mathbf{p} \times \mathbf{q} = -\mathbf{p} \times \mathbf{a}$ as

$$L = \mathbf{a}\,e_0 + \mathbf{m}\,I_3 \qquad\qquad (12.4)$$

The six coefficients $\{a_1, a_2, a_3, m_1, m_2, m_3\}$ of this line are on the bivector basis

$$\{\mathbf{e}_1\,e_0,\ \mathbf{e}_2\,e_0,\ \mathbf{e}_3\,e_0,\ \mathbf{e}_2\,\mathbf{e}_3,\ \mathbf{e}_3\,\mathbf{e}_1,\ \mathbf{e}_1\,\mathbf{e}_2\}. \qquad\qquad (12.5)$$

In Table 12.1, we follow [57] in using curly brackets for this data type. In derivations with lines, we obviously need to revert to the fully expressed form.

The dual representation of such a line is

$$L^* = L\,\mathbf{I}_3^{-1}\,e_0^{-1} = \mathbf{a}\,\mathbf{I}_3 + \mathbf{m}\,e_0^{-1}, \qquad (12.6)$$

so it has the coefficients $\{\pm m_1, \pm m_2, \pm m_3, a_1, a_2, a_3\}$ on the bivector basis, the $\pm$ signs are again determined by the metric chosen. Such a dual line is not in the usual Plücker coordinate tables, though we will soon see that it is convenient to have.

It should be clear that a mere coefficient-based representation is dangerous and needs to be accompanied by some bookkeeping, at the very least by maintaining separate data structures (and this is presumably what the brackets are intended to remind us of). We would be in favor of letting that bookkeeping be done by the algebraic tags on the coefficients; it is unambiguous, nonarbitrary, and computational. It is not necessarily more expensive, for the multiplications of the basis blades used as tags are essentially Boolean operations. We will expand on this in Chapter 19 for general geometric algebras.

12.1.3 COMBINING ELEMENTS

In interactions of elements, the explicit basis specification takes care of the computational rules for the Plücker formula table. We give some examples to explain entries in the table, and some that go beyond it, to show how the algebra empowers you to extend it if your application so requires.

1. **Plane through Point and Line.** The formula for augmenting a point p by a line L to make a plane is of course simply the outer product of point and the direct line representation. Rewriting the result in terms of a 3-D cross product, we get

$$p \wedge L = (e_0 + \mathbf{p}) \wedge (\mathbf{a}\,e_0 + \mathbf{m}\,\mathbf{I}_3) = e_0 \wedge (\mathbf{m}\,\mathbf{I}_3 + \mathbf{p} \wedge \mathbf{a}) + \mathbf{p} \wedge (\mathbf{m}\,\mathbf{I}_3)$$
$$= -(\mathbf{m} + \mathbf{p} \times \mathbf{a})\,e_0\,\mathbf{I}_3 + (\mathbf{p} \cdot \mathbf{m})\,\mathbf{I}_3.$$

   So this is indeed the plane $[\mathbf{a} \times \mathbf{p} - \mathbf{m} : \mathbf{m} \cdot \mathbf{p}]$ listed in Table 12.1. That this should be the natural extension of the construction of a line from two points is hard to guess from the Plücker table—yet it is a straightforward computation when one remembers their embedding in the full algebra.

2. **Point and Plane.** To compute the `meet` of a hyperplane Π and a point p in general position, we need to compute by (5.8)

$$\Pi \cap p = -p \cap \Pi = -\Pi^* \rfloor p = (\mathbf{n} - \delta e_0^{-1}) \rfloor (e_0 + \mathbf{p}) = \mathbf{n} \cdot \mathbf{p} - \delta.$$

   We recognize a *scalar* outcome, of which the geometry is the oriented distance between point and hyperplane. This imbues scalars (grade 0 blades) with a geometrical meaning: they are distances, and distances are apparently legitimate outcomes of a coincidence operation.

   This is of course the standard formula in the homogeneous model (calling this a Plücker coordinate formula is overdoing it), but is nice to see it as a standard `meet` rather than a special (albeit trivial) construct.

3. **Line and Plane**. The `meet` of a plane dually represented as $\Pi^* = -\mathbf{n} + \delta e_0^{-1}$ and a line L, in general position, is computed as

$$-\Pi \cap L = -L \cap \Pi = -\Pi^* \rfloor L = (\mathbf{n} - \delta e_0^{-1}) \rfloor (\mathbf{a}e_0 + \mathbf{m}\mathbf{I}_3)$$

$$= (\mathbf{n} \cdot \mathbf{a})e_0 + \mathbf{n} \rfloor (\mathbf{m}\mathbf{I}_3) + \delta \mathbf{a} + 0 = (\mathbf{n} \cdot \mathbf{a})e_0 + (\mathbf{n} \wedge \mathbf{m})\mathbf{I}_3 + \delta \mathbf{a}$$

$$= (\mathbf{n} \cdot \mathbf{a})e_0 + (\mathbf{m} \times \mathbf{n} + \delta \mathbf{a})$$

Note how the orthogonality relationships between the basis elements automatically kill the potential term involving δ and $\mathbf{m}$.

This is the correct result, representing a point at the location $(\mathbf{m} \times \mathbf{n} + \delta \mathbf{a})/(\mathbf{n} \cdot \mathbf{a})$ in its homogeneous coordinates. It corresponds to the result in Table 12.1.

4. **Skew Lines**. The `meet` of two lines L_1 and L_2 in general position (skew) is a measure of their signed distance. We have seen this in algebraic form in (11.9). Here we redo the computation compactly using the specific Plücker coordinates with their algebraic tags:

$$L_1 \cap L_2 = L_2^* \rfloor L_1 = (\mathbf{a}_2 \mathbf{I}_3 + \mathbf{m}_2 e_0^{-1}) \rfloor (\mathbf{a}_1 e_0 + \mathbf{m}_1 \mathbf{I}_3) = -\mathbf{m}_1 \cdot \mathbf{a}_2 - \mathbf{m}_2 \cdot \mathbf{a}_1.$$

Note that we used the dual line representation for one of the lines to derive this compact expression quickly. Again, the basis orthogonality relationships have prevented terms containing $\mathbf{m}_1 \cdot \mathbf{m}_2$ or $\mathbf{a}_1 \cdot \mathbf{a}_2$ from appearing (as they should!). If you had any problems evaluating the expression $(\mathbf{a}_2 \mathbf{I}_3) \rfloor (\mathbf{m}_1 \mathbf{I}_3)$, use duality (twice) and the commutation of the outer product of a vector and a bivector: $(\mathbf{a}_2 \mathbf{I}_3) \rfloor (\mathbf{m}_1 \mathbf{I}_3) = ((\mathbf{a}_2 \mathbf{I}_3) \wedge \mathbf{m}_1) \mathbf{I}_3 = (\mathbf{m}_1 \wedge (\mathbf{a}_2 \mathbf{I}_3)) \mathbf{I}_3 = (\mathbf{m}_1 \rfloor \mathbf{a}_2) \mathbf{I}_3 \mathbf{I}_3 = -\mathbf{m}_1 \cdot \mathbf{a}_2$.

The scalar result above retrieves the well-known compact Plücker-based method of determining how lines pass each other in space; three tests on the signs of such quantities representing the edges of a triangle determine efficiently whether a ray hits the inside of the triangle.

5. **Intersecting Lines**. When the parameters of the two lines $L_1 = p_1 \wedge \mathbf{a}_1$ and $L_2 = p_2 \wedge \mathbf{a}_2$ satisfy $\mathbf{a}_1 \wedge \mathbf{a}_2 \neq 0$ but $(p_1 \wedge \mathbf{a}_1) \wedge (p_2 \wedge \mathbf{a}_2) = 0$, then lines are not skew but intersect (the latter condition implies $(\mathbf{p}_1 - \mathbf{p}_2) \wedge \mathbf{a}_1 \wedge \mathbf{a}_2 = 0$). The resulting `meet` was given in (11.20); this is easily translated into Plücker format as

$$L_1 \cap L_2 = \|\mathbf{a}_1 \times \mathbf{a}_2\|^2 e_0 + \mathbf{a}_1 \mathbf{m}_2 \cdot (\mathbf{a}_1 \times \mathbf{a}_2) + \mathbf{a}_2 \mathbf{m}_1 \cdot (\mathbf{a}_2 \times \mathbf{a}_1)$$

$$+ \tfrac{1}{2} (\mathbf{a}_2 \cdot \mathbf{m}_1 - \mathbf{a}_1 \cdot \mathbf{m}_2)(\mathbf{a}_1 \times \mathbf{a}_2).$$

This outcome is immediately interpretable as a point; it is not one of the formulas one usually finds in the Plücker tables.

The other relationships of the table can be derived in a similar manner. If you need one that is not in the table, you can now derive it yourself.

12.1.4 MATRICES OF MOTIONS IN PLÜCKER COORDINATES

An affine transformation A transforms a vector $\mathbf{x}$ to

$$A[\mathbf{x}] = f[\mathbf{x}] + \mathbf{t},$$

and it has to satisfy $\overline{A}[e_0] = e_0$ to be affine. We found in (11.16) that a general affine transformation of the base space $\mathbb{R}^n$ transforms a blade in the representation space $\mathbb{R}^{n+1}$ as

$$A[X] = f[X] + \mathbf{t} \wedge (e_0^{-1} \rfloor f[X]),$$

where we conveniently define $f[e_0] = e_0$ so that f can be extended to any blade.

We can now spell out the effects on the different elements of their Plücker bases and write the result again on that basis. If we then represent the elements by Plücker coordinates, the affine transformation is representable as a matrix. By this transference principle, we can easily compute the affine matrices of some common transformations. We explain the derivations, which you can relate to the result given in Table 12.2.

The point transformation is familiar, and a direct consequence of the definition. You can compute its columns simply:

$$A[e_0] = f[e_0] + \mathbf{t} = e_0 + \mathbf{t}, \text{ and } A[e_i] = f[e_i].$$

The line transformation follows from the transformation of the elements of the line basis, and is more involved:

$$
\begin{aligned}
A[\mathbf{e}_i \wedge e_0] = A[\mathbf{e}_i] \wedge A[e_0] &= f[\mathbf{e}_i] \wedge (e_0 + \mathbf{t}) \\
&= f[\mathbf{e}_i \wedge e_0] - \mathbf{t} \wedge f[\mathbf{e}_i] \\
&= f[\mathbf{e}_i \wedge e_0] - (\mathbf{t} \times f[\mathbf{e}_i]) \, \mathbf{I}_3,
\end{aligned}
$$

which in its matrix form involves the cross product matrix $[\![\mathbf{t}^\times]\!]$. The other part of the line transformation follows from results such as

$$A[\mathbf{e}_2 \mathbf{e}_3] = f[\mathbf{e}_2 \mathbf{e}_3] = f[\mathbf{e}_1 \rfloor \mathbf{I}_3] = \overline{f}^{-1}[\mathbf{e}_1] \rfloor f[\mathbf{I}_3] = \det(f) \, \overline{f}^{-1}[\mathbf{e}_1] \rfloor \mathbf{I}_3.$$

The transformations for a dual line have not been specified in the table, but they are easily made. Comparing (12.4) and (12.6), line dualization is achieved by the matrix

$$
\begin{bmatrix}
[\![0]\!] & \pm[\![1]\!] \\
[\![1]\!] & [\![0]\!]
\end{bmatrix},
$$

the sign choice again depending on the metric of the representation space. Using this dualization matrix, any matrix for an operation on a dual line can be obtained from

Table 12.2: Transformation of the flats in the homogeneous model through matrices constructed from the outermorphisms acting on their Plücker coordinates. The same algebraic formula leads to quite different matrices. $[\![\mathbf{t}^\times]\!]$ is the matrix of a cross product, defined through $[\![\mathbf{t}^\times]\!][\![\mathbf{x}]\!] = [\![\mathbf{t} \times \mathbf{x}]\!]$. $[\![1]\!]$ is an identity matrix of appropriate size, $[\![0]\!]$ a zero matrix.

Element	Point	Line	Plane
Basis	$(\mathbf{e}_1, \mathbf{e}_2, \mathbf{e}_3, \mathbf{e}_0)$	$(\mathbf{e}_1\, e_0, \mathbf{e}_2\, e_0, \mathbf{e}_3\, e_0,$ $\mathbf{e}_2\, \mathbf{e}_3, \mathbf{e}_3\, \mathbf{e}_1, \mathbf{e}_1\, \mathbf{e}_2)$	$(\mathbf{e}_2\, \mathbf{e}_3\, e_0, \mathbf{e}_3\, \mathbf{e}_1\, e_0,$ $\mathbf{e}_1\, \mathbf{e}_2\, e_0, -\mathbf{e}_1\, \mathbf{e}_2\, \mathbf{e}_3)$
Affine	$\begin{bmatrix} [\![\mathbf{f}]\!] & [\![\mathbf{t}]\!] \\ [\![0]\!]^T & 1 \end{bmatrix}$	$\begin{bmatrix} [\![\mathbf{f}]\!] & [\![0]\!] \\ -[\![\mathbf{t}^\times]\!][\![\mathbf{f}]\!] & \det(\mathbf{f})\,[\![\mathbf{f}]\!]^{-T} \end{bmatrix}$	$\begin{bmatrix} \det(\mathbf{f})\,[\![\mathbf{f}]\!]^{-T} & [\![0]\!] \\ -[\mathbf{t}]^T \det(\mathbf{f})[\mathbf{f}]^{-T} & \det(\mathbf{f}) \end{bmatrix}$
Translation $\mathbf{t}$	$\begin{bmatrix} [\![1]\!] & [\![\mathbf{t}]\!] \\ [\![0]\!]^T & 1 \end{bmatrix}$	$\begin{bmatrix} [\![1]\!] & [\![0]\!] \\ -[\![\mathbf{t}^\times]\!] & [\![1]\!] \end{bmatrix}$	$\begin{bmatrix} [\![1]\!] & [\![0]\!] \\ -[\![\mathbf{t}]\!]^T & 1 \end{bmatrix}$
Rotation R	$\begin{bmatrix} [\![R]\!] & [\![0]\!] \\ [\![0]\!]^T & 1 \end{bmatrix}$	$\begin{bmatrix} [\![R]\!] & [\![0]\!] \\ [\![0]\!] & [\![R]\!] \end{bmatrix}$	$\begin{bmatrix} [\![R]\!] & [\![0]\!] \\ [\![0]\!]^T & 1 \end{bmatrix}$

the operation A on a direct line by undualization, performing A, and dualization. This transforms a matrix from acting on a direct line to acting on a dual line, according to

$$\begin{bmatrix} [\![A_{11}]\!] & [\![A_{12}]\!] \\ [\![A_{21}]\!] & [\![A_{22}]\!] \end{bmatrix} \mapsto \begin{bmatrix} \pm[\![A_{22}]\!] & [\![A_{21}]\!] \\ [\![A_{12}]\!] & \pm[\![A_{11}]\!] \end{bmatrix}.$$

In the plane transformation matrix acting on elements of the form $[\mathbf{n}, -\delta]$, you may recognize the 3×3 upper left submatrix as the transformation of the normal vector from (4.15). This also occurs in the direct representation of the plane, for the resulting coordinates should undergo the same transformation whether they represent a direct plane or a dual plane. You may derive it in the same manner as in its occurrence for line transformations, above.

12.1.5   SPARSE USAGE OF THE 2^4 DIMENSIONS

The Plücker coordinate representation and its associated matrices begin to show how an implementation of geometric algebra can make use of the sparseness of the geometrically significant structures. In principle, the Clifford algebra of the 4-D representation space has $2^4 = 16$ dimensions, and arbitrary linear transformations on this algebra would therefore require 16×16 matrices, to be applied to a 16×1 vector, for some $2 \times 16^2 = 512$ operations per transformation.

We have seen how the basic elements of 4-D geometry form subspaces of specific grades 1, 4, 6, 4, and 1. A linear transformation of the representation vector space (which is 4-D) leads to smaller matrices on each of those. This is already a reduction from $4,096$ operations to $1^3 + 4^3 + 6^3 + 4^3 + 1^3 = 346$, saving a factor of 12. Of course, one invokes only the transformation for the element at hand rather than transforming the whole ladder of subspaces the worst case among these are the middle grades, which require some 216 operations for a general linear transformation.

Even among those linear transformations on the representational spaces, not all are equal, for the extra dimension and the base space $\mathbb{R}^3$ have different semantics. Therefore, the useful transformations on the base space often have a special form. In the affine matrices above, this is reflected by their sparseness; one block is always zero. This leads to some additional reduction of the computational load.

We will go into the implementational issues in detail in Part III, after we have a more complete view of how geometric algebra performs its Euclidean computations. The matrices above will be found to play a role in fast implementations, and the consistent structure of geometric algebra permits them to be constructed implicitly by an automatic code generator rather than having to be coded by hand.

12.2   IMAGING BY MULTIPLE CAMERAS

When you have multiple cameras observing the geometry of the world, the same points or lines may be visible from several of them. If you know the relative positions and orientations of the cameras, this allows a reconstruction of the 3-D event; alternatively, the observed consistency can be used to estimate the camera parameters.

This geometrical situation is well suited to analysis by the homogeneous model, since it involves general points, lines, and planes. We do so in this section, retrieving classical results in a coordinate-free and highly geometrical manner. (Some indices on symbols are unavoidable, but they refer to the different cameras rather than to coordinates.)

As is usual when treating these issues, we simplify the situation geometrically by considering only ideal pinhole cameras, which perform an ideal central projection.

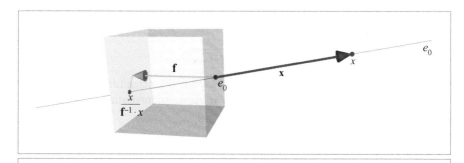

Figure 12.2: A pinhole camera.

12.2.1 THE PINHOLE CAMERA

Let us derive the formulas for the projection in a pinhole camera model using the homogeneous model. We choose to denote the vectors from the pinhole, so the pinhole itself is e_0, the point at the origin.[1]

The imaging plane of the pinhole is in a direction $\mathbf{e}$, at a distance f. We can merge these parameters f and $\mathbf{e}$ by saying that the support vector of the imaging plane is the vector $\mathbf{f} = f\mathbf{e}$. The imaging plane Π itself is then dually denoted by $\Pi^* \equiv \pi = e_0^{-1} - \mathbf{f}^{-1}$. Verify this: the equation $x \cdot \pi = 0$ for a probe point $x = e_0 + \mathbf{x}$ should lead to the Euclidean equation $\mathbf{x} \cdot \mathbf{f} = \mathbf{f} \cdot \mathbf{f}$, showing that the $\mathbf{f}$-component of the vector $\mathbf{x}$ has the correct magnitude to be in the image plane.

If we now take an arbitrary unit point x in 3-D, the light ray from the pinhole to that point is

$$L = e_0 \wedge x = e_0 \wedge \mathbf{x}. \tag{12.7}$$

This defines $\mathbf{x}$ fully as the Euclidean coordinate of x relative to $\mathbf{x}$, since the equation can be solved as $\mathbf{x} = e_0^{-1} (e_0 \wedge x)$. (This solution holds even when x has been given in homogeneous coordinates relative to a different origin; see structural exercise 2.) This shows clearly that the point at x plays the role of the Euclidean direction $\mathbf{x}$ of the ray line. To find the projection formula of the vector $\mathbf{x}$, we need to intersect this line with the imaging plane. This is an instance of the meet formula, and since we have the plane given dually, it is easiest to use the inner product form of the meet:

$$\Pi \cap L = L \cap \Pi = \Pi^* \rfloor L = \pi \rfloor L = (e_0^{-1} - \mathbf{f}^{-1}) \rfloor (e_0 \wedge \mathbf{x}) = e_0 (\mathbf{f}^{-1} \cdot \mathbf{x}) + \mathbf{x}. \tag{12.8}$$

---

1   The usual texts on imaging prevent awkward signs by imagining the imaging plane to be *in front of the pinhole*, at a distance f in the direction denoted by a unit vector $\mathbf{e}$. We do not need to do so in our formulas, since the intrinsic treatment of oriented elements in geometric algebra takes care of the signs automatically. You can take f positive or negative, and the formulas still work to produce the projection onto the corresponding image plane.

This is a point at the 3-D location

$$x' \equiv \mathsf{P}[\mathbf{x}] = \frac{\mathbf{x}}{\mathbf{f}^{-1} \cdot \mathbf{x}}, \tag{12.9}$$

that is precisely the rescaling of $\mathbf{x}$ you would expect: the magnitude guarantees that $\mathbf{x}' \cdot \mathbf{f} = \mathbf{f} \cdot \mathbf{f}$, so that $\mathbf{x}'$ indeed ends on the imaging plane. An interesting way of rewriting this is explored in structural exercise 3.

The mapping that we have produced is nonlinear—already, the test of the linear scaling property (which requires that $\alpha\mathbf{x}$ should be mapped onto $\alpha\mathbf{x}'$) fails. However, if we stay within the homogeneous domain and do not do the rescaling, the mapping *is* linear:

$$\mathbf{x} \mapsto e_0 \, (\mathbf{f}^{-1} \cdot \mathbf{x}) + \mathbf{x}.$$

To get the mapping in terms of points rather than vectors, we substitute $\mathbf{x} = e_0^{-1} \, (e_0 \wedge x)$, and obtain

$$x \mapsto e_0 \left((\mathbf{f}^{-1} - e_0^{-1}) \cdot x \right) + x. \tag{12.10}$$

Alternatively, you can derive this by using $L = e_0 \wedge x$ instead of $L = e_0 \wedge \mathbf{x}$ in (12.8). The matrix of this linear transformation on the basis $\{e_1, e_2, e_3, e_0\}$ is obtained by determining the action on each of the basis vectors. For instance, $e_0 \mapsto e_0 - e_0 = 0$; and $\mathbf{e}_1 \mapsto e_0 \, (\mathbf{f}^{-1} \cdot \mathbf{e}_1) + \mathbf{e}_1$, so that the e_0 coordinate of the image of $\mathbf{e}_1$ is $\mathbf{f}^{-1} \cdot \mathbf{e}_1$. The other components are similar, and the full matrix becomes

$$\begin{bmatrix} [\![1]\!] & [\![0]\!] \\ [\![\mathbf{f}^{-1}]\!]^T & 0 \end{bmatrix},$$

familiar from the classical treatment (although that usually gives the matrix multiplied by $\|\mathbf{f}\|$, with $\mathbf{e}_3$ as axis direction). In implementations, this is what you would use to act on the homogeneous vectors, but the explicit algebraic form is more convenient for symbolic simplifications.

In the classical way of using homogeneous coordinates, the linearity of the mapping is already an advantage, certainly combined with the representation of rigid body motions as linear transformations in the homogeneous coordinate representation. But the geometric algebra treatment goes beyond this: not only is the transformation linear, it is also an outermorphism. Therefore, combinations of the original objects using outer products transform nicely (i.e., covariantly). In particular, spanning behaves well under the central projection. The general projection of a homogeneously represented object X to the imaging plane is simply always the meet of $e_0 \wedge X$ with the image plane Π:

$$X \mapsto (e_0^{-1} - \mathbf{f}^{-1}) \rfloor (e_0 \wedge X) = e_0 \wedge \left((\mathbf{f}^{-1} - e_0^{-1}) \rfloor X \right) + X.$$

You can show, by analogy to the above, that this results in a flat with direction $\mathbf{D} = e_0^{-1} \rfloor (e_0 \wedge (\mathbf{f}^{-1} \rfloor X))$ and support point $(e_0^{-1} \rfloor e_0 \wedge X)/\mathbf{D}$.

Lines can now be projected immediately. On the Plücker line basis of (12.5), the projection is represented by the matrix

$$\left[\!\!\left[\begin{array}{cc} [\![0]\!] & [\![\mathbf{f}^{-1^\times}]\!] \\ [\![0]\!] & [\![1]\!] \end{array}\right]\!\!\right], \tag{12.11}$$

which is derived by techniques similar to our derivation of the matrices in Table 12.2. The projection of a general plane is not very interesting, since all planes project to the image plane by the matrix

$$\left[\!\!\left[\begin{array}{cc} [\![0]\!] & -[\![\mathbf{f}^{-1}]\!] \\ [\![0]\!]^T & [\![1]\!] \end{array}\right]\!\!\right].$$

Of course, in the homogeneous model of geometric algebra, we also have the rigid body motions present as linear transformations and their outermorphisms, so the whole framework now contains everything of interest in a consistent manner.

12.2.2 HOMOGENEOUS COORDINATES AS IMAGING

Two points spanning a line in the world generate a projected line in the image plane spanned by two projected points. It is an amusing (but rather confusing) property of the homogeneous model that homogeneous coordinates can be introduced in the image plane using the support vector $\mathbf{f}$ as the extra homogeneous embedding dimension. Let us consider a camera or focal length equal to 1, and focal vector $\mathbf{f}$. We will denote points in the image plane relative to the optical center (the location $\mathbf{f}$) by an underscore. Then a point at image location $\underline{\mathbf{x}}$ is seen from the pinhole as $\mathbf{f} + \underline{\mathbf{x}}$. It generates a ray of possible real-world positions $e_0 \wedge (\mathbf{f} + \underline{\mathbf{x}})$. The line spanned by two image points could be constructed in two ways, which are indeed consistent:

- The two rays $e_0 \wedge (\mathbf{f}+\underline{\mathbf{x}})$ and $e_0 \wedge (\mathbf{f}+\underline{\mathbf{y}})$ together span a world plane $e_0 \wedge (\mathbf{f}+\underline{\mathbf{x}}) \wedge (\mathbf{f}+\underline{\mathbf{y}})$, and this plane cuts the image plane in the world line

$$(e_0^{-1} - \mathbf{f}^{-1}) \rfloor (e_0 \wedge (\mathbf{f} + \underline{\mathbf{x}}) \wedge (\mathbf{f} + \underline{\mathbf{y}})) =$$
$$= e_0 \wedge (\underline{\mathbf{y}} - \underline{\mathbf{x}}) + (\mathbf{f} + \underline{\mathbf{x}}) \wedge (\mathbf{f} + \underline{\mathbf{y}})$$
$$= (e_0 + \mathbf{f} + \underline{\mathbf{x}}) \wedge (\underline{\mathbf{y}} - \underline{\mathbf{x}}).$$

We see from the final form (compare (11.3)) that this is a line in the direction $(\underline{\mathbf{y}} - \underline{\mathbf{x}})$ passing through the point $\mathbf{f}+\underline{\mathbf{x}}$ of the world (which is the point $\underline{\mathbf{x}}$ on the image plane). This is of course what we would expect.

- Directly taking the outer product of the two points in the image plane gives

$$(\mathbf{f} + \underline{\mathbf{x}}) \wedge (\mathbf{f} + \underline{\mathbf{y}}) = (\mathbf{f} + \underline{\mathbf{x}}) \wedge (\underline{\mathbf{y}} - \underline{\mathbf{x}}).$$

If we are to interpret this as the line through the image points at $\underline{\mathbf{x}}$ and $\underline{\mathbf{y}}$, then clearly $\mathbf{f}$ plays the role of the homogeneous coordinate in (11.3).

The similarity of the two approaches is apparent and generalizes to higher dimensions. Taking a slightly philosophical excursion, working in homogeneous coordinates is like viewing the 3-D world through a pinhole in a 4-D camera. Homogeneous coordinates are a useful version of Plato's cave metaphor, where the 3-D shadows of a 4-D higher reality are what we experience, but in which there is a clear mathematical advantage to use the 4-D "ideal world" as our spatial representation. With homogeneous coordinates, this metaphor can be taken literally as a projection from 4-D to 3-D.

12.2.3 CAMERAS AND STEREO VISION

When working with cameras, one often measures positions in the local coordinate system of the camera. In that case, the use of coordinates is essential for the conversions to objective world coordinates for comparison with other camera measurements. So even though geometric algebra is coordinate-free in a way that linear algebra is not, we still need coordinate transformations for common geometrical tasks.

Let us look at a single camera. Its position and orientation can be characterized relative to a standard position and orientation, at the origin e_0 looking in the fe_3-direction (where f is the focal length). Any general rigid body transformation can be performed as a pure rotation around the origin, followed by a translation. Let us denote the composite transformation by $\mathsf{A}[\cdot]$. In the homogeneous coordinate representation, it is linear and an outermorphism.

To be specific, consider a camera oriented by R_A (so that $\mathsf{R}_A[fe_3] = \mathbf{f}$) and its pinhole at the point $a = e_0 + \mathbf{a}$. This defines A as

$$X \mapsto \mathsf{A}[X] = \mathsf{R}_A[X] + \mathbf{a} \wedge (e_0^{-1} \rfloor \mathsf{R}_A[X]). \tag{12.12}$$

If we measure a point in the image of camera A at location $\underline{\mathbf{x}}$ (which for now we take to be converted to millimeters on the image plane after processing using the internal camera parameters), then this is a vector having the direction $\mathbf{x} = fe_3 + \underline{\mathbf{x}}$ in the local coordinate system. Also in that system, it determines the line $e_0 \wedge \mathbf{x}$ emanating from the pinhole. This line can be converted to world coordinates by applying A, which you may either see as a coordinate transformation or as moving the camera to its actual configuration:

$$\mathsf{A}[e_0 \wedge \mathbf{x}] = \mathsf{A}[e_0] \wedge \mathsf{A}[\mathbf{x}] = (e_0 + \mathbf{a}) \wedge \mathsf{R}_A[\mathbf{x}].$$

Here we applied the outermorphism A to each of the factors in the outer product. We wrote the result on e_0 as a translation characterized by the position vector $\mathbf{a}$ and its effect on the purely directional element $\mathbf{x}$ as a rotation R_A. Both are easy consequences of (12.12).

Introducing a second camera B, we have a similar equation generating a line $\mathsf{B}[e_0 \wedge \mathbf{x}_B]$ from the observation $\mathbf{x}_B$ (or $\underline{\mathbf{x}}_B$) and the rigid body motion B. If those two cameras are looking at the same point x and we know their relative positions and attitudes, then the measurements $\underline{\mathbf{x}}_A$ and $\underline{\mathbf{x}}_B$ (or rather $\mathbf{x}_A$ and $\mathbf{x}_B$) are not geometrically independent. There is a linear constraint relating them, known as the *epipolar constraint*, which arises from the fact that the two 3-D lines we have constructed should intersect in the point x.

For this to happen, the `meet` of the two lines should be degenerate, which happens when their naïve `join` (computed as an outer product) is zero:

$$0 = \mathsf{A}[e_0 \wedge \mathbf{x}_A] \wedge \mathsf{B}[e_0 \wedge \mathbf{x}_B].$$

Because A and B are outermorphisms, we can manipulate this equation to a more familiar form. First distribute the mappings over the outer products:

$$0 = \mathsf{A}[e_0] \wedge \mathsf{A}[\mathbf{x}_A] \wedge \mathsf{B}[e_0] \wedge \mathsf{B}[\mathbf{x}_B] = (e_0 + \mathbf{a}) \wedge \mathsf{R}_A[\mathbf{x}_A] \wedge (e_0 + \mathbf{b}) \wedge \mathsf{R}_B[\mathbf{x}_B].$$

Taking the inner product of this equation with e_0^{-1}, and noting that the second and fourth factor are purely Euclidean (so that they do not contribute), we get

$$0 = \mathsf{R}_A[\mathbf{x}_A] \wedge (e_0 + \mathbf{b}) \wedge \mathsf{R}_B[\mathbf{x}_B] + (e_0 + \mathbf{a}) \wedge \mathsf{R}_A[\mathbf{x}_A] \wedge \mathsf{R}_B[\mathbf{x}_B]$$
$$= \mathsf{R}_A[\mathbf{x}_A] \wedge (\mathbf{b} - \mathbf{a}) \wedge \mathsf{R}_B[\mathbf{x}_B].$$

The meaning of the resulting equation is sketched in Figure 12.3: the observed vectors and the relative camera position are in one plane. (If in the derivation the idea of taking the inner product with e_0^{-1} strikes you as unnatural, you can also write out the above equation in its Euclidean and non-Euclidean components, and focus on the non-Euclidean part being zero—that is, the same. The Euclidean part is then automatically zero, so we have not lost generality.)

The equation above is often rewritten in terms of the relative rotation of B relative to A, which is $\mathsf{R}_A^B \equiv \mathsf{R}_A^{-1} \mathsf{R}_B$. This gives, applying R_A^{-1} to the whole blade equation:

$$0 = \mathbf{x}_A \wedge \mathsf{R}_A^{-1}[\mathbf{b} - \mathbf{a}] \wedge \mathsf{R}_A^B[\mathbf{x}_B].$$

We recognize in $\mathsf{R}_A^{-1}[\mathbf{b} - \mathbf{a}]$ the position of the pinhole of B measured in the A-frame. Denoting that as the translation vector $\mathbf{t}_A$, we get

$$0 = \mathbf{x}_A \wedge \mathbf{t}_A \wedge \mathsf{R}_A^B[\mathbf{x}_B]. \tag{12.13}$$

This is the *epipolar constraint* on the point $\mathbf{x}_A$ given the point $\mathbf{x}_B$ (in B coordinates) and the relative poses of the cameras.

Note that this equation states that the measured $\mathbf{x}_A$ should lie on the 2-blade $\mathbf{t}_A \wedge \mathsf{R}_A^B[\mathbf{x}_B]$ parameterized by $\mathbf{x}_B$. In the image plane, this implies that the point $\underline{\mathbf{x}}_A/f_A$ lies on the line $\mathbf{t}_A \wedge \mathsf{R}_A^B[e_3 + \underline{\mathbf{x}}_B/f_B]$—note now $\mathbf{e}_3$ plays the role of a homogeneous model dimension for the image plane, just as we showed before.

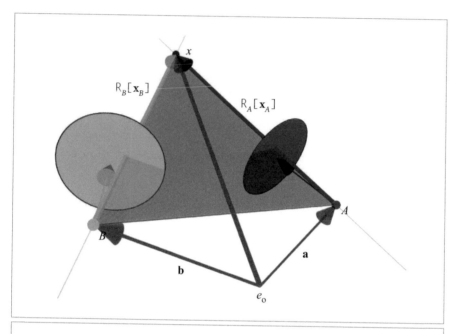

Figure 12.3: The epipolar constraint. We show two cameras A (in red) and B (in green) with their observations of the same point X. In the world, these give the vectors $R_A[\mathbf{x}_A]$ from the location $\mathbf{a}$ of camera A and $R_B[\mathbf{x}_B]$ from the location $\mathbf{b}$ of camera B. Since the cameras are looking at the same point, these ray vectors must be in one plane with the relative position vector $\mathbf{b} - \mathbf{a}$, and that is the epipolar condition. For clarity in visualization, we have followed the habit of most literature to draw the image planes in front of the pinholes.

In the classical treatment of the epipolar constraint, there are of course no trivector equations. We get to a scalar equation by taking the Euclidean dual of (12.13), which gives

$$0 = \mathbf{x}_A \cdot (\mathbf{t}_A \times R_A^B[\mathbf{x}_B]).$$

This is the form in which you find the epipolar constraint in texts on stereo vision, although it is often formulated in matrix form

$$0 = [\![\mathbf{x}_A]\!]^T [\![\mathbf{t}_A^\times]\!] [\![R_A^B]\!] [\![\mathbf{x}_B]\!]$$

by introducing the matrix $[\![\mathbf{t}_A^\times]\!]$ that performs the cross product with $\mathbf{t}_A$. The combination $[\![E_A^B]\!] = [\![\mathbf{t}_A^\times]\!] [\![R_A^B]\!]$ is known as the *essential matrix* of the stereo vision problem.

12.2.4 LINE-BASED STEREO VISION

In more advanced stereo vision, one does not use only potential point matches to reconstruct the depth image of reality, but also line matches. In the usual formulation, the representation of such lines and the matches that are their consequences is not always straightforward. The reason is that you require Plücker coordinates to represent lines, and this is an extra representational step in the classical approaches. In the geometric algebra approach, lines are natural elements of the algebra, just like points are, and the formulation of the line-based stereo matching is more clearly analogous to that for points. We investigate that briefly in this section.

Consider a camera A as above, with rigid body motion $A[\cdot]$, but now look at the interpretation of an observed line in the image plane. First, place a camera in the origin, in standard orientation. A line in the image plane at location $\underline{\mathbf{x}}$ in the image and direction $\underline{\mathbf{u}}$ (parallel to the image plane) is

$$L = (e_0 + \mathbf{f} + \underline{\mathbf{x}}) \wedge \underline{\mathbf{u}} = e_0\,\underline{\mathbf{u}} + \mathbf{M},$$

so it is characterized by a direction vector $\underline{\mathbf{u}}$ and a 3-D moment 2-blade $\mathbf{M} = (\mathbf{f} + \underline{\mathbf{x}}) \wedge \underline{\mathbf{u}}$. This line generates the plane of rays from the pinhole e_0:

$$e_0 \wedge L = e_0 \wedge \mathbf{M}, \tag{12.14}$$

which only depends on the moment of the line; see Figure 12.4. The 3-D Euclidean 2-blade $\mathbf{M}$ thus completely characterizes the observed line (obviously, you can retrieve the direction by intersection with the image plane), and we will denote the observed line by this 2-blade. (This is also shown by the matrix of the line projection in (12.11), which only involves the coordinates of the moment.) Note the subtle difference between the

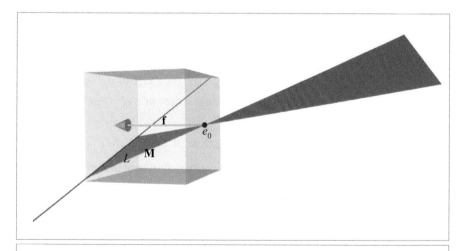

Figure 12.4: The plane of rays generated by a line observation L is characterized by its moment 2-blade $\mathbf{M}$.

effectively observed line $\mathbf{M}$ and the real-world 3-D line L, and how (12.14) expresses that they are geometrically and algebraically identical when seen from the pinhole.

Moving the camera to its general position and orientation by its rigid body motion A, we find that an observed line L_A in the image corresponds to a potential plane

$$A[e_0 \wedge L_A] = A[e_0 \wedge \mathbf{M}_A] = (e_0 + \mathbf{a}) \wedge R_A[\mathbf{M}_A].$$

Now consider three cameras A, B, C, each observing the same line L as L_A, L_B, and L_C, respectively, in their own local frames of reference. The world planes of the observed lines should intersect in line L. Algebraically, we can express this by stating that their meet naïvely computed with the pseudoscalar $e_0\,I_3$ as a join is zero (since they are not in general position). This gives the equation

$$0 = (A[e_0 \wedge L_A])^* \wedge (B[e_0 \wedge L_B])^* \wedge (C[e_0 \wedge L_C])^*.$$

To get more specific in the consequences, we expand the rigid body motion in its translational and rotational parts and write the full dual in terms of the Euclidean dual (since $I_4 = e_0\,I_3$, we have $X^* = X^\star\,e_0^{-1}$). As shorthand, we define Euclidean dual moment vectors $\mathbf{m}_A \equiv \mathbf{M}_A{}^\star$, and so on, in local camera coordinates, and their rotated versions $\mathbf{m}'_A$, $\mathbf{m}'_B$, $\mathbf{m}'_C$ to world coordinates:

$$\mathbf{m}'_A \equiv R_A[\mathbf{m}_A] = R_A[\mathbf{M}_A]^\star,$$
$$\mathbf{m}'_B \equiv R_B[\mathbf{m}_B] = R_B[\mathbf{M}_B]^\star,$$
$$\mathbf{m}'_C \equiv R_C[\mathbf{m}_C] = R_C[\mathbf{M}_C]^\star.$$

The $\mathbf{m}_A$, $\mathbf{m}_B$, $\mathbf{m}_C$ are dual representations of the planes of the rays for each of the lines and therefore simply dual representations of the observed lines in all their projective essence. You may think of them as the observed lines. The $\mathbf{m}'_A$, and so on, are those directions transferred to world coordinates. Then we obtain

$$
\begin{aligned}
0 &= \big((e_0 + \mathbf{a}) \wedge R_A[\mathbf{M}_A]\big)^* \wedge \big((e_0 + \mathbf{b}) \wedge R_B[\mathbf{M}_B]\big)^* \wedge \big((e_0 + \mathbf{c}) \wedge R_C[\mathbf{M}_C]\big)^* \\
&= ((e_0 + \mathbf{a})\rfloor(\mathbf{m}'_A\,e_0^{-1})) \wedge ((e_0 + \mathbf{b})\rfloor(\mathbf{m}'_B\,e_0^{-1})) \wedge ((e_0 + \mathbf{c})\rfloor(\mathbf{m}'_C\,e_0^{-1})) \\
&= \big((\mathbf{a}\cdot\mathbf{m}'_A)\,e_0^{-1} - \mathbf{m}'_A\big) \wedge \big((\mathbf{b}\cdot\mathbf{m}'_B)\,e_0^{-1} - \mathbf{m}'_B\big) \wedge \big((\mathbf{c}\cdot\mathbf{m}'_C)\,e_0^{-1} - \mathbf{m}'_C\big) \\
&= -\mathbf{m}'_A \wedge \mathbf{m}'_B \wedge \mathbf{m}'_C \\
&\quad + \big((\mathbf{a}\cdot\mathbf{m}'_A)\wedge\mathbf{m}'_B\wedge\mathbf{m}'_C - \mathbf{m}'_A\wedge(\mathbf{b}\cdot\mathbf{m}'_B)\wedge\mathbf{m}'_C + \mathbf{m}'_A\wedge\mathbf{m}'_B\wedge(\mathbf{c}\cdot\mathbf{m}'_C)\big)\,e_0^{-1}.
\end{aligned}
$$

Therefore, there are two equations that follow from the triviality of the meet:

$$0 = \mathbf{m}'_A \wedge \mathbf{m}'_B \wedge \mathbf{m}'_C$$

and

$$0 = (\mathbf{a}\cdot\mathbf{m}'_A)\,\mathbf{m}'_B \wedge \mathbf{m}'_C - (\mathbf{b}\cdot\mathbf{m}'_B)\,\mathbf{m}'_A \wedge \mathbf{m}'_C + (\mathbf{c}\cdot\mathbf{m}'_C)\,\mathbf{m}'_A \wedge \mathbf{m}'_B.$$

The first equation has a simple geometrical interpretation: the $\mathbf{m}'_i$ are the normal vectors of planes through a common line L, and these are of course coplanar. The second involves the positional aspects, and is more quantitative.

We can interpret these equations as a condition on any of the observed lines, given the other two lines. Let us develop this for $\mathbf{m}'_A$. To factorize $\mathbf{m}'_A$ out of the second equation (in terms of the outer product), we need to get rid of the first term. We can do so by taking the inner product with $\mathbf{a}$ of the first equation, which gives

$$0 = (\mathbf{a} \cdot \mathbf{m}'_A) \wedge \mathbf{m}'_B \wedge \mathbf{m}'_C - (\mathbf{a} \cdot \mathbf{m}'_B) \wedge \mathbf{m}'_A \wedge \mathbf{m}'_C + (\mathbf{a} \cdot \mathbf{m}'_C)\mathbf{m}'_A \wedge \mathbf{m}'_B.$$

Inserting this into the second equation yields

$$0 = \mathbf{m}'_A \wedge \left(\left((\mathbf{c} - \mathbf{a}) \cdot \mathbf{m}'_C\right) \mathbf{m}'_B - \left((\mathbf{b} - \mathbf{a}) \cdot \mathbf{m}'_B\right) \mathbf{m}'_C \right). \tag{12.15}$$

Given the observed lines $\mathbf{m}_B$ and $\mathbf{m}_C$ in their local coordinates, plus the relative poses of the cameras as given by A, B, C, this specifies what $\mathbf{m}_A$ should be (modulo the usual homogeneous scale factor). This is a linear relationship of two variables $\mathbf{m}'_B$ and $\mathbf{m}'_C$ to provide a third: it is a *tensor*, and we can rewrite it symbolically as a mapping T with two arguments:

$$0 = \mathbf{m}_A \wedge \mathsf{T}(\mathbf{m}_B, \mathbf{m}_C).$$

T is called the *trifocal tensor*. One can derive similar trifocal tensors by factoring out $\mathbf{m}_B$ or $\mathbf{m}_C$, so there are three of them in this situation.

Converting this into the more classical coordinate form, we consider everything relative to camera A. This means that we should apply A^{-1} to the tensor equation. All elements involved are Euclidean direction elements, so they simply transform by R_A^{-1}. Let us abbreviate $\mathsf{R}_A^B \equiv \mathsf{R}_A^{-1}\mathsf{R}_B$ and $\mathsf{R}_A^C \equiv \mathsf{R}_A^{-1}\mathsf{R}_C$, and the relative position of the cameras B and C in terms of A as $\mathbf{t}_B \equiv \mathsf{R}_A^{-1}[\mathbf{b} - \mathbf{a}]$ and $\mathbf{t}_C \equiv \mathsf{R}_A^{-1}[\mathbf{c} - \mathbf{a}]$. [2]

$$0 = \mathbf{m}_A \wedge \left(\mathsf{R}_A^B[\mathbf{m}_B] \, (\mathbf{t}_C \cdot \mathsf{R}_A^C[\mathbf{m}_C]) - (\mathsf{R}_A^B[\mathbf{m}_B] \cdot \mathbf{t}_B) \, \mathsf{R}_A^C[\mathbf{m}_C]\right)$$

The result is then written in terms of matrix operations on vectors (since that is all you have classically), or in tensor notation. The ith coordinate of the line $\mathbf{m}_A$ is proportional to

$$[\![\mathbf{m}_A]\!]_i \propto [\![\mathbf{m}_B]\!]_j \left([\![\mathsf{R}_A^B]\!]_i^j [\![\mathbf{t}_C]\!]^l [\![\mathsf{R}_A^C]\!]_l^k - [\![\mathsf{R}_A^B]\!]_m^j [\![\mathbf{t}_B]\!]^m [\![\mathsf{R}_A^C]\!]_i^k \right) [\![\mathbf{m}_C]\!]_k$$

$$= [\![\mathsf{T}]\!]_i^{jk} [\![\mathbf{m}_B]\!]_j [\![\mathbf{m}_C]\!]_k,$$

so that gives the trifocal tensor in tensorial notation (including the usual summation convention over repeated upper and lower index pairs).

---

2   In the other literature on multicamera treatments, it is customary to parameterize the rigid body motion inversely (not where the camera frame is in the world frame, but vice versa). Replace our rotations like R with R^{-1} and $\mathbf{t}$ with $-\mathsf{R}^{-1}\mathbf{t}$ to get that parameterization. It simplifies the appearance of the final results somewhat, since an R gets absorbed into the new $\mathbf{t}$ twice.

The trifocal constraint of (12.15) can be used to derive constraints on points as well as lines. For example, if we know the two observed lines $\mathbf{m}_B$ and $\mathbf{m}_C$, and wonder what the constraint might be for a point $\mathbf{x}_A$ to be the projection of the line Λ, then this follows immediately from the demand that it should lie on the line L_A, which is $\mathbf{x}_A \wedge e_0 \wedge \mathbf{M}_A = 0$, so $\mathbf{x}_A \wedge \mathbf{M}_A = 0$. Introducing $\mathbf{x}'_A = \mathsf{A}[\mathbf{x}_A] = \mathbf{a} + \mathsf{R}_A[\mathbf{x}_A]$ as the same point in world coordinates, we have $\mathbf{x}'_A \cdot \mathbf{m}'_A = 0$, so that the constraint on the point is obtained by taking the inner product of the trifocal constraint with $\mathbf{x}'_A$:

$$0 = (\mathbf{x}'_A \cdot \mathbf{m}'_B) \left((\mathbf{c} - \mathbf{a}) \cdot \mathbf{m}'_C \right) - \left(\mathbf{m}'_B \cdot (\mathbf{b} - \mathbf{a}) \right) (\mathbf{x}'_A \cdot \mathbf{m}'_C).$$

Therefore, $\mathbf{x}'_A$ lies on the dual line

$$\mathbf{m}'_B \left((\mathbf{c} - \mathbf{a}) \cdot \mathbf{m}'_C \right) - \left(\mathbf{m}'_B \cdot (\mathbf{b} - \mathbf{a}) \right) \mathbf{m}'_C.$$

Another way of looking at the trifocal tensor is as a line-parameterized *homography* (i.e., a a projective mapping between spaces). Suppose we fix $\mathbf{m}_C$; then this determines a spatial plane, which can be used to transfer points in B that are observations from the same world line to points in A (project the ray of observation $\mathbf{x}_B$ to meet the plane in a point, then observe this point as $\mathbf{x}_A$). This mapping can be extended uniquely to the whole image plane as a homography. We can give an explicit formula for this homography between the planes A and B: it should be proportional to the inverse adjoint of the mapping for lines $\mathbf{m}_A$ and $\mathbf{m}_B$, given $\mathbf{m}_C$. The adjoint of

$$\mathbf{m} \mapsto \left((\mathbf{c} - \mathbf{a}) \cdot \mathbf{m}'_C \right) \mathsf{R}_B[\mathbf{m}] - \left((\mathbf{b} - \mathbf{a}) \cdot \mathsf{R}_B[\mathbf{m}] \right) \mathbf{m}'_C$$

is easily found as

$$\mathbf{x} \mapsto \left((\mathbf{c} - \mathbf{a}) \cdot \mathbf{m}'_C \right) \mathsf{R}_B^{-1}[\mathbf{x}] - (\mathbf{m}'_C \cdot \mathbf{x}) \mathsf{R}_B^{-1}[\mathbf{b} - \mathbf{a}],$$

and we take the inverse by simply considering this not as a mapping from $\mathbf{x}_B$ to $\mathbf{x}_A$, but from $\mathbf{x}_A$ to $\mathbf{x}_B$. Therefore the homography between image A and image B is

$$\mathbf{x}_B \propto \left((\mathbf{c} - \mathbf{a}) \cdot \mathbf{m}'_C \right) \mathsf{R}_B^A[\mathbf{x}_A] - (\mathbf{m}'_C \cdot \mathbf{x}_A) \mathsf{R}_B^A[\mathbf{b} - \mathbf{a}].$$

Similarly, you can derive where a point should be in image A once you have observed it in image B and C, just using the trifocal tensor.

If you therefore have a robotic environment in which you see enough lines for which you are able to do the correspondence in the three images, then you can estimate the trifocal tensor (using techniques from [25]), and employ this to compute other correspondences between the images as well.

12.3 FURTHER READING

The impressive book by Faugeras and Luong [21] deals with the geometry of multiple cameras. Like [60], it uses the Grassmann-Cayley algebra of the outer product (which they call join) and its adjoint (which they call meet). This provides a concept of duality, but no metric. The book contains a compact mathematical explanation of this structure, illustrated with geometrical usage. Unfortunately, the authors quickly revert to a matrix-based notation in the chapters that put the structure to practical use, and what could have been explicitly defined algebraic products become tricks in matrix manipulation.

Another modern reference on the geometry of imaging is [25], which dextrously avoids using even Grassmann algebra. All operations are expressed in terms of matrices or tensors defined in terms of Plücker coordinates. The powerful results can therefore be implemented directly, but the geometrical structure of the techniques is often hard to grasp.

We find that both these books have become much more accessible now that geometric algebra offers a structural, representation-independent insight to help one understand what is essentially going on. They can then be browsed quickly for useful techniques.

12.4 EXERCISES

12.4.1 STRUCTURAL EXERCISES

1. Table 12.1 contains the case in which a line $\{\mathbf{a}, \mathbf{m}\}$ is extended to a plane by an additional direction $\mathbf{n}$ to form the plane $[\mathbf{a} \times \mathbf{n} : \mathbf{n} \cdot \mathbf{m}]$. Demonstrate the correctness of this formula by representing the spanning $L \wedge \mathbf{n}$ in terms of the Plücker coordinates.

2. Show that the solution of $\mathbf{x}$ from the relationship $e_0 \wedge x = e_0 \wedge \mathbf{x}$ of (12.7) is indeed $\mathbf{x} = e_0^{-1}(e_0 \wedge x)$, and understand to your satisfaction that this still gives the correct vector $\mathbf{x}$ for the point x, even when the latter would have been given relative to another origin e_0' for the homogeneous coordinates.

3. Knowing some of the standard formulas in geometric algebra, you may recognize that the central projection formula (12.9) is not unlike the usual orthogonal projection formula onto a line with direction $\mathbf{a}$, which maps $\mathbf{x}$ to $(\mathbf{x} \cdot \mathbf{a})\,\mathbf{a}^{-1}$. Demonstrate that we can consider the central projection $\mathbf{x}' \mapsto \mathbf{x}/(\mathbf{f}^{-1} \cdot \mathbf{x})$ as *the fixed vector $\mathbf{f}$ gets inverted, projected onto the variable vector $\mathbf{x}$, and then reinverted, to produce $\mathbf{x}'$.* This interpretation of the formula generalizes to substituting $\mathbf{x}$ by a line or plane (just replace the inner product with the contraction); it then produces the support vector of the projected line or plane. (Hint: Show that $\mathbf{x}'$ satisfies: $\mathbf{x}'^{-1} = (\mathbf{f}^{-1} \cdot \mathbf{x})\,\mathbf{x}^{-1} = (\mathbf{f}^{-1} \cdot \mathbf{x}^{-1})\,\mathbf{x}$ and interpret.)

4. Show that the images of all $\mathbf{x}$ under the construction of the previous exercise before the final inversion (that is, $(\mathbf{f}^{-1} \cdot \mathbf{x})/\mathbf{x}$) lie on a sphere through the pinhole and the end of the vector $\mathbf{f}^{-1}$. This is sketched in Figure 12.5 for $f = |\mathbf{f}\| = 1$. It bears uncanny resemblance to an eyeball!

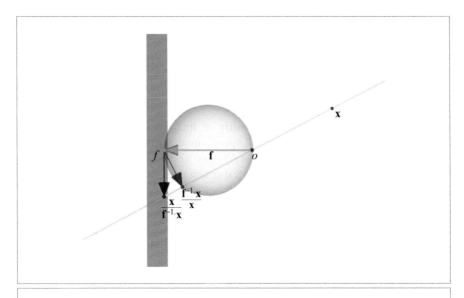

Figure 12.5: The projection of the optical center **f** onto all rays through the pinhole generates an eyeball.

5. Equation (12.9) gives the projection of a unit 3-D point at location **x** to become a point at location **x′**, which is on the image plane, but it is not expressed yet in image plane coordinates x̲. Show that the mapping from the 3-D point at **x** to the image point at $\underline{\mathbf{x}} = \mathbf{x}' - \mathbf{f}$ can be written as

$$\mathbf{x} \;\mapsto\; \frac{\mathbf{f}\,(\mathbf{f}^{-1} \wedge \mathbf{x})}{\mathbf{f}^{-1} \cdot \mathbf{x}}.$$

Interpret this expression geometrically, especially the numerator.

6. Represent the result of the previous problem in homogeneous coordinates, using the center of the image at **f** as origin. Before you do so, guess the answer!

12.5 PROGRAMMING EXAMPLES AND EXERCISES

12.5.1 LOADING TRANSFORMATIONS INTO OPENGL

This example illustrates the good interoperability between homogeneous coordinates-based software (such as OpenGL) and the homogeneous model.

Any outermorphism can be straightforwardly loaded into OpenGL. First, the images of the four basis vectors under the outermorphism are computed, and then the matrix representation of the outermorphism is constructed, whose coordinates are then loaded onto the modelview matrix of OpenGL.

We use a translation-rotation as an simple example. In our OpenGL examples so far, loading this transformation to the modelview matrix would be implemented as:

```
glTranslatef(0.0f, 0.0f, distance); // translate
rotor R = g_modelRotor;
rotorGLMult(R); // rotate (implemented in h3ga_util.cpp)
```

Using the matrix representation of the outermorphism, this becomes:

```
// get the translation vector & the rotor
h3ga::vector T = _vector(distance * e3);
rotor R = g_modelRotor;
rotor Ri = _rotor(inverse(R));

// compute images of basis vectors:
point imageOfE1 = _point(R * e1 * Ri + (T ^ (e0 << (R * e1 * Ri))));
point imageOfE2 = _point(R * e2 * Ri + (T ^ (e0 << (R * e2 * Ri))));
point imageOfE3 = _point(R * e3 * Ri + (T ^ (e0 << (R * e3 * Ri))));
point imageOfE0 = _point(R * e0 * Ri + (T ^ (e0 << (R * e0 * Ri))));

// create matrix representation:
omPoint M(imageOfE1, imageOfE2, imageOfE3, imageOfE0);

// load matrix representation into GL modelview matrix:
glLoadMatrixf(M.m_c);
```

For this simple example, it is of course counterproductive to replace the short OpenGL code with its geometric algebra equivalent: the example is intended to demonstrate the principles only.

12.5.2 TRANSFORMING PRIMITIVES WITH OPENGL MATRICES

This example performs the opposite of the previous example. Instead of setting the OpenGL modelview matrix using the matrix representation of an outermorphism, it reads out the modelview matrix and turns it into a matrix representation of an outermorphism. The actions are as follows:

- The code reads the OpenGL transformation matrix;
- It loads this into a matrix representation of an outermorphism;
- It resets the OpenGL modelview matrix;
- It applies the transform to the primitives, and then draws them.

We can then transform not only points, but also lines and planes, as this example is just the example from Section 11.13.2 with some modifications.

To initialize the matrix representation of the outermorphism, we use:

```
// extract OpenGL modelview matrix
float MVM[16];
glMatrixMode(GL_MODELVIEW);
glGetFloatv(GL_MODELVIEW_MATRIX, MVM);

// reset OpenGL modelview matrix(we will apply the transform ourselves)
glLoadIdentity();

// initialize the images of the basis vectors:
point images[4];
for (int i = 0; i < 4; i++) {
  images[i].set(point_e1_e2_e3_e0, &(MVM[i * 4]));
  /* or:
  images[i].set(point_e1_e2_e3_e0,
    MVM[i * 4 + 0],
    MVM[i * 4 + 1],
    MVM[i * 4 + 2],
    MVM[i * 4 + 3]);
  */
}

// initialized matrix representation of outer morphism
om M(images);
```

To draw any primitive (point, lines, plane):

```
void applyTransformAndDraw(const om &M, const mv &X) {
  // apply the outermorphism:
  mv MX = apply_om(M, X);
  draw(MX);
}
```

The actual code is slightly more complicated due to the use of a custom probing point; see the GA sandbox source code package.

In the graphics, there is a somewhat undesirable side effect due to the explicit use of a basis in a factorization algorithm that is used to draw the plane. It is best observed by starting the example and rotating the scene (drag the left mouse button along the edges of the viewport). You will notice that edges of the depicted plane stay in place while the rest of the scene rotates. This feels counterintuitive (compare this to Example 11.13.2 where the edges do rotate).

The cause of this is the following. To draw the plane, we need two vectors that span it. We obtain these by factorization, using the (basis-dependent) algorithm described in Section 21.6. In the original example (Section 11.13.2), we factor the plane, transform the factors, and then draw the plane using the rotated factors. Then the plane rotates visibly, because its vector factors do. In the present example, we first rotate the plane, then factor it, and then draw the rotated plane using its factors. When the rotation plane is the same as the plane itself, the rendering of the plane will not change at all. Intuitively, the former is better, but algebraically, the latter result is more correct, for it shows the true behavior of a plane under rotation.

12.5.3 MARKER RECONSTRUCTION IN OPTICAL MOTION CAPTURE

In Section 10.7.3 we implemented an algorithm for external camera calibration. That algorithm computed the position and orientation of any number of cameras. The current example builds upon that earlier example by using the calibrated cameras to perform measurements of the position of *markers*. We use the homogeneous model to reconstruct 3-D markers from actual capture data. Figure 12.6 shows the example in action.

The markers are little spheres (size from 0.5–3.0 cm) wrapped in retroreflective tape. Every time the camera records an image (between 20 to 500 times per second), an LED ring around the camera lens illuminates the scene, and the markers reflect this light back to the camera sensor. With proper exposure time (our system typically uses between 0.2 and 1ms), the result is that the markers are visible as bright blobs in the otherwise black images. The centers of the 2-D markers are easily retrieved from the images, since the center of a projected sphere is a good approximation to the projection of the center of that sphere when the markers are sufficiently far away.

When a marker is visible from multiple calibrated cameras, its position can be reconstructed by computing the intersection point of rays from the camera centers through

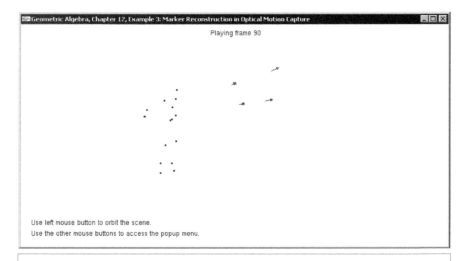

Figure 12.6: Example 3 in action. The four red arrows are the cameras and the black dots are the reconstructed markers. When the example is in motion, it is immediately clear that the black dots are attached to a human.

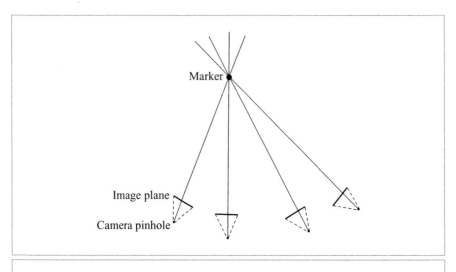

Figure 12.7: Reconstruction of markers.

the marker center. This is illustrated in 2-D in Figure 12.7. Markers can of course be occluded by the subject to which they are attached, so not all cameras can see every marker all the time. The more cameras that actually see a particular marker, the more precise the reconstruction can be. The simplistic marker reconstruction algorithm used in this example reconstructs the markers as follows: we iterate over every pair of 2-D markers from two different cameras. For each pair, we perform the following actions:

- First, we construct a plane through the two camera centers and the 2-D marker in the image plane of one camera. This plane is then intersected with the image plane of the other camera to compute the epipolar line:

```
// normalizedPoints P1, P2 are the camera positions.
// normalizedPoint M1 is the marker position, on image plane of
// camera 1.
// plane IP2 is the image plane of camera 2.

// compute the epipolar line in the image plane of camera 2:
line L2 = _line(unit_r(dual(P1 ^ M1 ^ P2) << IP2));
```

- Once we have the epipolar line, we search for 2-D markers sufficiently close to it (as the data is inherently noisy, we will rarely have a precise intersection of the epipolar line with a marker). We compute the distance of each candidate marker as $(M2*) \rfloor L2$:

```
// normalizedPoint M2 is the marker position, on image plane of
// camera 2
```

```
// compute distance in image plane:
mv::Float distance = fabs(_Float((M2 << IP2) << L2));
```

The dual is of course taken relative to the image plane of the second camera.

- If the distance is less than some epsilon value, we reconstruct a 3-D marker by computing the average of the closest points on the two lines P1 ∧ M1 and P2 ∧ M2. We collect all reconstructed points in an array. How these points are computed is described below.

After this loop, we iterate over all pairs of reconstructed 3-D markers to find the ones that are sufficiently close in 3-D space (where close is typically defined as within the distance of the radius of the marker). We merge close markers by summing them. We do not normalize the points, as we use the e0 coordinates as a rough counters for how many cameras contributed to a particular reconstruction. We return only those markers that have been seen by a sufficient number of cameras:

```
/*
R is the array of reconstructed 3D markers (candidates).
'reconstructedMarkers' is the final array of 3D markers
(normalizedPoints) that is returned to the caller.
*/
if ((int)R[i].e0() >= minNbCameras) {
  reconstructedMarkers.push_back(_normalizedPoint(R[i] /
    R[i].e0()));
}
```

When you read the full source code of the example in the GA sandbox source code package, you may encounter some use of the conformal model of the later chapters. Our actual motion capture system is based primarily on the conformal model, hence this example (which was derived from it to illustrate homogeneous model techniques) stores the camera transformations as conformal versors.

A more refined reconstruction algorithm would use some kind of data structure to quickly look up points close to particular epipolar lines. For each successful match in two cameras, an initial 3-D marker is then reconstructed. These 3-D markers can subsequently be projected onto the image planes of the other cameras, after which another data structure (e.g., a spatial hashing table) can be used to quickly find markers in the vicinity of the projected points. A least-squares-based solution (see Section 10.4.2) would then be used to find the optimal reconstruction of the 3-D marker given the 2-D measurements in each camera.

Computing Closest Points on Lines

Computing the closest points on two skew lines is an important operation that is performed in the inner loop of by the reconstruction algorithm. We provide an efficient geometric algebra implementation in Figure 12.8. The two lines are passed in factored form (i.e., as points and directions such that the actual lines are P1∧D1 and P2∧D2).

```
bool closestPointsOnCrossingLines(
  const normalizedPoint &P1, const h3ga::vector &D1,
  const normalizedPoint &P2, const h3ga::vector &D2,
  mv::Float &d1, mv::Float &d2) {

  // Compute difference between starting points
  h3ga::vector dif = _vector(P2 - P1);

  // compute inverse pseudoscalar of space spanned by D1 and D2
  bivector I = _bivector(D1 ^ D2);
  if (_Float(norm_e2(I)) == 0.0f) // check parallel
    return true;  // returning true means 'lines are parallel'
  bivector Ii = _bivector(inverse(I));

  // compute reciprocals:
  h3ga::vector rD1 = _vector(D2 << Ii);
  h3ga::vector rD2 = _vector(D1 << Ii);

  // solution:
  d1 = _Float(rD1 << dif);
  d2 = _Float(rD2 << dif);

  return false;
}
```

Figure 12.8: Crossing lines code.

The function returns scalars d1 and d2 such that P1 + d1 * D1 and P2 + d2 * D2 are the closest points on the two lines.

The function works by projecting the vector P2 − P1 onto the bivector D1∧D2. The geometric intuition behind this is that we are trying to minimize the distance P2 − P1 between the two points by moving along the lines. d1 and d2 are the coordinates of the projected vector relative to the basis spanned by D1 and D2. Because D1 and D2 are not orthonormal, we need to compute their reciprocals (rD1 and rD2) first.

13 THE CONFORMAL MODEL: OPERATIONAL EUCLIDEAN GEOMETRY

In the previous chapters, we studied the geometric algebra version of the homogeneous model. The homogeneous model of Euclidean geometry is reasonably effective since it linearizes Euclidean transformations, and geometric algebra extends the classical homogeneous coordinate techniques nicely through its outermorphisms. However, we remarked that we were not able to use the full metric products of geometric algebra, since the metric of the model was only indirectly related to the metric of the Euclidean space we had wanted to model.

Fortunately, we can do better. In the next few chapters, we present the new *conformal model* for Euclidean geometry, which can represent Euclidean transformations as orthogonal transformations. Encoding those as *versors*, we get the full power of geometric algebra, including the structure preservation of all constructions.

In this chapter, we start the exposition by defining the new representational space (which has two extra dimensions and an indefinite metric) and showing what its vectors represent. The two extra dimensions are geometrically interpretable as the point at the origin (as in the homogeneous model) and the point at infinity (which nicely closes Euclidean geometry, making translations into rotations around infinity).

We then focus on how to represent the familiar flats and directions already present in the homogeneous model and how to move them around. As first applications, we use the versor representation to provide straightforward closed-form interpolation of rigid body motions and universally valid constructions for the reflection of arbitrary elements.

The natural coordinate-free specification of elements and operations in the conformal model is best appreciated using the interactive illustrations. Even more than in the previous chapters, we encourage you to play around with the interactive software provided with this book.

13.1 THE CONFORMAL MODEL

The conformal model is especially designed for Euclidean geometry, which is the geometry of transformations preserving the Euclidean distances of and within objects. These *Euclidean transformations* are sometimes called *isometries*, and they include translations, rotations, reflections, and their compositions.

We will occasionally limit ourselves to *Euclidean motions*, which are the Euclidean transformations that can be performed a little at a time. These are also known as *rigid body motions*, and they consist of compositions of translations and rotations. They are *proper isometries*, i.e., they preserve the Euclidean distances and handedness.

To emphasize that the conformal model involves the representation of Euclidean geometry, we will denote the base space by $\mathbb{E}^n$ rather than $\mathbb{R}^{n,0}$. This notation is meant to suggest that the group of Euclidean transformations is to be represented in the model as well.

13.1.1 REPRESENTATIONAL SPACE AND METRIC

We start with some observations on points in a Euclidean space.

- A Euclidean space $\mathbb{E}^n$ has points at a well-defined distance from each other. If two points $\mathcal{P}$ and $\mathcal{Q}$ were created by a translation over the Euclidean direction vectors $\mathbf{p}$ and $\mathbf{q}$ from some origin $\mathcal{O}$, then their squared distance is $d_E^2(\mathcal{P}, \mathcal{Q}) = \|\mathbf{p} - \mathbf{q}\|^2 = (\mathbf{p} - \mathbf{q}) \cdot (\mathbf{p} - \mathbf{q})$. In many respects, these squared distances are more convenient to compute with than the actual distances; for instance, Pythagoras' theorem states that squared distances can be added under certain conditions.

- Euclidean spaces do not really have an origin: there is no special finite point that can be distinguished from other points. If an origin is used, it is for convenience, as a point to relate other points to. Similarly, there are no preferential directions, though it may be convenient to define an arbitrary standard frame at the equally arbitrary origin.

- Mathematicians have long known that it is convenient to close a Euclidean space by augmenting it with a *point at infinity*. (They call this compactification.) That point at infinity is a point in common to all lines and planes, and invariant under the Euclidean transformations. Having it explicit makes the algebraic patterns in geometrical statement more universal.[1]

When constructing a model for Euclidean geometry, we take these properties as central. The arbitrariness of the origin was something we wanted in the homogeneous model, and it was partly achieved by assigning an extra dimension to it in a representational space. So we need at least need an $(n+1)$-dimensional representational space. We are now also going to assign an extra dimension to the special point at infinity. In our model, we represent it as a vector denoted as ∞, to remind us of what it is. (There will be no confusion with the number infinity.) This turns our representational space for the n-dimensional Euclidean space $\mathbb{E}^n$ from $(n+1)$-dimensional to $(n+2)$-dimensional.

The interface between geometry and algebraic representation is that we represent *Euclidean points* in $\mathbb{E}^n$ by *representative vectors* in the $(n+2)$-dimensional representational vector space. The infinite point is represented by the vector ∞. A finite point $\mathcal{P}$ is represented by a vector p, with certain additional properties that we specify below. Now we still have the freedom to choose the metric of this representational space (and we must specify it if we want to endow it with a geometric algebra). We use this metric to encode the Euclidean distance in $\mathbb{E}^n$. In view of the observed linear behavior of square distances we do so as follows:

$$p \cdot q \sim d_E^2(\mathcal{P}, \mathcal{Q}). \tag{13.1}$$

On the left you see the inner product of two vectors of the representational space; on the right you see the squared distance of two points in the Euclidean space $\mathbb{E}^n$. They are *directly proportional!* (There are some scaling factors that we introduce below for backwards compatibility, but those are irrelevant to the principle of the setup.)

One immediate consequence of this definition of the metric is that a vector p representing a finite point must obey $p \cdot p = 0$, since the Euclidean distance of a point to itself is zero: $d_E^2(\mathcal{P}, \mathcal{P}) = 0$. Therefore, Euclidean points are represented by *null vectors* in the representational model. A null vector is a vector with norm zero; in a Euclidean space, such a vector would have to be the zero vector, and we could not use that to denote a great variety of points. So the $(n+2)$-dimensional representational space must be non-Euclidean.

Null vectors may seem strange, but it is fairly easy to make them in a manner that feels only moderately irregular. We may alternatively construct the $(n+2)$-dimensional representational space by augmenting the regular n Euclidean dimensions with two special dimensions, for which a basis is formed by two vectors e and $\bar{e}$ that square to $+1$ and -1, respectively, and that are orthogonal (so that $e \cdot \bar{e} = 0$). Having these, it is easy to make null

---

1   The homogeneous model has location-independent directions, which we identified as being characterized by improper points on the heavenly sphere, but this has a different flavor than a single point at infinity, which is a location that one can approach from *any* direction. We will not lose those extra elements representing directions for they are also found as elements in the new model.

vectors out of them: the vectors $(e + \bar{e})$ and $(\bar{e} - e)$ are both null without being zero, as you can check easily by squaring them. In this construction, the representational space has an orthonormal basis consisting of $(n + 1)$ positive dimensions (with basis vectors squaring to $+1$) and 1 negative dimension (with basis vector squaring to -1). We therefore denote it by $\mathbb{R}^{n+1,1}$. Such a space is called a *Minkowski space* in physics, where it has been well studied to represent space-time in relativity; the negative dimension is then employed to represent time. We will find a geometrically more natural basis than e and $\bar{e}$ for the conformal model, but the representational space is still the metric space $\mathbb{R}^{n+1,1}$.

This $(n + 2)$-dimensional space requires $n + 2$ coefficients to specify a vector. We should require only n coefficients to specify a point in Euclidean space, so there is more in this space than we appear to need. Since points are represented by null vectors, their representatives must obey $p \cdot p = 0$, and this is a condition on their coefficients that removes one degree of freedom, leaving $n + 1$. The remaining degree of freedom we use to denote the weight of the point, as in the homogeneous model. In that model, we retrieved the weight as $e_0^{-1} \cdot p$, since it was the coefficient of e_0; in the present model we extract it by the operation $-\infty \cdot p$. The relationship of inner product to distance of (13.1) should of course be defined on properly normalized points, since it is independent of the weight, so in that formula we should divide p by its weight, and q as well. We introduce an extra scaling factor of $-\frac{1}{2}$ for convenience later on, and actually define the inner product of vectors representing points through

$$\frac{p}{-\infty \cdot p} \cdot \frac{q}{-\infty \cdot q} \equiv -\tfrac{1}{2} d_E^2(\mathcal{P}, \mathcal{Q}). \qquad (13.2)$$

(The two minus signs on the left of course cancel each other, but we would like to get you used to viewing $-\infty \cdot p$ as a weight-computing expression.) Equation (13.2) is in fact the definition of the representational model; all the rest of the correspondence between elements of its geometric algebra and Euclidean geometry follows from it without additional assumptions.

The model thus constructed was called the conformal model in the literature, for the mathematical reason that it is capable of representing conformal transformations by versors. We will demonstrate that in Chapter 16, but prefer to focus on its use for Euclidean transformations first. We will show that it is an *operational model* for Euclidean geometry in the double sense that Euclidean transformations are represented as structure-preserving operators, which are moreover easy to put to operational use. In view of that, we would prefer to call it the operational model for Euclidean computations. This is too much of a mouthful, so we might as well conform to the term conformal model, despite its arcane origins.

Even though the conformal model hit computer science only a few years ago, when it was introduced by Hestenes et al. in 1999 [31], we are beginning to find that we could have had the pleasure of its use all along. Considerable elements of it are found in much older work, and it appears to have been reinvented several times. We do not know the whole story yet, but markers on the way are Wachter (a student of Gauss), Cox [9] and Forder 1941 [22] (who used Grassmann algebra to treat circles and their properties), Anglès 1980 [2] (who showed the crucial versor form of the conformal transformations) and Hestenes (who had already presented it in his 1984 book [33] but only later realized its true importance for Euclidean geometry). We are convinced that its time of general adaptation has finally arrived, since the conformal model is much more clearly useful to actually programming geometrical applications than it is to theorizing about geometry on paper.

13.1.2 POINTS AS NULL VECTORS

Let us be more precise about the point representation. As we realized before, the relationship of the metric to the Euclidean distance implies that finite points are represented by null vectors. What about the point at infinity?

The point at infinity is represented by ∞, and it should have infinite distance to all finite points. Substituting $q = \infty$ in (13.2), we find that an infinite result is reached only if we set $\infty \cdot \infty = 0$. This implies that the point at infinity is also represented by a null vector.

> *Null vectors in the conformal model $\mathbb{R}^{n+1,1}$ represent Euclidean points of $\mathbb{E}^n$ (both finite and infinite).*

After you have gotten used to this correspondence, it becomes natural to identify representative null vectors with points and talk about the null vector p as "the point p" rather than as "representing the point $\mathcal{P}$." We will gradually slip into that convenient usage as this chapter evolves.

We call a point p *normalized* or a *unit point* when its weight equals 1:

$$-\infty \cdot p = 1$$

We may be tempted to select one special unit point $\mathcal{O}$ as the origin of our space and represent it by a vector denoted o. It would then be a special vector in the conformal model with the property $o \cdot \infty = -1$, as well as $o \cdot o = 0$, by virtue of being a point. It therefore appears to bear a special relationship to the vector ∞ representing the point at infinity: though both are null vectors, they are (minus) each other's reciprocals in the sense that $o \cdot \infty = -1$. That simplifies the math of computing with them, and it reveals that the weight of a point is the coefficient of o of its representative vector (for a coefficient of a vector is retrieved by an inner product with its reciprocal, as we saw in Section 3.8—so the o coefficient is retrieved by an inner product with $-\infty$). The strong parallel with e_0 and e_0^{-1} in the homogeneous model suggests that an arbitrary unit point might be writable as $o + \mathbf{p}$, but this is not a null vector (it squares to $\mathbf{p}^2$). We should use the extra dimension to represent a unit point as $p = o + \mathbf{p} + \alpha \infty$, with an α that should be determined by the null vector condition $p \cdot p = 0$. You can easily verify that this gives

$$p = o + \mathbf{p} + \tfrac{1}{2}\mathbf{p}^2\,\infty \qquad\qquad (13.3)$$

as the representative null vector for a unit point relative to the origin o.[2]

Equation (13.3) resembles a coordinate representation of the vector p in terms of a purely Euclidean location vector $\mathbf{p}$ and two extra terms in a $\{o, \infty\}$-basis for the extra two dimensions of the conformal model. It is somewhat misleading: although ∞ indeed has special significance, the origin point o does *not*. Any unit point could have been used as origin; it would have had the same relationships with ∞ (since any finite point p obeys $-\infty \cdot p = 1$), and would merely have changed the value of the location vector $\mathbf{p}$ for the particular point $\mathcal{P}$ we are considering. So (13.3) is deceptive in that it is too specific. Still, it is convenient to have, since it gives a connection to the homogeneous representation (o is like e_0), and even to the vector space model (in exhibiting the Euclidean location vector $\mathbf{p}$). You can use it when you begin to work with the conformal model—as you advance, you will dare to work coordinate-free, merely using the algebraic properties of the point p, and that will make your formulas and algorithms more generally applicable.

But while we have this specific representation, let us use it to verify the consistency with the original definition of the metric:

$$
\begin{aligned}
p \cdot q &= (o + \mathbf{p} + \tfrac{1}{2}\mathbf{p}^2\infty) \cdot (o + \mathbf{q} + \tfrac{1}{2}\mathbf{q}^2\infty) \\
&= -\tfrac{1}{2}\mathbf{q}^2 + \mathbf{p} \cdot \mathbf{q} - \tfrac{1}{2}\mathbf{p}^2 \\
&= -\tfrac{1}{2}(\mathbf{q} - \mathbf{p})^2.
\end{aligned}
\qquad (13.4)
$$

Note how the metric of the conformal model plays precisely the right role to compose the correct terms in the squared difference required to express the Euclidean distance. So the conformal model indeed does what it was designed to do.

Letting $\mathbf{p}$ become large in (13.3), the dominant term is proportional to $+\infty$. This shows the reason for our choosing the normalization such that $\infty \cdot p$ is negative: now $+\infty$ is positively proportional to (the vector representing) the point at infinity.

We are especially interested in the Euclidean space $\mathbb{E}^3$, and when required, we use an orthonormal basis $\{\mathbf{e}_1, \mathbf{e}_2, \mathbf{e}_3\}$ for that part of the representation. That leaves the other two elements in the basis of $\mathbb{R}^{n+1,1}$. Earlier, we briefly introduced the basis $\{e, \bar{e}\}$ with their squares of $+1$ and -1, and that would work as basis for the part $\mathbb{R}^{1,1}$. But it is often geometrically more significant to make a change of basis to the two null vectors o and ∞, representing our arbitrary origin and the point at infinity. These null vectors can be defined in terms of e and $\bar{e}$ as

---

2   We should mention that the standards in this new model have not quite been established yet, and you may find other authors with different definitions and notations of the elements we have denoted ∞ and o, differing by a factor of ± 1 or ± 2 with our definition. For instance, in [15] our ∞ is their $-n$, and our o is their $-\tfrac{1}{2}\bar{n}$, so that their point representation reads $p = \mathbf{p}^2 n + 2\mathbf{p} - \bar{n}$, with normalization $p \cdot n = 2$. Any rescaling of ∞ is a trivial change to the conformal model, not affecting its structure in any way, though it is a bit annoying in cross-referencing texts from various sources. Always check the local definitions! We will give our reasons for our choices.

$$o = \tfrac{1}{2}(e + \bar{e}), \quad \infty = \bar{e} - e, \tag{13.5}$$

giving conversely

$$e = o - \tfrac{1}{2}\infty; \quad \bar{e} = o + \tfrac{1}{2}\infty. \tag{13.6}$$

So in terms of a basis, this is just a simple coordinate change in the representational subspace with pseudoscalar $e \wedge \bar{e} = o \wedge \infty$. We reiterate that *any* finite unit point could have been used instead of o without changing these defining equations (picking another point as origin merely changes where we attach the Euclidean part $\mathbb{E}^n$ to the basis for $\mathbb{R}^{1,1}$, and therefore just changes the Euclidean vectors by an offset). For $\mathbb{E}^3$, the basis elements have the inner product multiplication tables of Table 13.1. We can define the pseudoscalar by $I_3 = e_1 \wedge e_2 \wedge e_3$.

But throughout the remainder, we will avoid the use of the basis to specify elements. We shall hardly need it, for the Euclidean covariance properties of the conformal model will make computations in it delightfully coordinate-free.

13.1.3 GENERAL VECTORS REPRESENT DUAL PLANES AND SPHERES

By construction, some of the vectors in the representational space $\mathbb{R}^{n+1,1}$ represent weighted points of the Euclidean space; namely all null vectors, which must satisfy $p \cdot p = 0$. Yet there are many more vectors in $\mathbb{R}^{n+1,1}$. We should investigate whether those are also useful to represent elements of Euclidean geometry. And indeed they are: we will find that they are generally dual spheres (with points and dual planes included as special cases for radius zero and infinity, respectively).

To determine what such a nonnull vector v signifies, we should somehow use our fundamental definition of (13.2), since that is all that we have up to this point. Therefore we probe the vector v with a unit point x, solving the equation $x \cdot v = 0$ to find out what v is when considered as defining a Euclidean point set in this manner. Note that this already implies that we interpret a vector v as the *dual* representation of some Euclidean element.

Table 13.1: Multiplication table for the inner product of the conformal model of 3-D Euclidean geometry $\mathbb{E}^3$, for two choices of basis.

	e	e_1	e_2	e_3	$\bar{e}$
e	1	0	0	0	0
e_1	0	1	0	0	0
e_2	0	0	1	0	0
e_3	0	0	0	1	0
$\bar{e}$	0	0	0	0	−1

	o	e_1	e_2	e_3	∞
o	0	0	0	0	−1
e_1	0	1	0	0	0
e_2	0	0	1	0	0
e_3	0	0	0	1	0
∞	−1	0	0	0	0

It is perhaps a bit surprising that we meet these dual representations of elements before the direct interpretations. (Actually, something similar happened with the vectors of the homogeneous model. There, a vector of the form $e_0 + \mathbf{p}$ represented a point, a vector of the form $\mathbf{u}$ was a direction, and a vector $\mathbf{n} - \delta e_0^{-1} = \mathbf{n} \mp \delta e_0$ represented a *dual* plane).

- *Null Vector $p = \alpha (o + \mathbf{p} + \frac{1}{2} \mathbf{p}^2 \infty)$: Point.* We may as well first verify that this probing procedure gives the right interpretation for a general null vector of the form we saw before. Our earlier computation (13.4) gives $x \cdot p = -\frac{1}{2}\alpha(\mathbf{p} - \mathbf{x})^2$, so that this is zero if and only if $\mathbf{x} = \mathbf{p}$. Therefore x must be a point at the same location as p, albeit possibly with a different weight.[3]

  As we have seen, a vector p representing a point satisfies $p^2 = 0$, and $\infty \cdot p = -\alpha$. These properties define this class of vector of $\mathbb{R}^{n+1,1}$ representing a point of $\mathbb{E}^n$. The prototype of this class is the vector o representing the point $\mathcal{O}$ at the arbitrary origin.

- *Vector without o Component: $\pi = \mathbf{n} + \delta\infty$: Dual Plane.* A vector π without an o component has the general form $\pi = \mathbf{n} + \delta\infty$. It clearly does not represent a Euclidean point. We probe it to find out what it is in the Euclidean space:

$$x \cdot \pi = (o + \mathbf{x} + \tfrac{1}{2}\mathbf{x}^2 \infty) \cdot (\mathbf{n} + \delta\infty) = \mathbf{x} \cdot \mathbf{n} - \delta,$$

  and demanding this to be zero retrieves the familiar dual plane equation for a plane with normal vector $\mathbf{n}$ at a distance $\delta/\|\mathbf{n}\|$ from the origin. Therefore, the vector π dually represents a Euclidean plane. We will denote dual planes by π-based symbols (and direct planes by symbols based on Π).

  Such a dual plane vector satisfies $\infty \cdot \pi = 0$ and $\pi^2 = \mathbf{n}^2$, defining the type. The equation $\infty \cdot \pi = 0$ can be interpreted as: ∞ lies on the dual plane π. The prototypical member of this class is the vector $\mathbf{n}$ of $\mathbb{R}^{n+1,1}$ that lies completely in its subspace $\mathbb{E}^n$ (and hence is denoted in bold font).

- *General Vector $\sigma_{\pm} = \alpha (c \mp \frac{1}{2}\rho^2 \infty)$: Dual Sphere.* A general vector can be made as a scaled version of a null vector with an additional amount of ∞-component to make it nonnull. Let us write it as $\sigma = \alpha(c + \beta\infty)$, with c a unit point representative (so that $c^2 = 0$, and $-\infty \cdot c = 1$). Then we find

$$x \cdot \sigma = \alpha(x \cdot c + \beta x \cdot \infty) = \alpha\left(-\tfrac{1}{2}d_E^2(x,c) - \beta\right).$$

  Requiring this to be zero gives the equation $\|\mathbf{x} - \mathbf{c}\|^2 = -2\beta$, where we substituted the Euclidean distance between x and c as $\|\mathbf{x} - \mathbf{c}\|$. If β is negative, we redefine it as $\beta = -\frac{1}{2}\rho^2$, and obtain

---

3 Note however that there is something slightly uncomfortable going on: strictly speaking, if x is a point, then p must be the dual representation of a point (since it is the solution of $x \cdot p = 0$), rather than the direct representation (which would be the solution of $x \wedge p = 0$). Strangely enough, p is both, due to the special nature of null vectors. For $x \wedge p = 0$ is certainly valid for $x = p$, and computing: $x \wedge p = (o + \mathbf{x} + \frac{1}{2}\mathbf{x}^2 \infty) \wedge \alpha(o + \mathbf{p} + \frac{1}{2}\mathbf{p}^2 \infty) = \alpha o \wedge (\mathbf{p} - \mathbf{x}) + \cdots$, and already the first term shows that we can only make this zero by having $\mathbf{x} = \mathbf{p}$, so that $x = p$ is indeed the unique solution. We get back to the dual nature of the point representation when we treat the `plunge` in Section 15.1.

$$\|\mathbf{x} - \mathbf{c}\|^2 = \rho^2.$$

We recognize the equation of a sphere with center at the location $\mathbf{c}$, and radius ρ.

A general vector of the form $\sigma = \alpha(c - \frac{1}{2}\rho^2 \infty)$ *dually represents a Euclidean sphere with center c and radius ρ, and weight α.*

If β is positive, we can set it equal to $\frac{1}{2}\rho^2$, and we obtain the equation

$$\|\mathbf{x} - \mathbf{c}\|^2 = -\rho^2.$$

By analogy, this represents an imaginary sphere whose squared radius is negative. We do not need complex numbers to define it as long as we only represent squares of distances (which is indeed all that entered our model by its definition).

A general vector of the form $\sigma = \alpha(c + \frac{1}{2}\rho^2 \infty)$ *dually represents an imaginary sphere with center c and squared radius $-\rho^2$, and weight α.*

These imaginary spheres may appear useless, but they occur naturally as solutions to intersections and in duality computations. They keep our algebra consistent and closed, and we will try to develop a feeling for them (yes, you can learn to see imaginary spheres, in Section 15.1.3).

The dual sphere vectors satisfy $\sigma^2 = \pm\alpha^2\rho^2$ and $-\infty \cdot \sigma = \alpha$, so that a unit-weight sphere has the same square as its radius. The null vectors representing points can now be viewed as (dual) spheres of zero radius, which makes good geometric sense. Prototypical examples of this class are the vectors $e = o - \infty/2$ and $\bar{e} = o + \infty/2$, which represent the real and imaginary unit spheres at the origin.

This completes the Euclidean interpretation of the representational vectors in the conformal model; the results are collected in Table 13.2. The extension of the homogeneous

Table 13.2: The interpretation of vectors in the conformal model.

Element	Form	Characteristics	
Point	$p = \alpha(o + \mathbf{p} + \frac{1}{2}\mathbf{p}^2 \infty)$	$p^2 = 0$	$-\infty \cdot p \neq 0$
Dual plane	$\pi = \mathbf{n} + \delta\infty$	$\pi^2 \neq 0$	$-\infty \cdot \pi = 0$
Dual real sphere	$\sigma = \alpha(p - \frac{1}{2}\rho^2 \infty)$	$\sigma^2 = \rho^2 > 0$	$-\infty \cdot \sigma \neq 0$
Dual imaginary sphere	$\sigma = \alpha(p + \frac{1}{2}\rho^2 \infty)$	$\sigma^2 = \rho^2 < 0$	$-\infty \cdot \sigma \neq 0$

model by the extra dimension ∞, and the modification of the metric, have already given us (dual) spheres as typical Euclidean elements (with a point as a zero-radius sphere and a plane as an improper sphere that passes through infinity). This carries it beyond the homogeneous model, which could only represent flats. Clearly the set of transformations preserving properties of spheres are more limited than those preserving properties of flats, so the conformal model will be more specifically tailored to Euclidean geometry than the homogeneous model. The affine transformations in particular, which preserved the flats of the homogeneous model, are no longer all admissible since they may not preserve spheres. This loss is in fact a gain: the added precision makes the conformal model much more powerful for the treatment of Euclidean transformations than the homogeneous model.

We are obviously going to invoke the geometric algebra structure of the conformal model to make more involved elements from the dual planes and dual spheres at our disposal, but this will have to wait until the next chapter. In this chapter, we explore how the Euclidean transformations can be represented in the new model, and how that improves and extends the capabilities of the homogeneous model.

13.2 EUCLIDEAN TRANSFORMATIONS AS VERSORS

The crucial Euclidean property of point distance was embedded in terms of the inner product of the representational space $\mathbb{R}^{n+1,1}$ by (13.2). Euclidean transformations in $\mathbb{E}^n$ are *isometries*: they preserve distances of points, so they should be represented by transformations of $\mathbb{R}^{n+1,1}$ that preserve the inner product. Such transformations are therefore orthogonal transformations of the representational space. We know from Section 7.6 that geometric algebra can represent orthogonal transformations as versors. Therefore:

>   *Euclidean transformations are representable by versors in the conformal model.*

Since versors have structure-preserving properties, all constructions that we make using the geometric algebra of $\mathbb{R}^{n+1,1}$ will transform nicely (i.e., covariantly) in that algebra—which implies that they move properly with the Euclidean transformations. We never need to enforce that; it is intrinsically true due to the versor structure. This kind of *operational model* is very intuitive to use (for you can make a construction somewhere, and it will hold anywhere). The conformal model is the smallest known algebra that can model Euclidean transformations in this structure-preserving manner.

13.2.1 EUCLIDEAN VERSORS

Representing Euclidean transformations by versors is the plan, but when we put it into action we have to be bit more precise. Not just any versor in the algebra of $\mathbb{R}^{n+1,1}$ is a Euclidean motion: it also needs to preserve the vector ∞, since that is an essential part of the modeling interface between the inner product and its interpretation as Euclidean distance in (13.2). Such a versor induces an operation in the Euclidean base space that preserves the point at infinity, which is indeed a property of Euclidean transformations.

So in the conformal model, a Euclidean transformation is represented by a versor V that preserves ∞. Using the versor product formula of (7.17), this gives

$$\widehat{V} \infty \, V^{-1} = \infty.$$

It follows that $\infty V - \widehat{V} \infty = 0$, so that $\infty \rfloor V = 0$ is the condition on a versor V to be a Euclidean versor.

The simplest versor is a vector; and the most general vector that satisfies $\infty \rfloor V = 0$ is $\pi = \mathbf{n} + \delta \infty$. This represents a dual plane. All other Euclidean transformation versors should be products of such vectors; in the base space, we would phrase that as

> *All Euclidean transformations can be made by multiple reflections in well-chosen planes.*

This well-known fact from Euclidean geometry is the key to designing the Euclidean transformation versors in $\mathbb{R}^{n+1,1}$.

13.2.2 PROPER EUCLIDEAN MOTIONS AS EVEN VERSORS

We showed in Part I that an odd number of reflections will produce a transformation with determinant -1 (Section 7.6.2). The resulting versor represents an improper motion: it changes handedness, and therefore cannot be performed as a continuous motion. Proper, continuous Euclidean motions preserve handedness and are represented by even versors, which we can normalize to be rotors. There are two elementary proper motions: pure translations and pure rotations. They can of course be composed to make general proper motions. We construct their versors, and illustrate this in Figure 13.1.

- **Translations**. As a special case of versor composition, we compute the product of two reflecting parallel planes with the same direction but at a differing location, as in Figure 13.1(a). Let $\mathbf{n}$ be their unit normal vector, then we compute

  $$(\mathbf{n} + \delta_2 \, \infty)\,(\mathbf{n} + \delta_1 \, \infty) = 1 - (\delta_2 - \delta_1)\,\mathbf{n}\,\infty \equiv 1 - \mathbf{t}\infty/2 \equiv T_{\mathbf{t}},$$

  where we define the Euclidean *translation vector* $\mathbf{t} \equiv 2(\delta_2 - \delta_1)\mathbf{n}$. Geometrically, we would expect this double reflection to represent a translation in the direction $\mathbf{n}$, over twice the distance of the reflecting planes (i.e., over $\mathbf{t}$).

  For a point at the (arbitrary) origin o, this behavior is easily checked:

  $$\begin{aligned}
  T_{\mathbf{t}}\, o\, T_{\mathbf{t}}^{-1} &= (1 - \mathbf{t}\infty/2)\, o\, (1 + \mathbf{t}\infty/2) \\
  &= o - \tfrac{1}{2}(\mathbf{t}\infty\, o - o\,\mathbf{t}\infty) - \mathbf{t}\infty\, o\, \mathbf{t}\infty/4 \\
  &= o - \mathbf{t}\,(\infty \cdot o) - \infty\,(2o \cdot \infty - \infty\, o)\,\mathbf{t}^2/4 \\
  &= o + \mathbf{t} + \tfrac{1}{2}\mathbf{t}^2 \infty \equiv t.
  \end{aligned}$$

  (Verify this computation carefully, as it contains techniques you will frequently use.) This is the representative of the point t at location $\mathbf{t}$ relative to the origin o, so we

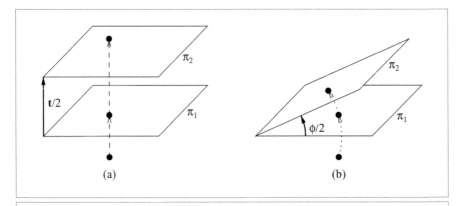

Figure 13.1: The versors for the Euclidean transformations can be constructed as multiple reflections in well-chosen planes. (a) Two parallel planes make a translation, and (b) two intersecting planes make a rotation.

have indeed found the translation versor. This versor is even a rotor, for $T_t \widetilde{T}_t = (1 - \mathbf{t}\infty/2)(1 + \mathbf{t}\infty/2) = 1 - \mathbf{t}\infty\mathbf{t}\infty/4 = 1 + \mathbf{t}^2\infty^2/4 = 1$.

Like all rotors in $\mathbb{R}^{n+1,1}$, the translation rotor has an exponential representation. This is easy to construct from the fact that ∞ squares to 0:

$$T_t = 1 - \mathbf{t}\infty/2 = 1 + (-\mathbf{t}\infty/2) + \tfrac{1}{2!}(-\mathbf{t}\infty/2)^2 + \cdots = e^{-\mathbf{t}\infty/2}.$$

The Taylor expansion of the exponential truncates by itself, after the first-order term. We foresaw such unusual rotors when discussing the exponential in Section 7.4.2.

- **Rotations in the Origin.** If we consider two planes in the origin as in Figure 13.1(b), these are dually represented as the purely Euclidean vectors $\pi_1 = \mathbf{n}_1$ and $\pi_2 = \mathbf{n}_2$. Viewed as reflection versors, we may take these as unit vectors without loss of generality. Their product is the versor

$$\pi_2\,\pi_1 = \mathbf{n}_2\,\mathbf{n}_1,$$

in which we recognize the purely Euclidean rotor R. Let us check what it does on a point, represented as a vector of $\mathbb{R}^{n+1,1}$:

$$R\,(o + \mathbf{p} + \tfrac{1}{2}\mathbf{p}^2\infty)R^{-1} = R\,o\,R^{-1} + R\,\mathbf{p}\,R^{-1} + \tfrac{1}{2}\mathbf{p}^2\,(R\infty R^{-1})$$
$$= o + (R\,\mathbf{p}\,R^{-1}) + \tfrac{1}{2}\mathbf{p}^2\infty$$
$$= o + (R\,\mathbf{p}\,R^{-1}) + \tfrac{1}{2}(R\,\mathbf{p}R^{-1})^2\infty,$$

which holds simply because the elements o and ∞ commute with the purely Euclidean elements. This is indeed the representation of the point at location

$R\,\mathbf{p}\,R^{-1}$, so successive reflections in planes through the origin are a rotation in a plane through the origin. This was also the case in the vector space model when we introduced its rotors in Chapter 7 (and therefore also in the homogeneous model). This rotor representation of rotations is therefore nicely backwards compatible: it is literally the same as in the vector space model contained in the space $\mathbb{R}^{n+1,1}$, and it merely acts on the whole space now.

Since this rotor is completely Euclidean, the earlier results of Chapter 7 and the vector space model in Chapter 10 apply, and we can write it as

$$R = e^{-\mathbf{I}\phi/2} = \cos(\phi/2) - \sin(\phi/2)\,\mathbf{I}$$

in terms of its rotation plane $\mathbf{I}$ and rotation angle ϕ.

- **General Rigid Body Motions**. A general rigid body motion can be constructed by first doing a rotation in the origin and following it by a translation. That gives a rotor of the form

$$(1 - \mathbf{t}\infty/2)\,R.$$

This generally contains terms of grades 0, 2, and 4. Being a rotor, the general rigid body motion can also be written in exponential form, but this is a bit involved: it requires viewing it as a screw motion (i.e., a rotation around a general axis combined with a translation along that axis). We will get back to analyzing and interpreting such motions in Section 13.5.2.

The above shows how the Euclidean proper motions are represented as rotors (i.e., as even unit versors). Improper Euclidean motions can be represented by odd unit versors (just multiply by a vector representing a dual reflection plane).

As a shorthand of the linear transformation produced by the application of $T_\mathbf{t}$ in a versor product, we will use $\mathsf{T}_\mathbf{t}[\cdot]$ (note the sans serif font). For the linear transformation of the rotor $R_{\mathbf{I}\phi}$, we may use the notation $\mathsf{R}_{\mathbf{I}\phi}[\cdot]$ (though we may conveniently drop the reference to the bivector angle).

13.2.3 COVARIANT PRESERVATION OF STRUCTURE

By representing Euclidean transformations as versors, we have achieved an important algebraic milestone in computing with Euclidean spaces, with far-reaching geometric consequences: automatic structure-preservation of all our constructions, in the sense of Section 7.6.3. By this we mean that a construction made anywhere, in any orientation, will automatically transform properly under a Euclidean motion.

As an example, we have just seen how we make a rotation by reflection in two planes passing through a point o as the product $R = \pi_2\,\pi_1$. If we now would want to have a similar rotation at a point p, we could define that new rotation by moving both planes to p by a translation versor $T_\mathbf{p}$ and making the computation $\pi_{1p}\,\pi_{2p}$ for the moved planes

$\pi_{1p} \equiv T_{\mathbf{p}} \, \pi_1 \, T_{\mathbf{p}}^{-1}$ and $\pi_{2p} \equiv T_{\mathbf{p}} \, \pi_2 \, T_{\mathbf{p}}^{-1}$. But when we do this, we find that this is the same as applying the translation $T_{\mathbf{p}}$ to the original rotor R, for $\pi_{1p} \, \pi_{2p} = T_{\mathbf{p}} \, \pi_1 \, T_{\mathbf{p}}^{-1} \, T_{\mathbf{p}} \, \pi_2 \, T_{\mathbf{p}}^{-1} = T_{\mathbf{p}} \pi_1 \, \pi_2 T_{\mathbf{p}}^{-1} = T_{\mathbf{p}} \, R \, T_{\mathbf{p}}^{-1}$. *The rotor of the moved planes is the moved rotor of the planes!* This is of course what we want for our original construction to be geometrically meaningful: it should transfer to another location in precisely this *covariant* manner (*co*-variant, implying: varying *with* the change) under the transformations of Euclidean geometry. We have effectively used similarity under Euclidean transformations as a generative principle rather than as a property to check afterwards.

It should be clear from this example that it is precisely the versor representation that makes this covariant structure preservation work so automatically, and that it should therefore work for any Euclidean transformation (since they are all represented by versors in the conformal model). We discuss the abstract principle involved in some detail, so that we can apply it throughout the remainder of this chapter. The most important structure-preservation properties of the versor product are easily proved.

- The versor product preserves the structure of a geometric product:
$$V[X\,Y] = V[X]\,V[Y].$$
  Proof:[4] Use associativity: $V\,(X\,Y)\,V^{-1} = V\,X\,Y\,V^{-1} = V\,X\,V^{-1}\,V\,Y\,V^{-1} = (V\,X\,V^{-1})\,(V\,Y\,V^{-1})$.

- The versor product is *linear*:
$$V[\alpha\,X + \beta\,Y] = \alpha\,V[X] + \beta\,V[Y].$$
  Proof: $V\,(\alpha\,X + \beta\,Y)\,V^{-1} = \alpha\,V\,X\,V^{-1} + \beta\,V\,Y\,V^{-1}$.

- The versor product is *grade-preserving*:
$$\langle V[X] \rangle_k = V[\langle X \rangle_k].$$
  Proof: Grades are defined in terms of the number of factors in an outer product. An outer product can be written as a sum of geometric products, and under transformation each of its terms transforms in a structure-preserving manner. The factorization of the transform is the transform of the factorization, and therefore the number of (transformed) vector factors in the outer product is preserved.

In Part I, we saw that we construct all objects and operators in geometric algebra using only these constructive elements (either as linear combinations of geometric products or by grade selection). Therefore the versors preserve all constructions, even including meet and join, inverses, and duality. So the universal structure preservation of constructions can be denoted compactly as

$$V(A \circ B)\,V^{-1} = (V\,A\,V^{-1}) \circ (V\,B\,V^{-1}),$$

where $\circ$ may be substituted by *any* of the products $\wedge, \rfloor, \lfloor, *, \cap, \cup$.

---

4   For structural clarity, we suppressed the signs in (7.18), which really should be there. You can easily verify that the result still holds when these are included.

This leads to the operational model principle, of which the conformal model is merely one manifestation:

If we have managed to construct a geometric algebra in which useful transformations (motions) in the base space are represented as versors, then all constructions in the model transform covariantly (i.e., with preservation of their structure).

We implicitly used this principle for rotations in the vector space model, which is why that is the operational model for directions in space. The homogeneous model augmented it with translations as linear transformations, but not as versors, which is why it had some problems preserving the structure of constructions (notably the metric properties). In the conformal model we finally have the Euclidean transformations as versors, so all constructions should be covariant under Euclidean transformations. It is the smallest known model that can do this. Operational models for other important geometries (such as projective geometry) have yet to be developed.

13.2.4 THE INVARIANCE OF PROPERTIES

Structure preservation is useful in analyzing and constructing all elements in the conformal model: it saves work. Since we have the Euclidean transformations available, we need to focus only on standardized elements at the origin (which is arbitrary anyway) to determine their form and defining equations. We can always move them to their desired location and orientation later. But are we guaranteed that their defining properties are preserved under such a transformation? For instance, could a point become a dual sphere or a dual plane? They are all represented as vectors, and could these not change into each other?

This brings us to how the characteristic properties change, that determine the interpretation of a representative element. For vectors, these properties were given in Table 13.2. They express that some combination of products is equal to a scalar. That algebraic form already guarantees that they are structurally preserved. To take an example, we have seen above how a unit point t at location $\mathbf{t}$ can be made by translation of a unit point at the origin: $t = T_{\mathbf{t}} o \, T_{\mathbf{t}}^{-1}$. But how do we know that it is still a unit point and not a dual sphere or dual plane? Of the original point o, we knew that $o^2 = 0$ and $-\infty \cdot o = 1$. Since these conditions are completely expressed in terms of the products of geometric algebra, they transform covariantly under the Euclidean transformation of translation, so we can relate the properties of t to those of o:

$$t\,t = T_{\mathbf{t}} o \, T_{\mathbf{t}}^{-1} \, T_{\mathbf{t}} o \, T_{\mathbf{t}}^{-1} = T_{\mathbf{t}} o\, o \, T_{\mathbf{t}}^{-1} = T_{\mathbf{t}} \, 0 \, T_{\mathbf{t}}^{-1} = 0,$$

so indeed, t is also a point. Moreover,

$$-\infty \cdot t = -\infty \cdot (T_{\mathbf{t}} o \, T_{\mathbf{t}}^{-1}) = -(T_{\mathbf{t}} \infty \, T_{\mathbf{t}}^{-1}) \cdot (T_{\mathbf{t}} o \, T_{\mathbf{t}}^{-1}) = -T_{\mathbf{t}} \, (\infty \cdot o) \, T_{\mathbf{t}}^{-1} = 1,$$

so t also has unit weight. Note that in the derivation of weight preservation, it is essential that ∞ is an *invariant* of the translation: it does not change, so the normalization equation preserves its form after transformation.

Characteristic properties expressed as scalar conditions in terms of the products and ∞ are automatically invariants of the Euclidean transformations.

In summary, as long as we construct elements and their defining properties in terms of the products of the geometric algebra of the operational model $\mathbb{R}^{n+1,1}$, we are automatically guaranteed of their covariance (for the elements) and invariance (for their properties) under Euclidean transformations. So we can construct everything around the origin, and yet get a full inventory of the general possibilities. With that labor-saving capability, we are ready to construct and interpret general blades of $\mathbb{R}^{n+1,1}$ effectively.

13.3 FLATS AND DIRECTIONS

Let us embed the geometrical elements of the homogeneous model into the conformal model, and investigate how the versor properties empower our geometric representation.

We originally started our exploration of the conformal model in Section 13.1.3 by interpreting vectors, which turned out to be the *dual* representations of spheres (real, imaginary, or zero radius), hyperplanes, and the point at infinity. To establish the correspondence with the vector space model and the homogeneous model, we now prefer instead to construct the *direct* representations of elements. The two representations are of course related by duality, so none is more fundamental than the other, but it is more natural to assign orientations to flats (and spheres) in the direct representation—and then dualization automatically introduces some grade-dependent orientation signs in their corresponding duals, which are good to get straight once and for all.

Constructing direct representations implies that we use the outer product $\wedge$, both for constructing the elements (as blades A) and for testing what Euclidean point set it represents (by solving $x \wedge A = 0$).

13.3.1 THE DIRECT REPRESENTATION OF FLATS

We start constructing an element of the form

$$X = \alpha \, (p_0 \wedge p_1 \wedge \cdots \wedge p_k \wedge \infty),$$

by which we mean the weighted outer product of $k+1$ unit points and the point at infinity ∞. This clearly contains the point at infinity (since $\infty \wedge X = 0$), and we will find out that such a blade is the direct representation of a k-*flat* in the conformal model (i.e., an offset k-dimensional linear subspace of $\mathbb{E}^n$). We do so by rewriting the blade to a form that we mostly recognize from the homogeneous model. This is done in three steps: standardization, interpretation, and generalization.

Standardization

By the structure preservation principle, we still have the general form if we take p_0 to be the arbitrary origin o (and we can view this either as moving the whole blade X back to the origin to make p_0 coincide with o, or as choosing p_0 as the origin of our representation). Note that the invariance of ∞ is essential (if it changed under translation, the form of the expression would not be the same at the origin). So we have reduced the element X to

$$X = \alpha\,(o \wedge p_1 \wedge \cdots \wedge p_k \wedge \infty).$$

Now because the outer product is antisymmetric, we can subtract one of the factors from each of the others without changing the value of the product (the extra terms would produce zero, $a \wedge b = a \wedge (b-a)$ being the prime example). We subtract o, and get the element X into the form:

$$X = \alpha\left(o \wedge (p_1 - o) \wedge \cdots \wedge (p_k - o) \wedge \infty\right).$$

We substitute the general form of the point representation of (13.3), setting $p_i = o + \mathbf{p}_i + \frac{1}{2}\mathbf{p}_i^2\,\infty$, to obtain:

$$X = \alpha\left(o \wedge (\mathbf{p}_1 + \tfrac{1}{2}\mathbf{p}_1^2\,\infty) \wedge \cdots \wedge (\mathbf{p}_k + \tfrac{1}{2}\mathbf{p}_k^2\,\infty) \wedge \infty\right).$$

The outer product with ∞ eliminates the extra terms $\frac{1}{2}\mathbf{p}_i^2\infty$. So we have:

$$X = \alpha\,(o \wedge \mathbf{p}_1 \wedge \cdots \wedge \mathbf{p}_k \wedge \infty).$$

We can now reduce the part involving the vectors $\mathbf{p}_i$ to a purely Euclidean k-blade $\mathbf{A}_k$, into which we also absorb the weight α, defining $\mathbf{A}_k = \alpha\,\mathbf{p}_1 \wedge \cdots \wedge \mathbf{p}_k$. We therefore ultimately find that this class of blade is equivalent to:

$$X = o \wedge \mathbf{A}_k \wedge \infty,$$

a much simplified form.

Interpretation

To find the Euclidean interpretation of $X = o \wedge \mathbf{A}_k \wedge \infty$ as a direct blade, we probe it with a point x and solve $x \wedge (o \wedge \mathbf{A}_k \wedge \infty) = 0$; that should give the set of Euclidean points it represents. Let us bring this into a more familiar form by expanding x as $o + \mathbf{x} + \frac{1}{2}\mathbf{x}^2\,\infty$. We immediately realize that the outer products with o and ∞ eliminate the first and last term of this point representation, so we effectively can substitute $x = \mathbf{x}$ without changing the solution:

$$0 = x \wedge o \wedge \mathbf{A}_k \wedge \infty = \mathbf{x} \wedge o \wedge \mathbf{A}_k \wedge \infty.$$

Therefore we need to solve $\mathbf{x} \wedge o \wedge \mathbf{A}_k \wedge \infty = 0$. This equation contains a purely Euclidean part ($\mathbf{x} \wedge \mathbf{A}_k$) and a non-Euclidean part ($o \wedge \infty$). These are orthogonal, so that their outer product is a geometric product, and we have $(o \wedge \infty)(\mathbf{x} \wedge \mathbf{A}_k) = 0$ (absorbing a sign into the zero). The model bivector $o \wedge \infty$ is not zero, so we must have $\mathbf{x} \wedge \mathbf{A}_k = 0$ (if you do not trust this reasoning, take the contraction of the original equation with $o \wedge \infty$ to eliminate the non-Euclidean part, and use $(o \wedge \infty) \rfloor (o \wedge \infty) = (\infty \cdot o)^2 = 1$). But the result

$$\mathbf{x} \wedge \mathbf{A}_k = 0$$

is a simple equation from the vector space model. It involves purely Euclidean vectors and implies that $\mathbf{x}$ is in the blade spanned by $\mathbf{A}_k$. Therefore the blade $X = o \wedge \mathbf{A}_k \wedge \infty$ represents a flat k-dimensional subspace through the origin o.

Generalization

We have found that flats through the origin are part of the conformal model, and since the Euclidean transformations are structure-preserving, all their translations and rotations should be in the model as well.

To find the general form of a k-dimensional flat through a point p, we need to translate $o \wedge \mathbf{A}_k \wedge \infty$ using the versor $T_{\mathbf{p}}$. Since this is a versor, we can distribute it over the terms o, $\mathbf{A}_k$, and ∞. The first and the last are easy: $T_{\mathbf{p}}$ applied to o is p, and ∞ is invariant, so the translation of X is $p \wedge T_{\mathbf{p}}[\mathbf{A}_k] \wedge \infty$. The translation of the Euclidean k-blade $\mathbf{A}_k$ requires some care:

$$\begin{aligned}
T_{\mathbf{p}}[\mathbf{A}_k] &= (1 - \mathbf{p}\infty/2)\,\mathbf{A}_k\,(1 + \mathbf{p}\infty/2) \\
&= \mathbf{A}_k - \tfrac{1}{2}(\mathbf{p}\infty\mathbf{A}_k - \mathbf{A}_k\mathbf{p}\infty) - \mathbf{p}\infty\mathbf{A}_k\mathbf{p}\infty/4 \\
&= \mathbf{A}_k + \tfrac{1}{2}\infty\,(\mathbf{p}\mathbf{A}_k - \widehat{\mathbf{A}_k\mathbf{p}}) + 0 \\
&= \mathbf{A}_k + \infty(\mathbf{p}\rfloor\mathbf{A}_k) = -p\rfloor(\infty\mathbf{A}_k).
\end{aligned} \tag{13.7}$$

where $p = T_{\mathbf{p}}[o]$. The final form is a multiplicative rewriting convenient for later computations. For our present purpose, the penultimate form is more convenient. Substituting it in $p \wedge T_{\mathbf{t}}[\mathbf{A}_k] \wedge \infty$, we find that its ∞-component is killed by the outer product with ∞, so that the general form of a flat k-dimensional offset subspace passing through p is

$$\textit{direct } \mathbf{A}_k\textit{-flat through } p: \quad p \wedge \mathbf{A}_k \wedge \infty. \tag{13.8}$$

Elements of this kind are depicted in Figure 13.2.

We see that a directly represented flat X in the conformal model always contains the point at infinity, so it satisfies $\infty \wedge X = 0$. In the origin, you can easily verify that $\|X\|^2 = X\widetilde{X} = -\|\mathbf{A}_k\|^2 \neq 0$, and since that is an invariant property, this holds anywhere. These two conditions characterize the blades representing a direct flat.

13.3.2 CORRESPONDENCE WITH THE HOMOGENEOUS MODEL

This expression $p \wedge \mathbf{A}_k \wedge \infty$ for a general direct flat in the conformal model is nicely backwards compatible with the homogeneous model. We can see this most clearly by

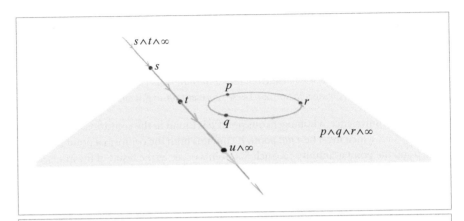

Figure 13.2: Flat elements in the conformal model: the plane $p \wedge q \wedge r \wedge \infty$ (with its orientation denoted by the circle), the line $s \wedge t \wedge \infty$, and the flat point $u \wedge \infty$ at their intersection.

trying to determine its meaning directly (without using the versor properties to bring things to the origin, as we did above). So we need to solve $0 = x \wedge (p \wedge \mathbf{A}_k \wedge \infty)$; it simplifies by the following steps:

$$
\begin{aligned}
0 &= x \wedge (p \wedge \mathbf{A}_k \wedge \infty) \\
&= (o + \mathbf{x} + \tfrac{1}{2}\mathbf{x}^2 \infty) \wedge (o + \mathbf{p} + \tfrac{1}{2}\mathbf{p}^2 \infty) \wedge \mathbf{A}_k \wedge \infty \\
&= (o + \mathbf{x}) \wedge (o + \mathbf{p}) \wedge \mathbf{A}_k \wedge \infty \\
&= \left(o \wedge (\mathbf{p} - \mathbf{x}) + \mathbf{x} \wedge \mathbf{p} \right) \wedge \mathbf{A}_k \wedge \infty \\
&= o \wedge (\mathbf{p} - \mathbf{x}) \wedge \mathbf{A}_k \wedge \infty + \mathbf{x} \wedge (\mathbf{p} - \mathbf{x}) \wedge \mathbf{A}_k \wedge \infty.
\end{aligned}
$$

The linear independence of the two terms implies that they should each be zero to make the total zero. The orthogonality of $o \wedge \infty$ to the Euclidean parts makes setting the first term equal to zero equivalent to solving the single equation $(\mathbf{p} - \mathbf{x}) \wedge \mathbf{A}_k = 0$. The second term is then also zero, so this is the general solution. We indeed have an offset flat with direction element $\mathbf{A}_k$ passing through the point at $\mathbf{p}$.

After the first two steps, this derivation is completely analogous to the computation in the homogeneous model:

$$
\begin{aligned}
0 &= (e_0 + \mathbf{x}) \wedge (e_0 + \mathbf{p}) \wedge \mathbf{A}_k \\
&= \left(e_0 \wedge (\mathbf{p} - \mathbf{x}) + \mathbf{x} \wedge \mathbf{p} \right) \wedge \mathbf{A}_k \\
&= \left(e_0 \wedge (\mathbf{p} - \mathbf{x}) + \mathbf{x} \wedge (\mathbf{p} - \mathbf{x}) \right) \wedge \mathbf{A}_k.
\end{aligned}
$$

The metric properties of o or e_0 (which are very different: $o^2 = 0$ while $e_0^2 = \pm 1$) do not enter the solution process, they only serve as bookkeeping devices for the outer product.

This equivalence of the derivations worked because the outer product with ∞ removed the terms involving the squared vectors from the point representation. We have demonstrated an important relationship between the two models:

The homogeneous model is embedded in the conformal model as governing the behavior of the blades involving a factor ∞. These represent offset flat subspaces.

Pushing this equivalence, a homogeneous point p is found in the conformal model as the element $p \wedge \infty$, which must be a *flat point*. It contains both the conformal point representative p and the point at infinity ∞. Such flat points occur as the result of the intersection of a line and a plane, which actually contains *two* common points: the finite intersection point and the point at infinity. An example is the point $u \wedge \infty$ in Figure 13.2. In containing this infinite aspect, they are subtly different from the point p itself, which is a dual sphere with zero radius, as we saw in Section 13.1.3. Separating these algebraically is natural in the conformal model and cleans up computational aspects. But we readily admit that these two conceptions of what still looks like a point in the Euclidean space $\mathbb{E}^n$ do take some getting used to.

Anyway, if you want the line through the points p and q, this is the 2-blade $p_H \wedge q_H$ in the homogeneous model (where p_H and q_H are the homogeneous representatives), and the 3-blade $\Lambda = p \wedge q \wedge \infty$ in the conformal model. It can be re-expressed as $p \wedge (q - p) \wedge \infty = p \wedge (\mathbf{q} - \mathbf{p}) \wedge \infty \equiv p \wedge \mathbf{a} \wedge \infty$, so the line passing through p with direction vector $\mathbf{a}$ is $p \wedge \mathbf{a} \wedge \infty$, just as it would have been $p_H \wedge \mathbf{a}$ in the homogeneous model. Our flexible computational rerepresentation techniques from the homogeneous model therefore still apply without essential change. In the conformal model, too, lines and planes can be represented by a mixture of locations and/or directions, with the outer product as universal construction operation.

13.3.3 DUAL REPRESENTATION OF FLATS

The dual representation of flats is simply found by dualization. As a pseudoscalar for the representational space $\mathbb{R}^{n+1,1}$ we use the blade representing the full Euclidean space as a flat, so

$$\text{pseudoscalar of conformal model: } I_{n+1,1} \equiv o \wedge \mathbf{I}_n \wedge \infty,$$

where $\mathbf{I}_n$ is the Euclidean unit pseudoscalar. As in the homogeneous model, we will denote the dualization in the full representational space by a six-pointed star (as X^*) and the dualization in its Euclidean part by a five-pointed star (as $\mathbf{X}^\star$, typically done on a purely Euclidean element).

Dualization in geometric algebra involves the inverse of the unit pseudoscalar. In the strange metric of the representational space $\mathbb{R}^{n+1,1}$, this is *not* equal to the reverse of $I_{n+1,1}$. The reason boils down to a property of the 2-blade $o \wedge \infty$. For we have

$$(o \wedge \infty)(o \wedge \infty) = (o \infty + 1)(o \infty + 1) = o \infty o \infty + 2 o \infty + 1 = 1$$

so that the $o \wedge \infty$ is its own inverse. (The final step involves a computation highlighted in structural exercise 2.) The inverse of the pseudoscalar is then easily verified to be

$$I_{n+1,1}^{-1} = o \wedge I_n^{-1} \wedge \infty,$$

so this is what we should use for dualization. It is satisfactory that the embedding of the Euclidean model is such that *the inverse of the representation of the Euclidean pseudoscalar is the representation of the inverse*. This is of course related to the general structure preserving properties of the conformal model. We did not have this property in the homogeneous model, where $(e_0 \, I_n)^{-1} = I_n^{-1} \, e_0^{-1} = (-1)^n \, e_0^{-1} \, I_n^{-1}$, involving an extra sign in odd dimensions. Because we have *two* extra representational dimensions in the conformal model (rather than just one), we can avoid those extraneous signs.

The structure-preserving property of the duality implies that we can make the dual of the standard flat through the origin and then translate it to the location of the desired flat to obtain its dual. This is permitted, for the structure preservation

$$V[X]^* = V[X^*]$$

holds for any even versor, especially for Euclidean motions. In the conformal model, we do not need to know whether an element is a direct or a dual representation to move it in a Euclidean manner by an even versor. This is an improvement over the homogeneous model. (However, in structural exercise 4, we ask you to show that there is an additional minus sign for an odd versor. So there is a small difference between reflections and motions.)

Dualizing the origin blade $o \wedge \mathbf{A}_k \wedge \infty$, we find

$$
\begin{aligned}
(o \wedge \mathbf{A}_k \wedge \infty)^* &= (o \wedge \mathbf{A}_k \wedge \infty)\rfloor(o \wedge I_n^{-1} \wedge \infty) \\
&= -(o \wedge \mathbf{A}_k)\rfloor(I_n^{-1} \wedge \infty) \\
&= -o\rfloor(\mathbf{A}_k^{\star} \wedge \infty) \\
&= \widehat{\mathbf{A}_k^{\star}}.
\end{aligned}
$$

Note that the grade inversion extends over the Euclidean dual, so it gives a sign of $(-1)^{n-k}$. The dual of a blade in the homogeneous model involved a similar sign change, so we are still backwards compatible in the dualization.

But in contrast to the homogeneous model (see Section 11.8.2), translations on dual blades are performed by the same operation as translations on direct blades: simply apply the translation versor. We have already computed the translation of a purely Euclidean element in (13.7), so the result is immediate:

$$\text{dual flat:} \quad \mathsf{T}_{\mathbf{p}}[\widehat{\mathbf{A}_k^{\star}}] = \widehat{\mathbf{A}_k^{\star}} + \infty(\mathbf{p}\rfloor\widehat{\mathbf{A}_k^{\star}}) = -p\rfloor(\mathbf{A}_k^{\star} \, \infty). \tag{13.9}$$

This makes it particularly easy to construct, say, the dual representation of a hyperplane with 2-blade $\mathbf{I}$ and associated normal $\mathbf{n} = \mathbf{I}^*$, passing through p: it is $\Pi^* = -p\rfloor(\mathbf{n}\,\infty) = -\mathbf{n} - (\mathbf{p} \cdot \mathbf{n})\,\infty$. This confirms our earlier result of Table 13.2, though the sign is now properly related to a properly oriented plane Π and a well-defined pseudoscalar in the dualization process.

Direct representations of flats contain a factor ∞ and are therefore characterized by $\infty \wedge X = 0$. Dualization of this condition yields that a blade X^* representing a dual flat is characterized by $\infty \rfloor X^* = 0$. This is of course interpreted to mean that the point at infinity is contained in the element dually represented by X. The condition itself is invariant under Euclidean transformations, because ∞ is. The other property of a direct flat, $\|X\|^2 < 0$, dualizes to $\|X^*\|^2 > 0$ for the dual representation.

13.3.4 DIRECTIONS

In the construction of the direct flat $p \wedge \mathbf{A}_k \wedge \infty$, we recognize the location p and the blade $\mathbf{A}_k \wedge \infty$. Bearing in mind how the conformal flats correspond to their representations in the homogeneous model, such an element $\mathbf{A}_k \wedge \infty$ must be the conformal representation of a k-dimensional *direction*:

$$\textit{directly represented direction:} \quad \mathbf{A}_k \wedge \infty.$$

To be a pure direction, it should only have directional properties and no locational aspects. It is easy to verify how it transforms under a general Euclidean motion consisting of a rotation R and a translation T:

$$T_t[R[\mathbf{A}_k \wedge \infty]] = \left(R[\mathbf{A}_k] - (\mathbf{t}\rfloor R[\mathbf{A}_k])\,\infty\right) \wedge \infty = R[\mathbf{A}_k] \wedge \infty.$$

So these elements of the form $\mathbf{A}_k \wedge \infty$ are rotation covariant, but translation invariant. That is precisely what you would expect from directions. The *dual directions* are simply their duals:

$$\textit{dual direction:} \quad (\mathbf{A}_k \wedge \infty)^* = (\mathbf{A}_k \wedge \infty)\rfloor(o \wedge \mathbf{I}_n^{-1} \wedge \infty) = -\mathbf{A}_k^\star \wedge \infty,$$

again involving signs that should be observed if you want to use them in consistently oriented computations. They are clearly also translation invariant.

You can make a flat by attaching a direction element to a point p using an outer product, as in $p \wedge (\mathbf{A}_k \wedge \infty)$. The corresponding dual flat is made by attaching the dual direction to the point p using the contraction, giving $p\rfloor(-\mathbf{A}_k^\star\,\infty)$ in agreement with (13.9).

A direction element X is characterized by containing ∞, so that $\infty \wedge X = 0$, and moreover by having a zero norm: $\|X\|^2 = 0$, as you can easily verify. A dual direction X^* has $\infty\rfloor X^* = 0$ and $\|X^*\|^2 = 0$.

13.4 APPLICATION: GENERAL PLANAR REFLECTION

In classical texts on subjects like computer graphics, many crucial formulas are given in a Euclidean or homogeneous-coordinate form. How are we to integrate those into the new conformal model? This is often surprisingly straightforward, since the more familiar vector space model and the homogeneous model are both naturally contained in the conformal model. All that is required is recasting of standard formulas into a geometric algebra format before you perform the embedding. You then find that the conformal model gives you the freedom to generalize the formulas, in a manner that the other models did not allow. The reason, as always, is the versor form of the Euclidean transformations and its associated structure-preserving properties.

As an application, let us consider the reflection of a general line Λ in a general plane π. See Figure 13.3.

- **Classical Linear Algebra.** In the classical way of doing this, you would have to treat positions and orientations separately. The intersection point q of Λ and π determines a point on the outgoing line. For its direction vector, you would reflect the direction **u** of Λ relative to the normal vector **n** of the plane. This would refer to a Euclidean formula such as

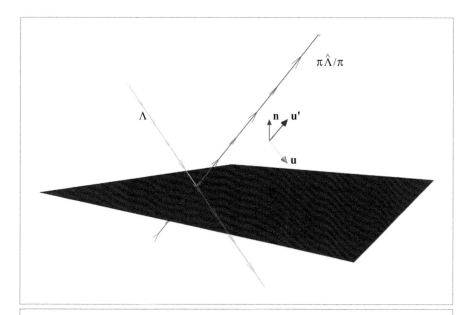

Figure 13.3:   Reflection of a line Λ in a (dual) plane π in the conformal model. This is derived from the reflection of a vector **u** in a normal vector **n** in the origin, according to the vector space model (also depicted).

$$\mathbf{u} \;\mapsto\; \mathbf{u}' = \mathbf{u} - 2(\mathbf{u} \cdot \mathbf{n})\mathbf{n}.$$

You would then reassemble the position q and the transformed direction $\mathbf{u}'$ and return that as the transformed line, in some data format.

- **Vector Space Model.** In the vector space model of geometric algebra, we can start from the same formula, since all quantities belong to it. We have seen in Section 7.1 how this can be reformulated to geometric algebra by introducing the geometric product, giving

$$\mathbf{u} \;\mapsto\; -\mathbf{n}\,\mathbf{u}\,\mathbf{n} = -\mathbf{n}\,\mathbf{u}/\mathbf{n} \tag{13.10}$$

(we introduced the division to waive normalization of $\mathbf{n}$.) This is clearly a versor product, and the embedding in the geometric algebra of the vector space model gives us the automatic extension to the reflection of any blade in the Euclidean model, not just a direction vector $\mathbf{u}$.

- **Homogeneous Model.** In the homogeneous model we can combine the directional and locational aspects of the line in one data structure, a 2-blade. We can use that blade to compute the intersection point, but we still need to isolate the direction vector to perform the reflection (by the same formula as the vector space model). We can then compose the resulting line from the intersection point and the new reflected direction as a resulting 2-blade in the homogeneous model.

- **Conformal Model.** In the conformal model, we can completely skip determining the intersection point. Instead, we extend the reflection formula (13.10) from the vector space model directly from merely working on directions to acting on the whole line, in the following manner.

We first note that the formula also holds in the conformal model, since the vector space model is contained in it. The direction vector $\mathbf{n}$ is now recognized as a dual plane through the origin, and this is the plane Π of the original problem, as long as we accept for the moment that it passes through the origin. We write this special plane as Π_o, dually represented by $\pi_o = \mathbf{n}$. We want the vector $\mathbf{u}$ to represent the direction of the line Λ_o also passing through the origin. This implies that the line should be represented as $\Lambda_o = o \wedge \mathbf{u} \wedge \infty$.

Let us now extend the reflection formula of (13.10) to this element Λ_o. This is straightforward, since o and ∞ are invariants of the reflection in π_o, and the versor structure of the conformal model does the rest.

$$o \wedge (-\mathbf{n}\,\mathbf{u}/\mathbf{n}) \wedge \infty = (-\mathbf{n}\,o/\mathbf{n}) \wedge (-\mathbf{n}\,\mathbf{u}/\mathbf{n}) \wedge (-\mathbf{n}\,\infty/\mathbf{n})$$

$$= \mathbf{n}\,(o \wedge \mathbf{u} \wedge \infty)\widehat{\;}/\mathbf{n}.$$

(We wrote the formula using the grade involution $\widehat{\;}$, since there are obviously as many signs as the grade of Λ_o and this will help the further generalization below.) So we now have a formula that reflects a line Λ_o through the origin in a dual plane π_o through the origin:

$$\Lambda_o \;\mapsto\; \pi_o \widehat{\Lambda_o}/\pi_o$$

Since this formula consists only of geometric products, it is clearly translation covariant. So moving both to a general position using the Euclidean rigid body motion V (first a rotation then a translation), we get a formula for a general line $\Lambda = \mathsf{V}[\Lambda_o]$ and a general dual plane $\pi = \mathsf{V}[\pi_o]$

$$\Lambda = \mathsf{V}[\Lambda_o] \;\mapsto\; \mathsf{V}[\pi_o \widehat{\Lambda}_o / \pi_o] = \mathsf{V}[\pi_o]\, \mathsf{V}[\widehat{\Lambda}_o] / \mathsf{V}[\pi_o] = \pi\, \widehat{\Lambda} / \pi.$$

Under this motion, the original intersection point o moves to $\mathsf{V}[o]$, but the final result holds whether we know the intersection point explicitly or not. So the reflection of a line Λ in a dual plane π in general is

$$\Lambda \;\mapsto\; \pi\, \widehat{\Lambda} / \pi,$$

in all its aspects of location and direction. It can do this *without actually computing the intersection point*, and that is an improvement over both the vector space model and the homogeneous model. (If you really want to compute the intersection point, the `meet` of the plane Π and the line Λ is the flat point $\Lambda \cap \Pi = \pi \rfloor \Lambda$.)

The result is clearly only the special case for a line of the general reflection formula that reflects any directly represented flat space in an arbitrary plane:

$$X \;\mapsto\; \pi\widehat{X}/\pi. \tag{13.11}$$

You see how much work the nice algebraic structure saves to transfer formulas, as well as providing clear insights in why they hold, and how to extend them. Classical geometrical results, typically formulated in the vector space model, are naturally embeddable within the conformal model. Moreover, we don't have to spell them all out: with a bit of practice and confidence, such transfers become one-liners. In the example above, you know that o is never special, so that (13.11) is the only actual computation to get to the general case. You may even learn to guess the result (which was clearly possible in the example above based on the pattern in reflection formulas in Table 7.1), and merely verify the correctness with interactive software (moving its constituents around to check its structure preservation, and the correctness of weights). That is how we ourselves typically design formulas for specific tasks, and the method is justified by virtue of the structure preservation properties of the conformal model.

Incidentally, (13.11) is the reflection formula used in the motivating example of Section 1.1.

13.5 RIGID BODY MOTIONS

We have seen in Section 13.2 how the conformal model represents the Euclidean transformations as versors. Of special interest are the Euclidean rigid body motions, represented by even versors that can be normalized to rotors. We explore their structure, culminating in a closed-form logarithm useful for interpolation of rigid body motions.

13.5.1 ALGEBRAIC PROPERTIES OF TRANSLATIONS AND ROTATIONS

The rotor $T_{\mathbf{t}} = \exp(-\mathbf{t}\infty/2) = (1 - \mathbf{t}\infty/2)$ has all the desirable properties of a translation versor. The occurrence of the null vector ∞ makes translation versors combine in an additive manner in terms of its vector parameter $\mathbf{t}$, since it kills any terms that might contain the combination of translations:

$$T_{\mathbf{s}}\, T_{\mathbf{t}} = (1 - \mathbf{s}\infty/2)\,(1 - \mathbf{t}\infty/2) = 1 - \mathbf{s}\infty/2 - \mathbf{t}\infty/2 + \mathbf{s}\infty\mathbf{t}\infty/4 = 1 - (\mathbf{s}+\mathbf{t})\infty/2 = T_{\mathbf{s}+\mathbf{t}}.$$

You may contrast this with the much more complicated composition of rotations, which we saw in Section 7.3. That was essentially the geometric algebra version of the quaternion product (in 3-D), extended to n-D. It, too, is automatic from the rotor representation of the rotation, but the result was in essence an addition of arcs on the rotation sphere, involving severe trigonometry in a coordinate representation.

The algebraic nature of the bivector in the exponent of these rotors is the crucial difference that makes us experience them so differently geometrically: when it has a negative square, it generates rotations, which do not commute and involve trigonometry, and when it has a zero square, it generates translations, which do commute and remain additive. We will meet the remaining possibility of a positive square later (in Section 16.3, where we show that they are scalings and involve hyperbolic functions).

The versor structure of the conformal model means that not only elements like flats and directions transform easily, but so do the versors themselves. A consequence of the properties of translations is that *translations are translation invariant*: a translated translation rotor acts in the same manner as the original rotor, for

$$T_{\mathbf{s}}[T_{\mathbf{t}}] = T_{\mathbf{s}}\, T_{\mathbf{t}}\, T_{\mathbf{s}}^{-1} = T_{\mathbf{s}}\, T_{\mathbf{t}}\, T_{-\mathbf{s}} = T_{\mathbf{s}+\mathbf{t}-\mathbf{s}} = T_{\mathbf{t}}.$$

(Note that T denotes the translation rotor, whereas $\mathsf{T}[\ ]$ denotes the translation operator!) This invariance property is why we feel so free to draw the translation vector anywhere, attached to any point we like in the same manner. By contrast, *rotations are not rotation invariant*:

$$R_{J\psi}[R_{I\phi}] = R_{R_{J\psi}[I\phi]},$$

and therefore the total rotation rotates in the $R_{J\psi}$-rotated plane $\mathbf{I}$, not in the original plane $\mathbf{I}$. (Again, R denotes the rotation rotor, while $\mathsf{R}[\ ]$ denotes the rotation operator.)

When we translate a rotation rotor $R_{I\phi} = \exp(-\mathbf{I}\pi/2)$, by a translation rotor $T_{\mathbf{t}}$, this has the same effect as shifting the rotation 2-blade by the translation $T_{\mathbf{t}}$:

$$T_{\mathbf{t}}[R_{I\phi}] = R_{T_{\mathbf{t}}[I\phi]}.$$

In 3-D space, this is particularly striking. The rotation 2-blade is the inverse dual of the rotation axis Λ_o also passing through the origin, since $\Lambda_o = \mathbf{I}^*$. Originally, the axis passes through the origin; after the translation the new axis Λ has become $T_\mathbf{t}[\Lambda_o]$. But the property that the rotation rotor is the exponent of the dual line is covariant, so the new rotation rotor is simply the exponent of the new dual axis Λ:

3-D rotation around line Λ over angle ϕ: $R = e^{-\Lambda^{-*}\phi/2} = e^{\Lambda^*\phi/2}$.

(Make sure you normalize the axis Λ so that the angle ϕ has the intended meaning.)

In Chapter 7, we showed that the 3-D rotation rotors around the origin were the geometric algebra way of doing quaternions, in what we now recognize was the vector space model. That had the advantage of embedding quaternions in a real computational framework, but did not add any computational power. In the conformal model, the concept of a quaternion is greatly extended. In 3-D, the extension encompasses the general rotation axis for an arbitrary rotation to encode the rotation. (In n-D, it still signifies the 2-blade of the rotational bivector, though this does then not dualize to a single axis.) This relationship to the rotation axis is very compact and convenient in applications. You have met it first in our motivating example in Section 1.1.

13.5.2 SCREW MOTIONS

A translated rotation is not yet the general rigid body motion, for it lacks a translation component along the rotation axis. So we would have to supply that separately, to obtain the general rigid body motion as

$$T_\mathbf{w}\, T_\mathbf{v}[R_{\mathbf{I}\phi}],$$

where $\mathbf{v}$ is a vector in the plane $\mathbf{I}$ (so that $\mathbf{v} \wedge \mathbf{I} = 0$), and $\mathbf{w}$ is vector perpendicular to it (so that $\mathbf{w}\rfloor\mathbf{I} = 0$).

Such a decomposition of a rigid body motion gives us a better understanding of what it does, and we can execute the two parts (translation along the axis and rotation around the axis) in any order or simultaneously. When we do them in similar amounts to reach a total resulting rigid body motion, we obtain a *screw motion*, as depicted in Figure 13.4. The simultaneous rotation and translation (around and along the spatial axis) are then related through the pitch of the screw. According to *Chasles' theorem*, an arbitrary rigid body motion can be represented in this manner. Given a general rigid body motion, we can compute the location and magnitude of the elements of the screw using the conformal model. Perhaps surprisingly, it completely avoids the rather involved trigonometry.

First, assume that we have a general rigid body motion $T_\mathbf{t}\, R_{\mathbf{I}\phi}$ composed of a standard rotation in a plane $\mathbf{I}$ at the origin, followed by a translation. This is the usual way to decompose rigid body motions, which we already encountered in the homogeneous model (in Section 11.8.5). It corresponds well to the homogeneous coordinate matrices for rigid body motions. Following Chasles, we attempt to rewrite this rigid body motion as a displaced rotation around a displaced axis parallel to $\mathbf{I}^*$, with its location characterized by

Figure 13.4: The computation of the screw corresponding to a rotor R followed by a translation over **t**. The rotor plane and angle are indicated by the blue disk. The translation **w** along the screw axis is the rejection of **t** by the rotation plane. The location vector **v** of the axis is found as the unique vector in the rotation plane whose chord under rotation is parallel to **w** − **t** (indicated by a thin blue line). The resulting screw motion around the axis (in black) is shown applied to a tangent bivector (in fading shades of green).

a vector **v** (which we can take to be in the **I**-plane without loss of generality), followed by a translation **w** in the direction of that axis. We saw this form of the motion above, and therefore find that we need to solve

$$T_t R_{I\phi} = T_w T_v[R_{I\phi}]$$

for **v** and **w**, where $\mathbf{v} \wedge \mathbf{I} = 0$ and $\mathbf{w}\rfloor\mathbf{I} = 0$. The two factors on the right-hand side should commute, since we can do the translation along the displaced axis before or after a rotation around it. Actually, that demand on commutation is an equivalent, more algebraic way of phrasing the problem, from which the conditions on **v** and **w** relative to **I** follow.

The Euclidean blade **I** characterizing the direction of the rotation plane can be computed as the normalization of $\langle R \rangle_2$. Since the only translation perpendicular to **I** is performed by **w**, we must have that **w** is the rejection of the translation vector **t** by **I**:

$$\mathbf{w} = (\mathbf{t} \wedge \mathbf{I})/\mathbf{I}.$$

That leaves the other part of the translation $(\mathbf{t}\rfloor\mathbf{I})/\mathbf{I}$. Call it $\mathbf{u}$ for the moment. The problem is now reduced to a 2-D problem in the $\mathbf{I}$-plane, namely that of solving for $\mathbf{v}$ in

$$T_\mathbf{u}\, R_{\mathbf{I}\phi} = T_\mathbf{v}\, R_{\mathbf{I}\phi}\, T_{-\mathbf{v}}.$$

We would like to swap the rightmost translation with $R_{\mathbf{I}\phi}$. We derive the following swapping rule (abbreviating $R_{\mathbf{I}\phi}$ to R for convenience):

$$R\, T_{-\mathbf{v}} = R\,(1+\mathbf{v}\infty/2) = (1+R\mathbf{v}\infty\widetilde{R}/2)\,R = (1+R\mathbf{v}\widetilde{R}\,\infty/2)\,R = T_{-R\mathbf{v}\widetilde{R}}\,R. \quad (13.12)$$

We also observe that since $\mathbf{v}$ is a vector in the R-plane, we have

$$\mathbf{v}\,\widetilde{R} = \mathbf{v}\,(\langle\widetilde{R}\rangle_0 + \langle\widetilde{R}\rangle_2) = (\langle\widetilde{R}\rangle_0 - \langle\widetilde{R}\rangle_2)\,\mathbf{v} = (\langle R\rangle_0 + \langle R\rangle_2)\,\mathbf{v} = R\,\mathbf{v}.$$

Therefore,

$$T_\mathbf{u}\, R = T_\mathbf{v} R\widetilde{T}_\mathbf{v} = T_\mathbf{v} T_{-R\mathbf{v}\widetilde{R}}\,R = T_{\mathbf{v}-R\mathbf{v}\widetilde{R}}\,R = T_{\mathbf{v}-R^2\mathbf{v}}R = T_{(1-R^2)\mathbf{v}}R.$$

It follows that

$$\mathbf{v} = (1-R^2)^{-1}\mathbf{u} = (1-R^2)^{-1}\,(\mathbf{t}\rfloor\mathbf{I})/\mathbf{I}. \quad (13.13)$$

Graphically, the vector $\mathbf{v} - R\mathbf{v}\widetilde{R} = 2(\mathbf{v}\rfloor R)/R$ occurring in this computation is the chord connecting the rotated $\mathbf{v}$ to the original $\mathbf{v}$. We should find $\mathbf{v}$ such that this equals $\mathbf{u}$, which gives the construction in Figure 13.4 as the geometric interpretation of this algebraic computation. To pull it all together, the final result giving the screw parameters is:

$$\textit{screw decomposition:}\quad T_\mathbf{t}\, R_{\mathbf{I}\phi} = T_{(\mathbf{t}\wedge\mathbf{I})/\mathbf{I}}\,{}^\mathsf{T}_{(1-R_{\mathbf{I}\phi}^2)^{-1}(\mathbf{t}\rfloor\mathbf{I})/\mathbf{I}}[R_{\mathbf{I}\phi}]$$

Realize again that T denotes a translation rotor, whereas $\mathsf{T}[\,]$ denotes a translation operator!

13.5.3 LOGARITHM OF A RIGID BODY MOTION

Using Chasles' theorem, we can determine the logarithm of a rigid body motion rotor V, which means that we can determine the bivector when we have been given the rotor. For such a rotor is, on the one hand,

$$V = T_\mathbf{t}\, R_{\mathbf{I}\phi} = (1-\mathbf{t}\,\infty/2)\,R_{\mathbf{I}\phi} = R_{\mathbf{I}\phi} - \tfrac{1}{2}\,\mathbf{t}\,R_{\mathbf{I}\phi}\infty, \quad (13.14)$$

and on the other hand,

$$V = T_\mathbf{w}\, R_{\mathsf{T}_\mathbf{v}[\mathbf{I}\phi]} = e^{-\mathbf{w}\infty/2}\, e^{-\mathsf{T}_\mathbf{v}[\mathbf{I}\phi]/2} = e^{-\mathbf{w}\infty/2 - \mathsf{T}_\mathbf{v}[\mathbf{I}\phi]/2},$$

in which the addition of the exponents is only permitted because the rotors commute. They were designed that way by Chasles' theorem, and it is the algebraic consequence of the possibility to execute the screw as one smooth motion. Substituting the values of $\mathbf{w}$ and $\mathbf{v}$ we found above, the requested rigid body motion logarithm is

$$\log(T_{\mathbf{t}}\,R_{\mathbf{I}\phi}) = -((\mathbf{t}\wedge\mathbf{I})/\mathbf{I})\,\infty/2 + {}^{\mathsf{T}}{}_{(1-R^2)^{-1}(\mathbf{t}\rfloor\mathbf{I})/\mathbf{I}}[-\mathbf{I}\phi/2]$$
$$= -((\mathbf{t}\wedge\mathbf{I})/\mathbf{I})\,\infty/2 + (1-R^2)^{-1}(\mathbf{t}\rfloor\mathbf{I}\phi)\,\infty/2 - \mathbf{I}\phi/2 \qquad (13.15)$$

The value of this expression can be determined via the computation of the screw parameters, provided that we can retrieve both $R_{\mathbf{I}\phi}$ (which gives $\mathbf{I}\phi$ by the logarithm of (10.14)) and $\mathbf{t}$ from the overall rotor V. But this retrieval is simple in (13.14):

$$R_{\mathbf{I}\phi} = -o\rfloor(V\infty), \quad \mathbf{t} = -2\,(o\rfloor V)/R_{\mathbf{I}\phi}.$$

Putting all formulas together, we get the pseudocode of Figure 13.5. Special cases happen when R equals $+1$ or -1. When R equals $+1$, the motion is a pure translation versor and we just return the logarithm of that (which is $-\mathbf{t}\,\infty/2$).When R equals -1, there is no unambigious logarithm, since a rotation of π within any plane could generate the -1. To prevent spurious terms in the product with the translation versor, one must choose a plane perpendicular to $\mathbf{t}$. In 2-D, this is impossible, and the logarithm does not exist (as we intimated in Section 7.4.4). Having the logarithm of a rigid body motion is a powerful result. It permits us to interpolate these motions in a total analogy to the rotation

```
log(V){
        R = -o⌋(V∞)
        t = -2 (o⌋V)/R
        if (R == -1) return ("no unique logarithm")
        if (R == 1)
                log = -t ∞/2
        else
                I = ⟨R⟩₂/√(-⟨R⟩₂²)
                φ = -2 atan2(⟨R⟩₂/I, ⟨R⟩₀)
                log = (- (t ∧ I)/I + 1/(1 - R²) t⌋Iφ) ∞/2 - Iφ/2
        endif
        }
```

Figure 13.5: Computation of the logarithm of a normalized rigid body motion rotor V. One may improve subsequent numerics by making sure to return a bivector by taking the grade-2 element by $\langle\log\rangle_2$.

interpolation procedure that we had in the vector space model. See Section 13.10.4 for an implementation.

13.6 APPLICATION: INTERPOLATION OF RIGID BODY MOTIONS

According to Chasles' theorem, a rigid body motion can be viewed as a screw motion. It is then natural to interpolate the original motion by performing this screw gradually. Figure 13.6 shows how simple this has become: the ratio of two unit lines L_2 and L_1 defines the square of the versor that transforms one into the other (see structural exercise 9). Performing this motion in N steps implies using the versor

$$V^{1/N} = e^{\log(L_2/L_1)/(2N)}.$$

Applying this versor repeatedly to the line L_1 gives the figure. It interpolates the transformation of L_1 into L_2 and extrapolates naturally. Note how all can be defined completely, and simply, in terms of the geometric elements involved. You do not need coordinates to specify how things move around. The same rotor $V^{1/N}$ can of course be applied to any element that should move similarly to L_1.

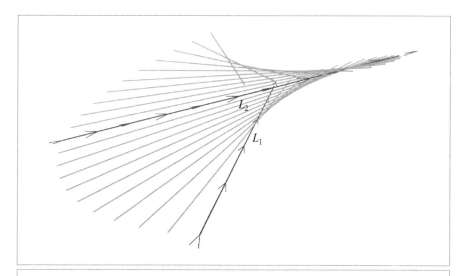

Figure 13.6: The interpolation and extrapolation of the rigid body motion transforming a spatial line L_1 into a line L_2 is done by repeated application of the versor $\exp\left(\log(L_2/L_1)/(2N)\right)$. The screw nature of the motion is apparent.

Since the resulting logarithms are bivectors, they can be interpolated naturally themselves. This allows one to estimate how various points on a rigid body move, establish their motion bivectors, and average those to get an improved estimate of the motion. Such numerical estimation techniques for motions are now being developed.

13.7 APPLICATION: DIFFERENTIAL PLANAR REFLECTIONS

Now that we have general planar reflections available in (13.11), we can redo the example of Section 8.5.2, in which we studied a rotating mirror and its reflected image using geometric differentiation. Before, we used what was effectively the vector space model and therefore could only work in the origin. Now we can treat the fully general planar rotation.

To restate the problem, we consider the reflection of an element X in the dual plane π (i.e., $X \mapsto \pi \widehat{X} \pi^{-1}$) in 3-D Euclidean space. We rotate the plane slightly around line Λ, and need to find out what this does to the reflection of X.

In Section 8.2.5, we performed the differential computation of such perturbed versors in completely general form, with as a result (8.6) for the bivector of the perturbation versor, given an original versor and a motion. We can apply this immediately, since the conformal model gives us these versors: the original versor is the reflection versor π, and the motion is $\exp(-\Lambda^{-*}\phi/2)$ with bivector $\Lambda^{-*}\phi$. Then according to application of (8.6), the versor displacing the reflection should have bivector

$$-2\phi\,(\pi \times \Lambda^{-*})/\pi = -2\phi\,(\pi\rfloor\Lambda^{-*})\,\pi = -2\phi\,(\pi\rfloor\Lambda^{-*}) \wedge \pi = 2\phi\,(\pi \wedge \Lambda)^* \wedge \pi \quad (13.16)$$

The geometric interpretation of $\pi \wedge \Lambda$ is the plane perpendicular to π and containing the line Λ; then the interpretation of the total formula is that it is dual of the meet of this plane with the dual plane π (i.e., it is dual of the orthogonal projection of the line Λ onto the mirror π). This line is indicated in Figure 13.7(a). Since the bivector of the resulting perturbing versor is this dual line, that versor is a rotation around this projected line when the reflection plane changes. This is clear geometry, easily verified. When it rotates around this line, the perturbed element follows a curved path: *first-order perturbations in versors can lead to element paths of second order.*

We would have to do a second-order Taylor series on the position to achieve a similar effect, so not only are the versor perturbations structure-preserving but they also give us a much better approximation for the same effort. You can work out the angle of rotation of the perturbation in a similar way as in the origin-based treatment, again resulting in (8.11).

Note that this computation is *not* valid for large perturbations: doing the same construction on a perturbed mirror shows that the projected line changes (since it now

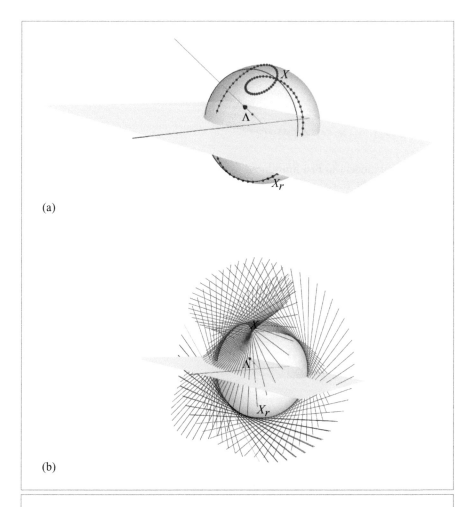

(a)

(b)

Figure 13.7: (a) The yellow planar mirror π reflects a point X to the point X_r. The green line Λ rotates the mirror; the red curve shows the orbit of the reflection in 100 steps. In black is the first-order perturbation description of the orbit around X_r. It is a rotation of X_r around the blue line, which is the projection of the line Λ onto the mirror π, which follows the tangent circle to the true orbit. The first few black points keep pace with the red points, showing that the velocity is correct. Note that a first-order perturbation of the versor generates a second-order curve. Both the black and red curves are on the sphere $X \wedge \pi^* \wedge \Lambda^*$ (perpendicular to both Λ and π), which has also been depicted, to make the spatial nature of the curves more obvious. (b) The same setup, merely replacing X by a line, to show the universality of constructions in geometric algebra. The black lines are again good first-order approximations to the exact results.

needs to be on the perturbed mirror), and therefore the arc changes as well. In fact, it becomes a caustic, of which the above effectively computes the local osculating cylinder. Because we have done the whole computation in the conformal model, we can substitute any of its other elements for X. Figure 13.7(b) shows the whole computation applied to a reflected line. After the next chapter you will not even hesitate to apply it to the reflection of a sphere, circle, or tangent blade.

13.8 FURTHER READING

The conformal model was first introduced to the engineering community in 1999 [31], but its roots can be traced to the 19th century. We will give more specific references in the following chapters, when we have more a complete view of its contents and capabilities.

Commercial applications of conformal geometric algebra to model statics, kinematics and dynamics of particles and linked rigid bodies are protected by U.S. Patent 6,853,964, "System for encoding and manipulating models of objects" [32]. Anyone contemplating such applications should contact Alyn Rockwood or David Hestenes. The patent does not restrict use of conformal geometric algebra for academic research and education or other worthy purposes that are not commercial.

13.9 EXERCISES

13.9.1 DRILLS

These drills intend to familiarize you with the form of common geometric elements and their parameters in the conformal model. We recommend doing them by hand first, and check them with interactive software later.

1. Give the representation of a point p_1 with weight 2 at location $\mathbf{e}_1 + \mathbf{e}_2$.
2. Give the representation of a point p_2 with weight -1 at location $\mathbf{e}_1 + \mathbf{e}_3$, and compute its distance to p_1.
3. Give the representation of the line L through p_1 and p_2.
4. Compute weight and direction of the line L.
5. Compute the support point on the line L.
6. Give the direct representation of the plane Π through L and the unit point at the origin.
7. Compute the direction and support of the plane Π.
8. Give the representation of the translation over $-\mathbf{e}_1$ of the plane Π.

9. Compute the dual π of the plane Π. Compute its dual direction and its moment.

10. Compute the dual of the line L.

13.9.2 STRUCTURAL EXERCISES

1. Show that on the $\{e, \bar{e}\}$-basis, the point p of (13.3) is represented as:

$$p = \mathbf{p} + \tfrac{1}{2}(1 - \mathbf{p}^2)e + \tfrac{1}{2}(1 + \mathbf{p}^2)\bar{e}.$$

   In [33] and [15], you find the close relationship of this formula with stereographic projection spelled out as a way of visualizing of the conformal model. Unfortunately, it needs the two extra dimensions, so you can only visualize the model for a 1-D Euclidean space. We will provide a better visualization in Section 14.3.

2. Show that $\infty\, o\, \infty = -2\,\infty$ and $o\,\infty\, o = -2\, o$.

3. In structural exercise 5 of Section 11.12.2, we introduced barycentric coordinates using the homogeneous model. Using the correspondence between homogeneous model and conformal model, give expressions for the barycentric coordinates in terms of conformal points.

4. When studying dualization in combination with versors in the main text, we were mostly interested in the even versors, for which $V[X]^* = V[X^*]$. For odd versors, you should use $V[X]^* = -V[X^*]$. Derive both simultaneously by using the general versor transformation of (7.18), which is is $V[X] = (-1)^{xv}\, V X V^{-1}$, to show that $V[X]^* = (-1)^v\, V[X^*]$. (Hint: What is the sign involved in swapping $V^{-1}\, I_{n+1,1}^{-1}$ to $I_{n+1,1}^{-1}\, V^{-1}$?)

5. For a pure translation versor T, the logarithm is easy to determine. Show that

$$\log(T) = \tfrac{1}{2}(T - \widetilde{T}).$$

6. Write the logarithm of the rigid body motion of (13.15) in terms of the logarithm of the rotor R introduced in (10.14). Your goal is to eliminate all $\mathbf{I}$ or ϕ from the formula.

7. Show that the ratio of two flat points $p \wedge \infty$ and $q \wedge \infty$ is a translation rotor. What is the corresponding translation vector?

8. Show that the ratio of two general planes passing though a common point p is a rotation versor. Do this in 3-D; you can represent the plane with normal $\mathbf{n}$ dually as $p\rfloor(\mathbf{n} \wedge \infty)$. The ratio of two elements is of course identical to the ratio of their duals. What is the bivector angle of the rotation?

9. Show that the ratio of two lines $p \wedge \mathbf{n} \wedge \infty$ and $q \wedge \mathbf{m} \wedge \infty$ is a screw motion (though not a general rigid body motion). What are the screw parameters?

10. To get back to an issue raised in Section 8.2.2, with translation represented as a rotor you can indeed change the position of a point x arbitrarily within a multiplicative framework. Show that the transformation by a translation rotor $T_t = \exp(-t\infty/2)$, when developed in the Taylor series expansion (8.3), generates an arbitrary additive change in the position of a point x.

13.10 PROGRAMMING EXAMPLES AND EXERCISES

The implementation of the conformal model of 3-D space is called c3ga. The basis of the representation space in this implementation is spanned by basis vectors no, e1, e2, e3, and ni, with the metric as in Table 13.1. As you notice, we write o and ∞ as no and ni in our code (short for null vector representing the origin and null vector representing infinity, respectively).[5] This, finally, is the implementation of geometric algebra that is behind the code of Figure 1.2 in the introductory example of Section 1.1.

The implementation contains a large number of specialized multivector types. Table 13.3 lists the most important ones. We have so far only treated the flats, the directions (labeled "free" in the table), and the rigid body motion operators (the table uses "rotor" exclusively for a rotation versor, calling a translation rotor a "translator"). The other elements will be introduced in the next chapter.

The specialized multivector types are complemented by the matrix representation of outermorphisms (om) and a specialized matrix representation that works well with OpenGL (omFlatPoint).

In addition to the basis vectors, which are present as constants by default, c3ga also contains some extra constants (see Table 13.4).

A 2-D version of the conformal model implementation is also provided. It is named c2ga and has most of the same specialized multivector types as its 3-D counterpart, but of course lacks spheres, freeTrivectors, and so on. We use c2ga in Sections 14.9.1 and 15.8.2.

13.10.1 METRIC MATTERS

As a first experiment to confirm that c3ga uses the strange metric of the conformal model, we print out the metric of the five basis vectors:

```
// get the basis vectors:
mv bv[5] = {no, e1, e2, e3, ni};
// ... (omitted)
for (int i = 0; i < 5; i++) {
```

---

5   When pronounced in a sentence, ni should be emphatic and slightly higher than the other words, whereas no should sound like a man imitating a woman imitating a man's voice.

```
// ... (omitted)
for (int j = 0; j < 5; j++) {
  printf(" % 1.1f", _Float(bv[i] << bv[j]));
}
printf("\n");
}
```

Table 13.3: A list of the most important specialized multivector types in c3ga.

Name	Sum of Basis Blades
vector	e1, e2, e3
point	no, e1, e2, e3, ni
normalizedPoint	no = 1, e1, e2, e3, ni
flatPoint	e1∧ni, e2∧ni, e3∧ni, no∧ni
pointPair	no∧e1, no∧e2, no∧e3, e1∧e2, e2∧e3,
	e3∧e1, e1∧ni, e2∧ni, e3∧ni, no∧ni
line	e1∧e2∧ni, e1∧e3∧ni, e2∧e3∧ni, e1∧no∧ni, e2∧no∧ni, e3∧no∧ni
dualLine	e1∧e2, e1∧e3, e2∧e3, e1∧ni, e2∧ni, e3∧ni
plane	e1∧e2∧e3∧ni, e1∧e2∧no∧ni, e1∧e3∧no∧ni, e2∧e3∧no∧ni
dualPlane	e1, e2, e3, ni
circle	e2∧e3∧ni, e3∧e1∧ni, e1∧e2∧ni, no∧e3∧ni, no∧e1∧ni,
	no∧e2∧ni, no∧e2∧e3, no∧e1∧e3, no∧e1∧e2, e1∧e2∧e3
sphere	e1∧e2∧e3∧ni, e1∧e2∧no∧ni, e1∧e3∧no∧ni, e2∧e3∧no∧ni, e1∧e2∧e3∧no
dualSphere	no, e1, e2, e3, ni
freeVector	e1∧ni, e2∧ni, e3∧ni
freeBivector	e1∧e2∧ni, e2∧e3∧ni, e3∧e1∧ni
freeTrivector	e1∧e2∧e3∧ni
tangentVector	no∧e1, no∧e2, no∧e3, e1∧e2, e2∧e3, e3∧e1, e1∧ni, e2∧ni, e3∧ni, no∧ni
tangentBivector	e1∧e2∧e3, e2∧e3∧ni, e3∧e1∧ni, e1∧e2∧ni, no∧e3∧ni,
	no∧e1∧ni, no∧e2∧ni, no∧e2∧e3, no∧e1∧e3, no∧e1∧e2
vectorE2GA	e1, e2
vectorE3GA	e1, e2, e3
bivectorE3GA	e1∧e2, e2∧e3, e3∧e1
translator	scalar, e1∧ni, e2∧ni, e3∧ni
normalizedTranslator	scalar=1, e1∧ni, e2∧ni, e3∧ni
rotor	scalar, e1∧e2, e2∧e3, e3∧e1
scalor	scalar, no ∧ ni

Name	Value
e1ni	$e_1 \wedge \infty$
e2ni	$e_2 \wedge \infty$
e3ni	$e_3 \wedge \infty$
noni	$o \wedge \infty$
I3	$e_1 \wedge e_2 \wedge e_3$
I5	$o \wedge e_1 \wedge e_2 \wedge e_3 \wedge \infty$
I5i	$(o \wedge e_1 \wedge e_2 \wedge e_3 \wedge \infty)^{-1}$

Table 13.4: Constants in c3ga.

(We omitted some code that prints out the names.) The output of this part of the example agrees with Table 13.1:

```
      no    e1    e2    e3    ni
no    0.0   0.0   0.0   0.0 -1.0
e1    0.0   1.0   0.0   0.0  0.0
e2    0.0   0.0   1.0   0.0  0.0
e3    0.0   0.0   0.0   1.0  0.0
ni   -1.0   0.0   0.0   0.0  0.0
```

The example also creates vectors e (denoted as ep) and $\bar{e}$ (denoted as em) from no and ni, according to (13.6):

```
// create 'e+' and 'e-'
dualSphere ep = _dualSphere(no - 0.5f * ni);
dualSphere em = _dualSphere(no + 0.5f * ni);
```

Note that dualSphere is now the specialized vector type that holds arbitrary 5-D vectors. The example continues to print out some information about e and $\bar{e}$. The output is

```
e+ = 1.00*no - 0.50*ni
e- = 1.00*no + 0.50*ni

The metric of e+ and e-:
e+  . e+ = 1.000000
e-  . e- = -1.000000
e+  . e- = 0.000000
```

13.10.2 EXERCISE: THE DISTANCE BETWEEN POINTS

This example draws five (draggable) points, connected by lines, and prints distance labels halfway between each pair of points. The exercise is to complete the following code:

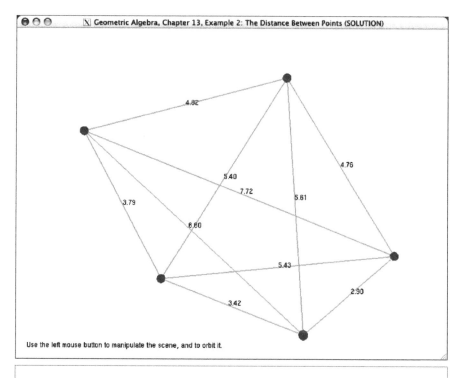

Figure 13.8: The output of the solution to Example 2.

```
const normalizedPoint &pt1 = g_points[i];
const normalizedPoint &pt2 = g_points[j];

// compute distance
// EXERCISE: fill in the code to compute the distance between pt1
// and pt2
float distance = 0.0;
```

Figure 13.8 shows sample output of the correct solution. If you want to write a robust solution, keep in mind that floating point roundoff error may cause values that should always be less than or equal to zero to turn up as a very small positive number. This may cause trouble when you hand this value (negated) to sqrt().

13.10.3 LOADING TRANSFORMATIONS INTO OPENGL, AGAIN

In this exercise we repeat the example from Section 12.5.1, but this time we use the conformal model. The goal is to build up the matrix representation of any outermorphism (in this case, a simple combination of translation and rotation) and to load that matrix into OpenGL.

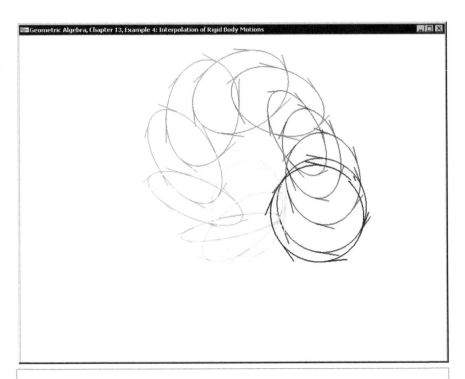

Figure 13.9: Example 4 in action. A trail of circles is drawn to visualize the interpolation of rigid body motions.

In the homogeneous version, we transformed the basis vectors $\mathbf{e}_1$, $\mathbf{e}_2$, $\mathbf{e}_3$, and e_0, and used their images to initialize the matrix representation of the transform. We then loaded this matrix directly into OpenGL.

In the conformal model, the blades $\mathbf{e}_1 \wedge \infty$, $\mathbf{e}_2 \wedge \infty$, $\mathbf{e}_3 \wedge \infty$, and $o \wedge \infty$ are the closest equivalent to the aforementioned basis vectors of the homogeneous model, since that is how the two models relate to each other. Geometrically, these are the blades representing the basis directions, just as $\mathbf{e}_1$, $\mathbf{e}_2$, $\mathbf{e}_3$, and $\mathbf{e}_0$ were in the homogeneous model. Using this, the code to load our conformal outermorphism into OpenGL is

```
// get translator and rotor
vectorE3GA t = _vectorE3GA(distance * e3);
normalizedTranslator T = exp(_freeVector(-0.5f * (t ^ ni)));
rotor &R = g_modelRotor;

// combine 'T' and 'R' to form translation-rotation versor:
TRversor TR = _TRversor(T * R);
TRversor TRi = _TRversor(inverse(TR)); // compute inverse
```

```
// compute images of basis blades e1^ni, e2^ni, e3^ni, no^ni:
flatPoint imageOfE1NI = _flatPoint(TR * e1ni * TRi);
flatPoint imageOfE2NI = _flatPoint(TR * e2ni * TRi);
flatPoint imageOfE3NI = _flatPoint(TR * e3ni * TRi);
flatPoint imageOfNONI = _flatPoint(TR * noni * TRi);

// create matrix representation:
omFlatPoint M(imageOfE1NI, imageOfE2NI, imageOfE3NI, imageOfNONI);

// load matrix representation into GL:
glLoadMatrixf(M.m_c);
```

In Section 16.10.1, we will perform the opposite operation of converting OpenGL matrices to conformal versors. We postpone this subject until we can properly handle (uniform) scaling, which is first introduced in Chapter 16.

13.10.4 INTERPOLATION OF RIGID BODY MOTIONS

Since we can now represent translation and rotation as exponentials of bivectors and have their logarithm available (see Section 13.5.3), we can interpolate them with ease. The example presented here is similar to the interpolation example treated in the context of the vector space model (Section 10.7.1), but the conformal model now allows us to do it properly.

First of all, we need an implementation of the logarithm of (normalized) translation-rotation versors. This is easily done given the pseudocode in Figure 13.5:

```
dualLine log(const TRversor &V) {
  // isolate rotation & translation part:
  rotor R = _rotor(-no << (V * ni));
  vectorE3GA t = _vectorE3GA(-2.0f * (no << V) * inverse(R));

  const float EPSILON = 1e-6f;

  if (_Float(norm_e2(_bivectorE3GA(R))) < EPSILON * EPSILON) {
    // special cases:
    if (_Float(R) < 0.0f)
    {// R = -1
      // Get a rotation plane 'I', perpendicular to 't'
      bivectorE3GA I;
      if (_Float(norm_e2(t)) > EPSILON * EPSILON)
        I = _bivectorE3GA(unit_e(t << I3));
      else {
        // when t = 0, any plane will do
        I = _bivectorE3GA(e1^e2);
      }
      // return translation plus 360 degree rotation:
      return _dualLine(0.5f *(I * 2.0f * (float)M_PI - (t^ni)));
    }
    else
```

```
  { // R = 1;
    // return translation :
      return _dualLine(-0.5f *(t^ni));
    }
  }
  else { // regular case
    // compute logarithm of rotation part
    bivectorE3GA Iphi = _bivectorE3GA(-2.0f * log(R));
    // determine rotation plane:
    rotor I = _rotor(unit_e(Iphi));

    // compose log of V:
    return _dualLine(
      0.5f * (
      -(t ^ I) * inverse(I) * ni +
      inverse(1.0f - R * R) * (t << Iphi) * ni -
      Iphi));
    }
  }
```

Note the slight misuse of the type system: the function returns a dualLine, which is offi-
cially a 2-blade. In general, the logarithm of a rigid body motion is not a 2-blade but
a bivector. It just so happens that this bivector can be represented employing the same
basis blades as for dual lines, so we can (ab-)use dualLine as the return type. That is
a consequence of the additive decomposition. But if you would pass the output of this
log() function to draw(), which can only draw blade, it will only draw a line if the screw
happens to be a pure rotation.

The example interpolates from one random versor to the next. These versors are computed as the product of a random translation and a random translated rotation:

```
void initRandomDest() {
  normalizedTranslator T1 = exp(_freeVector(randomBlade(2, 3.0f)));
  normalizedTranslator T2 = exp(_freeVector(randomBlade(2, 3.0f)));
  rotor R = exp(_bivectorE3GA(randomBlade(2, 100.0f)));

  g_destVersor = _TRversor(T1 * T2 * R * inverse(T2));
}
```

Interpolation between the versors is done using the following function, which is almost
identical to the one used in Section 10.7.1:

```
// interpolate between 'src' and 'dst', as determined by 'alpha'
TRversor interpolateTRversor(const TRversor &src,
  const TRversor &dst, mv::Float alpha) {
  // return src * exp(alpha * log(inverse(src) * dst));
  return _TRversor(src * exp(_dualLine(alpha *
    log(_TRversor(inverse(src) * dst)))));
}
```

14 NEW PRIMITIVES FOR EUCLIDEAN GEOMETRY

This chapter continues the development of the conformal model of Euclidean geometry. We have seen in the previous chapter how the model includes flats and directions as blades and Euclidean transformations on them as versors. That more or less copied the capabilities of the more familiar homogeneous model, though in a structure-preserving form, which permits metrically significant interpolation.

In this chapter, we show that the blades of the conformal model can represent many more elements that are useful in Euclidean geometry: They give us spheres, circles, point pairs, and tangents as direct elements of computation. Having those available will extend the range of computations that can be done by the basic products of geometric algebra, which is the subject of the next chapter. Here we carefully develop the representation of these new elements and show how to retrieve their parameters. You will for instance see that the sphere through the four points p, q, r, and s is the blade $p \wedge q \wedge r \wedge s$, and that you can immediately read off its center and radius from the dual sphere $(p \wedge q \wedge r \wedge s)^*$.

Again, we urge the use of interactive visualization to get the maximum enjoyment out of this chapter. The conformal model is not abstract mathematics, it is a practical tool!

14.1 ROUNDS

We have made the representation of flats by considering blades containing the point at infinity ∞. We can make other blades through the outer product of finite points only. It is a happy surprise that this gives us the direct representations of spheres, circles, and other related elements. For such elements we do not have the analogy of the homogeneous model to guide us, and it is actually somewhat easier to introduce them through their duals and then derive their direct representation from that. Once we have the results, either mode is easy to work in, it just depends on the manner in which your data has been given (we will find that if you know the center and radius of a sphere, you should use the dual to construct its blade; if you know four points on the sphere, you should use the direct representation).

14.1.1 DUAL ROUNDS

We start with the dual representation of a real sphere of radius ρ, located at the origin o (see Figure 14.1(a)). According to Table 13.2, this is

$$\textit{dual sphere at origin:}\quad \sigma = o - \tfrac{1}{2}\rho^2\,\infty.$$

Let us cut this dual origin sphere with a plane through the origin, dually represented as $\pi = \mathbf{n}$. Such a plane is perpendicular to the sphere in our Euclidean base space. In the representation, this perpendicularity of a plane and a sphere set corresponds to the statement

$$\pi \cdot \sigma = 0.$$

This is easily verified for the particular situation we have at the origin: $\mathbf{n} \cdot (o - \tfrac{1}{2}\rho^2\,\infty)$ $= 0$. The general result is achieved through the familiar argument on covariance: the

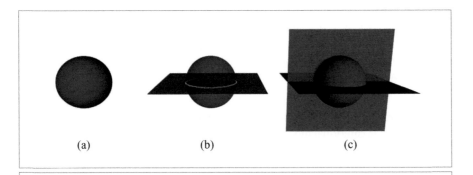

(a)   (b)   (c)

Figure 14.1: Dual rounds in the conformal model in 3-D, obtained by intersecting a sphere with planes. (a) The dual sphere σ, (b) the dual circle $\sigma \wedge \pi$ (in green), and (c) the dual point pair $\sigma \wedge \pi \wedge \pi'$ (in blue).

condition is formulated as a scalar-valued inner product; therefore, it is invariant under orthogonal transformations; therefore, it is invariant under versors; therefore, when both elements are rotated and translated it still holds; therefore, it holds universally as the condition for perpendicularity of a sphere and a plane through its center. Again, the structure preservation saves us proving statements in their most general form: one example suffices to make it true anywhere.

We can form the blade $\kappa = \sigma \wedge \pi$. Since this is the outer product of two duals, it must be the dual representation of the meet of plane and sphere (for $(A \cap B)^* = B^* \wedge A^*$, see (5.8)). In our Euclidean space, the meet is clearly a circle (see Figure 14.1(b)), and if everything is consistent, its dual representation, κ, must be a dual circle.

$$\text{dual circle at origin:} \quad \kappa = \sigma \wedge \pi = (o - \tfrac{1}{2}\rho^2 \infty) \wedge \mathbf{n},$$

where $\mathbf{n}$ is the dual representation of the carrier plane π of the circle. This is how simple it is to represent a dual circle in the conformal model: it is a 2-blade.

If you are still suspicious whether this really is a dual circle, probe this dual element κ with a point x and check for which points x this is zero. That gives the condition

$$0 = x \rfloor (\sigma \wedge \pi) = (x \cdot \sigma)\,\pi - (x \cdot \pi)\,\sigma.$$

Then take the inner product of this with π, which gives $0 = (x \cdot \sigma)\,\pi \cdot \pi$, and the inner product with σ, which gives $0 = (x \cdot \pi)\,\sigma \cdot \sigma$. Since both $\pi \cdot \pi$ and $\sigma \cdot \sigma$ are nonzero (at the origin, and therefore everywhere) this indeed retrieves the independent conditions $x \cdot \sigma = 0$ and $x \cdot \pi = 0$, so the point x must be both on the sphere and on the plane. In Euclidean terms, the former condition is $\mathbf{x}^2 = \rho^2$, and the second is $\mathbf{x} \cdot \mathbf{n} = 0$, which is clearly sufficient to construct the circle equation $x_1^2 + x_2^2 = \rho^2$ for coordinates of $\mathbf{x}$ in the $\mathbf{n}$-plane.

We can cut the circle with yet another plane π', perpendicular to both σ and π (see Figure 14.1(c)). In 3-D, that gives us a dual *point pair*, which is indeed a sphere on a line, the set of points with equal distance to the center o. It is dually represented as

$$\text{dual point pair at origin:} \quad \sigma \wedge \pi \wedge \pi'.$$

The Euclidean bivector $\pi' \wedge \pi$ is the dual meet of the two planes, and dually denotes the carrier line of the point pair. In an n-dimensional space, the process continues, and you can cut the original sphere with n hyperplanes before the outer product trivially returns zero (which makes geometric sense, since there are only n independent hyperplanes at the origin in n-dimensional space).

Let us call the elements we obtain in this way *dual rounds*. Their general form, when centered on the origin, is

$$\text{real dual round at origin:} \quad (o - \tfrac{1}{2}\rho^2 \infty)\,\mathbf{E}_k, \tag{14.1}$$

with $\mathbf{E}_k$ a purely Euclidean k-blade dually denoting the carrier flat of the round. Since the blades are generated in properly factored form, we will prefer to denote them using the geometric product; however, you could use the outer product instead (see structural exercise 8). When we translate them by a translation over $\mathbf{c}$, the dual sphere part moves to $c - \frac{1}{2}\rho^2 \infty$, whereas the purely Euclidean part changes according to (13.7) (i.e., to $-c \rfloor (\infty \mathbf{E}_k)$). Therefore, the general form for a round centered at c is

$$\textit{real dual round at c:} \quad (c - \tfrac{1}{2}\rho^2 \infty)\left(- c\rfloor (\infty \mathbf{E}_k)\right). \qquad (14.2)$$

We will later relate the Euclidean blade $\mathbf{E}_k$ to a specifically chosen orientation for the carrier flat $c \wedge \mathbf{A}_{n-k} \wedge \infty$, and then set $\mathbf{E}_k = (-1)^n \mathbf{A}^{\star}_{n-k}$ to represent this same orientation dually. For now, focus on the form of the expression rather than on such details.

We started from a real dual sphere to produce these elements; had we started from an imaginary sphere $o + \frac{1}{2}\rho^2 \infty$ (see Section 13.1.3), we would have produced dual *imaginary rounds*. These therefore only differ from the real rounds in replacing ρ^2 by $-\rho^2$. As we saw before, we can have representation of *imaginary* rounds in a *real* algebra, since only squared distances enter our computations.

You may wonder why we did not encounter *imaginary flats* in the previous chapters. But flats are always real, for they have no size measure like a radius. They merely have a weight, and when its sign changes, this simply changes their orientation. Of course, rounds also have a weight, which gives them an orientation and a density.

14.1.2 DIRECT ROUNDS

Now that we have made *dual* spheres, circles, and point pairs, real or imaginary, we also want to know their *direct* representation: they will give us yet another type of blade in the conformal model $\mathbb{R}^{n+1,1}$ with a clear geometric meaning in $\mathbb{E}^n$.

Instead of undualizing the duals, we now have enough experience with the conformal model to guess the direct representation and then relate it to the duals afterwards. The resulting computation is surprisingly easy and coordinate-free, and a good example of the kind of symbolic computational power the operational model of Euclidean geometry affords.

Let us consider the 3-D Euclidean space $\mathbb{E}^3$. A sphere in that space is determined by four points. We take four unit points p, q, r, s. We are bold and guess that the sphere Σ is represented by the blade

$$\Sigma = p \wedge q \wedge r \wedge s,$$

but of course we should prove that. First, we manipulate it using the antisymmetry of the outer product, subtracting p from all terms but the first, to obtain

$$\Sigma = p \wedge (q - p) \wedge (r - p) \wedge (s - p).$$

Then we dualize, producing

$$\Sigma^* = p \rfloor \big((q - p) \wedge (r - p) \wedge (s - p) \big)^*. \tag{14.3}$$

We suspect that this might be a dual sphere, but this is in a form we have not seen before: it is apparently characterized by a point on it contracted on something else. From Table (13.2), we only know it in the form $c - \frac{1}{2}\rho^2 \infty$, given its unit center point c and radius ρ. But if we know a unit point p on the sphere, then ρ is easily computed through $p \cdot c = -\frac{1}{2}\rho^2$, and using $p \cdot \infty = -1$, we can group these results in a compact form:

$$\textit{dual sphere around } c \textit{ through } p: \quad p \rfloor (c \wedge \infty). \tag{14.4}$$

We recognize $c \wedge \infty$ as the flat point at the center of the sphere.

Returning to (14.3), if Σ^* is supposed to be a dual sphere through p, then the expression under the dual should be a representation of the flat point at the center, possibly weighted. From its form, it is the $\mathtt{meet}$ of three elements that are dually represented by $(q-p)$, $(r-p)$, and $(s-p)$. What is $(q-p)$? Since it is a dual representation of some element, we probe it with a point x to find out: $0 = x \cdot (q - p)$, so $x \cdot p = x \cdot q$, so $d_E^2(x,p) = d_E^2(x,q)$. It follows that

$$q - p \ \textit{ is the dual midplane between the unit points } p \textit{ and } q. \tag{14.5}$$

So the expression $\big((q - p) \wedge (r - p) \wedge (s - p) \big)^*$ computes the intersection of the midplanes between the three point pairs, which is indeed a weighted copy of the flat point at the center $\alpha\,(c \wedge \infty)$. Therefore we indeed have a (weighted) dual sphere. It follows that the outer product of four points is a sphere.[1]

The relationship between dual and direct representation of the sphere,

$$\alpha\,(c - \tfrac{1}{2}\rho^2\,\infty) = (p \wedge q \wedge r \wedge s)^*,$$

implies that the (center, radius) formulation is exactly dual to the four-point definition. We can compute the weight α by taking the contraction with $-\infty$ on both sides, giving

$$\alpha = -\infty \rfloor (p \wedge q \wedge r \wedge s)^* = -(\infty \wedge p \wedge q \wedge r \wedge s)^*.$$

This give us an easy way to compute the squared radius of a sphere through four points as the square of the corresponding normalized dual sphere. Some duals and signs cancel, and we get

$$\rho^2 = (c - \tfrac{1}{2}\rho^2\infty)^2 = \big((p \wedge q \wedge r \wedge s)^* \big)^2 / \alpha^2 = \frac{(p \wedge q \wedge r \wedge s)^2}{(p \wedge q \wedge r \wedge s \wedge \infty)^2}.$$

---

1   To do things properly, we have to make sure the signs are consistent between the dual and the direct representation. You see that the intersection expression deviates from a $\mathtt{meet}$ in two aspects: it involves the wrong order (for a $\mathtt{meet}$ would normally swap the order of the arguments), and it uses the dual rather than the inverse dual. However, both involve a reversion of n elements in an n-dimensional base space, so the overall sign cancels. The result is therefore correct.

It is nice to have such a quantitative computational expression for the radius of the sphere through four points in such a coordinate-free form. In structural exercise 1, you derive a similar direct expression for the center of the sphere.

The computation we have just done was in 3-D space, which is why we needed to use four elements to span the sphere. Apart from that, nothing in the computation depended on the dimensionality. If we move into a plane, exactly the same construction can be used to show that three unit points $p \wedge q \wedge r$ span a circle, of which the radius squared is proportional to $(p \wedge q \wedge r)^2$. (If you are uncomfortable about whether this might be true in a general offset plane, just move the plane so that it contains the origin, perform the construction, and move it back to where it came from. The structure preservation of the versors then makes the result move with the motion and be valid in the offset plane as well.) Summarizing:

$$\text{oriented sphere through points } p, q, r, s: \quad p \wedge q \wedge r \wedge s$$
$$\text{oriented circle through points } p, q, r: \quad p \wedge q \wedge r$$
$$\text{oriented point pair through points } p, q: \quad p \wedge q$$

We emphasize that a directed point pair is a *single* element of conformal geometric algebra, just one 2-blade B. There is therefore no need to make a separate data structure for an edge, because the point pair contains all information. So in contrast to the homogeneous model, we can have a line segment as a single element of computation. It is even possible to retrieve the constituent points from the 2-blade B (see (14.13) in structural exercise 4). That is of course not possible for the other blades representing rounds; many triples of points determine the same circle.

As you play with such elements $p \wedge q \wedge r \wedge s$ in an interactive software package with visualization, it is pleasing to see how the independence of the result on motions of the points over the sphere is captured by the antisymmetry of the outer product. Switching on some display of the orientation of the sphere (as captured by the sign of its weight), you should see that orientation change in continuous and predictable manner, depending on whether p, q, r, s form a positively oriented tetrahedron or not. Can you determine the geometrical relationship between p, q, r, and s that makes the sphere become zero (see structural exercise 3)?

14.1.3 ORIENTED ROUNDS

We have seen above that a dual circle κ can be characterized as the dual meet $\sigma \wedge \pi$ of a sphere σ and a plane π through the sphere's center (so that the two are perpendicular and $\sigma \cdot \pi = 0$). This was a geometrical construction, and we were not too particular about the signs involved. But you normally want to use the capability of geometric algebra to represent *oriented* spheres, circles, and point pairs, so we should be more specific. The orientation of the direct circle K in direct representation is most easily characterized as the orientation of its carrier plane $K \wedge \infty$; proper dualization then gives the form we should use for the dual representation to get the desired matching orientation.

To avoid having to fix signs later, we work from a direct form with obviously correct orientation (though with a somewhat mysterious spherical part) to determine the correct dual. We claim that the direct form of a round at the origin is

$$\Sigma = (o + \tfrac{1}{2}\rho^2 \infty) \, \mathbf{A}_k, \tag{14.6}$$

with $\mathbf{A}_k$ purely Euclidean. The carrier flat of this round is $\Pi = \Sigma \wedge \infty = o \wedge \mathbf{A}_k \wedge \infty$, with the orientation of $\mathbf{A}_k$. We compute the corresponding dual:

$$
\begin{aligned}
\Sigma^* &= \big((o + \tfrac{1}{2}\rho^2 \infty) \, \mathbf{A}_k\big) \rfloor (o \wedge \mathbf{I}_n^{-1} \wedge \infty) \\
&= \big(\widehat{\mathbf{A}}_k \, (o + \tfrac{1}{2}\rho^2 \infty)\big) \rfloor (o \wedge \infty \wedge \widehat{\mathbf{I}}_n^{-1}) \\
&= \widehat{\mathbf{A}}_k \rfloor \big((o - \tfrac{1}{2}\rho^2 \infty) \wedge \widehat{\mathbf{I}}_n^{-1}\big) \\
&= (o - \tfrac{1}{2}\rho^2 \infty) \, (\mathbf{A}_k \rfloor \widehat{\mathbf{I}}_n^{-1}) \\
&= (o - \tfrac{1}{2}\rho^2 \infty) \, \mathbf{A}_k^\star \, (-1)^n. \tag{14.7}
\end{aligned}
$$

This shows both the correctness of the direct representation of the round (since it produces a dual of the correct form), and how we should choose the sign of the dual to represent a round with a particular orientation. You should really use this instead of the earlier expression (14.2), which was unspecific on the orientation.

For instance, to dually represent a circle with unit radius rotating positively (counterclockwise) in the $\mathbf{e}_1 \wedge \mathbf{e}_2$-plane, we have $\mathbf{A}_2 = \mathbf{e}_1 \wedge \mathbf{e}_2$. In 3-D space with the regular pseudoscalar, the dual circle is $(o - \tfrac{1}{2} \infty) \, (-\mathbf{e}_3)$; in a 4-D space, the dual circle would be $(o - \tfrac{1}{2} \infty) \, (-\mathbf{e}_3 \wedge \mathbf{e}_4)$.

Collectively, we call these elements derived from the intersection of a sphere with planes *direct rounds*, a term that includes oriented spheres, oriented circles, and oriented point pairs in 3-D, but clearly extends to any dimension. The above shows its representation to be

$$\textit{direct real round at origin:} \quad \Sigma = (o + \tfrac{1}{2}\rho^2\infty) \, \mathbf{A}_k. \tag{14.8}$$

Note the difference with the dual round representation of (14.1): for a real direct round, a factor appears that looks like an imaginary dual sphere. Direct imaginary rounds are of the form $(o - \tfrac{1}{2}\rho^2 \infty) \, \mathbf{A}_k$, simply changing the sign on ρ^2.

To correspond to mathematical tradition, we can call a round a $(k-1)$-sphere if its carrier has dimension k. The mathematical indication of dimensionality refers to the dimension of the manifold of its shell. What we call a sphere in 3-D is a 2-D curved surface, and therefore called a 2-sphere; we could call it a 2-round. A circle is a 1-D manifold and therefore a 1-sphere. A point pair is a 0-sphere; it is unique in having two separate components.

By translation using a rotor T_c, we can produce general rounds around any point c, as the elements

$$\textit{direct round:} \quad (c + \tfrac{1}{2}\rho^2 \infty) \wedge \big(-c \rfloor (\widehat{\mathbf{A}}_k \infty)\big), \tag{14.9}$$

where we used the translation formula (13.7). A k-sphere is therefore represented by a $(k+2)$-blade. The associated dual round with the same orientation is

$$\text{dual round: } (c - \tfrac{1}{2}\rho^2 \infty) \wedge \left(-c\rfloor(\widehat{\mathbf{A}}_k^{\star} \infty) \right), \qquad (14.10)$$

which is an $(n-k)$-blade. (To obtain (14.10), we applied (13.7) to (14.7), and canceled some signs.) These expressions have been collected in Table 14.1 for easy reference, though more compactly denoted through translation rotors applied to standard forms around the origin.

14.2 TANGENTS AS INTERSECTIONS OF TOUCHING ROUNDS

In the representation of a round the radius can be set to zero. This results in a blade that squares to zero. To find its geometric interpretation, let us construct a setup in which it naturally occurs. We put two (dual) spheres with equal radius ρ at opposite sides of the origin, in the unit $\mathbf{e}_1$-direction, as depicted in Figure 14.2. You can easily verify that these spheres are represented by the vectors $o \pm \mathbf{e}_1 + \tfrac{1}{2}(1 - \rho^2) \infty$. We compute the dual meet by the outer product of the duals and simplify the outcome:

$$\left(o - \mathbf{e}_1 + \tfrac{1}{2}(1 - \rho^2)\infty\right) \wedge \left(o + \mathbf{e}_1 + \tfrac{1}{2}(1 - \rho^2)\infty\right) = \left(o - \tfrac{1}{2}(\rho^2 - 1)\infty\right) \wedge (2\mathbf{e}_1) \quad (14.11)$$

There are three cases:

- When $\rho^2 > 1$ (as in Figure 14.2(a)) we get a real dual circle, nicely factored in the outer product as the intersection of a real dual sphere (with squared radius $\rho^2 - 1$) and the dual plane $2\mathbf{e}_1$ that denotes the flat carrier plane of the circle.
- But when we have $\rho^2 < 1$ (as in Figure 14.2(c)), the resulting intersection circle becomes imaginary and is factored as the intersection of a plane and an imaginary sphere (with the squared radius $\rho^2 - 1$ now being negative).
- When $\rho^2 = 1$, we get Figure 14.2(b). Geometrically, we would say that the spheres now have a tangent 2-blade in common. You can think of this tangent 2-blade as the intersection point of the two spheres with a tiny bit of plane attached. We have sketched it like that in Figure 14.2(b). This is apparently represented by the dual blade $o \wedge (2\mathbf{e}_1)$.

It is interesting that we obtain this kind of infinitesimal element through application of the incidence operator, not by any differentiation or limiting process, and that we therefore have them available as regular blades of the conformal model. The only unusual thing about them is that they square to zero.

Let us study this class of blade in its direct form, at the origin. We can make it by setting the radius of a direct round to zero, producing from (14.8) the element $o \wedge \mathbf{A}_k$ as the standard

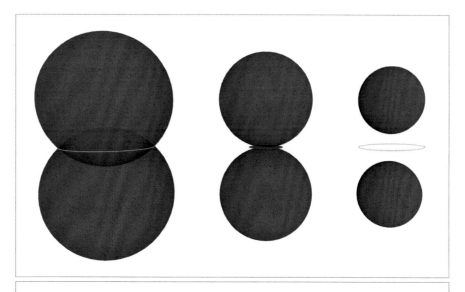

Figure 14.2: Intersection of two spheres of decreasing radii, leading to (a) a real circle, (b) a tangent 2-blade, and (c) an imaginary circle.

form at the origin. For $k = 2$, a tangent 2-blade results. These are new elements; we shall call them *(direct) tangents*.

$$\text{direct tangent at origin:}\quad o \wedge \mathbf{A}_k$$

They have a location (here o) and a direction, but they are not flats (since they lack the element ∞). A direct flat can be constructed from such an element by making it pass through infinity, as $(o \wedge \mathbf{A}_k) \wedge \infty$; this shows clearly that $\mathbf{A}_k$ is the directional element of the tangent.

If $\mathbf{A}_k$ is a vector $\mathbf{A}_1 = \mathbf{a}$, this tangent element looks like $o \wedge \mathbf{a}$. This may be pictured as a vector in direction $\mathbf{a}$ located at the origin. We can find out which point set it represents by solving $x \wedge (o \wedge \mathbf{a}) = 0$, and this yields that only $x = o$ is in it, but clearly $o \wedge \mathbf{a}$ is more than just o; it also has a directional part $\mathbf{a}$. It can be made as the meet of two circles (see structural exercise 7). Note that a tangent vector is *not* a small vector: the signed length of a vector is given by its weight, and that can be arbitrarily big. It is the *size* of the tangent vector element that is zero, as its zero square shows; you can interpret it as a pair of very close points, if that helps. The sharp distinction between size and weight occurs for the first time in the conformal model, which therefore appears as the natural framework to represent elements that are small in their 2-point geometric size, yet have a finite weight. These are finite elements of geometry, not infinitesimals!

The dual representation of the tangent is of course obtained from the representation of the dual round of (14.10), setting its radius to zero, giving

$$\text{dual tangent at origin:}\quad o \wedge A_k^\star \, (-1)^n.$$

As with the rounds, this involves a sign to relate the orientation to that of the direct representation. Apart from that, you see that the dualization is the Euclidean dual: *the tangents embed the vector space model into the conformal model, one copy at each location*. In mathematics, this is called the *tangent bundle*; the conformal model gives us its elements as regular elements of computation.

Displacing a direct tangent element using a translation rotor, it becomes

$$\text{direct tangent:}\quad p \wedge (-p \rfloor (\widehat{A}_k \, \infty)).$$

Note that you should not simply use $p \wedge A_k$ as the translated element!

A tangent X is a blade characterized by having zero square ($X^2 = 0$), since it is like a round with zero radius. But it does not contain ∞ (so that $\infty \wedge X \neq 0$). A dual tangent X^* satisfies the dual of these conditions: $(X^*)^2 = 0$, while $\infty \rfloor X^* \neq 0$. Because of these conditions, it is consistent to view a *point* as a special case of a tangent element, with a scalar directional part denoting its weight: $p \wedge \alpha$. In that view, a point is a localized scalar.

14.2.1 EUCLID'S ELEMENTS

We have now found all elements of Euclidean geometry as encoded in the conformal model. The basic objects are represented by the blades assembled in Table 14.1.

Note in the table that, apart from the flats, the elements retain their appearance after dualization: a dual direction may be confused with a directly represented direction, a tangent with a dual tangent, and a real dual round with an imaginary direct round, all of complimentary Euclidean properties of direction and/or size. Since we cannot tell them apart from their structure alone, we need to specify somehow whether they are intended for direct or dual usage.

This is a bit of a nuisance; we might have hoped that direct and dual representations had been more clearly separated in the conformal model. But at least direct and dual elements transform in precisely the same manner under even versors, so keeping track of their nature is only required for interpretation, not for correct algebraic manipulation. That in itself is an improvement over the homogeneous model in which direct and dual representations translate differently.

Table 14.1: All nonzero blades in the conformal model of Euclidean geometry and their parameters. Locations are denoted by dual spheres. The normalized points q are probes to give locations closest to q; one may just use $q = o$. The directions are computed in direct form, although the dualized version is also given. The weight in all cases is computed as $\sqrt{|q \rfloor \mathrm{direction}(X)|^2}$, orientation is relative to the chosen pseudoscalar for the directional subspace.

As for notation, $\hat{X} = (-1)^{\mathrm{grade}(X)} X$ is the grade involution, $\mathbf{E}^\star$ is a Euclidean dual, X^{-*} is the undualization in the full space. $\mathsf{T_p}$ denotes a translation to the position $\mathbf{p}$ relative to the origin, to signify the general form of the elements.

Standard Form X	Condition	Direction $\mathbf{E} \infty$	Location	Squared Size
direction $\mathbf{E}\infty$	$\infty \wedge X = 0$ $\infty \rfloor X = 0$	X	none	none
dual direction $-\mathbf{E}^\star \infty$	$\infty \wedge X = 0$ $\infty \rfloor X = 0$	X^{-*}	none	none
flat $\mathsf{T_p}[o \wedge (\mathbf{E}\infty)]$	$\infty \wedge X = 0$ $\infty \rfloor X \neq 0$	$-\infty \rfloor X$	$(q \rfloor X)/X$	none
dual flat $\mathsf{T_p}[\widehat{\mathbf{E}^\star}]$	$\infty \wedge X \neq 0$ $\infty \rfloor X = 0$	$-\infty \rfloor X^{-*}$	$(q \wedge X)/X$	none
tangent $\mathsf{T_p}[o\,\mathbf{E}]$	$\infty \wedge X \neq 0$ $\infty \rfloor X \neq 0$ $X^2 = 0$	$-(\infty \rfloor X) \wedge \infty$	$\dfrac{X}{-\infty \rfloor X}$	0
dual tangent $\mathsf{T_p}[o\,\mathbf{E}^\star(-1)^n]$	$\infty \wedge X \neq 0$ $\infty \rfloor X \neq 0$ $X^2 = 0$	$(-\infty \rfloor X^{-*}) \wedge \infty$	$\dfrac{X}{-\infty \rfloor X}$	0
round $\mathsf{T_p}[(o + \frac{1}{2}\rho^2\infty)\,\mathbf{E}]$	$\infty \wedge X \neq 0$ $\infty \rfloor X \neq 0$ $X^2 \neq 0$	$-(\infty \rfloor X) \wedge \infty$	$\dfrac{X}{-\infty \rfloor X}$ or $-\dfrac{1}{2}\dfrac{X\infty X}{(\infty \rfloor X)^2}$	$\rho^2 = +\dfrac{X\hat{X}}{(\infty \rfloor X)^2}$
dual round $\mathsf{T_p}[(o - \frac{1}{2}\rho^2\infty)\,\mathbf{E}^\star(-1)^n]$	$\infty \wedge X \neq 0$ $\infty \rfloor X \neq 0$ $X^2 \neq 0$	$(-\infty \rfloor X^{-*}) \wedge \infty$	$\dfrac{X}{-\infty \rfloor X}$ or $-\dfrac{1}{2}\dfrac{X\infty X}{(\infty \rfloor X)^2}$	$\rho^2 = -\dfrac{X\hat{X}}{(\infty \rfloor X)^2}$

14.2.2 FROM BLADES TO PARAMETERS

So far, we have constructed the elements from parameters such as centers, radii, and directions. But we also need the reverse process to find out what kind of element a computation has produced, so we need to be able to classify a given blade and retrieve its parameters. For this task, the intimidating Table 14.1 is intended as reference.

Computing these geometric parameters of the various kinds of blades is not too hard, and the various formulas are easily proved (or derived when you have forgotten them) from a blade in its standard form around the origin. This is allowed as long as all formulas are Euclidean covariant (i.e., they don't use position or orientation-dependent constructions). You know from Section 13.2 that a good way to guarantee that is to write them using only the standard products, and only ∞ as a special element (since the point at infinity is a Euclidean invariant).

We give the algebraic formulas for size, weight, location, and direction. In an implementation, you can often just read off these elements as the coefficients of specific components of the blade normalized by the overall weight (which may be found as another coefficient)—as long as you do not use different coordinate systems within a single application.

- **Squared Size.** The square of a normalized dual sphere gives you its radius squared, for

$$(o - \tfrac{1}{2}\infty\,\rho^2)^2 = -\tfrac{1}{2}(o\,\infty + \infty\,o)\,\rho^2 = \rho^2.$$

For a dual round in general, you need to do a normalization and mind some grade-dependent signs. An unnormalized dual round X has

$$X^2 = (o - \tfrac{1}{2}\rho^2\infty)\,\mathbf{E}^\star\,(o - \tfrac{1}{2}\rho^2\infty)\,\mathbf{E}^\star = \rho^2\,\mathbf{E}^\star\,\widehat{\mathbf{E}^\star}.$$

Since $\infty \rfloor X = \mathbf{E}^\star\,(-1)^n$, the squared size can now be computed by normalization and by a grade inversion to compensate for the extra sign in the Euclidean part. This gives

$$\rho^2 = \frac{X\widehat{X}}{(\infty \rfloor X)^2}.$$

For the direct representation of a round, the structure is similar. Therefore, the total formulas are as indicated in Table 14.1. It is good custom to denote the squared size rather than the size itself, since the square may be negative and we want to avoid introducing complex numbers.

All tangents and dual tangents have size 0. For the flats and directions, size is not an issue, since they do not have any (they only have a weight).

- **Direction, Weight, and Orientation.** A flat or round is in a translated linear subspace with a certain direction. This direction can be retrieved from the element by a specific formula for each category of element. The formulas are easily derived, they merely strip off the positional parts (if any) and supply a ∞-factor (if lacking).

All return a direction of the form $\mathbf{E} \wedge \infty$, with $\mathbf{E}$ purely Euclidean, and may be found in Table 14.1.

The Euclidean parts of these directions may not be of unit weight or positive orientation relative to the pseudoscalar of the Euclidean subspace they belong to. In that case, the magnitude of the direction is the *weight* of the element, and its sign the *orientation*. The orientation is somewhat arbitrary, since it must be relative to the orientation of the pseudoscalar, which is a matter of choice or convention. The weight of an element of the form $\mathbf{E}\infty$ can be computed by stripping off the ∞ (through a contraction with o or any other normalized point q), and computing the weight of the Euclidean part. That gives

$$\sqrt{|q\rfloor \text{direction}(X)|^2} = \sqrt{|q\rfloor (\mathbf{E}\infty)|^2}.$$

Since all elements have a direction, all elements have a weight and orientation. In some cases, these weights have a traditional way being displayed: a vector of weight 2 can be depicted as having length 2, and a tangent bivector of weight 2 as an area element of 2 area units. But a sphere of weight 2 is *not* a sphere of radius 2; rather it should be visualized as a heavier or denser spherical shell.

- **Location.** The location of a blade should be the Euclidean coordinates of some relevant point. For a round, this is naturally the center, but for flats such as lines and planes such a point is not uniquely indicated in a coordinate-free manner.

  For flats, we can either take the point of it that is closest to the origin (obviously not coordinate-free) or the point that is closest to some given point q. The formulas in the table actually produce a normalized dual sphere σ as the location (rather than a point, i.e., a dual sphere of radius zero), since that is a simple formula. This is often sufficient in intermediate computations, but you can take its Euclidean part as the Euclidean location vector (easy in most coordinate-based implementations, though algebraically it is $(o \wedge \infty)\rfloor(o \wedge \infty \wedge \sigma)$).

  For the rounds, we only need to strip off the carrier part to be left with a dual sphere around the center, and this is done by computing $X/(-\infty\rfloor X)$. Alternatively, we may compute the normalized center point directly as a dual sphere of radius zero through

$$c = -\tfrac{1}{2} \frac{X \infty X}{(\infty\rfloor X)^2}.$$

For now, just accept this formula. We will appreciate this properly as computing the reflection of ∞ in the normalized round when we treat the operators of the conformal model in Chapter 16.

The result of these considerations is Table 14.1, which computes all parameters of the elements of all classes.

14.3 A VISUAL EXPLANATION OF ROUNDS AS BLADES

The vector space model and the homogeneous model may have accustomed us to thinking of the blades in a representation space as always representing flat elements in the base space. Algebraically, this interpretation appears to be supported by the idea that they must represent subspaces and therefore be linear. However, this argument confuses the representational space with the base space that is being represented. We can show you more visually why the surprising characterization of Euclidean rounds by blades works, to convince you of its intuitive correctness. In this section we will "pop up" the ∞-dimension graphically, though we will necessarily show you only the conformal model for a 2-D Euclidean geometry $\mathbb{E}^2$.

14.3.1 POINT REPRESENTATION

We have seen that a Euclidean point at the location $\mathbf{x}$ is represented as the conformal vector

$$x = o + \mathbf{x} + \tfrac{1}{2}\mathbf{x}^2\,\infty.$$

In 2-D Euclidean space $\mathbb{E}^2$, this requires a 4-D representational space $\mathbb{R}^{3,1}$ with a basis like $\{o, \mathbf{e}_1, \mathbf{e}_2, \infty\}$. This would seem to be hard to visualize. However, the o-dimension works very much like the extra dimension in homogeneous coordinates: it allows us to represent offset linear subspaces (i.e., linear subspaces that are shifted out of the origin in a representational space with basis $\{\mathbf{e}_1, \mathbf{e}_2, \infty\}$). So because of the o-term, we are allowed to draw conformal planes, lines, and so on, that do not need to go through the origin. If you accept that, we do not need to draw this dimension explicitly. We just use this freedom and know that such elements are blades in the representational space because of the o-dimension.

The ∞-dimension is new relative to the homogeneous model, and much more interesting. If we draw the Euclidean 2-space as the $\mathbf{e}_1 \wedge \mathbf{e}_2$-plane, then there is apparently a *paraboloid* $(\mathbf{x}, \tfrac{1}{2}\mathbf{x}^2\,\infty)$ in the ∞-direction that we should get to know better, for the points of Euclidean space are represented on it. By studying the combinations of points on this paraboloid, we should be able to get an intuition on how the conformal model actually works. In Figure 14.3, you see the 2-D Euclidean space laid out in white, and the ∞-paraboloid indicated vertically above it.

A conformal null vector x representing a point ends on the paraboloid, just above the point at the location $\mathbf{x}$ in the Euclidean plane which it represents. We draw it as the parabola point with location vector $\mathbf{x} + \tfrac{1}{2}\mathbf{x}^2\,\infty$, letting the extra dimension o play the role of the homogeneous coordinate. In Figure 14.3, we have drawn a vertical connecting line between the point x and its representation; since that line runs in the ∞-direction, it is represented by the line $x \wedge \infty$ in this homogeneous depiction of the conformal model

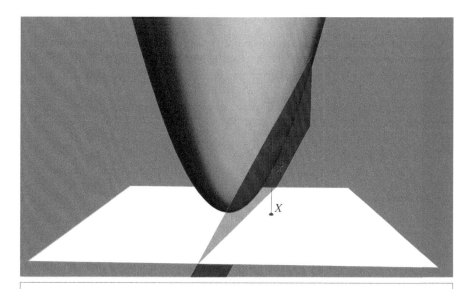

Figure 14.3: Visualization of a 2-D Euclidean point X as a conformal vector x on the representative paraboloid of the conformal model. The tangent plane at x has the dual representation x.

(with x on it and ∞ as its direction). That vertical line is therefore the visualization of the flat point $x \wedge \infty$ at the location of X.

We have also drawn a plane at x, tangent to the paraboloid. This is actually the representational plane x^*: it consists of all the vectors perpendicular to the vector x (in this metric). (It doesn't *look* perpendicular, but that is because we are interpreting with Euclidean eyes.) This is a direct consequence of the null-vector representation of Euclidean points. A conformal vector y is on a plane Π if and only if $y \wedge \Pi = 0$. Or, if we have a dual representation of the plane, $\pi = \Pi^*$, then y is in the plane if and only if $y \cdot \pi = 0$. You will remember that the metric of the conformal model is set up in such a way that $x \cdot x = 0$, and the motivation for that was that a point represented by x has distance 0 to itself in the Euclidean metric. We now see that we can also read this as

> *If the vector x represents a Euclidean point, then x is on the tangent plane to the paraboloid that is dually represented by x.*

Since $y = x$ is the only null vector that satisfies $y \cdot x = 0$, the plane dually represented by x cannot intersect the paraboloid anywhere else, so it must indeed be the tangent plane to the paraboloid.

14.3.2 CIRCLE REPRESENTATION

Let us find out what another representative vector σ on this tangent plane represents in the Euclidean base space. We can construct it from a point C in the Euclidean space, making a vertical line $c \wedge \infty$ out of its representative c, and intersect that with the tangent plane. That gives a representation point $\sigma = x \rfloor (c \wedge \infty)$. We have met this element σ as representing a dual circle through the point x with center c. Let us find out why this is the correct interpretation in this visualization as well. The position of σ relative to x is sketched in Figure 14.4. Since σ is on the plane with dual representation x, we must have $\sigma \cdot x = 0$. But this also implies that $x \cdot \sigma = 0$, which we would interpret as x being on the plane dually represented by σ. What is this dual plane of σ, and what other point representatives are on it? All such points y must satisfy $y \cdot \sigma = 0$, and therefore $\sigma \cdot y = 0$. So all these points must be such that σ lies on their tangent plane; that must imply that they are the contour of the parabola as seen from σ. Together, they lie on the plane dually represented by σ, which therefore passes through the parabola in the way indicated in Figure 14.4. That is then a geometrical way of constructing the dual of σ: as the connecting plane of tangent points to the paraboloid as seen from σ. In projective geometry, this construction is known as determining the polar of σ relative to the paraboloid. When you project down the resulting ellipse, you find a circle of Euclidean points, so *the plane dual to σ is the direct*

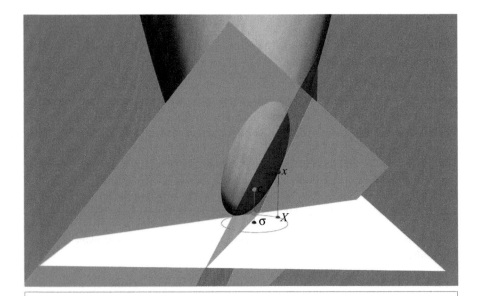

Figure 14.4: The representation of a circle through the point X is by a conformal point σ on the tangent plane of x. The plane dually represented by σ cuts the paraboloid in the representations of the points on the Euclidean circle.

representation of a circle. We show a cross section of this situation in Figure 14.5 for clarity (alternatively, you can see this as a depiction of the conformal model of a 1-D space).

> *A circle is represented by a plane that cuts the paraboloid in an ellipse, which projects down to a circle in the Euclidean space.*

The position of the plane relative to the point σ can be understood as follows. Since $\sigma = x \rfloor (c \wedge \infty) = (x \cdot c) \infty + c$, we have $\sigma^2 = -2(x \cdot c)$, so that σ can be written as $\sigma = c - \frac{1}{2}\sigma^2 \infty$. Therefore, the point σ is a distance of $\frac{1}{2}\sigma^2$ under the paraboloid at c. When we look for the intersections with a vertical line $x \wedge \infty$ at any location x with both the plane dually represented by σ and the tangent plane at c, we find that these are $x \cdot (c - \frac{1}{2}\sigma^2 \infty) = x \cdot c + \frac{1}{2}\sigma^2$ and $x \cdot c$, respectively. This shows that *the dual plane σ is parallel to and above the tangent plane at c, by an amount of $\frac{1}{2}\sigma^2$ in the ∞-direction.* So the dual plane σ is as far above the paraboloid as σ is below it. This computation of the relative position of σ and its dual plane σ holds independently of the value of σ^2. If it is positive, we have the construction depicted in Figure 14.4; if it is zero, we have the dual plane of a point, which is the tangent plane depicted in Figure 14.3; and if it is negative, the point σ is above the paraboloid, and the dual plane σ below it, still parallel to the tangent plane. Thus we find a sensible real construction of the dual of an imaginary circle.

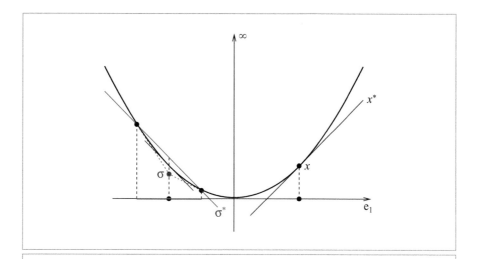

Figure 14.5: Cross section of the parabola of null vectors. On the right-hand side, a point x and its dual x^*, which is the direct representation of the tangent plane at x. On the left-hand side, the representation of a dual sphere σ. Its dual σ^* represents a plane as far above the parabola as σ is below it. It intersects the parabola in the points representing the sphere in the base space. These are on the contour of the parabola as seen from σ.

14.3.3 EUCLIDEAN CIRCLES INTERSECT AS PLANES

Figure 14.6 shows how to do the intersection of two Euclidean spheres (which in 2-D are of course circles) in the conformal model. The answer is a Euclidean point pair (i.e., a 0-sphere). We know that the 2-D conformal model would represent this as the outer product of two null vectors $x \wedge y$. That is represented as a line in this homogeneous depiction of the representational space, intersecting the paraboloid at the two representatives of the points x and y.

The figure shows clearly that this line representing the intersection of the two circles is the intersection of the two planes representing the circles. Those planes are in turn the duals of the circle representations σ and τ as homogeneous points under the paraboloid.

There is therefore nothing quadratic about intersecting two circles once we are in the conformal model; it is the same as intersecting planes. If the planes should intersect in a line that does not pass through the parabola, we would interpret this as an imaginary

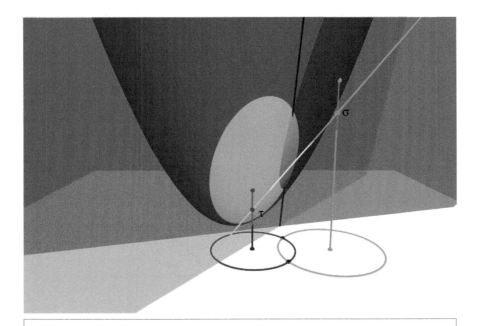

Figure 14.6: Visualization of the intersection of circles in the conformal model. The red circle in the 2-D Euclidean base space is represented by the red plane, and dually represented by the representative vector τ. The green circle is similarly represented by the green plane and the vector σ. The intersection of the two circles is the blue point pair, which is represented by the blue line $(\sigma \wedge \tau)^{-*}$, the intersection of the two planes. The dual of that line is the yellow line $\sigma \wedge \tau$ connecting the dual circle representatives.

point pair (it is a nice puzzle to construct its location), but there is nothing irregular about the intersection itself. Here is the take-home message of all this visualization:

> *In the conformal model of a 2-D Euclidean space, intersecting circles is identical to intersecting offset planes in a space of one more dimension (the ∞-dimension). In yet one more dimension (the o-dimension), this is identical to intersecting subspaces through the origin. Therefore, the* meet *of origin blades in the conformal model is effectively circle intersection.*

We hope this visualization helps you understand slightly better where those two extra dimensions come from, and why flat elements through the origin of the representational space $\mathbb{R}^{n+1,1}$ (its blades) can represent round elements in the Euclidean space $\mathbb{E}^n$.

14.4 APPLICATION: VORONOI DIAGRAMS

The *Voronoi diagram* of a set of points is a graph that partitions space into the parts closest to each of them, as in the base plane in Figure 14.7. In computational geometry literature [12], an interesting construction is made that turns the computation of a Voronoi diagram of a set of points into computing a convex hull of a properly constructed polyhedron in a space of one more dimension, projected down. The polyhedron is made up of parts of tangent planes of lifts of the original points to a paraboloid set up in the extra dimension. This paraboloid construction is usually presented as a special clever trick, and not used in other algorithms. We will show that it is essentially the conformal model, and therefore much more widely applicable than is usually appreciated. Let us analyze the Voronoi construction in more detail, making full use of the convenient metric that the conformal model dictates for the extension of the Euclidean base space $\mathbb{E}^n$ to $\mathbb{R}^{n+1,1}$.

We have just seen how points are mapped to the paraboloid in the conformal model construction and how we can use a homogeneous model in the resulting space to consider the tangent planes. Now consider two points $\mathcal{P}$ and $\mathcal{Q}$ of the Euclidean base space, lifted to the paraboloid as p and q. If we want to determine which points are closer to one than the other, the separation line between those is the *perpendicular bisector*, which is the line dually represented as $p - q$ in the conformal model. In the paraboloid representation, this line can be made by intersecting the tangent planes to p and q. We show this in two steps: first, since the tangent planes are dually represented by p and q, their meet is $\Lambda = (p \wedge q)^{-*}$. To find what this line in our model represents, we need to turn it into a plane stretching to infinity, and intersect that with the 2-D Euclidean base space. The plane is $\infty \wedge \Lambda = \infty \wedge (p \wedge q)^{-*} = (\infty \rfloor (p \wedge q))^{-*} = (p - q)^{-*}$. The intersection with the base plane gives $(p - q) \rfloor (o \wedge \mathbf{I}_2 \wedge \infty) = (p - q)^{-*}$, consistent with our assumption that this line was dually represented by $p - q$.

These perpendicular bisectors of point connections are the carriers of potential edges of the Voronoi diagrams in $\mathbb{E}^n$. We still need to select those among them that are actually

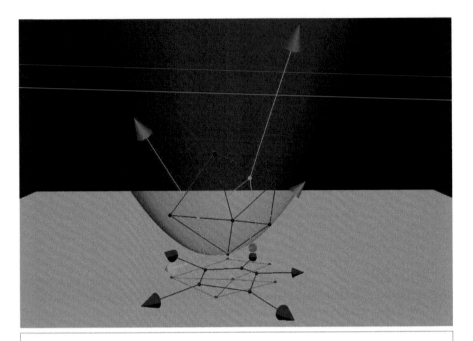

Figure 14.7: The Voronoi diagram of six red points in a 2-D Euclidean space is indicated in black, and their Delaunay triangulation in red. In the conformal model, the represented Delaunay triangulation (in blue) is obtained by making the convex hull of the represented points on the paraboloid, which is the five-sided pyramid. The representation of the Voronoi diagram is its dual in the conformal model, depicted in green, which is a planar pentagon with five rays (indicated by tangent vectors). The green points are below the paraboloid. They are the representations of the circumcircles of triangles, whose centers are the corresponding black points of the Voronoi diagram.

closest to the given points not overruled by other bisectors of even closer points. When we study this by entering a third point R, we have to decide among the bisectors dually represented as $p - q$, $q - r$, and $r - p$. These are lines in the intersection of the three tangent planes dually represented by p, q, and r, and they intersect in a single point in the representational space. This point is dually represented by $p \wedge q \wedge r$, and it represents the circumcircle of p, q, and r. Its projected location in the base space is the center of the circumcircle. It is clear that the parts of the bisectors that show up in the Voronoi diagram are the lines that are highest in the ∞-direction of the representation.

This holds for each triplet of points: the representations of the Voronoi lines connect the intersection points of each triplet of tangent planes, in the highest possible manner. Therefore we can determine the Voronoi diagram by constructing all such lines and determining their upper convex hull in the ∞-direction. It is illustrated as the green network in Figure 14.7, which projects down to the black Voronoi diagram in the 2-D Euclidean base space.

Using the principle of duality, this construction can be improved and made much more direct, for the intersection line $(p \wedge q)^{-*}$ of a tangent plane of point representatives p and q is exactly dual to the connection line $p \wedge q$ between the point representatives. When we take the dual of the complete network of line segments representing the Voronoi diagram, this generates a network that forms the convex hull of the point representatives themselves. It is illustrated as the blue network in Figure 14.7. It is in fact the representation of the *Delaunay triangulation* of the point set. That is the network of which the edges denote the closest point connections, indicated in red in Figure 14.7.

So the Delaunay triangulation of a set of points can be determined by lifting them all to a paraboloid, taking their convex hull, and projecting back. The corresponding Voronoi diagram is obtained from this by duality. That is precisely the convex hull algorithm from [12]. We have shown that this is actually a hidden application of the conformal model. You can play around with this algorithm in the programming example of Section 14.9.1.

14.5 APPLICATION: FITTING A SPHERE TO POINTS

14.5.1 THE INNER PRODUCT DISTANCE OF SPHERES

The inner product of the conformal model was defined to give a good correspondence to the squared Euclidean distance of normalized points in $\mathbb{E}^n$ represented as vectors in $\mathbb{R}^{n+1,1}$. Meanwhile, we have found that vectors of $\mathbb{R}^{n+1,1}$ can also represent dual spheres (and planes) of $\mathbb{E}^n$. We investigate what distance measure between spheres is defined by the inner product of vectors.

When we have two dual spheres $\sigma_1 = c_1 - \frac{1}{2}\rho_1^2 \infty$ and $\sigma_2 = c_2 - \frac{1}{2}\rho_2^2 \infty$, their inner product is

$$\sigma_1 \cdot \sigma_2 = c_1 \cdot c_2 + \tfrac{1}{2}(\rho_1^2 + \rho_2^2) = \tfrac{1}{2}\left(\rho_1^2 + \rho_2^2 - d_E^2(C_1, C_2)\right),$$

where $d_E^2(C_1, C_2)$ is the square of the Euclidean distance of their centers.

Figure 14.8 shows some situations for the geometry of the inner product. Let us first assume that one of the spheres is a point. In that case we find two situations, depending on whether the point is inside or outside the other sphere. If it is outside, as in Figure 14.8(a), the inner product denotes minus half the tangential distance of the point to the sphere, since its value becomes

$$\sigma \cdot p = -\tfrac{1}{2}\left(d_E^2(C, P) - \rho^2\right).$$

When the point is inside, the interpretation as a real distance changes to Figure 14.8(b). It is interesting to see what happens when the point gets close to the sphere at a distance

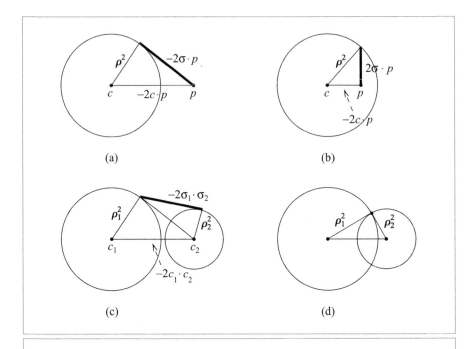

Figure 14.8: The inner product $\sigma_1 \cdot \sigma_2 = \frac{1}{2}\left(\rho_1^2 + \rho_2^2 - d_E^2(c_1, c_2)\right)$ can be interpreted as a squared distance measure between spheres. In the figure all distances are real and denoted by their squares. (a) When σ_2 is an outside point p, it is minus half the squared length of the tangent line segment from p to σ. (b) When σ_2 is an inside point p, it is half the squared length of the perpendicular arc from p to σ. (c) For two spheres, the construction is more involved. (d) The distance is zero when the two spheres intersect perpendicularly.

$\rho + \delta$ from the center: then the inner product is equal to $-\rho\delta$, to first order in δ. That implies that for points close to the sphere, it is linearly proportional to their distance to the sphere. This is a useful distance measure in optimization problems involving spheres, and we use it for sphere fitting in the next section.

For the general case of two spheres, the quantity representing the inner product can be constructed by a nested tangent construction, as illustrated in Figure 14.8(c). This does not immediately convey its usefulness. But Figure 14.8(d) shows an interesting special case: *when the inner product of two spheres is zero, they intersect orthogonally.* In this sense, the inner product of the conformal model is a measure of orthogonality.

By contrast, the outer product of two dual spheres is their dual intersection. When two spheres touch, this is a tangent element, characterized by having a zero square. So *when the outer product of two spheres has zero norm, they touch.*

14.5.2 FITTING A SPHERE TO POINTS

The representation of spheres as vectors leads to a simple method to fit the optimal sphere to a set of points, minimizing the sum of the squared distances (well, almost). This uses the structure of the conformal model to derive a procedure that can be executed classically. This section was inspired by [48].

In fitting a sphere to a set of points $\{p_i\}_{i=1}^N$, we can ask to minimize the sum of the squared distances. If a point p is at a close distance δ to a dual sphere σ with radius ρ, the inner product $p \cdot \sigma$ is $-\rho\delta$, to first order in δ (see previous section). Therefore for a close point, the squared distance is $(p \cdot \sigma)^2/\rho^2 = (p \cdot \sigma)^2/\sigma^2 = (p \cdot \sigma)(p \cdot \sigma^{-1})$, for a normalized sphere, σ, for which $-\infty \cdot \sigma = 1$.

So as a cost function for the fit we have

$$\Gamma \equiv \sum_i (p_i \cdot \sigma)(p_i \cdot \sigma^{-1}).$$

We ask for the σ that minimizes this. Differentiating with respect to σ and equating to 0 gives

$$0 = \frac{\partial \Gamma'}{\partial \sigma} = 2 \sum_i (p_i \cdot \sigma)(p_i \wedge \sigma)\sigma^{-3}.$$

Unfortunately, we do not know how to solve this equation, since it is strongly nonlinear in σ.

If we change the problem slightly, we can ask for the sphere that minimizes the summed squares of $\rho\delta$ rather than δ. This would minimize the distances to the points, but with a preference for small spheres. In practice, this solution will be close to the one we would have wanted. In this form it is solvable, for now the cost function is

$$\Gamma' \equiv \sum_i (p_i \cdot \sigma)(p_i \cdot \sigma),$$

and optimization of this gives

$$0 = \frac{\partial \Gamma'}{\partial \sigma} = 2 \sum_i p_i (p_i \cdot \sigma).$$

The single occurrence of σ makes this a linear equation that can be solved using standard linear algebra techniques, as follows. First, we realize that we can write the conformal space inner product in terms of a Euclidean inner product by the using metric matrix $[\![M]\!]$, which in 3-D is as given in Table 13.1:

$$[\![M]\!] = \begin{bmatrix} 0 & 0 & 0 & 0 & -1 \\ 0 & 1 & 0 & 0 & 0 \\ 0 & 0 & 1 & 0 & 0 \\ 0 & 0 & 0 & 1 & 0 \\ -1 & 0 & 0 & 0 & 0 \end{bmatrix}$$

for a conformal model with basis $\{o, \mathbf{e}_1, \mathbf{e}_2, \mathbf{e}_3, \infty\}$. Now $p \cdot \sigma$ is implemented on the vectors $[\![p]\!]$ and $[\![\sigma]\!]$ as an expression in the usual linear algebra of a Euclidean space:

$$p \cdot \sigma = [\![p]\!]^T \, [\![M]\!] \, [\![\sigma]\!],$$

where we should use the vector $[\![1, p_1, p_2, p_3, \frac{1}{2}(p_1^2 + p_2^2 + p_3^2)]\!]^T$ as a representation for each measured point p. We then need to solve, in least-squares fashion,

$$\sum_i [\![p_i]\!] \, [\![p_i]\!]^T \, [\![M]\!] \, [\![\sigma]\!] = 0.$$

To avoid the trivial solution $[\![\sigma]\!] = 0$, we use the scaling freedom in our representation to demand that $\sigma \cdot \sigma$ be constant and equal to 1. The nontrivial least-squares solution is then found using the standard singular value decomposition of the data matrix

$$[\![D]\!] \equiv \sum_i [\![p_i]\!] \, [\![p_i]\!]^T \, [\![M]\!] = [\![U]\!] \, [\![\Lambda]\!] \, [\![V]\!]^T.$$

The optimal σ is the last column of $[\![V]\!]$, which is the eigenvector corresponding to the smallest singular value. This is proportional to the dual sphere that minimizes Γ', and you can compute its center and radius using Table 14.1.

If you truly needed to minimize Γ rather than Γ', you can use the solution for Γ' as the seed for a nonlinear optimization method such as Levenberg-Marquardt.

14.6 APPLICATION: KINEMATICS

14.6.1 FORWARD KINEMATICS

In computer animation and robotics, *forward kinematics* is used to compute the location and change of geometrical objects given a change in their parameters. As a typical example of the kind of problem, we treat the forward kinematics of a humanoid arm, in which we need to compute where the limbs end up, given the angles of the various joints. With translations and rotations both available as versors, we can now write this in terms of the conformal model.

In the classical way of doing this, one uses the homogeneous model to compute the rigid body motion matrices for vectors, as in Table 12.2, and multiplies those to get a transformation matrices for points on each of the limbs. That chain structure is of course essential to the problem, and also found in the conformal model solution. The only difference is that the resulting rigid body motion versors can be applied to elements of any kind, not just vectors. When all you need to do is to control a robot, this is not that relevant, but if you want to render an animation of the arm, this is potentially useful. There is no longer a necessity to decompose the graphic elements of a limb down to the level of vectors or points; line elements, circles, or tangent bivectors are just as easily transferred and rendered at their proper location on the moved arm.

To take a specific example, let us take the structure of a Unimation Puma 560 robot. We start with the definition of the robot dimensions and the directions of their rotation planes from its manual. This gives the essential kinematics structure indicated in gray in Figure 14.9. It is encoded as translations (in meters) and rotation bivectors as:

$$
\begin{aligned}
\mathbf{t}_1 &= 6.60\,\mathbf{e}_3 & \mathbf{B}_1 &= \mathbf{e}_1 \wedge \mathbf{e}_2 \\
\mathbf{t}_2 &= 4.31\,\mathbf{e}_1 + 1.49\,\mathbf{e}_2 & \mathbf{B}_2 &= \mathbf{e}_3 \wedge \mathbf{e}_1 \\
\mathbf{t}_3 &= -0.20\,\mathbf{e}_3 & \mathbf{B}_3 &= \mathbf{e}_3 \wedge \mathbf{e}_1 \\
\mathbf{t}_4 &= 4.33\,\mathbf{e}_1 & \mathbf{B}_4 &= \mathbf{e}_2 \wedge \mathbf{e}_3 \\
\mathbf{t}_5 &= 0 & \mathbf{B}_5 &= \mathbf{e}_3 \wedge \mathbf{e}_1 \\
\mathbf{t}_6 &= 0.56\,\mathbf{e}_1 + 0.11\,\mathbf{e}_2 & \mathbf{B}_6 &= \mathbf{e}_2 \wedge \mathbf{e}_3
\end{aligned}
$$

We can introduce a set of graphic elements X_i^j to be drawn for limb i; they can be points, bivectors, circles, spheres, or whatever you need to draw the limb. You initially specify them in the coordinate frame of the corresponding limb i. In Figure 14.10, each limb is drawn by means of the point at its the origin o, the tangent blade of the rotation plane at the origin indicated as the tangent bivector $o \wedge \mathbf{B}_i$, and the point pair $o \wedge \mathsf{T}_{\mathbf{t}_i}[o]$ drawn as a line segment.

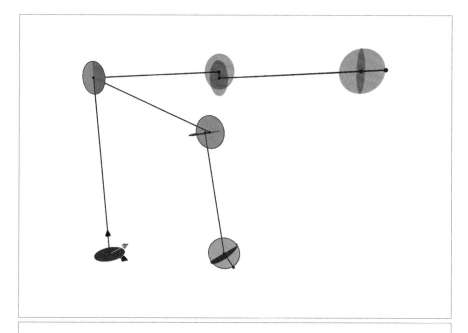

Figure 14.9: A Unimation Puma 560 arm, in its home position (gray) and during a motion. We used connected point pairs and tangent bivectors to denote the limbs. They transform in the same manner.

With that, we can initialize the translation versors for each limb L_i. We also compute where their rotation bivectors (the duals of the rotation axes) are in the home position when expressed in world coordinates relative to the base of the robot. This is done by simply translating the $\mathbf{B}_i$ by the successive translation of each joint. The representative rendering elements X_i^j are translated in precisely the same manner (we call the result Y_i^j).

$$T_0 = 1 \qquad T_i = T_{i-1}\, e^{-t_i \infty/2}$$
$$A_i = T_{i-1}\, \mathbf{B}_i\, \widetilde{T}_{i-1}$$
$$Y_i^j = T_{i-1}\, X_i^j\, \widetilde{T}_{i-1}$$

This concludes the setup of the robot kinematics parameterization, and allows us to draw the arm in the gray home position in Figure 14.9.

If the rotation angles ϕ_i are now specified at run time, we can compute the rigid body motion rotors M_i for each of the limbs. This merely involves rotating around each of the rotation bivectors A_i by the angle ϕ_i, and stacking the results for each successive joint.

You should realize in this that the extension of M_{i-1} to M_i should be done by a multiplication on the *right* by the next rotation around the A_i^*-axis, since that is a transformation given in the next object frame. The explanation is independent of rotors and holds in general for the composition of operators. To perform an operator Y (given in a local frame) when one is in the state X implies that one should undo the present state rotor X first (to get to the proper frame for application of Y), then do Y, and restore the frame by putting the result at X. Those operations are all performed by the customary left multiplication $(XYX^{-1})X = XY$, but the result is equivalent to the right multiplication of X by Y.

The resulting run-time algorithm is now straightforward:

$$M_0 = 1 \qquad M_i = M_{i-1}\, e^{-A_i \phi_i/2}$$
$$Z_i^j = M_i\, Y_i^j\, \widetilde{M}_i$$

We can use the rigid body motions M_i to draw the representative elements Z_i^j, giving a rendering of the moved arm as in Figure 14.9.

The universality of the transformation of the rendering elements X_i^j is the essential difference with the usual homogeneous coordinate solution. You could also have obtained this functionality by computing with the homogeneous coordinate matrices throughout and applying the result as an outermorphism matrix in the final equation. In that case, you would still need to employ the conformal model to make that work on rendering elements that are not the flats of the homogeneous model.

14.6.2 INVERSE KINEMATICS

In computer animation and robotics, *inverse kinematics* is the problem of finding parameters such as angles in a model to make it reach a certain specified position. Inverse kinematics is a notoriously hairy problem in geometrical computation and typically does not have a unique solution. Depending how it is hinged, each robot need its own specific solution to compute its joint angles. Closed solutions are rare, and typically numerical techniques are used.

The conformal model should be highly suited to represent this problem in Euclidean geometry, using its capability to compute directly with spatial quantities. Where a classically specified solution typically abounds in trigonometric functions because it tries to find a solution in terms of scalars (the angles), geometric algebra can avoid this almost completely. Instead, it computes spatial rotors by using intersections of spheres and ratios of lines. Trigonometry is only required if one wants angles as final results, and then involves a logarithm.

Having said that, these techniques have not yet been developed in great generality. To give a sample of what might be possible, we briefly discuss a very simple robot arm with a straightforward inverse kinematics algorithm, featuring some of the typical issues in such problems. The arm is of the humanoid type upper and lower limb, depicted in Figure 14.10, with limb lengths λ_1 and λ_2. The elbow is a simple planar hinge, but the shoulder is a spherical joint. With the shoulder fixed, we are asked to reach a target location p with the wrist from a standard pose in which the arm is horizontally stretched out in the $\mathbf{e}_1$ direction. The limb lengths are given. The principle of the solution we follow is partly inspired by [34].

We choose our coordinate origin at the shoulder, with $\mathbf{e}_3$ as the vertical direction.

1. ***The Tilt Plane.*** There is an obvious degree of freedom in this problem: we can freely choose the plane in which we bend the elbow. This tilt plane contains the upper and lower arm. It certainly needs to contain the line L through shoulder and wrist, which is

$$L = o \wedge p \wedge \infty.$$

Therefore it is natural to parameterize the tilt plane by a tilt angle of rotation around that line. We define the corresponding tilt rotor R_t as

$$R_t \equiv \exp(\check{L}^* \phi / 2),$$

where $\check{L}$ denotes the normalized L. To specify the direction of Π_t, we need a reference direction for the tilt plane. Let that be the vertical plane through L (i.e., the plane $\Pi_0 = \mathbf{e}_3 \wedge L$). Then the actual tilt plane is

$$\Pi_t = R_t \Pi_0 \widetilde{R}_t.$$

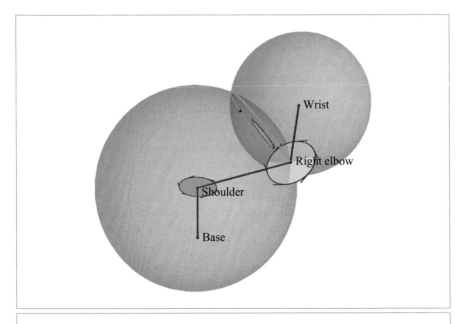

Figure 14.10: A humanoid arm with two movable limbs needs to reach a specified position of the wrist. With a choice for the free parameter determining the tilt of the elbow (that rotor is shown by the light blue tangent bivector), the location of the elbow and the rotors for the joint angles can be computed using the conformal model (see text).

2. ***The Elbow Location.*** The elbow needs to be in the tilt plane at the correct limb lengths from the shoulder and the wrist. We therefore set up two dual spheres: the shoulder sphere $\sigma_s \equiv o - \frac{1}{2}\lambda_1^2 \infty$ and the wrist sphere $\sigma_w \equiv p - \frac{1}{2}\lambda_2^2 \infty$. Their dual meet is the dual circle $\sigma_s \wedge \sigma_w$, and the intersection of that dual circle with the tilt plane is the elbow point pair P_e given by

$$P_e = (\sigma_s \wedge \sigma_w)\rfloor \Pi_s.$$

This gives two possibilities for the elbow. The point pair P_e can be split into its two components $P_e = p_- \wedge p_+$ by the technique in (14.13). We choose one of those as the elbow location q. The only concern here is consistency between subsequent solutions, and that is guaranteed by its automatic sign relationship to the orientation of the tilt plane.

The intersection may lead to an imaginary point pair. In that case there is no solution: the robot cannot reach the desired location. Whether this is the case is a simple test on the sign of P_e^2, which must be nonnegative for a real solution to exist.

3. ***The Elbow Angle Rotor.*** The elbow rotor R_e is the rotor required to rotate from the lower arm direction given by the line $A = o \wedge q \wedge \infty$ to the upper arm direction $B = q \wedge p \wedge \infty$.

We saw in the vector space model that there are several cheap ways of determining a rotor between two directions (Section 10.3). It is a happy surprise that these methods all transfer immediately to conformal lines through any point by literal substitution of the lines for the vectors in those equations. They then yield the conformal rigid body rotor (including translation) to turn one line to the other.

It is not hard to see why this should be so. We consider two lines U and V relative to their common point x. Then the lines may be represented as $U = x \wedge \mathbf{u} \wedge \infty$ and $V = x \wedge \mathbf{v} \wedge \infty$. The constructions to quickly produce a rotor from the two normalized Euclidean directions $\mathbf{u}$ and $\mathbf{v}$ in Section 10.3 were based on the geometric product $\mathbf{u}\,\mathbf{v}$ and on the squared sum $\mathbf{u}+\mathbf{v}$ of the two direction vectors $\mathbf{u}$ and $\mathbf{v}$. These quantities are easily computed from the two lines. The geometric product of lines through the origin is the same as of their directions, for we can factor out the flat point at the origin: $U V = \mathbf{u}\,(o \wedge \infty)\,(o \wedge \infty)\,\mathbf{v} = \mathbf{u}\,\mathbf{v}$. Also, the sum of the two lines conveys the sum of the directions $U + V = x \wedge (\mathbf{u} + \mathbf{v}) \wedge \infty$, and geometric ratios with the square of that quantity therefore also reproduce the vector space results.

We pick the method of the sum of the lines in (10.13) and compute the local elbow rotor as

$$R_e = \frac{1 + V U}{\sqrt{2(1 + V \rfloor U)}}. \tag{14.12}$$

This equation is translation-covariant and holds at $x = 0$, so it is valid everywhere. It is the rotor of a rotation in the tilt plane, at the elbow q, over the correct angle.

4. ***Shoulder Rotor.*** The rotor R_s for the shoulder angle of the spherical joint can be computed in a similar way. We use the two lines along the home position of the upper arm $A_0 = o \wedge \mathbf{e}_1 \wedge \infty$ and along its computed desired position $A = o \wedge q \wedge \infty$. That gives the spherical joint rotor R_s by applying a formula such as (14.12). This rotor resides at the shoulder.

5. ***Splitting the Spherical Joint.*** Not every spherical joint controller can handle a rotation in the combined form of simultaneous rotations over its two degrees of freedom. In that case, we need to split R_s into two rotors.

To be specific, let us assume that our arm needs to rotate first from the home position of the lower arm around the vertical axis $\mathbf{e}_3$, and then in a vertical plane.

To compute the first rotor, we simply compute the projected arm line onto the plane $\mathbf{e}_3{}^* = o \wedge \mathbf{e}_1 \wedge \mathbf{e}_2 \wedge \infty$ (call it Π_3):

$$A_1 = (A \rfloor \Pi_3)/\Pi_3.$$

This can also be computed more directly as the rejection $(A \wedge \mathbf{e}_3)/\mathbf{e}_3$. The rotor R_1 from A_0 to A_1 is then determined like (14.12). The remaining rotor is R_2 is determined by the condition that $R_1 R_2 = R_s$, so that $R_2 = \widetilde{R}_1 R_s$. This is a rotation in the plane of the original bivector $\mathbf{e}_1 \wedge \mathbf{e}_3$, but rotated by R_1.

6. **Determining the Angles.** All rotors we have determined are centered around their own points. The corresponding rotation angles follow from the Euclidean part of their logarithm (13.15), though it is actually more efficient to compute the vector space logarithm of (10.14) of their Euclidean part. (The outcome is the same, for taking the Euclidean part is the rejection of the rotor by $o \wedge \infty$, which is a covariant construction).

   If this is a robotics problem, you may indeed have to go down to scalar angles to feed them to the joint controllers. But if this were an animation problem, you should keep the results in rotor form, for they can be used directly to render the elements that characterize the limbs in their correct location and attitude.

You see how this solution uses a combination of the capabilities of the conformal model to compute with offset planes and spheres, and of its natural reduction to the vector space model at each fixed location.

The implementation of such an algorithm for this particular robot performs about 40 percent faster than the implementation of the classical, angle-based solution [34]. The main reason for that is partly the avoidance of trigonometry, and partly the avoidance of representational switching that is typically required within a classical solution between its several efficient internal representations (from homogeneous coordinate matrices to quaternions and back, for instance).

14.7 FURTHER READING

The use of circles as elements of computation is not new. A fairly complete treatment may already be found in 19th century works [9, 10] and a later summary of such techniques is in the 1941 book by Forder [22]. No one seems to have realized explicitly that the system can be extended with Euclidean versors, though they do make obviously covariant constructions. That breakthrough came in 1999 with [31].

With elements like lines, circles, and rigid body motion rotors as primitives, the conformal model can be used to describe arbitrary shapes in terms of such descriptors. Rosenhahn [53, 54] has used this to do model-based tracking. His work is especially interesting for explicit structural switching between the projective geometry of observation (naturally described as linear transformations in the homogeneous model) and the desired accurate Euclidean reconstruction (for which the conformal model is most suited). This stratification of geometries is important.

Conformal geometric algebra can be used to perform target calibration in a more direct way than by iteratively treating translational and rotational components separately as we did in the vector space model of Section 10.4.2. The first results may be found in the work of Valkenburg et al. [62, 64].

14.8 EXERCISES

14.8.1 DRILLS

These drills intend to familiarize you with the form of common geometric elements and their parameters in the conformal model. We recommend doing them by hand first, and check them with interactive software later.

1. Give the direct representation of the point pair (0-sphere) P spanned by the points p_1 and p_2 at location $\mathbf{e}_1$ and $\mathbf{e}_2$, with weights 2 and -1.
2. Compute center and radius of P.
3. Give the dual representation of P, and use it to compute radius and center.
4. Retrieve the locations of the original points from P (see (14.13) below).
5. Compute the carrier line (see Section 15.2.2) of P, both in direct and dual form.
6. Give the direct representation of the circle K through p_1, p_2, and the unit point at location $\mathbf{e}_3$.
7. Compute the squared radius and the center of the circle K.
8. Give the direct representation of the sphere Σ through K and the origin.
9. Compute the dual of Σ and read off its center and squared radius directly from that dual representation.

14.8.2 STRUCTURAL EXERCISES

1. The normalized sphere through four points p, q, r, s is: $\Sigma = (p \wedge q \wedge r \wedge s)/(p \wedge q \wedge r \wedge s \wedge \infty)^*$. Show that the Euclidean vector pointing to the center of this sphere is

$$\mathbf{c} = (o \wedge \infty)\rfloor \big(o \wedge \infty \wedge \Sigma^*\big) = \big((o \wedge \infty)\rfloor\Sigma\big)^\star,$$

   Note that the final rewriting involves the Euclidean dual. The first form is the rejection of the non-Euclidean parts from the dual. It is easily implemented as simply listing the Euclidean part of the normalized dual sphere.
2. The weight of a round spanned by four points is related to the volume of the simplex spanned by those points. Show that the weight of $p \wedge q \wedge r/2!$ is the area of the triangle pqr, and that the weight of the sphere $p \wedge q \wedge r \wedge s/3!$ is the volume of the tetrahedron $pqrs$.
3. The weight of a dual sphere σ is the weight of its center, and equal to $\infty \cdot \sigma$. Dualize this expression to discover when a sphere through the points p, q, r, s becomes zero.
4. In many computations resulting in a point pair P, you would like to have it in the factorized form $P = p_- \wedge p_+$, which is unique apart from scaling. Show that p_+ and p_- can be computed as:

$$\textit{point pair decomposition: } p_\pm = \frac{P \mp \sqrt{P^2}}{-\infty\rfloor P}. \tag{14.13}$$

(Hint: Simply substitute $P = p_- \wedge p_+$, and develop the terms in the formula; this shows why the formula works.)

5. For a flat point $P = p \wedge \infty$, (14.13) does not work, since it then requires division by a null vector. In that case, the simplest method is to retrieve the Euclidean position vector $\mathbf{p}$ and use that to make the point p. In an implementation, the coordinates of $\mathbf{p}$ are found as the coefficients of the basis blades $\mathbf{e}_1 \wedge \infty, \mathbf{e}_2 \wedge \infty$, and $\mathbf{e}_3 \wedge \infty$, divided by the coordinate of $o \wedge \infty$. Algebraically, show that

$$\mathbf{p} = -\frac{(o \wedge \infty) \rfloor (o \wedge P)}{(o \wedge \infty) \rfloor P}.$$

6. Show that the cosine of the angle between two lines L and M through a common point p can be computed as the usual formula (3.5) directly applied to the lines themselves rather than to their directions.

7. Compute the `meet` of the dual circles $\kappa_1 = \mathsf{T}_{\mathbf{e}_2}[(o - \frac{1}{2}\infty)(-\mathbf{e}_3)]$ and $\kappa_2 = \mathsf{T}_{-\mathbf{e}_2}[(o - \frac{1}{2}\infty)(-\mathbf{e}_3)]$, both residing in the $\mathbf{e}_1 \wedge \mathbf{e}_2$-plane. It is a tangent vector—what is its weight, and how is that related to the geometry of the situation?

8. In any of the expressions for the direct or dual directions $\mathbf{E} \wedge \infty$, you can replace the outer product with a geometric product to write $\mathbf{E}\infty$. Why is a similar substitution not true for general *tangents*?

9. In Figure 14.7, the green line segments are part of the Voronoi diagram. The points of these segments should represent Euclidean circles. Draw these circles in the Euclidean space. Similarly, the edges of the Delaunay triangulation represent circles, but they are imaginary. Draw some of those. (For a hint, see Figure 15.8.)

10. Use (3.6) and (3.18) to show that

$$\|\infty \wedge X\|^2 = -\|\infty \rfloor X\|^2$$

for general X. (Remember that the squared norm is defined through $\|Y\|^2 \equiv Y * \widetilde{Y}$.)

11. Extending Figure 14.8, draw pictures displaying the inner product of two spheres when the center of one is contained inside the other sphere, and when one sphere is fully contained inside the other sphere.

14.9 PROGRAMMING EXAMPLES AND EXERCISES

14.9.1 VORONOI DIAGRAMS AND DELAUNAY TRIANGULATIONS

This example uses `c2ga` (the 2-D conformal model implementation) to compute Voronoi diagrams and Delaunay triangulations. These geometric constructions were discussed in Section 14.4. The example lets the user drag 2-D points around, and create new ones.

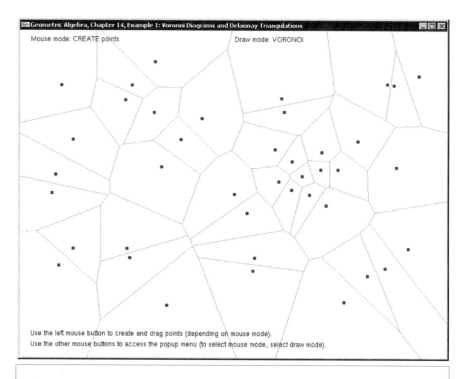

Figure 14.11: A Voronoi diagram of a set of points, as computed by Example 1.

The Voronoi diagrams and Delaunay triangulations are updated in real time, using the following steps:

- First, the 2-D points (including their ni-coordinate) are passed as a set of 3-D vectors to QHull, an existing library for computing convex hulls. The third dimension is of course the ∞-direction of the conformal model.
- The resulting convex hull is stored in a custom data structure called Delaunay-Triangulation.
- Backface culling is used to remove the unwanted part of the convex hull.
- The Voronoi diagram or the Delaunay triangulation is drawn.

We explain each of the steps in more detail below. (Using QHull is a bit backwards, since QHull itself is able to compute Voronoi diagrams and Delaunay triangulations directly. However, implementing our own convex hull algorithm just for this example would be overdoing things.)

Passing the Points to QHull

QHull does not accept 2-D conformal points as input. So we have to collect the e_1-, e_2-, and ∞- coordinates into an array that can be passed to QHull:

```
// the input points come from this array:
const std::vector<normalizedPoint> &points;

// We store the coordinates in the following array:
// 'coordT' is just a floating point type (e.g., float or double)
std::vector<coordT> qhCoord(points.size()*3);

// extract the e1-, e2-, ni-coordinates for each point:
for (unsigned int i = 0; i < points.size(); i++) {
  qhCoord[i * 3 + 0] = points[i].e1();
  qhCoord[i * 3 + 1] = points[i].e2();
  qhCoord[i * 3 + 2] = points[i].ni();
}

// pass 'qhCoord' to QHull
// ...
```

Note that p.ni() actually retrieves the ∞-coordinate of a point p, which would algebraically be denoted as the operation $-o \cdot p$, involving no rather than ni.

Storing the Hull in a DelaunayTriangulation

QHull returns the convex hull as a set of triangles. The vertices of the triangles are the same points that were originally passed to QHull (i.e., no new points are created). The example stores the convex hull in a class called DelaunayTriangulation. This class contains an array of DelaunayTriangles, and an array of DelaunayVertexes:

```
class DelaunayTriangulation {
public:
  // ... (Constructors, etc)

  std::vector<DelaunayVertex> m_vertices;
  std::vector<DelaunayTriangle> m_triangles;
};
```

Each DelaunayVertex contains a normalizedPoint that specifies its position. Each DelaunayTriangle contains an array of three vertices and an array of three neighboring triangles (both specified by their index). A DelaunayTriangle also contains a circle that passes through all three vertices, as described next.

Backface Culling

The convex hull is fully closed, but we do not need the triangles that close the top of the paraboloid, as these are not part of the Delaunay triangulation. Fortunately, QHull abides by the counterclockwise-orientation rule for triangles (see Section 2.13.2), so that we can use backface culling to remove the unwanted triangles. Here we compare the orientation of the circles spanned by the vertices of each triangle to the free bivector $\mathbf{e}_1 \wedge \mathbf{e}_2 \wedge \infty$:

```
// 'DT' is a DelaunayTriangulation
// get the vertices of the triangle:
DelaunayVertex &V1 = DT.m_vertices[vtxIdx[0]];
DelaunayVertex &V2 = DT.m_vertices[vtxIdx[1]];
DelaunayVertex &V3 = DT.m_vertices[vtxIdx[2]];

// Check if 'front-facing':
// 1: Compute the circle spanned by the three vertices:
circle C = _circle(V1.m_pt ^ V2.m_pt ^ V3.m_pt);
// 2: Compare orientation:
if (_Float(C << (e1 ^ e2 ^ ni)) < 0.0)
    continue; // do not use this triangle

// else: this triangle is valid
```

The circle C is stored along with each triangle for later use; the center point of the circle is a vertex of the Voronoi diagram.

Drawing the Result

Drawing the Delaunay triangulation from a `DelaunayTriangulation` is straight-forward. Drawing the Voronoi diagram is less so. First, we need to find the neighbors of every Delaunay triangle, on a per-edge basis. This is computed when the `DelaunayTriangulation` is initialized and stored in each `DelaunayTriangle`. With this information available, drawing the Voronoi diagram is done by connecting the center points of neighboring triangles with line segments. For each edge that does not have a neighboring triangle, we draw a (dashed) line from the center point through the midpoint of the edge.

14.9.2 EXERCISE: DRAWING EUCLID'S ELEMENTS

This example provides the bare-bones code to draw Euclid's elements (i.e., all flat, round, free, and tangents blades of the conformal model). It is your job to fill in the code that creates a typical element of each class and grade. You'll have to give each element the appropriate size and orientation, such that the end result looks somewhat like Figure 14.12. Note that you could copy the coordinates directly from the figure, but then you might as well go straight to the solution that is provided with the GA sandbox source code package. An example of the code you have to complete is

```
name = "flat point";
X = no; // <- EXERCISE: insert correct primitive (flat point) here
```

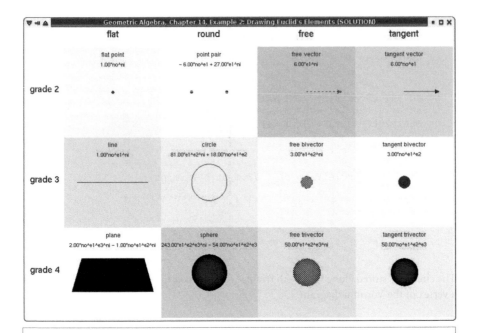

Figure 14.12: Euclid's elements (Example 2).

14.9.3 CONFORMAL PRIMITIVES AND INTERSECTIONS

In this example we repeat the blade visualization example of Section 11.13.2, only this time in the conformal setting that considerably extends the blades to be drawn. Again the user can create and drag points, and create primitives by selecting a number of these points. The primitives that can be constructed are lines, planes, circles, and spheres. The example draws all these primitives and their intersections. A screenshot is shown in Figure 14.13.

A circle is created by selecting three points. If one of these points is the point at infinity—which for this occasion is located conveniently in the upper-right corner of the viewport—then a line is created. Circles and lines can be computed using the same code, because a line is just a special type of circle (i.e., one that passes through infinity):

```
circle C = _circle(g_points[pointIdx1] ^ g_points[pointIdx2] ^
    g_points[pointIdx3]);
```

Spheres and planes are created by selecting four points, and computed as follows:

```
sphere S = _sphere(g_points[pointIdx1] ^ g_points[pointIdx2] ^
    g_points[pointIdx3] ^ g_points[pointIdx4]);
```

Again, a plane is a special type of sphere for which one of the selected points is infinity.

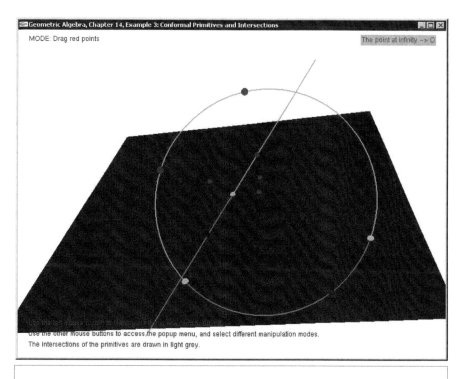

Figure 14.13: Example 3 in action.

The intersections of the primitives are computed by the following code, which spells out the `meet`:

```
// P1 and P2 are two multivectors (the primitives)
I = unit_e(join(P1, P2));
mv intersection = (P1 << I) << P2;
```

As in Section 11.13.2, we use orthogonal projection to enable the creation of degenerate situations.

```
// P1 and P2 are two multivectors (the primitives)
mv projection = (P1 << inverse(P2)) << P2;

// check if projection of P1 onto P2 is 'close' to P1
const float CLOSE = 0.02f;
if (_Float(norm_e(projection - P1)) < CLOSE) {
  // yes: P1 = projection of P1 onto P2
  P1 = projection;
}
```

This projection step is of course performed before the intersection test, not after. Note that orthogonal projections involving round conformal objects do not behave as you would expect, as this would result in conics (which the conformal model obviously cannot represent). For more details, see Section 15.3.

Creating Points

Just as in Section 11.13.2, new points can be created by a mouse-click. The user then expects a point to appear under the mouse cursor. The code to create a point at the correct location is very similar to that in Section 11.13.2, except that this time we use a versor to apply the inverse modelview transformation (in the homogeneous example, we constructed an outermorphism directly from the inverted modelview matrix):

```
// create point at required location and 'distance'
point pt = _point(c3gaPoint(_vectorE3GA(vectorAtDepth(distance,
  mousePos) − e3 * distance))));

// retrieve modelview matrix from OpenGL:
float modelviewMatrix[16];
glGetFloatv(GL_MODELVIEW_MATRIX, modelviewMatrix);

// convert modelview matrix to versor:
bool transpose = true;
TRversor V = _TRversor(matrix4x4ToVersor(modelviewMatrix, transpose));

// apply (inverse) OpenGL 'versor' to transform the point to the
// global frame
pt = inverse(V) * pt * V;
```

The function `matrix4x4ToVersor()` will be discussed in Section 16.10.1. It converts 4×4 translate-rotate-scale matrices into conformal versors.

14.9.4 FITTING A SPHERE TO A SET OF POINTS

This example implements the equations for fitting a sphere from Section 14.5. The user can drag and create points, as in the previous example. The program tries to fit a sphere to these points as well as it can under the least-squares condition. As noted in Section 14.5, the minimization criterion has a preference for small spheres.

The code, shown in Figure 14.14, is straightforward to implement given the algorithm that was spelled out in Section 14.5:

- The matrix $P = \sum_i \llbracket p_i \rrbracket \llbracket p_i \rrbracket^T$ is computed;
- Then, the metric matrix M is initialized;
- The product of these matrices is computed `PM = P * M`;
- The singular value decomposition (SVD) is computed, resulting in the matrices U, S, and V;
- The coordinates of the dual sphere DS are set to the last column of V.

```
dualSphere fitSphere(const std::vector<point> &points) {
  float P[5 * 5];

  // compute matrix P = sum_i (points[i] . points[i]^T)
  {
    // first clear all entries:
    for (int i = 0; i < 5 * 5; i++) P[i] = 0.0f;

    // fill the matrix:
    for (unsigned int p = 0; p < points.size(); p++) {
      // get coordinates of point 'p':
      const mv::Float *pc = points[p].getC(point_no_e1_e2_e3_ni);

      for (int i = 0; i < 5; i++)
        for (int j = i; j < 5; j++) {
          P[i * 5 + j] += pc[i] * pc[j];
          P[j * 5 + i] = P[i * 5 + j];
        }
    }
  }

  // initialize the metric matrix:
  float M[5 * 5] = {
  // no    e1    e2    e3    ni
    0.0f, 0.0f, 0.0f, 0.0f, -1.0f, // no
    0.0f, 1.0f, 0.0f, 0.0f,  0.0f, // e1
    0.0f, 0.0f, 1.0f, 0.0f,  0.0f, // e2
    0.0f, 0.0f, 0.0f, 1.0f,  0.0f, // e3
   -1.0f, 0.0f, 0.0f, 0.0f,  0.0f  // ni
  };

  // construct OpenCV matrices (on stack)
  CvMat matrixP = cvMat(5, 5, CV_32F, P);
  CvMat matrixM = cvMat(5, 5, CV_32F, M);

  // use OpenCV to multiply matrices
  float PM[5 * 5];
  CvMat matrixPM = cvMat(5, 5, CV_32F, PM); // create matrix P * M (on stack)
  cvMatMul(&matrixP, &matrixM, &matrixPM);

  // use OpenCV to compute SVD
  float S[5 * 5], U[5 * 5], V[5 * 5];
  CvMat matrixS = cvMat(5, 5, CV_32F, S); // create matrix S (on stack)
  CvMat matrixU = cvMat(5, 5, CV_32F, U); // create matrix U (on stack)
  CvMat matrixV = cvMat(5, 5, CV_32F, V); // create matrix V (on stack)
  int flags = 0;
  cvSVD(&matrixPM, &matrixS, &matrixU, &matrixV, flags);

  // extract last column of V (coordinates of dual sphere):
  dualSphere DS(dualSphere_no_e1_e2_e3_ni,
    V[0 * 5 + 4], V[1 * 5 + 4], V[2 * 5 + 4], V[3 * 5 + 4], V[4 * 5 + 4]);

  return DS;
}
```

Figure 14.14: Fitting-a-sphere code.

The OpenCV library is used to implement the linear algebra operations (matrix multiplication and SVD).

Interestingly, the sphere-fitting-algorithm will come up with a plane when this is the optimal solution! As an easy way to create this situation, we have provided an option in a popup menu that projects all points onto the $\mathbf{e}_3^*$ plane.

15 CONSTRUCTIONS IN EUCLIDEAN GEOMETRY

Now that the previous chapters have given us such a range of elements of Euclidean geometry as the blades in the conformal model, we want to combine them into useful constructions. As always, this is no more than the application of operations we introduced in Part I, but they take on intriguing and useful meanings.

The meet is greatly extended in its capabilities to intersect arbitrary flats and rounds (though always with the same formula $X \cap Y = Y^* \rfloor X$), or to compute other incidences. The results can be real or imaginary, or even the infinitesimal tangents. The dual of the meet provides a novel operation: the plunge, which constructs the simplest element intersecting a given group of elements perpendicularly. Once recognized, you will spot it in many constructions, and it enables you to quickly write down correct expressions for, say, the contour circle of a sphere seen from a given point.

We show how all the various concepts of a vector in classical geometry find their specific expression in the conformal model. Normal vector, position vector, free vector, line vector, and tangent vector now all automatically move in the correct way under the same Euclidean rotors. This demonstrates clearly that the conformal model performs both the geometrical computations and the "data type management" required in Euclidean geometry.

We conclude with a sample analysis of some involved planar geometry (the geometry of a Voronoi cell), and compare the coordinate-free conformal solution to the many parameters involved in even just specifying the classical solution.

15.1 EUCLIDEAN INCIDENCE AND COINCIDENCE

The basic constructions in geometric algebra are spanning and intersection. In the vector space model prevalent in Part I, these were fairly straightforward operations on subspaces through the origin. They still are in the representational space of the conformal model, but their interpreted Euclidean consequences deserve careful study. They offer a rich syntax for a constructive and consistent specification language for Euclidean geometry.

15.1.1 INCIDENCE REVISITED

We have encountered the `meet` in a general context in Chapter 5. Within the conformal model, we used it in the previous chapter to construct elements by intersection with other elements. Since the basic elements of point, sphere, and plane are all dually represented as vectors, it was easiest to study the `meet` in its dual form:

$$(A \cap B)^* = B^* \wedge A^*.$$

If your application involves a lot of intersections, it is often convenient to stay in the dual representation. However, you should still be careful: the duality in the `meet` needs to be taken relative to the `join`, which may change for successive `meet` operations, so you may not be able to use a single dual representation of an element for all its uses.

The principle of the `meet` is that it constructs the largest subblade common to the blades A and B. The duality relative to the `join` constructs the complements within the smallest blade containing both; these complements do not contain common factors, so that their outer product is nonzero. In that sense, they are independent blades. With that, the containment relationship that gives us the `meet` is easily verified in its dual form:

$$0 = x \rfloor (A \cap B)^* = x \rfloor (B^* \wedge A^*) = (x \rfloor B^*) \wedge A^* + \widehat{B^*} \wedge (x \rfloor A^*),$$

and the constructed independence of A^* and B^* then implies that x is contained in $(A \cap B)$ if and only if it is contained in both A and B. This is easily extended by associativity. Three intersecting spheres, dually represented by a, b, c, will `meet` in the point pair $(a \wedge b \wedge c)^{-*}$ (where the `join` is the full pseudoscalar); see Figure 15.1(a).

In the conformal model, the `meet` of two elements may be imaginary. For instance, the `meet` of three spheres that do not really intersect is an imaginary point pair. The dual of this imaginary point pair is a real circle; see Figure 15.1(b). When we draw that using visualization software, we find that this real circle intersects the original spheres perpendicularly. This bears investigating: can we construct elements that perpendicularly intersect other elements by taking the dual of the `meet`?

15.1.2 CO-INCIDENCE

Let us consider what kind of element is constructed by the dual operation to the `meet`. Its definition would be that its *direct* representation is $B^* \wedge A^*$. If we call the meet

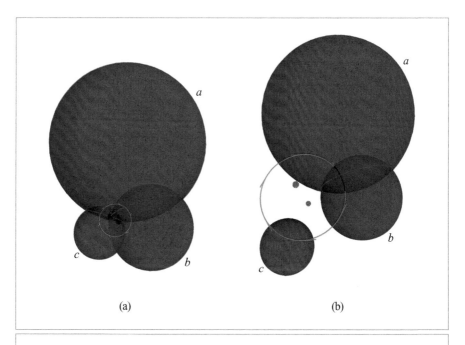

(a)   (b)

Figure 15.1: (a) The meet of three intersecting dual spheres a, b, c is the real point pair with dual representation $c \wedge b \wedge a$ (in blue), whereas their plunge is an imaginary (dashed) circle. (b) The plunge of three nonintersecting spheres is the circle with direct representation $c \wedge b \wedge a$ (in green), while their meet is an imaginary (speckled) point pair. To assist 3-D interpretation, some real intersection circles are indicated in red.

the incidence, we could call this "co-incidence", in the algebraic jargon that associates "co-" with "dual."

When we treated the scalar product in Section 3.1.4, we saw that two blades A and B of the same grade are orthogonal to each other when their scalar product is zero. This is equivalent to the scalar product of their duals being zero, since the scalar product of elements or their duals differ only by a sign (do structural exercise 1 if you want to know which sign). For blades of nonequal grades, this extends to their contractions, although the order now matters (since the contraction of a higher-grade blade onto a lower-grade blade is trivially zero anyway). For a blade X of grade smaller than A to be perpendicular to A, we must have

$$X \perp A \quad \Longleftrightarrow \quad X \rfloor A = 0.$$

Algebraically, when you write out X and A in their factors, the common parts of X and A produce a scalar, and the result then reduces to orthogonality of one of the remaining factors of X with all the remaining factors of A.

Suppose we have two blades A, B, and are looking for the Euclidean object X that is perpendicular to each. We therefore need to satisfy $X \rfloor A = 0$ and $X \rfloor B = 0$. Dualizing this we get $X \wedge A^* = 0$ and $X \wedge B^* = 0$. By choosing the duality relative to the `join` of the two blades A and B, we can make A^* and B^* to be independent blades, just as we did for the `meet` above. This prevents their outer product from being trivially zero. A blade X of lowest grade orthogonal to every non-scalar factor in A and B, and hence

$$X = B^* \wedge A^*,$$

where the order was chosen to correspond to the convention for the `meet`.

Whereas the `meet` constructs a representation of an object *in common* with given elements, this operation of plunging (our term) constructs the representation of an object that is *most unlike* other elements, in the orthogonal sense that it intersects them *perpendicularly*. We coined the term `plunge` (which according to Webster's dictionary may be etymologically related to plumb) to give the feeling of this perpendicular dive into its arguments. Since perpendicularity is a metric concept, the `plunge` is a truly metric operation, whereas the `meet` is not. The `plunge` is an elementary construction in Euclidean geometry that deserves to be better known. It is occasionally found in older works in the Grassmannian tradition [9, 22].

With the associativity of the outer product, the `plunge` easily extends to more elements, so that the general element perpendicular to A_1, A_2, and so on, has the direct representation $X = \cdots \wedge A_2^* \wedge A_1^*$. The `plunge` of three spheres, as in Figure 15.1(b), is therefore a circle:

$$X = C^* \wedge B^* \wedge A^*.$$

Shrinking the dual spheres to zero radius so that they become points, you see that the `plunge` gives a circle through those points, consistent with our earlier derivation of the interpretation of such an element $c \wedge b \wedge a$, with a, b, and c point representatives. We now see that to contain a point a is equivalent to plunging into the zero-radius sphere a^{-*} at that point. Letting the radius of a dual sphere go to infinity gives a dual plane; the `plunge` of such diverse elements can be mixed easily, as Figure 15.2 shows.

15.1.3 REAL MEET OR PLUNGE

The `meet` and the `plunge` are clearly each other's dual relative to the `join`. Moreover, should the result of one of them be imaginary, the other is real. This can only happen for rounds (since flats are always real). That principle helps with finding and visualizing an imaginary `meet`.

Take again the three spheres of Figure 15.1(b). They do not intersect and therefore have an imaginary point pair as their `meet`. This implies that their `plunge` is a real circle. This imaginary point pair and the real circle are algebraically each other's dual. Geometrically, this duality is a polar relationship on the unique smallest sphere containing either point

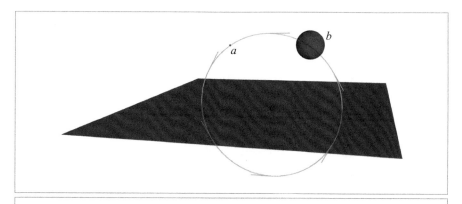

Figure 15.2: The plunge $c \wedge b \wedge a$ of dual plane c, dual sphere b, and point a.

pair or circle (or both, since they are on the same sphere). This is easily shown from the standard representation of such rounds:

$$\left(\left(o + \tfrac{1}{2}\rho^2 \infty \right) \mathbf{E} \right)^* = \left(o - \tfrac{1}{2}\rho^2 \infty \right) \mathbf{E}^\star (-1)^n.$$

If the former is a real circle on the sphere with squared radius $\rho^2 > 0$, then $\mathbf{E}$ is of grade 2; in a 3-D Euclidean space the latter is then a point pair on the imaginary sphere of the same radius (since $\mathbf{E}^\star$ is of grade 1). That locates the imaginary meet.

> *If a circle is considered as the real equator of a sphere, its dual is the imaginary point pair that form the poles. Vice versa, the dual of a real point pair is the imaginary equator of a sphere that has the point pair as its poles.*

Knowing this polar relationship of meet and plunge, we can also determine the location of the imaginary circle that is the meet of two nonintersecting spheres, as indicated in Figure 15.3(a). The plunge of the two spheres is a point pair (indicated in blue). The location of this point pair is found by realizing that any sphere that is perpendicular to both spheres should contain it (since the plunge is an associative operation), and so should every circle, plane, or line perpendicular to both spheres. The simplest way to localize it is then as the intersection of the line through the centers (which is clearly perpendicular to both spheres) and the unique perpendicular sphere with its center on this line, as indicated in Figure 15.3(a) (in white). That is precisely the sphere on which we should determine the meet as the dual to this point pair, which is the imaginary equator indicated as a dashed circle in Figure 15.3(a) (in green). We will give the formula for this sphere in Section 15.2.3.

Once you know how to find this round that contains both meet and plunge, their visualization is easy, whether they are imaginary or not. Verify that this indeed gives the localization of both in Figure 15.3.

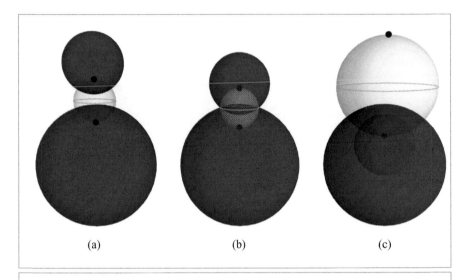

(a)   (b)   (c)

Figure 15.3: The meet and plunge of two spheres at increasing distances. When they do not intersect (a,c), the red dual spheres have a real plunge and an imaginary meet. When they do intersect (b), the meet is real and the plunge imaginary. In all cases meet and plunge are each other's dual (i.e., polars on the three white spheres). Imaginary elements are depicted as speckled.

15.1.4 THE PLUNGE OF FLATS

The construction of a general flat $p \wedge \mathbf{E} \wedge \infty$ can also be interpreted as a plunge. As a definite example, let us interpret the line $p \wedge \mathbf{u} \wedge \infty$. According to the plunge construction, this should perpendicularly intersect the points p and ∞ (so that it contains them), and it should moreover be perpendicular to $\mathbf{u}^*$, which is the plane through the origin with normal vector $\mathbf{u}$. Obviously, this plunge must then be the direct representation of the line through p in the $\mathbf{u}$ direction. This is illustrated in Figure 15.4(a). Including another Euclidean vector factor $\mathbf{v}$ gives an element that should meet the dual line $\mathbf{u} \wedge \mathbf{v}$ perpendicularly, and is therefore the direct representation of a plane. Removing the Euclidean factor (by setting it equal to the 0-blade 1) gives the representation of a flat of dimension zero (i.e., a direct flat point $p \wedge \infty$, which is the element that perpendicularly connects the small dual sphere p with the point at infinity ∞).

Now let us revisit the object $p \rfloor (c \wedge \infty)$ from (14.4), which contains such a flat point in its construction. It was a dual sphere through the point p, with center c. Undualizing this, we find that it is the direct object

$$S = p \wedge (c \wedge \infty)^{-*}.$$

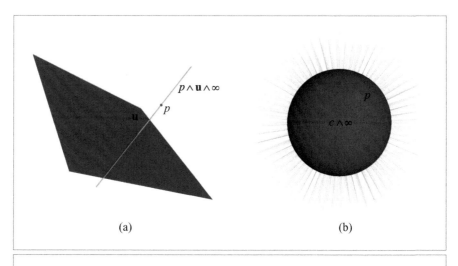

Figure 15.4: (a) The plunge construction of the line $p \wedge \mathbf{u} \wedge \infty$. (b) Visualization of the point $c \wedge \infty$ and its consistency with constructing the dual sphere $p \rfloor (c \wedge \infty)$, given a point p.

We see that this indeed contains p (it plunges into p since it is a direct representation of the form $p \wedge$), and that we should think of it as also plunging into the flat point $c \wedge \infty$. We have just seen that such a flat point in itself is the plunge of c and ∞, connecting those two dual elements. We observe that we get an intuitively satisfying picture like Figure 15.4(b), in which the flat point $c \wedge \infty$ has "hairs" extending from c to infinity. Intersecting all of those perpendicularly while also passing through the given point p must produce a sphere.

Carrying this type of thinking to its logical extreme, we can also interpret a purely Euclidean blade such as $\mathbf{v} \wedge \mathbf{u}$ as a direct element of our geometry. Since it is a purely Euclidean 2-blade, we have already met it as the *dual* representation of the line that is the meet of the two dual planes $\mathbf{u}$ and $\mathbf{v}$ passing through the origin. But if we interpret $\mathbf{v} \wedge \mathbf{u}$ directly, it should be the plunge of the two planes represented dually by $\mathbf{u}$ and $\mathbf{v}$ (i.e., the general element that is perpendicular to both).

Any cylinder with their common line as its axis is a solution, so this is like a nest of cylinders of arbitrary radius but fixed axis. However, as soon as we try to pick out one of the cylinders by specifying a point p on it, the element $p \wedge (\mathbf{v} \wedge \mathbf{u})$ is a unique *circle*. So a more accurate description is that $\mathbf{v} \wedge \mathbf{u}$ represents a stack of nests of circles. But that is also not quite right, since these elements are not circles yet (we have specified only a grade 2 object, not grade 3); rather, $\mathbf{v} \wedge \mathbf{u}$ is a stack of nests of potential circles. This may seem like an uncommon object, until you realize that a rotation rotor can be made from this dual line $L^* = \mathbf{v} \wedge \mathbf{u}$ as $\exp(\alpha \mathbf{v} \wedge \mathbf{u})$. Then the element $L^* = \mathbf{v} \wedge \mathbf{u}$ in fact describes a collection of

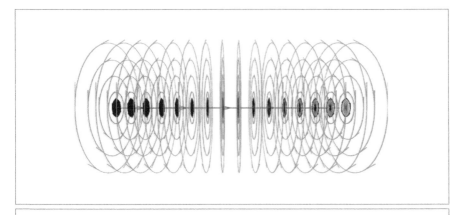

Figure 15.5: The circular orbits of a line L, which are $p \wedge L^*$ for various points p. The orbits of points on the line are tangent bivectors, depicted as small oriented disks.

orbits of the rotating points, as indicated in Figure 15.5 (in which the actual orbits $p \wedge L^*$ are drawn).

The point of all this discussion is that the conformal model suggests, in its algebraic coherence, a consistent and rich language for Euclidean geometry. It contains objects that we have not seen this explicitly before, but that on closer scrutiny represent familiar geometric concepts, and endows them with precise computational properties. Now that we have an algebraic way to combine them, it pays to try and get them into more practical form into our intuition, so that we can use them as active concepts in the solution of problems. This takes practice, and we recommend playing with interactive software packages like GAViewer to absorb this into your intuition—the old-fashioned derivations on paper are rather useless for this (though they should be used to check and/or prove the correctness of your new insights).

15.2 EUCLIDEAN NUGGETS

When you work with the conformal model in a practical application, you will find that you often need to convert elements, and that certain operations appear to be useful nuggets, occurring again and again. We provide some in this section, and try to discern some patterns. This is certainly not a complete inventory of such elementary constructions. We hope that the techniques we expose in deriving them will give you the skills to derive your own special expressions, should your application require them.

15.2.1 TANGENTS WITHOUT DIFFERENTIATING

When you need the tangent to a flat Π or to a round Σ passing through one of its points p, you can compute this without differentiation. Calling the element X (so that $p \wedge X = 0$), its tangent is

$$\text{tangent to } X \text{ at } p: \ p \rfloor \widehat{X} \qquad (15.1)$$

The tangents are illustrated in Figure 15.6.

The formula is easy to prove for a flat through the origin: $o \rfloor (o \wedge \mathbf{E} \wedge \infty)\widehat{} = o \wedge \mathbf{E}$, and because of Euclidean covariance, this holds everywhere for any flat. For a round the proof is also most simply done at the origin o. A direct round through o with center c and direction $\mathbf{E}\infty$ is $o \wedge (c \rfloor (-\widehat{\mathbf{E}}\infty))$, according to structural exercise 4. Then the proposed formula (15.1) gives, by straightforward computation, that the tangent equals $o \wedge (c \rfloor \mathbf{E})\widehat{}$, which is the correct result in its geometrical aspects: its directional part is an element of the carrier $\mathbf{E}$, perpendicular to the radial vector. Note that this local tangent blade is weighted by the magnitude of the radial vector c, which is the radius ρ. The occurrence of ρ (rather than ρ^2) may seem to require complex numbers for imaginary rounds. However, imaginary rounds contain no real points, so they have no real tangents either.

15.2.2 CARRIERS, TANGENT FLAT

Let us define the *carrier* of an element as the smallest grade flat that contains it. A flat is therefore its own carrier. The carrier of a round Σ (which may be a tangent) is

$$\text{carrier of round } \Sigma: \ \Sigma \wedge \infty.$$

This is easily proved: $\big((c + \frac{1}{2}\rho^2 \infty) \wedge \mathbf{E}\big) \wedge \infty = c \wedge \mathbf{E} \wedge \infty.$

We can use this to compute the *tangent flat* to an element X at one of its points p as $(p \rfloor \widehat{X}) \wedge \infty$. This applies to rounds and flats alike; though for a flat it is the identity. Tangents do not have tangent flats, though they do have a carrier.

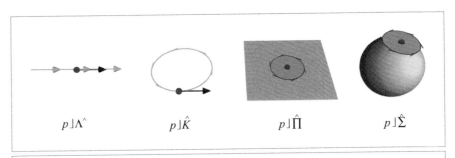

$$p \rfloor \Lambda\widehat{} \qquad\qquad p \rfloor \widehat{K} \qquad\qquad p \rfloor \widehat{\Pi} \qquad\qquad p \rfloor \widehat{\Sigma}$$

Figure 15.6: The tangents of flats and rounds at one of their points are simply computed by the contraction. No differentiation is required.

15.2.3 SURROUNDS, FACTORIZATION OF ROUNDS

A round Σ can be surrounded by the smallest sphere containing it. In its dual representation, that *surround* of Σ is

$$\text{surround of round } \Sigma: \quad \frac{\Sigma}{\Sigma \wedge \infty}, \tag{15.2}$$

in which the division denotes right division (i.e., $\Sigma(\Sigma \wedge \infty)^{-1}$). This even applies to tangents, but it then returns the locational point of the tangent, in agreement with the intuition that a tangent has zero size. Formula (15.2) can be interpreted as the dual of Σ relative to the carrier of dimension k; therefore the expression equals $(c - \frac{1}{2}\rho^2\infty)(-1)^k$, which is a dual sphere containing Σ. We prove the surround formula below.

If the round Σ is given in its dual form, you need the dual of (15.2):

$$\text{surround of dual round } \sigma: \quad (-1)^k \frac{\hat{\sigma}}{\infty \rfloor \sigma}. \tag{15.3}$$

The signs ensure that the computed surround is the same as before, but you could ignore them.

A round can be factored as the geometric product of its surround and its carrier:

$$\text{factorization of a round } \Sigma: \quad \Sigma = \left(\frac{\Sigma}{\Sigma \wedge \infty}\right)(\Sigma \wedge \infty), \tag{15.4}$$

and its dual is

$$\text{factorization of a dual round } \sigma: \quad \sigma = \left(\frac{\sigma}{\infty \rfloor \sigma}\right)(\infty \rfloor \sigma), \tag{15.5}$$

in which you need to include the $(-1)^k$ (as in (15.3)) if you need the proper orientation of the surround. These unique factorization formulas are the motivation behind the expression for the surrounding sphere. It is easy to show that the center of the surrounding sphere is in the carrier plane; and then (15.4) actually represents the perpendicular `meet` of carrier and surround. This makes the factorization geometrically very natural (see Figure 15.7). Algebraically, this factorization in orthogonal factors is important in symbolic simplification of formulas involving rounds. Given a round like a circle, such a factorization gives us a better understanding of it: the carrier plane gives us direction, weight, and orientation of the circle, and the sphere gives us its location (i.e., center) and size (i.e., radius squared).

A special case is a point pair, for which the same formulas hold, for it is clearly also a round. Now $\Sigma \wedge \infty$ represents the carrier line, and division again gives the containing sphere, as shown in Figure 15.7. The result is an expression for *the smallest sphere containing two points*:

$$\frac{p \wedge q}{p \wedge q \wedge \infty} = \frac{(p \wedge q)\rfloor(p \wedge q \wedge \infty)}{-2p \cdot q} = -\frac{1}{2}\left(p + q + (p \cdot q)\infty\right).$$

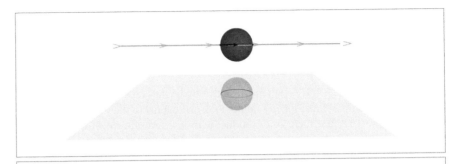

Figure 15.7: The factorization of rounds, using their carrier and surround. A point pair P is factorized as the intersection of its surrounding dual sphere $P/(P\wedge\infty)$ and the carrier line $P\wedge\infty$. A circle K is factorized as the intersection of the dual surrounding sphere $K/(K\wedge\infty)$ and the carrier plane $K\wedge\infty$.

The overall sign in this makes the `meet` correct, but you can ignore it if you are not interested in such matters of orientation.

For the proof of the correctness of the surround formula (15.2), we need to show that the center of the surround is in the carrier plane. So we reflect the surround in that plane and verify that this merely changes the surround by a sign. Using the reflection formula for a dual element into a directly represented mirror (from Table 7.1), we find that the derivation is quicker when we reflect in the inverse of the carrier (which makes no difference geometrically). We obtain

$$(\Sigma\wedge\infty)^{-1}\left(\Sigma(\Sigma\wedge\infty)^{-1}\right)(\Sigma\wedge\infty) = (\Sigma\wedge\infty)^{-1}\Sigma = \widehat{\Sigma}(\Sigma\wedge\infty)^{-1},$$

so that this dual sphere becomes a signed multiple of itself (in the final transition, we took the reverse and computed the resulting sign, after realizing that the reverse of a vector is the vector itself). It follows that the center of the sphere is on the mirroring carrier. As a consequence, the outer product of the dual surround sphere with the carrier must be zero; therefore you can replace the geometric product in (15.4) with a contraction, showing clearly that it expresses a `meet` of the carrier with the surround (in that order).

15.2.4 AFFINE COMBINATIONS

An affine combination of two normalized points gives a 1-parameter family of normalized dual spheres (so *not* of points!):

$$\lambda p + (1-\lambda)q. \tag{15.6}$$

The corresponding dual spheres are depicted in Figure 15.8 for the 2-D case. As λ becomes infinite, the dual sphere approaches the midplace $p - q$ (with enormous weight).

These dual spheres all intersect the smallest dual sphere through p and q orthogonally, as you may easily verify:

$$\left(\lambda p + (1 - \lambda) q \right) \rfloor \left(\tfrac{1}{2}(p + q) + \tfrac{1}{2}(p \cdot q) \infty \right) = 0.$$

Their radius squared is $-\lambda(1 - \lambda) d_E^2(p, q)$, so that imaginary dual spheres result for $0 < \lambda < 1$. (Those imaginary dual spheres do not look orthogonal to the smallest containing sphere in Figure 15.8, but their inner product with it is indeed zero.) Comparing this to the treatment in the homogeneous model where this affine combination is used in Section 11.2.3 to obtain the points between p and q, we find that the centers of the resulting dual spheres certainly behave exactly like the locations in the affine interpolation, though they are surrounded by imaginary dual spheres for the values of interest. To mimic that interpolation precisely within the conformal model, you should interpolate flat points rather than zero-radius dual spheres: $\lambda (p \wedge \infty) + (1 - \lambda) (q \wedge \infty)$ is the flat point at location $\lambda \mathbf{p} + (1 - \lambda) \mathbf{q}$.

Within the conformal model, other affine combinations become feasible. The affine combinations of flats are a simple extension of those seen in the homogeneous model in Section 11.2.3 and require little explanation. As there, 3-D lines can only be added if they have a finite point in common, which implies in the conformal model that they have two points in common, one of which is the point at infinity. Two circles K_1 and K_2 can now also be affinely combined, but like lines only if they have a point pair in common (which may be imaginary, as it usually is for two circles in a common plane); their sum

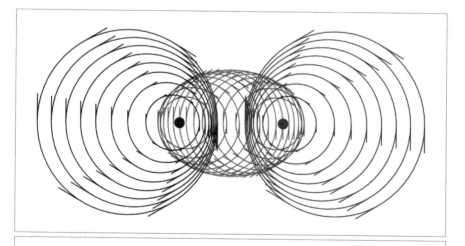

Figure 15.8: Affine combinations of the point p (red) and q (blue) in the plane are dual circles. The amount of each point is indicated by the color mixture. Imaginary circles result for some values; they have been dashed, and their centers follow the regular affine interpolation.

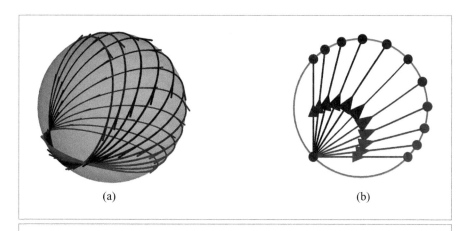

(a)   (b)

Figure 15.9: Affine combination of circles (a) and point pairs (b). These only generate other objects if they form a 1-parameter family, which implies that they should have a point pair or point in common, respectively. We show equal increments of the affine parameter to demonstrate its nonlinear relationship to angles or arcs.

then sweeps out circles on the common sphere $K_1 \cup K_2$ also containing the point pair, as illustrated in Figure 15.9(a). Similarly, two point pairs can be combined affinely if they have one point in common, and the results are point pairs that sweep out the circle that is their `join`, illustrated in Figure 15.9(b).

All these additive combinations transform covariantly, so they are geometrically sensible constructions. But you should avoid using them, for they parameterize the intermediate elements awkwardly. It is much more preferable to use the two generative elements to produce a *rotor* that affects the transformations; its parameters can be directly interpreted as angles and scalings of the intermediate elements. We develop that technique in the next chapter.

15.3 EUCLIDEAN PROJECTIONS

In geometric algebra, the projection of a blade X onto a blade P is another blade:

$$X \;\mapsto\; (X \rfloor P)/P, \tag{15.7}$$

where the division uses the geometric product. Applying this to the conformal model, we find the expected projection behavior if X is a flat (as we will show below when discussing Figure 15.10(b)).

But when X is a round, this is not the projection you might expect. For instance, consider Figure 15.10(a), the projection of a circle onto a plane. You may have hoped for an ellipse, but an ellipse is not represented by a blade in our model, and therefore cannot be the outcome of (15.7). Instead, the result of the projection of a circle onto a plane is a particular circle. To explain this effect, and derive precisely which circle, we recall (11.17):

$$A \cap (X \wedge A^{-*}) = (X \wedge A^{-*})^* \rfloor A = (X \rfloor A) \rfloor A.$$

So the projection is proportional to the meet of A with $X \wedge A^{-*}$ (by the magnitude A^{-2}). In our Euclidean situation, the latter is the plunge of X into A: it contains X and intersects A perpendicularly. In the case of projecting a circle onto a plane, we indeed get the construction of Figure 15.10(a), resulting in a circle that is the equator of the sphere that plunges into the plane and contains the circle C.

It is instructive to see what happens for flats, for instance when X is a line L and A a plane Π as in Figure 15.10(b). Now $L \wedge \Pi^*$ is the plane containing L perpendicular to Π, which is itself a flat. Therefore the meet with Π now produces the expected line onto the plane that is the orthogonal projection of the line onto the plane in the usual sense.

Figure 15.10(c) shows that when X is a line L and A is a sphere S, we get a great circle on the sphere, which is indeed a sensible interpretation of what it would be to project a line on a sphere. Obviously, these examples generalize to the other elements we have treated. You can even project a tangent vector onto a sphere (and the result is the point pair in which the plunging circle containing the tangent vector meets the sphere).

In summary, the operation of projection generalizes from flats in a sensible, but somewhat unusual manner, providing a fundamental operation that seems to be new to Euclidean geometry. We have yet to discover applications in which it might be useful to encode these constructions compactly.

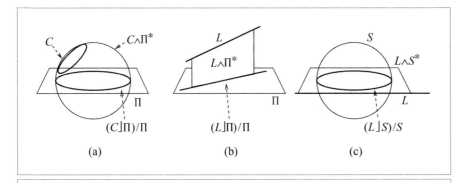

Figure 15.10: Orthogonal projections in the conformal model of Euclidean geometry.

15.4 APPLICATION: ALL KINDS OF VECTORS

In the classical way of computing with geometry, we have different needs for indicating 1-D directional aspects. They all use a vector $\mathbf{u}$, but in different ways, which should transform in different manners to be consistent with that usage.

We may just mean the direction $\mathbf{u}$ in general, which is a *free vector* that can occur at any location. Or we may attach a vector to a point p to make a *tangent vector p* at the location $\mathbf{u}$ (which feels more like two geometrical elements put together rather than as a unified concept). In 3-D, a *normal vector* can be used to denote a plane; it can be placed anywhere in that plane (this practice generalizes to hyperplanes in n-D). The *direction vector* of a line gives it an affine length measure that can be used anywhere along it; such a vector is free to slide along the line. Finally, a *position vector* is used to denote the location of a point relative to another point (the origin). This is actually a tangent vector that is always attached to the origin (though in coordinate-based approaches, the origin is often left implicitly understood).

Each of these concepts can be defined as an element in the conformal model, and this makes them have precisely the right transformation properties. Of course this is true for higher-dimensional directional elements as well, and we have treated them all above. But it pays to treat the vectors separately and explicitly. You know them intimately, and may have come across problems in modeling and coding in which the coordinate approach just was not specific enough to specify permitted transformations. The typical solution would be to define different data structures for each, with their own methods [24, 13]. The conformal model offers an alternative: use precise algebraic elements that have the correct transformational properties, and then use just general methods (versors) to transform them. Each object automatically transforms correctly. Moreover, its clear algebraic relationship to the other computational elements dispenses with the need for a profusion of new methods explicitly specifying the various interactions. The algebraic data structures automatically perform computation and administration at the same time.

So, let us make these types of vectors; they are illustrated in Figure 15.11.

- *Free Vector* $\mathbf{u} \wedge \infty$. A 1-D direction without location is represented by the element $\mathbf{u} \wedge \infty = \mathbf{u}\infty$. Its directional aspect is exhibited by applying a rotation rotor to it:

$$R[\mathbf{u} \wedge \infty] = R[\mathbf{u}] \wedge R[\infty] = R[\mathbf{u}] \wedge \infty,$$

which shows that $\mathbf{u}$ is the element denoting its direction. Applying a translation rotor gives

$$T_{\mathbf{p}}[\mathbf{u} \wedge \infty] = T_{\mathbf{p}}[\mathbf{u}] \wedge T_{\mathbf{p}}[\infty] = \big(\mathbf{u} + (\mathbf{p} \cdot \mathbf{u})\infty\big) \wedge \infty = \mathbf{u} \wedge \infty,$$

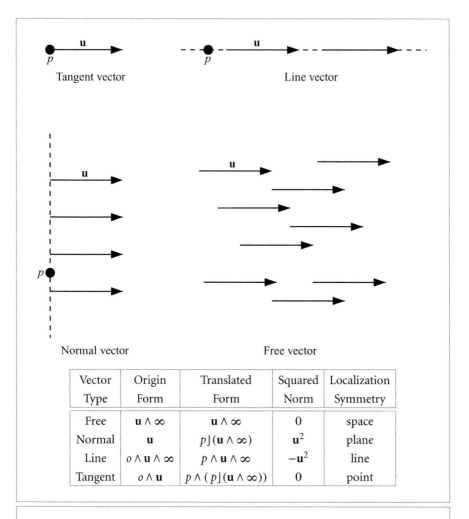

Vector Type	Origin Form	Translated Form	Squared Norm	Localization Symmetry
Free	$\mathbf{u} \wedge \infty$	$\mathbf{u} \wedge \infty$	0	space
Normal	$\mathbf{u}$	$p \rfloor (\mathbf{u} \wedge \infty)$	$\mathbf{u}^2$	plane
Line	$o \wedge \mathbf{u} \wedge \infty$	$p \wedge \mathbf{u} \wedge \infty$	$-\mathbf{u}^2$	line
Tangent	$o \wedge \mathbf{u}$	$p \wedge (p \rfloor (\mathbf{u} \wedge \infty))$	0	point

Figure 15.11: In the conformal model, the various kinds of vectorlike concepts of Euclidean geometry are clearly distinguished, each with its own form properly transforming under the permissible Euclidean transformations.

so that it is translation-invariant. It has no specific location, so this is truly a free vector. It corresponds to the entity $\mathbf{u}$ in the homogeneous model. In the conformal model, it is the parameter of the translation rotor $\exp(-\mathbf{u} \wedge \infty/2)$. That rotor can be applied anywhere, and we now see why we are allowed to draw its action as a vector at any location, as is customary.

- **Normal Vector u**. The normal vector of a plane in the origin is the purely Euclidean element **u**, just as in the vector space model. It clearly indicates a direction vector, for applying a rotation to it we get $\mathsf{R}[\mathbf{u}]$. However, there is also a translational element. Applying the translation rotor gives

$$\mathsf{T}_{\mathbf{p}}[\mathbf{u}] = p\rfloor(\mathbf{u} \wedge \infty) = \mathbf{u} + (\mathbf{p} \cdot \mathbf{u})\,\infty.$$

There are two consequences of this equation. First, it shows that the element **u** is insensitive to translations in the directions perpendicular to **u**. We can depict it freely anywhere in the dual plane **u**. It therefore truly is a normal vector of the plane; the classical liberty to draw such a normal vector anywhere on the plane is now fully part of the internal algebraic freedom of the computational element.

The second consequence is that this equation gives the normal vector representation of a plane with normal direction **u** passing through p instead of o. It is only the component $\mathbf{p} \cdot \mathbf{u}$ that affects the representation, and we recognize that as proportional to the oriented distance to the origin. This is the correct 1-D locational aspect of the plane: only its **u**-component matters.

- **Line Vector** $o \wedge \mathbf{u} \wedge \infty$. A concept that is less explicit in classical considerations in geometry is the direction vector of a line. It is important in classical mechanics, where it is used to denote a force. Such a force is free to act anywhere along its carrier line, and therefore is a vector-like concept that should be permitted to slide along a well-defined line.

An element with the correct symmetry is $o \wedge \mathbf{u} \wedge \infty$, which we are used to viewing as the line itself. It rotates to $o \wedge \mathsf{R}[\mathbf{u}] \wedge \infty$, which clearly indicates that it has a 1-D directional aspect. When we translate it, we obtain

$$\mathsf{T}_{\mathbf{p}}[o \wedge \mathbf{u} \wedge \infty] = p \wedge \mathbf{u} \wedge \infty = o \wedge \mathbf{u} \wedge \infty + (\mathbf{p} \wedge \mathbf{u}) \wedge \infty.$$

Again there are two consequences of this equation. First, it shows that the element is invariant under translations for which $\mathbf{p} \wedge \mathbf{u} = 0$ (i.e., motions along the line). Second, it shows how to make a general element with this property, passing through the point p. Only the part of **p** perpendicular to **u** affects the location, so a line in 3-D has a 2-D locational parameter complementary (i.e., dual) to its direction vector. This is effectively the "moment" we encountered in the homogeneous model.

It is perhaps surprising that a line vector should be a 3-blade, but it really unites three separate concepts: a location p, a direction **u**, and the straightness in its extension to the point at infinity ∞. All are required to define all aspects of its symmetry; the inclusion of ∞ allows the invariant shifting along the carrier.

- **Tangent Vector** $o \wedge \mathbf{u}$. A vector with direction **u** at the point location o is represented as $o \wedge \mathbf{u}$. A rotation changes this to $o \wedge \mathsf{R}[\mathbf{u}]$, so that **u** is its only rotational aspect. A translation carries the direction vector along with the point. This is perhaps best seen when we first rewrite to $o \wedge \mathbf{u} = o \wedge \left(o\rfloor(\mathbf{u} \wedge \infty)\right)$, and then translate:

$$\mathsf{T}_{\mathbf{p}}[o \wedge \mathbf{u}] = \mathsf{T}_{\mathbf{p}}[o \wedge \left(o\rfloor(\mathbf{u} \wedge \infty)\right)] = p \wedge \left(p\rfloor(\mathbf{u} \wedge \infty)\right).$$

The location p is explicitly present in the result; so the tangent vector has all the locational aspects of p, and moves with the point it is attached to.

In classical applications, you draw tangent vectors all the time, as velocities, directions of motions on a surface, and so on. Apparently we can do a lot without having them explicitly represented as an element of computation—but now that we have this representation, we can use it. We will see several examples of its potential use in the ray tracer application of Chapter 23.

- **Position Vector.** The most common informal use of a Euclidean vector $\mathbf{u}$ has no counterpart in the conformal model: to denote a point in space. That practice attempts to use an element of the vector space model (the algebra of directions) to denote a location. This is only possible by using some standard point, namely an origin o, as a reference. We need to make this algebraically explicit: if $\mathbf{u}$ is meant to emanate from o, the position vector should encode o in its definition. That makes it either the tangent vector $o \wedge \mathbf{u}$ or the line $o \wedge \mathbf{u} \wedge \infty$. It is not the line, for that loses essential aspects of the location o due to its linear translation symmetry. The position vector has some feature of the tangent vector $o \wedge \mathbf{u}$, and can be seen as an amount of travel along that vector. But that encodes one point (p) in terms of another (o), and therefore does not really resolve the issue.

Of course, in the conformal model we have a much better alternative: the point is represented as the representational null vector u. If we should want to know its location relative to any other point q, that is the Euclidean vector $\mathbf{u} - \mathbf{q} = u - q + \frac{1}{2}(u \cdot q) \infty$, but you do not need to know that to compute with the point u.

The representative vector for the point u may be obtained by using $\mathbf{u}$ in a translation rotor applied to q. Taking $q = o$, that gives us the correspondence to the classical characterization by a relative position vector:

$$u = \mathsf{T}_{\mathbf{u}-o}[o] = e^{-\mathbf{u}\infty/2} \, o \, e^{\mathbf{u}\infty/2}.$$

There is now a clear distinction between the vector parameter $\mathbf{u}$ characterizing the location and the standard element o to which it is to be applied. If o is chosen differently, $\mathbf{u}$ needs to change to reach the same point u. There is of course nothing geometrically intrinsic about the position vector $\mathbf{u}$ of a point u, and it is no wonder that the many tacit conventions in this position vector representation of a point are a source of errors and abuse. Homogeneous coordinates were the first step to a true point representation, but only the conformal model gives the full operational functionality expected from the representation of the most basic elements of geometry.

So, in summary: there is no need for separate data structures for the various kinds of vectors. As basic elements of computation, they are an integral element of the algebra, automatically inheriting their relationships with other elements through all of its products. Using this operational model of Euclidean geometry gives us more precision and flexibility, and simpler code. There is a small computational overhead to this, which we discuss in Part III; in a good implementation, this can be kept below 25 percent (which is actually also the cost of employing the structurally inferior homogeneous coordinates). Paradoxically, such efficiency is achieved by recognizing the class of an element and treating it with dedicated code; but the difference with the classical approach is that this special code is established by an automatic code generator rather than by the programmer herself.

15.5 APPLICATION: ANALYSIS OF A VORONOI CELL

To give an example of how the conformal model can be used in derivations of expressions, we take a detailed example from planar geometry involving distances in an essential manner. Given four points p, q, r, s in the plane, we will compute the potential edge of the Voronoi diagram across the common edge pr of the triangles prs and pqr. We compute all its parameters, its carrier line, an end point, and its edge length. In the classical approach, this would involve a lot of trigonometry, which in turn would require you to define a host of intermediate variables and edge length parameters, as in Figure 15.12(b). We will show how it can all be done based on the four points only, with seemingly no trigonometry at all. You should see this as an isolated exercise in uncovering some common computational techniques for this type of problem.

We will go slowly and point out useful techniques on the way. You can skip this section without missing much, though you may want to contrast the directness of the conformal model result (15.8) (fully expressible in the original four points) with the classical result (15.10) (which requires many derived quantities even to be stated).

15.5.1 EDGE LINES

To keep life simple, we start from normalized points p, q, r, s, so that $-\infty \cdot p = 1$, and so on.

The Voronoi edge element that interests us lies on the perpendicular bisector of the connection line of p and q, indicated in Figure 15.12(a). A point x on it therefore satisfies $x \cdot p = x \cdot q$, which we can rewrite to $x \cdot (q - p) = 0$. Therefore the perpendicular bisector is dually characterized by

$$\textit{dual perpendiculat bisector:} \quad q - p.$$

This element $q - p$ has a weight, a condition number that determines its significance. We can compute the norm squared as

$$(q - p) \cdot (q - p)^{\sim} = -2p \cdot q = d_E^2(p, q)$$

so the norm equals the squared length of the edge connecting p and q. But this poses the problem of whether to take the positive or negative square root, and thus threatens

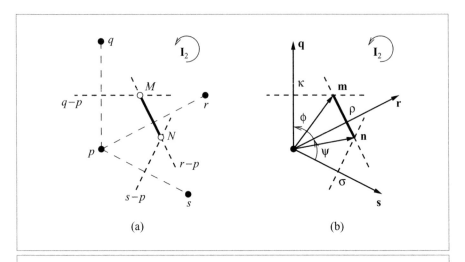

Figure 15.12: Definition of symbols for the Voronoi derivations, (a) for the conformal model derivation, and (b) for the classical, p-centered derivation.

to lose valuable orientation information. It is better to retain the directionality by computing the linear quantity

$$\infty \rfloor (q - p)^* = \left(\infty \wedge (\mathbf{q} - \mathbf{p})\right)^* = \left(\infty \wedge (\mathbf{q} - \mathbf{p})\right) \rfloor \left(o \wedge \mathbf{I}_2^{-1} \wedge \infty\right) = (\mathbf{q} - \mathbf{p})^\star \wedge \infty,$$

in which $(\ )^\star$ denotes the Euclidean dual in the plane. This direction is orthogonal in a consistent directed manner to the direction vector connecting $\mathbf{p}$ to $\mathbf{q}$, so this immediately gives both length and orientation. Consistency is obtained by relating all such signs and orientations to the same pseudoscalar of the plane (as implicit in the duality).

15.5.2 EDGE POINT

We form the dual midplane $r - q$, and compute the intersection of this with the earlier dual midplane $q - p$ to obtain a flat end point of our edge. This is, dually,

$$(q - p) \wedge (r - q) = p \wedge q + q \wedge r + r \wedge p,$$

which incidentally shows in its symmetry that only the order of p, q, r matters, and then only by changing the sign of the result. We would have obtained the same intersection by intersecting with the midplane $p - r$. You can confirm that the result is a dual flat by taking a contraction with ∞; that indeed gives zero.

This dual flat point is not normalized. (Note that a flat point $x \wedge \infty$ is normalized when its norm is 1, and a point x is normalized when its weight is 1 (so that $-\infty \rfloor x = 1$).) Its normalization factor may be useful in further computations of quantitative properties. If desired, we can normalize a flat point by the algebraic construction

$$ X \;\mapsto\; \frac{X}{(o \wedge \infty) \rfloor X}, $$

and this implies that a dual flat point is normalized by the substitution

$$ X^* \;\mapsto\; \frac{X^*}{o \wedge \infty \wedge X^*}. $$

We can take any point instead of the arbitrary origin o, and it is important to wield this capability to keep formulas simple. For the dual flat point we just computed, which is based on a q-centered computation, it seems indicated to take q for the arbitrary origin point o in the normalization (but we could take r or p as well). That yields as normalized flat intersection point the *circumcenter* of the triangle formed by p, q, r:

$$ \textit{flat circumcenter:} \quad M = \frac{p \wedge q + q \wedge r + r \wedge p}{p \wedge q \wedge r \wedge \infty}. $$

15.5.3 EDGE LENGTH

We can compute the other intersection in the quadrangle between the dual midlines $(r{-}p)$ and $(s-p)$ in the same manner, and normalize the resulting dual intersection point N^* by $(p \wedge r \wedge s \wedge \infty)$. From those two intersection points, we should be able to deduce the length of the potential edge of the Voronoi diagram.

It is possible to do this by computing their square distance, but this again gives consistency issues in choosing the square root. Let us instead derive this length as the weight of the line connecting the two intersection points. If we had been in the homogeneous model, this line would have been $M \wedge N$. Computations with flats in the conformal model do not work quite so simply. When we have the flat points $M = m \wedge \infty$ and $N = n \wedge \infty$ rather than the corresponding points m and n, we compute the desired line $m \wedge n \wedge \infty$ as $m \wedge N$. To do so, we should retrieve m from the dual flat point M^* as $o \rfloor M = (o \wedge M^*)^{-*}$. This also produces a term proportional to ∞, but the outer product in the line computation kills that anyway, which is a trick to remember! And again, since we can choose any normalized point as o, we pick q when we observe that doing so reduces the number of terms in the computation.

We now spell out the manipulation to obtain the connecting line; note that all we really do is substitution of known properties, perhaps guided by the idea that the result should be part of the dual line $(r - p)$. We denote the scalar normalizations of the two points by α_M and α_N, to unclutter the formulas.

$$ m \wedge n \wedge \infty = \alpha_M\, \alpha_N \left(q \wedge (q - p) \wedge (r - q) \right)^{-*} \wedge \left((s - r) \wedge (p - s) \right)^{-*} $$

$$
\begin{aligned}
&= -\alpha_M\,\alpha_N\left((r-p)\wedge p\wedge q\right)^{-*}\wedge\left((r-p)\wedge(p-s)\right)^{-*}\\
&= -\alpha_M\,\alpha_N\left((r-p)\wedge p\wedge q\right)^{*}\wedge\left((r-p)\wedge(p-s)\right)^{*}\\
&= -\alpha_M\,\alpha_N\left(\left((r-p)\wedge(p-s)\right)\cap\left((r-p)\wedge p\wedge q\right)\right)^{*}\\
&= -\alpha_M\,\alpha_N\,(r-p)^{*}\left((r-p)\wedge(p-s)\wedge p\wedge q\right)^{*}\\
&= \alpha_M\,\alpha_N\,(r-p)^{*}\,(p\wedge q\wedge r\wedge s)^{*}\\
&= \frac{(r-p)^{*}\,(p\wedge q\wedge r\wedge s)^{*}}{(p\wedge q\wedge r\wedge\infty)^{*}\,(p\wedge r\wedge s\wedge\infty)^{*}}.
\end{aligned}\tag{15.8}
$$

In the derivation we were striving towards a form that could use the meet identity (5.11). The final results confirms that we are on the line $(r-p)^{*}$, and it gives the length of the resulting segment between M and N. Don't forget that $(r-p)^{*}$ contributes a factor $d_E(r,p)$ to the length, as we derived above.

Taking q as variable for the moment, the length of the edge segment becomes zero when $p\wedge q\wedge r\wedge s = 0$, and this is when q lies on the circle $p\wedge r\wedge s$. Move it beyond the circle and it changes sign. Negative edge length is an indication that another edge should become the Voronoi edge, namely the part of $(s-q)^{*}$ computed completely analogously (most simply by a cyclic permutation of p, q, r, s). It is rather satisfying that this condition is completely embedded in the basic computation; nowhere did we put in the circle $p\wedge r\wedge s$ explicitly, but the rearrangements of the outer products just made it appear naturally. The two terms in the denominator are dual oriented planes describing the two triangles involved; each term is twice the signed area of its corresponding triangle.

15.5.4 CONVERSION TO CLASSICAL FORMULAS

The above computation using the conformal model was a fairly straightforward application of algebra on the combination of the basic points by standard operations. Where it may not have seemed straightforward, it was because this is the first time we encounter these techniques. The coordinate-free form of the final result is rather pleasing, and directly interpretable in terms of relative point positions.

By contrast, let us show how the results would look in classical form, to show that the conformal model result conveniently hides a lot of trigonometry. This requires the introduction of lots of symbols for local edge lengths and angles. For that, at least, we can use the natural correspondence to the geometric algebra of 2-D Euclidean space.

For the following computations, it is convenient to choose the (arbitrary) origin at p, denoting the Euclidean vectors to q, r, s by $\mathbf{q}, \mathbf{r}, \mathbf{s}$, respectively. Then the dual plane is

$$
r - p = \mathbf{r} + \tfrac{1}{2}\mathbf{r}^{2}\infty
$$

We will also use polar coordinates relative to p to convert the conformal expressions to more classical forms. That essentially means that we are using a vector space model based in p. We define the norms of $\mathbf{q}, \mathbf{r}, \mathbf{s}$ as κ, ρ, σ, and the angles through the implicit definitions also used in the vector space model treatment of a triangle (in (10.3)):

$$s/\mathbf{r} = \sigma/\rho\, e^{\mathbf{I}_2 \psi}, \quad \mathbf{r}/\mathbf{q} = \rho/\kappa\, e^{\mathbf{I}_2 \phi},$$

so that ψ is the angle from $\mathbf{s}$ to $\mathbf{r}$, and ϕ from $\mathbf{r}$ to $\mathbf{q}$ (see Figure 15.12(b)).

We prefer to redo parts of the computation rather than to perform an immediate substitution in the final conformal result (15.8). This shows the representation of intermediate quantities in an educational manner.

$$
\begin{aligned}
(q - p) \wedge (r - q) &= (q - p) \wedge (r - p) \\
&= (\mathbf{q} + \tfrac{1}{2}\mathbf{q}^2 \infty) \wedge (\mathbf{r} + \tfrac{1}{2}\mathbf{r}^2 \infty) \\
&= \mathbf{q} \wedge \mathbf{r} + \infty \wedge (\mathbf{r}\mathbf{q}^2 - \mathbf{q}\mathbf{r}^2)/2 \\
&= (1 - \frac{\mathbf{r}\mathbf{q}^2 - \mathbf{q}\mathbf{r}^2}{2\,\mathbf{q} \wedge \mathbf{r}} \infty)\,(\mathbf{q} \wedge \mathbf{r}).
\end{aligned}
$$

The final form can be interpreted using Table 14.1, or by realizing that it should be the dual of a flat point $x \wedge \infty$, and computing quickly what form such a dual should have:

$$(x \wedge \infty)\rfloor(o \wedge \widetilde{\mathbf{I}}_2 \wedge \infty) = ((o + \mathbf{x}) \wedge \infty)\rfloor(o \wedge \widetilde{\mathbf{I}}_2 \wedge \infty) = \widetilde{\mathbf{I}}_2 + \infty\, \mathbf{x}\rfloor\widetilde{\mathbf{I}}_2 = (1 - \mathbf{x}\infty)\widetilde{\mathbf{I}}_2.$$

Either way, it follows that the intersection point is located at

$$\mathbf{m} = \frac{\mathbf{r}\mathbf{q}^2 - \mathbf{q}\mathbf{r}^2}{2\mathbf{q} \wedge \mathbf{r}} = \frac{\rho\,\check{\mathbf{q}} - \kappa\,\check{\mathbf{r}}}{2\sin\phi}\widetilde{\mathbf{I}}_2, \tag{15.9}$$

where $\check{\mathbf{q}}, \check{\mathbf{r}}$ are unit vectors along $\mathbf{q}$ and $\mathbf{r}$, respectively. The intersection has a numerical strength measured by the norm of $\mathbf{q} \wedge \mathbf{r}$, which is $\kappa\rho\,\sin\phi$. Incidentally, this is the equation for the center of the circumcircle of the triangle $\mathcal{PQR}$ in its classical form.

The edge length computed before can be converted into a more classical form using the same polar coordinates relative to p (which is the chosen origin o):

$$
\begin{aligned}
(r - p)^* \frac{(p \wedge q \wedge r \wedge s)^*}{(p \wedge q \wedge r \wedge \infty)^* \,(p \wedge r \wedge s \wedge \infty)^*} &= \\
= (\mathbf{r} + \tfrac{1}{2}\rho^2 \infty)^* \frac{\left(o \wedge (\mathbf{q} + \tfrac{1}{2}\kappa^2 \infty) \wedge (\mathbf{r} + \tfrac{1}{2}\rho^2 \infty) \wedge (\mathbf{s} + \tfrac{1}{2}\sigma^2 \infty)\right)^*}{(o \wedge \mathbf{q} \wedge \mathbf{r} \wedge \infty)^* \,(o \wedge \mathbf{r} \wedge \mathbf{s} \wedge \infty)^*} & \\
= (\mathbf{r} + \tfrac{1}{2}\rho^2 \infty)^* \frac{\left(o \wedge \tfrac{1}{2}(\kappa^2\,\mathbf{r} \wedge \mathbf{s} + \rho^2\,\mathbf{s} \wedge \mathbf{q} + \sigma^2\,\mathbf{q} \wedge \mathbf{r}) \wedge \infty\right)^*}{(o \wedge \mathbf{q} \wedge \mathbf{r} \wedge \infty)^* \,(o \wedge \mathbf{r} \wedge \mathbf{s} \wedge \infty)^*} & \\
= (\mathbf{r} + \tfrac{1}{2}\rho^2 \infty)^* \frac{\kappa\rho\sigma\,(-\kappa\sin\psi + \rho\sin(\phi + \psi) - \sigma\sin\phi)}{2(-\kappa\rho\sin\psi)\,(-\rho\sigma\sin\phi)} & \\
= (\check{\mathbf{r}} + \tfrac{1}{2}\rho\infty)^* \frac{-\kappa\sin\psi + \rho\sin(\phi + \psi) - \sigma\sin\phi}{2\sin\phi\,\sin\psi}. &
\end{aligned}
$$

So this gives the length of the edge in completely classical terms as

$$\frac{-\kappa\sin\psi + \rho\sin(\phi + \psi) - \sigma\sin(\phi)}{2\sin\phi\,\sin\psi}, \tag{15.10}$$

and moreover that it is a part of the line orthogonal to $\check{r}$, passing midway through it (since the distance to p is clearly $\frac{1}{2}\rho$).

The derivation spelled out in this manner may look rather intimidating. Yet it is fully algebraic from the conformal model result, which itself was straightforward algebra (with a clear geometrical interpretation). There was no need for the usual careful checking of signs and symbols with messy figures to develop the result. And of course, one doesn't need to compute this result at all, because the original conformal expression of (15.8) has the same value. It is a disadvantage of the classical expression that the symmetry is hidden in the peculiar variables, and that the circumcircle as the essential curve for the change of sign is hidden in the final expression. All one should need to specify for a Voronoi edge are the original four points, and it is satisfying to see that the conformal model can express the total quantitative result in terms of those points only.

15.6 FURTHER READING

Geometry using `meet` and `plunge` was developed in a 1941 book by Forder [22], which is full to the brim with useful geometrical results. He uses Grassmann's notation, in which $a \wedge b$ is denoted as $[a\, b]$, and dualization is denoted by $|a$ rather than a^*.

Forder's book is one of several instances that demonstrate how far Grassmann's heritage was developed into quantitative geometry. He wrote it to "redress the balance" between the use of Grassmann algebra in physics (where he says "it has at last won an established place") and in geometry (where it is "less widely appreciated").

The book may look rather intimidating, but with interactive software should now have become quite readable: all equations can be drawn, for this is the algebra of geometry.

15.7 EXERCISES

15.7.1 DRILLS

1. Compute the tangent at the origin of the sphere Σ through the points at locations $0, \mathbf{e}_1, \mathbf{e}_2$, and $\mathbf{e}_3$ (you computed this sphere Σ in the drills of Chapter 14).
2. Factorize the circle K through the points at locations $\mathbf{e}_1, \mathbf{e}_2$, and $\mathbf{e}_3$ (you computed this circle K in the drills of Chapter 14).
3. Use that factorization of the circle K to spot its squared radius, center, carrier, and surround, by inspection.
4. Project the point at the origin onto the carrier plane of the circle K.
5. Make the free vector, tangent vector, line vector, and normal vector in the direction $\mathbf{e}_1$, at the origin (if a location is required).
6. Rotate each of the vectors of the previous exercise by $\pi/2$ in the $\mathbf{e}_1 \wedge \mathbf{e}_2$ plane. Explain the results.
7. Translate each of the vectors of the previous exercise by $\mathbf{e}_1 + \mathbf{e}_2$. Explain the results.

15.7.2 STRUCTURAL EXERCISES

1. Express the scalar product of two blades in terms of the scalar product of their duals. It should only differ by a sign, which you should express in terms of the grade of the blades and the space they reside in.

2. What is the geometry of the element $p \wedge \mathbf{u}$, where p is a point and $\mathbf{u}$ a Euclidean vector? (Hint: View it as a `plunge`. Counterhint: It is *not* the tangent vector $o \wedge \mathbf{u}$ moved to p.)

3. Show that the tangent of a tangent is zero. (Hint: Realize that a tangent is also a round; now use (15.1).)

4. On a circle K, find the closest and furthest points to a given point c (in n-D!). Hint 1: Set up a dual sphere with parametric radius around c, and compute the `meet` with K. Hint 2: The `meet` of parametric sphere and K should be a tangent vector at the desired points. Hint 3: Use the algebraic property of tangents, which is that they are rounds that square to zero, to get a second-order equation for the radius, and solve. Hint 4: Compute the locations of the tangent vectors.

5. Show that the dual sphere $s = r \rfloor (p \wedge q) = (r \rfloor p)\, q - (r \rfloor q)\, p$ is a member of the parameterized family (15.6), passing through r (but nonnormalized).

6. It is interesting to study the outcome of the construction $p \rfloor (q \wedge r)$, which crops up frequently in some form or other in computations. Depending on whether you let p, q, r be equal to points, dual spheres, or the point at infinity ∞, different constructions appear. Play with this construction, either by hand or (preferably) in an interactive visualization package. What is $\infty \rfloor (q \wedge r)$? What is $r \rfloor (q \wedge \infty)$?

7. Give the formula for the circle through a point pair $p \wedge q$ intersecting the dual plane π perpendicularly, and for the circle having a tangent vector in direction $\mathbf{u}$ at p and plunging into π.

8. Derive the factorization of a dual round (15.5) from the factorization of a direct round (15.4).

9. Construct the contour of a sphere Σ as seen from a point p (i.e., the circle K of points where the invisible part of the sphere borders the visible part, as in Figure 15.13). (Hint: The white sphere in the figure is a clue to the construction. Express it first, using the `plunge`. Then construct the circle as a `meet`.)

10. You can derive all types of vectors of Section 15.4 by relating them to two points a and b, which are combined using the various products of geometric algebra. Identify the types of the following elements and relate them geometrically to a and b: $\infty \rfloor (a \wedge b)$, $\infty \wedge (a \wedge b)$, $\infty \wedge (\infty \rfloor (a \wedge b))$, $\infty \rfloor (\infty \wedge (a \wedge b))$, $a \rfloor (\infty \wedge (a \wedge b))$, $a \wedge (\infty \rfloor (a \wedge b))$. How is their squared norm related to the oriented distance of a and b? For the null elements, can you retrieve this distance in another manner?

11. Show that the tangent with direction element $\mathbf{E}$ at p can be written in two equivalent forms:

$$\textit{tangent } \mathbf{E} \textit{ at } p: \quad p \wedge \left(-p \rfloor (\hat{\mathbf{E}}\infty) \right) = p \rfloor (p \wedge \hat{\mathbf{E}} \wedge \infty).$$

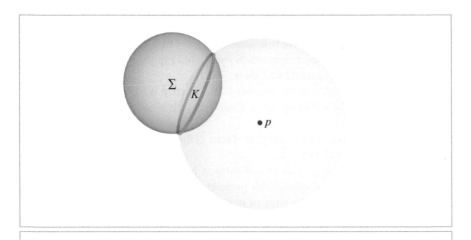

Figure 15.13: Construction of the contour circle K of a dual sphere $\sigma = \Sigma^*$ seen from a point p (see Exercise 9).

15.8 PROGRAMMING EXAMPLES AND EXERCISES

The three examples for this chapter are interactive versions of Figure 15.2 (the plunge), Figure 15.8 (affine combinations of points), and Figure 15.10 (Euclidean projections), respectively.

15.8.1 THE PLUNGE

This example draws a green circle that is the plunge of the following three primitives:

```
const int NB_PRIMITIVES = 3;
mv g_primitives[NB_PRIMITIVES] = {
    // point:
    c3gaPoint(-1.0f, 1.5f, 0.0f),
    // dual sphere:
    3.25f * e1 + 2.68f * e2 -0.34f * e3 + 1.00f * no + 8.43f * ni,
    // dual plane:
    -e2 + 2.0f * ni,
};
```

Computing and drawing the plunge is as simple as

```
draw(g_primitives[0] ^ g_primitives[1] ^ g_primitives[2]);
```

You can translate the sphere and the point. Notice how the circle always intersects the point, sphere, and plane orthogonally.

15.8.2 AFFINE COMBINATIONS OF POINTS

As described in Section 15.2.4, affine combinations of points are circles in $\mathbb{E}^2$. This example uses c2ga to implement an interactive version as shown in Figure 15.14. The code to draw the circles is:

```
const mv::Float STEP = 0.1f;
for (mv::Float lambda = -1.0f; lambda <= 2.0f; lambda += STEP) {
  // set color, turn GL stipple on when circle is imaginary.
  // ... (omitted)

  // draw the circle:
  draw(lambda * g_points[0] + (1.0f - lambda) * g_points[1]);
}
```

As an extra feature, the example also draws the bisecting line (in green) between the two points, using:

```
draw(g_points[0] - g_points[1]);
```

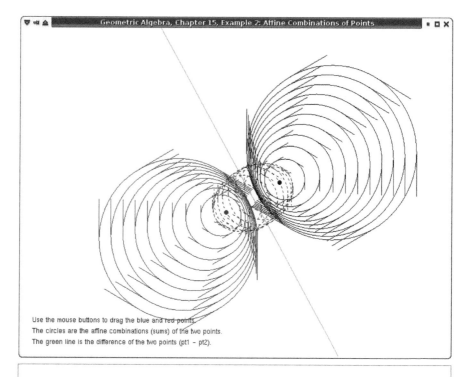

Use the mouse buttons to drag the blue and red points.
The circles are the affine combinations (sums) of the two points.
The green line is the difference of the two points (pt1 - pt2).

Figure 15.14: Screenshot of Example 2.

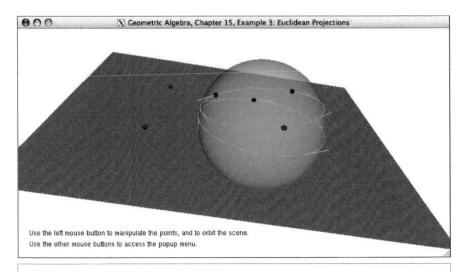

Figure 15.15: Screenshot of Example 3. The green circle is projected onto the blue plane. The projection is drawn in cyan. The sphere is the `plunge` of circle and plane, which explains the geometry of the projection.

15.8.3 EUCLIDEAN PROJECTIONS

As described in Section 15.3, orthogonal projection in the conformal model behaves somewhat unexpectedly. This example draws a line or circle, and its projection onto a plane. You can change the arguments interactively, and optionally cause the `plunge` to be drawn. Figure 15.15 shows a screenshot.

From formula to program is straightforward, for the projection itself is drawn using

```
// CL is circle or line
// PL is plane
draw((CL << inverse(PL)) << PL);
```

and the optional `plunge` is drawn as:

```
draw(CL ^ dual(PL));
```

16 CONFORMAL OPERATORS

Even with all the new techniques for Euclidean geometry in the previous three chapters, the possibilities of the conformal model are not exhausted. There are more versors in it, and they represent other useful transformations. Euclidean motions were just a special case of doing *conformal transformations*, which preserve angles. These also include reflection in a sphere and uniform scaling.

All conformal transformations are generated by versor products using the elementary vectors of the conformal model. Whereas the Euclidean motions of the previous chapter involved using the vectors representing dual planes, we now use the dual spheres. An important operation we can then put into rotor form is *uniform scaling*, and that permits a closed-form solution to the interpolation of rigid body motions with scaling.

The fact that the general conformal transformations can be represented as versors finally explains the name of the conformal model.

16.1 SPHERICAL INVERSION

The most elementary conformal transformation is the reflection in a unit sphere, called (spherical) inversion. As a versor, the spherical inversion in the unit sphere Σ around the origin involves the vector $\sigma = o - \infty/2$, representing this sphere dually. The spherical reflection is performed by the versor product:

$$X \mapsto \sigma \widehat{X} \sigma^{-1}.$$

A unit sphere of weight 1 (so that $-\infty \cdot \sigma = 1$) is equal to its own inverse, simplifying the equation slightly. We compute the effect on the basic elements:

$$o \mapsto -(o - \tfrac{1}{2}\infty)\, o\, (o - \tfrac{1}{2}\infty) = -\tfrac{1}{4}\infty\, o\, \infty = \infty/2$$

$$\infty \mapsto -(o - \tfrac{1}{2}\infty)\, \infty\, (o - \tfrac{1}{2}\infty) = -o\, \infty\, o = 2o$$

$$\mathbf{E} \mapsto (o - \tfrac{1}{2}\infty)\, \widehat{\mathbf{E}}\, (o - \tfrac{1}{2}\infty) = -\tfrac{1}{2} o\, \widehat{\mathbf{E}}\, \infty + \tfrac{1}{2}\infty\, \widehat{\mathbf{E}}\, o = -\mathbf{E}(o \cdot \infty) = \mathbf{E}.$$

It is clear that this is not a Euclidean transformation, for ∞ is not preserved but interchanged with o (and weighted). Geometrically, this is understandable: the point at infinity reflects to the center of the sphere, and vice versa. The total result on a point $x = \mathsf{T}_{\mathbf{x}}[o]$ is

$$\mathsf{T}_{\mathbf{x}}[o] = o + \mathbf{x} + \tfrac{1}{2}\mathbf{x}^2\infty \mapsto \mathbf{x}^2\left(o + \mathbf{x}^{-1} + \tfrac{1}{2}\mathbf{x}^{-2}\infty\right) = \mathbf{x}^2\, \mathsf{T}_{\mathbf{x}^{-1}}[o].$$

Not only does the point x end up at location $\mathbf{x}^{-1}$ (which is the inverse of $\mathbf{x}$, hence the name inversion), but its weight also changes by a factor $\mathbf{x}^2$. A real dual unit sphere gives a point in the same direction as $\mathbf{x}$; an imaginary dual unit sphere gives $\mathsf{T}_{\mathbf{x}}[o] \mapsto -\mathbf{x}^2\mathsf{T}_{-\mathbf{x}^{-1}}[o]$ (as you can derive in structural exercise 1). Therefore the latter performs an inversion, while simultaneously reflecting in the origin, which is somewhat surprising but pleasant to have available as a versor.

Figure 16.1 shows the inversion of various elements in a sphere. Containment relationships are of course preserved, but there is more. The spherical inversion is a *conformal transformation*: it preserves local angles. In the figure, the angle between the green line and blue circle is the same as between their images after reflection. This angle preservation property is best demonstrated by considering two tangents at the same location p, which are $p \wedge (p \rfloor (\mathbf{u}\,\infty))$ and $p \wedge (p \rfloor (\mathbf{v}\,\infty))$. Inverting the tangent $p \wedge (p \rfloor (\mathbf{u}\,\infty))$ using the equations above, we find it in a nonstandard form as

$$2\mathbf{p}^4 p' \wedge (p' \rfloor (\mathbf{u}\, o)),$$

where we used the notation p' for the unit point at location $\mathbf{p}^{-1}$. Using the direction formula from Table 14.1 and some algebraic simplification, you can establish that the direction of this tangent equals $-\mathbf{p}\,\mathbf{u}\,\mathbf{p}\,\infty$. Therefore the inversion of the tangent can be written in standard form as

$$p' \wedge (p' \rfloor (-\mathbf{p}\,\mathbf{u}\,\mathbf{p}\,\infty)).$$

The other tangent transforms in a similar way. If we now study the ratio of their Euclidean direction vectors at their common location, that transforms from $\mathbf{v}/\mathbf{u}$ to

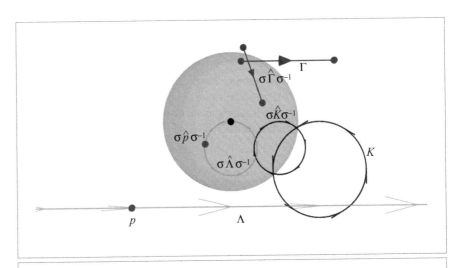

Figure 16.1: Inversion of various elements in a unit sphere (shaded in red) produces their reflection relative to that sphere. We show a point p at location $\mathbf{p}$ relative to the sphere's origin (denoted in black) that becomes a point at location $1/\mathbf{p}$; an oriented line Λ that becomes a circle through the center; a circle K that becomes a circle; and a point pair Γ that becomes a point pair. The black point at the center is the reflection of the point at infinity. Labels of the rounds are indicated at their centers. All containment and composition relationships are preserved, as are angles between the elements before and after reflection.

$(-\mathbf{p}\,\mathbf{v}\,\mathbf{p})/(-\mathbf{p}\,\mathbf{u}\,\mathbf{p}) = \mathbf{p}\,(\mathbf{v}/\mathbf{u})\,\mathbf{p}^{-1}$. This is a simple reflection in the plane with normal $\mathbf{p}$. For the two vectors $\mathbf{u}$ and $\mathbf{v}$, it affects the attitude of their plane $\mathbf{u} \wedge \mathbf{v}$, but not the magnitude of their relative sine or cosine. So although being in a different plane, the angle between $\mathbf{u}$ and $\mathbf{v}$ is preserved. It follows that reflection in a plane is a conformal transformation, and so is reflection in a sphere.

In contrast to the Euclidean operators we treated so far, the inversion in the sphere does not preserve the class of the element it acts on: the consequence of essentially swapping o and ∞ is that flats can become rounds and directions can become tangents at the origin. This is also clear from the characterization of the classes in Table 14.1, which worked by testing whether $\infty \rfloor X$ and/or $\infty \wedge X$ are zero: the nonpreservation of ∞ implies that such a condition holding for X may not hold for the transformed X.

Spherical inversions have interesting applications, and a well-chosen inversion can often transform a geometrical problem into a much simpler problem. Many examples of this may be found in [46], though in a classical form using complex numbers in the plane. The conformal model allows a much more powerful treatment of this principle in n-dimensional base spaces.

16.2 APPLICATIONS OF INVERSION

16.2.1 THE CENTER OF A ROUND

We have seen above how the point at infinity ∞ was reflected into (twice) the center o of the unit sphere at the origin that we were considering. Due to the versor nature, this result is translation-covariant, so that *the center of an arbitrary sphere can be obtained by reflection of ∞ in it.* There are some constants involved to obtain the center in normalized form, and the resulting formula is

$$\textit{center of sphere:} \quad c = -\tfrac{1}{2}\rho^2 \sigma \infty \sigma^{-1} = -\tfrac{1}{2}\frac{\sigma \infty \sigma}{(\infty \cdot \sigma)^2},$$

resulting in a normalized point. A general round can be factored into a sphere cut by planes through its center. Those subsequent reflections do not disturb the location of the center, so the general formula is as indicated in Table 14.1.

16.2.2 REFLECTION IN SPHERES AND CIRCLES

The straightforward application of spherical inversion is to compute the reflection in a sphere of arbitrary elements. We have used this above in Figure 16.1 for the reflection in a unit sphere around the origin. By now you should be confident that this may be extended to arbitrary spheres anywhere, because it is a versor product, and will therefore combine well with other versor products. We can move the unit sphere or the elements anywhere, in any orientation, and still $\sigma \widehat{X}/\sigma$ will perform the reflection, as illustrated in Figure 16.2. The only thing we cannot show now is that we are also allowed to change the radius of the sphere in a structure-preserving manner, although this may be obvious geometrically. We leave that until after we have the rotor for that scaling (already in the next section), and that will permit the rotor representation of reflection in spheres of arbitrary radius.

We can also compose the reflections, first reflecting in a plane π and then in a sphere σ cutting the plane orthogonally (so that $\pi \cdot \sigma = 0$). The total result is that we have used the versor $\pi \sigma$:

$$\textit{reflection in a circle:} \quad X \;\mapsto\; (\pi \sigma) \widehat{X} (\pi \sigma)^{-1}.$$

Since π was chosen through the center of the sphere σ, the versor $\pi \sigma$ is the dual representation of the circle that is their meet (as in the factorization of Section 15.2.3). This versor construction therefore gives a sensible meaning to *reflection in a circle.* In its plane, this acts like spherical inversion, but it can also be extended to act outside that plane. We found it hard to draw a convincing picture of this; because of its essential 3-D nature, it is best studied live in an interactive visualization. Structural exercise 3 explores reflection

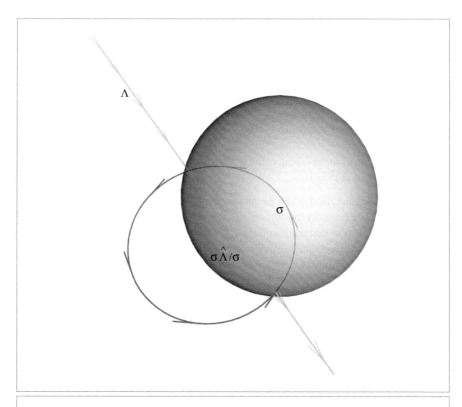

Figure 16.2: An oriented Euclidean line Λ is reflected in a sphere σ by the versor product $\sigma \hat{\Lambda} \sigma^{-1}$ to become a circle through the center.

in a point pair, which is interestingly different from spherical reflection, even in the plane.

16.3 SCALING

16.3.1 THE POSITIVE SCALING ROTOR

In Section 13.2.2, we generated a translation from reflections in two parallel dual planes. The natural question now is: What can we generate from the reflections in two "parallel" spheres (i.e., concentric spheres $o - \frac{1}{2}\rho_1^2 \infty$ and $o - \frac{1}{2}\rho_2^2 \infty$)?

We obtain a composite versor that is the product of the two inversions, and simplify:

$$(o - \tfrac{1}{2}\rho_2^2 \infty)\,(o - \tfrac{1}{2}\rho_1^2 \infty) = (\rho_1^2 + \rho_2^2) - \tfrac{1}{2}(\rho_1^2 - \rho_2^2)\,o \wedge \infty.$$

One would expect only the ratio of the sizes of the spheres to matter. We rescale the versor through dividing by $2\rho_1\rho_2$, and conveniently define a characteristic parameter γ through

$$\exp(\gamma/2) = \rho_2/\rho_1.$$

Now some algebraic manipulation gives a pleasant standard form,

$$S_\gamma = \cosh(\gamma/2) + \sinh(\gamma/2)\, o \wedge \infty = e^{\gamma o \wedge \infty/2},$$

where the rewriting of the resulting rotor as an exponential uses the results in Section 7.4 combined with $(o \wedge \infty)^2 = 1$.

You may guess what the effect of this rotor is, since you know that it is geometrically a double sphere inversion: it scales elements relative to the origin. We compute what happens to the basic components of our representation. For brevity define $O \equiv o \wedge \infty$ (it represents the unit flat point at the origin), and note that $O^2 = 1$, as well as $O\,o = -o = -o\,O$ and $O\infty = \infty = -\infty\,O$. With that, it is easy to derive that

$$S_\gamma[\infty] = \big(\cosh(\gamma/2) + \sinh(\gamma/2)\,O\big)\, \infty\, \big(\cosh(\gamma/2) - \sinh(\gamma/2)\,O\big)$$
$$= \big(\cosh(\gamma/2) + \sinh(\gamma/2)\,O\big)^2 \infty = e^\gamma\, \infty.$$

Using the commutation relationships, you should dare to abbreviate this derivation as

$$S_\gamma[\infty] = e^{\gamma O/2}\, \infty\, e^{-\gamma O/2} = e^{\gamma O}\, \infty = \big(\cosh(\gamma) + \sinh(\gamma)\,O\big)\, \infty = e^\gamma \infty. \qquad (16.1)$$

(Verify all steps in detail, if you need to be convinced.) Similarly,

$$S_\gamma[o] = e^{\gamma O/2}\, o\, e^{-\gamma O/2} = e^{\gamma O}\, o = \big(\cosh(\gamma) - \sinh(\gamma)\,O\big)\, o = e^{-\gamma}o. \qquad (16.2)$$

The bivector O commutes with a purely Euclidean blade $\mathbf{x}$, so

$$S_\gamma[\mathbf{E}] = \mathbf{E}.$$

In total, we find for the transform of a general point $x = T_\mathbf{x}[o]$

$$S_\gamma[x] = e^{-\gamma} \left(o + (e^\gamma \mathbf{x}) + \tfrac{1}{2}(e^\gamma \mathbf{x})^2 \infty \right) = e^{-\gamma} T_{e^\gamma \mathbf{x}}[o]. \qquad (16.3)$$

This clearly shows that the spatial aspects are scaled by e^γ (as well as weighted by $e^{-\gamma}$). So this is indeed a spatial scaling. You should realize that it has been characterized by an exponential parameter; so to obtain a scaling by a factor of 2, you need to use $\gamma = \log(2)$.

Note that this transformation is almost a Euclidean transformation: it does not quite preserve the point at infinity (as a Euclidean transformation would), but it merely weights it (which is not too bad). In particular, the type of an element is preserved in scaling, since the conditions of Table 14.1 are not affected by a weight on ∞. So rounds remain rounds, and flats remain flats. No Euclidean transformation can affect the size of an object, but once it has been determined by a scaling, it is nicely preserved.

There is a slight paradox in this use of the scaling rotor. The distance between Euclidean point representatives was originally defined through a ratio of inner products by (13.2). If the distance between the points should increase, so should the value of that inner product ratio. But the inner product is an invariant of the orthogonal transformation produced by the scaling rotor (by definition of an orthogonal transformation). When you check, you find that this indeed holds, because the scaling changes ∞ to $e^{\gamma}\infty$ and (16.3) trades spatial scale for point weight.

However, if we use the distance definition relative to the *original* point at infinity ∞, we find that the Euclidean distance of the points has indeed increased by a factor of e^{γ} after scaling by S_{γ}. Apparently, that is how we should use the definition of distance: always relative to the original point at infinity.

This interpretation afterwards of course does not affect the consistency of the conformal model computations. Having scaling available as a rotor means that all our constructions are automatically structurally preserved under a change of scale—we never need to check that they are. For instance, any construction made with a unit sphere using the products of our geometry can be simply rescaled to involve any sphere of any radius at any position without affecting its structure. This scaling does not even need to be done relative to the origin; by virtue of the structure preservation, a scaling around another location t is simply made by applying the translation rotor to the scaling rotor to obtain the translated scaling as $T_t[S_{\gamma}]$.

16.3.2 REFLECTION IN THE ORIGIN: NEGATIVE SCALING

When one of the parallel two reflecting spheres is imaginary, their product is a versor that is capable of reflecting in the origin. The prototypical example of this is the product of two unit spheres, one real, one imaginary. This gives the element

$$(o - \infty/2)\,(o + \infty/2) = o \wedge \infty.$$

This is the flat point in the origin, which makes perfect sense: negative scaling is like reflection in the origin. Its action on a conformal point is

$$T_{\mathbf{x}}[o] \quad \mapsto \quad -\,T_{-\mathbf{x}}[o].$$

Even though this is an even versor, it is not a rotor, for $(o \wedge \infty)\,(o \wedge \infty)\tilde{} = -1$. Therefore it cannot be performed in small amounts, and it cannot be expressed as the exponential of a bivector. It is a transformation, but not a motion.

You can combine this reflection in the origin with a scaling versor to effectively make a versor that can perform a negative scaling.

16.3.3 POSITIVELY SCALED RIGID BODY MOTIONS

The combination of a translation T, a rotation R, and a positive scaling S (all with arbitrary center) can always be brought in the standard form $T R S$. This is most easily shown by considering swapping the rotors pairwise, as follows:

- The swapping law for translation and rotation is

$$T_{\mathbf{t}} R = R\, T_{R^{-1}\,\mathbf{t}\,R}$$

  as we saw in (13.12).
- Rotation and scaling in the origin commute,

$$R_{\mathbf{I}\phi}\, S_\gamma = S_\gamma\, R_{\mathbf{I}\phi},$$

  since their generating bivectors do.
- Scaling and translation do not commute, but satisfy

$$T_{\mathbf{t}}\, S_\gamma = S_\gamma\, T_{e^{-\gamma}\mathbf{t}}.$$

You can derive this algebraically or confirm it by the simple sketch of Figure 16.4.

Because of these commutation properties, we can always convert any sequence of rotors S, R, and T into the standard order $T R S$, with suitable adaptation of the rotor parameters.

These rotors generate a logarithmic spiral; in 3-D. This can be used to generate a snail shell from a well-chosen circle, as in Figure 16.3. Note that this direct application to the 3-blade

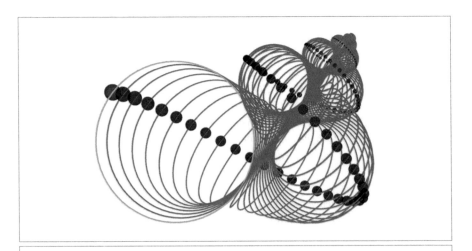

Figure 16.3: A positively scaled rigid body motion rotor repeatedly applied to a circle and a point (both displayed in light colors) generates an escargoid (snail shell).

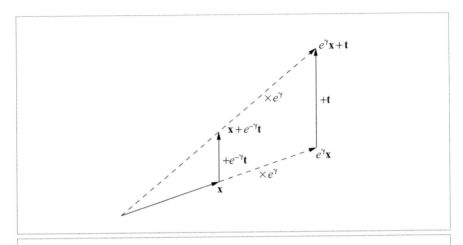

Figure 16.4: Swapping the order of a scaling rotor and a translation rotor implies that the translation needs to be rescaled.

of the circle implicitly uses the structure preservation of the rotor, and is therefore typical of the conformal model. In the figure, we have also applied the same rotor to a small sphere to give a better impression of the 3-D nature of the shell.

16.3.4 LOGARITHM OF A SCALED RIGID BODY MOTION

Refining the step size in the generation of transformations like Figure 16.3 should be done through interpolation of the rotor, completely analogous to the interpolation of rigid body motions in Section 13.6 and of rotations in Section 10.3.4. Correspondingly, we need a logarithm for this type of rotor to enable the computation of the required $V^{1/N}$ to perform the rotor V in N steps, as $\exp\big(\log(V)/N\big)$.

To determine the logarithm of the positively scaled rigid body motion $V = TRS$, we collate some partial results first.

- We have already determined the logarithm of the rigid body motion part TR, in (13.15).
- Scaling and rotation commute, so the logarithm of a rotor of the form RS is easy to determine:

$$\log(R_{\mathbf{I}\phi}\, S_\gamma) = \log(R_{\mathbf{I}\phi}) + \log(S_\gamma) = -\mathbf{I}\phi/2 + \gamma\, o \wedge \infty/2,$$

the sum of the logarithms of the commuting rotors.

- The really new combination is that of a rotor of the type $T S$. This can be determined to be

$$\log(T_{\mathbf{t}} S_\gamma) = \log(T_{\mathbf{t}\gamma/(e^\gamma-1)}) + \log(S_\gamma) = -\frac{\gamma}{e^\gamma - 1}\, \mathbf{t}\infty/2 + \gamma o \wedge \infty/2, \quad (16.4)$$

as you can verify by exponentiation and convenient grouping of terms into power series. If $\gamma = 0$, the fractional coefficient should take its limit value 1.

For the general positively scaled rigid body motion rotor in the form $V = TRS$, we decompose the rotor into known elements by applying the earlier factorization in the log for a rigid body motion of Section 12.5.3. We split off the part perpendicular to the plane of R, to write $V = T_{\mathbf{w}}\,(T_{\mathbf{v}} R_{I\phi} T_{-\mathbf{v}})\, S_\gamma$. The translation over $\mathbf{w} = (\mathbf{t} \wedge I)/I$ is not affected by the rotation and therefore interacts with the scaling as (16.4). The other part should be described as the translation of a rotor of the type $R S$, so we need to look for its center, as the location $\mathbf{c}$ satisfying

$$R_{I\phi}[e^\gamma \mathbf{c} - \mathbf{v}] + \mathbf{v} = \mathbf{c},$$

which solves to $\mathbf{c} = (1-R^2 e^\gamma)^{-1}(1-R^2)\,\mathbf{v} = (1-R^2 e^\gamma)^{-1}(\mathbf{t}\rfloor I)/I$ (where we used (13.13)). Relative to this center, the motion V is in standardized form $R S$, and its logarithm can be taken. That then needs to be translated to the correct in-plane location and composed with the remaining $\mathbf{w}$ component. Combining all these ingredients produces the rather intimidating

$$\log(T_{\mathbf{t}}\, R_{I\phi}\, S_\gamma) = \frac{\gamma}{1 - e^\gamma}\,((\mathbf{t} \wedge I)/I)\,\infty/2 + {}^\top_{(1-e^\gamma R^2)^{-1}(\mathbf{t}\rfloor I)/I}[-I\phi/2 + \gamma o \wedge \infty/2].$$

To apply this formula, you need to retrieve the parameters occurring in it from the total rotor V. This is done by observing that $-o\rfloor(V\infty) = -o\rfloor(R\,e^{\gamma O/2}\infty) = R\,e^{\gamma/2}$. Employing the additional demand on the normalization of the rotor $R\widetilde{R} = 1$, we find R and γ. Dividing out these rotors from V gives the rotor T. The parameters of R and T are computed as in Section 7.4.4. As we did there, if $R = 1$, you can set $I = 1$ and $\phi = 0$. If $\gamma = 0$ (so that $S = 1$), the fractional coefficient of $(\mathbf{t} \wedge I)/I$ should take its limit value for small γ, which is -1 (this is automatic if you rewrite it in terms of a sinch function). The resulting function is indicated in Figure 16.5. The exceptions for R are similar to the earlier rigid body motion logarithm of Figure 13.5.

Having the logarithm gives us the tool we need to interpolate these motions in the usual manner, and it was used to generate the stepwise transformation of the elements in Figure 16.3.

The versor of a *negatively* scaled rigid body motion is not a rotor. It cannot be written in exponential form and does not have a logarithm. Since you cannot do such a motion piecewise anyway, you should not want to interpolate it.

$$
\begin{aligned}
&\log(V)\{ \\
&\qquad X = -o\rfloor(V\infty) \\
&\qquad \gamma = \log(X\widetilde{X}) \\
&\qquad \text{if } (\gamma == 0) \text{ then} \quad (\textit{no scaling}) \\
&\qquad\qquad \gamma' = 1 \\
&\qquad \text{else} \\
&\qquad\qquad \gamma' = \gamma/(e^{\gamma} - 1) \\
&\qquad \text{endif} \\
&\qquad S = e^{\gamma o \wedge \infty/2} \\
&\qquad R = X e^{-\gamma/2} \\
&\qquad T = V\widetilde{S}\,\widetilde{R} \\
&\qquad \mathbf{t} = -2\,o\rfloor T \\
&\qquad \text{if } (\langle R\rangle_0 == -1) \quad \text{return (``no unique logarithm'')} \\
&\qquad \text{if } (\langle R\rangle_0 == 1) \quad (\textit{no rotation, so no } \mathbf{I}) \\
&\qquad\qquad \log = -\gamma'\,\mathbf{t}\,\infty/2 + \gamma\, o \wedge \infty/2 \\
&\qquad \text{else} \\
&\qquad\qquad \mathbf{I} = \langle R\rangle_2/\|\langle R\rangle_2\| \\
&\qquad\qquad \phi = -2\,\text{atan2}(\|\langle R\rangle_2\|,\langle R\rangle_0) \\
&\qquad\qquad T_{\mathbf{v}} = 1 - (1 - e^{\gamma}R^2)^{-1}\,(\mathbf{t}\rfloor\mathbf{I})/\mathbf{I}\,\infty/2 \\
&\qquad\qquad \log = -\gamma'\,(\mathbf{t}\wedge\mathbf{I})/\mathbf{I}\,\infty/2 + T_{\mathbf{v}}\,(-\mathbf{I}\phi/2 + \gamma o \wedge \infty/2)\widetilde{T}_{\mathbf{v}} \\
&\qquad \text{endif} \\
&\}
\end{aligned}
$$

Figure 16.5: Computation of the logarithm of a positively scaled rigid body motion rotor V. One may improve subsequent numerics by making sure to return a bivector through taking the grade-2 part $\langle\log\rangle_2$.

16.4 TRANSVERSIONS

In the classical literature on conformal operations, one introduces the operation of *transversion*. This may be defined as the triple composition of an inversion in the unit sphere, followed by a translation, followed by another inversion in the unit sphere. It is easy to compute its rotor, which simplifies nicely:

$$
(o - \infty/2)\,(1 - \mathbf{t}\infty/2)(o - \infty/2) = 1 + o\,\mathbf{t} = e^{o\mathbf{t}}.
$$

The two inversions make it a handedness-preserving operation, a proper conformal transformation represented by an even versor, which is even a rotor. You might prefer to use the analogy with the introduction of a Euclidean rotation as the reflection in two planes with a common line, and define the transversion as the reflection in two equal-radius spheres with a common point (use structural exercise 5 to show that this is essentially the same as the definition above).

From its form, the transversion clearly completes our types of primitive rotors in the conformal model using the last remaining factor of the bivector basis for its exponent (after $-\mathbf{t} \wedge \infty/2$ for translations, $\gamma\, o \wedge \infty/2$ for scaling, and a Euclidean $-\mathbf{I}\phi/2$ for rotations). The full list is in Table 16.1.

Table 16.1: Basic operations in the conformal model and their versors. The improper transformations have vector versors, and change handedness. The reflection in the origin is an even versor, but not a rotor, and it has no exponential form. The proper conformal transformations are composed from even unit versors and can be written as the exponentials of bivectors. The table shows that all elements in the bivector basis are associated with a specifically named operation.

Type of Operation	Explicit Form	Exponential Form
reflection in origin plane	$\mathbf{n}$	none
reflection in real unit sphere	$c - \infty/2$	none
reflection in origin	$o \wedge \infty$	none
rotation over ϕ in I-plane	$\cos(\phi/2) - \sin(\phi/2)\,\mathbf{I}$	$e^{-\phi\mathbf{I}/2}$
translation over $\mathbf{t}$	$1 - \mathbf{t}\infty/2$	$e^{-\mathbf{t}\wedge\infty/2}$
scaling by e^{γ}	$\cosh(\gamma/2) + \sinh(\gamma/2)\, o \wedge \infty$	$e^{\gamma o \wedge \infty/2}$
transversion over $\mathbf{t}$	$1 + o\,\mathbf{t}$	$e^{o \wedge \mathbf{t}}$

Table 16.2: Common proper transformations of some of the standard elements of the conformal model. The entry marked "Involved" is just the product of the transversions of ∞ and E.

Element	Translation $e^{-\mathbf{t}\infty/2}$	Rotation R	Scaling $e^{\gamma\, o\wedge\infty/2}$	Transversion $e^{o\mathbf{t}/2}$
o	$t = o + \mathbf{t} + \frac{1}{2}\mathbf{t}^2\infty$	o	$e^{-\gamma}o$	o
∞	∞	∞	$e^{\gamma}\infty$	$t' \equiv \infty + 2\mathbf{t} + 2\mathbf{t}^2 o$
E	$-(E\infty)\lfloor t$	$RE\widetilde{R}$	E	$E + o(2\mathbf{t}\rfloor E)$
$E\infty$	$E\infty$	$(RE\widetilde{R})\infty$	$e^{\gamma}E\infty$	Involved
$o \wedge E \wedge \infty$	$(o + \mathbf{t}) \wedge E \wedge \infty$	$o \wedge (RE\widetilde{R}) \wedge \infty$	$o \wedge E \wedge \infty$	$o \wedge E \wedge (\infty + 2\mathbf{t})$
$o - \infty/2$	$\mathbf{t} - \infty/2$	$o - \infty/2$	$e^{-\gamma}(o - e^{2\gamma}\infty/2)$	$o - t'/2$

A closed-from solution to the logarithm of a general conformal transformation also involving a transversion is not yet known.

16.5 TRANSFORMATIONS OF THE STANDARD BLADES

It is convenient to have a table with the most used transformations in standard form applied to the elements in standard form. This is Table 16.2.

Contemplating this table, it is striking how the same versor operation leads to very different explicit forms for the various elements. In actual computations, it is of course often simpler to compute with the versors as long as possible, converting to explicit form only at the end—if at all. This applies both to computations done by hand and by using a computer. The use of versor products keeps the software clear and reduces errors. Only in very time-critical applications would you write things out—and even then, geometric algebra is useful since it provides the code generator with the automatically correct way of transforming arbitrary elements.

16.6 GENERAL CONFORMAL TRANSFORMATIONS

With the reflections and the complete suite of rotors, we can make any conformal transformation. The usefulness of these still needs to be explored, for although they are familiar in mathematics, they are rather new to computer science. In the framework of geometric algebra, they can of course be applied to any element, and that leads to interesting operations.

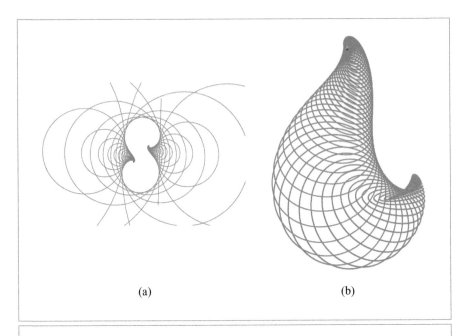

(a)   (b)

Figure 16.6: Loxodromes generated by applying the rotor $\exp\left((o \wedge \mathbf{e}_1 - \mathbf{e}_2 \wedge \infty)/10\right)$ multiple times to a circle: (a) in the $\mathbf{e}_1 \wedge \mathbf{e}_2$-plane, (b) in 3-D space.

16.6.1 LOXODROMES

Figure 16.6(a) shows what we get when applying the rotor

$$L \equiv e^{o \wedge \mathbf{e}_1 - \mathbf{e}_2 \wedge \infty} \qquad\qquad (16.5)$$

to a circle multiple times. (We actually applied $L^{1/N}$ to show its action piecewise.) This is a *loxodromic* transformation, known from the study of the conformal group in two dimensions. As the figure suggests, it has two special points, a source and a sink, and it scales, translates, and rotates elegantly between them.

The geometric algebra characterization of this transformation extends the classical treatment in two ways: it makes the transformation easily applicable to arbitrary basic elements such as flats and rounds (and the circle example gives a lot more insight in the structure of the mapping than merely transforming a point would have done). Also, there is no reason to limit the space to two dimensions; the same rotor works to transform n-dimensional space. The strange 3-D wormlike figures that result are best appreciated by an interactive software package such as GAViewer [18]; Figure 16.6(b) shows a rendering.

A rich class of shapes can be generated in this manner. Their simple expressibility means that they are easily refined, which is what one would wish from graphics primitives, so they may be useful to solid modeling. But this remains to be explored.

16.6.2 CIRCULAR ROTATIONS

More immediately useful are *circular rotations*, which have rotation orbits not around a line (as in Figure 15.5) but around a circle K. Such a circular rotation is of the form $\exp(K^*)$. Each element $p \wedge K^* = (p \rfloor K)^*$ is a circle, curving around the circle K, as in Figure 16.7. You can use this to your advantage to compactly generate a torus given its two generator curves (a dual line and a dual circle). You may explore this in structural exercise 7, which generates the torus in Figure 16.12.

The mere fact that circles can be used as basic descriptors to rule objects can already save a lot of memory and computation to manipulate them. This also remains to be explored.

16.6.3 MÖBIUS TRANSFORMATIONS

The usual way to study conformal transformations in mathematics is by means of Möbius transformations, so we mention that briefly. This technique, developed for 2-D, represents a vector as a complex number z and a conformal transformation as the mapping

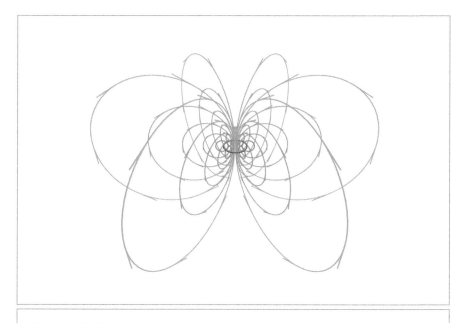

Figure 16.7: The circular orbits of a circle K, of the form $p \wedge K^*$.

$$z \mapsto \frac{az+b}{cz+d}.$$

By choosing the complex numbers a, b, c, and d appropriately, one can generate inversions, translations, rotations, and the other conformal transformations.

Since ratios are important, it makes sense to introduce complex homogeneous coordinates in the complex plane. Then the Möbius transformations admit a matrix representation: multiplication of the transformations is the same as multiplying the complex-valued matrices

$$\left[\!\!\left[\begin{array}{cc} a & b \\ c & d \end{array}\right]\!\!\right].$$

For instance, the conjugated inversion in the unit circle at the origin is represented by the matrix $\left[\!\!\left[\begin{smallmatrix} 0 & 1 \\ 1 & 0 \end{smallmatrix}\right]\!\!\right]$, and a translation over $\mathbf{t}$ by $\left[\!\!\left[\begin{smallmatrix} 1 & t \\ 0 & 1 \end{smallmatrix}\right]\!\!\right]$ (where t is the complex number corresponding to $\mathbf{t}$). Such representations allow one to compute products of conformal transformations easily.

All this is very clever, but also very limited. The matrix description of conformal transformations is very much point-based and, because of its use of complex numbers, tied to 2-D space. No such restrictions are felt when treating conformal transformations using the rotors of the conformal model. As we have seen before when discussing rotations and quaternions, there is no need to go to complex numbers to get an algebra that properly encodes the geometry.

16.7 NON-EUCLIDEAN GEOMETRIES

The conformal model can also be used for the description of other geometries. Hyperbolic and elliptic geometry find a natural home. We briefly mention the connection; more may be found in [15].

16.7.1 HYPERBOLIC GEOMETRY

In Euclidean geometry, we kept the null vector ∞ invariant, because we wanted it to represent the point at infinity. We can model *hyperbolic geometry* in the conformal model by keeping the vector $e = o - \infty/2$ invariant instead. This is the (dual) unit sphere, and we obtain in this manner the Poincaré disk model of hyperbolic geometry. Or rather, a Poincaré hyperball, for the conformal model is not limited to 2-D. By letting the spherical border play the role of infinity, the whole metric of the space must adapt in precisely the right manner.

We briefly indicate the parallel with Euclidean geometry, for convenience of terminology and depiction taking the 2-D case as in Figure 16.8. We are used to our lines of $\mathbb{E}^2$ in direct representation to be represented as 3-blades with a factor ∞; in 2-D hyperbolic geometry, the corresponding special elements must contain a factor e. As a consequence, they do not plunge into the point at infinity ∞, but into the unit sphere e. In the Poincaré depiction, they look like circles, as depicted in Figure 16.8.

Translations took place along Euclidean straight lines; in hyperbolic geometry, they must go along the hyperbolic lines. The translation along the line L in Euclidean space is by a rotor with exponent $-\infty\rfloor L/2$; in hyperbolic space, this becomes the rotor $\exp(-e\rfloor L/2)$.

That has a strange consequence. Because of the null nature of the translation bivector in the Euclidean case, distance was clearly additive in the usual Euclidean sense. In the hyperbolic case, additivity of the translation bivector generates a metric that looks rather distorted on the Poincaré disk. In Figure 16.8 we have indicated circles with the same center and increasing radii. Near the circle at infinity e, distances apparently become very large, and the points inside cannot escape the disk.

Hyperbolic motions can be made with versors that preserve e, and these can be generated by vectors of the form $\mathbf{n} + \delta\bar{e}$. The conformal representation for this geometry can then run completely parallel to how we handled Euclidean geometry in Chapter 13.

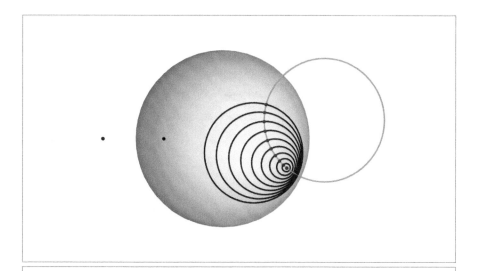

Figure 16.8: 2-D hyperbolic geometry in the conformal model. By considering the real dual unit circle $e = o - \infty/2$ as infinity, the hyperbolic lines become circles in the plane that meet the unit circle perpendicularly. (One such line is indicated as the green circle.) Translations are defined as moving elements along spherical lines. We show translated points at equal increments of hyperbolic distance, and the corresponding hyperbolic circles. This gives an impression of the hyperbolic metric on the Poincaré disk. A flat point in the hyperbolic plane consists of a point and its (unreachable) reflected twin outside the disk, indicated in blue.

16.7.2 SPHERICAL GEOMETRY

By taking the imaginary unit sphere $\bar{e} \equiv o + \infty/2$ as the preserved vector of all transformations, we get a model of *spherical geometry*. This is actually the geometry on an n-dimensional sphere, in which the role of the flats is played by the great circles.

We consider this model again in 2-D, for convenience in terminology and depiction (see Figure 16.9). The flats are now blades that contain $\bar{e}$. These are all circles that cut the unit circle in a point pair centered around the origin (i.e., in two diametrically opposite points). They are the images of the great circles on a sphere under stereographic projection from the south pole onto the plane of the equator. Such a projection is conformal, for it is the inversion in a sphere of radius $\sqrt{2}$ with its center at the south pole. That is how the conformal model contains a faithful representation of the geometry of the sphere into the plane. A flat point $p \wedge \bar{e}$ is a point pair (representing diametrically opposite points on the sphere).

The conformal translations are now effectively rotations on the sphere, and the corresponding distance measure is the angle between two great circles. The versors are

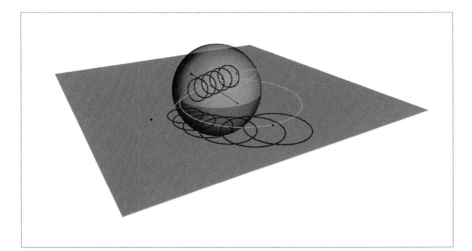

Figure 16.9: 2-D spherical geometry in the conformal model. By considering the unit dual imaginary circle $\bar{e} = o + \infty/2$ as infinity, the spherical lines become circles in the plane that meet the unit circle in two opposite points. (A spherical line through two red points is indicated as the green circle.) Translations are defined as moving elements along spherical lines. We show the translation of a black circle by equal increments of spherical distance.
Also shown is the spherical image of this planar geometry (by a 3-D inversion in the real dual sphere $T_{e_3}[o] - \infty$, which is not depicted). In that spherical image, the spherical lines are great circles, and the translation is a slide along such a great circle. A flat point of the plane becomes a pair of diametrically opposite points on the spherical image.

generated by subsequent reflections in vectors of the form $\mathbf{n} + \delta e$, and represent the expected sliding over the spherical space.

By the same principle, you can now make arbitrary conformal versors act on elements residing on a sphere. It is very satisfying that we do not need to write special software to deal with this geometry (or hyperbolic geometry): its elements and operators are all part of the conformal model, which we originally implemented to do Euclidean geometry.

16.8 FURTHER READING

The use of the conformal model to actually do conformal geometry is still very recent. The original successful attempt to use rotors for their representation may have been [2], but this reference is highly mathematical.

The richly illustrated book by Needham [46] helps develop your geometrical intuition on conformal mappings. Even though it is mostly restricted to the complex plane and the algebraic language of Möbius transformations, he works hard at getting the geometry up front. You can now reproduce most illustrations in his book through the conformal model in `GAViewer`.

Our usual sources [33,15] have additional material on the rotor method of representing conformal transformations. They explicitly spell out the relationship of the conformal model with stereographic projection.

A good source for the other geometries within the conformal model is the work of A. Lasenby, starting with the treatment in [15], which derives explicit expressions for the distances along the lines of the geometries.

16.9 EXERCISES

16.9.1 DRILLS

1. Reflect the line L through locations $\mathbf{e}_1$ and $\mathbf{e}_2$ in the unit sphere at the origin.
2. Factorize the result of the previous exercise to determine its center and squared radius.
3. Reflect the tangent vector at $\mathbf{e}_1 + \mathbf{e}_2$ in the direction $2\mathbf{e}_3$ in the unit sphere at the origin. Notice especially the weight of the result!
4. Scale the line L by a factor of e^2 from the origin.
5. Scale the line L by a factor of e^2 from the point $\mathbf{e}_1$.
6. Reflect the line L in the origin.
7. Reflect the line L in the point $\mathbf{e}_1$.

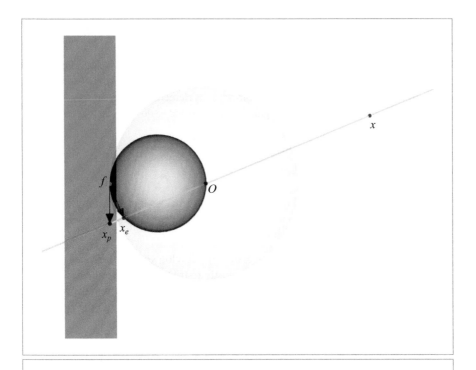

Figure 16.10: Imaging by the eye (see structural exercise 2).

16.9.2 STRUCTURAL EXERCISES

1. Show that the general inversion formula for a point in a dual sphere at the origin $o - \frac{1}{2}\rho^2\infty$ is

$$\mathsf{T_x}[o] \;\mapsto\; \frac{\mathbf{x}^2}{\rho^2}\, \mathsf{T}_{\rho^2\mathbf{x}^{-1}}[o].$$

   Note that this implies that imaginary spheres involve a central reflection in the origin (as well as the inversion of the distances to the origin).

2. Now that we have conformal model and spherical inversion, we can look at the pinhole camera with different eyes (see Figure 16.10). In that figure, an eyeball is defined by its pinhole o and the focal point f on the optical axis. A point x images on the eyeball as x_e. In the pinhole box camera of Figure 12.2, this point would have been imaged as x_p. Show that the two are related by inversion in the indicated sphere $f \rfloor (o \wedge \infty)$, which also transforms the eyeball as whole into the blue imaging plane.

3. Figure 16.11 shows the reflection of various elements in the brown point pair; all elements reside in the plane of the drawing. Compare this to the spherical reflection

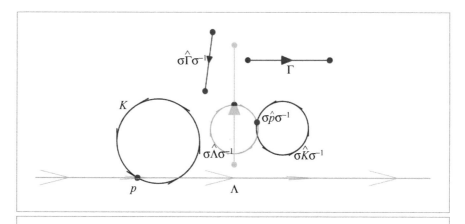

Figure 16.11: Reflection in a point pair (see structural exercise 3).

Figure 16.1 and understand the differences. (Hint: Use the factorization of the point pair by a sphere and well-chosen planes.)

4. Prove that flats through the origin are invariant under scaling.

5. We claimed in Section 16.4 that a transversion can also be constructed as the reflection in two touching equal-radius spheres. To determine the standard form of such a transversion, put the spheres symmetrically around the origin, with their centers at $\pm\mathbf{a}$. Show that this gives the transversion versor as $-\mathbf{a}^2\,(1 - 2o \wedge \mathbf{a}^{-1})$. What should you take as the distance of the touching spheres to obtain the standard transversion rotor $\exp(o \wedge \mathbf{t})$?

6. Show that the loxodromic rotor of (16.5) can be written as the commuting product of two more elementary rotors:

$$e^{(o-\infty)(e_1-e_2)/2}\, e^{(o+\infty)(e_1+e_2)/2}.$$

Analyze what these do using software.

7. In Figure 16.12, we have generated a torus by generating a few circles. The inversion in a sphere of a torus is called *Dupin's cyclide*. In the conformal model, its circles are simply the torus circles, inverted in the sphere. Write pseudocode to generate this figure with just a few parameterized operations.

8. The invariance of a vector a under the versors places the demand $a = -\widehat{V}\,a\,V^{-1}$ on the versor, which implies that $a \rfloor V = 0$. If the versor V is a rotor, it can be written as the exponent of a bivector. Show that any bivector of the form $a \rfloor T$ with T an arbitrary trivector leaves a invariant. Relate this to the translations of the various geometries we discussed (i.e., what trivector should you choose to get a translation?).

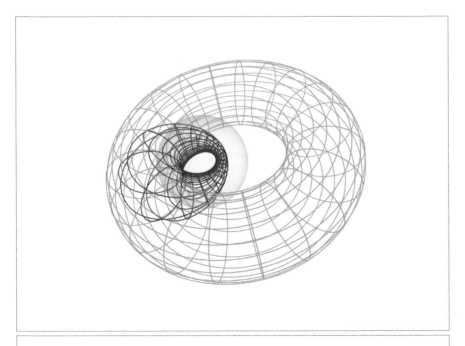

Figure 16.12: The Dupin cyclide as the inversion of a torus into a sphere. All elements in this sketch are primitives of computation in the conformal model (see structural exercise 7).

9. Figure 16.13 depicts the same situation, at the same scale, of a green line $L = p \wedge q \wedge i$ through two points p and q (one of which is the center of the black circles), and a flat point $r \wedge i$ (in blue). In all three figures, the line is used in a translation versor $\exp(i \rfloor L/2)$ to translate the point p over equal distances, and dual circles are made with the original point as its center as $t \rfloor (p \wedge i)$. The only difference is the element used for the infinity i (in light red). Identify the metrics and explain the differences as quantitatively as you can. You can pick i from $\mathbf{e}_1$, o, ∞, $o - \infty/2$, $o + \infty/2$, or $o + \mathbf{e}_2 + \infty/2$.

16.10 PROGRAMMING EXAMPLES AND EXERCISES

16.10.1 HOMOGENEOUS 4×4 MATRICES TO CONFORMAL VERSORS

This example draws the same GLUT models that we have used in many other examples before. The user can translate, rotate, and scale the model. The example intends

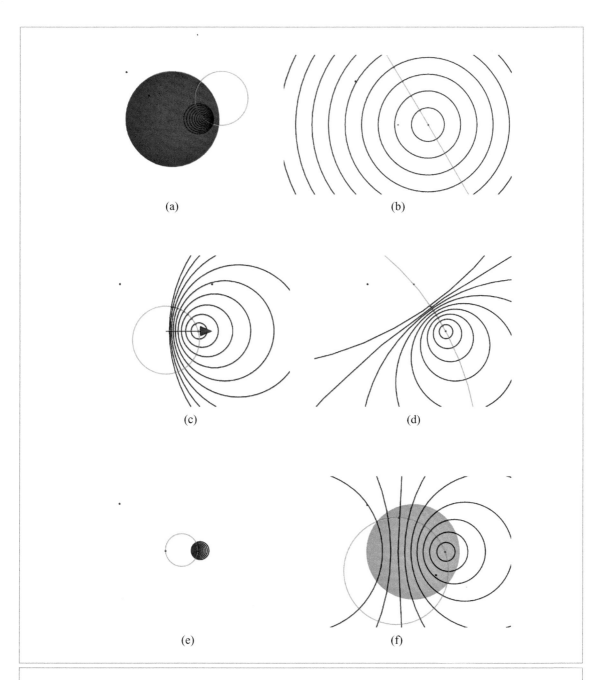

(a)

(b)

(c)

(d)

(e)

(f)

Figure 16.13: Metrical Mystery Tour: The conformal model can incorporate different conformal geometries, only differing by their concepts of infinity.

to demonstrate how to convert 4 × 4 homogeneous matrices (e.g., from OpenGL) to conformal versors: We construct a particular transformation on the OpenGL modelview matrix stack, read out the matrix, and convert this to a versor. Then we reset the modelview matrix to the identity and use the versor to apply the transformation.

The full matrix-to-versor conversion function is shown in Figure 16.14. The function is called `matrix4x4ToVersor()` and resides in `c3ga_util.cpp`, not in the source file of

```
TRSversor matrix4X4ToVersor(const mv::Float _M[4 * 4], bool transpose /*= false*/) {
  mv::Float M[4 * 4];
  if (transpose) {
    // transpose & normalize
    for (int i = 0; i < 4; i++)
      for (int j = 0; j < 4; j++)
        M[i * 4 + j] = _M[j * 4 + i] / _M[3 * 4 + 3];
  }
  else {
    // copy & normalize
    for (int i = 0; i < 4; i++)
      for (int j = 0; j < 4; j++)
        M[i * 4 + j] = _M[i * 4 + j] / _M[3 * 4 + 3];
  }

  // extract translation:
  vectorE3GA t(vectorE3GA_e1_e2_e3, M[0 * 4 + 3], M[1 * 4 + 3], M[2 * 4 + 3]);

  // initialize images of Euclidean basis vectors (the columns of the matrix)
  vectorE3GA imageOfE1(vectorE3GA_e1_e2_e3, M[0 * 4 + 0], M[1 * 4 + 0], M[2 * 4 + 0]);
  vectorE3GA imageOfE2(vectorE3GA_e1_e2_e3, M[0 * 4 + 1], M[1 * 4 + 1], M[2 * 4 + 1]);
  vectorE3GA imageOfE3(vectorE3GA_e1_e2_e3, M[0 * 4 + 2], M[1 * 4 + 2], M[2 * 4 + 2]);

  // get scale of the 3x3 part (e1, e2, e3)
  mv::Float scale = _Float(norm_e(imageOfE1) + norm_e(imageOfE2) + norm_e(imageOfE3))
          / 3.0f;

  // compute determinant of matrix
  // (if negative, negate all matrix elements)
  mv::Float n = 1.0f; // used to negate the matrix
  scalor negScale = _scalor(1.0f);
  if ((imageOfE1 ^ imageOfE2 ^ imageOfE3).e1e2e3() < 0.0f) {
    n = - 1.0f;
    // no^ni provides the negative scaling in the final versor:
    negScale = noni;
  }

                                                                        Continued
```

```
Continued

  // initialize 3x3 'rotation' matrix RM, call e3ga::matrixToRotor
  mv::Float si = n / scale;
  mv::Float RM[3 * 3] = {
    M[0 * 4 + 0] * si, M[0 * 4 + 1] * si, M[0 * 4 + 2] * si,
    M[1 * 4 + 0] * si, M[1 * 4 + 1] * si, M[1 * 4 + 2] * si,
    M[2 * 4 + 0] * si, M[2 * 4 + 1] * si, M[2 * 4 + 2] * si
  };
  e3ga::rotor tmpR = e3ga::matrixToRotor(RM);

  // convert e3ga rotor to c3ga rotor:
  c3ga::rotor R(rotor_scalar_e1e2_e2e3_e3e1,
    tmpR.getC(e3ga::rotor_scalar_e1e2_e2e3_e3e1));

  // get log of scale:
  mv::Float logScale = (mv::Float) ::log(scale);

  // return full versor:
  return _TRSversor(
    exp(_freeVector(- 0.5f * (t ^ ni))) *  // translation
    R *                                     // rotation
    exp(_noni_t(0.5f * logScale * noni)) *  // scaling
    negScale                                // negative scaling
    );
}
```

Figure 16.14: This function converts 4 × 4 homogeneous matrices to conformal versors (Example 1). The 4 × 4 matrix may only contain translation, rotation, and uniform scaling.

the example itself. We discuss it in detail below. Due to limitations on the transformations that the conformal model can represent, there are certain restrictions on the type of matrices we can handle. The matrix should only contain translation, rotation, and uniform scaling (i.e., the same along all axes). Such a matrix has the form

$$\begin{bmatrix} s_1 s_2 \, [\![R]\!] & s_2 \, [\![\mathbf{t}]\!] \\ 0^T & s_2 \end{bmatrix},$$

where $[\![R]\!]$ is a 3×3 rotation matrix, $[\![\mathbf{t}]\!]$ is a 3×1 translation vector, and s_1 and s_2 are scalars. s_1 is the actual scaling, while s_2 is just a homogeneous scaling factor that can in principle be removed without side effects (i.e., it does not affect our interpretation of transformed objects, only their weight).

As can be seen in Figure 16.14, the first step is to get rid of the s_2 factor by normalizing the matrix. Optionally, the matrix is transposed (OpenGL represents its matrices in column-major order, while we use row-major order). Once normalized, the translation part of the matrix is easily extracted as vector t.

To separate the rotation from the scale, we initialize three vectors that are the images of the basis vectors imageOfE1, imageOfE2, and imageOfE3. In principle, the overall scale is the norm of any of these three vectors. However, to increase precision slightly we set scale to the average of the norm all three vectors.

Once s_1 is found, we can normalize the rotation matrix and convert it to a rotor R using the existing matrixToRotor() function from Section 7.10.3. There is one problem, however; the scaling may be negative. In Section 16.3.2, we have described this as the application of an extra versor $o \wedge \infty$. The algorithm needs to detect and handle this, for the (Euclidean) norm of a vector is always positive, hence so is s_1. Negative scaling can be detected by computing the determinant of the rotation matrix. We implement this as follows:

```
if ((imageOfE1 ^ imageOfE2 ^ imageOfE3).e1e2e3() < 0.0f) {
    // negative scaling detected ...
}
```

When negative scaling has been detected, a negative scaling versor $o \wedge \infty$ is appended to the final versor. Hence, the code that handles the scaling and extracts the rotation is:

```
// get scale of the 3x3 part (e1, e2, e3)
mv::Float scale = _Float(norm_e(imageOfE1) + norm_e(imageOfE2) +
        norm_e(imageOfE3))  / 3.0f;

// compute determinant of matrix
// (if negative, negate all matrix elements)
mv::Float n = 1.0f; // used to negate the matrix
scalor negScale = _scalor(1.0f);
if ((imageOfE1 ^ imageOfE2 ^ imageOfE3).e1e2e3() < 0.0f) {
    n = -1.0f;
    // no^ni provides the negative scaling in the final versor
    negScale = noni;
}

// initialize 3x3 'rotation' matrix RM, call e3ga::matrixToRotor
mv::Float si = n / scale;
mv::Float RM[3 * 3] = {
    M[0 * 4 + 0] * si, M[0 * 4 + 1] * si, M[0 * 4 + 2] * si,
    M[1 * 4 + 0] * si, M[1 * 4 + 1] * si, M[1 * 4 + 2] * si,
    M[2 * 4 + 0] * si, M[2 * 4 + 1] * si, M[2 * 4 + 2] * si
};
e3ga::rotor tmpR = e3ga::matrixToRotor(RM);
```

Note that the variable negScale will either hold the value 1 or no∧ni to handle absence or presence of negative scaling, respectively. The final versor is returned as

```
return _TRSversor(
  exp(_freeVector(-0.5f * (t ^ ni))) *    // translation
  R *                                      // rotation
  exp(_noni_t(0.5f * logScale * noni)) *  // scaling
  negScale                                 // negative scaling
);
```

Negative scaling can still be represented by a translate-rotate-scale (TRS) versor, so the return type is TRSversor. Do note that the negative scaling versor $o \wedge \infty$ does not have a logarithm.

While we are on the subject, we can be more specific about the flexibility and use of the geometric algebra conformal versors versus the more common 4×4 homogeneous coordinate matrices. This refines the general observations made in Section 7.7.3.

- **Storage**. Storage is similar; to (densely) store a translation-rotation-scaling (TRS) versor, 12 floats are required, while the corresponding 4×4 matrix requires 13 floats.

- **Speed**. Application of a versor to a vector is about 30 percent slower than applying a 4×4 matrix to a vector. And because 4×4 matrices are so commonly used, mainstream CPUs have special hardware that can efficiently apply 4×4 matrix transformations. Hardware specialized for geometric algebra is not yet mainstream.

- **Universality**. Transforming blades that are not versors is straightforward with versors, since the same method is used to transform *any* blade or versor. The 4×4 matrices can be applied directly only to vectors. In practice, one uses outermorphisms to transform the higher-grade blades. These can be generated easily with the versors, but need to be hand-coded in the matrix approach.

- **Inverses**. Computing the inverse of a versor V is trivial ($V^{-1} = \widetilde{V}/(V\,\widetilde{V})$). Inverting a 4×4 matrix is harder, even when you use the tricks that speed up inversion of TRS transformations.

- **Interpolation**. Interpolation is more natural for versors, since for many rotors we have a logarithm in closed form. Logarithms of matrices are notoriously expensive to compute (although there are some special closed form solutions known here, too).

- **Conversion**. Versors can be converted to 4×4 matrices trivially (the columns of the matrix are the images of the four basis blades $\mathbf{e}_1 \wedge \infty$, $\mathbf{e}_2 \wedge \infty$, $\mathbf{e}_3 \wedge \infty$ and $o \wedge \infty$ under the versor). From matrix to versor is more involved.

- **Geometries**. The 4×4 matrices can also be used to do nonuniform scaling (and thus skewing) and (perspective) projection of vectors. The versors in the conformal model cannot, though other geometric algebras may be developed in which the versors have such actions. On the other hand, conformal versors can also represent inversions in a sphere and other conformal transformations that enable them to be used directly for spherical geometry.

```
TRSversorLog log(const TRSversor &V) {
  // get rotor part:
  rotor X = _rotor(- no << (V * ni));

  const float EPSILON = 1e-6f;

  // get scaling part
  mv::Float gamma = (mv::Float)::log(_Float(X * reverse(X)));
  mv::Float gammaPrime;
  if (fabs(gamma) < EPSILON) gammaPrime = 1.0f;
  else gammaPrime = gamma / (::exp(gamma) - 1);
  scalor S = exp(_noni_t(0.5f * gamma * noni));

  // get rotation part
  rotor R = _rotor(::exp(-0.5f * gamma) * X);

  // get translation part:
  translator T = _translator(V * reverse(S) * reverse(R));
  vectorE3GA t = _vectorE3GA(-2.0f * (no << T));

  if (_Float(norm_e2(_bivector E3GA (R))) < EPSILON * EPSILON) {
    if (_Float(R) > 0.0f) { // R = 1
      // no rotation, so no rotation plane
      return _TRSversorLog((0.5f * (- gammaPrime * (t ^ ni) + gamma * noni));
    }
    else { // R = - 1
      // We need to add a 360 degree rotation to the result.
      // Take it perpendicular to 't':
      bivectorE3GA I; // Get a rotation plane 'I', depending on 't'
      if (_Float(norm_e_2(t)) > EPSILON * EPSILON)
        I = _bivectorE3GA(unit_e(t << I3));
      else I = _bivectorE3GA(e1^e2); // when t = 0, any plane will do

      return _TRSvectorLog(0.5f * (
        -gammaPrime * (t ^ ni) +
        gamma * noni +
        2.0f * (float) M_PI * I));
    }
  }
  else {
    // get rotation plane, angle
    bivectorE3GA I = _bivectorE3GA(R);
    mv::Float sR2 = _Float(norm_e(I));
    I = _bivectorE3GA(I * (1.0f / sR2));
    mv::Float phi = -2.0f * (mv::Float)atan2(sR2, _Float(R));
```

Continued

```
(continued)

    // form bivector log of versor:
    normalizedTranslator Tv = _normalizedTranslator(
      1.0f - 0.5f * inverse(1.0f - (mv::Float)::exp(gamma) * R * R) *
      (t << I) * reverse(I) * ni);

    return _TRSversorLog(0.5f * (- gammaPrime * (t^I) * reverse(I) * ni +
      (- gammaPrime * (t^I) * reverse(I) * ni +
      Tv * (-phi * I + gamma * noni) * reverse(Tv)));
    }
}
```

Figure 16.15: Logarithm of scaled rigid body motion (Example 2). This function returns the logarithm of normalized translate-rotate-scale (TRS) versors.

We will show in Chapter 22 that for a ray-tracing application, the speed of an efficiently implemented conformal model is the same as for the homogeneous coordinate approach. Therefore, criteria such as expressibility of the operations or ease of programming can be the deciding factors.

16.10.2 LOGARITHM OF SCALED RIGID BODY MOTION

The code for computing the logarithm of a TRSversor (translate-rotate-scale-versor) is shown in Figure 16.15. It is a straightforward implementation of the pseudocode in Figure 16.5 in Section 16.3.4. The function resides in c3ga_util.cpp. Note that the function returns the TRSversorLog type, which is a specialized multivector type (the basis blades are e1∧e2, e1∧e3, e2∧e3, e1∧ni, e2∧ni, e3∧ni, and no∧ni). The example itself just tests the log() code by repeatedly generating random TRS versors V and checking that V = exp(log(V)).

16.10.3 INTERPOLATION OF SCALED RIGID BODY MOTIONS

We now revisit the interpolation examples from Sections 10.7.1 and 13.10.4, this time adding the scalability. We interpolate from one random TRS versor to the next. The random versors are created as follows:

```
void initRandomDest() {
    // get two random translators, a random rotor and a random scalor:
    normalizedTranslator T1 =  exp(_freeVector(randomBlade(2, 3.0f)));
    normalizedTranslator T2 =  exp(_freeVector(randomBlade(2, 3.0f)));
    rotor R = exp(_bivectorE3GA(randomBlade(2, 100.0f)));
    mv::Float s1 = (mv::Float)(1 + rand()) / (mv::Float)(RAND_MAX/2);
    scalor S1 = exp(_noni_t(0.5f * log(s1) * noni));
```

```
// return a random TRS versor:
g_destVersor = _TRSversor(T1 * S1 * T2 * R * inverse(T2));
}
```

Using the `log()` function from the previous example, the interpolation code requires little change compared to previous version:

```
TRSversor interpolateTRSversor(const TRSversor &src, const
  TRSversor &dst, mv::Float alpha) {
  // return src * exp(alpha * log(inverse(src) * dst));
  return _TRSversor(src * exp(_TRSversor(alpha *
    log(_TRSversor(inverse(src) * dst)))));
}
```

16.10.4 THE SEASHELL

This example code reproduces Figure 16.3. You can drag the outermost circle and sphere around to produce variations of the figure. A screenshot is shown in Figure 16.16. The code to draw the shell is

```
// Create versor that generates the sea shell:
TRSversor V = _TRSversor((1.0f - 0.25f * e3ni) *
  exp(_bivectorE3GA((e1^e2) * 0.4f)) *
  exp(_noni_t(-0.05f * noni)));
// Take 1/5st of the versor:
V = exp(0.2f * log(V)); // take only 1/5st of the versor

// precompute inverse of the versor:
TRSversor Vi = _TRSversor(inverse(V));

// get the circle:
circle C = g_circle;

// draw the circles:
const int NB_ITER = 200;
for (int i = 0; i < NB_ITER; i++) {
  draw(C);
  // update circle such that we draw a 'trail' of circles
  C = V * C * Vi;
}

// get the sphere:
sphere S = g_sphere;

// draw spheres:
for (int i = 0; i < NB_ITER; i++) {
  draw(S);
  // update sphere such that we draw a 'trail' of spheres
  S = V * S * Vi;
}
```

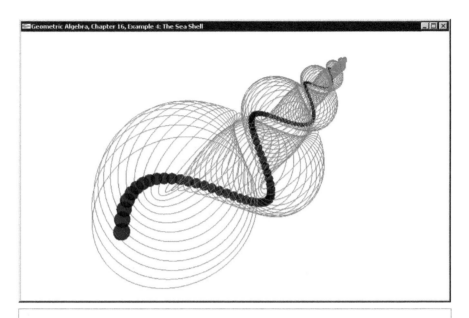

Figure 16.16: Screenshot of Example 4.

17 OPERATIONAL MODELS FOR GEOMETRIES

Now that we have the conformal model of Euclidean geometry and have seen its effectiveness, it is useful to take a step back and look in a more abstract manner at what we have actually done and why it works so well. That will provide a tentative glimpse of future developments in this way of encoding geometries.

17.1 ALGEBRAS FOR GEOMETRIES

A geometry (affine, Euclidean, conformal, projective, or any kind) is characterized by certain operators that act on the objects of the geometry. These objects can range from geometrical entities such as a triangle to properties such as length, so they may have various dimensionalities (which we call grades). The operators in the geometry change these objects in a *covariant* manner, preserving their structure in the following sense. The transformation $V[A \circ B]$ of an object constructed from A and B as $A \circ B$ (using some constructor product $\circ$) should transform to $V[A] \circ V[B]$:

$$\text{covariance: } V[A \circ B] = V[A] \circ V[B]. \tag{17.1}$$

These essential operators defining the geometry are sometimes called *symmetries*.

A convenient algebraic system encodes the operators so deeply into its framework that such covariant identities hold trivially. Conversely, only elements that transform in this manner deserve to be considered as objects in the geometry. Other elements fall apart under the transformations, and thus have no permanence of their defining properties.

Geometric algebra offers a method to produce such an automatically covariant representation system. Two features are essential for this:

- The two-sided versor product preserves the geometric product structure

$$V(XY)V^{-1} = (VXV^{-1})(VYV^{-1})^{-1}.$$

  Since the geometric product encodes the metric, a versor represents an orthogonal transformation.

- All geometrical constructions can be expressed in terms of the geometric product, both for objects and operators. (This can be done as linear combinations, or by grade selection; we have used both.)

In order to use this structural capability for a given geometry, we need to find a good representational space of which we then use the geometric algebra. In this representational space, the *symmetries should become isometries*. Isometries are distance-preserving transformations; they are the orthogonal transformations in the representational space that can be represented by versors in the corresponding geometric algebra.

We can call such a model of geometry an *operational model*, since it is fully designed around the operational symmetry that defines the geometry. An operational model is automatically structure-preserving, so the quest for a good representation of a geometry amounts to setting up an appropriate representational space with the proper metric to represent the symmetry invariants of the geometry.

We have seen some examples of this process in this book, and some more are known.

- *Algebra of Directions*. In the vector space model of Chapter 10, we considered the geometry of oriented and weighted directions. The basic operations in this geometry are the rotations, which can transform between directions. The metric of the directional space is Euclidean, so that different directions can be compared in weight (interpretable as length, velocity, area, volume, etc.). The versors we found are reflections and rotors (the latter are *n*-dimensional quaternions).

- *Euclidean Geometry*. Euclidean symmetries preserve Euclidean distances of points and the point at infinity. The operational model takes this to define the metric of the representational space. As we have seen, an isometric representation of the Euclidean geometry is reached by embedding in a space of two more dimensions and a Minkowski metric. One of the dimensions is used to encode the point at infinity. The other is used to represent the weight of elements, a somewhat unexpected aspect of geometry necessary for linearity of the basic operations in their algebraic representation.

- *Conformal Geometry*. In conformal geometry in general, the invariance of some set representing the points at infinity is important. In Euclidean geometry, this is a point; in hyperbolic geometry, a real unit sphere; and in spherical geometry, an imaginary unit sphere. The representational space that (dually) represents these sets as vectors is the conformal model. Its distance measures are related to the inner product through its translation versor.

- **Image Geometry**. Koenderink [35] has recently proposed an image algebra for the geometrical symmetries of images. These involve not only the spatial geometry of the image plane, but also the transformations on the value domain, and the interaction between the two. This induces a combined model with a mixed geometry, for 2-D gray value images the representation space is $\mathbb{R}^{4,2}$, stratified as a conformal model $\mathbb{R}^{2+1,1}$ for the Euclidean geometry in the base plane and a conformal model $\mathbb{R}^{1,1}$ for the scalar value dimension. The versors of this model give a basis for developing spatial smoothing operators.

- **Projective Geometry**. Unfortunately, there is not yet an operational model for projective geometry. That would have projective transformations as versors (rather than as linear transformations, as in the homogeneous coordinate approach). The metric of the representation space should probably be based on the cross ratio. Its blades would naturally represent the conic sections. Initial attempts [15, 48] do not quite have this structure, but we hope an operational projective model will be developed soon.

- **Contact Geometry**. There are more special geometries that might interest us in computer science. For instance, collision detection requires an efficient representation of contact. In classical literature, symplectic geometries have been developed for this based on canonical transformations (or contact transformations) [15]. It would be very interesting to compute with those by means of their own operational models, hopefully enabling more efficient treatment of problems like collision detection and path planning.

- **And More**. The more we delve into the mathematical literature of the 19th century, the more geometries we find. They all make some practical sense, but only projective geometry (and its subgeometries such as affine and Euclidean geometry) appears to have become part of mainstream knowledge. With the common representational framework of geometric algebra, we find that this obscure literature has become quite readable. As we begin to read it, we find it disconcerting that many aspects of the conformal model of Euclidean geometry pop up regularly (for instance, in [9, 10]). We could have had this all along, and it makes us wonder what else is already out there ...

We hope to have convinced you in Part II that the structural properties of an operational model are nice to have. Not only do they permit universal constructions and operators, but they also make available the quantitative techniques of interpolation and estimation. Linear techniques applied at the right level of the model (for instance, on the bivectors that are the logarithms of the rotors) provide more powerful results than the same techniques applied to the classical vectors or the matrix representations acting upon them.

It remains to show that such models are not only structurally desirable, but also efficient in computation. We do this in Part III.

PART III
IMPLEMENTING GEOMETRIC ALGEBRA

18 IMPLEMENTATION ISSUES

In the first two parts of this book, we have given an abstract description of geometric algebra that is mostly free of coordinates and other low-level implementation details. Even though we have provided many programming examples, this may still have left you with a somewhat unreal feeling about geometric algebra, and an impression that any implementation of geometric algebra would be computationally prohibitive in a practical application. To address these concerns, Part III gives details on how to create an efficient numerical implementation of geometric algebra.

We describe the implementation of all products, operations, and models that were used in the preceding parts. The description is from the viewpoint of representing multivectors as a weighted sum of basis blades, an approach that was already hinted at in Section 2.9.4. This is by far the most common way of implementing geometric algebra, although other approaches are also possible. We briefly mention alternatives in Section 18.3.

Our ultimate goal is efficient numerical implementation, not symbolic computer algebra. An efficient geometric algebra implementation uses symbolic manipulations only to bootstrap the implementation, and not during actual run-time computations. We have tried to keep the implementation description independent of any particular implementation, although Chapter 22 is basically a high-level description of how `Gaigen 2` works. `Gaigen 2` is the implementation behind the GA sandbox source code package that you have used in the programming examples and exercises.

18.1 THE LEVELS OF GEOMETRIC ALGEBRA IMPLEMENTATION

All multivectors can be decomposed as a sum of basis blades. An example of a basis for a 3-D geometric algebra is

$$\{ \underbrace{1}_{scalars}, \underbrace{\mathbf{e}_1, \mathbf{e}_2, \mathbf{e}_3}_{vector\ space}, \underbrace{\mathbf{e}_1 \wedge \mathbf{e}_2,\ \mathbf{e}_2 \wedge \mathbf{e}_3,\ \mathbf{e}_3 \wedge \mathbf{e}_1}_{bivector\ space}, \underbrace{\mathbf{e}_1 \wedge \mathbf{e}_2 \wedge \mathbf{e}_3}_{trivector\ space} \}.$$

The number of k-blades in a geometric algebra over an n-dimensional space is $\binom{n}{k}$, so there are a total of

$$\sum_{k=0}^{n} \binom{n}{k} = 2^n$$

basis elements required to span the entire geometric algebra. On such a basis we can represent any multivector $\mathbf{A}$ as a column vector $[\![\mathbf{A}]\!]$ with 2^n elements, containing the coefficients of $\mathbf{A}$ on the basis. $\mathbf{A}$ can be retrieved by a matrix multiply of a symbolic row vector containing the basis elements. For the 3-D example above, this would be

$$\mathbf{A} = [\![1, \mathbf{e}_1, \mathbf{e}_2, \mathbf{e}_3, \mathbf{e}_1 \wedge \mathbf{e}_2, \mathbf{e}_2 \wedge \mathbf{e}_3, \mathbf{e}_1 \wedge \mathbf{e}_3, \mathbf{e}_1 \wedge \mathbf{e}_2 \wedge \mathbf{e}_3]\!] \, [\![\mathbf{A}]\!].$$

A k-blade only contains elements of grade k, and therefore has many zero entries in its coefficient vector $[\![\mathbf{A}]\!]$, making the representation by 2^n elements rather wasteful. This effect becomes stronger in higher dimensions.

We should therefore explore the idea to make a more sparse representation of the elements of geometric algebra. It can still be based in this sum of basis blades principle, just executed with more sensitivity to the essential structure of geometric algebra.

With the sum of basis blades implementation as our core approach, a geometric algebra implementation naturally splits into four levels:

1. Selecting the basis blades and implementing the basic operations on them;
2. Implementing linear operations for multivectors;
3. Implementing nonlinear operations on multivectors;
4. The application level.

The following chapters first describe the first three levels, then continue to provide additional detail on efficiency and implementation, and finally talk about the fourth level, as follows:

- ***Implementing Basic Operations for Basis Blades (Chapter 19).*** Clearly, it pays to consider the lowest implementation level in detail. We introduce a convenient representation of basis blades and show how the elementary operations have a satisfyingly Boolean nature when considered on the basis. This will deepen your understanding of them and leads to efficient algorithms for the most basic computations.

- ***The Linear Middle Level (Chapter 20)***. We then use those basic capabilities to establish the middle level. Many operations of geometric algebra are linear and distributive. This property makes their implementation quite simple, given that we can already compute them for the basis blades. We consider implementation of this level through matrices (most directly exploiting the linear nature of the operators) or by looping over lists of basis blades (which more naturally leads to an efficient implementation).

- ***The Nonlinear Level (Chapter 21)***. Geometric algebra contains some important operations that are elementary, but not linear or distributive (examples are inversion, `meet` and `join`, and exponentiation). As a consequence, the approach used in the middle-level implementation cannot be used here. These operations are implemented using specialized algorithms, which are actually largely independent of the middle-level implementation. Only for efficiency reasons do we sometimes need direct access to the multivector representation (e.g., to find the largest coordinate in a numerically stable factorization algorithm).

- ***Our Reference Implementation (Online)***. To better illustrate the implementation ideas, we have written an accompanying reference implementation in Java based on the list of basis blades approach. It is available at *http://www.geometricalgebra.net*. This reference implementation was written for educational purposes, and it is not the same as the GA sandbox source code package. The difference is that in our reference implementation we have favored simplicity and readability over efficiency, so we do not recommend using it for computationally intensive applications.

- ***Efficient Implementation (Chapter 22)***. We show how to specialize the implementation according to the structure of geometric algebra to obtain high run-time efficiency. We also present benchmarks that illustrate that the performance of geometric algebra can be close to traditional (linear-algebra-based) geometry implementations, despite the much higher dimensionality of 2^n for its internal algebra. It describes techniques applied in `Gaigen 2`, which is the efficient software behind the GA sandbox source code package, and what we use in our applications.

- ***The Application Level (Chapter 23)***. You could consider the actual use of geometric algebra in your own application the fourth implementation level. In Chapter 23 we give an example of such practical use, in the form of the description of a ray tracer that was implemented using the conformal model. We describe the equations of the ray tracer in actual code and highlight decisions such as picking the right conformal primitive to represent a particular concept. The benchmarks of Chapter 22 are based on this ray tracer.

18.2 WHO SHOULD READ WHAT

The implementation description is quite detailed and not everyone will want to read all of it. We envision three types of audiences:

- If you are new to geometric algebra, the first two parts of the book may leave you wondering how this is ever going to work in an actual implementation (with those weird metrics, as a graded algebra, with so many diverse products, etc.). The following chapters should remove the fuzziness and magic. First, you will want to read Chapter 20 on linear operations. This should give you the comfortable feeling that geometric algebra is fully consistent, since it has the structure of the linear algebra of selected matrices. After that, you may want to read the first part of Chapter 19 on basis blades as it provides another viewpoint for understanding the basic products. Then give a cursory look to the rest of the chapters, but be sure to read the benchmarks that compare geometric algebra to traditional methods in Section 22.5.

- If you are considering using geometric algebra in one of your programs, you might want some background information that helps you pick a particular implementation, especially the appropriate representation of the elements of geometry. In that case we recommend skipping Chapter 19 on basis blades, but reading all of the linear Chapter 20. Then give a cursory look at the nonlinear algorithms and efficiency (Chapter 21 and 22). Finally, read the ray-tracing Chapter 23 in detail, as it will show you actual geometric algebra code that might be similar to what you're going to write yourself, regardless of what implementation you pick.

- If you are a hard-core coder and want to write your own implementation, or simply want to know all the details, we suggest you read all chapters in full detail and the proper order.

18.3 ALTERNATIVE IMPLEMENTATION APPROACHES

In the coming chapters, we will concentrate on one specific implementation approach, where multivectors are represented as a sum of basis blades. It is what we found worked best for low-dimensional geometric algebras useful to computer science. It is also the most commonly used approach by others. But before we explore the detailed consequences, we use the remainder of this chapter to mention some different approaches, just to widen your field of view. Some of these methods affect only the middle (linear) implementation level, while others are so radically different that they affect every level.

18.3.1 ISOMORPHIC MATRIX ALGEBRAS

The linearity and associativity of geometric algebra suggests representing the whole algebra with its geometric product by a single matrix algebra: all elements become represented as $2^n \times 2^n$ matrices, and the geometric product is represented as the matrix product. This is structurally pretty and quite well known among mathematicians as a desirable representation technique, but it is computationally very expensive. It is in fact a representation of Clifford algebra rather than geometric algebra (see Section 7.7.2), and as such rather wasteful: in a true geometric usage of properly constructed elements such as blades and versors, the corresponding matrices tend to be sparse, but this implementation does not make use of that property. Also, it cannot implement the contraction and outer product as easily as the geometric product.

We will use this idea briefly to implement inversion of general multivectors in Section 21.2.

18.3.2 IRREDUCIBLE MATRIX IMPLEMENTATIONS

The $2^n \times 2^n$ matrices do not form the smallest matrix representation of the algebra, if we follow common practice in mathematics and allow the field of the linear mappings they represent to include not only $\mathbb{R}$ (the real numbers), but also $\mathbb{C}$ (the complex numbers) and $\mathbb{H}$ (the quaternions).

Such linear matrix representations have long been known for Clifford algebras of spaces $\mathbb{R}^{p,q}$ with arbitrary signature (p, q) (which means $p + q$ spatial dimensions, of which p basis vectors have a positive square, and q have a negative square; see Appendix A). Again, each element of the algebra is represented as a matrix, and the matrix product of two such representatives is precisely the representation of the element that is their geometric product. We repeat part of the table of such representations in Table 18.1 (see [49]), and offer some structural exercises at the end of this chapter to familiarize yourself with them.

The structural advantage of such representations is that the geometric product becomes a simple matrix multiply (although the elements of the matrix may be reals, complex numbers, or quaternions). However, this is mostly a mathematical curiosity that will not

Table 18.1:   Matrix representations of Clifford algebras of signatures (p,q). Notation: $\mathbb{R}(n)$ are $n \times n$ real matrices, $\mathbb{C}(n)$ are $n \times n$ complex-valued matrices, and $\mathbb{H}(n)$ are $n \times n$ quaternion-valued matrices. The notation $^2\mathbb{R}(n)$ is used for ordered pairs of $n \times n$ real matrices (which you may think of as a block-diagonal $2n \times 2n$ matrix containing two real $n \times n$ real matrices on its diagonal and zeros elsewhere), and similarly for the other number systems.

	$q = 0$	$q = 1$	$q = 2$	$q = 3$
$p = 0$	$\mathbb{R}(1)$	$\mathbb{C}(1)$	$\mathbb{H}(1)$	$^2\mathbb{H}(1)$
$p = 1$	$^2\mathbb{R}(1)$	$\mathbb{R}(2)$	$\mathbb{C}(2)$	$\mathbb{H}(2)$
$p = 2$	$\mathbb{R}(2)$	$^2\mathbb{R}(2)$	$\mathbb{R}(4)$	$\mathbb{C}(4)$
$p = 3$	$\mathbb{C}(2)$	$\mathbb{R}(4)$	$^2\mathbb{R}(4)$	$\mathbb{R}(8)$
$p = 4$	$\mathbb{H}(2)$	$\mathbb{C}(4)$	$\mathbb{R}(8)$	$\mathbb{C}(8)$

save processing time in practice, since approximately the same number of operations have to be performed as in the basis-of-blades representation that is our reference implementation. (Only if special hardware were present to handle complex number and quaternion multiplications more efficiently than arbitrary multiplications, could these methods become more efficient than the multiplication of real valued matrices.)

Moreover, the matrix representations have the disturbing disadvantage that they only work for the geometric product. As long as they are used for the Clifford algebras for which they were developed, this is not a problem—but it makes this representation cumbersome for geometric algebra in general, where derived products such as the outer product and contractions are also important. True, a similar representation of outer products can be easily established, but one would have to switch between representations to perform one product or the other, as both may be needed in a single application. And unfortunately, since the contraction inner products are not associative, they are not isomorphic to matrix algebras, so they cannot be implemented in the same framework. (Actually, this nonassociativity of the contraction will require special measures in any implementation scheme, including ours.)

18.3.3 FACTORED REPRESENTATIONS

The mostly multiplicative structure of geometric algebra (with versors as geometric products of vectors, and blades as outer products of vectors) has recently been explored by the authors to produce factorized representations. This seems a viable implementation method for high-dimensional algebras, but more research is needed.

A k-blade can be stored as a list of k vectors. The outer product of these vectors is the value of the blade. Likewise, a versor can be stored as a list of vectors whose geometric product is the value of the versor. The storage requirement of blades and versors becomes $O(n^2)$ compared to $O(2^n)$ for the basis-of-blades method. Multivectors that are not blades or versors can only be represented as a sum of multiple blades or versors, and thus require more storage.

This multiplicative representation is radically different from the usual additive representation. As a consequence, the implementation of products and operations is also very different. What may be a difficult problem in one implementation approach can be trivial in the other, and vice versa. For example, addition is simple in the basis of blades approach, but one of the hardest problems in a factored representation. The `meet` and `join` are trivial with factored representation (given that you already have a linear algebra library like LAPACK), while they lead to a rather involved algorithm in the sum-of-basis-blades approach.

We do not discuss this approach further because it is best suited for high-dimensional ($n > 10$) geometric algebras, which are of less interest in this book. The details will appear in Daniel Fontijne's Ph.D. thesis.

18.4 STRUCTURAL EXERCISES

1. In a 2-D Euclidean geometric algebra on an orthonormal basis $\{\mathbf{e}_1, \mathbf{e}_2\}$, the general element X can be written as:

$$X = x_0 + x_1\,\mathbf{e}_1 + x_2\,\mathbf{e}_2 + x_{12}\,\mathbf{e}_{12}.$$

According to Table 18.1, this algebra should be representable as the matrix algebra $\mathbb{R}(2)$ (i.e., we should be able to find a 2×2 matrix representing the general element so that the geometric product of elements is represented as the matrix product). Show that the following works:

$$[\![\mathbf{x}]\!] = \begin{bmatrix} x_0 + x_2 & x_1 - x_{12} \\ x_1 + x_{12} & x_0 - x_2 \end{bmatrix}$$

This is not unique; some permutations of the same principles work as well, but not all. Why must the scalar part always be on the diagonal?

2. For a 3-D Euclidean vector space, show that you can generate a matrix algebra from the following representation of a vector $\mathbf{x} = x\,\mathbf{e}_1 + y\,\mathbf{e}_2 + z\,\mathbf{e}_3$:

$$[\![\mathbf{x}]\!] = \begin{bmatrix} z & 0 & x & -y \\ 0 & z & y & x \\ x & y & -z & 0 \\ -y & x & 0 & -z \end{bmatrix}$$

Compute the representation of a general element of this algebra. (This representation is not in Table 18.1 because it is not the smallest matrix algebra; see the following exercise).

3. For a 3-D Euclidean vector space, the matrix representation $\mathbb{C}(2)$ from Table 18.1 may be generated by:

$$[\![\mathbf{x}]\!] = \begin{bmatrix} z & x + iy \\ x - iy & -z \end{bmatrix},$$

where i is the complex imaginary (so $i^2 = -1$). Verify this and compute the representation of a general element of this algebra.

19 BASIS BLADES AND OPERATIONS

All geometric algebra implementations that represent multivectors as a weighted sum of basis blades will, at some point, have to compute products of basis blades. Among the various products, the capability to compute the geometric product (also for non-Euclidean metrics) is clearly the minimum requirement, as the outer product and various flavors of the inner products can be derived from it using the (anti-)symmetry or grade selection techniques of Chapter 6.

This chapter describes how to implement this geometric product of basis blades through a convenient representation of basis blades. The algorithms described are quite simple, yet intricate enough that we need (pseudo)code to describe them precisely. Instead of introducing our own pseudocode language, we opted to simply use Java. The code in this chapter therefore corresponds well to our reference implementation at our web site *http://www.geometricalgebra.net*, although it was sometimes slightly polished for presentation.

Conceptually, our basis blades representation uses a list of booleans and applies Boolean logic to the elements of that list. But in practice, the list of booleans is implemented using bitmaps and the bitwise boolean operators. To correspond more directly to the actual implementation, that is how we present the representation and its operators in this chapter.

19.1 REPRESENTING UNIT BASIS BLADES WITH BITMAPS

When we want to represent the geometric algebra of an n-dimensional space, we start with a set of n independent basis vectors $\{\mathbf{e}_i\}_{i=1}^{n}$. They need not be orthonormal or even orthogonal; the geometric product of the space will essentially define their relevant geometrical relationships. A *basis blade* is a nonzero outer product of a number of these basis vectors (or a unit scalar). Hence, a basis blade either contains a specific basis vector or not. Thus, each basis blade in an algebra can be represented by a list of boolean values, where each boolean indicates the presence or absence of a basis vector. The list of booleans is naturally implemented using a bitmap, which is why we call this the bitmap representation of basis blades.

The bits in the bitmap are assigned in the logical order: bit 0 stands for $\mathbf{e}_1$, bit 1 stands for $\mathbf{e}_2$, bit 2 stands for $\mathbf{e}_3$, and so on. Table 19.1 shows how this works out. The unit scalar does not contain any basis vectors, so it is represented by 0_b (in this chapter, the postfix, subscript b indicates a binary number). A blade such as $\mathbf{e}_1 \wedge \mathbf{e}_3$ contains $\mathbf{e}_1$ and $\mathbf{e}_3$, so it is represented by $001_b + 100_b = 101_b$.

Visually, the bits are stored in reversed order relative to how we would write them as basis blades. In a basis blade, $\mathbf{e}_1$ is the left-most basis vector, while in bitmap representation it is the right-most bit. While this is slightly inconvenient for reasoning about them, it is more consistent for implementation. If an extra dimension is required, the next most significant bit in the bitmap is used and previous results are automatically absorbed.

Table 19.1: The bitmap representation of basis blades. Note how the representation scales up with the dimension of the algebra.

Basis Blade	Bitmap Representation
1	0_b
$\mathbf{e}_1$	1_b
$\mathbf{e}_2$	10_b
$\mathbf{e}_1 \wedge \mathbf{e}_2$	11_b
$\mathbf{e}_3$	100_b
$\mathbf{e}_1 \wedge \mathbf{e}_3$	101_b
$\mathbf{e}_2 \wedge \mathbf{e}_3$	110_b
$\mathbf{e}_1 \wedge \mathbf{e}_2 \wedge \mathbf{e}_3$	111_b
$\mathbf{e}_4$	1000_b
...	...

Symbol	Operation
&	Bitwise boolean "and"
∧	Bitwise boolean "exclusive or"
>>>	Unsigned bitwise "shift right"

Table 19.2: Bitwise boolean operators used in Java code examples.

In our reference implementation, we use 32-bit integers as our bitmaps. This allows us to represent all basis blades of the geometric algebra of a 32-D space. This should be more than enough for practical usage in computer science applications. Further, this basis-of-blades method of implementing geometric algebra is only practical up to around 10-D spaces, since the number of basis blades required to represent an arbitrary multivector grows quickly, as 2^n. The products have to process combinations of all basis blades from both arguments, so you typically will run out of processing time before you run out of memory. But should you ever need more than 32 dimensions, extending the range of the bitmaps is straightforward.

The reference implementation stores basis blades in a class named `BasisBlade`. The class has only two member variables, `bitmap` (an integer) and `scale` (a floating point number). `scale` allows the basis blade to have an arbitrary sign or scale; this is what we called the *weight* in Part I. The name `ScaledBasisBlade` might have been a more accurate class description, but we found that a bit too verbose.

In Table 19.2, we show the symbols used for the bitwise boolean operations. These symbols are identical to the ones Java uses. Note that in code listings, ∧ may appear as ^.

19.2 THE OUTER PRODUCT OF BASIS BLADES

We now show how to compute the products on the basis blades using the bitmap representation. To compute the outer product of two basis blades, we first check whether they are dependent (i.e., whether the two blades have a common basis vector factor). If they do, they have a bit in common, so dependence of blades is checked simply with a binary *and*. If the blades are dependent, then the outcome of the outer product is 0 and the algorithm described below does not need to be executed.

If the blades are independent, computing their outer product is done (up to a scale factor) by taking the bitwise *exclusive or* of the bitmaps. For example,

$$(\mathbf{e}_2 \wedge \mathbf{e}_3) \wedge \mathbf{e}_1 = \mathbf{e}_1 \wedge \mathbf{e}_2 \wedge \mathbf{e}_3$$

is equivalent to

$$110_b \wedge 001_b = 111_b.$$

The hardest part in implementing the outer product is computing the correct sign for the result. The bitmap representation assumes that the basis vectors are in canonical order (e.g., using just the bitmap you cannot represent $e_3 \wedge e_1$, only $e_1 \wedge e_3$). We should employ the `scale` member variable to represent $e_3 \wedge e_1$ as $-1.0 * (e_1 \wedge e_3)$.

When you compute the sign of the outer product result by hand, you count how many basis vectors have to swap positions to get the result into the canonical order. Each swap of position causes a flip of sign. In our example, all that is required is that e_3 and e_1 swap positions, so we get a negative sign.

To compute this sign algorithmically, we have to compute, for each 1-bit in the first operand, the number of less significant 1-bits in the second operand. This is implemented somewhat like a convolution, in that the bitmaps are slid over each other bit by bit. For each position, the number of matching 1-bits counted. Figure 19.1 shows the implementation of this idea (the `bitCount()` function counts the number of nonzero bits in a word). The implementation of the outer product itself is in Figure 19.2; this function actually also implements the geometric product, as described in the next section.

```
// Arguments 'a' and 'b' are both bitmaps representing
// basis blades.
double canonicalReorderingSign(int a, int b) {
    // Count the number of basis vector swaps required to
    // get 'a' and 'b' into canonical order.
    a = a >>> 1;
    int sum = 0;
    while (a != 0) {
        // the function bitCount() counts the number of
        // 1-bits in the argument
        sum = sum + subspace.util.Bits.bitCount(a & b);
        a = a >>> 1;
    }

    // even number of swaps -> return 1
    // odd number of swaps -> return -1
    return ((sum & 1) == 0) ? 1.0 : -1.0;
}
```

Figure 19.1: This function computes the sign change due to the reordering of two basis blades into canonical order.

```
BasisBlade gp_op(BasisBlade a, BasisBlade b, boolean outer) {

  // if outer product: check for independence
  if (outer && ((a.bitmap & b.bitmap) != 0))
    return new BasisBlade(0.0);

  // compute the bitmap:
  int bitmap = a.bitmap ^ b.bitmap;

  // compute the sign change due to reordering:
  double sign = canonicalReorderingSign(a.bitmap, b.bitmap);

  // return result:
  return new BasisBlade(bitmap, sign * a.scale * b.scale);
}
```

Figure 19.2: This function will compute either the geometric product (in Euclidean metric only) or the outer product, depending on the value of argument `outer`.

19.3 THE GEOMETRIC PRODUCT OF BASIS BLADES IN AN ORTHOGONAL METRIC

The implementation of the geometric product is similar to that of the outer product as long as we stay in an *orthogonal metric* with respect to the orthonormal basis $\{\mathbf{e}_i\}_{i=1}^{n}$. Such a metric has $\mathbf{e}_i \cdot \mathbf{e}_j = 0$ for $i \neq j$, and $\mathbf{e}_i \cdot \mathbf{e}_i = m_i$; therefore, its *metric matrix* is diagonal:

$$\mathbf{e}_i \cdot \mathbf{e}_j = m_i\, \delta_j^i, \tag{19.1}$$

where δ_j^i is the *Kronecker delta function*, returning 1 when i equals j, and zero otherwise. A particular example is the Euclidean metric, for which $m_i = 1$ for all i, so that the metric matrix is the identity matrix. In general, the m_i can have any real value, but -1, 0, and 1 will be the most prevalent in many applications.

There are now two cases:

- For blades consisting of *different* orthogonal factors, we have the usual equivalence of the outer product and the geometric product:

$$\mathbf{e}_1 \wedge \mathbf{e}_2 \wedge \cdots \wedge \mathbf{e}_k = \mathbf{e}_1\, \mathbf{e}_2\, \cdots\, \mathbf{e}_k.$$

- However, when two factors are in common between the multiplicands, the outer product produces a zero result, but the geometric product does not. Instead, it annihilates the dependent-basis vectors, effectively replacing them by metric factors. For example,

$$(e_1 \wedge e_2)(e_2 \wedge e_3) = m_2\, e_1 \wedge e_3.$$

In the bitmap representation, these two cases are easily merged. Both results may be computed as the bitwise *exclusive or* operation (i.e., $011_b \wedge 110_b = 101_b$). In this sense, the geometric product acts as a "spatial exclusive *or*" on basic blades.

As for the outer product, the result of the geometric product "exclusive *or*" should be given the correct sign to distinguish between the outcomes $e_1 \wedge e_3$ and $e_3 \wedge e_1$. We therefore also have to perform reordering techniques to establish the sign of the result on the standard basis. Those sign changes are identical to those for the outer product: we count the number of basis vector swaps required to get the result into canonical order and use this to determine the sign.

In a Euclidean metric, all diagonal metric factors are 1, so this bit pattern and sign computation is all there is to the geometric product. The function that computes the geometric product in this simple Euclidean case is therefore virtually the same as the function for computing the outer product (see Figure 19.2), the only difference being the dependence check required for the outer product. That close and convenient similarity is due to the use of the orthonormal basis in the representation.

When the metric is not Euclidean but still diagonal, we need to incorporate the metric coefficients of the annihilated basis vectors into the scale of the resulting blade. Which basis vectors are annihilated is determined with a bitwise *and*. We do not show the resulting code, which simply invokes the Euclidean code with this extension; you can consult the reference implementation on our web site for the details.

19.4 THE GEOMETRIC PRODUCT OF BASIS BLADES IN NONORTHOGONAL METRICS

In practical geometric algebra we naturally encounter nonorthonormal bases. For example, in the conformal model we may want to represent o and ∞ directly as basis vectors. This leads to a nondiagonal multiplication table (and therefore nondiagonal metric matrix):

	o	e_1	e_2	e_3	∞
o	0	0	0	0	−1
e_1	0	1	0	0	0
e_2	0	0	1	0	0
e_3	0	0	0	1	0
∞	−1	0	0	0	0

The orthogonal metric method of the previous section does not seem to apply to this case. Yet the methods employed for the orthogonal metrics are more general than they seem. By the spectral theorem from linear algebra (a matrix is orthogonally diagonalizable if and only if it is symmetric), an arbitrary metric matrix can always be brought into diagonal form by an appropriate coordinate transformation to an orthonormal basis.

Therefore we can compute geometric products in the conformal model by temporarily switching to a new basis that is orthonormal. Such a basis is computed by finding the eigenvalue decomposition of the metric matrix. For our conformal model example, such a basis would be

$$\mathbf{f}_1 = \mathbf{e}_1$$
$$\mathbf{f}_2 = \mathbf{e}_2$$
$$\mathbf{f}_3 = \mathbf{e}_3$$
$$\mathbf{f}_4 = \tfrac{1}{2}\sqrt{2}(o - \infty)$$
$$\mathbf{f}_5 = \tfrac{1}{2}\sqrt{2}(o + \infty).$$

The $\mathbf{f}_i$ are all basis vectors with $\mathbf{f}_i \cdot \mathbf{f}_i = 1$, except for $\mathbf{f}_5 \cdot \mathbf{f}_5 = -1$. (Alternatively, we could use $\mathbf{f}_4 = e = o - \infty/2$ and $\mathbf{f}_5 = \bar{e} = o + \infty/2$, as in (13.6).) The new basis is therefore one of the orthogonal metrics of the previous section, and we can revert to the previous methods. So, to compute the geometric product in an arbitrary metric:

1. Compute the eigenvectors and eigenvalues of the metric matrix. This has to be done once, when the object that represents the metric is initialized.

2. Apply a change-of-basis to the input such that it is represented with respect to the eigenbasis.

3. Compute the geometric product on this new orthogonal basis (the eigenvalues specify the metric).

4. Apply another change of basis to the result, to get back to the original basis.

In practical implementations, the computation of the geometric product of the basis blades is done beforehand, so this way of implementing it does not slow down the application at run-time. The code for this algorithm is rather involved, so we do not show the implementation here. If you are interested, you can find it in the `subspace.basis` package at *http://www.geometricalgebra.net*.

One detail to note is that the result of a geometric product is not always a single basis blade anymore. When switching back and forth between one basis and another, a single basis blade can convert into a sum of multiple basis blades. This is in agreement with what one would expect. An example is easily given: in the conformal model, $o \infty = -1 + o \wedge \infty$.

19.5 THE METRIC PRODUCTS OF BASIS BLADES

We use the term *metric product* collectively to denote the scalar product, left and right contraction, and the inner products used by others and exposed in Appendix B. From the algorithmic viewpoint, they are all similar, for they can all be derived from the geometric product by extracting the appropriate grade parts of the result (the same could be done for the outer product, if you wish).

Let $\mathbf{A}$ and $\mathbf{B}$ be two basis blades of grade a and b, respectively. Then the rules for deriving a particular metric product from the geometric product of two basis blades are:

- **Left contraction.** If $a \leq b$, $\mathbf{A} \rfloor \mathbf{B} = \langle \mathbf{A}\,\mathbf{B} \rangle_{b-a}$, otherwise 0.
- **Right contraction.** If $a \geq b$, $\mathbf{A} \lfloor \mathbf{B} = \langle \mathbf{A}\,\mathbf{B} \rangle_{a-b}$, otherwise 0.
- **Dot product.** If $a \leq b$, then the dot product is equal to the left contraction. Otherwise, it is equal to the right contraction.
- **Hestenes metric product.** Like the dot product, but 0 when either $\mathbf{A}$ or $\mathbf{B}$ is a scalar.
- **Scalar product.** $\mathbf{A} * \mathbf{B} = \langle \mathbf{A}\,\mathbf{B} \rangle_0$.

The grade of a basis blade is determined simply by counting the number of 1-bits in the bitmap. The reference implementation can therefore implement the inner products exactly as defined. For example, to compute the left contraction of a grade-1 blade and a grade-3 blade, it extracts the grade $3 - 1 = 2$ part from their geometric product.

If one were to write a dedicated version of the metric product implementation, an optimization would be to check whether one lower-grade basis blade is fully contained in the higher-grade basis blade before performing the actual product. If this condition is not satisfied, the metric product (of whatever flavor) will always be 0. This containment is tested by checking whether all set bits in the lowergrade blade are also set in the higher-grade blade. For nonorthogonal metrics, this check should be done after the change to the eigenbasis.

19.6 COMMUTATOR PRODUCT OF BASIS BLADES

The commutator product of basis blades is easily derived from the geometric product using the following equation from Section 8.2.1: $\mathbf{A} \times \mathbf{B} \equiv \frac{1}{2}(\mathbf{A}\,\mathbf{B} - \mathbf{B}\,\mathbf{A})$.

19.7 GRADE-DEPENDENT SIGNS ON BASIS BLADES

Implementing the reversion, grade involution (Section 2.9.5), and Clifford conjugation (structural exercise 8, Section 2.12.2) is straightforward for basis blades. These grade-dependent sign operators toggle the sign of basis blades according to a specific multiplier based on the grade a of the blade (see Table 19.3).

Table 19.3: Reversion, grade involution, and Clifford conjugate for basis blades.

Operation	Multiplier for Grade a	Pattern
Reversion	$(-1)^{a(a-1)/2}$	$+ + - - + + - -$
Grade involution	$(-1)^a$	$+ - + - + - + -$
Clifford conjugation	$(-1)^{a(a+1)/2}$	$+ - - + + - - +$

As an example, the reverse of $\mathbf{e}_1 \wedge \mathbf{e}_2$ is computed by first determining the grade (which is 2) and then applying the correct multiplier (in this case $(-1)^{2(2+1)/2} = -1$) to the `scale` of the blade.

The last column of the Table 19.3 shows the repetitive pattern over the varying grades that describes the behavior of the each operation: "$-$" stands for multiplication by -1, while "$+$" stands for multiplication by $+1$. It shows, for instance, that the grade involution toggles the sign of all odd-grade parts while it leaves the even grade parts unchanged.

20 THE LINEAR PRODUCTS AND OPERATIONS

The linear products we use in this book are the geometric product, the outer product, the contraction inner products, the scalar product, and the commutator product. They are all linear in their arguments. Examples of unary linear operations that are discussed in this chapter are addition, reversion, grade involution, and grade extraction of multivectors. This chapter presents two ways to implement the linear products and operations of geometric algebra.

Both implementation approaches are based on the linearity and distributivity of these products and operations. The first approach uses linear algebra to encode the multiplying element as a square matrix acting on the multiplied element, which is encoded as a column matrix. We present this approach because the matrix ideas are familiar to many people, because it is convenient, and because it works for general Clifford algebras. However, it does not exploit the sparseness of most elements in geometric algebra, and is not used much in practice.

The second approach is effectively a sparse matrix approach and uses the basis blades idea of the previous chapter, storing multivectors as lists of weighted basis blades and literally distributing the work of computing the products and operations to the level of basis blades. This automatically employs the sparseness of geometric algebra and provides a more natural path towards the optimization of Chapter 22. This is also the approach used in our reference implementation.

20.1 A LINEAR ALGEBRA APPROACH

A multivector from an n-dimensional geometric algebra can be stored as a $2^n \times 1$ column matrix. Each element in the matrix is a coordinate that refers to a specific basis blade. To formalize this, let us define a row matrix $\mathbb{L}$ whose elements are the basis blades of the basis, listed in order. For example, in 3-D:

$$\mathbb{L} = [\![1,\ e_1,\ e_2,\ e_3,\ e_1 \wedge e_2,\ e_2 \wedge e_3,\ e_1 \wedge e_3,\ e_1 \wedge e_2 \wedge e_3]\!]$$

Using $\mathbb{L}$ we can represent a multivector A by a $2^n \times 1$ matrix $[\![A]\!]$ with elements

$$A = \mathbb{L}\,[\![A]\!].$$

We use the following notation to indicate equivalence between the actual geometric algebra operations and their implementation in linear algebra:

$$A \rightleftharpoons [\![A]\!].$$

20.1.1 IMPLEMENTING THE LINEAR OPERATIONS

The elements of geometric algebra form a linear space, and these linear operations are implemented trivially in the matrix approach:

- Addition of elements is performed by adding the matrices:

$$A + B \rightleftharpoons [\![A]\!] + [\![B]\!].$$

  Scalar multiplication is implemented as multiplication of the matrix $[\![A]\!]$ by the scalar:

$$\alpha A \rightleftharpoons \alpha\,[\![A]\!].$$

- The unary linear operations of reversion, grade involution, and Clifford conjugation can also be implemented as matrices. The entries of these matrices need to be set according to the corresponding operations on the basis blades in $\mathbb{L}$.

  For instance, for the reversion we define the (constant) diagonal matrix $[\![R]\!]$, whose entries are defined as

$$[\![R]\!]_{i,i} = (-1)^{\mathrm{grade}(\mathbb{L}_i)(\mathrm{grade}(\mathbb{L}_i)-1)/2}.$$

  Then reversion is implemented as

$$\widetilde{A} \rightleftharpoons [\![R]\!]\,[\![A]\!].$$

- Extracting the kth grade part of a multivector is also a unary linear operator. For each grade we need to construct a diagonal selection matrix $[\![S^k]\!]$, so that the operation can be performed as the matrix operator:

$$\langle A \rangle_k \rightleftharpoons [\![S^k]\!]\,[\![A]\!].$$

The entries of the matrix $[\![S^k]\!]$ must be defined as

$$[\![S^k]\!]_{i,j} = \begin{cases} 1 & \text{if grade}(\mathbb{L}_i) = k \text{ and } i = j \\ 0 & \text{otherwise.} \end{cases},$$

which may be summarized as $[\![S^k]\!]_{i,j} = \delta_i^j\,\delta_{\text{grade}(\mathbb{L}_i)}^k$.

20.1.2 IMPLEMENTING THE LINEAR PRODUCTS

The linear products can all be implemented using matrix multiplication. If we consider the geometric product $A\,B$, the result is linear in B, so A is like a linear operator acting on B. In the $\mathbb{L}$-based representation, that A-operator can be represented by matrix $[\![A^G]\!]$ acting on column matrix $[\![B]\!]$ (the superscript G denotes that A acts on B by the geometric product). That gives

$$A\,B \rightleftharpoons [\![A^G]\!][\![B]\!].$$

Here $[\![A^G]\!]$ is a $2^n \times 2^n$ matrix; we need to construct it so that it corresponds to the geometric product. As we do so, we find that its entries are certain linear combinations of the coefficients of A on the $\mathbb{L}$-basis. There is therefore not a single representation of the geometric product: each element acts through its own matrix. By the same reasoning, each element A also has an associated outer product matrix $[\![A^O]\!]$, which describes the action of the operation $A\wedge$, and a left contraction matrix $[\![A^L]\!]$ for the operation $A\rfloor$, and so on.

Whenever we want to compute a product, we therefore need to construct the corresponding matrix. Let us describe how that is done for the geometric product matrix $[\![A^G]\!]$.

For simplicity, we temporarily assume that the algebra has a diagonal metric matrix (e.g., a Euclidean metric). To devise a rule on how to fill in the entries of $[\![A^G]\!]$, we consider the geometric product of two basis blades weighted by their respective coordinates from $[\![A]\!]$ and $[\![B]\!]$. These are scalars, and they get multiplied by the geometric algebra elements from $\mathbb{L}$—that is where the structure of geometric algebra enters the multiplication. The product of two such elements $\mathbb{L}_k$ and $\mathbb{L}_j$ is a third element $\mathbb{L}_i$, with a scalar $s_i^{k,j}$ determined by the metric (in a Euclidean metric, $s_i^{k,j} = \pm 1$, with the minus sign occurring for instance for basis 2-blades, so these are not simply the m_j of the metric matrix in (19.1)).

$$\mathbb{L}_k\,\mathbb{L}_j = s_i^{k,j}\,\mathbb{L}_i \tag{20.1}$$

Think of the scalars $s_i^{k,j}$ as the "structure coefficients" of the algebra. As their definition shows, they involve the products of basis blades (the components of $\mathbb{L}$), so they can be determined efficiently by the methods of the previous chapter.

We can now specify what the product of the matrices $[\![A]\!]$ and $[\![B]\!]$ should satisfy, coefficient by coefficient:

$$([\![A]\!]_k \mathbb{L}_k)\,([\![B]\!]_j \mathbb{L}_j) = s_i^{k,j}\,[\![A]\!]_k [\![B]\!]_j\,\mathbb{L}_i. \tag{20.2}$$

To achieve the equivalent of this equation through matrix multiplication (i.e., $[\![A^G]\!]$ $[\![B]\!] = [\![C]\!]$), it is clear that $s_i^{k,j}[\![A]\!]_k[\![B]\!]_j$ should end up in row i of $[\![C]\!]$, so $s_i^{k,j}[\![A]\!]_k$ should be on row i of $[\![A^G]\!]$. The fact that $s_i^{k,j}[\![A]\!]_k$ should combine with $[\![B]\!]_j$ means that $s_i^{k,j}[\![A]\!]_k$ should be in column j. In summary,

$$([\![A]\!]_k \mathbb{L}_k)\,([\![B]\!]_j \mathbb{L}_j) = [\![A]\!]_k [\![B]\!]_j\, s_i^{k,j} \mathbb{L}_i \quad \rightarrow \quad [\![A^G]\!]_{i,j} = \sum_k s_i^{k,j}[\![A]\!]_k. \tag{20.3}$$

This final rule does not involve the symbolic basis list $\mathbb{L}$ anymore, yet is based fully on the geometric algebra structure it contributes. By executing this rule for all indices k, j, we obtain, by linearity, a full geometric product matrix $[\![A^G]\!]$. An example of the geometric product matrix $[\![A^G]\!]$ is shown in Figure 20.1 for a 3-D Euclidean metric.

When the metric matrix is nondiagonal, things get slightly more complicated, because the geometric product of two basis blades can result in a sum of basis blades:

$$\mathbb{L}_k \mathbb{L}_j = \sum_i s_i^{k,j}\, \mathbb{L}_i.$$

It is clear that we should propagate this change to the computation of the matrix $[\![A^G]\!]$:

$$([\![A]\!]_k \mathbb{L}_k)\,([\![B]\!]_j \mathbb{L}_j) = [\![A]\!]_k [\![B]\!]_j \sum_i s_i^{k,j} \mathbb{L}_i. \tag{20.4}$$

Thus we have to execute the rule in (20.3) for each $s_i^{k,j}[\![A]\!]_k$.

Summarizing, the following algorithm computes the geometric product matrix for a multivector A:

1. Initialize the matrix $[\![A^G]\!]$ to a $2^n \times 2^n$ null matrix.

2. Loop over all indices k, j to compute the geometric product of all basis blades as in (20.1). How to compute the products of basis blades is described in Chapter 19.

3. Add each $s_i^{k,j}[\![A]\!]_k$ to the correct element of $[\![A^G]\!]$ according to the rule on the right-hand side of (20.3).

$$[\![A^G]\!] = \begin{bmatrix} +A_0 & +A_1 & +A_2 & +A_3 & -A_{12} & -A_{23} & -A_{13} & -A_{123} \\ +A_1 & +A_0 & +A_{12} & +A_{13} & -A_2 & -A_{123} & -A_3 & -A_{23} \\ +A_2 & -A_{12} & +A_0 & +A_{23} & +A_1 & -A_3 & +A_{123} & +A_{13} \\ +A_3 & -A_{13} & -A_{23} & +A_0 & -A_{123} & +A_2 & +A_1 & -A_{12} \\ +A_{12} & -A_2 & +A_1 & +A_{123} & +A_0 & -A_{13} & +A_{23} & +A_3 \\ +A_{23} & +A_{123} & -A_3 & +A_2 & +A_{13} & +A_0 & -A_{12} & +A_1 \\ +A_{13} & -A_3 & -A_{123} & +A_1 & -A_{23} & +A_{12} & +A_0 & -A_2 \\ +A_{123} & +A_{23} & -A_{13} & +A_{12} & +A_3 & +A_1 & -A_2 & +A_0 \end{bmatrix}$$

$$[\![A^O]\!] = \begin{bmatrix} +A_0 & 0 & 0 & 0 & 0 & 0 & 0 & 0 \\ +A_1 & +A_0 & 0 & 0 & 0 & 0 & 0 & 0 \\ +A_2 & 0 & +A_0 & 0 & 0 & 0 & 0 & 0 \\ +A_3 & 0 & 0 & +A_0 & 0 & 0 & 0 & 0 \\ +A_{12} & -A_2 & +A_1 & 0 & +A_0 & 0 & 0 & 0 \\ +A_{23} & 0 & -A_3 & +A_2 & 0 & +A_0 & 0 & 0 \\ +A_{13} & -A_3 & 0 & +A_1 & 0 & 0 & +A_0 & 0 \\ +A_{123} & +A_{23} & -A_{13} & +A_{12} & +A_3 & +A_1 & -A_2 & +A_0 \end{bmatrix}$$

$$[\![A^L]\!] = \begin{bmatrix} +A_0 & +A_1 & +A_2 & +A_3 & -A_{12} & -A_{23} & -A_{13} & -A_{123} \\ 0 & +A_0 & 0 & 0 & -A_2 & 0 & -A_3 & -A_{23} \\ 0 & 0 & +A_0 & 0 & +A_1 & -A_3 & 0 & +A_{13} \\ 0 & 0 & 0 & +A_0 & 0 & +A_2 & +A_1 & -A_{12} \\ 0 & 0 & 0 & 0 & +A_0 & 0 & 0 & +A_3 \\ 0 & 0 & 0 & 0 & 0 & +A_0 & 0 & +A_1 \\ 0 & 0 & 0 & 0 & 0 & 0 & +A_0 & -A_2 \\ 0 & 0 & 0 & 0 & 0 & 0 & 0 & +A_0 \end{bmatrix}$$

Figure 20.1: Symbolic matrices for computing the geometric product ($[\![A^G]\!]$), outer product ($[\![A^O]\!]$) and left contraction ($[\![A^L]\!]$) in a 3-D Euclidean geometric algebra. A notational shorthand is used for readability, i.e., A_0 is the scalar coordinate of $[\![A]\!]$, A_1 is the coordinate from $[\![A]\!]$ that refers to $\mathbf{e}_1$; A_{123} is the $\mathbf{e}_1 \wedge \mathbf{e}_2 \wedge \mathbf{e}_3$-coordinate, etc.

The product matrices need to be computed only once, to bootstrap the implementation. The results can be stored symbolically, leading to matrices such as $[\![A^G]\!]$ in Figure 20.1.

The same algorithm can compute the matrix for any product derived from the geometric product. Figure 20.1 also shows matrices for the outer product ($[\![A^O]\!]$) and the left contraction ($[\![A^L]\!]$). Note how these matrices are mostly identical to the geometric product matrix $[\![A^G]\!]$, but with zeros at specific entries. The reason is of course that these products are merely selections of certain grade parts of the more encompassing geometric product.

After the symbolic matrices have been initialized, computing an actual product $C = A\,B$ is reduced to creating a real matrix $[\![A^G]\!]$ according to the appropriate symbolic matrix and the coordinates of $[\![A]\!]$, and computing $[\![C]\!] = [\![A^G]\!][\![B]\!]$. So computing the products of basis blades as in (20.1) is only required during the initialization step, during which symbolic matrices are computed. This is fine in a general Clifford algebra, in which A may be an arbitrary element of many grades. However, in geometric algebra proper elements tend to be rather sparse on the $\mathbb{L}$ basis, typically being restricted to a single grade for objects (which are represented by blades), or only odd or even grades for operators (which are represented by versors). Then the resulting matrices are sparse, and some kind of optimization in their computation should prevent the needless and costly evaluation of many zero results in their multiplication.

20.2 THE LIST OF BASIS BLADES APPROACH

Instead of representing multivectors as 2^n vectors and using matrices to implement the products, we can represent a multivector by a list of basis blades. That is how our reference implementation works. The linear products and operations are distributive over addition; for example, for the left contraction and the reverse we have

$$(\mathbf{a}_1 + \mathbf{a}_2 + \cdots + \mathbf{a}_n)\rfloor(\mathbf{b}_1 + \mathbf{b}_2 + \cdots + \mathbf{b}_m) = \sum_{i=1}^{n}\sum_{j=1}^{m} \mathbf{a}_i\rfloor\mathbf{b}_j,$$

$$(\mathbf{b}_1 + \mathbf{b}_2 + \cdots + \mathbf{b}_n)^\sim = \sum_{i=1}^{n} \widetilde{\mathbf{b}}_i.$$

Therefore, we can straightforwardly implement any linear product or operation for multivectors using their implementations of basis blades of the previous chapter. As an example, Figure 20.2 gives Java code from the reference implementation that computes the outer product.

The explicit loops over basis blades and the actual basis blade product evaluations are quite expensive computationally, and implementations based on this principle are about one or two orders of magnitude slower than implementations that expand the loops and optimize them. We will get back to this in Chapter 22.

```
Multivector op(Multivector x) {
    ArrayList result = new ArrayList(blades.size() * x.blades.size());

    // loop over basis blade of 'this'
    for (int i = 0; i < blades.size(); i++) {
        BasisBlade B1 = (BasisBlade)blades.get(i);

        // loop over basis blade of 'x'
        for (int j = 0; j < x.blades.size(); j++) {
            BasisBlade B2 = (BasisBlade)x.blades.get(j);
            // compute actual outer product of the basis blades ...
            // ... and add to result:
            result.add(BasisBlade.op(B1, B2));
        }
    }
    return new Multivector(simplify(result));
}
```

Figure 20.2: Implementation of the outer product of multivectors based in the list of blades approach. The `Multivector` class has a member variable `blades`, which is an `ArrayList` of `BasisBlades`. The function `simplify()` simplifies a list of `BasisBlades` by adding those blades that are equal up to scale.

20.3 STRUCTURAL EXERCISES

1. Why are the first columns of $[\![A^G]\!]$ and $[\![A^O]\!]$ equal to $[\![A]\!]$? Why does this not hold for $[\![A^L]\!]$?

2. Why is $[\![A^O]\!]$ lower triangular?

3. You know that $\mathbf{a}\,\mathbf{B} = \mathbf{a}\rfloor\mathbf{B} + \mathbf{a}\wedge\mathbf{B}$ for a vector $\mathbf{a}$ and a blade $\mathbf{B}$. How do you recognize this fact in the relationship of $[\![A^G]\!]$, $[\![A^L]\!]$ and $[\![A^O]\!]$? Why do we not have $[\![A^G]\!] = [\![A^L]\!] + [\![A^O]\!]$?

4. Compute the matrix for the right contraction.

21 FUNDAMENTAL ALGORITHMS FOR NONLINEAR PRODUCTS

In the previous chapter, we looked at how to implement the *linear* products in geometric algebra. The linearity of these products allowed us to implement them using linear algebra or through a simple double loop. However, there are other operations in geometric algebra that are *nonlinear* (such as inverse, `meet`, `join`, and factorization). These cannot be implemented in the same way.

In this chapter, we discuss the implementation of such nonlinear geometric algebra operations. The nonlinearity results in more complex algorithms, still reasonably efficient but typically an order of magnitude more time-consuming than linear operations.

We give algorithms for the inverse, for exponentiation, for testing whether a multivector is a blade or a versor, for blade factorization, and for the efficient computation of `meet` and `join`.

21.1 INVERSE OF VERSORS (AND BLADES)

We need to compute the inverse of the elements we construct in geometric algebra. Those are almost exclusively blades or versors (the only exceptions were the bivectors in an exponent, and we have no need to invert those). Invertible blades are always versors, since a k-blade (the outer product of k vectors) can always be written as the geometric product of k vectors (i.e., a k-versor); see Section 6.4.1. So if we can invert a versor V, we can also invert blades and thus most of the elements we construct.

There is a very efficient way to invert versors, which we call the *versor inverse* method (because, in general, it works only on versors). This inversion is almost trivial. It starts with the observation that the quantity $V\widetilde{V}$ is always scalar for any versor V. This is easily seen by factorizing the versor:

$$V\widetilde{V} = (\mathbf{v}_k \cdots \mathbf{v}_2\,\mathbf{v}_1)\,(\mathbf{v}_1\,\mathbf{v}_2\cdots\mathbf{v}_k) = (\mathbf{v}_k\cdots(\mathbf{v}_2\,(\mathbf{v}_1\,\mathbf{v}_1)\,\mathbf{v}_2)\cdots\mathbf{v}_k) = \mathbf{v}_k^2 \cdots \mathbf{v}_2^2\,\mathbf{v}_1^2.$$

The inverse of the versor is then simply

$$V^{-1} = \widetilde{V}/(V\widetilde{V}). \tag{21.1}$$

This method would fail if makes $V\widetilde{V} = 0$. But in that case (which may occur when one of the vector factors is a null vector), the multivector V is actually not a versor (since all versors are invertible by definition). V is then not actually a versor since versors are always invertible.

21.2 INVERSE OF MULTIVECTORS

As long as you stay within the bounds of geometric algebra (as opposed to Clifford algebra; see Section 7.7.2 for our take on the difference), the versor inverse method should cover all of your inversion needs. But should the need arise to invert a general multivector, you can use the following technique to invert *any* invertible multivector. It is based on the linear algebra implementation approach described in Section 20.1. Since

$$A\,B \rightleftharpoons [\![A^G]\!][\![B]\!],$$

it follows that

$$A^{-1}\,B \rightleftharpoons [\![A^G]\!]^{-1}[\![B]\!]$$

(just left-multiply both sides by A and $[\![A^G]\!]$). So the matrix of the inverse of A is the inverse of the matrix of A. This insight can be used to invert any multivector A as long as the matrix $[\![A^G]\!]$ is invertible, using the following steps:

- Compute the geometric product matrix $[\![A^G]\!]$;
- Invert this matrix;
- Extract $[\![A^{-1}]\!]$ as the first column of $[\![A^G]\!]^{-1}$.

To understand why the last step is correct, inspect the algorithm for computing the geometric product matrix. With the scalar 1 as the first element of $\mathbb{L}$, the first column of the matrix $[\![A^G]\!]$ is constructed from the geometric product of A with the first element of $\mathbb{L}$, which is 1. Therefore the first column of a geometric product matrix $[\![A^G]\!]$ is $[\![A]\!]$ itself (this may also be observed in Figure 20.1). Hence we find the coefficients of A^{-1} as the first column of $[\![(A^{-1})^G]\!] = [\![A^G]\!]^{-1}$.

This inversion method has two main disadvantages: it is both slow and numerically imprecise due to floating point round-off errors. Therefore, you should not use it to invert a blade or a versor—use the method of Section 21.1 for those.

21.3 EXPONENTIAL, SINE, AND COSINE OF MULTIVECTORS

The exponential and trigonometric functions were introduced for blades in Section 7.4.2. For easy reference, we repeat the equations for the polynomial expansion of the exponential:

$$\exp(\mathbf{A}) = 1 + \frac{\mathbf{A}}{1!} + \frac{\mathbf{A}^2}{2!} + \cdots$$

If the square of $\mathbf{A}$ is a scalar, this series can be easily computed using standard trigonometric or hyperbolic functions:

$$\exp(\mathbf{A}) = \begin{cases} \cos\alpha + \mathbf{A}\,\frac{\sin\alpha}{\alpha} & \text{if } \mathbf{A}^2 = -\alpha^2 \\ 1 + \mathbf{A} & \text{if } \mathbf{A}^2 = 0 \\ \cosh\alpha + \mathbf{A}\,\frac{\sinh\alpha}{\alpha} & \text{if } \mathbf{A}^2 = \alpha^2 \end{cases},$$

where α is a real scalar. Since both a versor and a blade have scalar squares, their exponentials are easy to compute by directly applying these formulas.

Unfortunately, we will need to take exponentials of nonblade elements. The exponentials of general bivectors generate the continuous motions in a geometry, and only in fewer than four dimensions are bivectors always 2-blades. In the important conformal model, rigid body motions are exponentials of bivectors that are not 2-blades, as we saw in Section 13.5. So we need techniques to process them.

When the special cases do not apply, the series for the exponential can be evaluated explicitly up to a certain order. This tends to be slower and less precise. Experience shows that evaluating the polynomial series up to order 10 to 12 gives the best results for 64 bit doubles. For the exponential, a rescaling technique is possible that will increase accuracy, as follows.

Suppose you want to compute $\exp(\mathbf{A})$. If $\mathbf{A}$ is a lot larger than unity, $\mathbf{A}^k$ can be so large that the series overflows the accuracy of the floating-point representation before it converges. To prevent this problem, we can scale $\mathbf{A}$ to near unity before evaluating the series, because of the identity

$$\exp(\mathbf{A}) = \left(\exp\!\left(\frac{\mathbf{A}}{s}\right)\right)^s.$$

For our purposes, any $s \approx \|\mathbf{A}\|$ will do. In a practical implementation, we choose s to be a power of two, so that we can efficiently compute $\left(\exp\frac{\mathbf{A}}{s}\right)^s$ by repeatedly squaring $\exp(\frac{\mathbf{A}}{s})$.

21.4 LOGARITHM OF VERSORS

We do not know of a general algorithm for computing the logarithm of arbitrary versors. However, for many useful cases a closed-form solution was found; see, for example, Section 10.3.3 (rotation), Section 13.5.3 (rigid body motion), and Section 16.3.4 (positively scaled rigid body motion). We have not yet included these logarithms in our reference implementation, but you can find a C++ implementation of these logarithms in the GA sandbox source code package.

21.5 MULTIVECTOR CLASSIFICATION

At times, it may be useful to have an algorithm that classifies a multivector as either a blade, a versor, or a nonversor. For instance, classification is useful as a sanity check of interactive input, or to verify whether and how a result could be displayed geometrically.

Yet testing whether a multivector is a versor or a blade is nontrivial in our additive representation. A blade test requiring that the multivector is of a single grade is insufficient (for example, $e_1 \wedge e_2 + e_3 \wedge e_4$ is of uniform grade 2, but not a blade). Adding the rule that the square of the multivector must be a scalar does not help (since $e_1 \wedge e_2 \wedge e_3 + e_4 \wedge e_5 \wedge e_6$ squares to -2 in a Euclidean metric, but it is not a blade).

The classification algorithm below (from [7]) performs this test correctly. It is one algorithm that can be used either for the versor test or for the blade test. It is natural that these tests should structurally be very similar, for all invertible blades are also versors. As you glance through it, you notice that the algorithm uses inverses. This is fine for a versor test, since versors need to be invertible by definition. However, a blade need not be invertible, and yet null blades are still blades.

In fact, "being a blade" is defined as "factorizable by the outer product", and that does not depend on a metric at all. Our reference [7] makes good use of this freedom: if you are testing whether V is a blade, you must perform the geometric products in the algorithm using a Euclidean metric. This choice of convenience eliminates all null blades that would have had to be taken into account without actually affecting factorizability, and that simplifies the algorithm. (As an example, the multivector $e_1 \wedge \infty$ is a blade, but in the conformal metric contains a null factor ∞. This makes the blade noninvertible, but it is of course still a blade. By using a Euclidean metric, ∞ is treated as a regular vector, the blade becomes invertible, and the test can run as it would for $e_1 \wedge e_2$).

On the other hand, determining whether a multivector V is a versor (i.e., a geometric product of invertible vectors) clearly depends on the precise properties of the metric. So for a versor test, you have to run this algorithm in the actual metric of the algebra of the versor.

The classification process for V consists of three parts:

1. Test if the versor inverse $\widetilde{V}/(V\widetilde{V})$ is truly the inverse of the multivector V. This involves the following tests:

$$\text{grade}(\widehat{V} V^{-1}) \stackrel{?}{=} 0,$$

$$\widehat{V} V^{-1} \stackrel{?}{=} V^{-1}\widehat{V}.$$

If either of these test fails, we can report that the multivector is a nonversor, and hence a nonblade.

The use of grade involution in the two equations above is an effective trick from [9]. It prevents multivectors that have both odd and even grade parts from sneaking through the test. If V is an even versor, then $\widehat{V} = V$. If V is an odd versor, $\widehat{V} = -V$. But if V has both odd and even grade parts, the grade involution prevents odd and even parts from recombining in such a way that they cancel each other out. See structural exercise 21.8 for an illuminating example.

2. The second test is on the grade preservation properties of the versor: applying a versor (and hence an invertible blade) to a vector should not change the grade. So for each basis vector $\mathbf{e}_i$ of the vector space $\mathbb{R}^n$, the following should hold if V is a versor:

$$\text{grade}(\widehat{V} \mathbf{e}_i \widetilde{V}) \stackrel{?}{=} 1.$$

When the multivector does not pass this test, we report a nonversor (and hence a nonblade); otherwise we know that it is either a versor or a blade.

3. The final part makes the distinction between blades and versors by simply checking whether the multivector is of a single grade. If so, it is a blade; otherwise it is a versor.

Unfortunately, we currently have no general solution to the problem of determining the type of multivectors that are not quite blades or not quite versors in a numerically informed manner. We would like to correct any numerical drift in these fundamental properties due to repeated floating point round-off errors, but we do not know how to make versors or blades out of almost-versors and almost-blades in an optimal way in general metrics.

21.6 BLADE FACTORIZATION

This section deals with generic *factorization of blades*. That is the problem to find, for a given blade $\mathbf{B}$ of grade k, a set of k vectors $\mathbf{b}_i$ such that

$$\mathbf{B} = \mathbf{b}_1 \wedge \mathbf{b}_2 \wedge \cdots \mathbf{b}_k.$$

You may want to factorize a blade because you want to use it as input to libraries that cannot handle blades (such as standard libraries for linear algebra or computer graphics), so that you are required to process their vector factors separately. Or you may need a factorization to implement another low-level algorithm such as the computation of `meet` and `join` below.

We are only concerned with the outer product and consequently (as in the previous section) are allowed to choose any convenient metric. To avoid any problems with null vectors, we use the Euclidean metric for all metric products in the algorithm that follows.

A first step towards a useful factorization algorithm is finding potential factors of a blade **B**. This can be done using projection. Take any candidate vector **c** (for example, a basis vector), and project it onto **B**:

$$\mathbf{f} = (\mathbf{c}\rfloor\mathbf{B})\,\mathbf{B}^{-1}.$$

If **f** is 0, then another candidate vector should be tried. For better numerical stability, we should also try another candidate when **f** is close to 0 (or better yet, try many different candidate vectors and use the one that results in the largest **f**). In any case, when **f** is a usable factor we can remove it from **B** simply by dividing it out:

$$\mathbf{B}_f = \mathbf{f}^{-1}\rfloor\mathbf{B}. \tag{21.2}$$

Then we can write (see structural exercise 2)

$$\mathbf{B} = \mathbf{f} \wedge \mathbf{B}_f,$$

so that we have found our first factor of **B**. We now repeat this process on the blade $\mathbf{B}_f$, iteratively, until we are left with a final vector factor.[1]

This basic idea works, but the procedure may be inefficient in high-dimensional spaces. Many candidate vectors may result in a zero value for **f**, which is a waste of effort. It is better to use the structure of **B** itself to limit the search for its factors in an efficient, stable technique, as follows. The blade **B**, as given, is represented as coordinates relative to a basis of blades. We take the basis blade with the absolute largest coordinate on this basis. Let that be $\mathbf{E} = \mathbf{e}_{i_1} \wedge \mathbf{e}_{i_2} \wedge \ldots \wedge \mathbf{e}_{i_k}$. We then use the basis vectors that make up **E** as candidate vectors for projection. This selection procedure guarantees that the projection of each of the candidate vectors is nonzero (we show this at the end of this section).

One final issue is the scaling of each of the factors. Due to the projection, the factors do not have a predictable scale, which is an awkward property. Our implementation of the factorization algorithm normalizes the factors and returns the scale of the blade as a separate scalar.[2]

---

1   We remove the factor **f** from **B**, because we want the factors to be orthogonal. If you just want factors that make up **B** and don't care about orthogonality or scale, you may adjust the algorithm to use the original **B** in each loop of the final algorithm below. But you must then make sure that your factors are linearly independent.

2   Other solutions are to premultiply the first factor with the scale, or to apportion the scale evenly over each factor. Which method is most convenient may partly depend on the subsequent use of the factorization. In any case, the unit factors with a separate scale can be transformed into any of the other representations easily.

The final algorithm [6, 8] for factorizing a blade becomes:

1. Input: a nonzero blade $\mathbf{B}$ of grade k.

2. Determine the norm of $\mathbf{B}$: $s = \|\mathbf{B}\|$.

3. Find the basis blade $\mathbf{E}$ in the representation of $\mathbf{B}$ with the largest coordinate; determine the k basis vectors $\mathbf{e}_i$ that span $\mathbf{E}$.

4. Let the current input blade be $\mathbf{B}_c \leftarrow \mathbf{B}/s$.

5. For all but one of the basis vectors $\mathbf{e}_i$ of $\mathbf{E}$:

   (a) Project $\mathbf{e}_i$ onto $\mathbf{B}_c$: $\mathbf{f}_i = (\mathbf{e}_i \rfloor \mathbf{B}_c) \, \mathbf{B}_c^{-1}$.
   (b) Normalize $\mathbf{f}_i$. Add it to the list of factors.
   (c) Update $\mathbf{B}_c$: $\mathbf{B}_c \leftarrow \mathbf{f}_i^{-1} \rfloor \mathbf{B}_c$.

6. Obtain the last factor: $\mathbf{f}_k = \mathbf{B}_c$. Normalize it.

7. Output: the factors $\mathbf{f}_i$ and the scale s.

Some notes on metric are in order. First, the algorithm also works for null blades, since no blade is actually null in the Euclidean metric that is used during the factorization algorithm itself. Second, the output of the algorithm is a set of orthonormal factors in the Euclidean metric that was used within the algorithm. That may not be the metric of the space of interest. If you desire orthonormality in some other metric, construct the metric matrix of the factors, perform an eigenvalue decomposition, and use this to construct an orthonormal set of factors (see also Section 19.4).

To see why the projection of $\mathbf{e}_i$ on $\mathbf{B}_c$ is never zero (step 5a), note that there always exists a rotor R that turns $\mathbf{E}$ to the original $\mathbf{B}$. This rotor will never be over 90 degrees (for that would imply that $\mathbf{E} \rfloor \mathbf{B} = 0$, yet we know that $\mathbf{E} \rfloor \mathbf{B}$ must be nonzero to be the basis blade with the largest coordinate in $\mathbf{B}$). We may not be able to compute the rotation R easily, but we can find $R^2 = R\,R = \mathbf{B}\mathbf{E}^{-1}$. Since R is never over 90 degrees, R^2 will never be over 180 degrees. Because of this, the quantity $\frac{1}{2}(\mathbf{e}_i + R^2 \, \mathbf{e}_i/R^2)$ must be nonzero: no $\mathbf{e}_i$ is rotated far enough by R^2 to become its own opposite. We rewrite this and find:

$$0 \neq \tfrac{1}{2}(\mathbf{e}_i + R^2 \, \mathbf{e}_i/R^2)$$
$$= \tfrac{1}{2}(\mathbf{e}_i + \mathbf{B}\,\mathbf{E}^{-1} \, \mathbf{e}_i \mathbf{E}\mathbf{B}^{-1})$$
$$= \tfrac{1}{2}(\mathbf{e}_i - \mathbf{B}\,\mathbf{E}^{-1}\widehat{\mathbf{E}} \, \mathbf{e}_i \mathbf{B}^{-1})$$
$$= \tfrac{1}{2}(\mathbf{e}_i - \widehat{\mathbf{B}} \, \mathbf{e}_i \, \mathbf{B}^{-1})$$
$$= (\mathbf{e}_i \rfloor \mathbf{B}) \, \mathbf{B}^{-1}.$$

Therefore none of the $\mathbf{e}_i$ from $\mathbf{E}$ projects to zero on $\mathbf{B}$, so $\mathbf{f}_1$ is nonzero in the first pass of the algorithm. After removal of $\mathbf{f}_1$, the same argument can be applied to the next $\mathbf{e}_i$ on $\mathbf{B}_c$, and so on. Hence none of the $\mathbf{f}_i$ are zero.

21.7 THE MEET AND JOIN OF BLADES

When we introduced the meet and join of blades in Chapter 5, we saw that they are in some sense the geometrical versions of intersection and union from set theory. We can use this correspondence to compute them. To simplify the description of the algorithm, we first illustrate some of the basic ideas with regular set theory and Venn diagrams and then transfer them to meet and join.

Figure 21.1(a) shows a Venn diagram of two nondisjoint sets $\mathbf{A}$ and $\mathbf{B}$, their union, $\mathbf{A} \cup \mathbf{B}$, and their intersection, $\mathbf{A} \cap \mathbf{B}$. We introduce a symmetric set difference through a *delta product*, defining $\mathbf{A} \bigtriangleup \mathbf{B}$ as $\mathbf{A} \cup \mathbf{B}$ minus $\mathbf{A} \cap \mathbf{B}$. This is illustrated in Figure 21.1(d). We also use the complement $(\mathbf{A} \bigtriangleup \mathbf{B})^*$ as illustrated in Figure 21.1(e). The dashed line along the border of Figure 21.1(e) indicates that $(\mathbf{A} \bigtriangleup \mathbf{B})^*$ extends to include all elements that are not in $\mathbf{A}$ or $\mathbf{B}$.

Suppose that we have a function s() that determines the size of a set. We can relate the size of the various sets through

$$s(\mathbf{A} \cup \mathbf{B}) = \frac{s(\mathbf{A}) + s(\mathbf{B}) + s(\mathbf{A} \bigtriangleup \mathbf{B})}{2}. \qquad (21.3)$$

To see why this holds, first convince yourself that

$$s(\mathbf{A}) + s(\mathbf{B}) = s(\mathbf{A} \cup \mathbf{B}) + s(\mathbf{A} \cap \mathbf{B}),$$

and then superimpose Figure 21.1(b), Figure 21.1(c), and Figure 21.1(d). Note that every area is covered twice, hence the division by 2 in (21.3). Likewise,

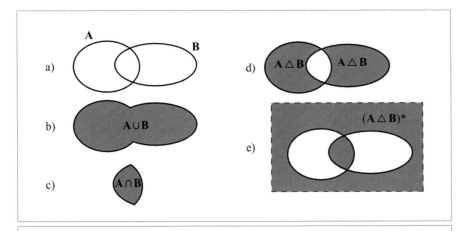

Figure 21.1: Venn diagrams illustrating union, intersection, and the delta product of two sets. (a) Two sets A and B, (b) the union $A \cup B$, (c) the intersection $A \cap B$, (d) the delta product $A \bigtriangleup B$, and (e) The dual of the delta product $(A \bigtriangleup B)^*$.

$$s(\mathbf{A} \cap \mathbf{B}) = \frac{s(\mathbf{A}) + s(\mathbf{B}) - s(\mathbf{A} \triangle \mathbf{B})}{2}. \tag{21.4}$$

The meet and join products are the geometrical versions of set intersection and union, respectively, applied to blades rather than sets. They really work on the bases spanning the blades, and are in that sense discrete. We want to compute the join of two blades $\mathbf{A}$ and $\mathbf{B}$. Instead of thinking of $\mathbf{A}$ and $\mathbf{B}$ as continuous oval blobs, think of them of as discrete factors, united by a boundary. That converts Figure 21.1 into Figure 21.2. Let us assume that $\mathbf{A}$ and $\mathbf{B}$ can be factored as

$$\mathbf{A} = \mathbf{a}_1 \, \mathbf{a}_2 \, \mathbf{c},$$

$$\mathbf{B} = \mathbf{c} \, \mathbf{b}_1.$$

This means that $\mathbf{A}$ and $\mathbf{B}$ have one factor $\mathbf{c}$ in common, and $\mathbf{a}_1, \mathbf{a}_2$, and $\mathbf{b}_1$ are independent (how we arrive at this factorization is discussed later on). Figure 21.2(a) illustrates this basic setting. Obviously, $\mathbf{A} \cap \mathbf{B}$ is proportional to $\mathbf{c}$ (Figure 21.2(c)), and the $\mathbf{A} \cup \mathbf{B}$ is proportional to $\mathbf{a}_1 \wedge \mathbf{a}_2 \wedge \mathbf{c} \wedge \mathbf{b}_1$ (Figure 21.2(b)). Figure 21.2(d) illustrates $\mathbf{A} \triangle \mathbf{B}$, and Figure 21.2(e) illustrates $(\mathbf{A} \triangle \mathbf{B})^*$.

The algorithm for computing the meet and join [6, 8] presented in this section first computes the required grade of the meet and join, for which we need the delta product for blades (from [8]). Then it works towards constructing either the meet or the join.

As we observed in Section 19.3, the geometric product acts like a spatial exclusive *or* on basis blades. The same is true for regular blades when we focus on highest nonzero grade part of the geometric product: it is the spatial exclusive or of the blades since it contains all the factors that do not occur in both arguments. In our example,

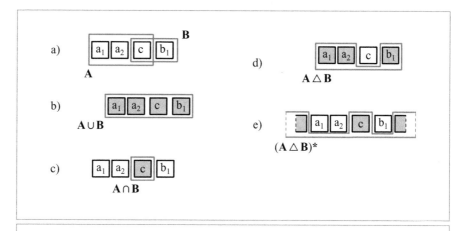

Figure 21.2: Venn diagrams illustrating Meet, Join, and the delta product of two blades. $\mathbf{A} = \mathbf{a}_1 \wedge \mathbf{a}_2 \wedge \mathbf{c}$, $\mathbf{B} = \mathbf{c} \wedge \mathbf{b}_1$ (see also Figure 21.1).

$$\mathbf{A}\,\mathbf{B} = (\mathbf{a_1}\,\mathbf{a_2}\,\mathbf{c})\,(\mathbf{c}\,\mathbf{b_1}) = (\mathbf{c} \cdot \mathbf{c})\,\mathbf{a_1}\,\mathbf{a_2}\,\mathbf{b_1}.$$

Note that we do not need to factorize the blades explicitly to get this result. The geometric product automatically eliminates the dependent factors. So the top grade part of a geometric product $\mathbf{A}\,\mathbf{B}$ is a blade that contains all factors that were present in either $\mathbf{A}$ or $\mathbf{B}$. We call this the *delta product of blades*, and define it as follows:

$$\mathbf{A}\,\triangle\,\mathbf{B} \equiv \langle \mathbf{A}\,\mathbf{B} \rangle_{\max}$$

where max is the largest grade such that $\mathbf{A}\,\triangle\,\mathbf{B}$ is not zero. The delta refers to the property of the delta product to compute the blade that contains factors that are present in either of the arguments, which makes it like the symmetric difference of the factors in $\mathbf{A}\,\mathbf{B}$ which we discussed above.

Once we have the delta product of blades, we can derive the following useful relations by analogy to (21.3) and (21.4):

$$\text{grade}(\mathbf{A} \cup \mathbf{B}) = \frac{\text{grade}(\mathbf{A}) + \text{grade}(\mathbf{B}) + \text{grade}(\mathbf{A}\,\triangle\,\mathbf{B})}{2} \tag{21.5}$$

$$\text{grade}(\mathbf{A} \cap \mathbf{B}) = \frac{\text{grade}(\mathbf{A}) + \text{grade}(\mathbf{B}) - \text{grade}(\mathbf{A}\,\triangle\,\mathbf{B})}{2} \tag{21.6}$$

Our `meet` algorithm starts with a scalar, and expands it by the outer product with new vectors until it arrives at the true `meet`. Potential factors of the `meet` satisfy the following conditions, which you may verify using Figure 21.2.

- They are not factors of the delta product;
- They are factors of **A** and **B**.

Likewise, the algorithm initially assumes that the `join` is the pseudoscalar, and removes factors from it until the true `join` is obtained. Factors that should not be in the `join` satisfy the following conditions:

- They are factors of the dual of the delta product;
- They are not factors of **A** and **B**.

Before we can formulate the algorithm, we repeat two relations from Section 5.3:

$$\mathbf{A} \cup \mathbf{B} = \mathbf{A} \wedge ((\mathbf{A} \cap \mathbf{B})^{-1} \rfloor \mathbf{B}) \tag{21.7}$$

$$\mathbf{A} \cap \mathbf{B} = (\mathbf{B} \rfloor (\mathbf{A} \cup \mathbf{B})^{-1}) \rfloor \mathbf{A} \tag{21.8}$$

These equations can be used to obtain the `join` from the `meet`, and vice versa. We now have enough building blocks to construct an algorithm to compute the `meet` and `join`:

1. Input: two blades **A**, **B**, and possibly a threshold ϵ for the delta product.

2. If grade(**A**) > grade(**B**), swap **A** and **B**. This may engender an extra sign, so be careful when you need to interpret the results (see Section 5.6).

3. Compute the dual of the delta product: $S = (A \triangle B)^*$. A threshold ϵ may be used to suppress floating-point noise when computing the delta product (specifically, when determining what is the top grade part of $A B$ that is not zero).

4. Factorize S into factors s_i.

5. Compute the required grade of the meet and join ((21.5) and (21.6)).

6. Set $M \leftarrow 1, J \leftarrow I_n$ (I_n denotes the pseudoscalar of the total space).

7. For each of the factors s_i:
   (a) Compute the projection p_i and rejection r_i of s_i and A:

$$p_i = (s_i \rfloor A) \rfloor A^{-1}$$
$$r_i = s_i - p_i$$

   (b) If the projection is not zero, then wedge it to the meet: $M \leftarrow M \wedge p_i$. If the new grade of M is the required grade of the meet, then compute the join using (21.7), and break the loop. Otherwise continue with s_{i+1}.
   (c) If the rejection is not zero, then remove it from the join: $J \leftarrow r_i \rfloor J$. If the new grade of J is the required grade of the join, then compute the meet using (21.8), and break the loop. Otherwise continue with s_{i+1}.

8. Output: $A \cap B$ and $A \cup B$.

For added efficiency, step 4 could be integrated into the main loop of the algorithm (i.e., only find factors of S as required).

Note that the algorithm returns both $A \cap B$ and $A \cup B$ at the same time, while only one of them may be required. However, benchmarks have shown that this algorithm is more efficient than an algorithm that searches specifically for either the meet or the join, as it will terminate as soon as it finds either of them. The returned join and meet are based on the same factorization, so we can use relationships of (21.7) and (21.8) to compute one given the other.

The number of loop cycles in the algorithm is $n-1$ in the worst case, where n is the dimension of the vector space. This can be understood by analyzing the following worst-case scenario,

$$A = e_n, \quad B = e_n,$$

in which case

$$A \triangle B = 1, \quad S = (A \triangle B)^* = e_1 \wedge e_2 \wedge \cdots \wedge e_n.$$

The algorithm will start by projecting and rejecting e_1. The projection will be zero, so the M will not grow towards the actual meet (likewise for $e_2 \cdots e_{n-1}$). When the projection is zero, the rejection is obviously nonzero, so each of the rejections is removed from the J. While this brings J closer and closer to the actual join, J has to shrink all the way until it is of grade 1, which will not happen until all $e_1 \cdots e_{n-1}$ have been processed, leading to $O(n)$ cycles of the loop.

We should note, however, that when the inputs $\mathbf{A}$ and $\mathbf{B}$ are in general position, the projection and rejection are likely to be nonzero in each cycle of the loop. In that case the required number of cycles is $\min(\text{grade}(\texttt{meet}), \text{grade}(\texttt{join})) \leq n/2$. For numerical stability, we should require that the projections and rejections have some minimum weight, which will increase the number of cycles (because some projections and rejections will not be used for computation).

21.8 STRUCTURAL EXERCISES

1. For $A = \mathbf{e}_1 + \mathbf{e}_2 \wedge \mathbf{e}_3$, compute A^{-1}, then verify that $\widehat{A} \, A^{-1}$ and $A \, A^{-1}$ indeed behave differently under the test in the algorithm of Section 21.5.

2. Prove that $\mathbf{f} \equiv (\mathbf{c} \rfloor \mathbf{B})/\mathbf{B}$ is a factor of $\mathbf{B}$ (if nonzero) so that $\mathbf{B} = \mathbf{f} \wedge \mathbf{B}_f$ for some blade $\mathbf{B}_f$. Give an expression for a possible $\mathbf{B}_f$ (it is not unique).

22 SPECIALIZING THE STRUCTURE FOR EFFICIENCY

This chapter describes an approach to efficiently implementing geometric algebra. By "efficient" we mean that if you write a program that uses geometric algebra to do some geometry, it should be about as fast as a similar program that uses a traditional optimized method to do the same geometry. For example, the homogeneous model in geometric algebra (Chapter 11) should perform on par with a traditional implementation of homogeneous coordinates in linear algebra. We will demonstrate that this is possible with benchmarks of a ray tracer. The ray tracer itself is described in the next chapter.

We assume that you have read the preceding implementation chapters, as our goal here is to implement efficiently what is described there. The approach we take is largely independent of a specific programming language, as long as the language has some concepts of object orientation (i.e., classes). Some examples of actual C++ code generated by `Gaigen 2` are given to make the approach more tangible.

22.1 ISSUES IN EFFICIENT IMPLEMENTATION

By now, we hope we have convinced you that geometric algebra is a structured and elegant mathematical framework for geometry. In computer science, better structure and greater elegance often help to simplify the implementation of problems and to avoid errors.

However, when geometric algebra is first presented to a computer scientist or programmer, the question arises how an implementation of geometric algebra can ever be as efficient as the current way of implementing geometry (based in linear algebra with various extensions). Many features of geometric algebra seem to work against efficiency: 2^n coordinates are required to represent a multivector for an n-dimensional space, the conformal model of 3-D space requires a 5-D algebra (with its 32-D basis), there appear to be many products and operations, and so on.

All initial geometric algebra implementations (like [43, 47]) seemed to confirm this, for they were structurally pretty but unacceptably slow for practical applications on high volume data. They were not designed with efficiency in mind. This is a pity, for the structure of geometric algebra actually offers many opportunities to home in on the essential computations that need to be done in an application. Effectively, these permit its practical use as a high-level programming language that computes efficiently.

Three issues will recur in the attempt to make an efficient implementation: multivectors, metrics, and operations.

- Multivectors are the fundamental elements of computation in geometric algebra, but they are big (2^n coordinates for n-D). So although it is mathematically attractive that all elements can be considered as multivectors, you would rather not base an implementation on it.

  Fortunately, most geometrically sensible elements of computation use only limited grades, and in a particular manner. This suggests defining a fixed, specialized multivector type for a variable whenever possible, to reduce storage and processing. Such a multivector type is an indication of its geometrical meaning, immediately implying a more limited use of grades or basis blades. Examples of specialized multivector types in the programming examples of Part I and II abound: `vector`, `translator`, and `circle`, to name a few.

  In a large group of geometric algorithms and applications, we can naturally impose such fixed multivector types on variables without compromising our solutions. (And if in some application we cannot, we can always revert to the multivectors.)

- The metric is a very fundamental feature of a geometric algebra, affecting the most basic products. Many different metrics are useful and need to be allowed, yet looking up the metric at run-time (e.g., from a table) is too costly. We should therefore not just implement a general geometric algebra and use it in a particular situation; for efficiency, we need a special implementation for every metrically different geometric algebra.

- The number of basic operations on multivectors is quite large, and every operation can be applied on every element. Written out in coordinates, the operations are not complicated, but precisely because the execution of each individual product or operation requires relatively few computations, any overhead imposed by the implementation (such as a conditional branch due to looping) will result in a significant degradation of performance.

As a consequence, expressions consisting of multiple products and operations can often be executed much more efficiently by folding them into one calculation, rather than executing them one by one through a series of generic function calls.

This suggests programming things out explicitly for each operation and each multivector type, with the danger of a combinatorial explosion in the volume of the code.

These considerations lead to the conclusion that for efficient performance, the elegant universal and unifying structure of geometric algebra needs to be broken up. Multivectors are general but too big, metrics are unifying but too basic to be variables, and the operations are simple and universal but too slow when not specialized and optimized.

Building such an optimized implementation of a particular geometric algebra is a lot of administrative work. It may be possible to do it by hand in limited, low-dimensional cases, but even then it is tedious and error-prone.

Fortunately, we do not need to do it ourselves. We should use the actual algebra to convert its coordinate-free expressions into low-level, coordinate-based code that can be directly executed by a processor. Used in that way, geometric algebra can be the engine of an *automatic code generator*, which can take a high-level specification of a solution for a geometric problem and automatically generate efficient implementations.

With such a tool, we never again have to write the error-prone, low-level code that directly manipulates coordinates. Instead, we just write our algorithms in the high-level language, confident that this will automatically lead to correct and efficient code.

22.2 GENERATIVE PROGRAMMING

Generative programming [11] is the craft of developing programs that synthesize other programs. One of its uses is to connect software components (such as algebra implementations) without loss of performance. Ideally, each component adapts dynamically to its context. In our case, we would like to use generative programming to transform the specification of a geometric algebra into an optimized implementation that is tailored to the needs of the rest of the program.

There are several points in the tool-chain (illustrated in Figure 22.1) from source code to fully linked and running program where this transformation can take place. Here we list three obvious examples, but other approaches (or hybrids) are also possible:

1. The most explicit approach is generating an implementation of the algebra before the actual compilation of the program takes place. This is the way classical code generators like `lex` and `yacc` work. Advantages are that generated code is directly available to the user as a library, and that this method does not cause interference with the rest of the chain. The main disadvantage is that it does not integrate well, since a separate program is required to generate the code. Another disadvantage is that

Figure 22.1: Basic tool-chain from source code to running application with three possible points where code generation can take place (see text).

some form of feedback from the final, running program (i.e., profiling) is required to allow the implementation to adapt itself to the context, and this may require an extra code generation pass. The geometric algebra implementations `Gaigen` and `Gaigen 2` take this approach.

2. The transformation can also take place at compile-time, if the programming language permits this. This is called meta-programming, where the algebra implementation is set up such that the compiler generates parts of it at compile time. For example, in C++ this is possible through the use of the *templates* feature that was originally added to the language for generic programming. The definite advantage of this method is the good integration with the language. A disadvantage is the limited number of mainstream programming languages that support meta-programming. Disadvantages specific to the C++ programming language are the complicated template syntax, hard-to-decipher compiler error messages (for both user of the library and its developer), and the long compile times. The `boost::math::clifford` library [61] uses this approach.

3. A third option is to delay code generation until the program is up and running. The program would generate the code as soon as it is required (e.g., the first time a function of the algebra is called with specific arguments). The advantages are that the algebra implementation can adapt itself to the actual input and the actual hardware that it runs on (e.g., available instruction set extensions, the speed of registers compared to cache compared to main memory, etc. [5]). Disadvantages are the slower startup time of the final program (due to the run-time code generation) and the fact that the method is rather nonconventional and hard to implement. At the time of writing, we are not aware of such a geometric algebra implementation.

The geometric algebra implementation described below can in principle be realized through each of these generative programming methods, so below we simply call it the code generator, regardless of the precise details.

22.3 RESOLVING THE ISSUES

The overall goals of our implementation can be stated as follows:

1. Waste as little memory as possible (i.e., store each variable as compactly as possible).

2. Implement functions over the algebra most efficiently. To do this you need several things:

    (a) Process as few zero coordinates as possible (which coincides with Goal 1),

    (b) Minimize (unpredictable) memory access. This coincides with Goal 1 and Goal 2a, but also demands avoiding lookup from tables, and so on.

    (c) Avoid conditional branches. When a modern processor mispredicts a conditional branch, a large number of processor cycles is lost.

    (d) Unroll loops (e.g., over the dimension of the algebra or the grade of a blade) whenever possible. This avoid branches and also allows you to apply further optimizations.

    (e) Optimize nontrivial expressions. Avoid implementing them as a series of function calls.

When we hold our (purposely naive) reference implementation based on Chapters 19 to 21 up to this list of demands, we see that it violates all but one of them. It wastes memory due to the bookkeeping required for per-coordinate compression, looks up the metric from tables, uses many conditional branches due to looping, does not unroll loops, and processes operations one by one. The only thing that it does right is minimizing the processing of zero coordinates.

22.3.1 THE APPROACH

Through generative programming, we can resolve each of the issues listed in Section 22.1 in a way that satisfies these goals. We first sketch how the issues can be resolved, followed by a more detailed description in the next section.

- As we indicated, the problem that multivectors are too general can be resolved by generating classes for specific multivector types. These classes store only the nonzero coordinates for that specific type (e.g., the 3-D vector class only stores the e_1-, e_2- and e_3-coordinates, as all other coordinates are always 0). We can also generate classes that represent useful constants (such as e_1, I, and $o \wedge \infty$). These classes are used mostly as symbolic tokens, as they will be optimized away when they are used in expressions. The combination of these two ideas leads to a hybrid where only some coordinates are constant. For example, the o-coordinate of a normalized homogeneous point is always 1.

- Using generative programming, the metric of the algebra can now be hard-coded directly into the implementation by the code generator, taking very little effort from the user. Likewise, the large implementation effort due to the combinatorial explosion caused by the large number of products/operations and multivector types is no longer an issue, as the code generator does the tedious work for us (note that not every function should be generated for every combination of arguments, but only those that are actually used by the program).

- Finally, by optimizing functions over the algebra, the code generator can fully optimize equations consisting of multiple products and operations, because it can oversee the whole context.

The approach is built on top of Chapters 19 and 20. First, it can be seen as an optimization of the list of basis blades approach from Section 20.2. The optimization is to generate a separate class for each type of list (i.e., multivector type) that we expect to encounter in our program, thus avoiding the need for explicit bookkeeping of the list. Second, to write out expressions (involving products and operations) on a basis, we need Chapter 19. The nonlinear functions from Chapter 21 are not essential to the approach, but most of them are suitable candidates for optimization, as described in Section 22.4.6.

22.4 IMPLEMENTATION

We now describe the implementation in some more detail, starting with the specification of the algebra, followed by a description of the classes and functions that should be generated.

22.4.1 ALGEBRA SPECIFICATION

The specification of the algebra is the starting point for the code generator. It specifies at least the following:

- The dimension of the algebra;
- The metric of the algebra;
- Definition of the specialized types;
- Definition of constants.

We briefly discuss these terms before moving on to their implementation.

Dimension of the Algebra

The geometric algebra of $\mathbb{R}^n$ has the dimensionality n, which should be specified. Within $\mathbb{R}^n$ it is convenient to specify a basis (not necessarily orthonormal; see below). Since geometric algebra is too new to have a standardized notation, it is sensible to permit the user to choose the names of the basis vectors. This results in more readable code, and output in a recognizable format.

Metric

The metric should allow for a nondiagonal metric matrix, even though an equivalent metric with a diagonal metric matrix always exists. The reason for this is that in certain models, it may be more intuitive and efficient to define the metric in an off-diagonal manner. For example, in the conformal model of Euclidean geometry, ∞ and o are more fundamental than the vectors e and $\bar{e}$ (see (13.5) for the relationships of these bases.

The increased efficiency of this basis is obvious when we write out the coordinates of a conformal line A relative to each basis. Using the o,∞-basis we get

$$A = A_{1o\infty}\, \mathbf{e}_{1o\infty} + A_{2o\infty}\, \mathbf{e}_{2o\infty} + A_{3o\infty}\, \mathbf{e}_{3o\infty} +$$
$$A_{12o\infty}\, \mathbf{e}_{12o\infty} + A_{13o\infty}\, \mathbf{e}_{13o\infty} + A_{23o\infty}\, \mathbf{e}_{23o\infty},$$

while the e, $\bar{e}$-basis results in

$$A = A_{1e\bar{e}}\, \mathbf{e}_{1e\bar{e}} + A_{2e\bar{e}}\, \mathbf{e}_{2e\bar{e}} + A_{3e\bar{e}}\, \mathbf{e}_{3e\bar{e}} +$$
$$A_{12\bar{e}}\, \mathbf{e}_{12\bar{e}} - A_{12e}\, \mathbf{e}_{12e} + A_{13\bar{e}}\, \mathbf{e}_{13\bar{e}} - A_{13e}\, \mathbf{e}_{13e} + A_{23\bar{e}}\, \mathbf{e}_{23\bar{e}} - A_{23e}\, \mathbf{e}_{23e}$$

(here we use shorthand to denote the coordinates $A_{\ldots}$ and the basis elements $\mathbf{e}_{\ldots}$, i.e., $\mathbf{e}_{13\infty} = \mathbf{e}_1 \wedge \mathbf{e}_3 \wedge \mathbf{e}_\infty$, etc.). So, storing a line requires six coordinates on the o,∞-basis compared to nine coordinates on the e, $\bar{e}$-basis, three of which are duplicates ($A_{12\bar{e}} = A_{12e}$, $A_{13\bar{e}} = A_{13e}$, and $A_{23\bar{e}} = A_{23e}$).

Specialized Types

The definition of each specialized type should list the basis blades required to represent it. The definition could also include whether the type is a blade or a versor, so that run-time checks can be made (e.g., in debug mode) to verify the validity of the value.

Preferably, the definitions should also allow certain coordinates to be defined as constant. For example, a normalized point has a constant o-coordinate, so there is no need to store it. It may also be useful to have the ability to specify the order in which coordinates are stored. This can simplify and optimize connections to other libraries, which may expect coordinates to be in a certain order.

Specialized types for matrix representations of outermorphisms are also an option. For example, in connection with OpenGL it is useful to have a special type to contain the outermorphism matrix representation that can be applied directly to a grade 1 blade (a 4×4 matrix). In our programming examples, we have used such a type in Sections 11.13.2, 12.5.1, and 13.10.3.

Constants

Important constant multivector values inside a program can be encoded as constants. Examples are basis vectors such as $\mathbf{e}_1$ and o or the pseudoscalar $\mathbf{I}$. We should include the constants in the algebra specification and generate special types for them, in such a way that their use in expressions can be optimized away by the code generator or the compiler.

22.4.2 IMPLEMENTATION OF THE GENERAL MULTIVECTOR CLASS

In our approach, an implementation of the general (nonspecialized) multivector is always required. The specialized types approach works best when the multivector type of each variable in a program is fixed. This assumption often holds, but also cripples one of the amazing abilities of geometric algebra: many equations work regardless of the multivector type of their input. For some programs (e.g., those dealing with noisy input or interactive input), the multivector type of variables may change at run-time. This prevents the use of specialized multivector types at compile-time, and hence for such programs we should provide a fallback option that may not be efficient, but at least works.

Coordinate Compression

An important issue when implementing the general multivector class is compression of the coordinates. The coordinates of the general multivector could be stored naively (all 2^n of them), but to increase performance, compression can be used to profit from the inherent sparseness of geometric algebra. By compression, we mean reducing the number of zero coordinates that are stored.

- **Per-Coordinate Compression**. The most straightforward compression method is per-coordinate compression. Each coordinate is tagged with the basis blade it refers to, and the coordinates are stored in a list. Zero coordinates are not stored in this list. We already encountered this type of compression in Section 20.2.

- **Per-Grade Compression**. Multivectors are typically sparse in a per-grade fashion: blades, by definition, have only one nonzero grade part. Likewise, versors are either even or odd, implying that at least half their grade parts are null. This suggests grouping coordinates that belong to the same grade part: when a grade part is not used by a multivector, the entire group of coordinates for that grade is not stored. Instead of signaling the presence of each individual coordinate, we signal the presence of the entire group.

    For low-dimensional spaces, per-grade compression is a good balance between per-coordinate compression and no compression at all. Per-coordinate compression is slow because each coordinate has to be processed individually, leading to excessive conditional branching, which slows down modern processors. No compression at all is slow because lots of zero coordinates are processed needlessly. Per-grade compression reduces the number of zero coordinates while still allowing sizable groups of coordinates to be processed without conditional branching.

- **Per-Group Compression**. Per-grade compression does miss some significant compression opportunities. For example, a flat point in the conformal model (a 2-blade) requires only four coordinates, while the grade 2 part of 5-D multivectors requires 10 coordinates in general. Thus per-grade compression stores at least six zero coordinates for each flat point. As the dimension and structure of the algebra increases, this storage of zero coordinates becomes more of an issue. A more refined grouping method may alleviate this problem, but what the groups are depends heavily on the specific use of the algebra. For example, in the conformal model for Euclidean geometry, one could group coordinates based on the presence of ∞ in the basis blades, in addition to grouping them based on grade; but if the conformal model would be used for hyperbolic geometry (see Section 16.7.1), a grouping based on the presence of e would be better.

Example of a Multivector Class

In C++, the general multivector class that uses per-grade compression could look like

```cpp
// C++ code:
class multivector {
public:
   /* ... constructors, compression functions, pretty printing,
      etc ... */


   /**
    * Bitmap that keeps track of grade usage (gu).
    * When bit 'i' of 'gu' is 1, then grade 'i' is present in the
      coordinates.
    */
   unsigned int gu;

   /** dynamically allocated memory that holds coordinates */
   double *c;
};
```

22.4.3 IMPLEMENTATION OF THE SPECIALIZED MULTIVECTOR CLASSES

Specialized multivector classes should be generated for each multivector type used in the program. The implementation of these classes can be straightforward. The class should provide storage for the nonconstant coordinates and some functionality is required to convert back and forth between the specialized multivectors classes and the general multivector class. The rest of the functionality is provided by the functions over the algebra.

As an example in C++, the specialized multivector class for a normalized flat point in the conformal model for Euclidean geometry could look like

```cpp
// C++ code:
class normalizedFlatPoint {
public:
   /* ... constructors, converters, pretty printing, etc ... */

   /** the no^ni coordinate is constant: */
   static const double noni = 1.0;

   /** holds the e1^ni, e2^ni and e3^ni coordinates: */
   double c[3];
};
```

22.4.4 OPTIMIZING FUNCTIONS OVER THE ALGEBRA

Optimizing functions over elements of the algebra is the key to achieving high performance. As described so far, we only have classes that can efficiently store general and specialized multivectors. Optimized functions over these algebra elements should also be generated to do something with the multivectors. Preferably, we generate these functions from their high-level definitions.

An example of the definition of such a function (in an imaginary language) is

```
// code in imaginary language:
// applies normalized versor 'V' to multivector'X'
function multivector applyVersor(multivector V, multivector X)
{
  return V X reverse(V);
}
```

Such a function would be used to apply any versor to a blade or another versor. The precise syntax of the definition depends on the generative programming method used. For example, in the meta-programming approach, they appear as a C++ template functions, while Gaigen 2 uses a domain-specific language to define them.

The point is that such functions should be instantiated with specific types of multivectors. For example, when applyVersor() is called with rotor and flatPoint arguments, the code generator should take the applyVersor() function definition and specialize it for those arguments. Figure 22.2 shows the resulting C++ code generated by Gaigen 2, which should serve as a convincing example that this type of code should indeed be automatically generated and not written by hand.

The process of generating code like Figure 22.2 from the high-level definition of the function like applyVersor() is as follows:

1. The types of the arguments are replaced with specializations.
2. The expressions in the function are written out on a basis.
3. The expressions are simplified symbolically: products and operations are executed at the basis level (Chapter 19), identical terms are added, unnecessary computations removed, and so on.
4. The return type of the function is determined.
5. The code is emitted.

This process is similar to how one would go about when writing such an optimized function by hand. As a result of the design (i.e., the use of specialized multivectors instead of other methods of compression) and code generation approach, the code in Figure 22.2 contains no conditional branches. These branches would otherwise be present due to coordinate compression and/or looping.

```
// generated C++ code:
flatPoint applyVersor(const rotor& V, const normalizedFlatPoint& X) {
  return flatPoint(
    // e1 ^ ni coordinate:
    2.0 * (
    V.c[0] * X.c[0] * V.c[0] - V.c[1] * X.c[0] * V.c[1] +
    V.c[1] * X.c[1] * V.c[0] + V.c[3] * X.c[1] * V.c[2] -
    V.c[0] * X.c[2] * V.c[3] + V.c[2] * X.c[2] * V.c[1])
    , // e2 ^ ni coordinate:
    2.0 * (
    V.c[3] * X.c[0] * V.c[2] - V.c[1] * X.c[0] * V.c[0] +
    V.c[0] * X.c[1] * V.c[0] - V.c[1] * X.c[1] * V.c[1] +
    V.c[2] * X.c[2] * V.c[0] + V.c[3] * X.c[2] * V.c[1])
    , // e3 ^ ni coordinate:
    2.0 * (
    V.c[0] * X.c[0] * V.c[3] + V.c[2] * X.c[0] * V.c[1] +
    V.c[3] * X.c[1] * V.c[1] - V.c[2] * X.c[1] * V.c[0] +
    V.c[1] * X.c[2] * V.c[1] - V.c[3] * X.c[2] * V.c[3])
    , // no ^ ni coordinate:
    V.c[0] * V.c[0] + V.c[1] * V.c[1] + V.c[2] * V.c[2] +V.c[3] * V.c[3]
  );
}
```

Figure 22.2: Code generated by Gaigen 2. The function applyVersor() (see text) was instantiated with a rotor and a flatPoint. Both variables V and X contain an array of floating-point coordinates named c. Note that Gaigen 2 does not know that the rotor is normalized, hence the needless computations in the last line of code, which always results in '1' for such normalized versors.

To illustrate the use of constants and constant coordinates, we present two more examples. Again in the conformal model of Euclidean geometry, let us instantiate the following function with a normalized flat point and a dual plane as arguments:

```
// code in imaginary language:
function multivector outerProduct(multivector a, multivector b)
{
     return a ^ b;
}
```

The o-coordinate of the normalized flat point is a constant 1. We can specialize our outer product code for when the operands are a flat point and a dual plane, which reduces the necessary computation by about 25 percent, as we have indicated in the following code generated by Gaigen 2:

```
// generated C++ code:
line outerProduct(const normalizedFlatPoint& x, const dualPlane& y)
```

```
{
  return line(
    x.c[1] * y.c[0] - x.c[0] * y.c[1],
    x.c[2] * y.c[0] - x.c[0] * y.c[2],
    x.c[2] * y.c[1] - x.c[1] * y.c[2],
    y.c[0],    // thanks to normalized flat point,
    y.c[1],    // no 'x' coordinates are used
    y.c[2]);   // on these lines
}
```

The constant coordinates save three multiplies in the last three lines of the function. Also, compare this optimized function to its (generic) equivalent from the reference implementation (Figure 20.2). The optimized function contains no loops or conditionals, which can make a difference of as much as two orders of magnitude in performance.

As a final example, let us instantiate a function that uses a constant (the unit pseudoscalar **I**) to dualize a multivector in the vector space model:

```
// code in imaginary language:
function multivector dual(multivector a)
{
    return a . inverse(I);
}
```

We instantiate this function using an ordinary vector. Because **I** is a constant, the function reduces to initializing a bivector with the (shuffled, negated) coordinates of the a:

```
// generated C++ code:
bivector dual(const vector& a)
{
    return bivector(-a.c[2], -a.c[0], -a.c[1]);
}
```

The inversion and the inner product have disappeared after this automatic optimization by the code generator. Compare this to the reference implementation, which would explicitly compute the inverse of I and then perform the actual inner product, only to achieve the same final result.

22.4.5 OUTERMORPHISMS

When an outermorphism has to be applied many times, it is often more efficient to compute a matrix representation of the outermorphism and use the matrix to apply the outermorphism.

Consider a situation where a lot of points have to be rotated. We could do this using a rotor R directly ($R X \widetilde{R}$) by calling the function in Figure 22.2. However, because $R X \widetilde{R}$ is an outermorphism, the same effect can be achieved by initializing an outermorphism

```
// generated C++ code:
flatPoint apply_om(const omFlatPoint& M, const normalizedFlatPoint& a)
{
  return flatPoint(
    M.c[0] * a.c[0] + M.c[1] * a.c[1] + M.c[2] * a.c[2] + M.c[3],
    M.c[4] * a.c[0] + M.c[5] * a.c[1] + M.c[6] * a.c[2] + M.c[7],
    M.c[8] * a.c[0] + M.c[9] * a.c[1] + M.c[10] * a.c[2] + M.c[11],
    M.c[12] * a.c[0] + M.c[13] * a.c[1] + M.c[14] * a.c[2] + M.c[15]);
}
```

Figure 22.3: Code generated for transforming a flat homogeneous point according to a 4×4 matrix. Note that the flat point has a constant $ni \wedge no$ coordinate, which saves four multiplies. Also note that the return type is deduced to be a `flatPoint` instead of a `normalizedFlatPoint`, because the outermorphism may undo the normalization of the point.

matrix (see Section 4.2) and then using this matrix to transform the points. The transform of points using such a matrix can be done using the automaticaly generated code in Figure 22.3, which is obviously more efficient.

The code in Figure 22.3 was generated from the following definition:

```
// code in imaginary language:
function multivector apply_om(outermorphism M, multivector a)
{
  return M * a;
}
```

Whether it is more efficient to use the matrix representation or apply versors directly depends on how much time it takes to initialize the matrix, how often you are going to apply it, and how much more efficient the matrix is compared to the straightforward geometric algebra implementation.

22.4.6 OPTIMIZING THE NONLINEAR FUNCTIONS

One might wonder if it is possible to use these optimization techniques to generate efficient implementations of the nonlinear functions in Chapter 21. The answer depends on whether the multivector types of intermediate and output variables of the algorithms are fixed for a specific instantiation of the algorithm. For most of the algorithms, it can be done, and leads to significant performance gains.

Each algorithm from Chapter 21 should be defined by function that—in order to generate an optimized version of it—should be instantiated with specialized multivector argument(s). For some of the algorithms (specifically multivector classification and factorization) it is necessary to unroll loops to make sure that each variable has a fixed multivector type. For example, the variable $\mathbf{B}_c$ in the blade factorization algorithm

(Section 21.6) changes type on each iteration, but by unrolling the loop this can be avoided (assuming that a new variable $\mathbf{B}_c$ is introduced for each iteration of the loop).

We briefly discuss how optimization would proceed for each nonlinear function from Chapter 21:

- *Inverse of versors (and blades)*. This is a simple equation that is very suitable for optimization.

- *Inverse of multivectors*. The initialization of the geometric product matrix can be done more efficiently because it is known that many coordinates are zero. The rest of the algorithm does not benefit from the technique described here. The return type of the function cannot be determined at code generation time, since it depends on the outcome of the matrix inversion, which is not predictable based solely on the specialized multivector type of the input argument.

- *Exponential, sine, and cosine of multivectors*. These functions can be highly optimized when the code generator is able to determine that the square of the specialized multivector argument is a scalar. Then the straightforward equations for the special cases can be used (see Section 21.3). Otherwise, the test for whether $\mathbf{A}^2$ is a scalar should be performed at run-time, and the outcome decides whether the special cases or the generic series evaluation algorithm should be used (neither of which can be now optimized).

- *Multivector classification*. This algorithm can be highly optimized. The first step (testing if the versor inverse is the true inverse) is straightforward. To optimize the second step (testing for grade preservation), the loop over all basis vectors should be unrolled. Then each test can be optimized for the individual basis vectors.

- *Blade factorization*. Optimizing blade factorization is similar to the multivector classification. As soon as the code generator is able to unroll loops, all variables in algorithm get a fixed type, and most conditional branches are removed (finding the largest basis blade of the blade and its basis vectors must be done at run-time, which requires some looping).

- meet *and* join *of blades*. The meet and join algorithms cannot be optimized by using specialized multivectors, because the multivector type of several variables in the algorithm is not fixed. For example, the type of the outcome of the delta product is by definition not predictable from just the multivector type of the arguments (the actual values of the arguments have to be known). There are several other unavoidable conditional branches in the algorithm.

22.5 BENCHMARKS

To get an idea of the relative performance of each model of 3-D Euclidean geometry and implementations thereof, we have written multiple implementations of a ray tracer.

The code for each implementation is basically the same, except each time we use a different model of the Euclidean geometry. This allows us to measure the relative performance of the models in a realistic context.

The models we used are:

- **3D LA**. The vector space model implemented with 3-D linear algebra. 3×3 matrices are used to represent rotations.

- **3D GA**. The vector space model implemented with 3-D geometric algebra. Rotors are used to represent rotations.

- **3D GA-OM**. The above, but using outermorphism matrices to represent rotations.

- **4D LA**. Homogeneous coordinates, with Plücker coordinates for lines and planes. 4×4 matrices are used to implement rotations, scaling, and translations.

- **4D GA**. The homogeneous model. Outermorphism matrices are used to implement rotations, scaling, and translations.

- **5D GA**. The conformal model. All transformations implemented using versors. This version of the ray tracer is described in detail in Chapter 23.

The geometric algebra models were implemented using Gaigen 2, which is basically an implementation of the ideas in this chapter (the GA sandbox source code package is also based on it). The linear algebra models use typical handwritten libraries (all code in-lined as much as possible). SIMD instructions (i.e., Intel's SSE instruction set) were used only as a result of optimizations performed by the Intel C++ compiler. You can download the source of each version of the ray tracer at *http://www.geometricalgebra.net*.

Table 22.1 shows the relative performance of the models. The 3D GA-OM model is most efficient by a small margin. This slight performance edge is due to the use of constants where the 3D LA model uses nonconstant vectors. The cost of using the conformal model

Table 22.1: Performance benchmarks run on a Pentium 4, 3GHz notebook, with 1GB memory, running Windows XP. Programs were compiled using Intel Compiler 9.0.

Model	Implementation	Rendering Time Relative to 3D LA
3D LA	Standard	1.00×
4D LA	Standard	1.22×
3D GA-OM	Gaigen 2	0.98×
3D GA	Gaigen 2	1.05×
4D GA	Gaigen 2	1.2×
5D GA	Gaigen 2	1.26×

5D GA (about 25 percent) is approximately the same as the cost of using the homogeneous model 4D GA or homogeneous coordinates 4D LA.

22.6 A SMALL PRICE TO PAY

In this chapter we presented a method that brings the performance of geometry implemented through geometric algebra close to that of geometry implemented through linear algebra. The core ideas of the method are specialization of multivectors and optimization of functions, implemented through generative programming. The benchmarks for the ray tracer example show that for the vector space model, performance is equal to that of linear algebra. In our benchmarks, the cost of using the conformal model is a drop of about 25 percent in performance, similar to using the customary homogeneous coordinates in such an application.

Of course, your mileage may vary. As described in detail in Chapter 23, the ray tracer was designed with performance in mind. When multiple representations were available for concepts (such as a ray), the alternative that performed the best was selected even though that representation may not have been the most elegant one. We have also seen applications where using the conformal model was *more* efficient than the traditional approach (40 percent in the inverse kinematics of a simple robot [34]). Different choices and different applications may perform better or worse, but we don't believe any well-designed conformal model application should be more than 50 percent slower than the best traditional implementation.

Besides achieving good performance, specializing multivectors also has the advantage that it makes your source code more readable. Instead of every geometric variable being of the `multivector` type, variables can have more informative types such as `line`, `circle`, and `tangentVector`, as can be seen in many programming examples in Part I and Part II. The disadvantage, of course, is that you lose the ability to store other types of multivectors in those variables.

The bottom line is that geometric algebra is competitive with classical approaches in computation speed, with the benefit of more readable high-level code and error-free low-level code.

22.7 EXERCISES

1. The weighted sum of which basis blades is required to represent a plane in general position on the o, ∞-basis? Which on the e, $\bar{e}$-basis?

2. What is the minimum number of coordinates required to represent any rigid body motion rotor versor in the conformal model?

3. Is there a difference in the basis for rotors and versors?

4. Write out the following equation in terms of coordinates: $\mathbf{n} = (\mathbf{a} \wedge \mathbf{b})^*$.

23 USING THE GEOMETRY IN A RAY-TRACING APPLICATION

In this chapter, we put the algebra to work by using it to implement a ray tracer in the C++ programming language. A ray tracer is a program that takes as input the description of a scene (including light sources, models, and camera) and produces an image of that scene as seen from the camera. We show how the conformal model of Euclidean geometry is very useful for the specification and computation of the basic operations. All geometrical elements occurring in this problem have a natural representation in that model.

When we really try to optimize for the speed of the ray tracer, we argue that one may want to deviate from the natural all-purpose representation to a more dedicated choice, which anticipates the typical ray tracing operations that are going to be performed on the data. That is, we trade some of the elegance of the conformal model for added efficiency. Even then, each of these choices can be described and compared within the conformal model, which moreover always supplies the necessary transformations in its versor representation. The resulting compromise between algebraic elegance and practical performance is probably typical of many applications.

In the process, we write a graphical user interface to manipulate objects in a viewport, as well as the camera that produces the view. This is another application in which the conformal model shows its power by giving directly executable expressions connecting mouse coordinates to the simulated motions. Here practical performance and algebraic elegance coincide in a satisfying manner.

23.1 RAY-TRACING BASICS

When you view a scene in the real world, rays originating from light sources interact with objects, picking up their colors before they reach your eye to form an image. In classical ray tracing, this process is reversed: starting from a pixel of the image, a ray is cast through the optical center of the camera and traced into the world. It encounters objects, reflects, refracts, and ultimately may reach one or more light sources. The product of all these interactions with the material and spectral properties of the objects and light sources determines the color, shading, and intensity that need to be rendered in the original pixel. This is called the *shading computation*. To do this physically exactly is impossible or slow, so many convenient shortcuts and simplifications have been introduced. Often, performing the shading computation requires new rays to be spawned and traced. Among those are the *shadow rays* that are used to check whether a direct path exists between some intersection point and a light source. New rays must also be spawned if the material at the intersection point is reflective or if it is transparent (which requires a new ray to be created after refraction, according to Snell's law).

From the algebraic point of view, the representation of rays and their intersections with objects is central to a ray tracer. The possibilities the conformal model offers for these elements will be carefully considered in their advantages and disadvantages, both algebraically and implementationally. Most of the time, algebraic elegance and computational efficiency go hand-in-hand, but sometimes we will be forced to choose, and then we will opt for efficiency. Yet we will always remain within the framework of the conformal model and benefit fully from its capability of specifying all operations in terms of the geometrical elements rather than their coordinates.

As we have seen in the previous chapter, using geometric algebra does not need to go at the expense of efficiency. A good code generator can convert the CGA specifications to executable code. (We will sometimes use CGA as the abbreviation for conformal geometric algebra in this chapter.) The code generator should certainly allow the creation of new types for specific conformal primitives (e.g., points, lines). But a good package goes further than merely creating the data structures. With a proper implementation, some seemingly expensive algebraic operations can be surprisingly cheap. For instance, to extract the Euclidean direction $\mathbf{u}$ from a line L, we can write $\mathbf{u} = (\infty \wedge o)\rfloor L$. In an efficient geometric algebra package, this computation requires very little effort, since it can be simplified to just assembling the coefficients L has on the basis blades $o \wedge \mathbf{e}_1 \wedge \infty$, $o \wedge \mathbf{e}_2 \wedge \infty$, and $o \wedge \mathbf{e}_3 \wedge \infty$ into a vector with basis $\mathbf{e}_1, \mathbf{e}_2, \mathbf{e}_3$. Dualization with respect to basis elements, in particular the pseudoscalar, can also be implemented by coordinate manipulation rather than true inner products. Such implementational considerations will affect our representational choices in the ray tracer considerably.

We will implement a complete but limited ray tracer. It includes a simple modeler with OpenGL visualization, in which one can interactively create the scene to be rendered. It can render only still frames (no animations). It can render only polygonal meshes (no curved surfaces). No kinematic chains are possible. It has support for texturing and bump mapping. Shading is adopted from the standard OpenGL shading model (ambient, diffuse, specular; see [58], Chapter 5). It supports reflection and refraction. We apply no optimizations to improve ray-model intersection test efficiency except simple bounding spheres and BSP trees.

The equations in this chapter are directly shown as C++ code, with occasional reference to their occurrence in the earlier chapters. Since the code is usually close to the mathematical notation, it should be easily readable even for non-C++ literate. The only catches are:

- The " . " denotes not the inner product, but access to a member variable in C++ .
- Since the " . " is already taken, we use the $\ll$ symbol to denote the inner product.
- We write `ni` for ∞ and `no` for *o*.

The code examples we show are sometimes polished or slightly changed for presentation. The full, unedited source code can be inspected at *http://www.geometricalgebra.net*. We use `Gaigen 2` as our conformal model implementation in this chapter. It is not required to have read the programming examples in Parts I and II to understand this chapter. (The ray-tracer code is not built on top of the GA sandbox source code package because the ray tracer was created before we had that available.)

23.2 THE RAY-TRACING ALGORITHM

We present the basic outline of the ray-tracing algorithm to define our terms and geometrical subproblems.

To render an image, the ray tracer spawns rays through the optical center of the camera position and each pixel of the image. (For antialiasing, multiple rays per pixel can be spawned, each with a small offset.)

For each spawned ray we want to find the closest intersection with the surface of a model. A quick intersection test is performed by first checking the intersection with a bounding sphere that encloses the entire model. If that test turns out positive, a more complex intersection test is done. The polygons of the models are stored in binary space partition (BSP) trees. We descend down the BSP tree until we discover what polygons the ray intersects. We pick the intersection that was the closest to the start of the ray and return it as the result of the intersection test.

Once the closest intersection point is known, we query the appropriate polygon to supply information about the material at the intersection point. Material properties include color, reflectivity, transparency, refractive index, and surface attitude (classically called surface normal, which is of course the dual of the direction 2-blade).

Using the material properties and light source information, local shading computations are made. The shading equations that we use are the same as those employed for the fixed function pipeline of OpenGL, except that we do shadow checks: for a light source to illuminate the intersection point, there must be a clear path between them. So a *shadow ray* is spawned from the intersection point towards each of the light sources to check the path. If the shadow ray cannot reach a light source unobstructed, that light source does not contribute to diffuse and specular lighting for the intersection point. Note that transparent objects are also treated as obstructions, because refraction causes our shadow ray to bend away from the light source, and we cannot easily compensate for that.

The outcome of the shading computations contributes to the final color of the pixel from which the ray originated. If the material is reflective or refractive, new rays are spawned appropriately. The colors returned by these rays also contribute to the final color of the pixel that the ray originated from.

In the next three sections we discuss the low-level geometric details of each part of the ray tracer. We start with the representation and interactive modeling of the scene, followed by the tracing of the rays, and end with shading computations.

23.3 REPRESENTING MESHES

To represent the shape of objects in the scene we need to model them. We will use a triangular mesh, consisting of a set of vertices and triangular faces, bounded by edges. Figure 23.1 shows an example of such a mesh in both solid and wireframe rendering.

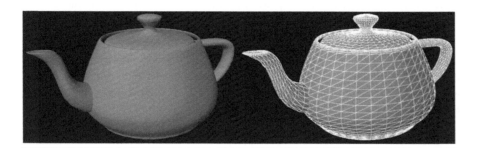

Figure 23.1: A polygonal mesh rendered in solid and with its wireframe model superimposed.

Meshes are read from .rtm files, which describes the faces and vertices of the mesh. The description of a face is simply a list of indices of the three vertices. The vertex description is more involved, specifying position, local surface attitude, and texture coordinates. Finally, the mesh file contains material properties such as reflectivity, transparency, and optional texture information.

Vertices

Vertices are stored in the vertex class. Vertices naturally require a 3-D position $\mathbf{x}$. We choose to use a normalized conformal point (of the form $o + \mathbf{x} + \frac{1}{2}\mathbf{x}^2\infty$) to represent the position of a vertex. Such a normalized conformal point requires four coordinates for storage (the constant o does not have to be stored). Alternatively, we could have used a flat point (of the form $(o+\mathbf{x}) \wedge \infty$) that would require one less coordinate, but this would be annoying in the algebra of the spanning operation (in which the plane determined by three points is the outer product of three regular points with infinity, not flat points) and during barycentric interpolation (see equation (11.19)).

To initialize the points, we read the Euclidean x, y, and z coordinates from the .rtm and construct a point from them:

```
normalizedPoint pt = cgaPoint(x * e1 + y * e2 + z * e3);
```

cgaPoint() is a function that constructs a point that looks like the familiar (13.3):

```
// Returns a point at the location specified by 3D Euclidean vector p
normalizedPoint cgaPoint(const vectorE3GA &p)
{
    return p + no + 0.5 * norm_e2(p) * ni;
}
```

The norm_e2() function computes the squared Euclidean norm (i.e., the sum of the squares of all coordinates).

It is common practice to approximate the look of a smooth surface by interpolating the local surface attitude at the vertices across each faces. This affects only the shading computations and does not change the actual shape of the model (hence the contours of a model show that it is actually a polygonal mesh). The mesh files specify the *normals* for each vertex. We construct a free vector from these coordinates (to refresh your memory on the kinds of vectors in the conformal model, review Section 15.4). This free vector is then dualized into a free 2-blade that is used to represent the surface 2-direction (which we call its *attitude*). Conceptually, attitude and normal represent the same geometrical feature, either by a 2-blade or by a perpendicular vector.

```
// nx, ny and nz are floats.
// They represent the classical surface normal as an attitude.
freeBivector att = dual((nx * e1 + ny * e2 + nz * e3) ^ ni);
```

Each vertex also has an associated 2-D point that is used for *texture mapping*, which is a common way of coloring the surface of a 3-D model by wrapping a 2-D image around it. In computer graphics, texture mapping is also used to apply bumps to the surface of a model (bump mapping, or displacement mapping), to specify the surface transparency, and so on. Because we won't be doing a lot of geometry on the texture coordinates, we use *normalized flat 2-D points* to represent them (this requires only two coordinates). That is effectively equivalent to representing them by homogeneous coordinates (see Section 13.3.2). Note that these points live in a different space than the rest of our geometry, namely in the 2-D carrier space of the texture image.

So putting it all together, the storage part of the `vertex` class looks like

```
class vertex {
    normalizedPoint pt;                 // position of the vertex
    freeBivector att;                   // local surface attitude
    normalizedFlatPoint2d texPoint;     // position in 2D texture image(s)
};
```

Representing Vertices with Tangent 2-Blades

An alternative way to represent the position and the local surface direction of a vertex is to use a tangent 2-blade. We can combine point p and attitude $\mathbf{A}$ into one vertex primitive with their combined properties and the clearer semantics of a tangent 2-blade: $V = p \rfloor (p \wedge \mathbf{A} \wedge \infty)$, see (15.1).

However, there are some efficiency concerns with this tangent 2-blade representation, for though its attitude can be extracted for free, its position cannot. The attitude of a tangent 2-blade V is the free 2-blade algebraically given by $-(\infty \rfloor V) \wedge \infty$, and this can be extracted by taking the $o \wedge \mathbf{e}_i \wedge \mathbf{e}_j$ coordinates of V and making them into $\mathbf{e}_i \wedge \mathbf{e}_j \wedge \infty$-coordinates of an attitude element $\mathbf{A} \wedge \infty$. But the position of a tangent 2-blade cannot be extracted in this manner from V. A method to get the flat point at the location of V is $(\infty \rfloor V) \rfloor (V \wedge \infty)$, which you may recognize as the `meet` of the line perpendicularly through V with the plane that contains V. On the coordinate level, implementing this equation involves some multiplication and addition rather than mere coordinate transfer. Since the position of the vertices is required often during rendering to do interpolation, the cost of its extraction made us decide against using the tangent 2-blade representation of vertices for the ray tracer. This is an example where a more algebraically elegant possibility is overruled by efficiency considerations.

Faces

The `face` class describes the faces of the polygonal model. The main data in this class are the indices of the vertices that make up the face. These indices are read from the `.rtm` file, stored in an integer array `vtxIdx`, and accessed through a function `vtx()` that returns the vertex.

The ray tracer then precomputes some blades to speed up later computations using standard conformal operations:

- The plane that contains the face:

```
// wedge together the position of vertex 0, 1, 2 and infinity
pl = vtx(0).pt ^ vtx(1).pt ^ vtx(2).pt ^ ni;
```

We also check whether the face actually spans surface area:

```
if (norm_e2(pl) == 0.0) // -> no surface area
```

- The three conformal lines along the three edges of the polygon:

```
// each edge line is its (start vertex)^(end vertex)^infinity
edgeLine[0] = vtx(0).pt ^ vtx(1).pt ^ ni;
edgeLine[1] = vtx(1).pt ^ vtx(2).pt ^ ni;
edgeLine[2] = vtx(2).pt ^ vtx(0).pt ^ ni;
```

The edge lines are used to compute whether a specific line intersects the face, and they come in handy to compute the direction of edges. For example,

```
((ni ^ no) << edgeLine[0]) ^ ni}
```

is the free vector along the direction of edge 0. With these details, the storage part of the `face` class looks like:

```
class face {
    // indices of the three vertices
    int vtxIdx[3];

    // lines along the edges of the face
    line edgeLine[3];

    // the plane of the face
    plane pl;
}
```

Computing the Bounding Sphere

For each mesh, a bounding sphere is computed that contains all the vertices of the mesh. As we mentioned before, this is a common trick to speed up intersection computations: when we need to test for the intersection of some primitive with a model, we do not do a detailed intersection test straight away, but instead first check if it intersects the bounding sphere. If so, then we proceed with the detailed intersection test, otherwise we can report that no intersection was found. The bounding sphere does not need to be tight (although the tighter the better). For simplicity we compute a sphere of which the center is at half the extent of the data set in the directions of the coordinate axes, and with an appropriate radius. This sphere is stored as a regular conformal sphere.

```
/*
Compute the center of the bounding sphere:
-create three orthogonal planes through the origin:
-compute minimal and maximal signed distance of each
vertex to these planes
*/
```

```
dualPlane pld[3] = {
  dual(no ^ e2 ^ e3 ^ ni),
  dual(no ^ e3 ^ e1 ^ ni),
  dual(no ^ e1 ^ e2 ^ ni)
};
mv::Float minDist[3] = {1e10, 1e10, 1e10};
mv::Float maxDist[3] = {-1e10, -1e10, -1e10};
for (int i = 0; i < vertices.size(); i++) {
  for (int j = 0; j < 3; j++) }
    minDist[j] = min(Float(pld[j] << vertices[i].pt)), minDist[j]);
    maxDist[j] = max(Float(pld[j] << vertices[i].pt)), maxDist[j]);
  }
}

/*
Compute the center 'c' of the bounding sphere.
*/
normalizedPoint c(cgaPoint(
  0.5 * (minDist[0] + maxDist[0]) * pld[0] +
  0.5 * (minDist[1] + maxDist[1]) * pld[1] +
  0.5 * (minDist[2] + maxDist[2]) * pld[2]));

/*
Compute the required radius 'r' (actually, -radius^2/2) of the
bounding sphere:
*/
mv::Float r = 0.0;
for (int i = 0; i < vertices.size(); i++)
  r = min(vertices[i].pt << center, r)

/*
Construct the bounding dual sphere:
*/
boundingSphere = c + r * ni;
```

Constructing the BSP Tree

To do a detailed intersection test of a model with, say, a line, we could simply check the intersection of the line with every face of the model. For complex models with many faces, this can become computationally expensive.

To speed up the intersection test, we apply a standard technique that is similar to bounding spheres but employed recursively. First we search a plane that has about half of the vertices of the model at its back side and the other half of the vertices in front. We sort the vertices according to the side of the plane on which they lie, and then recurse: for each of the halves of the vertex sets, we search for another plane that splits these halves in half. And so on, until we have partitioned the space into chunks that each contain a reasonable number of vertices. This is called a *binary space partition tree* (BSP tree). The intersection test of a line with such a BSP tree is discussed in Section 23.5.3.

During the construction of the BSP tree, the main geometric computation is the selection of the plane to partition the space. We use the following approach: for the subdivision planes, we cycle, modulo 3, through translated versions of the three orthogonal planes $o \wedge e_2 \wedge e_3 \wedge \infty$, $o \wedge e_3 \wedge e_1 \wedge \infty$, and $o \wedge e_1 \wedge e_2 \wedge \infty$. The remaining difficulty is to compute the translation that halves the vertex set in the chosen direction. This is done by applying the Quicksort algorithm. All of these computations are just standard techniques cast into a conformal notation.

```
/*
Get the 'base plane' we are going to use for partitioning the space:
The integer 'basePlaneIdx' tells us what plane to use:
*/
dualPlane dualBasePlane;
if (basePlaneIdx == 0) dualBasePlane = dual(no ^ e2 ^ e3 ^ ni);
else if (basePlaneIdx == 1) dualBasePlane = dual(no ^ e3 ^ e1 ^ ni);
else dualBasePlane = dual(no ^ e1 ^ e2 ^ ni);

/*
Compute the signed distance of each vertex to the plane.
This distance is stored inside the vertices,
so we can use it to sort them:
*/
for (i = 0; i < nbVertices; i++)
  vtx(i).setSignedDistance(
    dualBasePlane << getVertex(vertex[i].idx).pt;

/*
Use quicksort to sort the vertices.
The function 'constructBSPTreeSort()' compares the
signed_distance fields of the vertices.
*/
qsort(vertex, nbVertices, sizeof(bspSortData), constructBSPTreeSort);

/*
The required translation of the base plane is now simply the
average of the signed distance of the two vertices that came out
in the middle after sorting:
*/
splitIdx = nbVertices / 2;
mv::Float d = 0.5 * (vertex[splitIdx -1].signedDistance +
        vertex[splitIdx].signedDistance);

/*
We now translate the dual base plane, (un)dualize it and store
it in the pl of the BSP node:
*/
pl = dual(dualBasePlane - d * ni);
```

23.4 MODELING THE SCENE

This section treats the geometry used for modeling the scene. We need some way to place our light sources, cameras, and polygon models. So we need both methods to represent and apply Euclidean transformations, and methods to interactively manipulate these transformations.[1]

In the conformal model, the natural choice to represent Euclidean transformations is to use rotors. In Chapter 16 we showed how conformal rotors can be used to represent and apply rotations, translations, reflections, and uniform scaling. These are the most basic transformations used in computer graphics, and they would classically be represented with 4×4 matrices.

To apply a conformal rotor to a blade, we use code like:

```
/*
xf is the rotor (xf is short for 'transform').
ptLocal is a point that we want to take from a
'local' coordinate frame to the global coordinate frame.
*/
pt = xf * ptLocal * reverse(xf);
```

23.4.1 SCENE TRANSFORMATIONS

Concatenating Transformation Rotors

During interactive modeling, we want to concatenate small, interactively specified transformations to existing transformations. This can be done by either pre- or postmultiplying the existing rotor with the new rotor. As explained in Section 14.6.1, premultiplying applies the transformation in the global frame; postmultiplying applies the transformation in the frame represented by the current rotor. Which method is most practical depends on which frame gives the parameters that specify the small transformation.

Knowing how to apply rotors to blades and how to combine rotors, the only problem left is to construct the required transformation rotors in response to interactive user input. Figure 23.2 shows a screenshot of the user interface of the modeler. The caption of the figure lists all the available manipulations; most of them we treat below.

Mouse Input

All manipulations are done in response to mouse motions that arrive from the user interface toolkit as (x, y) coordinates. We immediately transfer these coordinates to a 2-D Euclidean vector:

---

1   We will not treat interactive modeling of the polygonal models themselves (e.g., manipulating the individual vertices of a model), although the geometry required to do that is similar to the techniques exposed here.

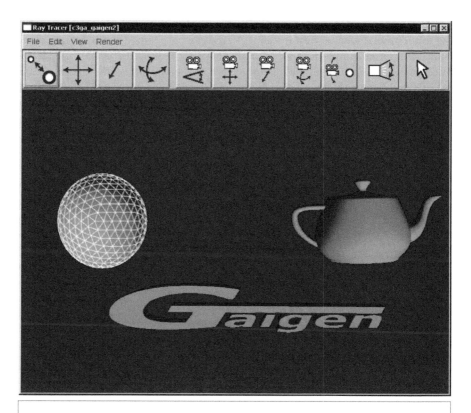

Figure 23.2: Screenshot of the user interface of the modeler. In the main viewport cameras, models, and light sources can be manipulated using the mouse. The available options are denoted by the buttons: scaling, translation (parallel to viewport), translation (orthogonal to viewport), rotation, camera field of view (active in this screenshot), camera translation (parallel to viewport), camera translation (orthogonal to viewport), camera rotation, camera orbit, size of spotlight, and selection. The selected model or light source is indicated by superimposing its wireframe mesh on top of its solid rendering.

```
vectorE2GA modelingWindow::getFLTKmousePosition() const {
  /*
  The calls to Fl::event_x() and Fl::event_y() return the coordinates
  of the mouse in pixels. We reflect the y-coordinate by subtracting
  it from the height h() of the window.
  */
  return Fl::event_x() * e1 + (h() - Fl::event_y()) * e2;
}
```

By subtracting mouse positions from two consecutive mouse events, we get the mouseMotion, which is used in the code below.

Scaling an Object

The first transformation we treat is scaling. To scale an object (model or light source) in a way that is intuitive to the user, we need a scaling rotor at a specific location; that is, at the center of the model. By *post*multiplying by the current transform of the object, we can apply the scaling rotor at the location of the model with no extra effort. Therefore, it is enough to specify the scaling rotor at the origin:

```
scalor createScalingVersor(const vectorE2GA &mouseMotion) {
  /*
  We use the 'vertical' part of the mouse motion to
  indicate the amount of scaling:
  */
  return exp((0.01 * (e2 << mouseMotion) * (no ^ ni));
}
```

Note that we extract the vertical part of the motion (i.e., dragging the mouse up and down will grow or shrink the object), and automatically uses a logarithmic scale (see Section 16.3.1).

Translating an Object

Translation is only slightly harder. The direction of the translation is specified in the camera frame though `mouseMotion`. We need to transform it into the global frame before we can *pre*multiply it with the object transform. Here is the code that computes a rotor for translation parallel to the camera:

```
/*
'cameraXF' is the current transform of the camera.
*/
translator createTranslationVersorParallel(
        const TRSversor &cameraXF,
        const vectorE2GA &scaledMouseMotionInCameraFrame) {
  /*
  Compute translator in camera frame.
  scaledMouseMotionInCameraFrame is
  'ready for use' except that it is in the wrong frame.

  So we compute the free vector that represents the
  mouse motion, transform that to global coordinates,
  and then compute the exponent of that:
  */
  return exp(0.5 * cameraXF *
    (ni ^ scaledMouseMotionInCameraFrame) *
    reverse(cameraXF)));
}
```

Note that the mouse motion must be scaled such that the selected model appears to stick to the cursor. For this to work, the mouse motion is scaled according to the distance of the object to the camera and the field of view of the camera:

```
/*
getSelectionDepth() returns the distance of the selected
object to the camera.
C.getFOVwidth() returns the field of view of the camera at
distance 1 to the camera.
w() returns the width of the viewport.
*/
vectorE2GA scaledMouseMotionInCameraFrame =
    (mouseMotion * getSelectionDepth() * C.getFOVwidth() / w();
```

Object translation orthogonal to the camera is similar to parallel translation. One difference is that we scale the mouse motion according to the current distance of the object, so that the further away the object is, the faster it will translate in response to vertical mouse motion.

Rotating an Object

Rotation is the most involved object transformation. When the user selects the rotation mode, a transparent bounding sphere is drawn around the model (this is indeed the bounding sphere that we computed earlier). The user can grab any point on the sphere and drag it around. The object will follow the motion as indicated by the user.

Figure 23.3 shows how we compute the required rotation. First, we construct two lines into the scene and compute where they intersect the bounding sphere of the object. The lines go from camera position through the previous and current mouse positions, respectively.

When we have computed the intersection points, we construct two lines from the center of the bounding sphere through the two points. The rotor turning one into the other can be constructed from their halfway line, which is computed from the point average of the two conformal points. This is essentially the trick of (10.13) in the vector space model that we extended to the conformal model in Section 14.6.2.

Here is the code that computes the rotor based on the two intersection points:

```
sphere s = dual(getSelectedObjectBoundingSphere());

// Get the intersection points:
point ptCurrent = getRotatePoint();
point ptPrevious = getPreviousRotatePoint();

// Compute the two lines by wedging the points/spheres together:
line L1 = s ^ ptCurrent ^ ni;
line L2 = s ^ (ptCurrent + ptPrevious) ^ ni;

// Compute the (normalized) versor, and apply to object transform:
L.preMulXF(unit_r(L1 * L2));
```

The function unit_r() computes the unit line under the reverse norm $\left(\text{i.e., } L/\sqrt{-L \cdot \tilde{L}}\right)$.

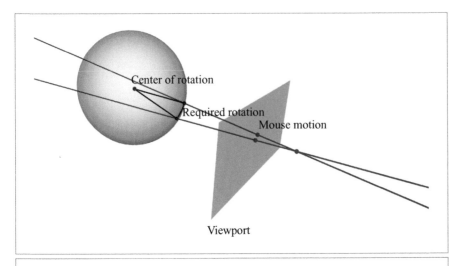

Figure 23.3: Rotating an object. On the right is the camera center (a red point). The user specifies the rotation by dragging the mouse in the viewport plane. This results in two points in the image plane (the current and previous mouse position). The two lines from the camera center through the mouse locations intersect with the bounding sphere in the blue points. The ratio of the lines from the center of the bounding sphere through both intersection points gives (twice) the required rotation.

Translating the Camera

Translating the camera is simpler than translating an object. Because the `mouseMotion` is specified in the camera frame, we can directly form a translation rotor:

```
TRSversor createCameraTranslationVersorParallel
    (const vectorE2GA &mouseMotion) {
  // Exponentiate the free vector to get the translation versor:
  return exp((0.5 * (mouseMotion ^ ni));
}
```

For the depth translation, we use only the vertical part of the mouse motion:

```
TRSversor createCameraTranslationVersorOrthogonal
    (const vectorE2GA &mouseMotion)
{
    /*
    Same code as in create_translation_versor_parallel,
    except we replaced mouseMotion with
    ((e2 << mouseMotion) * e3):
    */
    vectorE3GA mouseMotionOrthogonalCameraPlane =
        (((e2 << mouseMotion) * e3) ^ ni);
```

```
/*
We can now directly exponentiate this free vector
to get our translation versor:
*/
return exp(0.5 * mouseMotionOrthogonalCameraPlane);
}
```

Rotating the Camera

Camera rotation is done through a spaceball interface (see Figure 23.4). If the user drags the mouse on the outside of the (invisible) circle, the camera rotates about the e_3 vector: a rotation in the screen plane. If the user drags the mouse inside the dashed circle, the camera rotates about the local e_1 and e_2 vectors.

Construction of the rotation rotors is satisfyingly straightforward. For a rotation in the screen plane, we compute the 2-blade spanned by the mouse motion and mouse position:

```
TRSversor createCameraRotationVersorInScreenPlane(
        const vectorE2GA &mousePosition,
        const vectorE2GA &mouseMotion)
{
    return exp(mouseMotion ^ mousePosition);
}
```

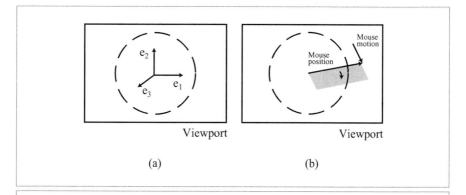

$$\text{Viewport} \qquad\qquad \text{Viewport}$$

(a) (b)

Figure 23.4: The spaceball interface. (a) When the user drags the mouse on the inside of the dashed circle the camera rotates about e_1 and e_2; outside of the circle, the camera rotates about e_3. (b) Computing the rotation when the user drags the mouse on the outside of the circle, We compute the 2-blade (drawn in gray) spanned by the mouse motion and the mouse position (drawn as vectors). The exponential of that 2-blade is the required rotation rotor.

For a rotation outside the screen plane, we compute exponential of the 2-blade $e_3 \wedge \mathtt{mouseMotion}$:

```
versor createCameraRotationVersorOutsideScreenPlane(
        const vectorE2GA &mouseMotion)
{
  return exp(e3 ^ mouseMotion);
}
```

Orbiting the Camera around a Selected Object

Orbiting the camera around a selected object is the most complicated transformation. The difficulty is that we want to rotate about a point in the world frame (the center of the selected object), with the rotation angle and attitude specified in the camera frame.

The standard way to construct a rotation about an axis through an arbitrary point is to first create a rotation rotor about the origin, and then to translate that rotor. So, first we compute a translator for the offset from the camera to the rotation rotor we want to use for orbiting. We do this by transforming the flat point at the origin according to the orbit rotor and transforming that point into the camera frame:

```
// Get the offset from the camera to the orbit versor.
// C.getXF() returns the camera transform.
// getOrbitVersor() returns the transform of the object we want
// to orbit
flatPoint t =
  reverse(C.getXF()) * getOrbitVersor() *
  (no ^ ni) *
  reverse(getOrbitVersor()) * C.getXF();
```

Once we have the location of the orbit rotor relative to the camera frame as a normalized flat point, we create a translator to that location:

```
// compute the translation versor from the origin to 't'
translator T = exp(- 0.5 * (t - (no ^ ni)));
```

Note that we cannot apply `reverse(C.getXF()) * getOrbitVersor()` directly to our translator `T`, because it may contain unwanted scaling and rotation; hence the trick of transforming a flat origin.

The actual rotation is computed from the mouse parameters in the same way as with a regular camera rotation (see above), but it now needs to be performed at the location specified by the translation rotor `T`:

```
// post multiply the camera transform with the rotation 'xf'
C.postMulXF(T * xf * reverse(T));
```

23.5 TRACING THE RAYS

With our scenes (interactively) modeled and represented, we arrive at the core of the ray tracer. In this section we discuss how rays are represented, how they are spawned, and how the ray tracer computes the intersection of rays with the models.

23.5.1 THE REPRESENTATION OF RAYS

Since rays interact with almost every part of our ray tracer, we should choose their representation carefully. We consider several alternatives and pick the one that we think works best in the context of the operations that need to be done.

Classical Representation

A ray is a directed half-line, with a position (where it was spawned) and a direction. Classically, we would represent this as the point (requiring three or four coordinates) and a direction vector (requiring three coordinates).

Point-Line Representation

A sensible generalization of this classical representation in CGA would be to use a conformal point and a conformal line through it. This is, of course, somewhat redundant, since the line also contains partial position information. But it would seem convenient to have the ray in line representation so that it can be used directly in intersection testing.

However, this representation requires a lot of coordinates for storage (five for the point, six for the line). A regular conformal point might seem a good way to represent position, but it is awkward in a ray tracer. When we intersect the ray (line) with a plane (the most common intersection operation in the ray tracer), we get a flat point as a result. To compute the representation of a spawned ray at that location, that flat point would have to be converted to a regular point, which is extra work not warranted by the geometry of the situation. Also, the point/line combination is expensive to transform to and from coordinate frames. Finally, even though it is algebraically possible to reflect and refract the lines by substituting them in the classical equations for reflection and refraction of directions (illustrated for reflection in Section 13.4), this is rather computationally expensive.

Tangent Vector Representation

A third idea is to use a tangent vector to represent rays. It seems perfect: a tangent vector has position and direction all in one. This would representing a ray as a single blade, conforming to the geometric algebra philosophy of representing one object by one blade, whenever possible. Besides, a tangent vector can be turned into its carrier line at any time (just add infinity using the outer product).

Yet this representation suffers from the same problems as the point-line representation. A tangent vector requires 10 coordinates for storage, and this makes it expensive in operations. Moreover, extraction of the locational part of a tangent vector cannot be done by a computationally free coordinate transfer (this problem was already described for tangent bivectors in Section 23.3).

Rotor Representation

Another possible ray representation is as a rotor, since a translation/rotation rotor naturally represents a position and a direction (the axis of a general rotation). It requires just seven coordinates for storage. But the position and direction information cannot be extracted for free from a rotor V by coordinate extraction. Instead, we have to apply the rotor to some blade to determine such properties: $V o V^{-1}$ is the position of the ray (as a conformal point), and $V(\mathbf{e}_3 \wedge \infty) V^{-1}$ is the direction of the ray. Both these equations are relatively expensive to evaluate.

More seriously, we need to intersect rays, but there is no `meet` operation for rotors (since they are not blades).

Flat Point-Free Vector Representation

In the end, we settled for the less elegant but more efficient representation of a ray as a flat point with a free vector. The flat point gives the position, the free vector gives the direction. The choice of a flat point may not seem natural. You cannot directly create a line from a flat point $p \wedge \infty$ and a free vector $\mathbf{u} \wedge \infty$, since they both contain an infinity factor. This is solved by removing the infinity factor from the free vector before wedging them together:

```
// fp is a flat point, fv is a free vector
line l = fp ^ (- no << fv)
```

The number of coordinates of a (flat point, free vector)-pair is as low as the classical representation. But, unlike the classical representation, the flat point and free vector automatically have clear semantics in operations. The flat point responds to both translation and rotation, the free vector only to rotation (it is translationally invariant).

Still, the disadvantage of this mixed CGA representation is that we miss out on automatic algebraic properties for the ray as a whole; the programmer has to remember that the flat point and the free vector belong together and what their combination means. Had we used a tangent vector, there would be no possibility of getting its raylike properties wrong in transformations. Now the programmer must specify semantics that are not intrinsic to the algebra by putting the blades together in one class and adding comments that give detailed information on what each blade represents. For instance:

```
class ray {
public:
  flatPoint pos;        // position of the ray
  freeVector direction; // direction of the ray
};
```

Another downside of the (flat point, free vector)-pair representation is that distance computations become less elegant. In ray tracing, when we search along the ray for the first intersection we need to compute the distance of candidate intersections with the start of the ray. If we had the ray position as a regular conformal point, we could do that with a simple inner product. With the ray position as a flat point, we have to use the classical approach of subtracting the two points and computing the Euclidean norm (squared).

23.5.2 SPAWNING RAYS

There are four occasions where we have to spawn new rays:

- The initial spawning of a ray through camera center and image plane;
- For shadow rays that test whether a light source is visible from a specific point;
- When reflecting a ray;
- When refracting a ray.

The geometry involved in doing reflections and refractions is treated in separate sections. Here we just show how camera rays and shadow rays are spawned.

Camera Rays

To render an image, we trace rays through each pixel in the image. If antialiasing is required, we spawn multiple rays through each pixel, each ray offset by a small random vector. The next block of code spawns the initial rays:

```
mv::Float pixelWidth = cameraFOVwidth / imageWidth;
mv::Float pixelHeight = cameraFOVheight / imageHeight;

// for each pixel in the image:
for (int y = 0; y < imageHeight; y++ {
  for (int x = 0; x < imageWidth; x++ {
    /*
    Compute the direction of the ray if it has to
    got to go through sensor pixel [x, y]:
    */
    freeVector rayDirection =
      ((x * pixelWidth - 0.5 * cameraFOVwidth) * e1 +
      (y * pixelHeight - 0.5 * cameraFOVheight) * e2 +
      e3) ^ ni;

    // sample multiple times, accumulate the result in pixelColor:
    color pixelColor;
    for (int s = 0; s < multisample; s++){
```

```
/*
Add a small perturbation within the square.
To generate random numbers, we call mt_rand(),
the Mersenne twister random number generator.
*/
freeVector perturbedRayDirection =
  rayDirection +
  (((mt_rand() - 0.5) * e1 * pixelWidth
  (mt_rand() - 0.5) * e2 * pixelHeight) ^ ni);

// make the direction unit:
freeVector unitRayDirection = unit_e(perturbedRayDirection);.

// trace the ray from camera position towards 'unitRayDirection'
ray R((C.getPosition(), unitRayDirection);
pixelColor += trace(R, maxRecursion);
  }
}
```

Note that one could say that we make direct use of (image) coordinates to specify the
directions of the rays ($x\,\mathbf{e}_1$, $y\,\mathbf{e}_2$), which is a somewhat against our philosophy, but we
believe this is the most sensible way to start the rays. Trying to avoid coordinate usage
here would make things more awkward instead of simpler.

Shadow Rays

Shadow rays are spawned to check whether there is a clear path between some surface
point and a light source. The surface point is naturally represented as a flat point, since it
comes from an intersection computation.

The light source position is encoded in a rotor. We can extract a point from that rotor by
using it to transform the flat point at the origin:

```
flatPoint lightPoint =
    light->getXF() * (no ^ ni) * reverse(light->getXF());
```

This computation can be done before we start rendering the scene, since the position of
the light does not change during the rendering of a single frame.

We now have two flat points that we can simply subtract to get a free vector that points
from the surface point to the light:

```
/*
'surfacePoint' is a flat point at the surface of some model.
'lightPoint' is a flat point at the position of the light for
which we have to check visibility:

We can subtract the points since they are both normalized.
*/
```

```
freeVector shadowRayDirection =
    unit_e(lightPoint - surfacePoint);
```

This gets us the representation of the shadow ray as `surfacePoint` and `shadowRayDirection`.

23.5.3 RAY-MODEL INTERSECTION

A typical ray tracer spends most of its time finding if and where rays intersect models. Our ray tracer does this in two steps. First, it checks if the ray intersects the bounding sphere. If so, a descent down the BSP tree follows, until it eventually finds actual intersections with faces of the model.

Bounding Sphere Test

The bounding sphere computation is quite simple:

```
// compute the line representation of the ray:
line rayLine = ray.pos ^ (- no << ray.direction);

// intersect the line and the sphere
pointPair intersection = dual(rayLine) << boundingSphere;

/*
Check if intersection is a real circle by computing
the radius squared:
*/
if ((intersection << intersection) > 0) {
  /*
  intersection with bounding sphere detected:
  . . .
  */
}
```

A Trip Down the BSP Tree

The descent down the BSP tree is more involved. Remember that the BSP tree recursively splits space into halves. The leaves of the tree contain the model faces that lie in that particular partition of space. During an intersection test, the goal of the BSP tree is to quickly arrive in those faces (leaves of the BSP tree) that the ray actually intersects. For instance, the first partition plane of the BSP tree divides space into two halves. Unless the ray happens to be parallel to this plane, it always intersects both halves, so we split the line in two, and each line segment goes down to the appropriate half of the BSP tree to be tested for intersections. Of course, we can take advantage of the fact that no faces lie outside of the bounding sphere: before we start our descent, we clip the ray against the bounding sphere, resulting in the initial line segment.

On the geometric algebra side of things, what this all boils down to is that we require a representation for line segments. This specific representation must be picked so that it works well while descending down a BSP tree.

We decided to use a pair of flat points to represent such line segments. One flat point represents the start and the other the end of the line segment. We now show how the BSP descent algorithm looks with this choice of representation. Afterwards, we briefly discuss the alternative representation of CGA point pairs to represent line segments.

To bootstrap the BSP descent algorithm, we need to have a line segment that represents the ray as clipped against the bounding sphere. As input for the recursive descent, we take the point pair named `intersection` from the previous code fragment and dissect that into two flat points using (14.13):

```
/*
 Dissects a point pair 'pp' into two flat points 'fpt1' and 'fpt2'
*/
void dissectPointPair(const pointPair &pp,
    normalizedFlatPoint &fpt1, normalizedFlatPoint &fpt2) {
  mv::Float n = sqrt(pp << pp);
  dualPlane v1 = ni << pp;
  dualPlane v2 = n * v1;
  dualSphere v3 = v1 << pp;

  mv::Float scale = 1.0 /(v3 << ni);
  fpt1 = (scale * ni) ^ (v3 + v2);
  fpt2 = (scale * ni) ^ (v3 - v2);
}
```

We sort the two points according to distance from the start of the ray. This is useful because it allows us to prune part of the BSP tree descent: we want to find the closest intersection of the ray with the model. So we first check the parts of the BSP tree closest to the start of the ray to see if we find any intersection there.

We determine what part of the tree to descend in first by checking on what side of the plane the start of the ray lies:

```
// partitionPlane is the plane that splits the space in two halves.
if ((ray.pos << dual(partitionPlane)) > 0.0) {
  // descend front side first
}
else if ((ray.pos << dual(partitionPlane)) < 0.0) {
  // descend back side first
}
else {
  // _always_ descend both sides
}
```

For the actual descent of the tree, we first check on what side(s) of the partition plane our line segment lies (this is not necessarily the same side as the start of the ray):

```
/*
partitionPlane is the plane that splits the space in two halves.
We compute the signed distance of the two flat points to the plane:
*/
mv::Float sd1 = no << (dual(partitionPlane) << fp1);
mv::Float sd2 = no << (dual(partitionPlane) << fp2);
```

If the line segment lies on both sides of the partition plane, we have to compute the intersection point of the line segment and the plane. Otherwise, we can simply descend down the appropriate node in the BSP tree.

The following code checks if we have to compute the intersection point, and computes it:

```
/*
If one signed distance is negative, and the other is positive,
then the line segment must lie on both sides of the plane:
*/
if (sd1 * sd2 < 0.0) {
  /*
  Compute the intersection point:
  -compute the line representation of the ray
  -intersect rayLine line and the partitionPlane
  */
  line rayLine = ray.pos ^ (- no << ray.direction);
  flatPoint fpI = unit_r(dual(partitionPlane) << line);

  // descend further down both sides of the BSP tree
  // . . .
}
```

A degenerate case occurs when the line segment lies inside the partition plane. We then simply descend to both children of the current BSP node.

Representing Line Segments with Point Pairs

Instead of pairs of flat points, we could use point pairs (i.e., 0-spheres) to represent line segments. But at some point we will have to dissect the point pair into two points to form two new point pairs (each on a different side of the partition plane), and this would then require the rather expensive (14.13). It is computationally more efficient to have the ends of the segments readily available.

23.5.4 REFLECTION

Reflections are computed as part of the shading computation treated in the next section. Reflection of any blade in any other blade is quite simple in the conformal model, though there are some signs depending on whether they are represented directly or dually (see Table 7.1). The following function implements the reflection. It forms a plane from the surface attitude (a free bivector), and reflects the ray direction (a free vector) in it:

```
color scene::reflect(const ray &R, int maxRecursion, const
        surfacePoint &spt) const {
```

```
/*
Reflect the ray direction in the surface attitude,
and spawn a new ray with that direction at the surface point.
*/
ray reflectedRay(
  spt.getPt(),                  // starting position of new ray
  -((spt.getAtt() ^ no) *
  R.get_direction() *
  reverse(spt.getAtt() ^ no))   // direction of new ray
  );

  return trace(reflectedRay, maxRecursion -1);
}
```

23.5.5 REFRACTION

Like reflections, refractions are also computed as part of the shading computation. The classical equation for refraction of direction vectors is:

$$\mathbf{u}' = \left(\text{sign} \, (\mathbf{n} \cdot \mathbf{u}) \, \sqrt{1 - \eta^2 + (\mathbf{n} \cdot \mathbf{u})^2 \, \eta^2} - (\mathbf{n} \cdot \mathbf{u}) \, \eta \right) \mathbf{n} + \eta \mathbf{u} \qquad (23.1)$$

where $\mathbf{n}$ is the surface normal, $\mathbf{u}$ is the ray direction, and η is the refractive constant. In our conformal implementation, both $\mathbf{n}$ and $\mathbf{u}$ are free vectors that can be plugged directly into the equation. As an interesting side note we mention that the same equation holds at the level of intersecting lines, by the extension principles of Section 13.4.

23.6 SHADING

The ray tracer performs shading computations for every spawned ray that intersects a model (unless it is a shadow ray). The shading computations should approximate how much light the material reflects along direction of the ray (and thus towards the camera). We do not discuss the shading computations in detail because they do not differ much from the classical computations. Even in the conformal model, shading computations are most conveniently performed using 3-D Euclidean vectors. For completeness, we briefly list the steps involved in shading:

- Before the ray tracer can perform any shading computations, it must interpolate the surface properties defined at the vertices for the intersection point. This is done by barycentric interpolation (see (11.19))
- If the model has a bump map applied to it, we have to modulate the surface attitude according to that bump map. This again involves interpolation.

- Finally, we have to perform the actual shading computations. Our ray tracer mimics the simple fixed function pipeline shading model of OpenGL, although we perform shading per intersection point instead of per vertex.

23.7 EVALUATION

We have shown most of the geometry involved in implementing a classical ray tracer. We have emphasized that there are usually several ways to represent computer graphics primitives in the conformal model of Euclidean geometry. For the ray tracer, we picked the most efficient representations we could find, but in other, less time-sensitive applications, we may pick the more elegant universal representations from the conformal model, since we do think it is ultimately beneficial to represent each geometric concept with a single type of blade. It would permit one to build the elements of a geometry library independent of the particular application they originated in.

Most interactive modeling transformations have been reduced to simple one-liners, direct conformal model formulas involving the mouse parameters. This is a part of the code where the conformal model shows off some of its power. The directness of the approach helped make the actual writing of the code a quick and enjoyable exercise.

The direct use of coordinates in the equations and the source code has been eliminated, except for places where they were actually useful (e.g., systematically generating position of all pixels in a 2-D image). This would not be possible with classical homogeneous coordinates. Just pick up any computer graphics text that treats construction of Euclidean transformations, and the pages will be littered with 4×4 matrices filled with coordinates.

As we already saw in the benchmarks in Chapter 22, there is a performance penalty for using the 5-D conformal model over the 3-D vector space model: our conformal ray tracer is about 25 percent slower than the vector space version. The difference compared to the 4-D homogeneous version is less than 5 percent.

In the graphical user interface of Section 23.4, we demonstrated the most pure manifestation of the conformal model. There, the correspondence to the algebraic elegance of the model versors and the simplicity of the code was most striking. We view it as an ideal example of what can be achieved by the methods for object-oriented programming of geometry exposed in this book.

PART IV
APPENDICES

A METRICS AND NULL VECTORS

A vector space $\mathbb{R}^n$ does not necessarily have metric properties: it may not be possible to measure and compare arbitary lengths and angles of vectors. For that, we need an *inner product*. This does not need to be Euclidean.

A.1 THE BILINEAR FORM

In mathematics, the metric properties are usually introduced through a *bilinear form* that encapsulates the measurement of all lengths in that space in a concise manner. As its name implies, a bilinear form has two arguments and is linear in each of them; the word "form" is mathematical jargon for "scalar-valued function." The bilinear form Q of the vector space $\mathbb{R}^n$ is therefore a bilinear function $Q : \mathbb{R}^n \times \mathbb{R}^n \to \mathbb{R}$, mapping pairs of vectors to a scalar.

If we assume that the bilinear form is given (with the original definition of the metric vector space), then we can define the inner product in terms of it as

$$\mathbf{x} \cdot \mathbf{y} \equiv Q[\mathbf{x}, \mathbf{y}].$$

Since we define our scalar product in terms of the inner product and the contraction in terms of the scalar product (in Chapter 3), all these subspace products pull in the same metric as characterized by Q.

A.2 DIAGONALIZATION TO ORTHONORMAL BASIS

Elementary linear algebra shows that any bilinear form in $\mathbb{R}^n$ can be brought into a standard diagonalized form; we would say that we can choose a certain basis $\{e_i\}_{i=1}^n$ in $\mathbb{R}^n$ so that for that basis the inner product becomes a particularly simple expression. This basis is *orthonormal*, which means that it satisfies

$$e_i \cdot e_j = \pm \delta_{ij}$$

(where δ_{ij} is 1 when $i = j$, and 0 when $i \neq j$). For each basis vector, the inner product with all other basis vectors is equal to 0, and with itself equal to 1 or -1. If all are $+1$, this is obviously equivalent to a Euclidean space, and it implies that on that basis, the inner product of two vectors $\mathbf{x} = \sum_{i=1}^n x_i e_i$ and $\mathbf{y} = \sum_{j=1}^n y_j e_j$ can be written as

$$\mathbf{x} \cdot \mathbf{y} = \sum_{i=1}^n x_i y_i.$$

In particular, $\mathbf{x} \cdot \mathbf{x} = \sum_{i=1}^n x_i^2$, which is positive, and which we interpret as the norm squared.

But the general definition also allows spaces in which some of the coefficients are negative. In physics, an important space is the *Minkowski space*, in which one of the vectors, say e_1, has $e_1 \cdot e_1 = -1$. Then we obtain, for the inner product of $\mathbf{x} = \sum_{i=1}^n x_i e_i$ and $\mathbf{y} = \sum_{j=1}^n y_j e_j$,

$$\mathbf{x} \cdot \mathbf{y} = -x_1 y_1 + x_2 y_2 + \cdots + x_n y_n. \tag{A.1}$$

Now $\mathbf{x} \cdot \mathbf{x}$ is no longer guaranteed to be positive, and strange effects occur with rotation-like operations in this space. Four-dimensional space-time has this structure; it is the framework of relativity. Spaces like this are surprisingly useful for computer science as well, even when you only want to do Euclidean geometry. They are the basis of the conformal model of Euclidean geometry in Chapters 13 through 16.

A.3 GENERAL METRICS

We develop some terminology for general metrics. If there are p independent unit vectors satisfying $e_i \cdot e_i = 1$, and q independent unit vectors satisfying $e_i \cdot e_i = -1$, then it is customary to say that the space has p "positive dimensions" and q "negative dimensions" (with $n = p + q$). Rather than $\mathbb{R}^n$, we then write $\mathbb{R}^{p,q}$ and call (p, q) the *signature* of the space. An n-dimensional Euclidean space is then written as $\mathbb{R}^{n,0}$, the representational space for its conformal model as $\mathbb{R}^{n+1,1}$, and an n-dimensional Minkowski space as $\mathbb{R}^{n-1,1}$. When we write $\mathbb{R}^n$, we mean to be nonspecific on the metric (don't confuse this with $\mathbb{R}^{n,0}$, which has a Euclidean metric!).

A.4 NULL VECTORS AND NULL BLADES

Mixed signature spaces contain vectors that have an inner product equal to 0 with themselves. These are called *null vectors*. In some sense, such a null vector is perpendicular to itself, which is somewhat hard to imagine.

We can construct null vectors easily in the Minkowski space above: take $\mathbf{n}_+ = (\mathbf{e}_1 + \mathbf{e}_2)/\sqrt{2}$ and $\mathbf{n}_- = (\mathbf{e}_2 - \mathbf{e}_1)/\sqrt{2}$. Then

$$\mathbf{n}_+ \cdot \mathbf{n}_+ = \tfrac{1}{2}(\mathbf{e}_1 + \mathbf{e}_2) \cdot (\mathbf{e}_1 + \mathbf{e}_2) = \tfrac{1}{2}(\mathbf{e}_1 \cdot \mathbf{e}_1 + \mathbf{e}_2 \cdot \mathbf{e}_2) = \tfrac{1}{2}(-1 + 1) = 0,$$

and similarly $\mathbf{n}_- \cdot \mathbf{n}_- = 0$. The inner product of $\mathbf{n}_+$ and $\mathbf{n}_-$ equals 1:

$$\mathbf{n}_+ \cdot \mathbf{n}_- = \tfrac{1}{2}(\mathbf{e}_1 + \mathbf{e}_2) \cdot (\mathbf{e}_2 - \mathbf{e}_1) = \tfrac{1}{2}(-\mathbf{e}_1 \cdot \mathbf{e}_1 + \mathbf{e}_2 \cdot \mathbf{e}_2) = 1.$$

If the space has a basis with p "positive" basis vectors and q "negative" basis vectors, the basis can contain at most $\min(p, q)$ null vectors. They then come in pairs, as in the example above, and for every null vector we can find another null vector such that their mutual inner product equals 1. Such vectors are each other's inverses relative to the inner product: they are *reciprocal* null vectors.

By using null vectors in composite constructions, we can make elements of higher grade that square to zero. A *null blade* must contain at least one null vector to square to zero. Yet beware that this is not a sufficient condition, for if the blade contains two reciprocal null vectors, it may not be null. An example is the 2-blade $\mathbf{n}_+ \wedge \mathbf{n}_-$, which squares to 1 (it is identical to $\mathbf{e}_1 \wedge \mathbf{e}_2$).

A.5 ROTORS IN GENERAL METRICS

The treatment of rotors in general metrics leads into subtle issues for which the understanding of the much simpler Euclidean case does not prepare. They include the explicit demand that the reverse of a rotor should be its inverse; that not all rotors are continuously connected to the identity; and that only in Euclidean and Minkowski metrics rotors can be written as the exponentials of bivectors. These matters are touched upon in Section 7.4; for a more precise treatment we recommend [52].

B CONTRACTIONS AND OTHER INNER PRODUCTS

This appendix collects some additional facts about contractions and inner products to give the correspondence with much other literature, as well as three long proofs for statements in Chapter 3.

B.1 OTHER INNER PRODUCTS

When you study the applied literature on geometric algebra, you will find an inner product used that is similar to the contraction, but not identical to it. It was originally introduced by Hestenes in [33], so we refer to it as the *Hestenes inner product* in this text. We define it here, and relate it to the contraction. It is most convenient to do so via an intermediate construction, the "dot product." More detail about these issues may be found in [17], which defends the contraction in detail.

B.1.1 THE DOT PRODUCT

Let $\mathbf{A}_k$ and $\mathbf{B}_l$ be blades of grade k and l, respectively. Then the *dot product* $\mathbf{A}_k \bullet \mathbf{B}_l$ is defined as

$$dot\ product: \quad \mathbf{A}_k \bullet \mathbf{B}_l = \begin{cases} \mathbf{A}_k \rfloor \mathbf{B}_l & \text{if } k \leq l \\ \mathbf{A}_k \lfloor \mathbf{B}_l = (-1)^{l(k-l)}\, \mathbf{B}_l \rfloor \mathbf{A}_k & \text{if } k \geq l \end{cases} \tag{B.1}$$

The dot product thus switches between left and right contraction depending on the grades. As a consequence of the definition, the dot product has a different and somewhat more symmetrical grade than the contraction:

$$\text{grade}(\mathbf{A}_k \bullet \mathbf{B}_l) = |k - l|.$$

It is therefore a mapping $\bullet : \bigwedge^k \mathbb{R}^n \times \bigwedge^l \mathbb{R}^n \to \bigwedge^{|k-l|} \mathbb{R}^n$. It is less often zero than the contraction (when the grade of the first arguments gets too high to have a non-zero outcome for the contraction, the second argument "takes over"). This may seem to be an advantage, but remember from Section 3.3 that a zero contraction has a geometric significance: it means that not all of $\mathbf{A}_k$ is contained in $\mathbf{B}_l$. Since that concept is of geometrical importance it needs to be encoded somewhere in the framework. When you use the dot product, you get a lot of conditional statements to do this, which could have been avoided by using the natural asymmetry of the contraction.

B.1.2 HESTENES' INNER PRODUCT

The literature on applied geometric algebra almost exclusively contains another inner product, which is like the dot product, *except that it is zero whenever one of the arguments is a scalar*. We refer to it here as the *Hestenes inner product*, and denote it $\bullet_H$. This is used in [33], and most texts derived from his approach, such as [15], denote it as $\cdot$ since it is the only inner product they use. Like the dot product, it leads to many special cases in general derivations (even more so, to treat the scalar cases).

B.1.3 NEAR EQUIVALENCE OF INNER PRODUCTS

Whenever the first argument has a lower or equal grade to the second, and there is no possibility of any of them being scalar, the contraction and inner products are all identical:

$$\mathbf{A} \bullet \mathbf{B} = \mathbf{A} \rfloor \mathbf{B} = \mathbf{A} \bullet_H \mathbf{B}, \quad \text{when } 0 \neq \text{grade}(\mathbf{A}) \leq \text{grade}(\mathbf{B}).$$

With careful writing, that limits the consequences of the choice of inner product, so that results can be compared and transferred with ease.

We have made an effort to make most equations in this book of the type "lowest grade first," so that one can easily substitute one's favorite inner product and follow the computation, but in some results of general validity (extending to scalars and arbitrary order of arguments), the use of the contraction is essential. Therefore we have kept the asymmetric notation $\rfloor$ for this inner product.

Other texts often have the same tendency in their ordering of the arguments, so the issue is not as serious as it might appear, at least algebraically.

In our GAviewer software, which was used to render most of the figures in this book, a preferred interpretation of the inner product can be set, so that formulas from this or any other book may be entered by using a dot notation. Its default setting is the left contraction. In the GA sandbox of the programming exercise, we only provide compact operators for the left contraction $\ll$ and the right contraction $\gg$ (with the arrows pointing to the argument that should have the lower grade for the result to be nonzero).

B.1.4 GEOMETRIC INTERPRETATION AND USAGE

The interpretation of the contraction $\mathbf{A} \rfloor \mathbf{B}$ is clear: it is "the part of $\mathbf{B}$ most unlike $\mathbf{A}$" (in linear fashion proportional to both, and with a specific sign, see Section 3.3). Because of the above, the dot product $\mathbf{A} \bullet \mathbf{B}$ and the Hestenes inner product $\bullet_H$ have a slightly more involved interpretation: they are the part of the larger grade blade most unlike the lower grade blade (in linear fashion, and with a different sign). This causes tests on the relative grade in many derivations and statements, complicating their range of validity and proofs considerably.

If you follow good programming practice and let the interpretation of your computational results be ruled automatically by the grades appearing in it, this is a nuisance. For computer vision in which data can accidentally degenerate (4 points may determine a plane rather than a polyhedron, somewhere deep inside a program), you may not be able to predict the grades of the result. The contraction then appears to be the more convenient inner product in the design of algorithms, and that is the reason we use it in this book.

On the other hand, if you know beforehand what grades to expect from your computation, and how to interpret them, then the dot product or the similar Hestenes inner product may be easier to work with in compact computations, especially in combination with the geometric product of Chapter 6. Apparently, this is the case in the applications to physics covered in [15], and in many engineering applications by the same school. If they had a strong need to switch to the mathematically tidier contraction, they would have done so by now.

If you are interested in more detail on these issues, [17] attempts to give a fair overview of pros and cons (and of course ends up preferring the contraction).

B.2 EQUIVALENCE OF THE IMPLICIT AND EXPLICIT CONTRACTION DEFINITIONS

We show that the contraction product explicitly defined by (3.7) through (3.11) is consistent with the implicit definition (3.6). That is a bit of work, but it contains useful techniques, and we prove a useful recursive formula in the process.

Some remarks before we start. In a degenerate metric, the implicit definition is not specific to define the contraction for all arguments, whereas the explicit definition can do that. The strongest we can do in this case is show that in all cases where the implicit definition does define an outcome, this is *consistent* with the explicit result. In nondegenerate metrics, we have a unique outcome even with the implicit definition, so we can even show *equivalence*. In that case:

$$\mathbf{X} * \mathbf{A} = \mathbf{X} * \mathbf{B} \text{ for all } \mathbf{X} \iff \mathbf{A} = \mathbf{B}. \tag{B.2}$$

We omit the proof of this statement, which is similar to the corresponding statement for the inner product of vectors in linear algebra; it is most simply done by developing all blades on a basis.

It is bothersome to point out the different behavior for the two kinds of metric with each statement we derive; the nondegenerate case is "typical" in the text that follows.

Proof of consistency of (3.7): $\mathbf{X} * (\alpha \rfloor \mathbf{B}) = (\mathbf{X} \wedge \alpha) * \mathbf{B} = (\alpha \mathbf{X}) * \mathbf{B} = \alpha(\mathbf{X} * \mathbf{B}) = \mathbf{X} * (\alpha \mathbf{B})$, so $\alpha \rfloor \mathbf{B} = \alpha \mathbf{B}$.

Proof of consistency of (3.8): $\mathbf{X} * (\mathbf{B} \rfloor \alpha) = (\mathbf{X} \wedge \mathbf{B}) * \alpha$ and the latter can only be nonzero if $(\mathbf{X} \wedge \mathbf{B})$ is a scalar. Under the assumption of grade$(\mathbf{B}) > 0$, therefore, $\mathbf{B} \rfloor \alpha = 0$.

Proof of consistency of (3.9): $\mathbf{X} * (\mathbf{a} \rfloor \mathbf{b}) = (\mathbf{X} \wedge \mathbf{a}) * \mathbf{b} = \mathbf{X}(\mathbf{a} * \mathbf{b}) = \mathbf{X}(\mathbf{a} \cdot \mathbf{b})$, since $\mathbf{X}$ is scalar if the result is to be nonzero.

Proof of consistency of (3.10): We show how a contraction of a vector with a blade can be computed by passing $\mathbf{x}\rfloor$ through the outer product, with appropriately alternating signs:

$$\mathbf{x} \rfloor (\mathbf{a}_1 \wedge \mathbf{a}_2 \wedge \cdots \wedge \mathbf{a}_k) = \sum_{i=1}^{k} (-1)^{i-1} \mathbf{a}_1 \wedge \mathbf{a}_2 \wedge \cdots \wedge (\mathbf{x} \rfloor \mathbf{a}_i) \wedge \cdots \wedge \mathbf{a}_k. \tag{B.3}$$

Since the term $\mathbf{x} \rfloor \mathbf{a}_i$ is simply the inner product $\mathbf{x} \cdot \mathbf{a}_i$, this allows rapid, one-line evaluation of an inner product of a vector on a blade as a simple sum of terms. That makes it an important formula to remember.

When we want to derive (3.16) from (3.6) for the $(k-1)$-blade $\mathbf{x} \rfloor (\mathbf{a}_1 \wedge \cdots \wedge \mathbf{a}_k)$, we have to take its scalar product with an arbitrary $(k-1)$-blade $(\mathbf{y}_{k-1} \wedge \cdots \wedge \mathbf{y}_1)$. We then write the scalar product as a determinant, and use the properties of the determinant in a Laplace expansion procedure:

$$(\mathbf{y}_{k-1} \wedge \cdots \wedge \mathbf{y}_1) * (\mathbf{x} \rfloor (\mathbf{a}_1 \wedge \cdots \wedge \mathbf{a}_k))$$
$$= (\mathbf{y}_{k-1} \wedge \cdots \wedge \mathbf{y}_1 \wedge \mathbf{x}) * (\mathbf{a}_1 \wedge \cdots \wedge \mathbf{a}_k)$$
$$= (\mathbf{a}_1 \wedge \cdots \wedge \mathbf{a}_k) * (\mathbf{y}_{k-1} \wedge \cdots \wedge \mathbf{y}_1 \wedge \mathbf{x})$$
$$= \begin{vmatrix} \mathbf{x} \cdot \mathbf{a}_1 & \mathbf{x} \cdot \mathbf{a}_2 & \cdots & \mathbf{x} \cdot \mathbf{a}_k \\ \mathbf{y}_1 \cdot \mathbf{a}_1 & \mathbf{y}_1 \cdot \mathbf{a}_2 & \cdots & \mathbf{y}_1 \cdot \mathbf{a}_k \\ \vdots & & \ddots & \vdots \\ \mathbf{y}_{k-1} \cdot \mathbf{a}_1 & \mathbf{y}_{k-1} \cdot \mathbf{a}_2 & \cdots & \mathbf{y}_{k-1} \cdot \mathbf{a}_k \end{vmatrix}$$

$$= \sum_{i=1}^{k}(-1)^{i-1}(\mathbf{x}\cdot\mathbf{a}_i)\begin{vmatrix} \mathbf{y}_1\cdot\mathbf{a}_1 & \cdots & \mathbf{y}_1\cdot\mathbf{a}_{i-1} & \mathbf{y}_1\cdot\mathbf{a}_{i+1} & \cdots & \mathbf{y}_1\cdot\mathbf{a}_k \\ \vdots & & \vdots & \vdots & & \vdots \\ \mathbf{y}_{k-1}\cdot\mathbf{a}_1 & \cdots & \mathbf{y}_{k-1}\cdot\mathbf{a}_{i-1} & \mathbf{y}_{k-1}\cdot\mathbf{a}_{i+1} & \cdots & \mathbf{y}_{k-1}\cdot\mathbf{a}_k \end{vmatrix}$$

$$= \sum_{i=1}^{k}(-1)^{i-1}(\mathbf{x}\cdot\mathbf{a}_i)\,(\mathbf{y}_{k-1}\wedge\cdots\wedge\mathbf{y}_1)*(\mathbf{a}_1\wedge\mathbf{a}_2\wedge\cdots\wedge\breve{\mathbf{a}}_i\wedge\cdots\wedge\mathbf{a}_k)$$

$$= (\mathbf{y}_{k-1}\wedge\cdots\wedge\mathbf{y}_1)*\left(\sum_{i=1}^{k}(-1)^{i-1}(\mathbf{a}_1\wedge\mathbf{a}_2\wedge\cdots\wedge(\mathbf{x}\cdot\mathbf{a}_i)\wedge\cdots\wedge\mathbf{a}_k)\right)$$

where the notation $\breve{\mathbf{a}}_i$ means that the term $\mathbf{a}_i$ is omitted from the outer product. Since this must hold for any blade $(\mathbf{y}_{k-1}\wedge\cdots\wedge\mathbf{y}_1)$, we apply (B.2), and we have proved (3.16) when we write the inner product between vectors as contractions.

It is now easy to establish a recursive formula for a k-blade $\mathbf{A}_k$:

$$\mathbf{x}\rfloor\mathbf{A}_k = \mathbf{x}\rfloor(\mathbf{a}\wedge\mathbf{A}_{k-1}) \equiv (\mathbf{x}\rfloor\mathbf{a})\wedge\mathbf{A}_{k-1} - \mathbf{a}\wedge(\mathbf{x}\rfloor\mathbf{A}_{k-1}). \tag{B.4}$$

Note that this formula even works for $k = 1$. Applying this repeatedly, we get (3.10), which completes its proof.

Proof of consistency of (3.11): The proof of (3.11) shows how nicely all our products intertwine:

$$\mathbf{X}*((\mathbf{A}\wedge\mathbf{B})\rfloor\mathbf{C}) = (\mathbf{X}\wedge(\mathbf{A}\wedge\mathbf{B}))*\mathbf{C}$$
$$= ((\mathbf{X}\wedge\mathbf{A})\wedge\mathbf{B})*\mathbf{C}$$
$$= (\mathbf{X}\wedge\mathbf{A})*(\mathbf{B}\rfloor\mathbf{C})$$
$$= \mathbf{X}*(\mathbf{A}\rfloor(\mathbf{B}\rfloor\mathbf{C}))$$

This is valid for arbitrary $\mathbf{X}$; therefore, by the orthogonality property of the scalar product (B.2), we must have (3.11).

All the above shows that the contraction operation defined by equations (3.7) through (3.11) is consistent with (3.6). For a space with a nondegenerate metric, in which both of these definitions define the contraction, they are therefore equivalent.

The explicit 5-part definition allows us to also compute the contraction for a space with a degenerate metric, without getting a contradiction with the now-incomplete implicit definition. The explicit definition is therefore the one we should use in general. But of course, (3.6) still holds for any $\mathbf{X}$ in any metric. That property, which shows how the contraction interleaves with the outer product and the scalar product, remains the main algebraic motivation for introducing the contraction product. (Mathematically, it makes the contraction the adjoint of the outer product relative to the scalar product a natural construction in the treatment of algebras.)

B.3 PROOF OF THE SECOND DUALITY

Since the second duality relationship is so important, we prove that it follows from our definitions. We repeat the statement of the second duality (from (3.21)):

$$(A \rfloor B) \rfloor C = A \wedge (B \rfloor C) \text{ when } A \subseteq C$$

Its proof is straightforward and rather instructive if you are interested in the mathematics of using the algebra; but you will not miss anything essentially geometrical if you skip it.

We prove (3.21) by a four-step induction, each result forming the basis for the next.

0. As *step 0* of the induction, if at least one of A, B or C is a scalar, this is covered by either of the cases "A is a scalar" or "A is not a scalar, but B is." Both are trivial.

1. As *step 1* of the induction, we have the straightforward point of departure for vectors:

$$c \wedge (b \rfloor c) = (c \rfloor b) \rfloor c, \tag{B.5}$$

which is basically two ways of writing the scalar product of the scalar $b * c = b \rfloor c = b \cdot c$ with the vector c.

2. Next, we prove from this *step 2*:

$$a \wedge (b \rfloor C) = (a \rfloor b) \rfloor C \text{ if } a \subseteq C. \tag{B.6}$$

This is done by induction: assume that $a \wedge (b \rfloor C_{k-1}) = (a \rfloor b) \rfloor C_{k-1}$ if $a \subseteq C_{k-1}$. Then form $C_k = c_k \wedge C_{k-1}$ and try to prove validity for that. First, note that $a \subseteq C_k$ is equivalent to $a \wedge C_{k-1} = C_k$ for some blade C_{k-1}. Then we obtain:

$$\begin{aligned}
a \wedge (b \rfloor C_k) &= a \wedge (b \rfloor (a \wedge C_{k-1})) \\
&= a \wedge ((b \rfloor a) \wedge C_{k-1} - a \wedge (b \rfloor C_{k-1})) \\
&= a \wedge (b \rfloor a) \wedge C_{k-1} \\
&= (a \rfloor b) \rfloor C_k.
\end{aligned}$$

Induction from (B.5) then proves (B.6).

3. Then we prove *step 3*:

$$a \wedge (B \rfloor C) = (a \rfloor B) \rfloor C \text{ for } a \subseteq C, \tag{B.7}$$

through assuming $\mathbf{a} \wedge (\mathbf{B}_{l-1}\rfloor\mathbf{C}) = (\mathbf{a}\rfloor\mathbf{B}_{l-1})\rfloor\mathbf{C}$ for $\mathbf{a} \wedge \mathbf{C} = 0$, and deriving for $\mathbf{B}_l = \mathbf{b}_l \wedge \mathbf{B}_{l-1}$:

$$
\begin{aligned}
(\mathbf{a}\rfloor\mathbf{B}_l)\rfloor\mathbf{C} &= \\
&= (\mathbf{a}\rfloor(\mathbf{b}_l \wedge \mathbf{B}_{l-1}))\rfloor\mathbf{C} \\
&= (\mathbf{a}\rfloor\mathbf{b}_l)(\mathbf{B}_{l-1}\rfloor\mathbf{C}) - (\mathbf{b}_l \wedge (\mathbf{a}\rfloor\mathbf{B}_{l-1}))\rfloor\mathbf{C} \\
&= (\mathbf{a}\rfloor\mathbf{b}_l)(\mathbf{B}_{l-1}\rfloor\mathbf{C}) - \mathbf{b}_l\rfloor((\mathbf{a}\rfloor\mathbf{B}_{l-1})\rfloor\mathbf{C}) \\
&= (\mathbf{a}\rfloor\mathbf{b}_l)(\mathbf{B}_{l-1}\rfloor\mathbf{C}) - \mathbf{b}_l\rfloor(\mathbf{a} \wedge (\mathbf{B}_{l-1}\rfloor\mathbf{C})) \\
&= \mathbf{a} \wedge (\mathbf{b}_l\rfloor(\mathbf{B}_{l-1}\rfloor\mathbf{C})) \quad \text{[by developing the last term]} \\
&= \mathbf{a} \wedge ((\mathbf{b}_l \wedge \mathbf{B}_{l-1})\rfloor\mathbf{C}) \\
&= \mathbf{a} \wedge (\mathbf{B}_l\rfloor\mathbf{C}),
\end{aligned}
$$

and induction from (B.6) now proves (B.7).

4. Finally, we prove the actual second duality in *step 4*:

$$
\mathbf{A} \wedge (\mathbf{B}\rfloor\mathbf{C}) = (\mathbf{A}\rfloor\mathbf{B})\rfloor\mathbf{C} \quad \text{for } \mathbf{A} \subseteq \mathbf{C}, \tag{B.8}
$$

through assuming $\mathbf{A}_{m-1} \wedge (\mathbf{B}\rfloor\mathbf{C}) = (\mathbf{A}_{m-1}\rfloor\mathbf{B})\rfloor\mathbf{C}$ for $\mathbf{A}_{m-1} \subseteq \mathbf{C}$, and deriving for $\mathbf{A}_m = \mathbf{a}_m \wedge \mathbf{A}_{m-1}$:

$$
\begin{aligned}
\mathbf{A}_m \wedge (\mathbf{B}\rfloor\mathbf{C}) &= (\mathbf{a}_m \wedge \mathbf{A}_{m-1}) \wedge (\mathbf{B}\rfloor\mathbf{C}) \\
&= \mathbf{a}_m \wedge ((\mathbf{A}_{m-1}\rfloor\mathbf{B})\rfloor\mathbf{C}) \\
&= ((\mathbf{a}_m\rfloor(\mathbf{A}_{m-1}\rfloor\mathbf{B}))\rfloor\mathbf{C} \quad \text{by (B.7)} \\
&= (\mathbf{a}_m \wedge \mathbf{A}_{m-1})\rfloor\mathbf{B})\rfloor\mathbf{C} \\
&= (\mathbf{A}_m\rfloor\mathbf{B})\rfloor\mathbf{C}
\end{aligned}
$$

and induction from (B.6) proves the result.

The required containment of $\mathbf{A}$ into $\mathbf{C}$ enters this proof already in step 2, to prevent a proliferation of extra terms that would make the statement less simple and therefore less usable.

B.4 PROJECTION AND THE NORM OF THE CONTRACTION

We claimed in Section 3.3 that the norm of the contraction $\mathbf{A}\rfloor\mathbf{B}$ is proportional to the norm of $\mathbf{B}$ times the norm of the projection of $\mathbf{A}$ onto $\mathbf{B}$. Let us combine the proof of that statement with a proof of the projection formula $P_{\mathbf{B}}[\mathbf{A}] = (\mathbf{A}\rfloor\mathbf{B})\rfloor\mathbf{B}^{-1}$ itself. It is all based on the scalar product definitions of the norm (3.4), the angle (3.5), the left contraction (3.6) and the right contraction (3.18), plus usage of duality and the reversion properties of the scalar product in (3.3).

We first express the norm of $\mathbf{A} \rfloor \mathbf{B}$ in something resembling the projection:

$$
\begin{aligned}
\|\mathbf{A} \rfloor \mathbf{B}\|^2 &= (\mathbf{A} \rfloor \mathbf{B}) * (\mathbf{A} \rfloor \mathbf{B})^\sim = (\mathbf{A} \rfloor \mathbf{B}) * (\widetilde{\mathbf{B}} \lfloor \widetilde{\mathbf{A}}) \\
&= \mathbf{B} * \left((\mathbf{A} \rfloor \mathbf{B})^\sim \wedge \mathbf{A} \right) = \left(\mathbf{B} \lfloor (\mathbf{A} \rfloor \mathbf{B})^\sim \right) * \mathbf{A} \\
&= \left((\mathbf{A} \rfloor \mathbf{B}) \rfloor \widetilde{\mathbf{B}} \right) * \widetilde{\mathbf{A}}.
\end{aligned}
$$

We can express the norm of $(\mathbf{A} \rfloor \mathbf{B}) \rfloor \widetilde{\mathbf{B}}$ back in terms of the norm of $\mathbf{A} \rfloor \mathbf{B}$:

$$
\begin{aligned}
\|(\mathbf{A} \rfloor \mathbf{B}) \rfloor \widetilde{\mathbf{B}}\|^2 &= \left((\mathbf{A} \rfloor \mathbf{B}) \rfloor \widetilde{\mathbf{B}} \right) * \left((\mathbf{A} \rfloor \mathbf{B}) \rfloor \widetilde{\mathbf{B}} \right)^\sim = \left((\mathbf{A} \rfloor \mathbf{B}) \rfloor \widetilde{\mathbf{B}} \right) * \left(\mathbf{B} \lfloor (\mathbf{A} \rfloor \mathbf{B})^\sim \right) \\
&= \mathbf{B} * \left((\mathbf{A} \rfloor \mathbf{B})^\sim \wedge \left((\mathbf{A} \rfloor \mathbf{B}) \rfloor \widetilde{\mathbf{B}} \right) \right) = \mathbf{B} * \left(\left((\mathbf{A} \rfloor \mathbf{B})^\sim \rfloor (\mathbf{A} \rfloor \mathbf{B}) \right) \rfloor \widetilde{\mathbf{B}} \right) \\
&= \|\mathbf{A} \rfloor \mathbf{B}\|^2 \, \|\mathbf{B}\|^2.
\end{aligned}
$$

Now, defining $\mathbf{P} \equiv \left((\mathbf{A} \rfloor \mathbf{B}) \rfloor \widetilde{\mathbf{B}} \right) / (\mathbf{B} * \widetilde{\mathbf{B}})$ and denoting the cosine of $\mathbf{P}$ and $\mathbf{A}$ as $\cos(\mathbf{P}, \mathbf{A})$ we rewrite these results as

$$
\|\mathbf{A} \rfloor \mathbf{B}\|^2 = \|\mathbf{P}\| \, \|\mathbf{A}\| \, \|\mathbf{B}\|^2 \cos(\mathbf{P}, \mathbf{A})
$$

and

$$
\|\mathbf{P}\|^2 = \|\mathbf{A} \rfloor \mathbf{B}\|^2 / \|\mathbf{B}\|^2.
$$

It follows that

$$
\|\mathbf{P}\| = \|\mathbf{A}\| \, \cos(\mathbf{P}, \mathbf{A}),
$$

and since $\mathbf{P}$ was proved to be contained in $\mathbf{B}$ in Section 3.3, it is the projection of $\mathbf{A}$ onto $\mathbf{B}$. (Its idempotence is proved in Section 3.6.) With that meaning of $\mathbf{P}$, we interpret the second result as the desired expression for the norm of $\mathbf{A} \rfloor \mathbf{B}$.

C SUBSPACE PRODUCTS RETRIEVED

C.1 OUTER PRODUCT FROM GEOMETRIC PRODUCT

We prove that the outer product defined by (6.10) and (6.11) is consistent with the familiar outer product we had before. We repeat the statements for convenience:

$$\mathbf{a} \wedge \mathbf{B} = \tfrac{1}{2}(\mathbf{a}\,\mathbf{B} + \widehat{\mathbf{B}}\,\mathbf{a}), \tag{C.1}$$

$$\mathbf{B} \wedge \mathbf{a} = \tfrac{1}{2}(\mathbf{B}\,\mathbf{a} + \mathbf{a}\,\widehat{\mathbf{B}}), \tag{C.2}$$

where $\widehat{\mathbf{B}} = (-1)^{\text{grade}(\mathbf{B})}\,\mathbf{B}$ is the grade involution of $\mathbf{B}$ introduced in Section 2.9.5. (Writing the equations in this form makes it easier to lift them to general multivectors B.)

We check the earlier defining properties of Section 2.10.

- If $\mathbf{B}$ is a scalar β, we get $\mathbf{a} \wedge \beta = \beta\,\mathbf{a} = \beta \wedge \mathbf{a}$.
- If $\mathbf{B}$ is a vector, we obtain antisymmetry, as we saw above.
- To demonstrate associativity, let us develop $(\mathbf{a} \wedge \mathbf{B}) \wedge \mathbf{c}$ and $\mathbf{a} \wedge (\mathbf{B} \wedge \mathbf{c})$.

$$(\mathbf{a} \wedge \mathbf{B}) \wedge \mathbf{c} = \tfrac{1}{4}\,(\mathbf{a}\,\mathbf{B}\,\mathbf{c} + \mathbf{a}\,\mathbf{c}\,\widehat{\mathbf{B}} + \widehat{\mathbf{B}}\,\widehat{\mathbf{c}}\,\mathbf{a} + \widehat{\mathbf{c}}\,\mathbf{B}\,\mathbf{a})$$

$$\mathbf{a} \wedge (\mathbf{B} \wedge \mathbf{c}) = \tfrac{1}{4}\,(\mathbf{a}\,\mathbf{B}\,\mathbf{c} + \widehat{\mathbf{B}}\,\mathbf{a}\,\mathbf{c} + \mathbf{c}\,\widehat{\mathbf{a}}\,\widehat{\mathbf{B}} + \mathbf{c}\,\mathbf{B}\,\widehat{\mathbf{a}})$$

The two are equivalent if $a c \widehat{B} - c \widehat{a} \widehat{B} = \widehat{B} a c - \widehat{B} \widehat{c} a$. But this can be simplified using (6.9) to $(a \cdot c) \widehat{B} = \widehat{B} (a \cdot c)$. That holds by the commutativity of scalars under the geometric product, so the two expressions are indeed equivalent. In particular, this demonstrates the associativity law for vectors:

$$(a \wedge b) \wedge c = a \wedge (b \wedge c).$$

By applying it recursively we get general associativity for blades.

- Linearity and distributivity over addition trivially hold by the corresponding properties of the geometric product. They permit extending the results to general multivectors B.

This demonstrates that the two equations above indeed identify the same outer product structure that we had before, at least when one of the factors is a vector. Since we have associativity, this can be extended to general blades, and by linearity to general multivectors. Only the case of two scalars is formally not yet included.

C.2 CONTRACTIONS FROM GEOMETRIC PRODUCT

We prove (6.12) and (6.13), repeated for convenience:

$$a \rfloor B = \tfrac{1}{2}(a B - \widehat{B} a), \qquad\qquad (C.3)$$

$$B \lfloor a = \tfrac{1}{2}(B a - a \widehat{B}). \qquad\qquad (C.4)$$

We focus on the formula for the left contraction $a \rfloor B$, and show consistency with the earlier definition of (3.7) through (3.11). The right contraction $B \lfloor a$ is completely analogous.

- If B is a scalar β, we get $a \rfloor \beta = 0$.
- If B is a vector b, we obtain symmetry and equivalence to the classical inner product $a \cdot b$ of (6.9).
- To demonstrate the distribution of the contraction over an outer product, we first show

$$a \rfloor (b C) = \tfrac{1}{2}(a b C + b \widehat{C} a) = \tfrac{1}{2}(a b C + b a C - b a C + b \widehat{C} a)$$
$$= (a \cdot b) C - b (a \rfloor C). \qquad\qquad (C.5)$$

Similarly, you may derive that

$$a \rfloor (\widehat{C} b) = (a \rfloor \widehat{C}) b + C (a \cdot b). \qquad\qquad (C.6)$$

These equations are important by themselves, for they specify the distributivity of the contraction over the geometric product. Adding them produces $a \rfloor (b \wedge C) = (a \cdot b) \wedge C - b \wedge (a \rfloor C)$. This is essentially (B.4), and we can pick up the rest of the proof from there: applied repeatedly, this equation gives us the desired result (3.10) for blades.

- If we follow the same procedure we used to demonstrate associativity of the outer product, we get a surprise. That derivation does not show the associativity of the contraction (fortunately!), but the associativity of a "sandwich" combination of the two contractions:

$$(\mathbf{a} \rfloor \mathbf{B}) \lfloor \mathbf{c} = \mathbf{a} \rfloor (\mathbf{B} \lfloor \mathbf{c}).$$

For vectors, this becomes the rather trivial $(\mathbf{a} \rfloor \mathbf{b}) \lfloor \mathbf{c} = \mathbf{a} \rfloor (\mathbf{b} \lfloor \mathbf{c})$, which holds because both sides are zero.

- That leaves us with no tools to prove the remaining defining equation (3.11) for the earlier contraction: $\mathbf{A} \rfloor (\mathbf{B} \rfloor \mathbf{C}) = (\mathbf{A} \wedge \mathbf{B}) \rfloor \mathbf{C}$. In fact, we cannot even compute both sides simultaneously, since we always need a vector as an argument to use (6.12).

As with the outer product, the proposed equations do not define the contraction, and do therefore not prove it to be identical with the earlier contraction from Chapter 3; but we have at least shown that they are consistent with it.

C.3 PROOF OF THE GRADE APPROACH

We prove that the subspace products of blades can be retrieved as certain selected grades of the geometric product, as stated in (6.19) through (6.22). We repeat those statements for convenience:

$$\mathbf{A}_k \wedge \mathbf{B}_l \equiv \langle \mathbf{A}_k \, \mathbf{B}_l \rangle_{l+k} \tag{C.7}$$

$$\mathbf{A}_k \rfloor \mathbf{B}_l \equiv \langle \mathbf{A}_k \, \mathbf{B}_l \rangle_{l-k} \tag{C.8}$$

$$\mathbf{A}_k \lfloor \mathbf{B}_l \equiv \langle \mathbf{A}_k \, \mathbf{B}_l \rangle_{k-l} \tag{C.9}$$

$$\mathbf{A}_k * \mathbf{B}_l \equiv \langle \mathbf{A}_k \, \mathbf{B}_l \rangle_0 \tag{C.10}$$

We show the correctness of the equation for the outer product using the earlier derived (6.10). Write $\mathbf{A}_k$ as $\mathbf{A}_k = \mathbf{a}_k \wedge \mathbf{A}_{k-1}$. Then:

$$
\begin{aligned}
\langle \mathbf{A}_k \, \mathbf{B}_l \rangle_{l+k} &= \langle (\mathbf{a}_k \wedge \mathbf{A}_{k-1}) \, \mathbf{B}_l \rangle_{l+k} \\
&= \tfrac{1}{2} \langle \mathbf{a}_k \mathbf{A}_{k-1} \mathbf{B}_l + \widehat{\mathbf{A}}_{k-1} \mathbf{a}_k \mathbf{B}_l \rangle_{l+k} \\
&= \tfrac{1}{2} \langle \mathbf{a}_k \mathbf{A}_{k-1} \mathbf{B}_l + \widehat{\mathbf{A}}_{k-1} \mathbf{a}_k \mathbf{B}_l + \widehat{\mathbf{A}}_{k-1} \widehat{\mathbf{B}}_l \mathbf{a}_k - \widehat{\mathbf{A}}_{k-1} \widehat{\mathbf{B}}_l \mathbf{a}_k \rangle_{l+k} \\
&= \langle \mathbf{a}_k \wedge (\mathbf{A}_{k-1} \mathbf{B}_l) + \widehat{\mathbf{A}}_{k-1} (\mathbf{a}_k \rfloor \mathbf{B}_l) \rangle_{l+k} \\
&= \langle \mathbf{a}_k \wedge (\mathbf{A}_{k-1} \mathbf{B}_l) \rangle_{l+k} \quad \text{[other term at most of grade } l+k-2] \\
&= \mathbf{a}_k \wedge \langle \mathbf{A}_{k-1} \mathbf{B}_l \rangle_{l+k-1} \\
&\quad \vdots \\
&= \mathbf{A}_k \wedge \langle \mathbf{B}_l \rangle_l \\
&= \mathbf{A}_k \wedge \mathbf{B}_l
\end{aligned}
$$

This proof does not work when $\mathbf{A}_k$ is a scalar α of grade 0. But then $\langle \alpha \mathbf{B}_l \rangle_l = \alpha \mathbf{B}_l = \alpha \wedge \mathbf{B}_l$, so the result still holds.

The definition (C.8) of the left contraction, in terms of geometric product and grade operator, is also consistent with the earlier definition of (3.7) through (3.11), as we show one by one.

- The scalar properties (3.7) and (3.8) are straightforward:

$$\mathbf{A}_k \rfloor \alpha = \langle \mathbf{A}_k \alpha \rangle_{-k} = 0 \ \ \text{if} \ \ k \neq 0$$

$$\alpha \rfloor \mathbf{A}_k = \langle \alpha \mathbf{A}_k \rangle_k = \alpha \mathbf{A}_k$$

- The inner product correspondence property (3.9) is rather obvious: $\langle \mathbf{a}\,\mathbf{b} \rangle_0 = \mathbf{a} * \mathbf{b} = \mathbf{a} \cdot \mathbf{b}$.
- The property (3.10), which would read $\mathbf{a} \rfloor (\mathbf{b} \wedge \mathbf{B}_l) = (\mathbf{a} \rfloor \mathbf{b})\mathbf{B}_l - \mathbf{b} \wedge (\mathbf{a} \rfloor \mathbf{B}_l)$ for a blade $\mathbf{B}_l$ of grade l, takes some more work:

$$
\begin{aligned}
(\mathbf{a} \rfloor \mathbf{b})\mathbf{B}_l - \mathbf{b} \wedge (\mathbf{a} \rfloor \mathbf{B}_l) &= \langle (\mathbf{a} \cdot \mathbf{b})\mathbf{B}_l - \mathbf{b}(\mathbf{a} \rfloor \mathbf{B}_l) \rangle_l \\
&= \langle (\mathbf{a} \cdot \mathbf{b})\mathbf{B}_l - \mathbf{b} \wedge (\mathbf{a} \rfloor \mathbf{B}_l) - \mathbf{b} \rfloor (\mathbf{a} \rfloor \mathbf{B}_l) \rangle_l \\
&= \langle (\mathbf{a} \cdot \mathbf{b})\mathbf{B}_l - \mathbf{b} \wedge (\mathbf{a} \rfloor \mathbf{B}_l) \rangle_l \quad \text{[last term of too low grade]} \\
&= \langle \mathbf{a} \rfloor (\mathbf{b} \wedge \mathbf{B}_l) \rangle_l \\
&= \langle \mathbf{a}\,(\mathbf{b} \wedge \mathbf{B}_l) - \mathbf{a} \wedge (\mathbf{b} \wedge \mathbf{B}_l) \rangle_l \\
&= \langle \mathbf{a}\,(\mathbf{b} \wedge \mathbf{B}_l) \rangle_l \quad \text{[last term of too high grade]} \\
&= \mathbf{a} \rfloor (\mathbf{b} \wedge \mathbf{B}_l)
\end{aligned}
$$

- In contrast to the algebraic methods of Section 6.3.1, we now have in (C.8) a complete definition of the contraction and are therefore also able to prove the final contraction property (3.11).

$$
\begin{aligned}
(\mathbf{A}_k \wedge \mathbf{a}) \rfloor \mathbf{B}_l &= \langle (\mathbf{a} \wedge \widehat{\mathbf{A}}_k)\mathbf{B}_l \rangle_{l-k-1} \\
&= \tfrac{1}{2} \langle (\mathbf{a}\widehat{\mathbf{A}}_k)\mathbf{B}_l + \mathbf{A}_k \mathbf{a} \mathbf{B}_l \rangle_{l-k-1} \\
&= \tfrac{1}{2} \langle \mathbf{A}_k \mathbf{a} \mathbf{B}_l - \mathbf{A}_k \widehat{\mathbf{B}}_l \mathbf{a} + \mathbf{A}_k \widehat{\mathbf{B}}_l \mathbf{a} + \mathbf{a}\widehat{\mathbf{A}}_k \mathbf{B}_l \rangle_{l-k-1} \\
&= \langle \mathbf{A}_k (\mathbf{a} \rfloor \mathbf{B}_l) + \mathbf{a} \wedge (\widehat{\mathbf{A}}_k \mathbf{B}_l) \rangle_{l-k-1} \\
&= \langle \mathbf{A}_k (\mathbf{a} \rfloor \mathbf{B}_l) \rangle_{l-k-1} \quad \text{(lowest grade last term $|l-k|+1$, too high)} \\
&= \mathbf{A}_k \rfloor (\mathbf{a} \rfloor \mathbf{B}_l),
\end{aligned}
$$

and recursion over the factors of $\mathbf{A}_k$ then proves the general result $(\mathbf{A} \wedge \mathbf{A}') \rfloor \mathbf{B} = \mathbf{A} \rfloor (\mathbf{A}' \rfloor \mathbf{B})$.

So we see that the symbol $\langle \mathbf{A}_k \mathbf{B}_l \rangle_{l-k}$, which we proposed as the grade-based alternative definition of the left contraction inner product, has all the desired properties. Therefore it is algebraically identical to the contraction of Chapter 3.

For the right contraction we could repeat this proof, or just show that its correspondence (3.19) with the left contraction holds:

$$\mathbf{A}_k \lfloor \mathbf{B}_l = \langle \mathbf{A}_k\, \mathbf{B}_l \rangle_{k-l} = \langle (\widetilde{\mathbf{B}}_l\, \widetilde{\mathbf{A}}_k)^{\widetilde{}} \rangle_{k-l} = \langle \widetilde{\mathbf{B}}_l\, \widetilde{\mathbf{A}}_k \rangle_{k-l}^{\widetilde{}} = (\widetilde{\mathbf{B}}_l \rfloor \widetilde{\mathbf{A}}_k)^{\widetilde{}}.$$

Since that is universal for all blades (and can be extended for multivectors), the right contraction defined by (C.9) must be identical to our earlier right contraction from Chapter 3.

The scalar product of blades of equal grade should be consistent with how it can be defined as a special case of the contraction: $\mathbf{A}_k * \mathbf{B}_k = \mathbf{A}_k \rfloor \mathbf{B}_k = \langle \mathbf{A}_k\, \mathbf{B}_k \rangle_0$. For nonequal grades, the scalar part of $\mathbf{A}_k\, \mathbf{B}_l$ is zero (as it should be), since the lowest grade present always exceeds zero when $k \neq l$. Therefore, (C.10) for the scalar product is correct.

D COMMON EQUATIONS

Not all equations are equal, and some occur more often than others in the practice of geometric algebra. We have collected the equations we use most for convenient reference. We specify their sources so that you may find the context or a similar equation if the one we give does not quite satsify your needs.

Some equations have been made somewhat more robust for practical use (such as using an absolute value under a square root to guarantee validity even with round-off errors).

Table D.1: Notation of products, operators, and possible definitions. For alternative definitions, disambiguation of `meet` and `join`, and other details, we refer to the chapters indicated. In the formulas, $\mathbf{I}_n$ is the unit pseudoscalar for the total space, and a and b are the grades of the blades $\mathbf{A}$ and $\mathbf{B}$. All definitions except `meet` and `join` can be extended from blades' general multivectors. The `C++` column presents the operators and function names used in the programming examples in Part I and Part II.

Product or Operation	Notation	Definition	Reference	C++
geometric product	$\mathbf{A}\,\mathbf{B}$	fundamental	Chapter 6	`*`
outer product	$\mathbf{A} \wedge \mathbf{B}$	$\langle \mathbf{A}\,\mathbf{B} \rangle_{a+b}$	Chapter 2	`^`
scalar product	$\mathbf{A} * \mathbf{B}$	$\langle \mathbf{A}\,\mathbf{B} \rangle_0$	Chapter 3	`%`
left contraction	$\mathbf{A} \rfloor \mathbf{B}$	$\langle \mathbf{A}\,\mathbf{B} \rangle_{b-a}$	Chapter 3	`<<`
right contraction	$\mathbf{A} \lfloor \mathbf{B}$	$\langle \mathbf{A}\,\mathbf{B} \rangle_{a-b}$	Chapter 3	`>>`
commutator product	$\mathbf{A} \times \langle B \rangle_2$	$\langle \mathbf{A}\,\langle B \rangle_2 \rangle_a$	Chapter 8	N.A.
cross product ($\mathbb{R}^{3,0}$ only)	$\mathbf{a} \times \mathbf{b}$	$(\mathbf{a} \wedge \mathbf{b}) \rfloor \mathbf{I}_3^{-1}$	Chapter 3	N.A.
meet $\mathbf{M}$	$\mathbf{A} \cap \mathbf{B}$	$(\mathbf{B} \rfloor \mathbf{J}^{-1}) \rfloor \mathbf{A}$	Chapter 5, 21	`meet(A, B)`
join $\mathbf{J}$	$\mathbf{A} \cup \mathbf{B}$	$\mathbf{A} \wedge (\mathbf{M}^{-1} \rfloor \mathbf{B})$	Chapter 5, 21	`join(A, B)`
extraction of grade i	$\langle A \rangle_i$		Chapter 2	`takeGrade(A, 1 << i)`
reverse	$\widetilde{\mathbf{A}}$	$(-1)^{a(a-1)/2}\,\mathbf{A}$	Chapter 2	`reverse(A)`
grade involution	$\widehat{\mathbf{A}}$	$(-1)^a\,\mathbf{A}$	Chapter 2	`gradeInvolution(A)`
inverse	$\mathbf{A}^{-1}$	$\widetilde{\mathbf{A}}/(\mathbf{A}\,\widetilde{\mathbf{A}})$	Chapter 3, 6	`inverse(A)`
dual	$\mathbf{A}^*$	$\mathbf{A}/\mathbf{I}_n$	Chapter 3, 6	`dual(A)`
undual	$\mathbf{A}^{-*}$	$\mathbf{A}\,\mathbf{I}_n$	Chapter 3, 6	`undual(A)`
squared norm	$\|\mathbf{A}\|^2$	$\mathbf{A}\,\widetilde{\mathbf{A}}$	Chapter 3	`norm_r2(A)`

Table D.2: Sign change per grade for reversion, grade involution, and Clifford conjugation.

Operation	Notation	Multiplier for grade x	Pattern
reversion	$\widetilde{X}$	$(-1)^{x(x-1)/2}$	+ + − − + + − −
grade involution	$\widehat{X}$	$(-1)^x$	+ − + − + − + −
Clifford conjugation	$\overline{X}$	$(-1)^{x(x+1)/2}$	+ − − + + − − +

Table D.3: Linear transformation f applied to common products.

Product	Linear transformation	Reference
outer product	$f[\mathbf{A} \wedge \mathbf{B}] = f[\mathbf{A}] \wedge f[\mathbf{B}]$	(4.3)
scalar product	$f[\mathbf{A} * \mathbf{B}] = \mathbf{A} * \mathbf{B}$	(4.9)
left contraction	$f[\mathbf{A} \rfloor \mathbf{B}] = \overline{f}^{-1}[\mathbf{A}] \rfloor f[\mathbf{B}]$	(4.13)
dual	$f^*[\mathbf{A}^*] = \det(f)\,\overline{f}^{-1}[\mathbf{A}^*]$	(4.14)
cross product	$f[\mathbf{a} \times \mathbf{b}] = \det(f)\,\overline{f}^{-1}[\mathbf{a} \times \mathbf{b}]$	(4.15)
pseudoscalar	$f[\mathbf{I}_n] = \det(f)\,\mathbf{I}_n$	(4.7)
meet	$f[\mathbf{A} \cap \mathbf{B}] = f[\mathbf{A}] \cap f[\mathbf{B}]$ (or with $\det(f)$)	Section 5.7
join	$f[\mathbf{A} \cup \mathbf{B}] = f[\mathbf{A}] \cup f[\mathbf{B}]$ (or with $\det(f)$)	Section 5.7

Table D.4: Basic equations in geometric algebra.

Operation	Formula	Reference
symmetry of outer product	$\mathbf{a} \wedge \mathbf{B} = \frac{1}{2}(\mathbf{a}\mathbf{B} + \widehat{\mathbf{B}}\mathbf{a})$	(6.10)
symmetry of outer product	$\mathbf{B} \wedge \mathbf{a} = \frac{1}{2}(\mathbf{B}\mathbf{a} + \mathbf{a}\widehat{\mathbf{B}})$	(6.11)
symmetry of left contraction	$\mathbf{a} \rfloor \mathbf{B} = \frac{1}{2}(\mathbf{a}\mathbf{B} - \widehat{\mathbf{B}}\mathbf{a})$	(6.12)
symmetry of right contraction	$\mathbf{B} \lfloor \mathbf{a} = \frac{1}{2}(\mathbf{B}\mathbf{a} - \mathbf{a}\widehat{\mathbf{B}})$	(6.13)
contraction vector on blade	$\mathbf{a} \rfloor (\mathbf{b} \wedge \mathbf{C}) = (\mathbf{a} \cdot \mathbf{b}) \wedge \mathbf{C} - \mathbf{b} \wedge (\mathbf{a} \rfloor \mathbf{C})$	from (3.10)
contraction vector on versor	$\mathbf{a} \rfloor (\mathbf{b}\,\mathbf{C}) = (\mathbf{a} \cdot \mathbf{b})\,\mathbf{C} - \mathbf{b}\,(\mathbf{a} \rfloor \mathbf{C})$	(C.5)
contraction vector on blade	$\mathbf{x} \rfloor (\mathbf{a}_1 \wedge \cdots \wedge \mathbf{a}_k) =$ $\sum_{i=1}^{k}(-1)^{i-1}\,\mathbf{a}_1 \wedge \cdots \wedge (\mathbf{x} \rfloor \mathbf{a}_i) \wedge \cdots \mathbf{a}_k$	(3.16)
reciprocal frame	$\mathbf{b}^i = (-1)^{i-1}(\mathbf{b}_1 \wedge \cdots \breve{\mathbf{b}}_i \wedge \cdots \wedge \mathbf{b}_n) \rfloor \mathbf{I}_n^{-1}$	(3.31)
dual	$\mathbf{X}^* = \mathbf{X} \rfloor \mathbf{I}_n^{-1}$	Section 3.5.3
undual in $\mathbb{R}^n$ (and $\mathbb{R}^{n+1,1}$)	$\mathbf{X}^{-*} = (-1)^{n(n-1)/2}\mathbf{X}^*$	Section 3.5.3
duality relationship	$(\mathbf{A} \wedge \mathbf{B})^* = \mathbf{A} \rfloor (\mathbf{B}^*)$	(3.24)
duality relationship	$(\mathbf{A} \rfloor \mathbf{B})^* = \mathbf{A} \wedge (\mathbf{B}^*)$, for $\mathbf{A} \subseteq \mathbf{I}$	(3.24)
projection of $\mathbf{X}$ onto $\mathbf{B}$	$\mathbf{X} \mapsto (\mathbf{X} \rfloor \mathbf{B}) \rfloor \mathbf{B}^{-1} = (\mathbf{X} \rfloor \mathbf{B})/\mathbf{B}$	Section 3.6, 6.4.2
rejection of vector $\mathbf{x}$ by $\mathbf{B}$	$\mathbf{x} \mapsto (\mathbf{x} \wedge \mathbf{B}) \rfloor \mathbf{B}^{-1} = (\mathbf{x} \wedge \mathbf{B})/\mathbf{B}$	Section 3.6, 6.4.2
reflection of direct $\mathbf{X}$ in dual $\mathbf{D}$	$\mathbf{X} \mapsto (-1)^{xd}\mathbf{D}\,\mathbf{X}\,\mathbf{D}$	Table 7.1
applying an even versor V	$\mathbf{X} \mapsto V\mathbf{X}V^{-1}$	(7.19)
applying an odd versor V	$\mathbf{X} \mapsto V\widehat{\mathbf{X}}V^{-1}$	(7.20)
inverse of a versor V	$V^{-1} = \widetilde{V}/(V\widetilde{V})$	(21.1)

Table D.5: Exponential, cosine, and sine of blades. See Section 7.4.

Taylor series	Special cases
$\exp\mathbf{A} \;=\; 1 + \frac{\mathbf{A}}{1!} + \frac{\mathbf{A}^2}{2!} + \cdots$	$\exp\mathbf{A} \;=\; \begin{cases} \cos\alpha + \mathbf{A}\,\frac{\sin\alpha}{\alpha} & \text{if} \quad \mathbf{A}^2 = -\alpha^2 \\ 1 + \mathbf{A} & \text{if} \quad \mathbf{A}^2 = 0 \\ \cosh\alpha + \mathbf{A}\,\frac{\sinh\alpha}{\alpha} & \text{if} \quad \mathbf{A}^2 = \alpha^2 \end{cases}$
$\sin\mathbf{A} \;=\; \mathbf{A} - \frac{\mathbf{A}^3}{3!} + \frac{\mathbf{A}^5}{5!} - \cdots$	$\sin\mathbf{A} \;=\; \begin{cases} \mathbf{A}\,\frac{\sin\alpha}{\alpha} & \text{if} \quad \mathbf{A}^2 = \alpha^2 \\ \mathbf{A} & \text{if} \quad \mathbf{A}^2 = 0 \\ \mathbf{A}\,\frac{\sinh\alpha}{\alpha} & \text{if} \quad \mathbf{A}^2 = -\alpha^2 \end{cases}$
$\cos\mathbf{A} \;=\; 1 - \frac{\mathbf{A}^2}{2!} + \frac{\mathbf{A}^4}{4!} - \cdots$	$\cos\mathbf{A} \;=\; \begin{cases} \cos\alpha & \text{if} \quad \mathbf{A}^2 = \alpha^2 \\ 1 & \text{if} \quad \mathbf{A}^2 = 0 \\ \cosh\alpha & \text{if} \quad \mathbf{A}^2 = -\alpha^2 \end{cases}$

Table D.6: Working with rotors.

Requirement	Formula	Reference
log of Euclidean rotor R	$\log(R) = \big(\langle R\rangle_2/\|\langle R\rangle_2\|\big)\,\mathrm{atan2}(\|\langle R\rangle_2\|,\langle R\rangle_0)$	(10.14)
shortest rotor from unit $\mathbf{a}$ to unit $\mathbf{b}$	$R = (1 + \mathbf{b}\,\mathbf{a})/\sqrt{2\,|1 + \mathbf{a}\cdot\mathbf{b}|}$	(10.13)
rotor/quaternion relationship		Section 7.3.5
matrix/rotor conversion		Section 7.10.4

Table D.7: Convenient equations from the homogeneous model in Chapter 11.

Requirement	Formula	Reference
parameters of all blades	flats, directions	Table 11.3
intersection and spanning	Plücker formulas	Table 12.1
rigid body motions	transformation matrices	Table 12.2
translation direct blade	$X \mapsto X + \mathbf{t} \wedge (e_0^{-1}\rfloor X)$	(11.13)
translation dual blade	$X^* \mapsto X^* - e_0^{-1} \wedge (\mathbf{t}\rfloor X^*)$	(11.14)

Table D.8: Convenient equations from the conformal model.

Requirement	Formula	Reference
Euclidean distance of normalized points p, q	$d_E(p, q) \;=\; \sqrt{\lvert -2p \cdot q \rvert}$	from (13.2)
normalized point from Euclidean vector p	$p \;=\; o + \mathbf{p} + \frac{1}{2}\mathbf{p}^2 \infty$	(13.3)
normalized point from normalized sphere Σ	$p \;=\; \Sigma^* - \frac{1}{2}\Sigma^2 \infty$	
normalized point from normalized flat point Φ	$p \;=\; o \rfloor \Phi + \frac{1}{2}(o \rfloor \Phi)^2 \infty$	
two normalized points from point pair P	$p_\pm \;=\; (P \mp \sqrt{\lvert P^2 \rvert})/(-\infty \rfloor P)$	(14.13)
normalize point or dual sphere p	$p \;\mapsto\; p/(-\infty \cdot p)$	Section 13.1.3
normalize flat F	$F \;\mapsto\; \widehat{F}/((o \wedge \infty) \rfloor F)$	
dual sphere of radius ρ at normalized point c	$\Sigma^* \;=\; c - \frac{1}{2}\rho^2 \infty$	Table 13.2
dual sphere through p, center c	$\Sigma^* \;=\; p \rfloor (c \wedge \infty)$	(14.4)
direct round at c, radius ρ	$\Sigma \;=\; (c + \frac{1}{2}\rho^2 \infty) \wedge \left(-c \rfloor (\widehat{\mathbf{A}}_k \infty) \right)$	(14.9)
dual round at c, radius ρ	$\Sigma^* \;=\; (c - \frac{1}{2}\rho^2 \infty) \wedge \left(-c \rfloor (\widehat{\mathbf{A}}_k^{\star} \infty) \right)$	(14.10)
dual plane normal vector $\mathbf{n}$, support δ	$\Pi^* \;=\; \mathbf{n} - \delta \lVert \mathbf{n} \rVert \infty$	Table 13.2
dual plane through p, normal $\mathbf{n}$	$\Pi^* \;=\; p \rfloor (\mathbf{n} \wedge \infty)$	
dual midplane normalized points p, q	$\Pi^* \;=\; p - q$	(14.5)
direct flat at p, direction $\mathbf{A}_k$	$\Pi \;=\; p \wedge \mathbf{A}_k \wedge \infty$	(13.8)
dual flat at p, direction $\mathbf{A}_k$	$\Pi^* \;=\; -p \rfloor (\mathbf{A}_k^{\star} \infty)$	(13.9)
parameters of all blades	flats, rounds, tangents, directions	Table 14.1
translation of Euclidean element	$T_{\mathbf{p}}[\mathbf{A}_k] \;=\; -p \rfloor (\infty \mathbf{A}_k) = \mathbf{A}_k + \infty (\mathbf{p} \rfloor \mathbf{A}_k)$	(13.7)
logarithm rigid body motion	algorithm based on formula	Figure 13.5
logarithm scaled rigid body motion	algorithm based on formula	Figure 16.5
conformal versors	complete list	Table 16.1

Bibliography

[1]   Ablamowicz, R., P. Lounesto, and J. Parra (Eds). *Clifford Algebras with Numeric and Symbolic Computations*. Boston: Birkhäuser, 1996.

[2]   Anglès, P. Construction de revêtements du group conform d'un espace vectorial muni d'une "métrique" de type (p,q). *Annales de l'Institut Henri Poincaré*, Section A, Vol. XXXIII:33–51, 1980.

[3]   Barnabei, M., A. Brini, and G. C. Rota. On the exterior calculus of invariant theory. *Journal of Algebra*, 96:120–160, 1985.

[4]   Bayro-Corrochano, E., and J. Lasenby. A unified language for computer vision and robotics. In G. Sommer and J. J. Koenderink (Eds), *Algebraic Frames for the Perception-Action Cycle. Lecture Notes in Computer Science*, 1315, Springer 1997.

[5]   Beckmann, O., A. Houghton, P. Kelly, and M. Mellor. Run-time code generation in C++ as a foundation for domain-specific optimisation. *Lecture Notes in Computer Science*, 3016:291–306, 2004.

[6]   Bell, I., Ian Bell's web pages on geometric algebra. *http://www.iancgbell.clara.net/maths/geoalg1.htm*, 2005.

[7]   Bouma, T. A., and G. Memowich. Invertible homogeneous versors are blades. *http://www.science.uva.nl/ga/publications*, 2001.

[8]   Bouma, T., L. Dorst, and H. G. J. Pijls. Geometric algebra for subspace operations. *Acta Mathematicae Applicandae*, pp. 73:285–300, 2002.

[9]   Cox, H. On systems of circles and bicircular quartics. *Quaterly Journal of Pure and Applied Mathematics*, 19:74–124, 1883.

[10] Cox, H. Application of Grassmann's Ausdehnungslehre to properties of circles. *Quarterly Journal of Pure and Applied Mathematics*, 25:1–71, 1891.

[11] Czarnecki, K., and U. W. Eisenecker. *Generative Programming—Methods, Tools, and Applications*. Addison-Wesley, 2000.

[12] de Berg, M., M. van Kreveld, M. Overmars, and O. Schwarzkopf. *Computational Geometry: Algorithms and Applications*, 2nd edition. Springer Verlag, 1998.

[13] DeRose, T. Coordinate-free geometric programming. Technical Report 89-09-16, University of Washington, September 1989.

[14] Doran, C., and A. Lasenby. Physical applications of geometric algebra. *http://www.mrao.cam.ac.uk/~ clifford/ptIIIcourse/*, 1999.

[15]   Doran, C., and A. Lasenby. *Geometric Algebra for Physicists*. Cambridge University Press, 2003.

[16]   Dorst, L. Analyzing the behaviors of a car: a study in the abstraction of goal-directed motion. *IEEE Systems, Man and Cybernetics A*, 28(6):811–822, 1998.

[17]   Dorst, L. The inner products of geometric algebra. In J. Lasenby, L. Dorst, and C. Doran (Eds), *Applications of Geometric Algebra in Computer Science and Engineering*, pp. 35–46. Boston: Birkhäuser, 2002.

[18]   Dorst, L., and D. Fontijne. 3D Euclidean geometry through conformal geometric algebra (a `GAViewer` tutorial). *http://www.science.uva.nl/ga/*.

[19]   Dorst, L., and S. Mann. Geometric algebra: a computation framework for geometrical applications: part I. *Computer Graphics and Applications*, 22(3):24–31, May/June 2002.

[20]   Eastwood, M. G., and P. W. Michor. Some remarks on the Plücker relations. *Rendiconti del Circolo Matematico di Palermo*, II-63:85–88, 2000.

[21]   Faugeras, O., and Q.-T. Luong. *The Geometry of Multiple Views*. MIT Press, 2001.

[22]   Forder, H. G. *The Calculus of Extension*. Cambridge University Press, 1941.

[23]   Goldman, R. Illicit expressions in vector algebra. *ACM Transactions on Graphics*, 4(3), July 1985.

[24]   Goldman, R. On the algebraic and geometric foundations of computer graphics. *ACM Transactions on Graphics*, 21(1), January 2002.

[25]   Hartley, R. I., and A. Zisserman. *Multiple View Geometry in Computer Vision,* 2nd edition. Cambridge University Press, 2003.

[26]   Hestenes, D. Geometric calculus. A chapter of an unpublished book, available at *http://modelingnts.la.asu.edu/html/NFMP.html*.

[27]   Hestenes, D. The design of linear algebra and geometry. *Acta Appl. Math*, 23:65–93, 1991.

[28]   Hestenes, D. Grassmann's vision. In *Hermann Günther Grassmann (1809–1877): Visionary Mathematician, Scientist and Neohumanist Scholar*, pp. 243–254. Dordrecht: Kluwer Academic Publishers Group, 1996.

[29]   Hestenes, D. *New Foundations for Classical Mechanics*, 2nd edition. Reidel, 2000.

[30]   Hestenes, D. Point groups and space groups. In J. Lasenby, L. Dorst, and C. Doran (Eds), *Applications of Geometric Algebra in Computer Science and Engineering*, pp. 3–34. Boston: Birkhäuser, 2002.

[31]   Hestenes, D., H. Li, and A. Rockwood. A unified algebraic framework for classical geometry. In G. Sommer (Ed), *Geometric Computing with Clifford Algebra*. Springer, 1999.

[32] Hestenes, D., A. Rockwood, and H. Li. *System for encoding and manipulating models of objects*, U.S. Patent 6,853,964, granted February 8, 2005.

[33] Hestenes, D., and G. Sobczyk. *Clifford Algebra to Geometric Calculus*. Reidel, 1984.

[34] Hildenbrand, D., D. Fontijne, W. Yusheng, M. Alexa, and L. Dorst. Competitive run-time performance for inverse kinematics algorithms using conformal geometric algebra. In *Proceedings of Eurographics 2006, Vienna*, 2006.

[35] Koenderink, J. J. A Generic Framework for Image Geometry. In J. Lasenby, L. Dorst, and C. Doran, (Ed), *Applications of Geometric Algebra in Computer Science and Engineering*, pp. 319–332. Boston: Birkhäuser, 2002.

[36] Lasenby, J., W. J. Fitzgerald, C. J. L. Doran, and A. N. Lasenby. New geometric methods for computer vision. *International Journal of Computer Vision*, 36(3):191–213, 1998.

[37] Lasenby, J., R. Lasenby, A. Lasenby, and R. Wareham. Higher dimensional fractals in geometric algebra. Technical Report CUED/F-INFENG/TR.556, Cambridge University Engineering Department, 2006.

[38] Lasenby, J., and A. Stevenson. Using geometric algebra for optical motion capture. In E. Bayro-Corrochano and G. Sobcyzk, (Eds), *Applied Clifford Algebras in Computer Science and Engineering*. Birkhäuser, 2000.

[39] Lounesto, P. Marcel Riesz's work on Clifford algebras. In E. F. Bolinder and P. Lounesto (Eds), *Clifford Numbers and Spinors*, pp. 119–241. Kluwer Academic, 1993.

[40] Lounesto, P., R. Mikkola, and V. Vierros. CLICAL user manual: complex number, vector space, and Clifford algebra calculator for MS-DOS personal computers. Technical Report Institute of Mathematics Research Reports A248, Helsinki University of Technology, 1987.

[41] Lounesto, P. *Clifford algebras and spinors*. London Mathematical Society Lecture Note Series 239, 1997.

[42] Mann, S., and L. Dorst. Geometric algebra: a computation framework for geometrical applications: part II. *Computer Graphics and Applications*, 22(4):58–67, July/August 2002.

[43] Mann, S., L. Dorst, and T. Bouma. The making of GABLE, a geometric algebra learning environment in Matlab. In E. Bayro-Corrochano and G. Sobczyk (Eds), *Geometric Algebra with Applications in Science and Engineering*, chapter 24, pp. 491–511. Birkhäuser, 2001.

[44] Mann, S., N. Litke, and T. DeRose. A coordinate free geometry ADT. Technical Report CS-97-15, University of Waterloo, 1997.

[45] Mann, S., and A. Rockwood. Using geometric algebra to compute singularities in 3D vector fields. In *IEEE Visualization 2002 Proceedings*, pp. 283–289, 2002.

[46] Needham, T. *Visual Complex Analysis*. Oxford: Clarendon Press, 1997.

[47]   Perwass, C., CLU: Clifford algebra library and utilities.
       http://www.perwass.de/cbup/clu.html.

[48]   Perwass, C., and W. Förstner. Uncertain Geometry with Circles, Spheres and
       Conics, volume 31 of *Computational Imaging and Vision*, pp. 23–41.
       Springer-Verlag, 2006.

[49]   Porteous, I. R. *Topological Geometry*. Cambridge: Cambridge University Press,
       1981.

[50]   Porteous, I. R. *Geometric Differentiation for the Intelligence of Curves and Surfaces.*
       Cambridge: Cambridge University Press, 1994.

[51]   Porteous, I. R. *Clifford Algebras and the Classical Groups.* Cambridge: Cambridge
       University Press, 1995.

[52]   Riesz, M. *Clifford Numbers and Spinors.* Kluwer Academic, 1993.

[53]   Rosenhahn, B., and G. Sommer. Pose estimation in conformal geometric algebra,
       part I: the stratification of mathematical spaces. *Journal of Mathematical Imaging
       and Vision (JMIV)*, 22:27–48, 2005.

[54]   Rosenhahn, B., and G. Sommer. Pose estimation in conformal geometric algebra,
       part II: real-time pose estimation using extended feature concepts. *Journal of
       Mathematical Imaging and Vision (JMIV)*, 22:49–70, 2005.

[55]   Semple, J. G., and G. T. Kneebone. *Algebraic Projective Geometry*. Oxford
       University Press, 1952.

[56]   Shoemake, K. Animating rotation with quaternion curves. In *Proceedings of ACM
       SIGGRAPH*, volume 19:3, pp. 245–254, 1985.

[57]   Shoemake, K. Plücker coordinate tutorial. *Ray Tracing News*, 11(1), July 1998.

[58]   Shreiner, D., M. Woo, J. Neider, T. Davis, and the OpenGL Architecture Review
       Board. *OpenGL Programming Guide: The Official Guide to Learning OpenGL,
       Version 1.4, Fourth Edition.* Addison-Wesley, 2003.

[59]   Stewart. I. Hermann Grassmann was right. *Nature*, 321:6065, 1986.

[60]   Stolfi, J. *Oriented Projective Geometry*. Academic Press, 1991.

[61]   Suter, J. Clifford (software), 2003. Used to be available at
       http://www.jaapsuter.com.

[62]   Valkenburg, R. J., and N. S. Alwesh, Calibration of Target Positions using the
       Conformal Model and Geometric Algebra, *Image and Vision Computing, New
       Zealand*, 241–246, 2005.

[63]   Zhang, Z. A flexible new technique for camera calibration. *IEEE Transactions on
       Pattern Analysis and Machine Intelligence*, 22(11):1330–1334, 2000.

[64]   Zhao, Y., R. J. Valkenburg, R. Klette, and B. Rosenhahn, Target Calibration and
       Tracking using Conformal Geometric Algebra, *IEEE Pacific-Rim Symposium on
       Image and Video Technology*, 2006.

Index

Printed and bound by CPI Group (UK) Ltd, Croydon, CR0 4YY

03/10/2024

01040309-0001